DEPARTMENT OF THE INTERIOR
Hubert Work, Secretary

U. S. GEOLOGICAL SURVEY
George Otis Smith, Director

Professional Paper 138

MINING IN COLORADO

A HISTORY OF DISCOVERY, DEVELOPMENT AND PRODUCTION

BY

CHARLES W. HENDERSON

WASHINGTON
GOVERNMENT PRINTING OFFICE
1926

CONTENTS

ILLUSTRATIONS

MINING IN COLORADO

By Charles W. Henderson [1]

HISTORICAL SKETCH

THE FIRST DISCOVERIES OF GOLD

Spanish explorers [1a] appear not to have visited the region that is now eastern Colorado until the middle of the seventeenth century, but traversed southern, central, and western Colorado in 1706, 1719, 1720, 1765, 1776, and 1779. A small party of Frenchmen, headed by two brothers named Mallet, crossed the eastern part of the State from north to south in 1739. The American explorers that left records of their journeys in the region were Pike, in 1805 and 1806; Long, in 1820; Bonneville, in 1832; Dodge, in 1835; Frémont, in 1842, 1843, 1845, 1848, and 1853; Gunnison and Beckwith, in 1853; Macomb and Newberry, in 1859 and 1860. Neither Pike, Long, Bonneville, nor Dodge seem to have reported the discovery of any minerals. In 1833 a party of trappers from Taos, N. Mex., camped on Dolores River near Trout Lake, about 13 miles northeast of the present site of Rico [2] but apparently paid no attention to mineral deposits. Gold is said to have been discovered near the site of Lake City about 1848 by a member of the Frémont party, but no one has been able to identify the place or even the stream from which the first gold was panned. [3]

The explorations of the geologists of the Hayden, Wheeler, King, and Powell surveys, and later of the United States Geological Survey, made chiefly by King, Hague, Emmons, Hayden, Powell, Gilbert, Walcott, Marvine, Howell, Holmes, Peale, and Dutton, early gave a general knowledge of the geography and geology of the State, which has later been studied in greater detail in many areas by members of the United States Geological Survey and others.

The great transcontinental migration to the California gold mines that began in 1849 found the high mountains in Colorado a barrier to travel, so that the main available avenues to the Far West were the Overland route, through Wyoming, on the north, and the Santa Fe route, through New Mexico, on the south.

It is the purpose here not to relate completely the history of Colorado, but to give as far as it is clearly known only the history of mining, with which the history of the State runs parallel after 1858. If hunters, trappers, and Indians knew of gold in Colorado before 1849 their knowledge occasioned no immigration; and even the knowledge possessed by those who may have known of gold in the period between 1849 and 1858 did nothing to stimulate the opening and settlement of Colorado other than to supplement the leadership of the Russell brothers, to whom nearly all historians now credit the rush to the Pikes Peak region in 1858, a rush that was fraught with much hardship, danger, and disappointment but that brought to the successful miners handsome rewards and hastened by many years the development of the State. There were then no roads through the mountains of Colorado to California, and even if there had been such routes it is quite possible that the "forty-niners" might have passed by the bonanza fields of Colorado as they did those of Nevada. The plains of Colorado were occupied by Indians, chiefly the Arapahoes and Cheyennes. The Russell brothers, who were originally from Georgia but who had gained experience in mining in California, led a party of Georgians, Cherokee Indians, and others to Cherry Creek in 1858 in search of gold in what was then western Kansas.

James H. Pierce, a member of the expedition under William Green Russell to Cherry Creek in 1858, wrote as follows in a manuscript in the files of the State Historical and Natural History Society of Colorado:

In April, 1858, in company of William Green Russell, Joseph O. Russell, Dr. Levi J. Russell, R. J. Pierce, Solomon Roe, I, with some six or eight more, all from the State of Georgia, outfitted in Leavenworth City, Kans., with the object of prospecting for gold in the Rocky Mountains. The place in view was South Platte, Cherry Creek, and Ralston Creek, by a former understanding between William Green Russell, the leader of our company, and John Beck, then a citizen of the Cherokee Nation, a man of experience in mining, and formerly from Georgia.

Beck, in company with Louis Ralston, in crossing the plains [of Kansas] in the year 1849, had found gold on Cherry Creek,

[1] In the preparation of the statistics the writer has had the assistance of Virginia E. Knowles, Bessie M. Peterson, John H. Winchell, Sue D. Boot, and Max J. Gleissner, of the United States Geological Survey.

[1a] Thomas, A. B., Spanish expeditions into Colorado: Colorado Mag., vol. 1, No. 7, November, 1924. Thomas disproves the entrance of the Spaniards into Colorado in 1540-1600.

[2] Ransome, F. L., The ore deposits of the Rico Mountains, Colo.: U. S. Geol. Survey Twenty-second Ann. Rept., pt. 2, p. 239, 1901.

[3] Irving, J. D., and Bancroft, Howland, Geology and ore deposits near Lake City Colo.: U. S. Geol. Survey Bull. 478, p. 13, 1911.

and also on Ralston Creek, in small quantities. Russell had also found gold on the North Platte and Sweetwater rivers, [*] while he and a small company of Georgians were on their way to California. Beck and Russell had been acquainted in Georgia before going to California. They met by chance in California, and in talking of their trip they discovered that both had found gold in the Rocky Mountains.

After [William] Green Russell had made two trips home from California he accumulated some means—he and his [two] brothers. His youngest brother, Dr. Levi John Russell, had gone to Philadelphia and there graduated at the Pennsylvania Medical College, after which he had returned to Georgia, married, and settled down to the practice of medicine. [William] Green, [Levi] John, and [Joseph] Oliver Russell had bought farms and a few slaves, and were comfortably situated in Auraria, Lumpkin County, in northeast Georgia, but all of them had been miners from boyhood. Green had kept up a correspondence with John Beck, of the Cherokee Nation, from the time he had last met him in California, and in the winter of 1857 they entered into an agreement to form parties each in their respective countries and to meet on the plains at the Big Bend of the Arkansas River.

The meeting at the Big Bend was accomplished on April 25, 1858. Beck had 78 men and Russell 26, making in all 104 men, composed of Georgians, Cherokees, Arkansans, Missourians, and some who joined us in Kansas. All or most of them were outfitted with ox teams and six months' provisions. We then proceeded on our journey, halting only on Sundays, until we reached the head of Cherry Creek. We there found gold as Beck had represented, generally in light sand and in small quantities, but in no place [did we try to] work. We stayed there only one or two days.

We then went on to the Platte [and] camped at the mouth of Cherry Creek a day or two. We got there May 23, 1858. We prospected up and down the Platte but found nothing of value. We then crossed the Platte, went to Ralston Creek, and camped there, where Beck and Ralston had found gold in 1849. The place I judge to be about 1 mile above the junction of Ralston and Clear creeks and perhaps a quarter of a mile west of the old Cherokee trail. We prospected up and down Ralston and Clear creeks and back into the mountains about 5 miles. We were there several days but found nothing of much importance; but we got gold every pan. * * *

As I said we had got gold in small quantities both on Clear Creek and Ralston. We had not at that time been in the mountains exceeding 5 miles, but part of the men began to get disheartened, and about one-half of the Cherokee party wanted to stay. They wanted us to unite and build a fort or stockade, but the majority of the party could see no use in such work. So they concluded to bust up, and about the 10th of June the whole of the Cherokee party left us. There were, however, a few that had fallen in with them who remained with us. We were now reduced to less than 50 men.

We spent a few days in the same camp, still prospecting. We went as far north as Boulder and Big Thompson, and south to Bear Creek and into the mountains a short distance. Finding nothing of any importance we concluded to recross the Platte, which we did about the 24th of June. [Note date.] In crossing the Platte we came near losing the wagon and our money and the summer's provisions, which were in the wagon, where I was myself, Doctor Russell, Bates, Roe, and McAfee being with me. We managed to get the wagon and most of our rations out, but our rations were damaged considerably by the water. We then camped there for a day or two and there had another split-up in the company. All of our original Georgian party but seven and six of those that had fallen in with us from Kansas left. The balance of the 104 who came

with us were all gone on the back track, leaving only 13 men to prospect the Rocky Mountains.

* * * I shall never as long as life lasts forget the last split in the company. [I and] Twelve [others] of us agreed to stay, and all the balance went home or somewhere else. * * * I have only seen a few of them since that day. We had in one month's time been reduced from the formidable number of 104 to only 13. * * *

On the very day that we were left alone we started up the Platte to make a camp at the canyon, some 12 or 15 miles from the mouth of Cherry Creek. As fortune would have it, some 3½ miles up the river, while the wagons were ahead of me some one hundred or two hundred yards, I discovered on the bank of the river a bed of alluvial gravel, and under it was a conglomerate or cement bedrock. I ran ahead to the wagons and got a pan, pick, and shovel and had taken out a handful of the gravel and had it about two-thirds panned when Green Russell came up to me and finished washing it. We had about 6 or 7 cents' worth of nice scale gold. He then says in rather an excited tone: "Our fortune is made. Run and stop the wagons and tell them to come back here." This I did. We panned several pans that evening and really the prospect was flattering. But it only proved to be a small deposit.

We camped there several days and made a hand rocker out of a cottonwood log and mined out something over $200 in thin scale gold. In the meantime Green Russell and Sam Bates found another little deposit in a bar up Dry Creek, some 2 miles from the river, that was richer than the one on the Platte. They got $2.50 out of one panful of the dirt. We sank a hole some 4 feet deep in the sand and got water to rock our dirt. We took out $50 in one day. This buoyed us up considerable. We felt that our trip, though hazardous, was not a failure, and another thing that helped us out was that those discoveries were made in so short a time after our comrades left us. But this also proved, like the prospect on the Platte, of but little extent.

The gold obtained here was scale gold, about as coarse as wheat bran, though coarser than any that had been found anywhere on the plains, and was evidently a float gold and had a source somewhere that we thought we could trace out. Our party, though, was weak, and we could not go into the mountains with our wagons. We had only two little mules, so that it was impossible for us to explore the mountains to any great extent, though at one time six of our little party started up the Platte and were gone six or seven days, carrying their provisions on the mules. There were no roads, and it is natural for men always to find the hardest places to get over first. That proved to be the case with them. They returned without making any discovery, tired, hungry, and disappointed.

We had now been something like a month during which time we had not seen a human being of any kind. We saw a man on horseback coming toward our camp. We soon saw that it was a white man, and what could a lone white man be doing in this part of the country? It was not any of our party or anyone that had been with us at any time but an entire stranger. He came up to us. He was a man that had evidently traveled east as well as west. He seemed to be at home among the western hills. It was a man by the name of Cantrell, from Kansas City, Mo. He had been to Fort Laramie in the Black Hills trading with the Government troops and had heard about our party starting out from Leavenworth and had concluded to come by and see what we were doing.

He happened to strike us at the right time to raise an excitement in the East. He had been to California and was a practical miner. We had not got that little bar on Dry Creek worked out. So he got me and one of the others to go with him and get a sackful of the dirt out of that place. We got it out,

[*] At the site of Atlantic City, Wyo., gold was found as early as 1842.—C. W. H.

and I suppose that it would have prospected 25 cents to the panful. He took it with him to Kansas City, and I learned afterward that he panned it out there and published it in the papers with an affidavit that it came from the Rocky Mountains and it was just as it came out of the mines, which I have every reason to believe was the case. He then left us and went all the way down the Arkansas River home by himself.

During all these times the Indians were entirely friendly to the whites in all this country, from the Rocky Mountains to the Arkansas. We never met up with any Utes. I rather suppose they might have been of a different temperament from developments a year or so after that time.

In a few days after Cantrell had left three more white men and one or two Mexicans came through here. It was Captain Dice and some men with him from New Mexico. They had been to Fort Bridger in Utah. They had traveled alone all the way from Fort Bridger, just these three white men by themselves. They stopped a couple of days with us and proceeded on their journey to Taos. Well, then we were alone but a short time when we were joined by an old trader by the name of John Smith. He had with him two or three white men and one or two families of Indians, and an Indian wife. I am now lost for dates. We did not take any papers, as our nearest post office was Fort Laramie, and we had not been there to get our mail and had not heard from home since leaving.

Captain Dice, as he went home, met up with another party of prospectors as he was on the way to Taos, which had been out all summer. They had come from Lawrence, Kans., and were known as the Lawrence party. They had started with the intention of joining us, but met those going back that left us in the early part of the summer, and from that, instead of coming on, they turned southwest along the Raton Mountains in New Mexico, coming hence north toward Pikes Peak and camping a month near the Peak. Hence the name Pikes Peak was given to the new gold mines. It was the Lawrence party that prospected around Pikes Peak, not the Georgia party, as many suppose. I have never been any nearer Pikes Peak than Colorado Springs, and not then until last year, that being the year 1884. None of the Georgians ever prospected in that place.

Sometime in August, 1858, our little party left the South Platte and went north as far as the Medicine Bow Mountains. On the 22d day of September, 1858, we camped on Medicine Bow Creek, at the foot of the mountains. While there a snow fell about 6 inches deep. We got some small gold on that creek and on the West Laramie River but only light, fine colors. That was a great country for game. We killed three bears one day and saw many more. In fact, game was plentiful everywhere than in all the country we had been in. This trip was more a hunting and fishing trip than prospecting. While camped there we saw a regiment of soldiers coming in from Salt Lake, who were the first men we had seen on that trip.

The snow coming made us think that perhaps we had better get out of there. So we turned south, and in about ten days we were again on the Platte where Denver City now stands. That place seemed to have a natural attraction for us; so we camped at the same place. We found there about 35 or 40 men. The Lawrence party had got in. So we were not so lonesome any more. But what was to be done for rations? There we were 700 miles from any depot of supplies. William G. Russell and J. O. Russell wanted to go home and return the next spring and get a few of their friends to come out with them. We had then only about $500 in the company and that was not enough to buy provisions and then go home on. Valverias Young, who had been with us all the time as one of the thirteen pioneers, had come from Kansas but was a native of Iowa. He owned the two little mules and he wanted to go also. So they three went together. Green Russell bought a pony from

one of our mountaineer acquaintances on time, and they went on horseback to Leavenworth City. They took with them most of the gold that had been taken out by all of us and left us ten to remain in the mountains. I will here give the names of our little band of pioneers:

William Green Russell, Dr. L. J. Russell, Joseph O. Russell, three brothers; James H. Pierce and R. J. Pierce, cousins to each other and cousins to the Russell brothers; Solomon Roe, brother to Doctor Russell's wife; Samuel Bates, no connection to any of us. These were all of our little party who came from Georgia. The others had joined us in Kansas: W. A. McPhadden, William McKimmons, Theodore Herring, Lute Tierney, and J. L. Hastison and Valverias Young. These were the names of our little band.

As I have said, two of the Russells and Young went home. Doctor Russell, McPhadden and McKimmons, and R. J. Pierce took the teams and went to Fort Garland, then in New Mexico, now Colorado, to get supplies. They had but little money to buy with, but Doctor Russell sold his watch to an officer for provisions. With that and what little they had, they procured enough to make out on until spring. * * *

While Doctor Russell, McPhadden, and the others were gone after provisions, myself and the others who were left were preparing winter quarters. We built a cabin near the Platte about 100 yards south of the mouth of Cherry Creek, and it was joined on to by John Smith and others, making a long row of log cabins, afterward called the Indian Row, and familiar to all old-timers. McPhadden's party had built a cabin on Ferry Street.

About the time that our party got back with provisions, the emigration began to pour in from the Missouri River. The sack of dirt that Cantrell had taken to Kansas City had raised an excitement, and men were coming in from St. Louis and Leavenworth and all the points along the Missouri River so late in the fall they could only build cabins and fit temporary winter quarters. By Christmas there must have been a thousand men on the South Platte. The excitement spread like wildfire. The presses in the towns on the river were heralding to the world that great and fabulous quantities of gold had been found near Pikes Peak, thus getting up a rush and stampede permanently. As yet nothing worthy of the excitement had even been found; in fact, nothing but some little deposits near the Platte and the light float gold on Cherry Creek and a little bar on Ralston and gold on Clear Creek— these were all that had been found up to that time.

The emigrants, having gotten here too late to prospect for themselves, were disposed to discredit what we told them and believed that we really had a big bonanza awaiting the return of our comrades in the spring of 1859. So, many of them sat in their cabins writing big yarns, and many of them drawing largely on their prolific brains and writing back for truth what was nothing but a hallucination of the brain. By this time all had gone into town-site speculations. The first of these [was] Auraria, and the next Denver, [now] the first town in the great State of Colorado. We had one hundred shareholders, and I was one of them. I helped to chain the town. There were 1,280 acres included in the site. Each shareholder had 26 lots of 68 feet front, and a solid block in the bottom of low land not surveyed into lots. William H. Foster was the surveyor. About the time that we had got Auraria laid out, there came a new outfit of men from Kansas and with a full board of county officers appointed by Governor Denver, of Kansas, who had authority to organize the county and to call it Arapahoe. They being officials and old town speculators could not recognize our town, but they went to work and laid out a town of their own and called it Denver City, after His Excellency, Governor Denver. They took 1,280 acres more just east of Cherry Creek and on the hill back of McPhadden's ranch on the Platte; McPhadden had a claim in the bottom.

Their site was all beautiful table-land and really a better town-site than ours; but it was useless to grieve after spilt milk.
* * *

The winter fortunately was one of the mildest I ever saw in the mountains and everybody was busy building towns to sell in the spring. All had their pockets full of town papers. Before spring there were perhaps 20 cities in the country as large as New York, minus the wealth, population, and buildings. That is, they had the ground to build on, but nothing to build with.

Well, by this time the winter was getting pretty well past the date of 1858, which was a thing of the past, and our town shares all bore the date of 1859. Time was flying by. We had lots of big log houses built. A. J. Williams and Charles Blake had the Denver Hotel, a large building of cottonwood logs, dirt floor, and canvas roof. Old Dick Wooton had his saloon of fine logs up and the upper story rented to William N. Byers as the Rocky Mountain News Office. Auraria was ahead with the first newspaper in the country, and everybody must be there to see the first sheet in press. Denver had lots of big log houses waiting for men in the spring. Levi, St. James & Co., were selling goods on the bank of Cherry Creek on Larimer Street, while John Ming was selling them near the foot of Ferry Street. Everything was booming.

So Auraria has a paper now and the town and every city are full of literary men. So I leave the history of the towns to them, and I go to prospecting again. It is now February, and men are spending their time in hunting and prospecting a little, but nothing is discovered to justify all this town business. What was to support it all? No mines that would pay wages. Thousands of men back on the Missouri waiting for the ice to give away so that they could come to the great bonanza at Pikes Peak, as all the Eastern papers called it. We were getting our mail once a month from Fort Laramie, carried by Old Big Phil, an old mountain man. * * *

In our papers coming from the East we could get all the lies that were being published about this country and knew that they were nothing but lies. We were not surprised to see the big rush in the spring of 1859, though our little prospecting party was not by any means responsible for it.

It was now well on to spring. We had in all this addition to our company added last year's miners. John Gregory, [who shortly afterward] discovere[d] the noted Gregory lode near Mountain City, came in the early part of the spring. He was an old Georgia miner. He had been spending the winter at Fort Laramie and came here early in the spring. * * * I was told by him that he was living on venison at the time he discovered the famous lode. It was truly a fortunate and timely discovery, and it caused many other discoveries in the same neighborhood.

About this time George Jackson discovered Idaho bar on South Clear Creek. Those two discoveries caused an immediate rush into the mountains.

All this time our own little party had not left their winter quarters, and those discoveries do not belong properly to us. All the honor that we can claim is that John H. Gregory was another Georgian added to the list.

How it happened that Gregory thus dropped in at this particular crisis I never was able to account for, but his discovery was really the most important of the early days of Colorado.

Soon after some one discovered the diggings on the Boulder. I do not know who the discoverer there was, but those discoveries were all made before Green Russell came back from Georgia. There was considerable anxiety by us regarding his return, and the people of Leavenworth City had induced him to come by the Smoky Hill Route, a route which up to that time was unexplored and unknown and was supposed to be a great part of the way a desert without water and inhabited by Indians.

In the meantime the great rush of immigrants had set in, and the cry of humbug was heard all over the land. Never did I see as many disappointed men in all my life as there were at that time coming into Colorado. One day the news was that men were traveling for months on the Smoky Hill; another that the Indians had killed all of them; another, that Green Russell had been hung for getting out the great humbug. In fact all the malicious lies that could be invented were circulated, but in a few days in came Green Russell and with some 30 or 40 men, all in perfect health and fine spirits. They only camped at Denver a day or two and then went into the mountains and in a short time they had found Russell Gulch, about 2½ miles west of the Gregory Road.

Russell Gulch proved to be one of the best towns in Colorado for some time after his coming.

On the Smoky Hill route in a short time the Holaday stage line was established and made regular trips for that year. Denver seemed to flourish and was at once the metropolis of the Rocky Mountains and ever since has been the leading town of Colorado.

According to the story given to Rickard [5] by James Andrew O'Farrell, who says that he was a member of the Russell expedition of 1858,

a party of six Georgians, in the summer of 1849, were taking a herd of thoroughbred horses across the continent. This party consisted of Green Russell [William Green Russell], and his younger brother Dr. Levi J. Russell, A. T. Lloyd, G. W. Kiker, Charles Kiker, and P. H. Clark. They reached Camp Lyon, on the Arkansas, in October, and were there persuaded by James Dempsey, a Government guide, that it was too late in the season to cross the mountains. Moving northward they established a winter camp at the junction of Cherry Creek and the Platte. There they built two cabins on a sand bar on the south side of the river, and during the rest of the year they prospected the alluvial banks of Cherry Creek, but they did not penetrate the mountains for fear of the Indians. They found gold at several places along the creek, particularly at a point 16 miles upstream from its mouth [probably in Newlin Gulch in alluvial gravels derived from the Dawson arkose beds or the Castle Rock conglomerate beds],[6] and they preserved small quantities of the gold dust in quills made of feathers of wild geese that they had shot.

Early in 1850 the party crossed the range by Bridger Pass, in Wyoming (then in Utah), and went on to California. They occasionally mentioned to others the gold they had found in "western Kansas," and in proof of their story they showed to them the gold-filled goose quills. They mined gold successfully near Downieville, Calif. In the spring of 1857 this party and others sold out their interests in California and returned to Georgia, but before they separated they agreed that in the near future they would form a prospecting party to go to "western Kansas" and search for gold.

In May, 1858, the six men who had mined gold on Cherry Creek and five others met at the Planter's House, in St. Louis, to carry out the agreement thus made. The five new members of the party were J. A. O'Farrell, three men named Chastine, and a man named Fields. All except two were old Californians. Having organized, they went to Leavenworth by water and thence to Camp Harney [obviously Camp Kearney, Nebr.], along the military road. Late in July they left this

[5] Rickard, T. A., The development of Colorado's mining industry: Am. Inst. Min. Eng. Trans., vol. 26, pp. 834-848, 1896.

[6] See also description of localities of discoveries on Cherry Creek in 1849 and 1858 in Richardson, G. B., U. S. Geol. Survey Geol. Atlas, Castle Rock folio (No. 198), p. 12, 1915.

frontier post, accompanied by an escort of 20 men under the command of Captain Lyon. In August [7] the party reached the log cabins at the junction of the Platte and Cherry Creek, where the banks of the creek were covered with the wild cherry trees that had suggested its name. As soon as they had established camp they went to the places where Russell and his friends had found sufficient gold to encourage them in 1849 [at Russellville Gulch, 5 miles southeast of Franktown; Ronk Gulch and Gold Run, southeast of Elizabeth; and Newlin Gulch, near Parker].

Hollister says: [8]

It was the commercial collapse of 1857 that set many adventurous spirits in the then West peering into the obscurity beyond them for a new field of enterprise.[8a] A party of Cherokee Indians, traveling overland to California in 1852, via the Arkansas River and along the base of the Sierra Madre to the North Platte at Fort Laramie, by some means found gold in the banks of Ralston Creek, a small affluent of the South Platte, emptying into it near its mouth; and each year thereafter parties of Cherokees had gone out and prospected the streams in the vicinity of what is now Denver City. At last they were successful; they obtained a few dollars worth of the glittering dust, which they carried home late in 1857, exhibiting it freely as they passed through Nebraska and Kansas.

The report of a new land of gold in the West spread like an epidemic through the country drained by the Missouri River, and soon traveled far beyond. These Indians appear to have gone home and told their story on the confines of the Gulf of Mexico, for Georgians were among the first to seek the new gold country.

On the 9th of February, 1858, W. G. Russell, with a party of nine men left the State of Georgia with a view of prospecting the eastern slope of the Sierra Madre along the heads of the South Platte, from Pikes Peak to the Black Hills. They arrived on the head of Cherry Creek about the first of June. They prospected Cherry Creek, the Platte, and the affluents of the Platte as far north as Cache la Poudre Creek without finding attractive gold placers. They returned to the Platte, and about 5 miles up a small dry creek that joins the Platte from the east 7 miles south of the mouth of Cherry Creek they found gold and panned several hundred dollars' worth. As soon as the work was well begun some of the party [including William Green Russell], returned to Kansas with the news. As the nearest notable object was Pikes Peak the new gold region was named from that mountain.

In the last days of the summer of 1858 several small parties left Lawrence, Plattsmouth, Omaha, Florence, Bellevue, Council Bluffs, and other towns on Missouri River for the new Pikes Peak district. On the 23d of October several parties, including Dr. [Levi John] Russell's, were encamped at the mouth of Cherry Creek. Three miles farther up the Platte, on its west [?] bank, were the "Mexican diggings" [9]—gold diggings worked by Mexicans. Parties were constantly arriving and were vigorously prospecting, though they were not very successful, only Dry Creek yielding gold in quantities large enough to be encouraging. The camp at the mouth of Dry Creek was called Placer Camp. [Placering was done on the Platte from what is now Alameda Avenue and the Platte, in Denver, where a dry creek entered the Platte, to the mouth

of the dry creek which flows into the Platte north of Petersburg. Placering also had some success up both these dry creeks, the most southerly one passing through the town of Englewood.—C. W. H.] The divide between the Platte and the Arkansas was then covered with snow, and on the 31st of October 10 inches of snow fell at the mouth of Platte Canyon. By the 4th of November a town plat had been surveyed, on the west side [of Cherry Creek near its junction with the Platte] by William Foster and christened Auraria by Doctor Russell, whose party had come from a town of that name in Lumpkin County, Ga., where gold is still found. This region was then within the bounds of Kansas, and a county was defined and called Arapahoe, after the neighboring tribe of Indians. Arrivals from the States continued. Early in the winter a town called St. Charles was laid out [on the east bank of Cherry Creek] opposite Auraria, by a party from Lawrence, most of whom returned home. The governor of Kansas Territory at that time was James William Denver, and the interests of St. Charles were left in the hands of a man who sold out to the Denver Town Co. So, through a natural process of aggregation, Denver was slowly established.

In August, 1858, George A. Jackson did some prospecting about Vasquez Fork (Clear Creek), and in the winter of that year reached the site of Idaho Springs, where, on January 7, 1859,[10] according to his diary on file in the State Historical and Natural History Society of Colorado, he found colors and one nugget at the mouth of a branch of South Clear Creek. Because of a heavy storm, he marked the spot and returned to Golden and in April returned with friends who named the branch Chicago Creek, after their home city. The gravel deposits became known as Jackson diggings.

The most complete story of the earliest days of mining in Colorado is given by Smiley,[11] whose details, obtained from Dr. Levi J. Russell, check well with the story told in Mr. James H. Pierce's manuscript, as follows:

When the extravagant accounts, based on practically nothing—because nothing beyond the old, time-worn stories we have recounted, was actually known—of the alleged "new gold field at Pikes Peak" were again put into circulation by newspapers and word-of-mouth late in the autumn of 1857, W[illiam] Green Russell, who was then at his home in Georgia, resolved to go to the Peak on a prospecting tour. After some correspondence with the Indian Territory people it was decided jointly to organize such an expedition, a part of the men to be from Lumpkin County, Ga., and others from the Cherokee Nation. This was the earliest movement toward making an organized attempt by experienced men to search systematically for gold in this Rocky Mountain country. Of the ensuing proceedings we quote from Doctor Russell's statement made for this History [of Denver]:

"A party was finally made up in Lumpkin County, consisting of the following-named persons: W. G. Russell, J. O. Russell, L. J. Russell, Lewis Ralston, William Anderson, Joseph McAfee, Solomon Roe, Samuel Bates, and John Hampton. This party left home on the 17th day of February, 1858, and expected to be joined by the Cherokee party at the Cherokee town of Maysville, near the northeast corner of the Cherokee Nation's territory. When we reached Maysville we found that the other party was not ready to move. After a conference with the

[7] The Russell party arrived about June 1, according to Hollister. On June 24, according to Smiley. Arrived here May 23, left for Ralston Creek, returned to mouth of Cherry Creek June 24, according to James H. Pierce.

[8] Hollister, O. J., The mines of Colorado, Springfield, Mass., 1867.

[8a] Confirmed in a personal interview May 22, 1924, with Mr. Hal Sayre, of Central City and Denver, who found himself, a civil engineer, out of work as the result of the panic of 1857 and started from Dubuque, Iowa, in 1859 for Pikes Peak and arrived with many others at Auraria June 1, 1859.

[9] Smiley locates the "Mexican or Spanish diggings" on the east bank, at West Virginia Avenue, six blocks south of Alameda Avenue.

[10] Hollister says that the discovery was made on Apr. 1, 1859.

[11] Smiley, J. C., History of Denver, with outlines of the earlier history of the Rocky Mountain country, pp. 184-197, 1903.

leader of the Cherokee party the Rev. John Beck, a Baptist preacher, and a future time and place of meeting agreed upon, our party moved on to Rock Creek, Kans., where W. G. Russell had taken up land in 1857, and where we went into camp and remained several days. J. H. and R. J. Pierce joined us at Rock Creek, and from that point W. G. Russell, J. O. Russell, William Anderson, Lewis Ralston, and Joseph McAfee went to Leavenworth and exchanged our mule teams for oxen, and procured a general outfit. Upon their return we left Rock Creek and moved to Manhattan, Kans., where we camped for two or three days. Here we were joined by William McFadding, William McKimmons, Jacob Masterson, Valarias Young, Theodore Herring, J. Brock, Luke Tierney, T. C. Dickson, George L. Howard, and a Frenchman whose first name, Henry, is the only one I remember. Leaving Manhattan we crossed the Kansas River near that place and struck out for the old Santa Fe trail on the Arkansas, reaching it near the Great Bend. Here we came up with Mr. Beck and some 30 other Cherokees, on June 3; they having arrived somewhat in advance of us and we having seen signs of their presence ahead of us for a couple of days."

Beck's contingent had been joined by several white people after it started. These were George McDougall, a brother of Senator McDougall, of California; Philander Simmons an experienced mountaineer, * * * a mountain and plains man of some experience. He had been attached to Bent's Fort as early as 1842, and was familiar with all the region hereabouts. * * *

The united companies made an imposing caravan, which consisted of 33 yoke of cattle, 14 wagons, 2 two-horse teams, a dozen or two of ponies. * * * But not all of the people were bound out to dig gold. * * *

On June 12 the expedition reached Fort Bent, where a brief stop was made. On the 16th of June the Russell-Cherokee part of it, including Simmons and the men who had in Kansas directly attached themselves to the expedition, left the Arkansas and struck across the country by the way of [Black] Squirrel Creek [which enters the Arkansas 12 miles east of Pueblo] to the South Platte-Arkansas divide. Beck claimed that the Cherokee party of 1850 had found gold in the hills down there while passing through them. Russell found nothing, and after a day or two of labor [on the divide], crossed over to Cherry Creek, and all the party began prospecting as soon as they struck it, and continued along its bed without finding anything of importance. Then it was decided to move down the creek to the South Platte. They came in here on June 24, 1858 [now the date of the Colorado Pioneers annual celebration of the arrival at the mouth of Cherry Creek], and crossing the creek at a point between the present Blake and Wazee streets, went into camp on the west side. * * * When the party went into camp on the 24th of June there was neither a white man nor the sign of one around the mouth of Cherry Creek; nor were there any Indians within sight.

The Russell party came nearly not being the first organized company of gold seekers in the Pikes Peak country. Not far behind it was another band, almost as large, winding its way across the plains. The new rumors of gold being in plenty at the Peak, or new versions of the old ones, had reached Lawrence, Kans., in the winter of 1857-58, as they had every other place along the Missouri border, and were so persistent as to excite great interest in the town. * * * A meeting of adventurous younger citizens was held, and a company organized by John Easter to go at once to Pikes Peak. These Lawrence men, who constituted what is known in the early annals of Denver as "the Lawrence party," * * * started on their long journey on May 19. This party comprised Albert W. Archibald, A. F. Bercaw, Giles Blood, Frank Bowen, Joseph Brown, W. J. Boyer, William Chadsey, John A. Churchill, Frank M. Cobb, William Copley, ——— Cross, John Easter, Adnah French, Peter Halsey, William Hartley, Josiah Hinman, Mrs. Anna A.

Holmes, Charles Holmes, George Howard, Howard Hunt, Roswell Hutchins, "Pap" Maywood, William McAllen, ——— McKay, John D. Miller, Austin R. Mills, Robert Middleton and his wife and child, Charles Nichols, William Parsons, George Peck, Robert Peebles, William Prentiss, William Regan, Charles Runyon, George W. Smith, Nicholas Smith, J. H. Tierney, Jack Turner, Augustus Voorhees, James H. White, A. C. Wright, and Jason T. Younker: with J. H. Tierney as the guide, or leader.

This Lawrence party also came out by the Arkansas River route. The Lawrence party followed the Arkansas to the mouth of the Fontaine-qui-Bouille [Fountain Creek] and then turned northward and on July 5 went into camp at Pikes Peak, which they fondly believed marked the location of the fortunes awaiting them. * * *

It is to be remembered that at this time this region here at the base of the mountains was not altogether unfrequented by white men. While the Russell and Lawrence parties were the only organized parties out here on the search for gold, there were several individuals and little parties of two or three persons in the country who knew nothing of the larger organizations until they fell in with them here along the South Platte. Furthermore, the first overland stage line was making occasional trips between the Missouri River and Salt Lake, by the way of Fort Laramie, and at longer intervals a wagon train would pass this way [by way of Denver] on a journey between New Mexico and Utah, or by the Arkansas River route to and from the Missouri River. * * *

The departure from the Missouri border of these two large companies bound for the alleged "new gold regions" was soon noised abroad. Wildly exaggerated reports of what was going on out at the base of the mountains were again in circulation that spring and summer [1858] along the Missouri and through the States and led to the organization of many other parties in the border towns to follow these pioneers. * * *

Philander Simmons, mentioned by Smiley as a member of Russell's party in 1858, says [12] that Russell did not come to Cherry Creek in 1849 and that 1858 marks his first trip to the Pikes Peak region.

The numerous histories seem to agree in the main in regard to the series of events that began in 1859. Rickard says that by the end of 1858 rumors of rich diggings had crossed the plains, the rush had set in, and crowds had begun to arrive. Hollister recounts in more detail some of the events that followed 1858, and the history given below is taken, with slight changes, from his work. [13]

Very early in 1859 the citizens of Auraria, in Arapahoe County, Kans., began to scatter out in a search for gold. After thoroughly prospecting all the streams in the region they decided that Clear Creek (Vasquez Fork of the South Platte), was the richest. Diggings were therefore begun on that stream 3 or 4 miles east of the gap between North and South Table Mountains, and at that place there soon sprang up a town called Arapahoe. At one time this town must have consisted of 50 houses, but in 1867 not one remained to mark its site. A little higher up on the stream, just where it fairly escapes from the foothills, another town was soon begun, called Golden City,

[12] Rockafellow, B. F., History of Fremont County, in History of the Arkansas Valley, Baskin & Co., 1881.
[13] Hollister, O. J., op. cit.

named for Tom Golden, its founder. In the later part of 1859 this town was built up rapidly, reaching, in the summer of 1860 the highest point of prosperity it attained during the years 1859–1867. Diggings were also opened and worked successfully on Ralston Creek, a small tributary of Clear Creek. The creek bars were a mass of boulders of all sizes, and it was soon observed that the gold there always occurred in scales, like flattened shot, which indicated to the prospectors that the gold had come from farther up the creeks.

Other placer gold had already been found in the mountains, too, at that time. About the end of January, 1859, B. F. Langley discovered some rich placers or bars in a gulch on South Boulder Creek, which was full of fallen timber, so that the place was called Deadwood diggings. This gulch is described by Hollister as at the junction of Gamble Gulch and its tributary, Lump Gulch, but the present Lump Gulch is parallel to Gamble Gulch; it is not a tributary of it. His location would appear to be near the present site of Rollinsville. Gold Dirt and Perigo are farther up Gamble Gulch. The map prepared in 1860 by S. W. Burt and E. L. Berthoud, mining and civil engineers, of Central City and Golden City, shows Deadwood diggings as on South Boulder Creek near the mouth of North Beaver Creek, and Jefferson diggings several miles up North Beaver Creek. These places could have been reached by starting from the camp of Ralston, following Ralston Creek to its most northerly source, and crossing the divide into Lump and Gamble gulches or by prospectors following Left-hand Creek and crossing the ridges into South Boulder Creek. By the end of March, 1859, a number of men were mining at Deadwood and took out considerable gold.

All the gold mined thus far had been found in stream sand and gravel, but on May 6, 1859, John Hamilton Gregory discovered the outcrop of the gold lode on North Clear Creek that was distinguished by his name. Gregory left his home in Gordon County, Ga., in 1857, and in 1858 he drove a Government wagon from Leavenworth to Fort Laramie, which stood at the junction of the Big Laramie and the North Platte, where he remained for several months. Meantime he heard that gold had been discovered on the South Platte, and early in January, 1859, he started alone on a prospecting tour southward along the base of the mountains. He found nothing of value until he reached the camp at Auraria and started prospecting the Vasquez Fork (Clear Creek) of the South Platte, which he followed upstream, his plan being to prospect thoroughly wherever the creek forked and to follow up the branch that gave the most promise. In this way he went up Clear Creek Canyon to the main forks, 14 miles above Golden, then up the north branch 7 miles to the gulch that bears his name, where he found promising indications, but a heavy snowstorm prevented further work. He returned to the valley for provisions and prevailed on Wilkes Defrees, of South Bend, Ind., to accompany him back to the mountains. The two reached Gregory Gulch after a tedious journey of three days and discovered what was soon after known as Gregory No. 5. They returned to the valley to get their friends, who with teams went laboriously (there was no road then) zigzagging up the north bank of Clear Creek over into Eightmile Canyon above Golden Gate, to Guy Hill, over to Ralston Creek, and down Dory Hill to North Clear Creek. Each member of this group of men except Gregory located a 100-foot claim on the Gregory Lode, Gregory being given two claims, Nos. 5 and 6. From the 16th to the 23d of May, Gregory worked five hands on one of the claims with a sluice and took out $972. He soon afterward optioned claims Nos. 5 and 6 to E. W. Henderson and A. Gridley for $21,000 and began to prospect for other properties at $200 a day and left in September, 1859, for the East with $30,000 in gold (so it is said), to return again in 1860.

Hollister's statements as to the quantity of gold washed and the results of hand labor alone show that in this district, as in other mining districts in the West, the rewards of many of the miners in those days were rich. Gregory advanced Henderson & Gridley $200 with which to commence operations, and their first four days' labor with a sluice brought them $607. During the summer they took out $17,000 and had besides a large pile of quartz and sluice headings and tailings, which they sold for $7,000 to Gregory & Reese, who were then operating a rude quartz mill on North Clear Creek.

In July, 1859, prospectors had found their way into the South Park, and in August extravagant stories were told of the richness of the newly discovered deposits in the South Park and in the mountains on the Cache la Poudre. The consequence was a stampede from the Gregory workings in both directions. Of the workings on the Cache la Poudre little has since been heard, but in South Park gulch and bar mining were carried on with fair success on all the affluents of the south fork of the South Platte. The towns of Montgomery, Buckskin [Laurette], Mosquito, Fairplay, Tarryall, Hamilton, and Jefferson sprang into being. At Tarryall, in one week in September, the Rocky Mountain Union Co., with four hands, took out $420. In one week Bowers & Co., with three hands, took out 57 ounces. In one week W. J. Holman, with five hands, took out $686. The discovery of placer gold was followed by the discovery of rich lodes from Mosquito Creek to the head of the Platte, where the town of Montgomery was located.

Many immigrants who arrived late in the season by way of Arkansas River went straight by either the northern or the southern base of Pikes Peak to the mines in South Park through a succession of delightful parks, which furnished a natural and easy route. Hence arose Canon City on the Arkansas, at the entrance to the mountains, and Colorado City on Fountain Creek, 40 miles north of the Arkansas, almost in the shadow of Pikes Peak.

In August and September, 1859, parties crossed into the Middle Park from Montgomery and Hamilton and discovered both placer and quartz gold on the upper reaches of Blue River and its tributaries. That year— in all the gulch and bar diggings in South Park; on South Clear Creek; on North Clear Creek, including Russell, Illinois, and Nevada gulches; on the several forks of Boulder Creek; on main Clear Creek below Golden City; on Ralston Creek; and on the Platte above Denver—miners at work were making from $3 to $5 a day to the hand. Others were finding gold on the Arkansas from Canon City to Tennessee Pass. Men were prospecting from the head of the Del Norte to the head of the Big Laramie, ranging west as far as the mouth of White River.

Gold was discovered in Hahns Peak district, Routt County, in 1865, and active work was begun there in 1866.

CHRONOLOGIC RECORD

1807

Pursley reports to Pike at Santa Fe that gold occurs on the South Platte, to which "he had refused to lead Spaniards." Nothing resulted from this report.

1848

Gold said to have been found near site of Lake City, Hinsdale County, by a member of an exploring party in command of J. C. Frémont, but the spot is unmarked and was unheralded.

1849, 1850, or 1852

Party on their way to California, which included men who were acquainted with gold placer mining and some Cherokee Indians, found gold on branches of the South Platte, Cherry Creek and Ralston Creek, a tributary of Clear Creek, and took away small quantities of it in goose quills.

1858

Members of party making discovery in 1849–1852 and others, including many with experience in placer mining in Georgia and California, under command of William Green Russell, of Auraria, Ga, having outfitted at Leavenworth, Kans., according to Levi J. Russell, James H. Pierce, and others of the party, go up the Arkansas to Fort Bent, and turn off the Arkansas east of the present site of Pueblo, up Black Squirrel Creek, over divide into Cherry Creek and begin prospecting for gold. Find colors all along Cherry Creek, with a slightly better showing about 16 miles above its mouth (probably at Newlin Gulch, near Parker) and colors at the mouth of Ralston Creek. Go as far north as Boulder and Big Thompson creeks and south to Bear Creek and find nothing of importance. Conclude to go back across the Platte. Start up the Platte on east bank for Platte Canyon and 3½ miles above mouth of Cherry Creek in a bed of alluvial gravel with a conglomerate bedrock, find gold and rock out about $200 in thin scale gold. In a bar up Dry Creek, about 2 miles from the river, find another little deposit, richer than the one on the Platte.

Traveler [Cantrell] from the west comes by and takes sackful of gravel from deposit on Dry Creek. He takes gold by way of Arkansas River to Kansas City (Westport) and has it panned before witnesses, helping the rush already started.

Captain Dice comes through, goes on to Taos, N. Mex., meets a party from Lawrence, Kans., which had left Lawrence intending to join Russell-Cherokee party but meeting those returning has gone southwest to Raton Mountains, thence north to Pikes Peak.

Russell party make a camp called Placer Camp, just north of Petersburg, back a little way from mouth of Dry Creek. In August, 1858, Russell party goes north as far as Medicine Bow Mountains. Turn back to mouth of Cherry Creek. William Green Russell, Joseph Oliver Russell, and Valerias Young return to Leavenworth for provisions and reinforcements. Others of party go to Fort Garland, then in New Mexico, for provisions. Remainder, joined by John Smith, Indian trader, build cabins near the Platte about 100 yards south of the mouth of Cherry Creek, joined by cabins of John Smith and others, later called Indian Row. About time party returned from Fort Garland with provisions immigrants began to pour in from Missouri River. By Christmas about 1,000 men on the South Platte. Town-site locations begin.

Montana City laid out in September by Lawrence party, between Platte River, West Evans, South Delaware, and West Iliff streets, south of Overland Park, Denver, and about 20 log houses built. Some men wintered here, but the place was overshadowed by Auraria City and Denver City and was abandoned in 1859.

Auraria City laid out in October by the Georgians (Levi J. Russell, secretary) on west bank of Cherry Creek near its junction with the Platte, at site of West Denver.

St. Charles laid out by Lawrence party, in September, on east bank of Cherry Creek, but no buildings erected. Most of Lawrence party return down the Platte route. Denver City laid out in November and December on St. Charles site by party of officials from Leavenworth sent out by Governor James William Denver, of Kansas. Streets paralleled approximately course of Cherry Creek.

The name of Auraria City (Russell had also presented name of Dahlonega, from the town of that name in Georgia, where there was once a branch of the United States Mint) was dropped upon completion of the first bridge across Cherry Creek at Larimer Street April 6, 1860.

The "City of Highland" or the "Highland Town Company," across the Platte, was organized in the summer or fall of 1859, and some part if not all of the site was surveyed and staked off. The village of Highlands was a later growth and a different enterprise from the pioneer Highland project.

Other parties arrive at Cherry Creek, some up the Platte route and others up the Arkansas and Fountain and down Cherry Creek, some of the latter perhaps crossing from Cherry Creek [Franktown] to Plum Creek and down the Platte.

1859

The rush to the Cherry Creek or Pikes Peak gold region continues.

Prospectors find gold along Platte from Alameda Avenue (Denver) to mouth of the dry creek that empties into the Platte six blocks south of Alameda Avenue and southward to the dry creek that empties into the Platte just north of Petersburg and up both dry creeks, some gold being found 12 miles up the most southerly creek (the one that flows through Englewood).

Gold Run, near Gold Hill, Boulder County, perhaps first discovered late in 1858, found to be rich early in 1859. The "Pikes Peak region," the destination in the minds of all, although Cherry Creek diggings was their objective point, become the lodestar of a large number of adventurous men, most of them ignorant of mining, but hopeful of improving their financial condition, hard hit by the depression of 1857–1859. Diggings begun on Clear Creek, just east of Table Mountains, and settlement of Arapahoe built. Golden City built higher up the stream. Diggings begun on Ralston Creek.

George A. Jackson, on January 7, finds gold on South Clear Creek, at the mouth of what is later called Chicago Creek.

Placer gold found at Deadwood diggings, at the mouth of North Beaver Creek and South Boulder Creek, near the present railroad station of Pactolus, on the Denver & Salt Lake Railroad (Moffat road), and in June in Lump Gulch and Gamble Gulch, tributaries of South Boulder, probably near site of present settlement of Rollinsville, Gilpin County.

Placer gold found on headwaters of Arkansas River, chiefly in Cache and Clear creeks, now in Chaffee County, early in 1859.

John Hamilton Gregory on May 7 finds rich, easily worked, decomposed, oxidized surface deposits of lode veins on North Clear Creek, soon called Gregory diggings, near the spot where later rose the town of Blackhawk.

William Green Russell returns with provisions for his party, coming by way of the old Cherokee trail, then known as Smoky Hill route.

William Green Russell discovers, in the first days of June, placer gravels and residual decomposed, oxidized surface outcrops of lode veins in what is now Russell Gulch.

Pattern of mining districts and miners' laws roughly outlined and adopted at Gregory diggings June 8.

Provisional local government formed at Gregory diggings July 9 and 16.

First arrastre built in Gregory diggings to crush quartz in July, 1859. Placer gold found at Buckskin, Mosquito, Hamilton, Tarryall, Montgomery, and Fairplay, on branches of the south fork of the South Platte, in the northwest part of South Park (July to August).

Placer gold found in Georgia Gulch and other gulches on the headwaters of the Blue and the Swan, over the mountain passes from the South Park, in August and September.

George Griffith discovers gold in place near Georgetown August 1, but does not pursue discovery. First steam quartz mill erected in Gregory diggings in September, 1859. September 29 a foot of snow in Gregory diggings. Stampede to the plains, but about 1,500 stay.

Placer gold found at junction of California Gulch and Arkansas River (now in Lake County) late in 1859.

1860

A second rush, larger than the one in 1859. Most of the immigrants head for Gregory diggings, which they find overcrowded, some obtaining work of some description there and others (with even earlier arrivals) spreading in every direction.

First stagecoach from Denver to Gregory district arrives over Golden Gate-Dory Hill road March 4.

Mining laws of Gregory district codified and amplified.

Horace Greeley, Albert D. Richardson, and Henry Villard, of the New York Tribune, the Boston Journal, and the Cincinnati Commercial, respectively, visit the Pikes Peak country.

On July 4 consolidated ditch completed, bringing water from Fall River to Russell Gulch.

California Gulch (now Lake County) scene of big rush to rich placer diggings. Placers found in McNulty Gulch, Tenmile district, Summit County.

Clark, Gruber & Co. build in Denver a banking, coining, and assay establishment, which was purchased by the United States in April, 1862.

Disastrous expedition of Baker up the Rio Grande into Bakers Park (now Silverton). Deposit at Phillips mine, Buckskin Gulch, Park County, discovered.

1860–1865

Much prosperity at Empire, owing to ease of sluicing gold of surface deposits.

1861

Only a slight immigration during the year.

Congress organizes Territory of Colorado February 26 with William Gilpin as governor.

First Territorial legislature of Colorado meets September 9. Creates seventeen counties, fixes the county seats, and upholds the mining laws of local mining districts.

Much experimentation in Gregory and other districts of Gilpin County in treatment of ores by various forms of crushers, by steam, by fire, and by chemicals,

Whale, Lincoln, and other lodes at Idaho Springs producing in a 20-stamp mill.

1862–63

Slight immigration. Effects of Civil War seen. Two regiments enlisted in Colorado and moved south. Some miners go home to join either the North or South. The members of the Russell party try to return to the South but are made political prisoners.

Mills of Gregory district work only half successfully. Losses of gold heavy, perhaps 75 per cent.

1863

Great speculation in Gregory mine by investors in New York and Boston.

Good gold recovery in small mill at Georgetown.

1864

A 40-stamp mill and reverberatory furnace for gold ores built at Georgetown.

Mining nearly at a standstill at Gregory.

Discovery of silver early in 1864 (Coaley claim, on Glacier Mountain, near Montezuma, Summit County) draws attention to headwaters of South Clear Creek. In September the Belmont lode, on Mount McClellan, 8 miles above Georgetown, is discovered—the first paying silver mine.

1865–66

Repetition of 1864 at Gregory.

1865

Placer gold found at Hahns Peak.

Rush to Argentine silver district, above Georgetown.

1866

Elizabethtown and Georgetown both in their making; later joined to form Georgetown. Idaho Springs mines mostly idle.

Machinery arrives at Florence to bore holes for oil. Plans for prospecting for oil near Golden.

1866–67

Gold ores of Griffith district and silver ores of Argentine draw attention to Georgetown.

1867

Placers in Gunnison County discovered.

Union Pacific Railroad completed from Omaha to Cheyenne.

Hill smelter, or Boston & Colorado Smelting Co.'s experimental plant, erected at Blackhawk in June.

Placers in California Gulch (Lake County) about exhausted.

1868

Easily worked rich placers and surface deposits of South Park mines about gone. Sulphides in lode mines giving trouble.

Dives-Pelican silver lode, Silver Plume, discovered.

Burleigh tunnel, Silver Plume, started by hand.

Burleigh drills introduced at Burleigh tunnel, Silver Plume, early in 1869. Power drills first used for driving a mine adit in the United States.

Deposit at Ward, Boulder County, discovered.

Gold lodes (leaf gold in nests of lead carbonate) worked at Leadville.

Hill smelter open for business in January; first shipment of matte made in June. This smelter revived the waning mining industry of Colorado.

1869

Silver at Caribou, Boulder County, recognized.

1870

Denver Pacific Railroad (Denver to Cheyenne, Wyo.) completed in summer. Kansas Pacific (now Union Pacific) Railroad reached Denver by Smoky Hill route in August.

Colorado Central Railroad between Denver and Golden completed in September, 1870.

Gold discovered in Wightman's Gulch, Summitville district, Rio Grande County.

Gold found in Arrastra Gulch, near Silverton.

1870–1880

Georgetown and Silver Plume grow fast.

1871

Lodes discovered in the Archean granite in Tenmile district, Summit County, but no development until 1879, and that in deposits in limestone.

Denver & Rio Grande Railroad (now Denver & Rio Grande Western) built south from Denver to Colorado Springs. Ultimate destination the City of Mexico.

Ute and Ulay deposits discovered at Lake City.

Hukill mine at Idaho Springs opened. Silver discovered on Lincoln and on Bross Mountains, Park County.

1872

Homestake mine, near Tennessee Pass, shipping.

In November lead blast furnace of Mount Lincoln Smelter Works (Dudley smelter) built. (Changed to copper reverberatory in August, 1873; ran until January 25, 1874.)

Small blast furnace in operation at Montezuma, Summit County.

Small matte smelter in operation at Whale mine near Idaho Springs.

Silver discovered in Elk Mountains, Gunnison County.

Golden City Smelting Works (Bagley & Sons) built at Golden (a blast furnace).

Small Mexican smelting furnace in operation at Rico.

Colorado Central extended from Golden to Blackhawk and branch from Forks to Floyd Hill.

Mount Lincoln Smelting works at Dudley, Park County, completed in November.

Petzite discovered at Gold Hill, Boulder County.

1872–73

Swansea lead smelter (8 tons in 24 hours) operating near Empire.

1873

Branch smelter of Boston & Colorado Smelting Works erected at Alma, Park County. Operated through the years 1873–1875.

Small reverberatory furnace built at Lincoln City, French Gulch, near Breckenridge, which did not prove successful. In 1874 a blast furnace substituted, not satisfactory, so two Drummond lead furnaces added; successful for four months of 1874 on ore from Cincinnati lead mine.

Denver Smelting Co.'s lead smelter built at Denver.

1874

Rush into San Juan country. Hotchkiss (Golden Fleece) deposits discovered.

Gold discovered at Sunshine, Boulder County. Silver at Humboldt-Pocahontas mine, Rosita, Custer County.

The South Park Railroad built from Denver to Morrison.

Two unsuccessful smelters built at Silverton.

Boyd smelter built at Boulder.

1874–75

Smelter at Halls Gulch, Park County, tried out.

1875

Grand View mine, half a mile north of Ouray, located.

Lake City, Hinsdale County, founded. Crooke smelter built.

Deposits at Salina, Magnolia, and Jamestown, Boulder County, discovered.

Malta smelter built at Malta, at mouth of California Gulch.

Mary Murphy mine, on Chalk Creek, Chaffee County, being opened.

Greene smelter built at Silverton. Active to 1880. Moved to Durango, where forerunner of present smelter was built later.

Collom lead smelter built at Golden.

Golden Smelting Co.'s smelter altered to a copper matting plant.

Smuggler deposit, above Telluride, discovered.

1876

Colorado Dressing & Smelting Co.'s matting plant began operation at Golden.

Virginia Canyon mines, between Russell Gulch and Idaho Springs, brought to notice by the opening of the Specie Payment mine.

South Park Railroad being pushed up north fork of the South Platte, with the San Juan country in view as its final destination.

Discovery made that heavy sands in sluice boxes, which had for years bothered placer miners in California Gulch, were rich lead carbonates.

1877

Una and Gertrude claims of Camp Bird, Ouray County, staked.

Discovery of lodes of lead carbonate leads to great rush to California Gulch, and the establishment of Leadville, which became the county seat of Lake County. As a consequence of this discovery Lake County has produced the greatest variety of metal in Colorado and has the record of the greatest total value of metal mined in all counties, amounting roughly to one-third of the grand total value for the State.

Leadville smelter built at Leadville.

Colorado Central Railroad reached Georgetown.

Bassick gold deposit, at Querida, Custer County, discovered.

1878

Grant built smelter at Leadville. Small smelter built at Crested Butte.

Harrison Reduction Works, of St. Louis, Mo., build smelter at Leadville.

La Plata smelter built at Leadville.

Great Monarch claim and Madonna deposit discovered at Monarch, Chaffee County.

Silver discovered at Robinson and Kokomo in Tenmile district, Summit County, during the summer.

Mines at Silver Cliff, Custer County, and Bull Domingo mine opened.

1878-1883

Crooke Smelter, at Lake City, in operation.

1879

Eilers & Billing built Arkansas Valley smelter at Leadville. Little Chief; Ohio & Missouri; Cumming & Finn; Gage, Hagaman & Co.; Raymond, Sherman & McKay; Elgin; and Adelaide smelters built at Leadville. Opening of mines at Rico, on deposits that had been known for several years.

Discovery of Belden lode in Eagle County, near Red Cliff.

Bonanza, Saguache County, opened by rush into Gunnison County. Gold discovered in Independence district and silver at Aspen, Pitkin County. Ore from Perigo mine, Gold Dirt district, Gilpin County, yielding well under stamps. Deposit at Terrible mine, Ilse, Custer County, discovered. Robinson-Kokomo district, Summit County, outcrops opened by shallow shafts.

1879-80

Ocean Wave smelter operated at Lake City.

1880

At Leadville 11 or 12 smelters operating, several ore-buying firms, and 4 stamp mills.

Robinson-Kokomo district, in Summit County, produces $200,000 in silver and lead.

Unsuccessful Eclipse smelter built at Animas Forks, above Silverton.

Grand View smelter built at Rico. Ran a few years.

Three smelters in operation at Golden.

Five small and inefficient smelters at Garfield, Maysville, Poncha Springs, and on Chalk Creek, Chaffee County.

Rich lodes discovered in French Gulch, near Breckenridge, in Summit County; the source of the placers that had produced in 1859 and later.

Two smelters built at Tincup (short-lived). Small smelter at Gothic.

Denver & Rio Grande Railroad reached Leadville.

Three smelting plants in operation at Golden.

1881

Virginius mines, at head of Canyon Creek, Ouray County, worked by three levels and two shafts.

Denver & Rio Grande Railroad reached Crested Butte.

Greene smelter moved from Silverton to Durango and operated.

Deposits at Red Mountain and Ironton, Ouray County, discovered.

Smelters in Summit County are White Quail lead smelter and Greer lead smelter at Kokomo; Summit two-blast lead furnace at Robinson; Wilson matte smelter, Boston & Colorado (two stacks), and Elyria (two stacks) at Breckenridge; Lincoln at Lincoln City, above Breckenridge; Sissapo, at Montezuma; Battle Mountain, two stacks, at Red Cliff [now Eagle County.]

1882

Denver & Rio Grande Railroad reached Silverton.

Present Durango smelter opened for business.

Grant smelter, at Denver, built. Permanently closed at time of strike in 1903.

Unsuccessful Martha Rose smelter built at Silverton.

Pueblo plant built at Pueblo by Mather & Giest.

Second smelter at Rico, later purchased by Pasadena Co., operated for two years.

Smelter built at Aspen. Ran until 1887, could not survive on ore of one district.

1883

Silver mines of Breckenridge opened up.

Denver, South Park & Pacific Railroad (later part of Colorado & Southern) reached Leadville by way of Como, Breckenridge, and Tenmile Creek.

Small smelter at Ames, San Miguel County, unsuccessful.

Colorado Smelting Co. built Eilers plant at Pueblo.

1884

Bonanza smelter, Saguache County, idle because of inability to treat ores.

Only one smelter in operation at Golden.

1885

Zinc first recovered from Colorado ores; long considered only as something to be avoided and cursed when found with other ores; often called "poison" by operators of amalgamating and concentrating mills.

1886

Yak tunnel, Leadville, started. From 1888 to 1898 operated intermittently; from 1898 to completion in 1910 driven practically continuously. Length 23,800 feet.

Globe smelter built near Denver by Holden & Chanute; later taken over by Dennis Sheedy and Charles Kountze, and later (1899) by American Smelting & Refining Co.

1887

Gilpin County tramway, 24-inch gage, completed between Blackhawk and Russell Gulch.

Denver & Rio Grande Railroad reached Aspen.

1888

Colorado Midland Railroad reached Aspen.

Philadelphia plant built at Pueblo by Holden, general manager for Guggenheim Bros.

Belgian retort zinc furnace built at Denver. Failure.

1890

Large copper deposits opened in Henriette and Maid of Erin mines at Leadville.

1891

Bartlett erected zinc oxide plant at Canon City, called the American Zinc-Lead Co.

Ores at Creede and Cripple Creek discovered. Denver & Rio Grande Railroad extended from Wagonwheel Gap to Creede.

Leadville finds gold in the Little Jonny and other mines on Breece Hill.

1892

Manganiferous iron ores shipped from Leadville, 3,100 tons, worth at mines $15,500. Manganiferous silver ores shipped from Leadville, 62,309 tons.

1893

Beginning of aerial tram transportation and extension of use of electricity for power.

Newhouse tunnel, Idaho Springs, started by hand in September. Machine drills installed January, 1894. Completed, 22,000 feet long, November 17, 1910.

1894

Labor trouble at Cripple Creek.

1895

Wilfley table invented at Kokomo, Summit County.
Gravels on Swan River tested with oil drill.

1896

Camp Bird mine relocated by Thomas F. Walsh.
Thomas F. Walsh running pyritic smelter at Ouray.
Labor strike at Leadville June 1.
New Colorado-Philadelphia chlorination mill at Colorado City to treat Cripple Creek ores.

1897

Labor strike ended at Leadville in March.
Eighteen stamp mills, aggregating 870 stamps, in operation in Gilpin County.
Fire in the carbonaceous shale in Smuggler mine, at Aspen.
Small smelter erected at Dunton, Dolores County.
Centrally located smelters of the State receive large quantities of ore from mines outside Colorado.
Carter tunnel, Gunnison County, started in July.
Camp Bird and Revenue mines, Sneffels, most prominent in Ouray County.
Silver Lake mill, San Juan County, operated by electric power generated on Animas River.
Smuggler Union mills, San Miguel County, formerly amalgamation mills, now entirely concentration. Tom Boy uses amalgamation-concentration. Liberty Bell mine examined and purchased. Telluride Power Transmission Co. transmits power 17 miles to Camp Bird mine, said to be the longest distance electric power had been transmitted in the United States up to that time.
Copper ore discovered in La Sal Mountains, Montrose County, near Utah line; also uranium minerals.

1898

Five sampling works in operation in Clear Creek County and two in Gilpin County.
Two New Zealand type gold dredges built on Swan River. Failed to excavate deep and coarse gravel.
Colorado & Northwestern Railroad (later Denver, Boulder & Western) reaches Ward from Boulder.

1899

Liberty Bell mine, Telluride, adds a 7-ton cyanide leaching plant to its 80-stamp amalgamation mill.
Zinc ore shipped from Georgetown.
Bismuth ore shipped from Leadville.
Smelters in Colorado shut down for two months in strike over eight-hour law.
Home Mining Co. unwaters Penrose shaft, Leadville, in May.
Certain smelters in Kansas received small shipments of zinc blende concentrate from Creede, the beginning of the zinc industry in the Rocky Mountains. Large shipments of zinc ore and concentrates made to Belgium from Leadville in 1899 to 1903 aided in starting zinc industry in Colorado.
Cripple Creek reached by two railroads—the Florence & Cripple Creek, from Florence, and the Midland Terminal from Divide.
American Smelting & Refining Co. organized.
Stratton's Independence mine at Victor sold to a British syndicate.
Four smelters at Leadville: Arkansas Valley, Union, Bimetallic, and Boston Gold-Copper.
Two orange-peel dredges built on Blue River. Failures.
Cripple Creek mills are the Metallic Extraction Co. (cyanide) near Florence, the Colorado-Philadelphia Reduction Co. (chlorination) at Colorado City, the El Paso Reduction Co. (chlorination) at Florence, the National Gold Extraction Co. (a part of the Colorado-Philadelphia Reduction Co.) (chlorination) at Florence, the Colorado Ore Reduction Co. (cyanide) at Elkton, the Gillette Reduction Co. (chlorination) at Gillette, and the Brodie Reduction Works, (cyanide) at Mound City. The Economic Gold Extraction Co. built a plant (chlorination with precipitation by hydrogen sulphide, also using the cyanide process) in Eclipse Gulch, on west side of Squaw Mountain.
One Risdon and one Bucyrus dredge built on Swan River.
Camp Bird and Revenue mines, Sneffels, two most prominent in Ouray County.

1900

Two public sampling works in operation at Boulder.
Mary Murphy mine, Chalk Creek, Chaffee County, under lease to Buena Vista Smelting & Refining Co., which has a matting plant at Buena Vista treating a mixture of siliceous ore and heavy iron sulphides carrying one-half to 4 per cent copper from Leadville, some Sedalia ore, and some Mary Murphy ore. The smelter did not operate long.
Two public samplers at Georgetown and three at Idaho Springs, Clear Creek County.
W. S. Stratton buys much property in the Cripple Creek district.
Three of the first mills erected in the Cripple Creek district, the Chlorination, at Gillette; the Brodie Cyanide mill, at Cripple Creek; and the Chlorination Cyanide mill in Arequa Gulch, find that they can not compete with the mills in the valley, leaving in operation in the district only the Economic Chlorination mill, on the west slope of Squaw Mountain, at the mouth of the Columbine-Victor tunnel.
Union Gold Extraction Co.'s Chlorination mill (with precipitation by hydrogen sulphide and with Wilfley tables for the slimes) built at Florence by the syndicate controlling the Vindicator-Golden Cycle-Anaconda mines.
Dorcas cyanide mill built at Florence.
Rocky Mountain Smelting Co.'s matting plant built at Florence.
Portland Gold Mining Co. starts construction at Colorado City of 350-ton chlorination mill.
The Colorado-Philadelphia and Standard Chlorination mills under one management, in operation at Colorado City.
Metallic Extraction Co. at Cyanide, near Florence, in operation. Heavy hoisting, pumping, and compressor plants placed at the Gold Coin and Portland shafts at Victor.
Building of roadbed into Cripple Creek of the Colorado Springs & Cripple Creek District Railway (the Short Line), 45 miles between the two towns, expected to be in operation by April.
Colorado Electric Power Co., of Canon City, furnishes power to Cripple Creek mines.
Three concentrating mills, none working at full capacity, in operation at Rico.

Agent of the Director of the Mint estimates that 68,689 tons of ore were mined in 1900 in Gilpin County. Many stamp mills in operation at Blackhawk.

In Hinsdale County the Ute and Ulay mine produces 5,000 tons of concentrates and the Golden Fleece mine ships about $100,000 worth of ore.

Leadville Pumping Association undertakes to unwater the down-town mines of Leadville.

Home Mining Co., at Leadville, operates through the Penrose, Star, and Bon Air shafts the Coronado, Midas, and Penrose, near the head of Seventh, Fifth, and Fourth streets, respectively.

Cloud City Mining Co. obtains mineral rights by purchase of city lots in Leadville, in what is known as the Stevens and Leiter addition.

Arkansas Valley smelter, Leadville, is enlarged.

A semipyritic smelter started at Leadville, in November 1899, by the Boston Gold-Copper Smelting Co. is operated continuously through 1900.

Shipments of iron-manganese ore to steel works—chiefly from the Catalpa-Crescent mines—cease in June.

Leasing system predominates at Leadville.

Treatment of lead-zinc ores receives attention at Leadville.

Nelson adit, at Creede, in over 9,000 feet and drains Amethyst and Last Chance mines.

Two sampling plants and three mills operate at Aspen, which together with the town are furnished with electric power by the Castle Creek Power Co.

The Baca grant, in Saguache County, in litigation for 25 years, passes to a syndicate which proposes to develop the mineral resources. This company erected a large stamp mill at Crestone. As depth is reached, the free-milling ores change to a sulphide carrying considerable copper.

Two concentrating mills built at Bonanza, Saguache County.

Kendrick-Gelder semipyritic smelter built at Silverton.

Silver Lake mine, San Juan County, installs electric drills. Sunnyside and Gold King continue to operate.

In San Miguel County, Tom Boy amalgamation-Frue Vanner concentration mill in operation. Liberty Bell completes 250-ton cyanide leaching plant. Smuggler-Union mill at Pandora concentrates only. Telluride Power Transmission Co. increases its capacity. Copper ore found in sandstone in western part of San Miguel County.

1901

Magnetic separating plants built at Canon City, Pueblo, and Denver.

Three sampling works and many mills in operation at Idaho Springs.

Humphreys concentration mill built at Creede.

Work in progress at Emma mine, at Dunton, Dolores County.

Stamp mills and two samplers active in Gilpin County.

Southwestern lead smelter "blown in" at Whitepine, Gunnison County, and operates for a short time.

The small Hoffman smelter, at Marble, in operation that year.

The Arkansas Valley smelter only smelter in operation at Leadville.

Home pyritic smelter at Ouray fails.

Cyanidation added to Camp Bird mill, Ouray County.

Silver Lakes group of mines, comprising 175 claims, mill sites, and placer claims, San Juan County, sold to American Smelting & Refining Co. by E. G. Stoiber.

Kendrick-Gelder smelter in operation at Silverton part of the year.

Smuggler-Union, Liberty Bell, and Tom Boy mines continue to operate at Telluride.

Smuggler-Union introduces cyanidation of tailings.

Semipyritic smelter built at Golden. Small smelter built at Rico.

Semipyritic smelter at Robinson, Summit County.

1902

Zinc mining assuming considerable importance in the State.

Picking and washing belts and machines enter Cripple Creek district metallurgy. Chlorination process treats most of the ore from Cripple Creek, at the Union mill of the United States Reduction & Refining Co., at the Economic mill at the portal of the United Mines Transportation Co.'s tunnel on the west slope of Squaw Mountain, at the Standard and Colorado mills of the United States Reduction & Refining Co., at Colorado City, and at the Portland, completed in early fall. Near Colorado City another mill, the Telluride, is completed at Colorado City to use bromine instead of chlorine as the solvent for the gold.

Fryer Hill Mines Co. unwaters Fryer Hill mines, Leadville; idle since 1896.

Improvements are made at Arkansas Valley smelter, Leadville.

Cyanide mill built at Leadville, to treat ore from the Ballard mine.

New zinc concentrating mills built at Leadville.

Smuggler-Union, Liberty Bell, and Tom Boy mines active at Telluride.

Kendrick & Gelder semipyritic smelter operated during the summer at Silverton.

Silver Lake, Gold King, Sunnyside, and other mines active in San Juan County.

Camp Bird and Revenue mines active in Ouray County.

Durango smelter is active.

Salida lead-bullion smelter built.

Consolidation of mines at Rico.

Good Hope, at Vulcan, Gunnison County, ships copper ore.

Ohio City district, Gunnison County, active.

The Smuggler Leasing Co. acquires by lease nearly all the well-known properties at Aspen.

Milling active at Kokomo, Summit County.

1903

Severe labor troubles in nearly all the mining camps. Chlorination plants at Florence and Colorado City in operation only part of the year. The Omaha & Grant smelter, at Denver permanently closed and the Globe smelter, near Denver, idle part of the year. Nearly the entire militia of the State was stationed for months at Cripple Creek and Telluride. Nevertheless, production continues fairly well under the conditions.

United States Zinc Co.'s retort furnaces completed at Pueblo.

1903-4

El Paso drainage adit, at Cripple Creek, is completed. About 6,000 feet long; starting a mile below the El Paso mine, tapping shaft of that mine at 600 feet, and continuing northeastward. Operators planning a longer adit to drain 740 feet below the El Paso adit.

1904

The Midas Mining & Leasing Co. sinks the Penrose, Coronado, and other shafts in the down-town part of the district; that is, within the town of Leadville.

San Miguel County one of the most prominent milling centers in the State.

Ouray semipyritic smelter runs for six weeks.

In this year 578 producing mines in Colorado.

Eight lead plants, two zinc smelters, and eight semipyritic plants operating. New semipyritic plants were erected at Grand Junction and at Pearl.

Denver, Northwestern & Pacific Railway (Moffat Road) constructed over main range to a point 70 miles west of Denver.

Colorado & Northwestern Railroad (beginning at Boulder) builds an extension to Eldora and Sunset.

Mineral County (Creede) ships large quantity of concentrates.

Mining in San Juan County very active.

Five companies operating through the Cowenhoven adit, at Aspen. A zinc concentrating mill erected at Aspen. Carbonaceous shale in Smuggler mine still burns.

Reliance Gold Dredging Co. constructing dredge at Breckenridge.

Twenty stamp mills, aggregating 780 stamps, in operation in Gilpin County most of the year.

The May Day, La Plata County, ships ore.

1905

Active smelters are lead-smelting plants at Globe, Pueblo, Eilers, Arkansas Valley, Durango, Salida; there is a copper smelter at Argo; zinc retorts at Pueblo; zinc oxide-copper matte plant at Canon City; semipyritic plants at Silverton, Grand Junction, and Pearl.

Iron-manganese ore shipped from Red Cliff to the steel works at Pueblo.

Two electrically driven Bucyrus dredges in operation during a part of 1905 near Golden, in the Clear Creek gravel beds below Table Mountain. The gold is fine and difficult to save. Operations not resumed in 1906 nor later.

Zinc industry at Leadville growing. Zinc ores shipped to zinc retort plants return precious metal content in residues shipped back to Colorado. Midas Co., at Leadville, sinks Penrose shaft to 920 feet. The Yak tunnel 10,800 feet in at the end of the year.

Reliance Gold Dredging Co. completes a double-lift dredge in French Creek, Summit County.

Four concentrating mills, having a combined capacity of 850 tons per 24 hours, including the Smuggler 400-ton mill, in operation on Hunter Creek and Roaring Fork Creek, at Aspen.

Electric power available in San Juan County by building of Animas Power Co.'s plant.

Smelters receive about one-quarter of ore mined in the Cripple Creek district, the remainder going to the chlorination mills of the Portland Gold Mining Co. and the United States Reduction & Refining Co. at Colorado City and to the Economic mill at Victor.

1906

Thirty-three counties in Colorado produce metals.

Active smelters are lead plants at Globe, near Denver; Arkansas Valley, at Leadville; the Pueblo and the Eilers, at Pueblo; Durango; and Salida; a copper smelter at Argo, near Denver (plant for refining of copper destroyed by fire this year); the zinc oxide and copper matte plant of the United States Smelting Co., at Canon City; and the zinc retort smelter at Pueblo. Grand Junction smelter made an unsuccessful run on ores from Eagle and Ouray counties. Kendrick & Gelder pyritic smelter at Silverton operated by the Ross Mining & Milling Co. on company and custom ore. Saratoga semipyritic smelter, in Ouray County, ran part of the summer. Small copper matte smelter at Pearl, Jackson County.

Camp Bird mill, Ouray County, destroyed by snowslides and fire March 17.

Bull Domingo mine near Silver Cliff, Custer County, flooded since 1899, unwatered. Bassick mine, near Querida, developed by an 1,800-foot shaft, working.

Australian flotation process tried out at Rico.

Fulford district, Eagle County, producing from 10-stamp amalgamation mill.

Blevin dredge, 8 miles north of Lay, Moffat County, produces most of gold credited to the county (then Routt).

Silver Lake mill near Silverton burned but replaced by a 300-ton concentration plant.

Zinc carbonate discovered at Madonna mine, Chaffee County.

Dorcas chlorination and cyanide mill, at Florence, destroyed by fire in March.

Construction of Golden Cycle roast-amalgamation-cyanidation mill started at Colorado Springs.

1907

Roosevelt drainage tunnel started May 11 at Cripple Creek to gain 740 feet of drainage.

Golden Cycle mill destroyed by fire in July. Reconstruction begun.

Portland Gold Mining Co. erects cyanidation annex to chlorination mill at Colorado Springs.

Reliance Gold Dredging Co. reports a successful season at Breckenridge.

The Colorado Gold Dredging Co. acquires extensive holdings of stream and bench gravels and begins the installation of two dredges near Breckenridge.

Increased building and extension of electric power plants.

Decline in prices of zinc and lead in July affects lead and zinc production heavily.

Money stringency of 1907 affects mine operations. Miners turn to gold mining.

Active smelters are lead smelters at Globe, near Denver; Pueblo and Eilers, at Pueblo; Arkansas Valley, at Leadville; Durango; Salida; a copper plant at Argo, near Denver; the zinc oxide and copper matte plant of the United States Smelting Co., at Canon City; the zinc retort smelter of the United States Zinc Co., at Pueblo, and the Kendrick & Gelder semipyritic smelter at Silverton (Ross Mining & Milling Co.). Rico smelter idle.

Production of lead-zinc ores from Leadville dropped from 250,734 tons in 1906 to 199,573 tons in 1907.

A small shipment of ore from Halls Valley district, Park County, where little work had been done for 25 years.

Three dredges at work near Breckenridge, and another moved there from Golden, Colo. Electric power made available by Central Colorado Power Co. and Summit County Power Co. A dredge operates at Hahns Peak, Routt County.

Wellington mine completes new 100-ton concentration mill at Breckenridge.

Ratification of contracts between Gilpin County mine operators and Newhouse Tunnel Co.

Laterals being driven to Clear Creek County and Gilpin County mines from Newhouse tunnel.

Large quantity of iron-manganese ore shipped from Red Cliff, Eagle County, to steel mill at Pueblo.

1908

Ore of Mary Murphy mine, Chalk Creek, Chaffee County, being treated by amalgamation and concentration in 20-stamp Pawnee mill.

In Clear Creek County 104 mines in operation.

Active smelters are lead smelting plants at Globe, near Denver; Pueblo and Eilers, at Pueblo; Arkansas Valley, at Leadville; Durango; and Salida; a copper plant at Argo, near Denver; zinc retort smelter (United States Zinc Co.) at Pueblo; and the zinc oxide and copper matte plant at Canon City (United States Smelting Co.). Several magnetic separating plants operating on lead-zinc-iron ores.

Milling rates are so low that $8 ore can be shipped from Cripple Creek at a profit. Some of the Cripple Creek local cyanide plants treat profitably dump ore (already mined) containing $2 to $4 in gold per ton. Smelters receive only about 8 per cent of the ore mined in the Cripple Creek district.

At the close of 1908 the Roosevelt drainage adit at Cripple Creek has 4,872 feet completed in three headings.

Golden Cycle roast-amalgamation-cyanidation plant, with eight Edwards roasters, again in operation, early in 1908.

From October 1, 1907, to March 1, 1908, the freight and treatment charges on Cripple Creek ores at Colorado Springs up to $10 gold content per ton reduced from $6.25 to $3.50 per ton.

Completion and extension of electric power facilities in Boulder, Clear Creek, Gilpin, Lake, and Summit counties.

1909

Stratton's Independence mill at Victor, started in April with a capacity of 4,500 tons a month; enlarged in December, 1909, to 7,000 tons, and in November, 1910, to 9,000 tons.

Eilers smelter, Pueblo, dismantled.

So-called Modern smelter at Utah Junction, near Denver, starts October 22.

United States Smelting Co. (formerly American Zinc-Lead Co., or Bartlett zinc oxide process) plant, at Canon City, closes in September.

Dump of old Bassick mill, at Querida, reworked by cyanidation.

Burleigh tunnel drains Seven-Thirty mine, Silver Plume, in July.

In Gilpin County 110 companies or individuals operating mines or groups of mines.

Placer gold produced at Radium, Grand County.

Large quantities of zinc mill dumps from A. Y. & Minnie and Colonel Sellers mines, at Leadville, shipped to magnetic separating plants and zinc retort smelters.

Yak tunnel, at Leadville, 3.32 miles long at end of year.

New camp of Goldstone, on East Elk Creek, 9 miles from Newcastle, Garfield County, opened. Shipments made from Gray Eagle mine 1910.

1910

A Hammond dredge, of close-connected type of 54 buckets, of 4 cubic feet capacity each, or 2,000 cubic yards a day, in operation from July to October at Russell, on Placer and Sangre de Cristo creeks, Costilla County. Operated again in 1911, when overturned and never righted again.

Furnaces of Boston & Colorado Smelting Co.'s reverberatory copper matting plant, at Argo, near Denver, were "out" on March 17. This plant had been started in Denver in 1878, being moved from Blackhawk, where started in 1868 as the first successful smelter in Colorado.

Modern smelter, Utah Junction, near Denver, closes April 9.

North American Co.'s semipyritic smelter at Golden starts old Carpenter smelter April 9.

New semipyritic smelter operates for a short time at Alma.

Five lead smelters—Globe, Pueblo, Arkansas Valley (at Leadville), Durango, and Salida—and one zinc retort smelter, at Pueblo, in operation throughout the year.

Zinc carbonate ore discovered at Leadville.

In June Portland Gold Mining Co. completes Victor nonroast concentration-cyanidation mill.

In December Roosevelt drainage tunnel, Cripple Creek, pierces Pike Peak granite and enters breccia, but no heavy flow of water ensues. Work at tunnel continued.

1911

Semipyritic smelter (North American Smelter & Mines Co.) at Golden operated until November. Five lead smelters and one zinc smelter in operation in the State.

Rawley tunnel, Saguache County, started May 27, 1911. Completed 6,235 feet in October, 1912.

Three of four dredges in operation at Breckenridge.

Chlorination process of Cripple Creek ores ceases for all time. The United States Reduction & Refining Co.'s chlorination and cyanide plants at Colorado City close permanently.

1911–1913

New 500-ton cyanidation mill worked tailings at Old Union chlorination mill at Florence.

1912

Roosevelt tunnel 16,857 feet long in March, 1912, but drainage unsatisfactory.

New mill built at Mary Murphy mine, Chalk Creek district, Chaffee County.

Argo amalgamation-concentration-cyanidation mill completed at mouth of the Argo (Newhouse) tunnel, Idaho Springs.

Three of four dredges in operation at Breckenridge.

Roasting and magnetic separation plant added to Wellington mill, at Breckenridge.

Shipments of ore from the Equity mine, 10 miles northwest of Creede, Mineral County, show that there are ore deposits in the county outside of Creede.

Revival of mining at Bonanza, Kerber Creek, Saguache County.

Zinc carbonate ore from Leadville amounted to 142,782 tons of 29.2 per cent zinc in 1912.

Year shows increase of 15 per cent in value of gold, silver, copper, lead, and zinc produced in Colorado.

1912–1914

Australian process of flotation first successfully used in Colorado, chiefly as accessory to other methods of concentration.

1913

Gold Links mine, at Ohio City, Gunnison County, closes after long period of production.

Argo amalgamation-concentration-cyanidation mill at mouth of Argo (Newhouse) tunnel, Idaho Springs, discards amalgamation.

Terrible mine, Ilse, Custer County, unwatered but allowed to refill.

Only three of the four dredges in the Breckenridge district operated.

New silver district found on Brush Creek, west of Eagle, Eagle County.

During the year five lead smelters in the State treated ore from British Columbia and Canada as well as from Colorado, Idaho, South Dakota, Utah, and other States, and a considerable quantity of zinc residues from Kansas-Oklahoma zinc retort smelters but, as for several years, these smelters (all occasionally making leady copper matte as well) were not operated at full capacity. Ouray copper matting plant operated during year.

1914

Small quantity of copper ore shipped from Blair Athol, 6 miles north of Colorado City, El Paso County.

Four dredges in operation at Breckenridge, Summit County.

Preparations made to unwater Penrose shaft, at Leadville.

A 50-ton zinc oxide plant to handle low-grade zinc carbonate ore built at Leadville.

Cripple Creek gold district made largest output since 1908.

Work resumed on Roosevelt tunnel August 4 but stopped November 4.

Decrease in value of metals produced in the State in 1914 amounted to 6 per cent. Production of gold, however, increased $1,736,189.

Five lead smelters in the State, one zinc retort smelter with 1,920 retorts at Pueblo, and 1 copper matting plant at Ouray in operation.

1915

Dredge removed from York, Mont., installed on Box Creek, west of Arkansas River, 12 miles below Leadville, ran 70 days, and produced gold bullion valued at $69,292.

Black Bear mine, at Telluride, begins to mine large quantities of complex lead-zinc-copper-iron-silver-gold ore.

Leadville ships 15,956 long tons of iron-manganese ore to steel plants.

Portland Gold Mining Co. buys Stratton Independence mine and mill.

Vindicator Consolidated Gold Mining Co. buys Golden Cycle mine, at Victor, Cripple Creek district.

Copper ore shipped from Parkdale, Fremont County.

Largest production in Clear Creek County since 1907.

Iron Mask mine and mill at Belden, Eagle County, purchased by Empire Zinc Co. in March.

Work resumed March 2 in driving Roosevelt tunnel, then 17,127 feet long.

Five lead smelters active in the State, treating much zinc residues from Oklahoma.

Production at Leadville in 1915 largest since 1883.

Increase in value of metals produced in 1915 is 30 per cent, due almost entirely to increase in production of zinc. Silver falls off 1,768,093 ounces. Number of producing mines decreases 45.

1916

Globe, Leadville, Pueblo, Durango, and Salida smelters were operated continuously on ore from Colorado and other States, including a considerable and increased quantity of zinc residues from Kansas and Oklahoma zinc retort smelters. The quantity of zinc residues sent to Colorado has been increasing for several years.

Copper matte smelter at Ouray revived and newly constructed copper matte plant at Vulcan operated for a short time.

Gold production of Cripple Creek decreases. Roosevelt tunnel 22,100 feet long.

San Juan, San Miguel, and Ouray counties combined increase production, but Camp Bird mine closes July 1, final clean-up being made August 1. The old base of operations at the Camp Bird mine was at the mouth of No. 3 tunnel, at an elevation of 11,300 feet. A new base was established at 9,700 feet, from which a crosscut tunnel is being driven to intersect the Camp Bird vein 11,000 feet away. The new crosscut was in 329 feet by March 31.

Four dredges in operation at Breckenridge.

Increased production of silver at Aspen and at Creede.

Lake County makes total gross production of $16,082,059 of gold, silver, copper, lead, and zinc—largest in history of mining in that county.

Leadville ships 90,600 long tons of iron-manganese ores to steel plants. Derry Ranch dredge, in Box Creek below Malta, produced $119,169 in 1916. Largest production (in value) at Red Cliff in history of the district. High price of copper caused considerable activity in Fremont County and revival at Pearl, Jackson County.

Increase in value of metals produced in 1916 is 13 per cent, due almost entirely to increase in production of zinc.

1917

Powder River Dredging Co. built new dredge in Breckenridge district; five dredges at work in that district.

Five lead-silver plants, 2 zinc oxide plants, 1 zinc retort smelter, two custom magnetic separation plants in operation in Colorado.

Gold production at Cripple Creek decreased. Roosevelt tunnel extended to 24,000 feet.

Gross production at Leadville decreased. Heavy shipments from Down Town Co.'s mines, unwatered in 1916. Dredge operated at Box Creek, below Malta. Leadville also ships 104,169 long tons of iron-manganese ore to steel mills.

Zinc output increased heavily at Breckenridge, but dredge gold output fell off.

San Juan and San Miguel counties maintained production; Ouray County showed effect of idleness of Camp Bird mine; Georgetown-Silver Plume district shows increased production.

Metal production decreased 14 per cent in value in 1917.

1918

Five lead-copper plants in operation—Salida, Globe, Pueblo, Durango, and Leadville. Quantity of zinc residues from Kansas-Oklahoma fell off heavily.

Zinc plants in operation are United States Zinc Co. at Pueblo, River Smelting & Refining Co. at Florence, Western Zinc Concentrating Co. at Leadville, and new Ohio Zinc Oxide Co. at Canon City.

Empire Zinc Co.'s magnetic separation plant, Canon City, active, but Western Chemical & Manufacturing Co.'s plant at Denver idle part of the year.

Cripple Creek gold decreases. Portland cyanide mill at Colorado Springs and Portland cyanide-concentration mill at Victor close. Portland Independence mill operated. Golden Cycle roast cyanidation-amalgamation mill continues. Roosevelt tunnel connected with Portland No. 2 shaft.

Fryer Hill, Leadville, unwatered in 1916, closed in June. Moyer-Tucson mines of Iron Silver Co., at Leadville, closed Company purchases mines in Graham Park, Leadville. Western Mining Co.'s properties (Wolftone shaft) closed.

Leadville also ships 99,100 long tons of iron-manganese ore on war contracts to steel mills.

Arduous situation in San Juan County owing to influenza and high cost of labor and mining, but production maintained fairly well.

San Miguel County maintains production.

Camp Bird treats no ore but does development.

Eagle County zinc mines very productive.

Summit County dredges and zinc mine active.

Chamberlain-Dillingham Ore Sampling Co. closes Blackhawk, Georgetown, and Idaho Springs samplers and dismantles them.

Clear Creek-Gilpin Ore Co. opens a sampler at Idaho Springs in November. Custom mills at Idaho Springs active.

Climax Molybdenum 400-ton mill starts at Climax, Lake County, in February.

Decrease in quantity of all metals produced and decrease in value of combined metals amounts to 19 per cent. Silver alone shows increase in value for the year due to increase in price.

1919

Graham Park mines, Leadville, closed in April.

Globe plant ceases in April to accept regular consignments of ore, but some special ores, flue dust, and other material received in order to continue plant as a producer of arsenic.

Wellington mill, Breckenridge, closes early in 1919 because of lack of market for zinc.

Leadville, Pueblo, Durango, and Salida lead plants operated at reduced capacity. Leadville ships 11,055 long tons of iron-manganese ore on unexpired war contracts to steel mills.

United States Zinc Co.'s retort furnaces at Pueblo operated at reduced capacity.

Zinc oxide plants of Western Zinc Concentrating Co., at Leadville, and of Ohio Zinc Co., at Canon City, operated at increased capacity on zinc carbonate ores.

Empire Zinc Co.'s magnetic separation plant at Canon City operated steadily.

Western Chemical & Manufacturing Co., of Denver, drops purchase of Leadville zinc-lead sulphide ores but continues using pyrite ores from Leadville and Kokomo.

Shipments of iron manganese ore from Leadville to steel mills cease July 1.

Cripple Creek gold production decreases heavily.

Down Town Mines Co. (Penrose shaft) continues operations.

Zinc mines at Red Cliff reduced to development only, after producing during early part of year.

Climax Molybdenum mill closes April 1.

New Colorado Central mill, Georgetown, makes large production of silver from old Equator dump.

Decrease in production of all metals for the year.

1920

Continuous operation to December of Sunnyside new 500-ton mill at Eureka.

Salida smelter closed in February, principally because of falling off of receipts of Montana zinc residues from Oklahoma zinc retort smelters, leaving three lead smelters in operation in the State—at Leadville, Durango, and Pueblo.

Empire Zinc Co. completes modern zinc oxide plant at Canon City but does not operate it.

Empire Zinc Co.'s magnetic separating plant, successfully operated since 1902, closes in November.

River Smelting & Refining Co.'s matting and fuming plant at Florence closed in December.

Cripple Creek gold production falls.

Leadville production increases after bad year of 1919 Arkansas Valley blows in fourth furnace. Derry Ranch dredge operates May to December.

Leadville also ships 11,056 long tons of iron-manganese ore to steel mills.

San Miguel County product is as usual. Tom Boy early in year set in motion its new oil-flotation mill. Smuggler Union new flotation mill burns but replaced by a concrete mill. Liberty Bell resumes after short idleness.

Camp Bird mine continues small development.

Colorado Central mill, Georgetown, closed in May.

1921

Burleigh mill, Silver Plume, makes good silver concentrates from old Dives-Pelican dump until destroyed by fire in December.

Pueblo flood of June 3, combined with decreased receipts of Colorado ores and Montana residues from Oklahoma zinc retort smelters, closed Pueblo lead smelter. San Juan mines made smallest production since 1882.

Almost complete cessation in marketing of zinc ores throughout year and decrease in production of lead.

Liberty Bell mine, Telluride, abandoned all operation in August. Active since 1898.

Leadville ships 924 long tons of iron-manganese ore to steel mills.

Gold production continued to fall. Three dredges idle at Breckenridge, at least two permanently.

1922

Pittman act maintained price of silver at $1 per ounce Remodeled Nashotah mill makes large production of silver concentrates from Dives-Pelican dump, at Silver Plume, from May throughout year.

South Park Dredging Co. commenced operating a 70-bucket dredge of 8 cubic feet capacity, capable of handling 4,680 cubic yards per 24 hours, below Fair Play, in May.

Only two dredges operate at Breckenridge.

Empire Zinc Co. opens large zinc oxide plant at Canon City in August.

Pueblo lead smelter sold and torn down.

Leadville ships no iron-manganese ore to steel mills.

Some zinc mines reopen in November and December.

1923

Production of gold at Cripple Creek increased during the year. Portland mine finds at 2,600-foot level some of the best ore found in the mine below the 1,600-foot level. Cresson produces 10,000 tons a month from above 1,700-foot level; completes installation of hoist capable of sinking to 3,000 feet. Stratton Estate mines optioned.

Sunnyside 500-ton lead-zinc concentration mill at Eureka reopened in January.

Pittman Act expires in June.

In July lead-silver concentrate at rate of 2,500 tons per month began from Rawley mill, Bonanza, Saguache County (closed in December).

Reed-Coolbaugh sulphating plant at Durango completed in August.

Unwatering begun at Pyrenees and Greenback shafts, Leadville, in August. Arsenic shipped from Madera, Gunnison County. Zinc ore in quantity shipped from Whitepine, Gunnison County.

Eagle mines, Red Cliff, shipped large quantity of zinc ore to Canon City.

Pumps pulled at Penrose mine, Leadville, in November.

Leadville again finds market for iron-manganese ore to steel mills.

Only two dredges in operation at Breckenridge.

Smelting and milling ore shipped from Gold Belt tunnel, above Silver Plume, making a new mine for this district.

Zinc concentrates shipped steadily from Wellington mine at Breckenridge.

At end of year it is found that total production of metal mines gained 21 per cent; gold increased, silver decreased, copper increased, lead nearly doubled, zinc more than doubled.

LOCATION, AREA, AND ORGANIZATION OF THE MINING COUNTIES OF COLORADO

When the first Territorial legislature of Colorado met, on September 9, 1861, it created 17 counties and fixed the county seats. These 17 counties have now expanded to 63 by the addition of the Arapahoe and Cheyenne Indian reservations and by the subdivision of counties, some of which were originally larger than several of the Eastern States. The area considered under each county name given here is the area comprised within the present county boundary.

The following details of the location, area, and organization of the mining counties in Colorado are transcribed, with slight modifications, from the Year Book of the State of Colorado for 1920, published by the State Board of Immigration.

ADAMS COUNTY

Adams County lies in the north-central part of the State. The city of Denver forms a part of its western boundary. It is an irregular rectangle, with an extreme length, from east to west, of 72 miles and a width of 18 miles. Its area is 807,680 acres, about 125,000 acres more than the area of Rhode Island.

Adams County was organized in 1902 from a part of Arapahoe County. Parts of it were annexed to Washington and Yuma counties in 1903, and a part of Denver County was added in 1909. Long's expedition crossed the northwestern corner of what is now Adams County in 1820. Other exploring and prospecting expeditions followed the same route along the South Platte prior to the discovery of gold, in 1859. The early gold seekers wasted little time in Adams County, though they did some prospecting for placer gold in the sands of the Platte north of the

Blanca, which stand on the boundary line between Alamosa and Costilla counties.

Early in 1807, after making unsuccessful attempts to scale Pikes Peak, Capt. Zebulon M. Pike's expedition crossed the Sangre de Cristo Range, skirted the base of Sierra Blanca, and camped on the banks of the Rio Grande, near the present site of the city of Alamosa. Captain Pike's diary contains the first authentic record made by any American traveler in this territory. Numerous exploring parties, including that led by John C. Frémont, followed the Rio Grande through

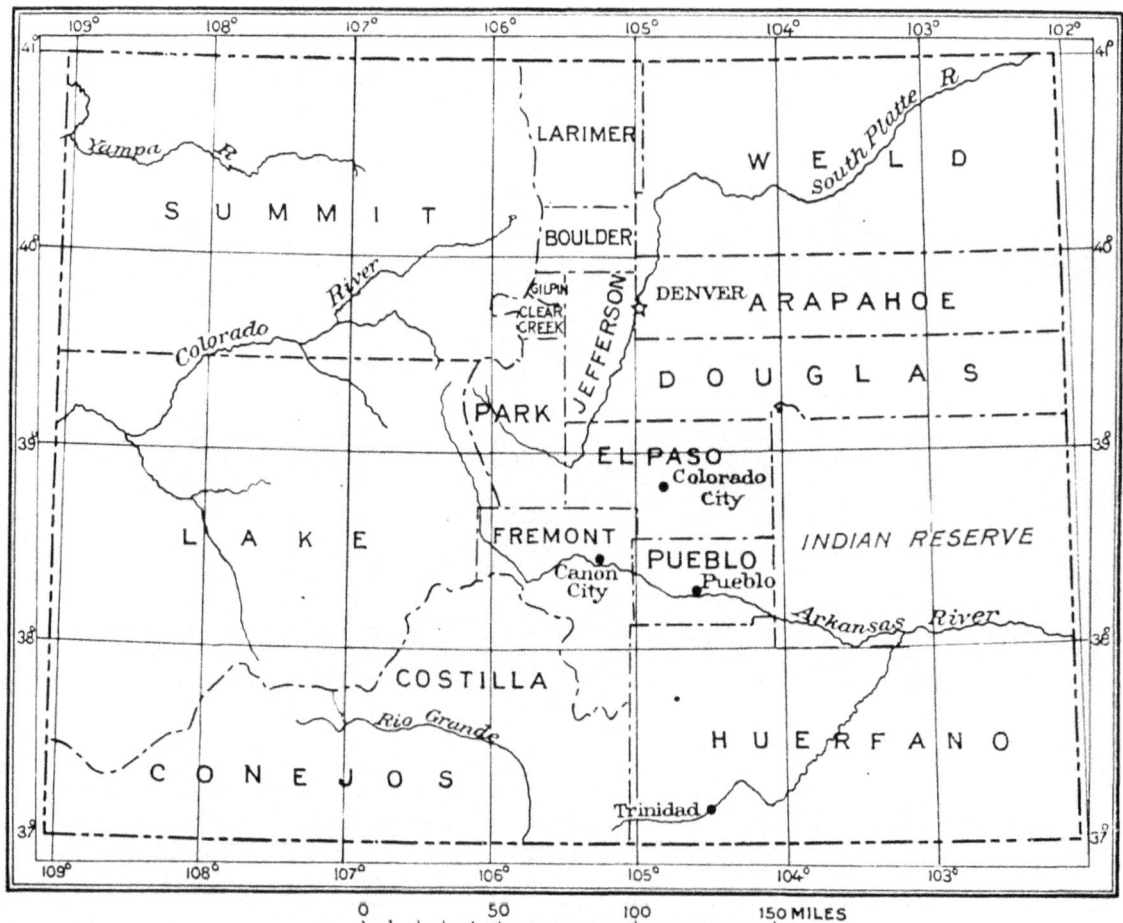

FIGURE 1.—Map showing original 17 counties created in 1861 by first Territorial legislature of Colorado. Colorado City was the first capital, in 1861; Golden was the capital in 1862–1867; Denver has been the capital since 1868

site of Denver. A few temporary camps were established by gold seekers south of the site of Brighton in the early sixties, but no permanent settlements were made.

ALAMOSA COUNTY

Alamosa County lies in the south-central part of the State, in the heart of the San Luis Valley. In outline it is an irregular pentagon having an extreme length, from east to west, of 30 miles, and an extreme width, from north to south, of 27 miles. Its area is 465,280 acres. The surface is generally level except in the northeast part, where it rises into broken hills, which culminate in two massive peaks, Old Baldy and Sierra

this country. The town of Alamosa was founded in 1878. Alamosa County is the youngest in the State; it was created by the State legislature in 1913 from parts of Conejos and Costilla counties.

ARAPAHOE COUNTY

Arapahoe County lies in the north-central part of the State, a part of its western boundary being formed by the city of Denver. It is an irregular rectangle, 72 miles long and 12 miles wide. Its area is 538,880 acres.

Arapahoe was one of the original 17 counties in Colorado Territory as organized in 1861. It was originally much larger than it is now, having at one time extended to the Kansas line. Parts of it were taken to form Adams and Denver counties in 1902 and Washington and Yuma counties in 1903.

ARCHULETA COUNTY

Archuleta County is in the southwestern part of the State. Its southern boundary is formed by the State of New Mexico, and its eastern boundary by the main range of the Rocky Mountains. It is rectangular in outline. Its extreme length, from east to west, is about 60 miles, and its extreme width is 33 miles. Its area is 780,800 acres, or about 100,000 acres greater than that of the State of Rhode Island.

When Colorado Territory was organized in 1861, the area that now forms Archuleta County was included in Conejos County. Archuleta County was organized in 1885 and was named in honor of J. M. Archuleta, then a prominent citizen of old Conejos County.

BACA COUNTY

Baca County lies in the extreme southeast corner of the State. It is bounded on the east by Kansas, and on the south by Oklahoma and New Mexico. It is a regular rectangle, 55 miles long from east to west, and 44 miles wide. Its area is 1,633,280 acres, or about 400,000 acres more than that of the State of Delaware. Its surface is a comparatively level plateau, broken by a low range of hills in the southwest. The altitude ranges from about 3,800 feet in the extreme east to 5,700 feet in the southwest.

It has long been believed that Coronado, in his explorations of the Southwest in 1540 (see p. 1), crossed the corner of what is now Baca County. He appears to have gone up Cimarron River over a part of what later became the Santa Fe Trail. Early in the last century there was considerable travel through this part of Colorado between Missouri River and Santa Fe. The mountain division of the Santa Fe Trail crossed what is now Baca County, along the north side of Cimarron River. Three granite markers now show the course of this historic trail through Baca County. Although the travel there was considerable no settlements were made until the early sixties. For about 20 years isolated ranchers made their homes in the valleys of streams in this territory. The actual settlement of the county, however, did not begin until 1887, when there was a considerable influx of stockmen and some farmers to this part of Colorado.

The county was created in 1889 from a part of Las Animas County.

BOULDER COUNTY

Boulder County lies in the north-central part of the State. The Continental Divide forms its western boundary. It is of somewhat rectangular outline, is 33 miles long east to west, and is 24 miles wide. Its area is 488,960 acres. The surface is extremely varied, being a rolling or broken valley in the east and rising to the summit of the Continental Divide on the west.

Boulder County was one of the original 17 counties included in Colorado Territory when it was organized in 1861. Its boundaries have never been changed.

CHAFFEE COUNTY

Chaffee County lies near the central part of the State. Its western boundary is formed by the Saguache Mountains, which here constitute the Continental Divide, and its eastern boundary by the Park Range. It has an extremely irregular outline, is about 45 miles long from north to south and about 25 miles wide near its central part. Its area is 693,120 acres, about 10,000 acres more than that of the State of Rhode Island. Its surface is principally mountainous and its altitude ranges from about 7,000 feet at the point where Arkansas River crosses its southern boundary to more than 14,000 feet at the summits of some of the peaks in the Saguache Range.

In 1879 Chaffee County was organized from a part of Lake County and was named in honor of Jerome B. Chaffee, one of Colorado's first United States Senators.

CLEAR CREEK COUNTY

Clear Creek County lies in the north-central part of the State. Its western boundary is formed by the Continental Divide. It is of irregular outline and has an extreme length, from east to west, of about 25 miles, near the central part, and an extreme width of about 20 miles. Its area is 249,600 acres. The surface is principally mountainous, and the altitude ranges from 6,880 feet, at the northeast corner, to more than 14,000 feet at the summits of some of the peaks in the western part.

The county was organized in 1861, soon after Colorado Territory had been formed. It was named for the stream along the course of which most of the early prospecting was done.

CONEJOS COUNTY

Conejos County lies in the south-central part of the State and contains a portion of the southern end of the San Luis Valley. The Rio Grande forms the eastern boundary and the main range of the Rocky Mountains forms the western. It is of rectangular outline and has an extreme length, from east to west, of 45 miles and an extreme width, from north to south, of 30 miles. Its area is 801,280 acres, or about 119,000 acres greater than the area of Rhode Island. The surface is level in the east but rises rather steeply in the west to the Continental Divide. The altitude ranges from 7,000 feet in the extreme southeast to more than 13,000 feet at the summits of some of the mountain peaks near the western border.

Conejos was one of the original 17 counties in Colorado Territory and was, when first organized, much larger than it is to-day.

COSTILLA COUNTY

Costilla County is in the south-central part of the State and includes a part of the south end of the San Luis Valley. The Rio Grande forms a part of its western boundary, the Sangre de Cristo Range its northern and eastern boundary, and the State of New Mexico its southern boundary. Its area is 758,400 acres, about 32,000 acres more than the combined areas of Rhode Island and the District of Columbia. The county is of irregularly rectangular shape and has an extreme length, from north to south, of about 54 miles and an extreme width, from east to west, of about 32 miles. The surface in the southwest is a level valley, which rises rather steeply toward the east and northeast, culminating in the high peaks of the Sangre de Cristo Range. The altitude ranges from about 7,500 feet in the southwest to more than 14,000 feet at the summit of Old Baldy and other peaks of the Sangre de Cristo Range.

Costilla County was organized as one of the original 17 counties of Colorado Territory in 1861 and was at that time considerably larger than now. A large part of it was included in old Spanish land grants.

CUSTER COUNTY

Custer County lies in the south-central part of the State, the Sangre de Cristo Range forming its western boundary. It is of irregularly triangular shape and has an extreme length at the base, which is the north boundary, of 38 miles and a width of 25 miles. Its area is 478,080 acres. Its surface forms a plateau that rises into a rugged range of hills near the eastern boundary and that culminates in the Sangre de Cristo Range on the west. The altitude ranges from about 6,700 feet at the northern boundary to more than 14,000 feet at the summits of some of the peaks of the Sangre de Cristo Range.

The county was organized in 1877 from a part of Fremont County.

DENVER COUNTY

Denver County is identical in its boundaries with the city of Denver. It lies near the foothills on the eastern side of the Rocky Mountains, in the north-central part of the State. It is the smallest county in Colorado, having an area of 37,120 acres. South Platte River flows north through the central part of the county, and Cherry Creek, which comes in from the southeast, enters the Platte near the center of the business part of the city. The valleys of these streams contain the lowest altitudes in the county, and the surface rises gradually to the east and west of these streams, being generally level or gently sloping. The altitude ranges from 5,180 feet to about 5,350 feet.

The area was originally included in Arapahoe County. Denver County was not organized until 1902. Arapahoe County was then much larger than it is at present, including all of what is now Adams County and extending east to the Kansas line. Denver County was larger when first created than it is now. In 1909 a part of its original territory was added to Adams County, leaving the boundaries of Denver County as they are at present.

DOLORES COUNTY

Dolores County, in the southwestern part of the State, is bounded on the south by Montezuma County and on the west by Utah. It is of rectangular outline and has an extreme length, from east to west, of 65 miles and an extreme width of 24 miles. Its area is 667,520 acres, slightly less than the area of the State of Rhode Island. The surface consists of broken table-land in the west, which rises to the summits of the La Plata and San Miguel mountains on the eastern border. The altitude ranges from about 5,900 feet in the extreme southwest to about 13,000 feet at the summits of some of the peaks on the eastern boundary.

The area now included in Dolores County was at first a part of La Plata County and was made a part of Ouray County in 1877. Dolores County as it now exists was created in 1881.

DOUGLAS COUNTY

Douglas County lies in the north-central part of the State. Its western boundary is formed by Platte River and the South Fork of the Platte. In outline it is a truncated triangle, the southern boundary forming the base. It is 30 miles long and is 30 miles wide at its southern boundary and about 20 miles at its northern boundary. Its area is 540,800 acres. The surface varies from level or gently rolling plains in the west and north to a rugged foothill district in the southwest. The altitude ranges from 5,400 feet in the northwest to about 7,600 feet in the extreme southwest.

The county was one of the original 17 counties of Colorado Territory as it was organized by the act of the first Colorado territorial legislature, in 1861. It was named for Stephen A. Douglas. At that time the county extended eastward to the Kansas line. A part of it was taken to form Elbert County in 1874.

EAGLE COUNTY

Eagle County lies in the west-central part of the State. Its surface is principally mountainous and its eastern boundary is formed by the Gore Range of mountains. Its area is 1,036,800 acres. It is rectangular in outline and has an extreme length, from east to west, of 48 miles and an extreme width of 38 miles. The altitude ranges from about 6,150 feet where Colorado River crosses the western boundary to over 13,000 feet at the summits of the mountain peaks in the east and southeast.

The county was organized in 1883 from a part of Summit County.

EL PASO COUNTY

El Paso County lies in the east-central part of the State and is, as its name implies, a sort of open door or "pass" between the Great Plains region of eastern Colorado and the picturesque mountain region beyond. It is an almost perfect rectangle, though it has some slight irregularities on the western boundary. Its extreme length from east to west is 55 miles, and its width is 42 miles. Its area is 1,357,440 acres, or a little more than one-third that of the State of New Jersey. The surface is principally a level or somewhat broken plain, but it rises steeply in the extreme west to the summit of Pikes Peak and other high mountains in the district immediately west of Colorado Springs. The altitude ranges from about 5,000 feet in the southwest to 14,110 feet at the summit of Pikes Peak, near the western boundary.

El Paso County was one of the original 17 counties included in Colorado Territory. A part of it was taken in 1899 to form Teller County.

FREMONT COUNTY

Fremont County lies in the south-central part of the State on the eastern boundary of the mineralized belt. A part of the western boundary is formed by the Sangre de Cristo mountain range. It is of rectangular outline and is about 60 miles long from east to west and about 30 miles wide. Its area is 996,480 acres, a little less than two-thirds that of the State of Connecticut. The surface is principally rolling or mountainous. The altitude ranges from about 5,000 feet at the point where Arkansas River crosses the eastern boundary to more than 12,000 feet at the summits of some of the peaks in the southwestern part.

Fremont is one of the original 17 counties in Colorado Territory, organized in 1861, and was named in honor of John C. Frémont, who crossed this territory several times in his efforts to discover a feasible railway route across the Rocky Mountains.

GARFIELD COUNTY

Garfield County lies in the western part of Colorado and includes a part of the Grand Valley, which is one of the best known agricultural and fruit-raising districts in the State. It forms an extremely irregular rectangle which is 110 miles long from east to west and about 50 miles wide at the eastern end. Its width at the west end, where it touches the State of Utah, is about 20 miles. Its area is 1,988,480 acres, a little more than the combined areas of the States of Delaware and Rhode Island. The surface is extremely irregular, ranging in altitude from about 4,700 feet at the western boundary to over 13,000 feet at the summits of some of the peaks in the northeastern part.

The territory now included in Garfield County was originally occupied by the Ute Indians. There was no development worthy of note until after 1881, when the Indians were by treaty removed from this part of Colorado to western Utah. Small prospecting parties explored the mountainous areas, both north and south of Colorado River, about 1879 and built a fort not far from the site of Glenwood Springs, which they called Fort Defiance. The county was organized in 1883 from a part of Summit County and was named in honor of President James A. Garfield. A part of it was taken to form Rio Blanco County in 1889.

GILPIN COUNTY

Gilpin County lies in the north-central part of the State. A part of its western boundary is formed by the Continental Divide. It is an irregular triangle, having an extreme length of about 16 miles near the center and an extreme width on the eastern boundary of 13 miles. It is the smallest county in Colorado except Denver, which includes only the city of Denver. Its area is 84,480 acres. The surface is almost all mountainous, and the altitude ranges from 6,880 feet at the southeast corner to about 14,000 feet at the summits of some of the peaks on the western boundary.

The county was one of the original 17 counties included in Colorado Territory as organized in 1861. It was named in honor of William Gilpin, the first governor of the Territory.

GRAND COUNTY

Grand County lies in the north-central part of the State. Its eastern boundary is formed by the Continental Divide, the northern boundary by the Rabbit Ears mountain range, and part of the southern boundary by the Williams Fork Mountains. It is made up principally of a mountain park known as Middle Park, which is surrounded by mountain ranges. Its outline is irregular. The greatest length, from north to south, is about 55 miles, and the greatest width is

22   MINING IN COLORADO

about 52 miles. Its area is 1,194,240 acres. The altitude ranges from about 7,800 feet in the extreme southwest to more than 13,000 feet at the summits of some of the peaks on the eastern boundary.

The county was organized in 1874 from a part of Summit County.

GUNNISON COUNTY

Gunnison County lies in the north-central part of the State. Its eastern boundary is formed principally by the Continental Divide. It is of very irregular triangular outline and has an extreme length from north to south of about 90 miles and an extreme width of 65 miles. Its area is 2,034,560 acres, or a little more than the combined areas of the States of Delaware and Rhode Island. The surface is extremely irregular and in most parts mountainous. The altitude ranges from about 6,875 feet at the place where the Gunnison crosses the western boundary to about 14,000 feet at the summits of some of the peaks in the north and east.

The county was organized in 1877 from a part of Lake County.

HINSDALE COUNTY

Hinsdale County lies in the southwestern part of the State, in what is known as the San Juan mining district. It is of irregularly rectangular outline, broadened at the north end. Its extreme length from north to south is about 52 miles and its extreme width from east to west is 26 miles. Its area is 621,440 acres. The surface is nearly all mountainous, the altitude ranging from about 8,500 feet where Lake Fork branch of Gunnison River crosses the north boundary to more than 14,000 feet at the summits of some of the peaks in the San Juan Range near the central part.

The county was organized in 1874 from parts of Conejos, Costilla, and Lake counties.

HUERFANO COUNTY

Huerfano County lies in the south-central part of the State. Its western boundary is formed by the Sangre de Cristo and Culebra mountain ranges, which are really but one range, though known by different names in different places. It has a more irregular outline than any other county in the State. Its extreme length, from east to west, is about 48 miles, and its width, from north to south, near the central part, is about 40 miles. Its area is 960,000 acres, or 300,000 acres more than that of the State of Rhode Island. The surface is an irregular plateau broken by many narrow valleys in the east and rising into a rugged mountainous area in the west. The altitude ranges from about 5,690 feet at the north boundary to more than 13,000 feet at the summits of the mountains on the west.

The county was organized in 1861 as one of the original 17 counties in Colorado Territory but was much larger at that time than it is at present.

JACKSON COUNTY

Jackson County lies in the north-central part of the State and includes the mountain valley known as North Park. The State of Wyoming forms its northern boundary. Mountain ranges bound it on the other sides—the Medicine Bow Range on the east, the Rabbit Ears Range on the south, and the Park Range on the west. It is very irregular in outline and has an extreme length, from north to south, of about 45 miles and an extreme width of 42 miles. Its area is 1,044,480 acres. The surface is principally rolling or level mountain valley, which rises gradually to mountain ranges on all sides except the north. The altitude ranges from about 7,800 feet at the point where the North Platte crosses the north boundary to more than 12,000 feet at the summits of the peaks in the bordering ranges.

The county was organized in 1909 from a part of Larimer County and named in honor of President Andrew Jackson.

JEFFERSON COUNTY

Jefferson County lies in the north-central part of the State, the city of Denver forming a part of its eastern boundary. It is an irregular triangle, with an extreme length of 72 miles from north to south. Its width is about 20 miles at the north boundary and a little more than 1 mile in the extreme south. Its area is 517,120 acres. Its surface is principally mountainous, though some level or rolling valley land lies along the courses of the streams. The altitude ranges from about 5,300 feet in the east to nearly 10,000 feet in the extreme west.

The early history of this county is closely linked with that of the city of Denver. The first settlements within the present limits of the county were made by gold seekers about the time the foundations of the city of Denver were being laid, in 1859. The city of Golden was founded in 1859 and was first called Golden City. For a number of years it rivaled Denver for the honor of being the first city in the State. It was made the capital of Colorado Territory in 1862 and retained the honor until 1867, when the seat of government was transferred to Denver. The Colorado School of Mines was opened at Golden in 1874. Jefferson County was one of the original 17 counties in Colorado and was named in honor of Thomas Jefferson. The Territory itself was first called Jefferson, but the name was afterward changed to Colorado, from the Spanish adjective meaning red, suggested by the abundance of red rock outcrops. A part of the territory of the county was added to that of Park County in 1908.

LAKE COUNTY

Lake County is an extremely rugged, mountainous area near the center of the State, at the very crest of the main range of the Rocky Mountains. It is relatively small but is famous the world over as one of the richest known mineral-producing districts. It is of an irregular rectangular shape, is 24 miles long from north to south, and is about 22 miles wide at its southern boundary. It is bounded on the east by the Park Range and on the west by the Saguache Range, which here forms the Continental Divide. Its area is 237,440 acres. The surface is nearly all mountainous. The altitude ranges from about 8,935 feet at the point where Arkansas River crosses the south boundary to 14,420 feet at the summit of Mount Elbert, the highest point in Colorado.

Lake County was organized in 1861 as one of the original 17 counties of Colorado, but at that time was much larger than it is at present.

LA PLATA COUNTY

La Plata County is in the southwestern part of the State and includes a considerable portion of the agricultural territory popularly known as the San Juan Basin. Its southern boundary is formed by the State of New Mexico. Its shape is that of a truncated triangle. It has an extreme length of about 40 miles, from north to south, and an extreme width of 38 miles near its southern end. Its area is 1,184,640 acres, or about 73,000 acres less than that of the State of Delaware. In its southern part its surface is divided into level table-lands interspersed with small timbered hills, but it rises steeply into a rugged mountainous region in its northern part. Its altitude ranges from about 5,900 feet at its southern boundary to more than 14,000 feet at the summits of some of the peaks in the north.

The county was organized in 1874, then comprising a territory nearly four times as large as the present county.

LARIMER COUNTY

Larimer County lies in the north-central part of the State. Its northern boundary is formed by the State of Wyoming and its western boundary by the Medicine Bow mountain range. It is of irregularly rectangular outline except along its western boundary. Its extreme length from east to west, along the northern boundary, is 64 miles and its width is about 50 miles. Its area is 1,682,560 acres. The surface ranges from level plains in the eastern part to an extremely rugged mountainous area in the west. The altitude ranges from about 4,800 feet in the east to more than 14,000 feet at the summits of some of the peaks near the western boundary.

Larimer County, one of the original counties of Colorado Territory, was named in honor of Gen. William Larimer, a well-known Colorado pioneer and one of the founders of the city of Denver. A part of its original territory was taken to form Jackson County in 1909.

LAS ANIMAS COUNTY

Las Animas County lies in the southeastern part of the State. Its southern boundary is formed by the State of New Mexico, and part of its eastern boundary by the Culebra Mountains. It is of irregularly rectangular outline and has an extreme length, from east to west, of 116 miles, and an extreme width, near its central part, of about 55 miles. It is the largest county in Colorado. Its area is 3,077,760 acres, or about 7,000 acres less than that of the State of Connecticut. The surface in the east is broken prairie and in the west a plateau, which rises into a mountainous district west of Trinidad. Its altitude ranges from about 5,300 feet in its northeastern part to more than 14,000 feet at the summits of the highest peaks in the Culebra Range.

The county was organized in 1866 from a part of Huerfano County.

MESA COUNTY

Mesa County is the center of the tier of western counties that border on the State of Utah. It is of irregularly triangular shape and has an extreme length, on the north, of about 84 miles from east to west and a width of 62 miles on the western boundary and of about 10 miles in the extreme northeast corner. Its area is 2,024,320 acres, or a little less than two-thirds that of the State of Connecticut. Its surface is extremely varied, and its altitude ranges from about 4,360 feet at the point where Colorado River crosses the western boundary to over 9,000 feet on the Uncompahgre Plateau, on the south, and about 10,000 feet on the Battlement Mesa, on the northeast.

The county was organized in 1883 from a part of Gunnison County and received its name from the great table-land on its eastern side called Battlement Mesa.

MINERAL COUNTY

Mineral County lies in the south-central part of the State, just west of the San Luis Valley and near the crest of the continent. It is of rectangular outline and has an extreme length, from north to south, of 40 miles and an extreme width of 24 miles. Its area is 554,240 acres. The surface is generally rugged and mountainous, and the altitude ranges from 8,250 feet where the Rio Grande crosses the eastern boundary to more than 13,000 feet at the summits of peaks in the San Juan Range.

The county was created in 1893 from parts of Hinsdale, Rio Grande, and Saguache counties.

MOFFAT COUNTY

Moffat County is in the extreme northwest corner of the State, the northern boundary being formed by the State of Wyoming and the western boundary by the State of Utah. It is a perfect rectangle except for slight irregularities on its eastern boundary. Its extreme length, from east to west, is about 91 miles, and its width is about 55 miles. Its area is 2,981,120 acres, or about 100,000 acres less than that of the State of Connecticut. It is the second county in size in Colorado, being surpassed only by Las Animas County. Its surface is a broken plateau, which becomes slightly mountainous in the northeast and in the extreme northwest. The altitude ranges from about 5,400 feet at the point where Yampa River crosses the western boundary to about 7,600 feet in the extreme northeast part. The county was organized in 1911 from the western part of Routt County and named in honor of David H. Moffat, builder of the "Moffat" railroad and one of the best known of Colorado's pioneers.

MONTEZUMA COUNTY

Montezuma County is in the extreme southwest corner of Colorado, the southern boundary being formed by New Mexico and the western boundary by Utah. It is of irregularly rectangular outline and has an extreme length, from east to west, of about 50 miles and an extreme width, from north to south, of about 38 miles. Its area is 1,312,640 acres, or about twice that of the State of Rhode Island. It is a broken table-land in the south and west and rises rather abruptly to the summits of the La Plata Mountains in the northeast. The altitude ranges from about 5,600 feet in the southeast to nearly 13,000 feet at the summits of some of the peaks in the northeast. The county, which was organized in 1889, was formed from the western part of La Plata County and was named for the famous ruler of the Aztecs.

MONTROSE COUNTY

Montrose County lies somewhat south of the west-central part of the State. Its western boundary is formed by the State of Utah. Its outline is that of a double rectangle having an extreme length, from east to west, of about 86 miles and an extreme width of 35 miles. Its area is 1,448,960 acres, or about one-fourth that of the State of New Hampshire. Its surface in general is a broken table-land crossed by numerous valleys that extend generally from the southeast to the northwest. The Uncompahgre Plateau extends northwest from the San Juan Mountains across its central part. Its altitude ranges from about 5,150 feet on the western boundary to about 9,600 feet in the most elevated points of the Uncompahgre Plateau.

The county was organized in 1883 from a part of Gunnison County.

OURAY COUNTY

Ouray County lies in the southwestern part of the State and includes a part of the rich mineral belt known as the San Juan district. It is of irregularly triangular outline, with the base to the north. Its extreme length, from north to south, is 33 miles, and its extreme width is about 29 miles. Its area is 332,160 acres, or about half that of the State of Rhode Island. Its southern part is mountainous, and its northern part is level or broken, including a portion of the Uncompahgre Valley. The altitude ranges from 6,300 feet at the north boundary to over 14,000 feet at the summits of some of the mountains in the southern part.

The territory was included in the tract of land ceded by the Southern Ute Indians to the United States in 1873. It had been but little explored prior to this time, but settlers and prospectors flocked into the entire territory immediately after the treaty was ratified, and rich mineral deposits were soon found in the district now included in Ouray County. In the summer of 1875 a permanent mining camp grew up in the heart of the mountains near the southern end of the Cimarron Range. This camp formed the nucleus of the town of Ouray, which was named in honor of a well-known Ute chief, whose services to the whites in this section were very great. Rich deposits of gold and silver were found in the Mount Sneffels district in 1875, and two years later the Virginius mine was opened. The county was organized in 1877. At that time it extended west to the State line and included the territory now embraced in Dolores and San Miguel counties.

PARK COUNTY

Park County lies almost exactly in the center of the State and includes the beautiful mountain-rimmed meadow known as South Park. Its western boundary is at the summit of the Park Range, which at some places forms the Continental Divide. The county is extremely irregular in outline. It is about 60 miles long from north to south and has an extreme width of about 45 miles. Its area is 1,434,880 acres. Its surface is principally hilly or mountainous except in South Park, which lies near its center and which is nearly 50 miles long and from 11 to 40 miles wide. In altitude the county ranges from about 7,200 feet at the point where Platte River crosses its eastern boundary to more than 14,000 feet at the summits of some of the peaks in its western part.

Park County was organized as one of the original 17 counties in Colorado Territory. It was named in honor of the beautiful valley, on the rim of which most of the prospect camps were located.

PITKIN COUNTY

Pitkin County is in the central part of the State, just west of the main range of the Rockies. It is of extremely irregular outline, is about 54 miles in length along its north boundary and about 30 miles in width north and south through its central part. Its area is 652,160 acres. The surface varies greatly, ranging from rugged mountains resplendent with natural grandeur to broad valleys in which agriculture is practiced profitably. Several mesas scattered through the county provide a considerable area of level, fertile, and productive farm land. The altitude ranges from about 6,625 feet in its northwestern part to more than 14,200 feet at the summits of some of the peaks in the east and south.

The county was organized in 1881 from a part of Gunnison County.

PUEBLO COUNTY

Pueblo County lies in the south-central part of the State and includes a portion of the Arkansas Valley, one of the best known agricultural areas in Colorado. It is of irregular outline and has an extreme length, from north to south, of 54 miles on its eastern boundary and an extreme width of 54 miles. Its area is 1,557,120 acres, a little more than half that of the State of Connecticut. The surface is principally a broken plain, through the central part of which passes the valley of Arkansas River. Toward the southwest it rises gradually into a rugged foothill district, its altitude ranging from about 4,350 feet at the point where Arkansas River crosses its eastern boundary to a little over 8,000 feet in the extreme southwest.

Pueblo County was one of the original 17 counties in Colorado Territory.

RIO BLANCO COUNTY

Rio Blanco County lies in the northwestern part of the State. Its western boundary is formed by the State of Utah, and it includes the northern part of the old Uinta Indian Reservation. It is of irregularly rectangular shape, and its area is 2,062,720 acres, about two-thirds of that of the State of Connecticut. It is the fourth county in Colorado in size, being surpassed only by Las Animas, Moffat, and Weld counties. Its extreme length, from east to west, is about 110 miles and its extreme width, along its western boundary, is about 40 miles. The surface in the west is a high, broken plateau, which rises rather steeply to the mountainous district known as the White River Plateau. The altitude ranges from about 5,800 feet at its western boundary to more than 12,000 feet at the summits of some of the peaks in the eastern part.

This region played a prominent part in the early history of Colorado, as it was the scene of encounters with the Ute Indians, which finally led to the removal of all the members of this tribe in Colorado to western Utah. In the spring of 1878 Nathan C. Meeker, for whom the town of Meeker, the county seat of Rio Blanco County, was named, was appointed Indian agent in this territory. He had trouble with the Indians from the first, and in the fall of 1879 he asked for troops to protect him and his associates. Major T. T. Thornburgh, with a company of 160 men, was ordered to assist Meeker, and started for the White River Agency in September, 1879. On the morning of September 29 Major Thornburgh and his men were ambushed in Red Canyon, a narrow ravine in the northern part of Rio Blanco County, where 15 soldiers were killed and 35 wounded. Major Thornburgh himself was killed and scalped. Meanwhile a party of Utes attacked Meeker and the employees at the Indian agency, killed most of them, and took the women prisoners. Immediately after these outrages there was a general demand for the removal of the Indians from this region, and in 1881 about 17,000 of them were placed on the Uinta Reservation in Utah. Rio Blanco County was organized in 1889 from the northern part of Garfield county.

RIO GRANDE COUNTY

Rio Grande County lies in the south-central part of the State and includes most of the western extension of the San Luis Valley. It is of irregularly rectangular outline and has an extreme length, from east to west, of 30 miles and an extreme width of 25 miles. The surface is generally level except in the southwest, where it rises steeply to form the San Juan Mountains. Its area is 574,720 acres. The altitude ranges from about 7,600 feet where the Rio Grande crosses the eastern boundary to about 13,000 feet at the summits of peaks of the San Juan Mountains in the southwest.

Early explorers frequently followed the Rio Grande across the territory now included in this county. John C. Frémont's fourth Rocky Mountain expedition crossed the area in 1848 and came to grief in the bleak San Juan Range, farther west. In 1860 a colony of Mexicans settled in the valley of the Rio Grande not far from the present site of Monte Vista. In 1870 gold was discovered in the western part of the county, and for several years mining development was rapid. For a short time in the early eighties this county ranked third in the State in production of gold. The county was organized in 1874 from parts of Conejos and Costilla counties and was named from the principal stream of the San Luis Valley.

ROUTT COUNTY

Routt County lies in the northwestern part of the State. Its northern boundary is formed by the State of Wyoming and a part of its eastern boundary by the Continental Divide. It is of extremely irregular rectangular shape, 75 miles long from north to south and about 42 miles wide. Its area is 1,477,760 acres, about 220,000 acres more than that of the State of Delaware. The surface is generally rough or mountainous, except in the valleys of Yampa River and its tributaries. The altitude ranges from about 6,230 feet at the point where Yampa River crosses the western boundary to about 12,000 feet at the summits of some of the peaks on the eastern boundary.

This part of Colorado was frequently visited by trappers, explorers, and prospectors before 1860, but no settlement was made until about 1866. In 1864 a prospector named Way discovered placer gold at the base of Hahns Peak while returning to Clear Creek County, from which he had started on his prospecting tour. He told the story of his discovery to Joseph Hahn, for whom the peak was later named. The two organized a party of miners and went to the Territory in 1866, establishing a small settlement near the present site of the town of Hahns Peak. They encountered many hardships in the severe winter that followed and finally gave up further efforts to develop the deposits they discovered. The county was organized in 1877 from a part of Grand County and was named in honor of John L. Routt, twice governor of Colorado.

SAGUACHE COUNTY

Saguache County is in the south-central part of the State and includes the north end of the San Luis Valley. It is of irregular shape and has an extreme length, from east to west, of about 85 miles and an extreme width, from north to south, of about 48 miles. Its area is 2,005,120 acres, or about 65,000 acres greater than the combined areas of the States of Rhode Island and Delaware. Its eastern boundary is formed by the Sangre de Cristo mountain range, and the Continental Divide passes across its northwestern corner. The San Luis Valley extends northward about 30 miles into the central part of the county. The surface here is level, forming a plain, which rises gradually to the Sangre de Cristo Range on the east. The altitude ranges from 7,500 feet in the southern part of the county to more than 14,000 feet at the summit of peaks of the Sangre de Cristo Range. For a distance of more than 50 miles every peak in this range rises to a height of 13,500 feet or more.

The first settlement was made in 1865 on Saguache River, near the present site of Saguache, by soldiers of the First Regiment of Colorado Volunteers. In 1867 Otto Mears, whose name is woven into the history of every county in southwestern Colorado, began his work of opening up wagon roads into the San Juan district, and for several years he did considerable work in Saguache County. The county itself was organized in 1867 from a part of Costilla County. The name is of Indian origin and is said

to be abridged from a Ute expression meaning "blue earth." The first settlers were chiefly miners, prospectors, and cattlemen.

SAN JUAN COUNTY

San Juan County is in the southwestern part of the State, in the heart of what is known as the San Juan mining district. This district takes its name from the San Juan Mountains, the principal range in this part of Colorado, whereas the agricultural district to the south, popularly known as the San Juan Basin, takes its name from San Juan River, which drains southwestern Colorado and northwestern New Mexico. The county is of triangular shape and has an extreme length, from north to south, of 30 miles and an extreme width, at the base of the triangle, of 25 miles. Its area is 289,920 acres. There are but four smaller counties in Colorado—Denver, Gilpin, Clear Creek, and Lake counties. The surface is extremely rugged, but there are a few small mountain valleys. The altitude ranges from about 8,500 feet at the point where Animas River crosses the southern boundary to more than 14,000 feet at the summits of some of the peaks in the north.

The early Spanish explorers penetrated the rugged area now included in San Juan County, where Spanish names have been given to numerous rivers and mountains. John C. Frémont's fourth expedition is supposed to have reached a point in this county late in 1848 before the severe winter forced the few remaining members of his party to make a painful journey back over the mountains into the San Luis Valley. John Baker's expedition passed through this region in 1860. Prospectors found pay ore here about 1870, but it was not until after the region had been bought from the Southern Ute Indians in 1873 that settlers began to come in. Mining development was rapid, for this is one of the richest gold and silver bearing areas in the State. The Durango & Southern Railroad, now a part of the Denver & Rio Grande Western system, was completed in 1882, and from that time on an immense store of wealth was poured out from the mines in the narrow canyons above Silverton. The county, which was organized in 1876, was taken from the northern part of La Plata County.

SAN MIGUEL COUNTY

San Miguel County lies in the southwestern part of the State. Its western boundary is formed by the State of Utah. It is of rectangular form, and its boundary lines are regular, except in its eastern and southeastern parts, where they lie along the summits of mountain ranges. Its extreme length, from east to west, is about 75 miles, and its extreme width is about 25 miles. The area of the county is 824,320 acres, or about 433,000 acres less than the area of the State of Delaware. The altitude ranges from about 5,000 feet in the west to nearly 14,000 feet at the summits of some of the peaks on the eastern boundary.

The county was organized in 1883 from a part of Ouray County.

SUMMIT COUNTY

Summit County is in the north-central part of the State. The Gore Range forms most of its western boundary, and its eastern boundary is formed by the Williams Fork Mountains and the Continental Divide, here called the Snowy Range. The county is very irregular in outline. It has an extreme length, from north to south, of about 48 miles and an extreme width of 38 miles. Its area is 415,360 acres. Most of its surface is mountainous. The altitude ranges from about 8,500 feet in the north to more than 14,000 feet at the summits of some of the peaks along the eastern and southern boundaries.

Summit County is one of the original 17 counties in Colorado Territory, organized in 1861. It was much larger then than now, including most of the area now divided into Eagle, Garfield, Grand, and Routt counties.

TELLER COUNTY

Teller County lies in the central part of the State directly west of Colorado Springs, and Pikes Peak, the best known mountain in Colorado, lies near its eastern boundary. It is an irregular rectangle in outline, about 27 miles long from north to south and 21 miles wide in its southern part. Its area is 550,080 acres, or a little less than half that of the State of Rhode Island. Most of its surface is mountainous, though there are a few tracts of rolling land in the mountain valleys. The altitude ranges from 7,600 feet, in the north, to about 13,000 feet at the summits of some of the mountain peaks in the southeast.

Teller County was organized in 1899 from parts of El Paso and Fremont counties and was named in honor of Henry M. Teller, for thirty years United States Senator from Colorado.

DEVELOPMENT BY COUNTIES

ARAPAHOE, DENVER, AND JEFFERSON COUNTIES

The city and county of Denver was separated in 1902 from Arapahoe County, one of the original 17 counties of the State, in 1861. Arapahoe County originally included Cherry Creek and the "dry" creeks flowing from the Cherry Creek divide to the Platte—the streams on which the gold was found that brought the miners to Colorado in 1858. The bars on these creeks yielded fine gold that had been reconcentrated from gold-bearing channels in the beds of the Tertiary Dawson arkose and of the Castle Rock conglomerate, the latest sediments that were deposited along the flanks of the Rocky Mountains from Colo-

rado Springs to Denver and that extended over the Arkansas-Platte divide, but the work of mining was not highly remunerative, although it was continued throughout 1859. In the geologically younger gravels of Clear Creek, which empties into the Platte north of Denver, the prospectors had better success at Arapahoe and Golden City, and from those places they followed Clear Creek to Jackson diggings and to Gregory diggings.

GILPIN COUNTY [14]

John H. Gregory, the discoverer of Gregory diggings, was allowed two claims, Nos. 5 and 6, in the newly formed Gregory district, but he very soon gave an option on his claims to E. W. Henderson and A. Gridley. Henderson and Gridley apparently paid Gregory for claims Nos. 5 and 6 out of the gold won from these claims during the season of 1859. Several groups of men leased parts of the mine and worked with varying failure and success until the property had been opened on the surface for a considerable length and to a depth of 130 feet at one end and 180 feet at the other, where the vein pinched—that is, "the cap was reached."

In 1860, according to Hollister, the east half of Gregory's Discovery claim was leased to an association called the American Mining Co. The leasing system was therefore introduced by Americans and not, as has been supposed, by Welshmen, who came later. Forty feet below the surface, as Hollister says,[15] the miners passed through "decomposed pyrites, quartz, dirt, and gossan and reached the undecomposed quartz, iron and copper pyrites, called the cap." This early local use of the term "cap" shows an interesting contrast with the present generally accepted use to indicate the decomposed vein matter or "gossan" near the surface. Instead of sinking a shaft they continued to work out the deposit through its whole length. The mixed decomposed quartz and "cap" paid them tolerably well for a few weeks, but they found that the amalgam would unite with only a small proportion of the gold in the "cap," and the mine soon ceased to pay, so that they were unable to make their payments. The original owners therefore took the property back and leased it successively to several others on the same terms, with the same result. All this time the mine was "in the cap," which was worked for its entire length.

Gridley finally sold his interest to Henderson. In June, 1862, Henderson put on a force that worked night and day up to February, 1863, when the vein was found to widen again. During these seven or eight months the ore taken out paid a little more than expenses, yielding $8 to $10 a ton. From February

[14] See also Bastin, E. S., Henderson, C. W., and Hill, J. M., Ore treatment, labor and royalties, freight rates: U. S. Geol. Survey Prof. Paper 94, pp. 153-170, 1917.
[15] Hollister, O. J., op. cit., p. 64.

to August, 1863, when work was stopped to permit a hoisting apparatus to be erected, ore yielding $60,000 was taken out. A 30-horsepower engine was set up, 40 hands were employed, and by the end of 1863 ore yielding $20,000 additional had been taken out. Negotiations had then been begun for the sale of the property, together with others on the Gregory vein, to persons in the East, and in the following spring it passed into the hands of the Consolidated Gregory Co. for the cash price of $1,000 a linear foot and a heavy consideration in the stock of the company. It lay idle for a year. In the beginning of 1865 work was begun in cutting down and timbering a working shaft, putting in a pump, and opening levels for stopes.

The sketch thus given is fairly typical of the general history of the gold quartz veins in Colorado. At first they yielded enormously in a simple sluice. It was not unusual for four or five men to wash out $150 a day for weeks at a time from such a lode as the Gregory, Bates, Bobtail, Mammoth, or Hunter. Single pans of "dirt" that would yield $5 could be taken up carefully from any of a dozen lodes. One group of men ran a sluice three weeks on the Gregory and cleaned up 3,000 pennyweights, a pennyweight being worth about 80 cents. Their highest day's work yielded $495, their lowest $21. Another group of four men that worked for four months on the Fisk lode made an average of $100 a day.

By July 1, 1859, a hundred of these sluices were running within a short distance of Gregory Point. A year later their owners were vainly trying to make a profit from "cap" and sulphides, but nearly all their gold was running off in the water used on the plates, leaving them a mixture that was worthless.

A sketch by Hollister of the developments on a claim on the Bobtail lode from its discovery to the beginning of 1863 should be of interest to the geologist. The first pay material was decomposed quartz, which was struck in a crevice 6 feet below the surface and which extended downward for 39 feet. The crevice was 4 feet wide and the ore averaged $41 a ton. Here a "cap"—an obstacle—was struck. The vein widened, its content of gold decreased, and its content of sulphide increased. It proved to be 26 feet thick and yielded an average of $20 a ton. A crevice then struck was found to be 3½ feet wide and to extend downward for 70 feet. The material in it yielded an average of $38 a ton. A second "cap" was here struck, which proved to be 35 feet thick and averaged $15 a ton. Next came a 3-foot crevice that extended downward for 40 feet and yielded $60 a ton. A third cap, which extended downward for 40 feet, yielded ore paying $42 a ton. The claim was the Bobtail No. 2 East. To a depth of 260 feet and a length of 100 feet along the vein it produced $204,000.

Meanwhile other diggings had been begun and hundreds of men were at work. Four or five months after gold was discovered in Deadwood Gulch, on South Boulder Creek, 300 men were employed there. Later in the season South Boulder Creek itself was flumed at several places to permit its bed to be worked out. The prospectors scattered, and mining was done in Twelvemile diggings (at the head of North Clear Creek), on Lefthand Creek, and on the smaller tributaries of North and South Boulder creeks. Very rich quartz veins were struck at Gold Hill, 12 miles west of Boulder City, and about October 1, 1859, a rude quartz mill was started there. All the Boulder diggings paid from $3 to $5 a day per hand.

Bar mining was also carried on energetically on the Platte above Denver, on Clear Creek near Arapahoe and Golden City, and on Ralston Creek near its mouth. Various contrivances were used to bring water to the mines. Wheels were put into the creeks to raise it out on the banks, and ditches from 3 to 10 miles long were dug. Towne & Patterson washed out $117 on the Platte in 15 days with a rocker. The Georgia Co., having brought water in a ditch 3 miles to the same diggings, washed out $54 in four days with a tom. Mining on the streams in the valley, however, was eclipsed by that in the mountains before the end of the season of 1859, and work on the valley gravels was abandoned and not resumed until 1904–5, when one futile attempt was made to dredge the gravels below Golden near the old workings at the town of Arapahoe.

In the first days of June, 1859, William Green Russell had found gold and had begun mining in the gulch that bears his name, a tributary of North Clear Creek, a little south of Gregory Gulch and parallel with it. A week's work with six men brought 76 ounces of gold worth $16 to $18 an ounce, according to its purity and fineness. Others had taken up claims above and below his, and toward the end of September 891 men were at work in the gulch, producing an average of $35,000 a week. At the same time 213 men were working in Nevada Gulch, in Illinois Gulch, on Missouri Flats, and along the upper tributaries of Gregory and Russell gulches, producing an average of $9,000 a week. As water became scarce for the gulches and lodes two large ditches were projected to bring it from the head of Fall River. These eventually became the Consolidated Ditch, which was 10 or 12 miles long and took in at its head 300 inches of water and delivered 150 inches on Quartz Hill. This ditch was completed during the fall, winter, and spring of 1859–60 at a cost of $100,000.

In the Gregory district several rude quartz mills and some arrastres, worked by water or teams, were in operation and returned handsome profits. Water soon began to be scarce, however, and work was stopped on many paying claims.

The pattern after which the mining districts and miners' laws were fashioned had been roughly outlined and adopted at a miners' meeting held in Gregory diggings June 8, 1859. The Gregory lode was discovered in May, 1859. For a few days the small party that went to the spot with Gregory on his second visit had a monopoly of the discovery and working of lodes. Many of the best lodes in the Gregory diggings were found and claimed before the end of May. The Gregory lode was divided into claims of 100 feet each, Gregory, as the discoverer, receiving two claims. This mode of division allotted the entire lode satisfactorily to all the men in the party.

By June Gregory Gulch from North Clear Creek to the confluence of Eureka, Nevada, and Spring gulches was crowded with canvas tents, log shanties, and bough houses, as thick as they could stand. William N. Byers had pitched his tent near what is now the corner of Main and Lawrence streets, Central City, and had suggested that name for the future city. It was estimated that there were 5,000 people in the gulch. The lower part of the gulch was swampy and overgrown with alders and willows.

The newcomers soon began to murmur about the privileges claimed by those who came first; none of the first comers had been there a month, yet they monopolized everything, so the newcomers contended that the Gregory, Mammoth, Hunter, Bates, Bobtail, Gregory Second, and other claims should be cut down to 25 feet each and redistributed. In consequence of this contention a great meeting, numbering some 3,000 men, was convened at Gregory Point. At this meeting resolutions were adopted defining the boundaries of the district and the conditions under which claims could be taken and held. One of the resolutions provided that lode claims were to be 100 feet long and 50 feet wide and that "creek" or placer claims were to extend 100 feet up or down the gulch from wall to wall—provisions that remained in force until the enactment of the Federal mining act of May 10, 1872. Subsequent meetings, held July 9 and 16, 1859, organized a provisional local government.

As early as the middle of July, 1859, Lehmer, Laughton & Peck started a Spanish arrastre [16] for grinding quartz near the mouth of Gregory Gulch.

On September 17, 1859, Prosser, Conklin & Co. began to build the first steam quartz mill erected in Colorado. This mill, which was small, was soon in successful operation.

By October 1, 1859, five arrastres and two small wooden stamp mills were running on North Clear Creek, all driven by water, and four arrastres were being built. Each of these was making about $200 a week by treating the headings of the sluices—that is, the quartz that was too coarse to pass through the screens used, which were pierced with about half-inch holes.

On October 7, 1859, Coleman, LeFevre & Co.'s steam mill was started, but it soon broke down. By November 4 it had been repaired and started again, and in the first seven days' run on quartz of the Gunnell lode it produced 1,442 pennyweights of gold. At this time the miners were 56 feet down on the Gunnell lode, and the decomposed pyrites or gossan of the surface was giving place to more solid material, which yielded $60 a ton and grew richer the deeper they went. At 76 feet from the surface 15 tons yielded $1,700. On the morning of September 29, 1859, the hills and gulches were covered by a foot of snow. The climate of the region was then unknown, and there was only a small stock of provisions in the country, especially in the mountains; so it is no wonder that, seeing so heavy a fall of snow early in the autumn, the miners, afraid of being snowed in and starved, left the place in a stampede. Probably not more than 1,500 men wintered in the mining region in 1859–60, but these were agreeably surprised at the mildness of the weather. The first snow melted and ran off in a few days, and the cold spell was followed by as lovely an Indian summer as was ever enjoyed anywhere. Good weather may almost be said to have lasted throughout the winter, as there were only three or four falls of snow that were accompanied by severe cold, and the cold spells were comparatively short. A few men stuck to their toms or to their hand rockers, burrowing on rich pay streaks in the gulches all winter. In the three months ending January 31, 1860, one man took out $2,400 in Nevada Gulch with a rocker.[17]

A rude quartz mill had been in operation for several months in 1859 at Gold Hill, between Lefthand and North Boulder creeks, and in the autumn a large water mill was being built. The owners of lodes in the Gregory district were busy during the winter getting out ore to supply the projected mills in the spring. From each of the lodes—the Cotton, Fisk, Bobtail,

---

16 An arrastre is a circular trough, generally 10 or 12 feet in diameter, made of stone, in which other heavier stones, called "mullers," are dragged around continuously, thus grinding up the quartz. They were generally run by water. The firm named above set up the first one run in Colorado.

17 A miner's rocker or cradle is a rectangular box provided with rockers and separated into an upper and lower part by a coarse iron screen. The miner sets it in an inclined position beside a spring or stream where a steady supply of water can be had and brings to it his "pay dirt." He places the dirt in shovelfuls on the screen and pours water on it with a pan or other vessel while he rocks the cradle. By this washing and rocking the finer material in the "pay dirt" is carried through the screen into the lower part of the rocker; the coarse material remains on the screen and after it is thoroughly washed is removed from time to time by the miner. The finer material includes mud and fine sand, which is washed out of the rocker over a lip at its lower end, as well as the particles of gold, which, being heavier, are not washed over the lip but are held in the rocker.

A "tom" is a long box or trough containing a screen like that in a cradle, but having no rockers. It is placed near a creek, so set as to slope downstream. The "pay dirt" is shoveled on the screen, and the water of the creek is led to and over it in a continuous stream. To the lower end of the tom are attached in sequence two or three sluice boxes—troughs having square cross sections, open at the top and at both ends—in which cross slats ("riffles") are nailed at intervals. Above the slats (that is, on their upstream sides) quicksilver is placed to catch and hold the particles of gold washed out of the pay dirt. A dozen or a score of men shovel the dirt into the tom and keep it loose, so that the water will flow freely through it, and another man keeps it free from large stones with a sluice fork. As the fine dirt passes along the tom the gold is stopped at the riffles, where it unites with the quicksilver, forming an amalgam. At some places the bedrock may be used as a sluice.

Clay County, Gunnell, Maryland, Casto, Kansas, Burroughs, and others—there had been piled up 300 or 400 tons of ore, the success of the Coleman & Le Fevre mill with rock from the Gunnell lode greatly encouraging the miners.

Very fair roads had been made into the mountains— one by way of Golden Gate (built by Tom Golden), 2 miles north of Golden, and one by way of Bradford (almost identical with the site of the houses of the present Ken-Caryl ranch), south of the mouth of Turkey Creek. A road from Denver to South Park by way of Mount Vernon and Bergen's ranch had been projected, and work on it was vigorously pushed during the winter. So also was work on the St. Vrain, Golden City, and Colorado (?) wagon road, which avoided Denver entirely. Fair wagon roads ran from Canon City and from Colorado City to South Park. There were trails from South Park to Middle Park, which was reached also by a trail from Gregory. On March 4 1860, Kehler & Montgomery's express coach—the first coach ever run on the Denver-Golden Gate-Dory Hill line—arrived at the mines of the Gregory district from Denver.

A second rush to Colorado, larger than the first, took place in 1860. By the first of May immigrants were arriving from the States at the rate of a hundred a day. It was estimated that up to that time 11,000 wagons had passed Plum Creek, counted perhaps at some spot on Plum Creek between the present Palmer Lake and Littleton, bound for "the Pikes Peak region." The Platte River route may be said to have contained for a full month but a single train, which extended from the mountains to Missouri River. A great many came up the Arkansas and went directly into South Park. The Gregory district was the destination of most of the newcomers, but it soon became overcrowded, so that there were many hardships.

During 1860 the mining laws for the Gregory district were codified and amplified. The integrity of the local laws was upheld by one of the first acts of the Territorial Legislature of Colorado, in 1861, and was further supported by the act of the Congress of the United States approved July 26, 1866. This act contained the first statement of the famous "Apex law." On February 26, 1861, the region, which had up to this time formed a part of the Territory of Kansas, was organized by an act of Congress as the Territory of Colorado, and William Gilpin was appointed governor.

There was only slight immigration in 1861, but work went ahead. The lode mines were active, and the mills were generally successful. There was a great deal of experimentation in the treatment of ores by different forms of crushers, by the use of steam, fire, and chemicals. Doctor Burdsall, of Nevadaville, had for some time been experimenting with smelting and seemed to be getting encouraging results,

but his furnace was destroyed in the fire of November 4, 1861, when Nevadaville was almost destroyed.

During 1862 and 1863 the mills of Gilpin County had much the same experiences with sulphides but did very well. As the mines reached greater depth the problem of removing the water became serious. Toward the end of 1863 more mines were bought by New York and Boston investors. Gulch, bar, and placer mining had nearly ceased as an attractive business in the original Arapahoe County in 1859, in Boulder and Clear Creek counties in 1860, in South Park in 1861, and in Gilpin County in 1863.

Much of the oxidized free-milling gold ore had been exhausted in 1859, but during the next four years some had been found, and the difficulties of amalgamation of the sulphides had been overcome with varying success. In 1864 mining came to a standstill. The Civil War had drained much of the source of immigration from the East and had also taken two regiments from Colorado. The stamp mills were saving only about one-fourth of the gold and wasting all the other metals. The history of 1865 and 1866 was but a repetition of that of 1864. At the end of the Civil War, with a reduction in price of food and the return of labor to its pursuits, there was some improvement in the conditions at the mines. The completion of the Union Pacific Railroad from Omaha, Nebr., to Cheyenne, Wyo., in 1867 also stimulated mining, but the most effective stimulus was the successful smelting of the ores.

In 1864 James E. Lyon took some selected ore from the Gregory lode to furnaces near New York City and, finding that he could recover not only the gold but much of the silver and copper, devoted his attention entirely to the introduction of smelting in Colorado. In 1865 he erected at Blackhawk small furnaces that were operated with partial success for nearly a year. At first the gold and silver were obtained by making a lead bullion from which the lead was driven off by cupellation, but this process was abandoned for the matting process, with copper as the vehicle. By October, 1866, necessary alterations in the furnace of Lyon & Co. had been made. There were then three roasting and three smelting furnaces, each set having a capacity of 20 to 25 tons a day. Besides these there was a patent roaster, three American hearths for lead ores, and one cupel furnace. Later, in 1866, these works were sold to the Consolidated Gregory Co. In 1867 there was 100 to 200 tons of matte on hand from these works.

In 1867 Nathaniel P. Hill began to construct at Blackhawk a matting smelter consisting of a calcining furnace and a small reverberatory. The Hill smelter, or Boston & Colorado Smelting Co. was organized in 1867; the first experimental plant was erected at Blackhawk in June, 1867; the establishment was opened for business January, 1868; the first shipment of matte was made in June, 1868. The fire brick for this plant were shipped by rail from St. Louis to the terminus of the Kansas Pacific Railroad, and thence to Blackhawk by wagon, 600 miles. The matte was hauled to Missouri River and shipped by way of New York to Swansea, Wales, for resmelting and refining. This plant was successfully operated, and its capacity was enlarged from time to time until in 1878 it was removed to Argo, near Denver, where it was operated continuously until 1910. Associated with Mr. Hill from 1873 was Mr. Richard Pearce, who had experience in smelting ore from Gilpin and Clear Creek counties at the Swansea smelter near Empire in 1872–73.

In 1875 Mr. Pearce invented a process for the separation of the gold, silver, and copper at Blackhawk.[18] This process [19] was not made public until after the decision in 1908 to close and dismantle the smelter at Argo.

In 1860 the Independent or Gold Dirt district, the source of the placer gold of Rollinsville and perhaps of that of the Deadwood diggings as well, with its 10 or 12 stamp mills and its town of log houses along Gamble gulch, was a competitor with the Gregory district. In 1867 only four or five companies were operating here. After 1868 the camp seems to have been nearly deserted until 1879, when ore from the Perigo mine yielded well in an old 16-stamp mill at the town of Gold Dirt. Work was continued at the Perigo mine until 1888 and perhaps later, and some work has been done there in a desultory way every year since. In 1880 a large amount of capital was expended in the construction of a ditch and flume, 4½ miles in length, along the valley of South Boulder Creek and across Moon and Gamble gulches, preparatory to working the placer ground in the valley of South Boulder Creek near Rollinsville. There is a record of placer mining near Rollinsville in 1897. The Smuggler mine of Moon gulch has been an irregular producer from 1908 to 1924.

From 1875 to 1908 Gilpin County continued to make a steady output of gold. From 1909 to 1924 the output showed a slow decrease.

CLEAR CREEK COUNTY

As a result of Jackson's discovery in 1859 exploration and placer mining were begun immediately on South Clear Creek and extended, according to Spurr and Garrey,[19a] from the junction of Fall River and Clear Creek to the forks of North Clear Creek and South Clear Creek, on river bars, stream placers, and bench placers. The principal deposits were in the vicinity of Idaho Springs. The exact site of Jackson's discovery is said to be beneath one of the large willow trees along the road on the north side of Chicago Creek about halfway between the Jackson and Waltham concentrating mills. The Jackson diggings consisted of bars or flats in or adjacent to the streams and of deposits forming the stream bed itself. The low benches that occur along the sides and in places on the top of the low ridge that separates Chicago Creek from Clear Creek just above their union were also worked extensively. The three old river terraces from 25 to 180 feet above the present Clear Creek were operated profitably, particularly the higher bench southeast of the junction of Soda and Clear creeks. Spanish Bar, as the creek and bars from the mouth of Fall River to a point below the Stanley mine were called, produced much gold, although Hollister says that the boulders were so numerous and so heavy and the water so troublesome that little of the bedrock was uncovered.

Very rich quartz veins were early discovered on the bordering hills, and in 1861 a 20-stamp water mill was started. This mill was run pretty steadily for two years on surface material from the Whale (now known as the Stanley), Lincoln, and other lodes in the vicinity.

After 1859 both gulch and lode mine development was active from Idaho Springs (then Idaho City) to Dumont (then Mill City, at the mouth of Mill Creek, 7 miles above Idaho Springs) and beyond. About August 1, 1859, George Griffith discovered the Griffith lode, near the present town of Georgetown, from which he sluiced out $100 in two days. Other deposits were discovered in this vicinity, but richer diggings elsewhere drew the miners away from this locality. Placer mining in this county continued until 1913, although the greatest output from this source was made in 1859–1863. Spurr and Garrey estimate that South Clear Creek and its branches, including Chicago Creek, yielded $750,000 in gold dust from 1859 to 1880.

From 1860 to 1865 Empire was very prosperous, owing to the ease with which gold could be sluiced from decomposed quartz lodes treated in the same way as placer gravel. This superficial, oxidized portion of the lodes extended down to a depth of 40 feet or more, where sulphides were encountered, containing gold in a free state. The sulphides were not amenable to the same simple treatment as the oxidized deposits, but amalgamation in sluices and in stamp mills at Empire continued until 1875, and desultory production was made there until 1924. From 1859 to 1865 the Empire sluicing operations produced about $1,500,000 in gold.

[18] For description of the Blackhawk plant and process, smelting charges, capacity, etc., see Egleston, Thomas, The Boston & Colorado smelting works, in Raymond, R. W., Statistics of mines and mining in the States and Territories west of the Rocky Mountains for 1875, pp. 294–295, 379–394, 1877; Am. Inst. Min. Eng. Trans., vol. 4, pp. 276–298, 1876.

[19] Anonymous, The revelation of a metallurgical secret: Eng. and Min. Jour., vol. 87, pp. 464, 963, 1909. Pearce, H. V. (son of Richard), The Pearce gold-separation process: Am. Inst. Min. Eng. Trans., vol. 39, pp. 722–734, 1908.

[19a] Spurr, J. E., and Garrey, G. H., Geology of the Georgetown quadrangle, Colo.: U. S. Geol. Survey Prof. Paper 63, pp. 312–314, 1908.

In 1863 a small stamp mill at Georgetown made a few good runs. During the next winter, nearly all the lodes in this district passed into the hands of eastern men, and a great deal of work was done on the Griffith and other lodes. A 40-stamp mill and a reverbatory furnace were erected. In September, 1864, prospectors from Empire started out in search of silver, which they expected to find in and around the range near the headwaters of the southwestern branches of Clear Creek, where these branches interlock with the headwaters of Snake River. The Coaley claim, on Glacier Mountain, near Montezuma, Summit County, the deposit on which was discovered early in 1864, was the first silver-bearing claim in Colorado but never produced much, only leading the way for the discoveries in Clear Creek County. On September 14, 1864, the Belmont lode, on McClellan Mountain 8 miles above Georgetown, was discovered. In 1865 there was a rush to this district, then known as the Argentine. In 1866 some prospecting and development work was done. At that time there were in the Griffith district, at the forks of Clear Creek, two towns half a mile apart—Georgetown, just below the forks, and Elizabethtown, in a small park just above them. These towns were soon after merged into one, called Georgetown. The Argentine companies put up mills and smelters at Georgetown and Elizabethtown. In 1867 the active development of some of the richer lodes of both the Griffith and Argentine districts was begun, including the Equator, Terrible, Baker, Brown, Coin, Griffith, and others. In 1868 the lodes worked by the Dives and Pelican mines, on Republican Mountain, were discovered, but active production from them was not begun until 1871. From 1870 to 1880 the silver-mining industry of upper Clear Creek grew in magnitude.

In 1866 most of the lode mines at Idaho Springs were idle, although some development work was done and some ore was produced, particularly at the Seaton and Crystal mines; at Spanish Bar, where the Whale, Lincoln, and Edgar mines were active; and in the Trail district, where the Freeland and others were being worked. Up Fall River and between Fall River and Mill Creek prospecting and development was in progress, and there was considerable activity also at Empire.

In the Lamartine-Trail Creek district the Freeland mine, on a deposit discovered in 1861, yielded handsomely to 1888 but irregularly thereafter to 1924, and the Lamartine, located in 1867, produced regularly from 1898 to 1905, when it was turned over to lessees.

In 1871 operations were begun at the Hukill mine. The mines of Virginia Canyon were brought to notice by the opening of the Specie Payment in 1876. In 1871 gulch mining on the Chambers claim, on South

Clear Creek about 200 yards from the junction of North and South Clear creeks, in Jefferson County, was reported to yield $5 a day per hand, and several other claims half a mile to a mile above the forks were then producing and had been producing for five years.

In 1875 the greater part of the activity of the county was concentrated at Georgetown, where there was a competitive market for silver ores through several outside purchasers for eastern smelters and for smelters in Germany. The mines at Empire were not producing largely. Those near Idaho Springs yielded about $90,000, most of it in smelting ore from the Hukill, Victor, Veto, Queen, and Seaton mines, sold to the Boston & Colorado works, at Blackhawk. From 1876 to 1880 the Georgetown district continued to be the principal producer, and a little work was done at Empire, but the Idaho Springs and Freeland districts gradually increased their production of gold, and Dumont and Lawson each made a small output.

The activity of the mines in the vicinity of Lawson dates back to the discovery of the Free America vein on Red Elephant Hill, in 1876. The Commodore and others produced rich silver ores to 1889. On the south side of Clear Creek the Jo Reynolds was discovered in 1865 and was operated almost continuously from 1877 to 1907. South of Dumont the well-known properties are the Senator, Blue Ridge, and Syndicate; on the north, the Albro.

By September 24, 1870, the Colorado Central (now Colorado & Southern) Railway had been built between Denver and Golden, and in 1872 it was extended from Golden to Blackhawk and a branch was run up Clear Creek from the forks to Floyd Hill. In 1877 the railroad was completed from Floyd Hill to Georgetown.

The year 1880 held the county record for the production of silver until 1894, when, despite the gradually lowering price, the output of silver for the county was the highest ever made. In 1883 the most extensive placer operations were in the vicinity of Alice, on Silver Creek, a branch of Fall River. From 1904 through 1918 the silver mines on upper Clear Creek gave way to the gold-silver mines on lower Clear Creek, although after 1903 the mines on upper Clear Creek became large producers of zinc, in addition to lead and silver, and from 1919 through 1924 surpassed the lower creek mines in value of output. Clear Creek County ores have always carried much zinc, of which relatively little of the total content has been saved. Metallurgy that would save the zinc should prove an impetus to the revival of mining in this county.

SUMMIT COUNTY

In the summer of 1859 prospectors from South Park ascended Michigan Creek, a tributary of the Tarryall, and pushed over the Continental Divide at what is now known as Georgia Pass. On the north side of Farncomb Hill, in Georgia Gulch, a tributary of the Swan, east of the site of Breckenridge, they discovered rich placer ground. From 1859 to 1862 Georgia Gulch alone is said to have yielded $3,000,000. Soon after gold was discovered in Georgia Gulch it was found in paying quantity on Swan River—in Gold Run Gulch, Galena Gulch, American Gulch, Humbug Gulch, and Delaware Flats—and on Blue River, in French Gulch and its tributaries, Gibson, Nigger, Corkscrew, Illinois, and Hoosier gulches. All these gulches were worked more or less for two or three years after 1860, and at Gold Run Gulch and one or two other localities extensive operations were carried on for years. Munson, the assayer in charge of the Denver Mint in 1887, estimated the yield of placer gold in Summit County from 1860 to 1869 as $5,500,000. In 1870 the most productive localities were Illinois, Iowa, French, Gold Run, Galena, and Georgia gulches and Buffalo and Delaware flats.

As early as 1860 a ditch was constructed to carry water from Blue River to Gold Run Gulch, a distance of 6 miles. Two years later another ditch, 9 miles long, which brought 500 inches of water from Blue River, was constructed. Other ditches and bedrock flumes were constructed later. A creek claim in Gold Run embraced 100 feet of the stream and 75 feet on each side of it. A bank claim was 100 feet square. Anybody was entitled to hold a creek claim and a bank claim in each tier of claims on either or both sides of the creek. Buffalo Flats, to which water was brought in ditches from French Gulch, is said to have yielded more gold than Gold Run. In 1870 there were 100 miles of ditches and flumes in the Breckenridge region. At this time hand washing had generally given place to hydraulic methods, and booming—damming and flushing—afterward extensively practiced, was being introduced. The bed of French Gulch, especially in the neighborhood of Lincoln, was laboriously explored by drifting in those early days.

Silver-bearing lodes were opened on Glacier Mountain, near Montezuma, as early as 1864; and in 1866–67 about twenty lodes of gold and silver quartz, seamed with galena and pyrite, were recorded in Buffalo Flats. In French Gulch a good deal of work had been done. In 1865 a stamp mill was set up in French Gulch, but it was merely started and was then apparently idle for some time. In 1869 some argentiferous lead ore was taken from the Old Reliable vein, at Lincoln, and about the same time the Laurium (or Blue Flag), in Illinois Gulch, appears to have shipped argentiferous lead ore by wagon to Denver and Golden. The Cincinnati (Robley claim), on Mineral Hill, near Lincoln, was developed in the early seventies and for 10 years made shipments of high-grade galena and cerusite ore. As early as 1873 a reverberatory furnace was erected in French Gulch to treat this ore. Other lead-silver mines operated prior to 1880 were the Union, Lucky, and Minnie. Late in the seventies an unusually rich

mass of silver ore carrying both lead and copper was uncovered in the reddish sandstones northwest of Breckenridge Pass, at the Warriors Mark, but later developments there were disappointing until 1922–23, when the mine was reopened and shipments of ore made.

Raymond reports that as early as 1870 more than 4,000 lodes were listed on the records of Summit County, but very few of them were developed, as the owners of most of them were working placer mines. Most of the lodes then under development were at Montezuma and Sts. John, in the Snake River mining district. Montezuma was at that time reached by stage from Como to Breckenridge and thence by trail, or from Georgetown by way of the road crossing the range near Grays Peak. From 1870 to 1875 there appears to have been much development in this district and in the adjoining Peru district on the Comstock, Coaley, Sukey, Silver Wing, Napoleon, Chautauqua, and other lodes, and many mills and smelting plants of different kinds were erected, but little ore seems to have been treated and little shipped because of the difficulties of transportation and the consequent high cost. Neither the mills nor the smelters in this district appear to have been successful, owing, no doubt, to the heavy content of zinc in the ore, and in 1875 the St. Lawrence mill was moved to the Pelican lode, at Georgetown. In 1883 the district seems to have reawakened, for the railroad furnished better transportation than could be had before, and many shipments of silver-lead and silver-copper ores were made. In 1890 and 1891 some shipments were made, and in 1892, according to the report of the Director of the Mint, the Decatur Mines Syndicate produced $319,275 worth of silver (coinage value), and again in 1899 some work was evidently done. Production was resumed in 1906, but it fell off again until 1909, when there was a slight revival. In 1910 some fairly modern mills were erected and some mine and dump ore was treated. From 1911 to 1924 milling operations were spasmodic.

The following paragraphs are quoted from Ransome,[20] whose sketch of the history of mining at Breckenridge is here freely drawn upon:

Notwithstanding the fact that rich placers had been washed on the slopes of Farncomb Hill since 1860, it was not until the end of 1879 or the beginning of 1880 that gold was found in place on the Ontario claim; this event was rapidly followed by discoveries on the Elephant, Boss, Key West, Bondholder, Gold Flake, and other now well-known claims on the hill. In view of the extreme narrowness of these veins and their failure to outcrop above their covering of soil the comparatively late date at which they were discovered is not altogether surprising. For about 10 years these wonderful little veins were actively exploited, chiefly by lessees, who riddled the northeast side of Farncomb Hill with tunnels and drifts and broke into pocket after pocket of the beautifully crystallized wire and flake gold

[20] Ransome, F. L., Geology and ore deposits of the Breckenridge district, Colo.: U. S. Geol. Survey Prof. Paper 75, pp. 17-20, 1911.

for which this locality is justly famous. In 1885 there were over 100 men working on the hill; but by 1890 the search had lost some of its zest. * * * There was a period, between 1889 and 1898, when considerable work was done by the companies that successively controlled what was originally known as the Ware property, after Col. A. J. Ware, one of the first to operate in the district on an extensive scale. Thus late in 1888 the Victoria Mining Co. built a mill in American Gulch and this was run for a few years on such low-grade ore as could be gathered from the workings and dumps on the north slope of Farncomb Hill. Another mill, [later] known as the Gold Dust mill, was built by the same company in 1889 on the west side of the Blue near Breckenridge, and for a time 9 or 10 teams were kept busy hauling ore to it from the company's mines, in which numerous lessees were at work. The total capacity of the two mills was 120 tons, but they were not long in operation. * * * About the year 1894 the Victoria Mining Co. was succeeded by the Wapiti Mining Co., which built many miles of flume, bringing the water from the Middle Swan, and which hydraulicked many of the old dumps and much of the surface material on the north side of the hill.

Another part of the district that for a time rivaled Farncomb Hill as a source of gold, although never noted for such beautiful specimens, is Gibson Hill. Here the first event of note was the discovery of the Jumbo lode by E. C. Moody in the summer of 1884, this being followed by active prospecting all over the hill. A settlement known as Preston was established on the north slope, and for several years the Jumbo, Buffalo, Extension, and Little Corporal mines produced a large quantity of comparatively low-grade free-gold ore that was milled in part at Preston and in part at what was known as the Eureka mill, at the mouth of Cucumber Gulch, below Breckenridge. About the year 1886 Moody began work on the Seminole and other claims north of Gold Run, afterward developed into the Jessie mine. Before 1890 production had begun also from the Hamilton and Cashier mines.

Meantime, while the gold deposits were being developed, the silver-lead mines were not idle. In 1883 the Cincinnati was the largest mine on Mineral Hill, but it was soon surpassed by the Lucky. In the middle eighties Lincoln was a thriving town in which three small mills were active, treating about 60 tons of ore a day. In one of these was concentrated the first ore taken from the Oro mine in 1887. In the following year the owners of the Oro built their own mill at the mine, which soon became one of the most productive in the district. Another mine that came into prominence about this time is the Iron Mask, situated on the west side of the Blue, in Shock Hill. From this mine shipments of high-grade silver-lead ore began in 1888 and continued with few interruptions for about 10 years. Other mines shipping in 1889 or 1890 were the Ohio, on Shock Hill; the Kellogg and Sultana, on Gibson Hill; the Washington, Dunkin, and Juniata, on Nigger Hill; the Mountain Pride, at the head of Illinois Gulch; the Oro and Lucky, on Mineral Hill; the Victoria (Wapiti) group, on Farncomb Hill; and the I. X. L., on the Swan. Just beginning noteworthy development and production at this time were the Puzzle, Ouray, Country Boy, and Wellington mines. In 1891 the Boss and Gold Flake mines, on Farncomb Hill, yielded some rich masses of crystalline gold. Among he events of 1892 was the organization of the Jessie Gold Mining & Milling Co., which took over the property of the Gold Run Mining Co. and began the building of a new mill at the Jessie mine. The Extension Gold Mining & Milling Co. undertook in the same year the thorough development of what had hitherto been generally known as the "Fair property," near the Jumbo mine.

About the year 1896 shipments were resumed from the Mountain Pride, and this mine shortly afterward began extensive development and in 1898 was the leading producer in the district. This preeminence, however, was not long maintained, and the mine had been idle for a number of years when visited in 1909. The Cashier and Jessie mines were actively worked and produced large quantities of milling gold ore in the late nineties. In 1909 the only mines producing were the Wellington, Country Boy, and Sallie Barber. The Hamilton mine, after being productive for many years, was abandoned about the year 1902. * * * During [1909] also work was resumed in the Puzzle and Gold Dust workings, from which and from the Ouray mine large quantities of high-grade silver-lead ore with some gold were stoped during the 20 years following 1885.

In 1908 a new 100-ton wet concentration mill was completed for the lead-zinc ores of the Wellington, and beginning in 1909, the yield of zinc from this property has been large except during a period of idleness from December, 1920, to December, 1922. Early in 1912 a 50-ton roaster and magnetic separating mill was built to remove the iron from the zinc middlings. In 1913-1915 considerable quantities of metallic gold were found in pockets of the Dunkin mine on Nigger Hill.

The ordinary modes of placer mining, particularly hydraulic washing, booming, and bedrock drifting, continued to be actively practiced up to about the year 1900 and then gradually fell into disuse. In 1909 none of this work was in progress, except at one place on the upper Swan, where an attempt was being made to convert a bedrock drift into a sluice connected with an open pit; attention had been diverted from the high-level and superficial placers to those amenable to the modern method of dredging.

It was in 1895 that Mr. Ben Stanley Revett, recognizing the possibility of working the deep gravels along the main streams, began by attempting to sink a shaft to bedrock on the Swan, near the mouth of Galena Gulch. This shaft, owing to the large quantity of water present, was not successful. He then undertook to test the gravels with an oil drill, this probably being the first application of such an implement to prospecting in Colorado.

In 1898 the American Gold Dredging Co., organized in Boston under the laws of Michigan, built two dredges on the Swan, but these, planned in accordance with New Zealand experience, proved unable to excavate the deep and coarse gravel near the mouth of Galena Gulch. In this year also the same company set up two Evans hydraulic elevators on the same stream. On the Blue, near the mouth of Cucumber Gulch, Pence & Miller were trying to sink a placer pit, using first hydraulic elevators and then steam pumps. At 30 feet in depth the pit had not reached bedrock. In 1899 the Blue River Gold Excavating Co. began work on the Blue, about 2 miles north of Breckenridge, with two dredges of the orange-peel type. These were failures. Toward the end of the year the North American Gold Dredging Co. built one Risdon and one Bucyrus dredge on the Swan and dismantled the two first constructed. The new boats, the larger having a capacity of 2,500 cubic yards a day, were operated for a few years but were never fully successful and were finally abandoned. The gravels having been found difficult to handle with the lightly constructed dredges then in use, the Gold Pan Mining Co., organized in December, 1899, with a capital of $1,750,000, acquired 1,700 acres of placer ground and undertook to work the bed of the Blue, at the south end of the town of Breckenridge, by using hydraulic elevators. This company spent $750,000 in cash and gave 400,000 shares of stock, par value $1 each, to pay for the construction of 3 miles of 8-foot ditch and a connecting pipe line having a capacity of 6,000 miner's inches, the erection of machine shops, and actual excavation. Bedrock was reached in October, 1902, at 73 feet. But this ambitious project, like its lesser predecessors, failed to wrest riches from the river channel, although it has left an enduring monument to itself as well as an instructive warning to others in the huge pile of boulders, many of them over 6 feet in diameter, that now overlooks the town. In the prospectus issued by the company the productive life of the ground had been estimated at 40 years, the average value of the gravel at 60 cents a yard, the cost of working at 10 cents a yard, and the total profits at $80,000,000. The plant has not proved entirely useless; part of the capacity of the ditch and pipe line has been utilized for lighting the town, and the machine shops have proved a valuable adjunct to the gold-dredging industry.

In 1905 the American Gold Dredging Co. was operating one dredge on the Swan, but in 1906 it sold its property to Lewisohn Bros., of New York, who spent the year 1906 drilling and testing placer claims from Browns Gulch on Swan River down to the junction of the Swan with the Blue and ground between these two streams. In 1907 this company started building two 9 cubic foot open-connected Bucyrus dredges, to be operated by electricity. Under the name Colorado Dredging Co. these two dredges were operated for six months in 1908. In 1905 Mr. Revett, acting as trustee for the Reliance Gold Dredging Co., unincorporated, began the construction of a double-lift dredge of his own design on French Gulch. In 1908–9 the Reliance dredge was operated during the winter, work that had been thought impossible. In the summer of 1909 the Reliance dredge was overhauled and adapted to the use of electricity for power, the steam equipment being abandoned. The overhauling also included the change from an open-connected type with buckets of 9 cubic foot capacity to a close-connected type with buckets of 5 cubic foot capacity. The hull for a fourth dredge, known as the Reiling, operated under the name French Gulch Dredging Co., was built in 1908 in French Gulch. The machinery for this dredge was that removed from one of the two Reiling dredges that operated below Golden in 1905–6. All four dredges were operated in 1909 and 1910. One of the two dredges of the Colorado Gold Dredging Co. was idle in 1911, 1912, and 1913; in 1912 it dredged 1,270,476 cubic yards of gravel, with a yield of $208,248, an average of 16.39 cents a yard, at an operating cost of 5.296 cents, a general expense of 0.263 cent, and a profit of 10.83 cents.

In 1914 the Tonopah Placers Co. took over the two dredges of the Colorado Gold Dredging Co., and the dredge of the Reliance Co., and until 1918 operated all three steadily. To the end of 1915 this company had dredged 300 acres, and there remained 4,783 acres under company ownership to be dredged. The average value per yard dredged was $0.147 in 1916, $0.1344 in 1917, and $0.086 in 1918. At the end of 1918 Tonopah dredge No. 1 on the Magnum Bonum placer, on Blue River, had good ground ahead and gave encouraging promise of future earnings. In 1918 Tonopah dredge No. 2 was operating on Swan River on ground belonging to the Farncomb Hill Gold Dredging Co. under a contract by which the Tonopah Placers Co. received half the net profits. This ground proved to contain less gold than the ground dredged earlier. In 1918 Tonopah dredge No. 3, in French Gulch, reached the end of the property in the direction it was going and was turned back. In 1919 this dredge started work on ground that showed a fair content of gold and had at least two years' work ahead. The Powder River Dredging Co. installed a fifth dredge in the district on Blue River in 1917 and operated it 8 months in 1918 and 10 months in 1919. The dredge was then taken over by the Blue River Dredging Co., which operated it almost continuously in 1920, 1921, 1922, 1923, and 1924.

Dredge No. 1 of the Tonopah Placers Co.'s dredges continued work during 1919 on the Magnum Bonum placer; No. 2 continued on the ground of the Farncomb Hill Gold Dredging Co., but this land proved unprofitable, and the dredge was closed down in December; No. 3 was in continuous operation on ground that had been left by other dredging operations some years ago. In 1920 dredge No. 1 continued operations on the Magnum Bonum placer and had work ahead for probably four or five years more; No. 2 was idle; and No. 3 was operated on ground of the Long Island Mining Co. but was closed in December because it had reached a point where further operations became unprofitable. In 1921 dredge No. 1 only was operated; No. 2 was floated and dismantled; No. 3 stood where left in 1920. In 1922 and 1923 only dredge No. 1 was operated. In 1924 the Tonopah Dredging Co. sought unsuccessfully to obtain permission to dredge a passageway between the railway and rear of the town of Breckenridge to reach the ground of Illinois Gulch and the upper Blue.

Production of Tonopah Placers Co., Breckenridge, Summit County, Colo., 1914–1923

Year	Cubic yards dredged	Gold produced	Silver produced	Gross value of bullion shipped to Denver Mint	Value per yard	Cost of operations at dredges	Fineness of gold	Fineness of silver
		Ounces	*Ounces*					
1914	2,995,256	22,542.00	4,782	$467,204.00	$0.156	$290,527.00	0.807	0.174
1915	3,242,247	23,034.30	5,416	479,017.56	.148	267,324.57	.803	.189
1916	3,199,962	19,557.68	4,748	406,367.68	.127	257,138.76	.798	.194
1917	3,351,821	21,745.65	5,210	451,743.20	.135	285,104.75	.793	.194
1918	3,267,307	13,564.99	3,345	282,973.83	.087	332,902.26	.788	.194
1919	3,122,571	14,776.48	3,852	309,509.22	.099	293,255.27	.738	.193
1920	1,791,143	11,008.96	3,026	233,661.96	.130	181,007.61	.624	.169
1921	1,541,454	9,299.73	2,445	194,689.03	.126	174,012.44	.725	.196
1922	1,393,148	7,428.09	1,958	155,512.19	.112	140,888.07	.762	.201
1923	1,228,555	5,716.14	1,443	119,175.37	.097	117,333.68	.757	.191
	25,133,464	148,675.02	36,225	3,099,854.04	.123	2,339,494.41		

Improved facilities for handling ore and concentrates were afforded by the completion, in 1883, of the narrow-gage Denver, South Park & Pacific Railroad (now part of the Colorado & Southern) from Como to Leadville by way of Breckenridge, and in 1906 and 1907 advantage was given to all kinds of mining through the introduction into the district of electric power by the Central Colorado Power Co. and the Summit County Power Co.

According to S. F. Emmons,[21] gold-bearing placers were discovered in the Tenmile region in the early sixties by prospectors who came from the placer diggings around Breckenridge, and it was probably by them that the name Tenmile was given to the stream, this being about its distance from Breckenridge. A few vein deposits in the Archean rocks were afterward opened, but no important mining development took place until 1879–80, after the discovery in 1877–78 of rich silver deposits in stratified limestone around Leadville, which directed the attention of prospectors to this then somewhat novel class of ore deposits. A very considerable number of more or less oxidized bodies of pyrite, blende, and galena were opened along the eastern slopes of the Tenmile Valley, but the most productive ore bodies were found on the west side of the valley, from Robinson northward to Jacque Mountain. Most of the richer oxidized ores were exhausted during the decade from 1880 to 1890, and many of the mines were closed, the enormous quantities of unaltered pyritous ores being then too poor for profitable exploitation.

Since 1890 there have been in this district intermittent periods of activity of mining, each period being characterized by some single outstanding operation, stimulated by an increased demand by the smelters for pyritous ores, or by the attempts to treat the lead-zinc sulphide ores by magnetic separation, or by the sale of the iron-zinc sulphide ores to American zinc retort plants. It was in this district, at Kokomo, that A. R. Wilfley began to develop the Wilfley table, which was put into successful operation in 1895. McNulty Gulch, near Robinson, was the scene of placering in 1860, shortly after the discovery of gold in California Gulch, in Lake County, and some gold was recovered in McNulty Gulch in 1920–21. The molybdenum deposits of Bartlett Mountain, near Climax (Fremont Pass) were successfully operated in 1918–19 and again in 1924–25. The greater part of these deposits were decided in 1918 to be in Lake County.

PARK COUNTY

In 1859 some prospectors who ascended the North Fork of the Platte and others who arrived from the East by way of Arkansas River, passing by the sites of Canon City and Colorado City, found gold in the streams of South Park and later rich lodes at the headwaters of the tributaries of the Platte. The exploration of these placers and lodes in 1859–1861 caused settlements to rise overnight. Among these settlements were Montgomery, at the base of Mount Lincoln, on the headwaters of the Platte; Buckskin, 6 or 7 miles south of Montgomery; Fairplay, on Beaver Creek; and Tarryall and Hamilton, near the head and in the western edge of the park, on Tarryall Creek, at the point where it breaks away from the range. The richer placers were worked out feverishly, and by 1867 many of the settlements were deserted. Only Tarryall and Fairplay remained. From August, 1859, to 1872, the placers in the vicinity of Fairplay yielded about $1,000,000 and those in the vicinity of Hamilton and Tarryall yielded about $1,000,000.

After 1867 the placer ground became too poor to pay by washing in small claims, and the companies acquired sufficient territory to justify the construction of flumes and the purchase of hydraulic machinery. Hydraulicking and sluicing, including for many years the operations of Chinese, have been carried on more or less continuously since then, and the output from 1868 to 1918, inclusive, a period of 51 years, amounted to $1,518,924, or $29,783 a year; and as the output from 1859 to 1867, inclusive, was $1,780,000, the total output of the placers to 1919 was $3,298,924. In 1919–1921 the placer ground near Fairplay was drilled and sampled, and construction of a dredge was started in 1922 by the South Park Dredging Co.

[21] U. S. Geol. Survey Geol. Atlas, Tenmile district special folio (No. 48), p. 1, 1898.

This dredge was built by the Yuba Construction Co. of California and has 70 buckets of 9 cubic feet each, capable of handling 4,680 yards in 24 hours. It was set in operation in May, 1922, in the flood plain of the South Platte, between the high morainal banks below the town of Fairplay. Dredging was successfully continued in 1923–24, and two larger dredges may be installed later to dredge the deep morainal gravels. The gold that was received at the Denver Mint in 1923 showed an average fineness of 0.774 and 0.160 fineness silver.

One of the most valuable gold lodes found was the Phillips, in Buckskin Gulch, discovered in 1860 by Joseph Higginbotham, known as "Buckskin Joe." In June, 1861, twelve persons were working on this lode. In September, 1861, the town of Buckskin contained a thousand inhabitants. From June 18 to October 19 about $50,000 was taken out. The process employed was simple. The top quartz and dirt was run through sluices, and the headings were reworked in arrastres. The ground worked yielded about $350 per cord of ore, a cord of ore being equal in size to a cord of wood, measuring 128 cubic feet and weighing 8 to 9 tons. The retorted gold sold for $16 in coin an ounce.

During the same season $25,000 was taken out of the Phillips lode by other miners. The lode was worked until 1863, when the miners reached sulphides, which could not be treated by ordinary milling methods. The total output, according to Raymond, was about $250,000. The history of the work on this lode is identical with that of the work on all the gold lodes about Hamilton, Montgomery, and Mosquito. The top quartz was mined easily and was treated readily by amalgamation. Large companies were formed and extensive mills were built, but as soon as the sulphides were reached the mills were closed and operations were stopped, for the sulphides could not be treated by amalgamation.

The gold obtained from the lodes of Park County from 1859 to 1867, according to Raymond, amounted to about $710,000. In 1868 and 1869 probably only a little gold lode mining was done in Park County. In 1870 the Pioneer mine produced $40,000 in four months. In July or August, 1871, silver ore was discovered on Mount Bross and Mount Lincoln. The ore of the Moose mine is said to have averaged $460 a ton. During 1871 about 30 tons was shipped to Swansea, Wales. In 1872 prospecting was done from Ute Pass (at the head of Michigan Creek) to Buffalo Peaks. In that year about 1,500 tons of silver ore was sold for about $150,000. Most of this ore was purchased by a branch of the Boston & Colorado Smelting Co. (the Blackhawk smelter) and by the Mount Lincoln Smelting Works, a company that had built a 10-ton blast furnace and had begun operations

on December 1, 1872, producing lead bullion and copper matte.

In 1873 there was an increase in production, due chiefly to the operation of the smelters. The Mount Lincoln Smelting Works had been erected by E. D. Peters, who at first believed that large quantities of lead ores could be obtained in the vicinity, especially from the Horseshoe deposit, which had been discovered as early as 1867. As the rich silver ores from Mount Lincoln and Mount Bross contained a great deal of lime and heavy spar and considerable galena, Mr. Peters erected a blast furnace, but after operating it as a lead plant from December 1, 1872, to August, 1873, and finding that an adequate supply of lead could not be obtained, he decided to erect a reverberatory furnace to produce copper matte. The reverberatory furnace was completed September 10, 1873, and was operated for the rest of the year but was idle in 1874 and 1875. The branch of the Hill [Blackhawk] smelter at Alma was operated in 1873–74 but was idle in 1875, and no record has been found that either smelter was ever revived, for Hill no doubt consolidated his operations at Argo, near Denver, after he left Blackhawk in 1878. In 1909 the Colorado Gold Mining & Smelting Co. built a semipyritic matting plant, which was operated for a short time and then closed.

By 1881 the Moose mine is said to have produced in all $3,000,000, and the Fanny Barrett mine, on Loveland Mountain, was making a notable output. During that year some very rich free gold was taken from the London mine, on the ridge between Mosquito Creek and Leadville. In 1882 little work was done at the Moose mine, but the Dolly Varden mine, south of the Moose, was productive. This mine had been worked steadily since the deposit there was discovered. The output of the mine from 1872 to 1882 was 15,000 to 20,000 tons of ore, which averaged 150 ounces of silver per ton. In 1882 probably 1,000 tons was taken out of the Fanny Barrett mine, but only 200 tons was treated, the rest being kept for treatment at a smelter which the company had erected at Alma. The London mine in that year built 7 miles of railroad from the foot of London Mountain to the South Park Railroad near Alma. The U. P. and K. P. mines were actively operated, and the milling ore was treated in the 20-stamp amalgamation-concentration mill. The Last Chance mine, in the Horseshoe district, was also operated.

In 1883 the Montgomery district showed signs of new life, work having been done on the Nova Zembla vein and on the Harrington vein, which is said to have yielded $500,000 in 1863. Some ore produced in development was shipped from the Fanny Barrett mine. The new mill of the London mine (near Mosquito Pass) was in operation and produced $124,000 in gold bullion in seven months and 420 tons of concentrates worth $60 a ton. The mine also shipped 210 tons of high-grade ore, worth at least $21,000, and 600 tons averaging $18 a ton, to mills in Idaho Springs and Central City. In 1883 the Whale mine, in Hall Gulch, shipped silver-lead ore worth $24,516. The London mine has been worked almost continuously since 1875, although from 1911 to 1914 it made only a small output. The Moose, Dolly Varden, and Fanny Barrett mines have made but small output since 1884. The Hilltop made a large output in 1888 and 1889 and has been worked intermittently ever since. Each year from 1908 to 1915 it produced considerable zinc carbonate. The New York, Orphan Boy, and Atlantic & Pacific mines were formerly well-known producers.

BOULDER COUNTY

In 1858 a party of prospectors went up what is now known as Sunshine Canyon, followed the main ridge between Fourmile Creek and Lefthand Creek, and in December, 1858, camped on the west slope of Gold Hill, Boulder County. They found gold in a gulch called Gold Run, and the next spring they returned and discovered many veins. The Horsfal property in particular yielded much free gold on the surface that could be recovered in sluice boxes. Soon other veins were supplying surface ores to the stamp mills and arrastres on Lefthand Creek. When the oxidized ores failed mining was nearly stopped until the Blackhawk smelter was erected, in 1868. The few remaining miners continued to work until 1872, when gold telluride was found in the Red Cloud vein.

As early as 1860 placer gold was mined on Fourmile Creek from its mouth upstream, but active placer mining lasted only about a year in Boulder County, although a little gold was taken out every year for many years. According to reports, $6,000 was taken out in 1872 on Fourmile Creek at a point 8 miles west of Boulder, and from 1887 to 1889 placer mining was active near Sugar Loaf.

In 1859–60 prospectors from Gilpin County roamed over the hills near Caribou, but they did not recognize the silver ore there. It is said that a prospector named Conger saw some silver ore from Nevada in a railroad car and came back to Caribou and located the Poorman mine in 1869. In the same year William Martin and George Lythe located the Caribou mine.

According to Raymond,[22] 26 tons of ore from the Caribou mine containing $3,217 in silver were sold in 1869 to Professor Hill at Blackhawk; in 1870 about 425 tons of ore, worth about $175 a ton, were extracted. According to the report of H. C. Burchard, Director of the Mint,[23] Conger discovered the Poorman lode in

1869 and took out $15,000 at 70 feet and $100,000 in opening the mine, but Raymond, in his detailed reports from 1869 to 1875, does not mention the Poorman mine as producing; in fact, he does not mention it at all until 1875 and then as a small shipper only. Burchard reports that the Poorman was developed in 1882 by a 370-foot shaft and four levels aggregating 1,000 feet and in 1883 by a shaft 550 feet deep and says that it was producing regularly. In addition he reports that the Poorman had been from the first a steady and profitable producer. In his report for 1870 Raymond speaks of the Conger mine as one "concerning which the general opinion is favorable," but in his report for 1871 he says that the Caribou mine was the only mine developed and that the other deepest workings were on the Idaho and Boulder County, the shafts of which reached depths respectively of 45 and 50 feet. In 1871 the Caribou mine was opened by two shafts, one 205 feet, and the other, which was 110 feet east of the first, 115 feet deep. The two shafts were connected by drifts. In estimating the production of silver for 1869 to 1875 Raymond's figures (not Burchard's) have been used.

The deposits at the Idaho and Boulder County mines were discovered soon after the Caribou. In 1871 Raymond reports that the Caribou mine was either producing daily or was capable of producing 1 ton of high-grade ore, worth $500 to $700 a ton; 30 tons of second-grade ore, worth $150 to $200 per ton; and some third-grade ore, worth about $60 a ton. For the third-grade ore a mill for chloridizing, roasting in four Bruckner furnaces, and amalgamating in pans was being erected in 1871 on Middle Boulder Creek and was in operation in 1872, when it made large shipments of bullion. From September 21, 1870, to October 1, 1872, the Caribou mine produced 3,651 tons of both smelting and milling ore, on which the net profit was $90 a ton. At the end of 1872 the main working shaft was 329 feet deep, and 34,082 tons of ore was exposed. Other mines in the district were the Idaho, Perigo, Boulder County, Sherman, Seven-Thirty, Grand View, Bullion, No Name, Sovereign People, and Arlington. In 1872, at the town of Middle Boulder, Hetzer & McKenzie had a 7-ton 15-stamp mill, with stamps weighing 500 pounds each. In 1873 the bulk of the Caribou district ores was treated at the mills in the district, and in that year the Caribou mine was sold. In 1874 Caribou was perhaps the most prosperous mining camp in northern Colorado. The output from the Caribou mine alone in that year was 1,800 tons, which when milled produced $130,000 in silver bullion, and the ore shipped from the Sherman mine amounted to 220 tons and was said to have been worth $40,000. In that year work was begun on a mill to treat ore from the Sherman and No Name mines. In 1875 the Caribou district produced $450,000. At the end of

---

[22] Raymond, R. W., Statistics of mines and mining in the States and Territories west of the Rocky Mountains for 1870, p. 326, 1872.

[23] Report of the Director of the Mint upon the production of the precious metals in the United States during the calendar year 1882, p. 398, 1883.

that year the shaft in the Caribou mine was 500 feet deep and the output of that mine alone was $204,000. There are no detailed records of mining at Caribou during the next four years, but in 1881 the silver bricks shipped from Caribou through the express office at Boulder were worth $227,983, and the bricks shipped in 1882 were worth $192,881. In 1883 the Caribou mine was not worked, the pumps were taken out, and the mill was closed, but the Poorman mine was operated. In 1884 the Caribou was still closed, but the Poorman was worked to a depth of 600 feet, and indications favored the continuance of production at that mine.

From 1887 to 1890, inclusive, the Caribou mine was again worked, and from 1888 to 1893 the Poorman mine was worked, but the drop in the price of silver practically closed the silver mines of Caribou, although an output was made intermittently from the Boulder County, St. Louis, and other mines. In 1914 a 100-ton cyanidation mill was completed at Caribou to treat ores from the dumps of the Poorman and Caribou mines, and the old Boulder County mill was remodeled into an amalgamation-concentration mill and operated for six months on ore from the Boulder County mine. The new Caribou mill was started to work in 1915, the Boulder County mill continued in operation, and both mills were operated during the greater part of the year. The Caribou mill was changed to a flotation-concentration mill in 1916 and operated only in that year and then closed. Considerable smelting ore was shipped from the district from 1917 to 1923. The Boulder County mill was remodeled in 1920 with the addition of oil-flotation equipment and was operated for short periods in 1920 and 1921. In 1923, the Boulder County mill was completely overhauled and remodeled and was operated for a short period.

In 1896 the discovery of gold on Spencer Mountain led to the founding of Eldora, 2 miles south of Caribou. This camp, which has produced gold tellurides, was fairly active until 1904.

Free gold was found on the Columbia vein, at Ward, in the late sixties. After gold was discovered on the Columbia vein a stamp mill was freighted across the plains and installed on Niwot Hill to treat the oxidized surface ores, and for some time it was operated with success. The oxidized ores disappeared, however, as the depth of the workings increased, and the mill would no longer save the gold. In 1871 Raymond [24] reported that about 20 per cent of the assay is saved by the stamp mills, that the Columbia lode had yielded, to that date, not far from $250,000, and that while this amount was being saved probably not less than $750,000 had run down the creek and been lost. During 1871 Mitchell & Williams were repairing their

20-stamp mill, which was also equipped with "percussion" concentrating tables that were said to work well; Smith & Davidson were running their 50-stamp mill; and Richardson was constructing a chlorination mill, with four Bruckner cylinders for roasting.

In 1872, owing to the lack of success in recovering gold in stamp mills and the failure of the chlorination works, comparatively little work was done at Ward. In 1873 it appeared that the chlorination works would be successful, but in 1874 the mines seem to have been generally idle. In 1875 the Niwot was worked with some success after an idleness of several years. Some interesting experiments were made with the Pomeroy percussion concentration tables.

In 1871 the only mine operating at Gold Hill seems to have been the White Rock lode, the ore of which is said to have returned 14 ounces a cord (8 or 9 tons) in a stamp mill. At that time the bed of Fourmile Creek was being worked for gold with fair success, and some of the operators had worked well up toward the head of that creek. Their work resulted in the discovery of silver-lead ore at Garden Gulch and Williamsburgh in 1872 and in the establishment of the gold camp of Sunnyside in 1873.

In May, 1872, petzite, a gold-silver telluride, was discovered at the Red Cloud mine, at Gold Hill. Raymond reports that from August 1 to December 31, 1872, this mine yielded 40 tons of $900 ore and 250 tons of $100 ore. During the next two years this mine was sunk to a depth of 430 feet and 1,000 feet of levels were run. Ore was found almost continuously in all parts of the mine, but the wonderfully rich pocket of tellurides near the surface was never duplicated. In September, 1874, the mine was shut down. During 1873 and 1874 only one other mine, the Cold Spring, was found to contain tellurides, but the record of this mine equaled if it did not surpass that of the Red Cloud, and the two mines yielded in that time about $600,000 from about 400 tons of ore, which thus averaged $1,500 a ton. During this time the old workings at Gold Hill were yielding a little gold under stamps.

In March, 1874, gold was first discovered at Sunshine, only 2 miles east of Gold Hill, in a region frequently traversed by prospectors on their way to other camps. The first discovery did not cause much excitement, but later in the season, when the American mine was located and considerable quantities of free gold were found on the outcrop, prospectors hurried to the camp from every neighboring point, particularly from Gold Hill, so that with the closing down of the Red Cloud mine in September, the future of Gold Hill seemed dark. About 300 lodes were located at Sunshine, of which a few were profitable. Probably $20,000 to $30,000 was taken out at Sunshine that year, the largest part of which was produced from surface ores carrying free gold that were treated in the small works of J. Alden Smith, at Sunshine.

[24] Raymond, R. W., Statistics of mines and mining in the States and Territories west of the Rocky Mountains for 1871, p. 351, 1873.

Early in 1875 prospectors set out in every direction from Sunshine, and during that year there was a succession of notable discoveries. The first of these discoveries was made in the vicinity of Fourmile Creek, at Salina, Camp Tellurium, and along Gold Run Gulch. Then came reports from a new camp on South Boulder Creek, later called Magnolia, after which the Slide lode, on Gold Hill, was found. Prospectors also found ores in the valley of Jim Creek, a tributary of Lefthand Creek, where, in 1872, iron and copper sulphides, argentiferous galena, and antimonial silver ores had been discovered but not worked. Glendale, Springdale, and Providence were camps in 1874. The output from all these camps in 1874, in small lots of ore that assayed from $500 to $5,000 a ton, Raymond estimates as "telluride ore worth not less than $200,000." One lot of ore from the Malvina mine, amounting to 1,500 pounds, sold for $8,300. In 1875 at Magnolia, the Keystone lode, which was discovered in August, 1875, yielded more than $20,000 worth of telluride ore from a 75-foot shaft, the Little Dorrit yielded $4,000, and the Mountain Lion more than paid expenses from the surface to a depth of 50 feet. The Malvina lode, at Gold Hill, was discovered in June, and in sinking the first 25 feet ore worth nearly $20,000 was taken out, and from the time of discovery to the end of the year more than $37,000 worth of ore was taken out in development only, without stoping, from a shaft 100 feet deep. The Slide lode, which was discovered in July, yielded a combination of tellurides and native gold; the Cash, an old discovery, did not carry tellurium ores, but its ore yielded $20 a ton in a stamp mill; some remarkably rich telluride ores were found in the Sterling; the Red Cloud was idle; the Cold Spring continued to ship regularly as it had done for two years; the Horsfal and Alamakee (original discoveries at Gold Hill) were idle; the Washington Avenue, west of Gold Hill at Williamsburgh, developed deposits of galena, pyrites, chalcopyrite, and zinc blende. In Gold Run Gulch stood the camps of Salina and Camp Tellurium. The American mine, at Sunshine, produced about $125,000, and many others made a fair output. At Springdale a considerable quantity of telluride ore was taken out, and at Providence some exceedingly rich ore of the same kind.

From 1876 to 1881, inclusive, the gold mines of Boulder County made an average yearly output of about $340,000. The Ward district had by this time partly solved the problem of treating its sulphide gold ores by mills; Jamestown had been founded between Providence and Balarat; and some work had been done on the argentiferous lead mines of Albion, north of Caribou.

In 1882 many former producing mines were idle—the Niwot, at Ward; the John Jay, at Jamestown; the Malvina and Cold Spring, at Gold Hill; the Crater, at Magnolia; the Poorman, at Caribou; and the mines at Albion. During the year, however, new producing mines were brought in—the Senator Hill, at Magnolia; the Ingram, near Salina; and the Golden King, at Jamestown. The Golden Age free gold mine at Jamestown, however, surpassed any record of former years. Most of the high-grade ores of Boulder County were sent to the Argo or to the Golden smelter, but some were sent to the Grant smelter at Denver, and to Omaha. In 1882 very little seems to have been done at Ward, and Jamestown was the principal district, with the Golden Age mill operating 25 stamps and the Pell 6 stamps. At Balarat the Smuggler mine was developed only, and its output from 1877 to 1882 was said to be from $400,000 to $600,000 of tellurium ores. During 1882 placer mining was still continued in a small way at several places, particularly at Sugar Loaf.

From 1883 to 1912 Boulder County continued to produce gold, principally from Gold Hill, Jamestown, and Ward, and reached the height of its production in 1892 with a total value of $1,141,852, of which $982,988 was gold. In 1913 and 1914 the output o silver exceeded that of gold; in 1915 gold again took the lead; but from 1916 to 1924 the shipments of silver ore from the Yellow Pine, at Sugar Loaf; from the White Raven, at Ward (until it closed in 1921); and from the mines at Caribou placed silver in the lead.

The Colorado & Northwestern Railroad, later called the Denver, Boulder & Western Railroad, starting from Boulder, reached Sunset in 1897, Ward in 1898, and Eldora in 1902, but was dismantled in 1918.

LAKE COUNTY

In 1859 prospectors who had followed Arkansas River found gold along its course, particularly at Georgia Bar, 8 miles above the mouth of Clear Creek, in Lake County; at Kellys Bar and at Cache Creek, 3 miles above Clear Creek, both in Chaffee County; and at other places along Arkansas River for about 30 miles below Lake Creek.

Late in 1859 Slater & Co. discovered gold at the junction of California Gulch and the Arkansas, but not until April 6, 1860, were encouraging deposits found, and those were in a valley leading from Iowa Gulch. Upper California Gulch was soon found to contain the most plentiful supply of gold and was preempted in 100-foot claims for 7 miles. The town that grew up from the influx of miners was called Oro and stood near the head of California Gulch, just south of Iron Hill. In 1860 the Discovery claim, which was just above the site of the A. Y. and Minnie mine, produced $60,000; claims Nos. 5 and 6 produced $65,000, and other claims made about the same output. A quartz vein running through the Discovery claim and claim No. 1 below yielded $216 in half a day by sluicing by three men. For the first three years mining was carried on by sluices, long toms, Georgia and hand rockers, and pans. Then followed consolidation and mining by ground-sluicing and hydraulicking.

The most productive years of placer mining in California Gulch were from 1860 to 1867, the population meanwhile dwindling from 5,000 or 6,000 to a very small number. In the late sixties the Printer Boy, Pilot, and Five-Twenty lodes were producing gold found in nests of lead carbonate. Other less productive placer districts in the county have been Colorado Gulch (which opens into the Arkansas opposite the mouth of California Gulch), Iowa Gulch, and the Little Fryingpan, a tributary of Colorado Gulch. Other placer and lode districts located in the sixties but never very productive include the Westphalian, Pine Creek, La Plata, Hope, Lake Falls, and Red Mountain districts, of Chaffee and Lake counties. In 1871 placer mining was being continued in California Gulch with a decreased yield, and there was a 5-stamp mill on the Five-Twenty property at which ores from the Printer Boy, American Flag, Pilot, and Berry tunnel were treated. The total yield of the California Gulch placers to 1872 is estimated by Raymond at $3,000,000, but other estimates are higher than this. Burchard's figures in the report of the Director of the Mint for 1882, give the placer output of Lake County (including Chaffee County) for 1860–1869, as $5,812,000. The output of Chaffee and Lake counties has been separated by means of data furnished by Hollister. The Homestake mine, near Tennessee Pass, was worked extensively in 1872. In that year a ditch 14 miles long and 6 feet wide was under construction in Iowa Gulch. In 1873 only a few placers were worked. Several rich gold strikes were made on lodes in California Gulch, and the Homestake mine was reported to be shipping to Golden ore that carried 30 to 60 per cent lead and 200 to 500 ounces of silver. In 1874 there was little placer mining, and most of the old, lower ground was regarded as worked out, but many new ditches were being built to carry water to higher ground. The Homestake mine shipped some argentiferous galena containing nickel, but there was no market near at hand for silver ores.

In 1874 W. H. Stevens and A. B. Wood came over the range from Fairplay, where they were mining, to build the Oro ditch. In examining California Gulch Wood found float consisting of carbonate of lead and began digging on the south side of the gulch on Dome Ridge, now known as Rock Hill, on what was afterward called Rock claim. In the fall of 1875 he sank a little shaft through the drift that covered the outcrop, at a point subsequently worked by an open cut. He made arrangements to have some work done that winter, and in 1876 this led to the uncovering of the outcrop of the lode across California Gulch, up Iron Hill. Raymond received reports that silver-lead-copper ore worth $25,000 was produced from mines in California Gulch in 1875. He suspected, however, that these figures represent the value of ore which had been hoisted but not treated and which then lay at the smelting works of the Cincinnati Co., at Malta, at the mouth of California Gulch. These works, which had been erected but not put in operation that year, consisted of a long reverberatory furnace, for roasting, a shaft furnace, and a cupelling furnace. There was still no near market for silver ores, but Breece was building chlorination works in California Gulch for treating gold ores taken from the Berry tunnel, and in addition to the smelter at Malta a plant of 10 tons capacity, comprising two roasting furnaces and one blast furnace, was under construction in Chalk Creek, in Chaffee County. The Homestake mine was idle in 1875. The gold lodes of California Gulch were worked in a small way, and some work was done at the Yankee Blade, near Granite. The production of the placer mines decreased heavily.

In 1876 the series of outcrop claims running from Rock claim were located on the supposed vein. Ore from the Rock claim was taken in 1877 to the smelter at Malta. In 1877 Stevens persuaded the Harrison Reduction Co., of St. Louis, to erect smelting works (completed in 1878), and in 1878 James B. Grant put up the smelter which grew into the Omaha & Grant Smelting & Refining Co. In 1879 Anton Eilers and Gustav Billing erected the smelter which in later years became the property of the Arkansas Valley Smelting Co. In 1877–78 the greatest rush to any camp in the history of the State occurred, resulting in the building of a new town, called Leadville, 7 miles below the old town of Oro.

In 1878 George Fryer sunk a hole on a hill north of Stray Horse Gulch and found a deposit of carbonate ore that proved to be one of the most remarkable ore bodies ever discovered. A month later Rische & Hook happened to sink a hole where the "contact" or the mass of the ore approached the surface and found the ore body on which was developed the Little Pittsburgh mine, the foundation of the fortune of H. A. W. Tabor. That year Leadville's output was $2,490,000. In 1879 it was $11,285,276. In 1880 there were at Leadville several ore-buying firms, 11 or 12 smelters yielding bullion, and 4 stamp mills. Production increased materially in 1882, when the output was valued at $15,256,375.

In 1882 considerable attention was paid to Sugar Loaf Mountain, one of the foothills of Mount Massive, at the head of Little Fryingpan Gulch. In Little Fryingpan Gulch, the Shields, Venture, and Welsh mines had already been producers, but this new discovery on Sugar Loaf included the deposits worked by the Dinero, Birdie R., and other mines. A 10-

stamp mill was erected on the Shields property and operated on ore from the Shields and other mines.

In 1883 six large smelters were in operation at Leadville. The Breece Iron mine was the only paying mine on Breece Hill, although the St. Louis mine was worked and a concentrator was projected, and the Little Jonny shaft was down about 120 feet. In 1884, as formerly, the Leadville smelters obtained the largest part of the product of the camp, but the smelters at Denver and Pueblo were beginning to increase their purchases of Leadville ore. The Oro and Antioch stamp mills were in operation, a mill was erected on the Lilian property, a small concentrating mill was built for mine ore and operated successfully, and an amalgamation mill was leased to work the dumps of the Chrysolite mine. Iron Hill continued to be the largest producing district, from the Iron Silver, A. Y., Minnie, Colonel Sellers, and other mines. The Little Jonny mine was actively worked, the product being silver-lead ores carrying some gold.

In 1885 there was considerable competition between the Leadville and the valley smelters, and the Iron Hill mines shipped ore on a schedule providing for a penalty of 50 cents for every unit of zinc above 12 per cent, although on one contract a content above 20 per cent was accepted without penalty. It was in this year that F. L. Bartlett commenced his experiments in the East on Leadville zinc-lead sulphides, which in 1890–91 resulted in the erection of the zinc oxide plant at Canon City. In 1889, Messrs. Argall, Ingalls & Wood seriously considered the building of a plant for the recovery of zinc. In 1885 there was a concentrating plant at the mines of the Iron Silver Co. and at the Colonel Sellers mine for the treatment of zinc-lead sulphides. These mills and later mills were designed chiefly to treat the low-grade zinc-lead sulphide and to make a high-grade lead concentrate, the zinc concentrate or middling being discarded. Fortunately, a large part of these zinc concentrates was allowed to accumulate in piles, and large quantities were shipped in 1899–1901 to Belgium. In 1907–1914 large quantities were re-treated by magnetic separation at Canon City and Pueblo. In 1886 more concentrating plants for treating low-grade zinc-lead sulphides were built, and in 1887 concentrating equipment was further increased.

In 1886, lead bullion aggregating 25,963 tons was produced by smelters at Leadville, and 138,335 tons of ore containing 22,526 tons of lead was shipped from Leadville to outside smelters. Part of the product of the Leadville smelters, however, was derived from ores shipped from Red Cliff and Kokomo. From 1880 to 1889 the production of lead at Leadville was enormous, but after 1889 the average annual output was only about half the average reached during the 10-year period named, although from 1899 to 1901, inclusive, the output increased heavily.

In 1890, a large deposit of copper ore was found in the Henriette and Maid of Erin properties. In 1893 Lake County began to produce considerable gold and has continued to do so ever since, chiefly from properties on Breece Hill.

The Yak Mining, Milling & Tunnel Co. was incorporated in May, 1894, for the purpose of acquiring mines and constructing a drainage and transportation tunnel from California Gulch toward Breece Hill. Work was started in March, 1895, with a single-track tunnel about 4,000 feet long, originally driven by one of the older companies—the Silver Cord. The Yak company enlarged the first 3,000 feet to a double track, left the next 1,842 feet as a single track, and from there on drove a double track. By 1911 this drainage tunnel had been driven about 3½ miles in a general eastward direction toward the amphitheater of Big Evans Gulch and toward the Mosquito Range, and laterals were run to many of the large mines along its course. Very little additional work was done at the heading after 1911, the extremities of the tunnel and its laterals being marked by the Vega, Diamond, and Resurrection No. 2 mines.

In 1896–97 the effects of the labor strike are seen in the output.

In 1898–99 the driving of a drainage tunnel from Malta was debated but without results.

In 1899 the American Smelting & Refining Co. was organized, and took over nearly all the lead smelters of the Rocky Mountain States. The downtown mines, which had filled with water at the time of the miners' strike in 1896, were unwatered in May, 1899.

In 1901 the Colorado Zinc Ore Co., at Denver, and the Empire Zinc Co., at Canon City, were erecting magnetic separating plants, and the United States Zinc Co. was organized to build a zinc plant at Pueblo. During this year new zinc concentrating mills (the Minnie and the Resurrection) were built at Leadville, and Leadville's first and perhaps last cyanide mill was erected at the Ballard mine, unless the financial difficulties connected with the building of a cyanide plant for the treatment of the Garbutt-Little Jonny ore can be overcome.

In 1903 the shipments of zinc sulphide became large and continued so until 1908, when they were very small, because of the low price of zinc. In 1905 the Rho magnetic separation mill was erected at the mouth of the Yak tunnel, replacing the old Yak concentration mill. In 1906 the Damascus, A. Y. & Minnie, Adams, and Rho mills were in operation at Leadville. In 1907 the Adams and Rho mills were the only ones in operation, and in 1908 the Adams, Rho, and Leadville District (new in 1907). In 1909 the Adams mill closed, and in 1911 the Rho mill closed.

In 1910 large bodies of zinc carbonate ore were found at Leadville, and shipments were made. The next year the shipments increased heavily. The ore was first found in the Madonna mine, at Monarch, Chaffee County, and later at the Chance-Hilltop mines, in the Horseshoe district, Park County. It was first discovered at Leadville in the Robert E. Lee mine, but its grade was very low, and no attempt was made to search for more of it. In 1910 Mr. Howard E. Burton, associated with Messrs. H. K. White and Alfred Thielen, leasing on the Hayden shaft of the May Queen mine, was the first to find and ship high-grade zinc carbonate. The Western Mining Co. soon afterward found large bodies of zinc carbonate and silicate in the old workings of the Maid of Erin, Henriette, Waterloo, Adams, Mahala, Big Chief, and other mines. After this discovery, zinc carbonate was found in nearly all parts of the Leadville district, and the shipments of it were heavy until 1915, by which time the grade of ore remaining in some of the large properties had decreased materially.

During 1915 the output of lead and silver continued to decrease, but the output of gold from Breece Hill mines increased greatly, and the placer industry was revived, after years of nonexistence, by the installation of a dredge on Arkansas River at the mouth of Box Creek, 12 miles below Leadville. In June, 1916, the downtown mines, which had been allowed to fill with water in 1907, were again unwatered and from 1917 to 1923, when they were again closed, they produced large quantities of lead oxide, zinc carbonate, iron-manganese, and other ores. In August, 1923, unwatering by electric pumps was begun in the Carbonate Hill mines that had been closed in 1918-19. The water was not completely removed until the spring of 1925.

In 1910 interest was aroused by the development of gold ore in Lackawanna Gulch, and a small shipment was made that year from the Miller mine. A 30-stamp amalgamation mill was also in course of construction at this property and a 100-ton amalgamation-concentration mill at the Mount Champion mine, which became a rather large producer in 1913 and increased its production in 1914, 1915, and 1916. This mill was idle after 1917, but smelting ore was shipped in 1917. The mill was torn down in 1923, and parts of it were used in the new Griffin mill, in the St. Kevin district.

CHAFFEE COUNTY

The placer deposits in Chaffee County were worked contemporaneously with those of Lake County, and the gold-lode deposits at Granite were discovered and worked almost contemporaneously with the early work at the California Gulch gold deposits in Lake County.

According to Crawford,[25] ore was discovered in 1878 on the Great Monarch claim, at Monarch, and in 1878 or 1879 the Madonna deposit was discovered. Very little work was done until 1883, when the railroad was extended from Maysville to Monarch, and for 10 years the production was large. From 1893 until 1906 little was done in the district, but in 1906 shipments of zinc carbonate ore in considerable quantity were made and have been made ever since. The shipments of zinc carbonate from the Madonna mine in 1906 were made by Howard E. Burton, of Leadville, who in 1910 discovered zinc carbonate on the May Queen lode at Leadville.

In 1870 the lode mines at Granite, particularly the Yankee Blade, which was equipped with a 20-stamp mill, alone produced $60,000 in gold, but since that time they seem to have made little or no output. According to Burchard,[26] the ore above the water level in the veins of the Granite district carried free gold but became refractory when water was reached, so that the mines one after the other suspended operations. In 1879 one of the old stamp mills was remodeled without success. In 1889-1892, in 1900, and again in 1917, there was a small output of gold from the Belle of Granite mine, which is in Lake County, very near the line between Chaffee and Lake counties. The placer mines near Granite made a continuous output from 1860 until hydraulicking was discontinued, in 1911. The chief placer mine near Granite was that in Cache Creek, on a deposit discovered in 1860. Ground-sluicing by individuals was continuous from 1860 to 1883, when the Twin Lakes Hydraulic Gold Mining Syndicate (Ltd.), of London, began sluicing operations. In 1889 sluicing gave place to hydraulicking. The company owned what was known as the Cache Creek ditch, which brought water from Clear Creek by ditch and tunnels.

In 1875 a lead smelter plant was under construction at the Mary Murphy mine, on Chalk Creek. In 1881 about 30 tons a day was being mined from the Mary Murphy mine. In 1882 a shipment of 500 tons was made from drifts run for development on the Mary Murphy and 150 tons from the Iron Chest mine. In 1883 shipments were made from eight mines in this district, and in 1884 the ore shipped amounted to 3,010 tons, valued at $113,524. From 1887 to 1908 considerable quantities of gold-silver-lead ore were shipped from the Mary Murphy mine. A mill was built there in 1886, and at different times the property was leased with varying success. Lessees built mills at Romley and St. Elmo and a smelter at Buena Vista, all now destroyed or dismantled. In 1909 and 1910 the Chalk Creek district made very little output, but development was started and ship-

[25] Crawford, R. D., Geology and ore deposits of the Monarch and Tomichi districts, Colo.: Colorado Geol. Survey Bull. 4, pp. 195-196, 1913.
[26] Burchard, H. C., Report of the Director of the Mint on the production of the precious metals in the United States during the calendar year 1882, p. 407, 1883.

ments were increased in 1911 and 1912, and in 1913 a new mill was in operation. The output increased greatly from that year until 1918, when the ore reserves were reported as nearly exhausted. Lessees continued to ship smelting ore from 1918 to 1924.

The Sedalia mine, 4 miles north of Salida, is one of the few large copper mines in Colorado, having shipped from 1884 to 1908 probably 60,000 to 70,000 tons of ore containing 5 per cent or more of copper and $1 to $2.50 to the ton in gold and silver. In recent years it has also been a producer of zinc carbonate and lead-zinc sulphide ores. In 1900 the Buena Vista Smelting & Refining Co. was operating a matte smelter (short-lived) at Buena Vista and in 1902 the Ohio and Colorado plant was built at Salida (closed February, 1920). Some of the steel in this plant was used to build the Rawley mill at Bonanza, Saguache County, in 1923, and the stack and other parts of the smelter were purchased in 1924 for use in a plant employing the Gordon ammoniacal process for the treatment of complex lead-zinc ores.

GUNNISON COUNTY

In 1867 the route used by the Ute Indians from Arkansas River to the Gunnison went up Lake Creek, which empties into Twin Lakes, in Lake County. At that time placer deposits had been discovered and worked in Taylor, Kent, Union, Washington, and German gulches, in Gunnison County. Lodes also were known to exist, but the district was far from civilization. Gold was discovered in 1861 on Taylor River, in what has since been known as Tin Cup district, and almost simultaneously gold was discovered in Washington Gulch, in the northern part of Gunnison County. Little work was done until 1872, when notable discoveries of silver-bearing rock were made in the Elk Mountains. During the next five years there was a small increase in the number of settlers, but in 1878 Leadville drew off many of them, though others came in. In that year a smelter was being put up at Crested Butte, and in the fall mines were opened in the eastern part of the county and Hillerton, Virginia City, Ohio City, Pitkin, Gothic, and Irwin were laid out and in process of building. During this time the town of Gunnison was growing. In 1879 the gold-bearing veins of the Independence district were discovered and also the deposits of the Aspen silver-lead district on Roaring Fork River. Both districts were then in Gunnison County. In 1880 the Cochetopa, Ruby, Ohio City, and Elk Mountain districts were in existence.

In the fall of 1881 the Denver & Rio Grande Railroad was completed to Gunnison and later to Crested Butte, a few miles from Ruby Camp. Beyond Ruby in 1881 lay a vast unexplored region which until then had been occupied by the Ute Indians. Tin Cup was the most productive camp in the county. Two

smelters had been erected, the Virginia City and the Willow Creek. Gothic had a 15-ton smelter but had not attained any importance. In the Quartz Creek district there were several valuable properties. From 1,200 to 1,500 locations had been made at Ruby, but less than 50 were paying. In 1882 the Virginia City smelter, at Tin Cup, was being operated, and ore had been shipped from the Eureka mine, on Treasury Mountain, in the Gothic district, to the Argo works, near Denver, and to Kansas City. In 1883 high-grade lead-silver ore was shipped from the North Star mine, in the Tomichi district. The smelter in the Tin Cup district had not been operated, but a large quantity of gold-silver ore had been shipped from the Gold Cup mine. About $100,000 in gold and silver had been produced at Ruby, and concentrating mills had been erected there. Very little ore had been produced in the Elk Mountain district because the concentrating mills had not been started. In the Gothic district (including Washington Gulch) the Eureka mine continued to make shipments, and a concentrating mill was erected in Rock Creek. Some development work had been done in the Cochetopa district.

In 1884 little was done in the Ruby district. Most of the shipments were made from the Forest Queen mine, where a concentrating mill had been erected. In the Tomichi district considerable ore had been shipped from the North Star mine; Quartz Creek had not been very active; and the Tin Cup district had showed little activity outside of the placers. In 1885 in the Tomichi (White Pine) district the Eureka-Nest Egg was the most productive property, shipping lead carbonate to the Royal Gorge smelter at Canon City. Zinc carbonate appears to have been an undesirable constituent of the ores of this camp at this time.

From 1885 to 1893 the Tomichi district was fairly prosperous, but from 1893 to 1913 only a few mines produced much ore. In 1914 and 1915 a considerable quantity of lead-zinc-silver ore was shipped from the Morning Star mine. During 1916–1918 the Akron Mines Co. shipped lead-zinc sulphide ore direct to smelters and milled in the 60-ton concentration plant on the property considerable ore from the Akron group and from the Eureka-Nest Egg group of mines. There was much development and experimenting with the milling problem at the Akron mines, but not much ore was marketed until 1922, when a favorable contract allowed large monthly shipments to be made on a basis of the sum of the lead and zinc assay.

From 1914 to 1918, inclusive, a large quantity of zinc carbonate ore was shipped from the Doctor mine, on Spring Creek, in the Elk Mountain district.

Ohio City and Pitkin were towns in 1879. Both have survived to the present day, and much of the output of gold of Gunnison County has been mined in those two districts, particularly from the Ohio City or Gold Brick district, where the Gold Links mine was a notable producer of gold ore until 1913. The Carter mine has made intermittently considerable production up to and including 1924.

South of Iola, in the Domingo district, the Vulcan and Good Hope mines have produced 1,000 tons of ore carrying 10 ounces of silver to the ton and 6 per cent of copper.

PITKIN COUNTY

According to Rickard, ore was first discovered in the Roaring Fork district, of which Aspen is the center, on July 3, 1879, when Philip W. Pratt and Smith Steel, coming from Gothic by way of Maroon Pass, found the Galena lode on West Aspen Mountain. On the following day they located the Spar Claim on Aspen Mountain, and on July 5 Allbright and Fuller located, at the foot of Smuggler Mountain, the Little Rock claim, which covered a part of the property of the present Smuggler mine. The Smuggler claim itself was located August 30 by Charles Bennett. The first mineral survey was made on the Monarch mine on October 12, 1879, by John Christian, of Leadville. The report of Spurr[27] on the Aspen district has been drawn upon freely in the following pages.

Though it is doubtful whether the early prospectors possessed a sufficiently broad knowledge of geology to have observed the geologic conditions in the region, it is tolerably certain that those who first came to Aspen in 1879—men who had been working in Leadville—had observed on the maps of the Geological Atlas of Colorado that the Paleozoic rocks which carry the silver at Leadville nearly encircle the Sawatch uplift and that, with the keen observation of men of their profession, they selected limestone beds at the same horizon as the ore-bearing zone at Leadville in which to make their investigations.

In the summer of 1879 the Durant, Iron, Spar, Monarch, Late Acquisition, and Smuggler claims at Aspen were located. Work was suspended during the winter, chiefly because of the Indian revolt in the neighborhood. In the spring of 1880, however, the Emma, Aspen, Vallejo, Mollie Gibson, Argentum-Juniata, Della S., J. C. Johnston, Park-Regent, and other claims were located. The town, which had at first been called Ute, was rechristened Aspen, probably because the tree so named grew in abundance on the neighboring hills. Exploration along the strike of the limestone belt was continued, and claims were located along it for 30 to 45 miles—from the valley of Fryingpan Creek on the northeast to that of Taylor River on the south. Ashcroft, at the head of Castle Creek, was at first the largest town, but, although the geologic conditions around it are most promising, few considerable bodies of rich ore have been discovered in that region, the only mines producing in 1896 being

the Express mine, whose deposits lie at the Leadville horizon, and the Montezuma group, on Castle Peak, in the Maroon formation and diorite, about 13,500 feet above sea level. The rich deposits near Aspen itself made but little show upon the surface. On Smuggler Mountain and along the base of Aspen Mountain their outcrops are buried beneath glacial gravels; moreover, the ore contained much less iron and manganese than the Leadville deposits, and the outcrops of the ore bodies were therefore not so readily distinguishable from ordinary altered limestone or dolomite.

Thus, in 1881 and 1882, the prospects on the Castle Creek slope of Aspen Mountain were considered the more promising, and it was not until 1884 that the existence of the very rich ore bodies on Spar Ridge was disclosed by the workings of the Emma and Aspen mines. As a result of these discoveries the town of Ashcroft was moved almost bodily to Aspen, many houses having been dragged over the 12 miles that separate the two towns.

In the meantime, from January, 1881, to August, 1882, there had been running at Independence or Sparkill, at the headwaters of the Roaring Fork, in the eastern part of Pitkin County, a 15-stamp mill and later an additional 30-stamp mill on the gold ores of this district, which produced from beginning to end about $190,000. The mines were again operated in 1891, again in 1897–1899, and possibly for a short period in 1900 but not since 1900. In 1906–7 some work was done and some ore produced from silver-bearing copper-lead ores in the adjoining Lincoln district. The Montezuma mine, in the Ashcroft district, was worked from time to time until 1915.

For the first six years of its existence the great drawback to the development of the Aspen district was its inaccessibility. It could be reached from existing railroads only by crossing the summits of lofty ranges of mountains. The shortest and most generally traveled line of approach from the east left the railroad at Granite, 15 miles below Leadville, in the valley of the Arkansas, and after ascending the Lake Fork passed Twin Lakes, crossed the summit of the Sawatch by Hunter Pass and descended to Independence, and thence went down the Roaring Fork to Aspen, a distance of about 40 miles. A second line, 72 miles in length, left the railroad at Buena Vista, lower down the Arkansas Valley, crossed the Sawatch by Cottonwood Pass or Chalk Creek Pass, each about 11,000 feet high, into the valley of Taylor River, and after ascending that river crossed Taylor Pass to Ashcroft and thence followed Castle Creek down to Aspen.

The first lot of ore shipped from Aspen was taken from the Spar and Chloride mines, on Aspen Mountain. The ore was transported on the backs of burros or jackasses to Granite or Leadville to be smelted.

[27] Spurr, J. E., Geology of the Aspen mining district, Colo.: U. S. Geol. Survey Mon. 31, pp. xix-xx, 1898.

The cost of such transportation was at first $50 to $100 a ton, but as competition increased these rates were reduced, until near the time of the advent of the railroads they were $25 a ton.

In 1886 the Colorado Midland Railroad, which had built its line from Colorado Springs to Leadville in order to get part of the profitable ore-carrying business of Leadville, was induced by the promising developments of ore at Aspen to project a line to that point. This work had hardly been undertaken when the Denver & Rio Grande Railroad Co., whose line was already built down Eagle River to Red Cliff, felt obliged to enter into competition for the Aspen trade, and a railroad-building contest ensued, each road striving to reach the objective point first. The line of the Colorado Midland, which was a broad-gage road, ascended the Sawatch Range directly opposite Leadville, passed through its crest by a tunnel at Hagerman Pass, and descended Fryingpan Creek to the Roaring Fork. The route of the Denver & Rio Grande Railroad was longer, but it was then a narrow-gage line, and it followed valleys all the way, descending Eagle and Grand rivers to Glenwood Springs, and thence ascending the valley of the Roaring Fork. In spite of the difficult engineering and the many tunnels in the magnificent canyon of Colorado River above Glenwood, the Denver & Rio Grande reached Aspen first, in October, 1887, and the trains of the Colorado Midland did not actually reach the town limits until February, 1888. By the advent of the railroads the expense of transportation of ore to the smelters at Leadville, Pueblo, or Denver, was reduced to $10 or $15 a ton, and in later years this rate has been still further reduced, the charges being in a measure proportioned to the value of the ore.

Another cause besides the difficulty of transportation that retarded the development of the mines at Aspen was the many lawsuits in regard to the ownership of the most valuable ore bodies, which sprang up as a natural consequence of the peculiar unfitness of the United States mining laws to give a clear title, or even any title at all, to deposits of this kind.

In the exploitation of its mines and the reduction of its ores Aspen was unusually enterprising and led the way in many improvements in both these branches of mining. As early as 1882 smelting works were built at the north edge of the town and were run more or less continuously until 1887. That they should be financially successful when obliged to depend on the ores from a single district was hardly to be expected, and when by the advent of the railroad they were brought into competition with centrally situated works at Denver and Pueblo, which drew their ore supplies from all parts of the mountains, they were naturally closed down. Extensive lixiviation works, designed by C. A. Stetefeldt, were erected in 1891 on the north bank of Castle Creek and were operated until the financial crash of 1893. They employed a modification of the Russell process. The financial success of these works is also said to have been doubtful. There have been many sampling works in the district, the first of which was opened in 1883.

In 1897 there were at Aspen four concentrating plants, whose total capacity was 500 tons a day. In 1898 about 10 per cent of the shipments from the district was concentrate. The Smuggler mills have been the most active plants in the district, and one of them, the only one operated for many years, still survives. The equipment used in the mills has been a combination of Hallett tables, Wilfley tables, and Frue vanners. Hartz jigs were also used in the Smuggler mill. For several years after 1901 the Smuggler mill settled its zinc slimes in a reservoir, and in 1904–1908 these slimes were shipped to the Canon City lead-zinc oxide plant and to other zinc plants. In 1906 more than half of the crude ore mined was milled. In 1915 nearly 80 per cent was milled, and more than half of the quantity shipped from the district was concentrate, most of it lead concentrate. In recent years the zinc output has been derived mainly from lead-zinc carbonate ores taken from the Durant mine. In 1910 the Smuggler Leasing Co. unwatered the Mollie Gibson and other mines and kept them unwatered from 1910 to 1918 at great expense, but in 1919 it allowed the two lower levels to fill with water. Lessees continued to work in the district, and the 360-ton concentration mill continued to be operated, so that there was an increase in the production of silver in 1919 but a decrease in that of lead. About the usual production was maintained in 1920, and a considerable production of lead-zinc ores was made from a deposit found at Lenado, near Aspen. The Smuggler mill was idle in 1921, but there was much new development work in the district; the output decreased. The Smuggler mill was in operation again in May, 1922, and there was increased production of silver for the year. The mill closed again in 1923.

Many adits have been driven in this district, the principal one being the Cowenhoven, which has penetrated Smuggler Mountain 2½ miles.

EAGLE COUNTY

Hollister relates that in July, 1860, a party of 100 persons left Breckenridge to explore White River and its tributaries for mineral deposits. They went up Tenmile Creek for 10 miles, crossed over the divide southwestward to "Piney Creek" (now known as Eagle River), went down that creek to its canyon, and then over to the Roaring Fork, which they followed to its junction with the Colorado. Here they found the hot and cold springs of Glenwood Springs. The party crossed Colorado River and struck for the head of the South Fork of White River, which they followed down to its junction with the Green. Turning to the southeast they traveled "over ashy and sandy deserts and sedimentary rocks" to the sources of the Animas and the Rio Grande, in the San Juan Mountains. They went down the Rio Grande and returned by way of Fort Garland to Denver, where they finished their tour, and after that time the greater part of northwest Colorado, then within the bounds of Summit County, was regarded as destitute of minerals. Hollister fails to comment on the effect of the reports of this trip in regard to the thought of mineral deposits at the sources of the Animas and Rio Grande, but the impression of that region, then in Lake and Conejos counties, was probably likewise unfavorable.

This party obviously turned away from Eagle River Canyon and Battle Mountain, in Eagle County, where deposits of silver-lead carbonates were not discovered until 1879, the first claim being the Belden. In 1880 several promising deposits were found, notably on the Belden claim, and silver valued at $50,000 was produced. In 1881 the Belden furnished large quantities of lead-silver ore for the Battle Mountain smelter. In 1883 properties at Red Cliff produced 19,859 tons of ore carrying 232,031 ounces of silver and 8,000 tons of lead, the largest output of lead ever made in one year at that place. During the next two years the production fell off, but in 1886, owing to an extraordinary increase in the production of gold and silver, the total value for the year was $1,079,458, which was not again equaled until the high price of zinc in 1915 gave a value of $1,643,056. After 1886 the production at Red Cliff rose and fell, but apparently never again did the shipments show so high a tenor of lead as in that year. The large quantities of zinc in the ores were troublesome until 1905, when the Pittsburg Gold Zinc Co., which was succeeded by the Eagle Mining & Milling Co., erected a magnetic separation mill to handle the ores of the Iron Mask mine. In 1917 the Empire Zinc Co. acquired much of the territory. Except in 1907 and 1908 the output of zinc from this district to 1918 gradually increased, reaching in 1915 more than 11,000,000 pounds. In 1916 it was 28,438,052 pounds, and in 1917 it was 23,715,412 pounds. But in 1918 it was only 14,845,341 pounds, and in 1919 only 3,387,548 pounds. In 1920 it rose to 6,653,235 pounds, and 517,109 pounds of copper also were produced. In 1921 no zinc ores were marketed, but the output of copper amounted to 1,833,078 pounds. In 1922 the output of zinc was 11,000,000 pounds and that of copper was 1,330,296 pounds. In 1923 zinc reached 23,600,000 pounds. The Black Iron mine at Red Cliff has produced much iron-manganese ore, and shipments of ore of this class were made again in 1923–24.

In 1881 a large stamp mill was erected in the Holy Cross district to crush the surface gold ores, but the activity there did not continue long.

In 1913 silver ores were found in steeply dipping sedimentary rocks on Brush Creek, and 200,000 ounces of silver were produced from surface and shallow workings up to 1918, but work was at first handicapped by the fact that the principal property fell into litigation soon after its discovery, and later development has turned toward driving a long crosscut adit, with the object of striking the sedimentary beds at depth, a project not yet completed.

CUSTER COUNTY

In 1872 galena and rich silver glance had been discovered at Rosita, in Custer County, but the ore seemed to pinch out, and the deposit first discovered was for a time abandoned. In April, 1874, a thin seam of carbonate of copper, which was accompanied by native silver, was discovered on the south slopes of the hills back of Rosita. This seam was the outcrop of the famous Humboldt-Pocahontas vein, which was worked more or less continuously for 15 years and produced more than $900,000 worth of ore. In 1877 the Bassick deposit was discovered 2 miles north of Rosita and is said to have yielded ore worth $500,000 in gold and silver in the first year and a half. The Bassick mine was then sold and is said to have produced $1,500,000 more up to 1885. Since then it has had a checkered career, but it made an intermittent output up to and including 1923.

In 1878 Silver Cliff sprang into existence with the discovery of horn silver at the south end of the White Mountains. Soon afterward, in the Blue Mountains, about 2 miles north of Silver Cliff, the remarkable deposit was discovered that became known as the Bull Domingo mine, which by 1881 had produced $290,000 in silver in coinage value at $1.29 an ounce and considerable lead. In 1881 the Denver & Rio Grande Railroad, in order to reach the iron mines on Grape Creek as well as to serve the Wet Mountain Valley, built a narrow-gage line from Canon City up the winding valley of Grape Creek to Westcliffe, but the track was continually being washed out, and after a particularly extensive washout in 1888 the railroad was abandoned and the remaining tracks were removed. For 12 years the region was without railroad communication, but in 1900 the Denver & Rio Grande built a standard-gage road from Texas Creek to Westcliffe. The year 1885 saw the end of active production at Silver Cliff, although mining has been continued there intermittently to the present time, with a large output in 1917, 1918, and 1919, chiefly from the Passiflora mine. Outside of this area, at Ilse, on Oak Creek, the Terrible mine, which was located in 1879, produced, from 1884 to 1889, $500,000 to

$800,000 in lead. No output was made here again until 1897–1900, when about a thousand tons of ore carrying about 5 per cent of lead was taken out. The mine was idle until 1921–1924, when the ore was successfully concentrated.

RIO GRANDE COUNTY

In 1870 prospectors went south from Del Norte and in June of that year discovered gold in Wightmans Gulch, Summitville district (now Rio Grande County), which led to the opening, in 1872–73, of the Little Annie and other lodes, the erection of stamp mills, an output of $2,063,964 in gold and silver from 1873 to 1887, and a small yearly output to 1917, with a record value of $112,117 for 1909. From 1873 to 1923 a total gross calculated value of gold, silver, lead, and copper of $2,556,909 is recorded.

SAN JUAN COUNTY

The town of Del Norte was the natural provision point and gateway to the watershed of the Rio Grande. In 1860 Baker made his expedition from Del Norte into Bakers Park, where Silverton now stands, but found no profitable gulch mining. He was overtaken by the heavy winter snows and harassed by the Ute Indians, and many of his party perished miserably, a remnant escaping over the mountains only after suffering great hardships.

The following historical sketch of the Silverton district is taken without essential modification from Ransome.[28]

For several years the memory of [Baker's] unfortunate expedition seems to have discouraged further attempts at prospecting in the neighborhood of Bakers Park.

It was not until the early seventies that reports of mineral wealth again began to draw the more adventurous miners into the San Juan region. Some gold was early obtained by washing in Arrastra Gulch [near Silverton], and this led, in 1870, to the discovery by a party of prospectors sent out by Governor Pile, of New Mexico, of the first mine which was successfully operated, the Little Giant, on the north side of Arrastra Gulch. This produced a gold ore, of which some 27 tons were treated in arrastres, yielding $150 a ton. The first shipment of ore from the district is said to have been from this mine. In 1872 troops were sent into the region to keep out the miners, as their presence constituted a violation of the treaty of 1868, by which the Utes were secured in sole possession. In the same year a commission was appointed by Congress to negotiate a new treaty with the Indians to reduce the extent of their reservation. The Little Giant Co. was organized in Chicago in 1872, and in 1873 the arrastres were replaced by an amalgamating mill equipped with a Dodge crusher, a ball pulverizer, and five stamps. Power was furnished by a 12-horsepower engine. The mill was built 1,000 feet below the mine, and the ore was brought down on the first wire-rope tramway built in the region. This year the mine produced $12,000 out of a total of about $15,000 for the entire region. The pay shoot, however, began to diminish, and after the milling of a few hundred tons of ore mine and mill were

abandoned. Several lodes had by this time been opened in the region and some small amounts of rich ore had been taken out, but it was not until 1874 that the main rush to the country began. In September of the previous year a treaty, known as the Brunot treaty, had been drawn up with the Utes, whereby the San Juan Mountains were thrown open to settlement. The ratification of this treaty by the Senate in April, 1874, was followed by a sudden influx of miners, chiefly from the northern camps of Colorado, but including also a few from the south, and some even from the far West. It is estimated that about 2,000 men came into the district during the summer of 1874, and Endlich[29] reports that more than this number of lodes were then staked out.[30]

At that time La Plata, Hinsdale, and Rio Grande were the only counties into which the former reservation had been divided. The chief settlement and the county seat of La Plata County was Howardsville; but in the autumn of 1874 the county seat was moved to Silverton, then a growing town of some dozen houses, admirably situated in Bakers Park. The nearest post office at this time was Del Norte, about 125 miles distant. In 1876 San Juan County was formed from a portion of La Plata County, with Silverton as the county seat. At this time the town is said to have had a population of about 500 voters. Ouray, San Miguel, and Dolores counties were subsequently formed by legislative enactment from the territory originally included in La Plata County.

In 1874 real mining began, principally on Hazelton Mountain, and several hundred tons of gray copper and galena ore were taken out from the Aspen, Prospector, Susquehanna, and neighboring claims during this and the immediately succeeding years. This ore was treated chiefly in Greene & Co.'s smelter, which was erected just north of Silverton in 1874 but which was not successfully blown in until the following year. The machinery was brought in on burros from [Pueblo], then the terminus of the Denver & Rio Grande Railroad. The product of the entire quadrangle [the area shown on the U. S. Geol. Survey's Silverton map] for 1875 was about $35,000, and an estimate made in 1877 places the total product from the beginning of mining to the close of 1876 at a little over $1,000,000. The Greene smelter was in intermittent operation until 1879 and was the first successful water-jacket furnace in the State. Its daily capacity was about 12 tons, and it is said to have smelted nearly $400,000 worth of silver-lead bullion. The bullion was shipped by pack train and wagon to Pueblo. The cost of transporting it to the railway terminus was $60 per ton in 1876, $56 per ton in 1877, and $40 per ton in 1878. The average price for treatment was not far from $100 per ton. During the seventies the chief route into the Animas mining district was by the trail from Del Norte, on the Rio Grande, by way of Antelope Park and Cunningham Gulch. Over this route the first ore sold from the Pride of the West mine, in Cunningham Gulch, was taken out in 1874. It was not until 1879 that the wagon road from Antelope Park was completed by way of Stony Gulch, and ore could be hauled out to Del Norte by teams at $30 a ton.

The founding of Lake City, about the year 1875, and the establishment there by Crooke & Co. of a smelting plant, afforded a market for the ores of the northeastern portion of the quadrangle. The first ore shipped out from this part of the district was from the Mountain Queen mine, at the head of California Gulch, in 1877. It amounted to 370 tons, and contained 64 per cent of lead and 30 ounces of silver per ton. It was carried by pack animals to the end of the road at Rose's cabin, at a cost of $3 per ton. Crooke & Co., of Lake City,

[28] Ransome, F. L., A report on the economic geology of the Silverton quadrangle, Colo.: U. S. Geol. Survey Bull. 182, pp. 19–25, 1901.

[29] U. S. Geol. and Geog. Survey Terr. Ann. Rept. for 1874, pp. 120–121, 1876.

[30] According to Bancroft, "more than 1,000 lodes claimed." History of Nevada, Colorado, and Wyoming, p. 501, San Francisco, 1890.

and Mather & Geist, of Pueblo, both had ore-buying agencies in Silverton in 1879. During this year about 500 tons of ore, worth about $60,000, were sent to the Lake City smelter, and about 185 tons went to Pueblo. The value of the latter was probably about $25,000.

In 1879 a road was completed from Silverton up Cement Creek to the head of Poughkeepsie Gulch, where prospecting and mining was going on with great activity on the Old Lout, Alabama, Poughkeepsie, Red Roger, Saxon, Alaska, Bonanza, and other claims. Chlorination and lixiviation works were erected at Gladstone about this time, to treat these ores by the Augustin process. Their capacity was about 6 tons per day.

During the seventies the eastern and northeastern portions of the Silverton quadrangle were actively prospected, and nearly every lode which has subsequently proved valuable was then located. In some cases paying ore was taken out in large quantities, as from the North Star mine on Sultan Mountain and others. But this activity was in great part feverish and unwholesome. The success of a few encouraged extravagance in the incompetent, and opened a rich field to unscrupulous and dishonest promoters. Smelting plants and mills were erected before the presence of ore was ascertained. Reduction processes were installed without any pains having been taken to ascertain their applicability to the particular ores to be treated. Thus in 1876 Animas Forks was a lively town of some 30 houses and 2 mills and in 1883 boasted of a population of 450. * * * Built upon hopes never realized, its decline was almost as rapid as its rise, and the town is now ruined and desolate. Its principal mill was put up in 1875 or 1876 to treat ore from the Red Cloud mine but was never successful. The Eclipse smelter, erected by James Cherry as late as 1880, at the mouth of Grouse Gulch, ostensibly to run on lead ores from the Mountain Queen and other claims, was also a costly failure. The Bonanza tunnel, a mile and a half west of the town, was run 1,000 feet at the extravagant cost of $300,000 or $400,000, and then abandoned. Around Mineral Point probably $2,000,000 or $3,000,000 were squandered in mining operations which resulted in no permanent improvements or actual development. * * *

In 1881 the remarkable deposits between Red Mountain and Ironton were discovered, and in 1882 and 1883 prospectors swarmed into this new field * * * [See Ouray County, p. 182.]

Previous to the advent of the railroad in Silverton, ores running less than $100 per ton could seldom be handled with profit, but with the completion of the Silverton branch of the Denver & Rio Grande narrow-gage railroad in July, 1882, the rate of transportation on low-grade ores was much reduced, and many mines hitherto unavailable became productive. Freight charges, at first $16 per ton to Denver or Pueblo, were soon dropped to $12, at which high figure they stood for some time. Over 6,000 tons of ore were shipped from Silverton during the first six months after the advent of the railroad. The Greene smelter had some years previously (about 1880) come into the possession of the New York & San Juan Smelting Co., which in 1881 moved the Silverton plant to Durango and in 1882 started the present smelter in that town. In September, 1887, the name was changed to the Durango Smelting Co., which operated until April 1, 1888. From that date until May 1, 1895, business was carried on under the name of the San Juan Smelting & Mining Co., a corporation organized through the consolidation of the Durango Smelting Co., of Durango, and the Hazelton Mountain Mining Co., of Silverton, owners of the Aspen group of mines. The Martha Rose smelter, with a capacity of about 20 tons, began operations in Silverton in 1882, but after smelting about 11 tons of bullion shut down and was never successfully reopened.

The year 1883 was a busy one in Silverton, and the population of the town rose to over 1,500 inhabitants. Sampling works had previously been erected by E. T. Sweet and T. B. Comstock & Co. Late in the season a third plant was opened by Stoiber Brothers. The North Star on Sultan [Mountain], the Belcher, Aspen, Gray Eagle, North Star on Solomon [Mountain], the Green Mountain, as well as the Red Mountain mines were all actively producing, while great strikes were announced in the Ben Franklin and Sampson mines. The Silver Lake mine also came into prominence and shipped that year 72 tons of ore to Sweet's sampling works.

It was not until about 1890 that any real attempt was made to concentrate low-grade ores. The credit of thus initiating a procedure upon which largely depends the future of the whole district must be divided between J. H. Terry, of the Sunnyside mine, and E. G. Stoiber, of the Silver Lake mine. Both men have been successful, and Mr. Stoiber in particular has shown how low-grade veins may be worked successfully on a large scale with a modern plant. About this time the North Star mine, on Sultan Mountain, put up the present mill, run by water power, and in 1894 Thomas Walsh and others erected the matte smelter just west of Silverton. Walsh treated by the Austin process the low-grade Guston ore, and bought siliceous ores wherever he could obtain them. This smelter ran pretty steadily for three years and finally shut down. Its capacity was about 100 tons of ore a day for ten months in the year. In all, about 100,000 tons of low-grade siliceous and pyritiferous ores were treated. There was no further attempt made to smelt ores in Silverton until the construction in 1900 of the pyritic smelter, near the mouth of Cement Creek. The smelter at Durango, which was leased in 1895 by the Omaha & Grant Co., and which on May 1, 1899, became the property of the American Smelting & Refining Co., has continued to handle the bulk of the ore and concentrates from the Silverton region.

With a few notable exceptions, the mines of the Silverton quadrangle produced ores in which silver and lead are the predominant metals. Naturally the rapid decline in the value of silver in 1892 and succeeding years resulted in the closing of many mines hitherto productive and in a general decrease of mining activity. * * * The success of Messrs. Stoiber and Terry in handling low-grade ores has demonstrated that when wasteful and inadequate methods are replaced by modern appliances and shrewd management mines carrying abundant low-grade ore may be made profitable. * * *

Placer mining has never been extensively practiced within the Silverton quadrangle. In former years a little washing was done on the east side of California Mountain, in Picayune Gulch, and in Arrastra Gulch, but there are no extensive deposits of auriferous gravels in the district, and the total output from placer mining is probably insignificant.

In the summer of 1899 the 9-mile Gladstone, Silverton & Northern Railroad was completed from Silverton to the Gold King mine, near Gladstone. In the same year the Gold King mine was equipped with a new 100-ton amalgamation-concentration mill, and the Sunnyside mine, near Eureka, with a 100-ton amalgamation-concentration mill, and the American Smelting & Refining Co. acquired the lead-bullion copper-matte Durango smelter, which is still serving the San Juan region. In 1900 the Kendrick-Gelder pyritic smelter was built at Silverton, but like all other "pyritic" smelters of the United States it sought copper ore to make a matte. It operated part of each year from 1900 to 1905. In 1906 and 1907 this smelter was operated as a matting plant by the Ross Mining & Milling Co., chiefly on copper ores from its Champion, Silver Wing, St. Paul, and Galta Boy mines.

In 1901 the Silver Lake group of 175 claims was sold to the American Smelting & Refining Co. In 1901 the amalgamation and concentration mills reported in the county had a daily capacity of 1,470 tons. From 1902 to 1906 the 80-stamp 200-ton amalgamation-concentration mill of the Gold King mine, the combined 130-ton amalgamation-concentration mills of the Sunnyside mine, the combined 400-ton concentration mills of the Silver Lake mines, and many other mills contributed a heavy stream of amalgam bullion and concentrates to the output of the State. In 1904 magnetic separating machines were added to the Silver Ledge mill, at Chattanooga, to separate the iron from the lead-zinc-iron ore, making this mill the first in San Juan County in which zinc was recovered as a marketable product. As early as 1883 the Sunnyside ore was known by the smelters as one of the most desirable dry ores in the State, except for the large percentage of zinc contained. It then, as now, carried considerable gold and only about 5 to 10 per cent of lead. In 1905 there was a new mill on the Old Hundred mine, which was productive through 1910. In 1906 the Animas Power Co. completed its hydroelectric plant at Rockwood, and since that year it has furnished electric power to the mines. In 1906, at the Gold Prince mine, near the Gold King, was built a 500-ton stamp-amalgamation, wet-concentration, and magnetic-separation mill, which was operated from 1907 through 1910. In 1906 the Dives and Shenandoah mines are mentioned as producing. These mines have made a very creditable production to 1923. In 1907 the Iowa-Gold Tiger 150-ton concentration mill was added to the list of operating mills. Fire and snowslides cut the production for 1908. The Hercules mine became a producer in 1908 and operated until 1911. In 1909 a severe spring and no railroad service for 46 days in July and August and for 40 days in September and October, owing to washouts, cut the production in San Juan County again. In 1910 the production recovered with the addition of several new mills, although the Gold King mine was turned over to lessees and the Silver Lake mines showed signs of exhaustion. The renewed efforts made in 1910 to separate and market the zinc in the ores of the county resulted in a considerable production of that metal that year. In 1911 production again decreased. In 1912 an electrostatic mill was built at the Sunnyside mine to treat the zinc-iron middlings from the amalgamation-concentration mill.

In 1913 the production increased over that in 1912, although the Silver Lake mines were turned over to lessees. Production fell off in 1914. The Silver Lake mines continued to be operated by lessees. The Silver Lake 300-ton concentration mill was opened as a custom mill in May, and a 100-ton flotation unit was added to the mill. A flotation unit was added to the Gold King mill to treat the slimes. Flotation cells were also added to the Iowa-Gold Tiger mill. In 1915 there was an increase for all the metals except silver. The Iowa-Gold Tiger, Silver Lake, Intersection, Gold King, and Sunnyside mines were the chief producers. The years 1916 and 1917 showed further increases in production, but in 1918 the Gold King mine was closed down and the Sunnyside's new 600-ton flotation plant, built in 1917–18, was operated for only eight months. The Gold King was not reopened until 1924. In 1918 the Dives, Highland Mary, Mayflower, and Pride of the West mines, in addition to those mentioned in previous years, made a notable production. In 1919 the regular operations at the Sunnyside mine and mill were carried on for only four months, for a fire on April 26 made it necessary to rebuild the boarding house, bunk house, compressor plant, emergency hospital, and snowsheds. The production of San Juan County for 1920 was the largest in gross value in the history of the county, because of the continuous operations of the Sunnyside 500-ton gravity-concentration and selective-flotation mill, but in 1921, as a result of the low prices (beginning in November, 1920) for lead and zinc, the Sunnyside mill was idle all the year, as were nearly all the other mines, and the production of the county was the smallest since 1882. In 1922 most of the mines continued idle or were not reopened until late in the summer, but in 1923, with increased prices for lead and zinc, mining was resumed and continued, with the operation of the Sunnyside mill, for the full year. The Sunnyside mill increased its output of lead concentrate and zinc concentrate in 1924.

HINSDALE COUNTY

The following account of the Lake City district is taken, with slight changes, from Irving and Bancroft:[31]

Precious metal was probably first discovered in the Lake City area in 1848 by a member of the Frémont party, but the discovery apparently was not followed by any search for mineral deposits. On August 27, 1871, the Ute and Ulay veins were discovered. At that time all the land of the San Juan region belonged to the Ute Indians. The reports of mineral wealth brought many prospectors into the region, but the encroachment on their lands was resented by the Indians. In 1874, by the Brunot treaty, the San Juan Mountains were thrown open to settlement. In August, 1874, Hotchkiss,[32] the leader of the expedition that built a wagon road from Saguache to Lake City, discovered the rich vein now known as the Golden Fleece and named it the "Hotchkiss." News of the strike spread rapidly, and Lake City soon became the center of activity. In 1874 Hinsdale County, with

[31] Irving, J. D., and Bancroft, Howland, Geology and ore deposits near Lake City, Colo.: U. S. Geol. Survey Bull. 478, 1911.
[32] Raymond says Hotchkiss Finley; Rickard says Hotchkiss.

Lake City as the county seat, was established by Territorial legislation, but later legislative enactments have materially cut down the original area. During the same year reduction works were built at Lake City. In 1875 ore aggregating 18 tons, worth $1,319, was shipped from the Hotchkiss mine, and new discoveries were made almost daily. The continued production of the Hotchkiss and the Ute and Ulay mines and the opening up of the Ocean Wave group made the year 1876 seem very promising. During that year the Ute and Ulay mines were purchased by the Crooke Mining & Smelting Co., which erected a lead smelter, the Ocean Wave mine was opened, and in April, 1878, the Excelsior mine was located. By 1880 the Ocean Wave had yielded 110,000 ounces of silver at the Ocean Wave Works, but both mill and mine became idle in that year. The Hotchkiss, or Golden Fleece, did little between 1876 and 1878, and was then idle until 1883 and again until 1889, but from that time on it was very profitable until 1897 [33] and from 1900 until 1902 was worked with a 60-stamp mill. In 1880 a great deal of work was done on the Palmetto group, which lies at the head of Henson Creek, on Engineer Mountain, west of Capitol City, and a 15-stamp mill was built on this property. The St. Louis, Capitol, Czar, Silver Chord, Young America, Yellow Medicine, Pride of America, Vermont, Red Rover, and many other properties near Capitol City were also being worked. In 1881 the Crooke smelter was operating on ore from the Ute and Ulay mines and from the Polar Star, and 400 tons of ore from the Palmetto mine yielded $28,000 worth of silver in the stamp amalgamation mill. In January, 1881, the Frank Hough mine, on Engineer Mountain, was started, and late in the year shipped 60 tons of high-grade copper-silver ore having an average value of $125 a ton. In 1882 concentration works having a capacity of 150 tons a day were erected at the Ute mine, and the Palmetto mill was actively operated. In April, 1884, the Crooke smelter closed, but a large quantity of copper ore was shipped from the Frank Hough mine. The years 1885 and 1886 were dull in this district.

In 1887 considerable ore was shipped from the Ulay, Vermont, and Yellow Medicine properties. The shipments from the Yellow Medicine fell off perceptibly in 1888, but the Ulay and Vermont continued to ship.

In 1889 there was only a small output, but in that year the branch railroad to Lake City was completed, and soon afterward very rich ore was reported from the Golden Fleece. A single car of petzite ore from this mine is said to have yielded $50,000.

There was a revival in the district in 1891, and the period from that year to 1902, inclusive, was the most productive in the history of the county. The principal producing mines were the Golden Fleece, Ute, Ulay, Hidden Treasure, and Czar. The total output of the Golden Fleece was $1,400,000. From 1903 to 1915 little was done in the district. The Ute and Ulay mines were reopened and were productive for a short time in 1918 and were then again abandoned. There is much low-grade ore in this district, and associated with the low-grade ore is an enormous quantity of zinc blende, which should be amenable to proper treatment.

DOLORES COUNTY

The following account of the history of the Rico district up to 1900 is taken with slight modifications from Ransome,[34] who in turn has acknowledged his indebtedness to an article entitled "The early trail blazers," which was published in the Rico News of June, 1892.

In 1861 Lieutenant Howard and other members of John Baker's expedition into the San Juan region made their way over the mountains from the east and prospected Dolores River. In 1866 a party from Arizona reached the bend of Dolores River, where the town of Dolores now stands, and explored the river to its source. Thence they crossed the divide to Trout Lake and went down San Miguel River.

In 1869 two prospectors on their way to Montana from Santa Fe prospected at Rico and located the Pioneer claim, a name that afterward became the official designation of the district.

In 1870 R. C. Darling, a surveyor of the boundaries of the Ute Indian Reservation, passed up the Dolores on his way to Mount Sneffels. He located some claims at Rico and passed on his way. During the same year Gus Begole, John Echols, Dempsey Reese, and "Pony" Whittemore came into the district from New Mexico and discovered the Aztec and Yellow Jacket lodes. On the approach of winter all the prospectors left the district. Apparently none of the prospectors came back the following summer, but in 1872 Darling led a large party into the Pioneer district from Santa Fe, erected a Mexican smelting furnace, and produced three bars of bullion. The results were not encouraging, and the party returned to Santa Fe. In 1875 members of the Hayden survey mapped the region.

Prospecting was again resumed in 1877, and in 1878 it became active through the energy of John Glasgow, "Sandy" Campbell, David Swickhimer, and others, who located the Atlantic Cable, Grand View, and other claims but abandoned work in the winter. In the spring of 1879 rich oxidized silver ore was discovered, and in the summer there was a rush to the

33 See Rickard, T. A., Across the San Juan Mountains: Eng. and Min. Jour., vol. 76, pp. 307-308, 1903.

34 Ransome, F. L., The deposits of the Rico Mountains, Colo.: U. S. Geol. Survey Twenty-second Ann. Rept., pt. 2, pp. 240-242, 1902.

district from neighboring camps. Ore was also found in the Chestnut vein, on Newman Hill, and a small shipment was made to Swansea, Wales. A settlement was begun, a town site was surveyed, and a post office called Rico was opened.

In the fall of 1880 the Grand View smelter began operations, the machinery coming from the railway terminus at Alamosa by wagons to Mancos and thence by Bear Creek to Rico. In 1882 a second smelter was built, which was purchased by the Pasadena Co. in 1884 and operated for nearly two years as a custom plant. In 1887 Swickhimer struck ore in the Enterprise claim, infusing new life into the camp, and large bodies of ore were found in the Rico-Aspen and other claims. In 1890 the Rio Grande Southern Railroad reached Rico. From 1889 to 1894 the largest annual production of the district was made, particularly in 1893, when 2,675,238 ounces of silver, 4,500,000 pounds of lead, and $442,105 in gold were produced. After 1894 the production decreased heavily, though it recovered slightly in 1897–98. The district produced a large quantity of zinc in 1898. In 1901 a magnetic separator and wet concentration mill was erected at the Atlantic Cable mine for the treatment of lead-zinc sulphide ores, and test runs were made during that year and the next. In 1902 the Pro Patria magnetic-separation and jig-concentration mill was set in operation and produced a lead and a zinc concentrate. The milling results were not satisfactory, and production of zinc fell off until 1905 and 1906.

The year 1907 was a dull one for the district, but in 1908 the Pro Patria mill was again set in motion and made a considerable output of zinc and lead concentrates. This mill was destroyed by fire in October, 1908, and no milling was done again until 1913, when the mill was remodeled into a straight wet-concentration plant and operated during that year only. In 1912 the district took on new life from shipments of copper and lead-zinc smelting ore, and in 1913 the value of the silver, copper, lead, and zinc produced was larger than in 1894. In 1914 and 1915 the lead-zinc shipments decreased, and in 1914 the copper shipments, but in 1915 the largest output of copper in the history of the camp was made. In 1916 the production of copper declined, but in 1917 the production of lead and zinc increased heavily and that of copper also increased. In 1918 the production of copper increased somewhat but that of lead and zinc decreased. There are large reserves of lead, zinc, and pyrite ores in the district.

During 1900 the Emma mine, at Dunton, became a producer, and in 1901 an amalgamation-concentration mill was erected. The mine continued as an intermittent producer to 1924.

LA PLATA COUNTY

According to Cross,[35] it was not until 1878 that prospecting was begun in La Plata County. In that year the Comstock mine was opened, and work was begun at the Cumberland and Snowstorm properties, near the head of the La Plata Valley. By the end of 1881 many locations had been made, and the nature of the richest ores, tellurides of gold and silver, had become well known. In 1881 in addition to the activity in the La Plata district, there was some in the Needle Mountains and Vallecito districts, but neither of these two districts has since developed much ore. The La Plata (or California) district, in La Plata and Montezuma counties, embracing the La Plata Mountains and including the watersheds of La Plata River, Mancos River, Bear Creek, and Junction Creek, has been practically the sole producing district in La Plata County. The output was very small until 1894, for which year the report of the Director of the Mint shows a heavy increase in the output of gold and an amazing increase in that of silver, for which no explanation can be given. It would appear that the figures for silver are wrong. For 1895 and 1896 the production was nominal, but in 1897 there seems to have been more activity, and in 1902, owing to the output from the Neglected mine, at the head of Junction Creek, of $117,041 in gold and $1,682 in silver, the production took a very considerable leap forward, and afterward gradually increased year by year—in 1903 from the Neglected, in 1904 from the Neglected and May Day, in 1905 from the May Day—until in 1907 the total output, chiefly from the May Day and Valley View, exceeded $500,000. The production declined in 1908 and 1909, but in 1910 and 1911 the record for 1907 was again approached. Another decline followed in 1912, but an increase was made in 1913. The output decreased again in 1914 and still farther in 1915, 1916, 1917, and 1918. The total production of the La Plata district from 1878 to 1923 was $4,775,699.

SAN MIGUEL COUNTY

In 1870 Darling, the Government surveyor, on his way up Dolores River, passed through the area on which Rico now stands on his way to Mount Sneffels, but if he went as far as the mountains he apparently did not discover any valuable mineral deposits. In 1872 he returned to prospect at Rico. In 1874 prospectors who probably crossed the head of Dolores River from Baker Park into Marshall Basin were led by the placers of the San Miguel to search for lodes in

[35] Cross, Whitman. U. S. Geol. Survey Geol. Atlas, La Plata folio (No. 60), p. 12, 1899.

the mountains. In 1875 they made locations on what is now the Smuggler vein and shipped a ton of ore worth $2,000 to the smelter at Alamosa. Other claims were located, and in 1877 small shipments were made, principally to Silverton, the ores being mined for silver. In 1878 Marshall Basin produced 200 tons, and small lots were sent to the Silverton smelter from the mines about Ophir. In 1879 milling was first attempted by arrastres. In 1881 the Virginius mine, at the head of Canyon Creek (in Ouray County), was worked by three levels and two shafts, and in 1882 produced $75,000 in silver. In 1883 a small smelter was built at the old town of Ames, but apparently it did not prove successful, as it ran only a year. In 1883 a shipment of 4 tons of ore from the Smuggler vein gave a return of 800 ounces of silver and 18 ounces of gold to the ton.

Placer mining in San Miguel County has never been very successful, and by 1896, according to estimates, only $100,000 had been taken out. The auriferous gravels were not generally found at the level of the present streams but occur in small areas at a greater or less distance above them, and operations have been hampered because of the heavy weight of the gravel and because of the great size of the boulders that have to be handled. The total placer production for San Miguel County from 1878 to 1924 has been only $188,635.

The lode mines in the county continued to increase in production steadily from year to year. In 1881 the principal shipping mines were the Smuggler, Mendota, Cimarron, and Argentine. Among others developed were the Alta, Palmyra, and Silver Chief.

According to Rickard,[36] the Pandora mill was erected and the Pandora and Oriental mines were operated during 1877 and succeeding years. Among the early locations were the Belmont and the Tomboy, on the trail which crosses the range from Silverton and Red Mountain, but the value of the lode was not shown in the first workings, and the company organized in 1892 was unsuccessful. Later developments after a reorganization in 1894 resulted two years later in making the lode a large producer. In 1899 the Tomboy Gold Mines Co. was organized to take over the Tomboy mine. In 1901 the ore showed signs of approaching exhaustion, so the Argentine property, near by, was acquired. In 1911 the Argentine began to give poor results, and the Montana group of claims on the opposite side of the basin was purchased from the Revenue Tunnel Co. In 1915 the extension of the Montana group, as covered by the Sydney-White Cloud group of claims, was purchased. During all this period the operations of the company have been successful by the use of amalgamation and concentration, with the addition of cyanidation in December, 1914, processes used until January, 1920, when a new oil-flotation plant superseded all other plants, except that subsidiary amalgamation was installed in the new plant in 1921.

In 1882 the Pandora and Oriental had a 40-stamp mill in operation and a mill on the Gold King had been completed. Reports published in 1882 show that ores from the upper San Miguel district were shipped in 1881 by way of Gunnison, then the nearest railroad point. In 1882 there was a 10-stamp Frue vanner amalgamation-concentration mill on the N. W. H., jr., claim. In 1883 San Miguel County was created from a part of Ouray County. In 1884 the Smuggler Mendota, Sheridan, Union, Cleveland, Bullion, Hidden Treasure, Cimarron, and other mines in Marshall Basin were working, and the Ophir district was also active. During the next six years the Smuggler, Mendota, Sheridan, and Union continued to be large producers. In 1891, the Smuggler-Union Mining Co. was formed, including also the Sheridan-Mendota claims. That year the Carribeau, at Ophir, yielded large quantities of silver. The operations of the Smuggler Union Co., have been continuous.

In the five years from 1893 to 1898, according to Purington,[37] the increased facilities for the transportation of ore from the mines to the mills by means of wire-rope trams and the widespread use of electricity for power contributed greatly to the increase of production.

Since 1898 the large output of the Telluride district has come chiefly from the mines of three large companies—the Liberty Bell, the Smuggler Union, and the Tomboy. The last two have been already mentioned. According to Chase,[38] the Liberty Bell mine is on the western front of the San Juan Mountains, 2 miles north of Telluride. The vein was discovered in 1876 by W. L. Cornett, who, with subsequent locators, took up claims along the apex. A few hundred feet of development work was done and a few tons of ore were smelted or milled, but profitable working proved impossible, and the property lay idle until 1897, when Arthur Winslow acquired it for the United States & British Columbia Mining Co. After due investigation preliminary development, and the initial construction of mine buildings, tramway, and 10-stamp section of the proposed 80-stamp mill, the Liberty Bell Gold Mining Co., was organized and began operations in December, 1898. Since then, until 1920, there have been only two complete suspensions aggregating 10 months, for extensive additions and alterations to the mill; a suspension of three months in 1902 for reconstruction, following disastrous snowslides; and one for four months in 1903, by reason of labor troubles—a total of about a year and a half; otherwise, the mine

[36] Op. cit., p. 843.

[37] Purington, C. W., Preliminary report on the mining industries of the Telluride quadrangle, Colo.: U. S. Geol. Survey Eighteenth Ann. Rept , pp. 755-756, 1898.

[38] Chase, C. A., Notes on the Liberty Bell mine: Am. Inst. Min. Eng. Trans., vol. 42, pp. 694-741, 1911. See also Winslow, Arthur, Am. Inst. Min. Eng. Trans., vol. 29, pp. 285-307, 1900.

has been worked continuously and its output has expanded to 500 tons of ore daily.

Mr. Winslow revived the enterprise at the time when the treatment of raw mill-tailings by direct cyanidation was first shown to be profitable, within a year or two after the first successful long-distance transmission of electric power. Experiments in the cyaniding of tailings from amalgamation and concentration were begun almost immediately. In September, 1899, an experimental 7-ton leaching plant was installed, and in May, 1900, a 250-ton leaching plant of the South African type was ready for operation. It was evident from the outset that the mine could be made profitable, although this plant treated probably the lowest grade of material then handled in this country by this process. The mill was idle for four months in 1920. In September, 1920, the cyanide treatment was discontinued, and for the rest of the year concentration only was applied to the ore, the reserves of which were seen to be small. In 1921 the mill was operated in a final clean-up until August, 1921, when operations were discontinued and the company was dissolved.

From 1897 onward San Miguel County has been one of the principal milling centers of Colorado. Very little crude ore was shipped to smelters, and practically the entire output of ore was milled by amalgamation, concentration, cyanidation, and flotation, producing refined gold-silver bars and high-grade concentrates. From 1913 to 1918 the concentrates in the Liberty Bell mill were also given a rough treatment by cyanidation, thus recovering an additional percentage as bullion and reducing the freight charges on the concentrates by reducing their value. The county averaged annually over $2,000,000 in gold from 1897 to 1919 and over $1,000,000 from 1920 to 1924. From 1897 to 1924 it averaged more than 1,000,000 ounces of silver, and in 1923 it reached 1,982,007 ounces. The lead output averaged more than 5,000,000 pounds from 1897 to 1924 and reached 9,360,637 pounds in 1923. From 1908 zinc middlings were shipped from the Tomboy mill until 1914, when, because of the change in the ore milled, the zinc content was not saved. From 1915 to 1920 the zinc output of the county was obtained from concentrates from the Smuggler-Union mill, which were derived from ore taken from the Black Bear mine. The zinc concentrates from the Black Bear ore were of low grade, however, and the price received for that part of the mill product was nominal, so development only was done until August, 1923, when the Reed-Coolbaugh sulphating plant, which was built at Durango to remove the zinc, allowed large quantities of the complex Black Bear ore to be milled, first making a high lead concentrate on the tables and then floating all remaining sulphides as a lead-zinc-iron-copper concentrate.

OURAY COUNTY

In 1874 prospectors were at work at the headwater of the Uncompahgre and in Poughkeepsie Gulch within the present boundaries of Ouray County, and also just over the present county line, in what is now San Juan County. In 1875 the Grand View claim was located just below the town of Ouray, and in 1877 it was patented. In 1881 work was being done at the Belle of Ouray and Union mines, on Bear Creek, 3 miles from the town of Ouray; at the Silver Link and Silver Point mines, on the Uncompahgre, within a mile of Ouray; at the Mineral Farm mine, near the mouth of Canyon Creek, 2 miles from Ouray; and at the Virginius, Yankee Boy, Revenue, Governor, and other mines at the head of Canyon Creek, over the divide from the San Miguel drainage basin.

In 1882 miners from Silverton went across the range and found the Yankee Girl, National Belle, and other deposits at the headwaters of Red Mountain Creek, a tributary of the Uncompahgre. In September, 1882, the Denver & Rio Grande Railroad reached a point within 30 miles of Ouray. It is not positively known whether the Virginius deposit was discovered by prospectors who came from New Mexico by way of Dolores River, by prospectors who came from Silverton, or by prospectors who came from the Gunnison by way of Ouray.

At first the mines at the head of Canyon Creek were worked vigorously. Red Mountain and Ironton were very active from 1882 to 1893, when the fall in the price of silver and the exhaustion of the phenomenally rich portions of the ore body caused the boom to collapse. The railroad from Silverton to Ironton was completed in 1888. The Yankee Girl and Guston mines were worked until 1896, but the low price of silver, the increase in cost as the workings became deeper, the expensive trouble in handling the corrosive waters of the mines, and the low grade of the ore taken from the deep workings caused the mines to close down. These two mines alone have produced between $6,000,000 and $7,000,000. From 1896 to 1915 these camps were practically idle. Owing to the increase in the price of copper and lead, a revival began in 1915, but it continued only until 1918.

The Virginius mine was first worked at the upper level, at 12,500 feet, in the Virginius Basin, and all ore was taken out through this opening. About 1893 the Revenue adit was completed. This adit started at an altitude of 10,800 feet in the bed of Canyon Creek and ran 7,500 feet to cut the Virginius vein at a depth of 2,000 feet vertically below the surface. The Revenue tunnel is about 1,000 feet above the lower limit of the San Juan formation. Owing to the desultory manner in which the Virginius had been worked, it is impossible to estimate the output of the mine up to 1896. The ore is said to have carried about $14 in silver to the ton and varying quantities of gold. The value of the output of the mines of the Virginius Basin up to 1896 is estimated at $4,000,000, which includes the output of practically all the productive part of Ouray County that lies in the Telluride quadrangle as mapped by the United States Geological Survey.

Ransome,[39] writing in 1901, says of the Camp Bird mine:

Although the first mine to be worked in the district was a gold producer, yet it is an interesting fact that for many years prospecting was restricted to a search for silver and lead ores. It was apparently owing to this adherence to an established routine that the Una and Gertrude claims, in Imogene Basin, worked 20 years ago for silver and lead, were subsequently abandoned, with no knowledge of the remarkable gold ore which lay alongside the argentiferous streak and which was thrown out as waste. Masses of this rich ore were discovered by Thomas F. Walsh on the Camp Bird claim and subsequently on the dump of the Una and Gertrude and he purchased the latter in 1896 for $10,000.

In an article published in 1911 Rickard[40] agrees in the main with Ransome. An article[41] entitled "The true story of the Camp Bird discovery" describes the journey of William Weston in October, 1877, by railroad to La Veta, by wagon to Del Norte, by pack outfit over Cunningham Gulch to Silverton, thence up Mineral Creek and over the divide by Commodore Gulch into Imogene Basin, in the Sneffels district. Weston and his partner staked the Gertrude, Una, and other claims and worked them single-handed for four years. Weston assayed his own ores and obtained $12 to $20 worth of gold per ton from the outcrop of the Gertrude and Una, but at that time the smelters would not pay for less than 1 ounce of gold a ton, and it cost $35 a ton to pack the ore to the Greene smelter, at Silverton, and $45 a ton for treatment; so it would not pay to extract ore running less than $100 a ton. In 1881 the Weston & Barber group was sold for $50,000 and a mill was erected, but the company failed, and it was 14 years later, in September, 1896, when Thomas F. Walsh, in looking for siliceous ores to flux the basic ores of Red Mountain in his pyritic smelter at Silverton, sampled the Gertrude dump and found rich gold ore in the face of the Gertrude drift.

The Camp Bird mine was worked actively from 1896 to June 30, 1916, when it was closed to await the results of the new adit, which at 11,000 feet in June, 1918, cut the vein 450 feet below the lowest old workings. Development work on the vein failed to find sufficient ore to operate the 60-ton mill. From 1896 to June 30, 1916, this mine yielded $27,269,768 in value of recovered gold, silver, lead, and copper, and

the profit at the mines and mill, exclusive of depreciation, was $17,731,788, or 65 per cent.

SAGUACHE COUNTY

The town of Saguache, the county seat of Saguache County, had gained considerable prominence as a distributing point in 1867 and 1868, but the history of mining there practically begins with 1879-80, during the rush to the Gunnison country. The silver-lead-manganese veins of Kerber Creek attracted many prospectors, and Bonanza had grown rather large before the fall of 1880. Free gold in quartz was found on the western slopes of the Cochetopa Mountains, and the camp of Willard was established on Cochetopa Creek. From 1880 to 1923 the county produced $2,776,554, of which $1,626,385 represents the output of silver. The greater part of the output came from the Rawley, Antoro, Michigan, Paragon, Cocomongo, and Eagle mines, in the Kerber Creek district.

In July, 1923, monthly shipments of 2,500 tons of lead-silver concentrate to the Leadville smelter were begun as the result of a reorganization of the Rawley Co., with help from the Metals Exploration Co. and the American Smelting & Refining Co., whereby a 300-ton mill was erected at Bonanza and connected with a 7½-mile aerial tram to Shirley, on the Marshall Pass branch of the Denver & Rio Grande Western Railroad. Unfortunately the mine closed in December, 1923, because of certain financial and other difficulties.

The Bonanza district is interesting because of its zonal ore bodies, manganese-silver oxidized outcrops, silver-lead-iron zone, silver-iron-zinc zone, and silver-iron-copper zone. The manganese surface deposits are thought to be rich enough to ship as manganese ores.

On the eastern side of the San Luis Valley is the Orient mine, once a large producer of iron ores, which, with new development in 1923-24, promises to equal its former output.

The Crystal Hill and Embargo districts have produced good ore from time to time.

MINERAL COUNTY—THE CREEDE DISTRICT[42]

In the eighties a wagon route in the upper part of the valley of the Rio Grande extended between Wagonwheel Gap, Silverton, and Lake City. This route passed very near the present site of Creede and still nearer Sunnyside, a small camp about 2 miles west of Creede. Some of the prospectors on this route halted long enough to prospect the steep mountain slopes along the valley, and after finding encouraging indications they located several claims. J. C. McKenzie and H. M. Bennett located the Alpha, at Sunnyside,

39 Ransome, F. L., A report on the economic geology of the Silverton quadrangle, Colo.: U. S. Geol. Survey Bull. 182, p. 24, 1901.
40 Min. and Sci. Press, vol. 103, p. 827, 1911.
41 Denver correspondence, Eng. and Min. Jour., vol. 89, p. 1266, 1910.
42 Emmons, W. H., and Larsen, E. S., Geology and ore deposits of the Creede district, Colo.: U. S. Geol. Survey Bull. 718, 1923.

April 24, 1883, and with James A. Wilson staked out the Bachelor Claim, near the present site of Creede, July 1, 1884. Some prospecting was done in the middle eighties, principally at Sunnyside, and futile attempts were made to work the ores in arrastres. There is no record of any further discoveries from 1886 until August, 1889, when N. C. Creede, E. R. Naylor, and G. L. Smith located the Holy Moses mine on Campbell Mountain. The following summer Creede located the Ethel claim and C. F. Nelson located the Solomon claim. The mining district thus formed, which lies just east of the Sunnyside district, was called the King Solomon district. No valuable deposits were found until June, 1891, when D. H. Moffat and Capt. L. E. Campbell came to Wagonwheel Gap to visit the Holy Moses mine, on which they had obtained an option, and employed Creede to prospect for them. Soon afterward Theodore Renniger found rich float on Bachelor Mountain but could not find its source until Creede discovered the outcrop of a large lode at a point 200 feet above it on the hill slope. Creede accordingly located the lode as the Amethyst and informed Renniger, who on August 8, 1891, took up another claim, which he called the Last Chance, on the same lode. Three years earlier J. C. McKenzie had located several claims on a large vein of quartz about 150 feet above the Amethyst lode. These claims were, however, abandoned until the fall of 1891, when the Del Monte location covered them. Débris from the upper part of the mountain had so deeply covered the Amethyst lode as to prevent its recognition earlier. After Creede's discovery but little work was necessary to prove that the Amethyst lode was an immensely valuable deposit, and it was accordingly staked for a distance of nearly 2 miles along its strike.

The railroad was extended from Wagonwheel Gap to the Creede district on December 16, 1891. The district has been producing almost continuously since the advent of the railroad. Mining there was pushed actively in 1892 and culminated in the largest recorded annual production of the district—4,897,684 ounces of silver and quantities of lead and gold, all having a value of $4,150,946. The continued drop in the price of silver caused a collapse in 1894, but there was a recovery from 1898 to 1900, and a fairly steady output until the end of 1907. From 1901 to 1911, inclusive, the production of lead and zinc was large, though that of silver steadily declined. From 1912 to 1914 the production of silver increased, but it decreased greatly in 1915, 1916, and 1917. There was a large increase, however, in both quantity and value in 1918. From 1919 to 1923, inclusive, the output of silver steadily decreased.

TELLER COUNTY—THE CRIPPLE CREEK DISTRICT

GENERAL FEATURES

Creede and Cripple Creek were rivals in attracting attention in 1891. According to Lindgren and Ransome,[43] the historic rush of prospectors to the Pikes Peak region in 1859 did not result in discoveries in that immediate neighborhood.

It was not until 1874 that the region adjacent to Cripple Creek began to attract the attention of prospectors. The report that H. T. Wood, while connected with the Hayden Survey, had found gold ore near Mount Pisgah (west of the town of Cripple Creek) drew a number of men to that locality. A few loose fragments of ore were found on the surface and the Mount Pisgah mining district was organized, but no valuable deposits were uncovered, though in 1878 Henry Cocking is said to have driven a tunnel in Poverty Gulch near the point where the Gold King and C. O. D. mines were afterward developed, and openings were made by B. F. Requa and others in what is now the productive part of the district.

The district was then gradually deserted. There was a brief renewal of activity in 1884, caused by the reported rich placer deposits near Mount Pisgah. The alleged discovery, however, appears to have been fraudulent, and the grassy hills of the Cripple Creek region, now thoroughly discredited in the eyes of mining men, were given over to the grazing of cattle.

Rickard[44] writes:

Among the earliest of the gold seekers was Robert Womack, who once owned a small ranch in the district. He sold it to Bennett & Myers, the proprietors at that time of the cattle range, which covered a large part of the area now forming the environs of the town of Cripple Creek. For many years, between 1880 and 1890, Bob Womack lived in the district, doing occasional work for Bennett & Myers and spending his spare time in prospecting. He had previously had some experience in Gilpin County and knew gold ore when he saw it. In the course of desultory diggings he found several veins, and when he would turn up at intervals, at Colorado Springs, he exhibited pieces of float as evidence of his discovery; but * * * his statements made little impression. For many years he worked on a hole in Poverty Gulch without staking a claim in proper form. * * * In December, 1890, E. M. de la Vergne and F. F. Frisbee came up from Colorado Springs to prospect. * * * On Guyot Hill, in Eclipse and Poverty gulches [they] found evidences of gold veins, and samples were taken away. These averaged about 2 ounces of gold per ton. * * * [They] returned early in February, 1891. They found Bob Womack at work in Poverty Gulch. * * * [Frisbee] sent 1,100 pounds of surface ore by wagon to the Pueblo Smelting & Refining Co., who gave returns at the rate of $200 per ton. This was in August, 1891. * * * In May, 1891, Frisbee and De la Vergne happened to be at Colorado Springs and met W. S. Stratton, to whom they showed certain assays of ores brought down by them from Cripple Creek. * * * After the meeting with De la Vergne and Frisbee [Stratton went to] Cripple Creek and camped there * * * [and on the 4th of July, 1891] he made two locations, the Washington and the Independence.

43 Lindgren, Waldemar, and Ransome, F. L., Geology and ore deposits of the Cripple Creek district, Colo.: U. S. Geol. Survey Prof. Paper 54, pp. 130-131, 1906.
44 Quoted by Lindgren, Waldemar, and Ransome, F. L., op. cit., pp. 130-132.

The following account of the development of the district from 1892 to 1905 is taken without modification from Lindgren and Ransome: [45]

The development of the district, notwithstanding the fact that many mining men of capital and experience looked askance at what they regarded as another Cripple Creek bubble, was extraordinarily rapid. Before the opening of spring in 1892 the hills swarmed with prospectors, and on February 26 the town of Cripple Creek was incorporated. Adjoining it on the southwest sprang up the town of Fremont, afterward absorbed by Cripple Creek. The main route into the district at this time was by wagon road from Florissant.

In October the Anaconda, Arequa, Blue Bell, Buena Vista, Deerhorn, Eclipse, Gold King, Matoa, Mountain Boy, Ophir, Pharmacist, Plymouth, Strong, Summit, Sweet, Victor, and Work mines were shipping ore, and railroads were under construction from Canon City on the south and from Divide on the north.

In the autumn of 1893 the list of producing mines had become a long one and included the Blue Bird, C. O. D., Dead Pine, Doctor, Eclipse, Elkton, Gold Dollar, Granite, Ingham, Logan, Mary McKinney, Moose, Morning Glory, Portland, Raven, Stratton's Independence, Strong, Tornado, Zenobia, and many other well-known properties.

The Midland Terminal Railroad, connecting Cripple Creek with Colorado Springs by way of Divide, was completed December 16, 1893, and the Florence & Cripple Creek Railroad was opened to traffic July 2 of the following year.

The year 1894 is memorable on account of a strike during which the miners resorted to arms, property was destroyed, and lives were lost. A large force of deputy sheriffs was finally enrolled to restore order, but at this stage the governor of Colorado called out the militia and put a stop to what threatened to become a miniature war. The mine owners, by the "Waite agreement," consented to the establishment of a minimum wage, to the eight-hour day, and to the avoidance of all discrimination between union and nonunion men. In spite of these disturbances the development of the district made notable strides, and the Independence mine at this time was only 70 feet deep and was worked with a horse whim. It shipped in August 800 tons, of which the poorest carload averaged 3½ ounces of gold per ton. The Portland mine at this time was shipping about 60 tons of smelting ore daily. About 100 men were employed, and the mine produced more ore than any other property in the district. It was in the latter part of this year that Cross and Penrose investigated the district for the United States Geological Survey.

In 1895 the Portland mine had reached a depth of 600 feet and the Independence 470 feet. The latter was the most profitable mine in the district, and Stratton, now a rich man, began the purchase of outlying property. The Logan and American Eagle mines were bought by him this year and were consolidated as the American Eagle group. He acquired a number of other mines in succeeding years. The Vindicator, 60 feet deep; the Mary McKinney, 146 feet deep; the Anna Lee, 760 feet deep; and the Elkton, Pharmacist, Isabella, Victor, Last Dollar, Strong, Anchoria-Leland, Abe Lincoln, C. O. D., and Gold King were all shipping ore in this year, and considerable excitement was caused by the remarkably rich ore shoots in the Moose, Raven, and Doctor mines on Raven Hill. Several of the mines encountered water about this time and had to begin pumping.

During the next few years the number of producing mines continued to increase, and in 1900 the maximum output of $18,000,000 was obtained. Beacon Hill attracted much attention in consequence of rich ore found in the Prince Albert

and adjacent mines. The Victor and Isabella mines were highly productive up to 1898 and 1900, respectively, and shipped large quantities of very rich ore. Four long tunnels—the Chicago, Good Will, Ophelia, and Standard—were begun about this time. In 1899 the Standard tunnel encountered a flow of water 2,800 feet from the portal, which compelled a suspension of operations. Teller County, with Cripple Creek as its county seat, was formed from a portion of El Paso County. Another notable event of the year was the sale of Stratton's Independence, the most famous and profitable mine in the district, to the Venture Corporation (Ltd.), of London, for $10,000,000.

In 1901, the Colorado Springs & Cripple Creek District Railway was completed into the district. About this time many of the larger mines, having worked down to the water surface determined by the outflow through the Standard tunnel, were again compelled to face the question of deeper drainage. A drainage commission was formed, subscriptions were collected, and the El Paso tunnel was begun in 1903. Connection was made with the El Paso mine, under Beacon Hill, in the autumn of the same year.

The year 1902 is noteworthy chiefly on account of the discovery of remarkably rich ore in the recently opened C. K. & N. mine on Beacon Hill and the coming into prominence of the El Paso and Golden Cycle mines as large producers.

Early in 1903 a strike was ordered by the Western Federation of Miners in all mines shipping ore to certain reduction works in Colorado City. The difficulty was adjusted for the time, but after some months of agitation and uncertainty another strike was called, on August 10, which resulted in the closing of nearly all the mines in the district except the Portland. The mine owners organized and took active steps to reopen the mines with nonunion labor. Work was first resumed at the El Paso under strong guard, and some of the other mines were soon afterward reopened under similar conditions. It soon became evident, however, that any general attempt on the part of the mine owners to work the mines would be the signal for violence. Governor Peabody accordingly ordered the militia into the district, and under their protection all of the mines gradually resumed operations with nonunion miners. A number of dastardly outrages, such as the murder of the superintendent of the Vindicator mine and the blowing up of a station platform at Independence at a time when it was crowded with nonunion miners, were perpetrated about this time, and, being generally charged to the union men, led to coercion and deportations.

The general depression caused by the labor difficulties of 1903 and 1904 was partly relieved by a number of discoveries of new pay shoots, particularly of a body of remarkably rich ore in the W. P. H. claim, on Ironclad Hill. Ore of very high grade was found also in the El Paso mine and new pay shoots opened in the Gold Coin and Granite mines. In the Portland mine a number of new ore bodies were discovered, indicating that the ore reserves on several of the levels were larger than had been generally supposed.

THE ROOSEVELT OR CRIPPLE CREEK DRAINAGE TUNNEL

During 1905 and 1906 deeper drainage to remove the water from the Cripple Creek volcanic vent was seriously considered. The result was the choice of the Gatch Park plan, according to which a tunnel was to be started at a distance of 14,550 feet from the El Paso shaft, at an elevation of 8,020 feet, at a point 770 feet below the El Paso tunnel. The depth gained was 740 feet. After much desultory work had been done by the tunnel company and by contractors in 1907, when the tunnel had been driven 1,500 feet, the contract for its completion was awarded, in January, 1908, to A. E. Carlton. From that time the work progressed steadily from the portal and from headings that lay north and south of a shaft at a point 7,975 feet from the portal. The elevation of the collar of this shaft was 8,743 feet and its depth to the grade of the tunnel was 685 feet. This shaft was started August 16, 1907, and was finished September 16, 1908. At the end of 1908 three headings in all had been completed for a distance of 4,872 feet, all in Pikes Peak granite. In 1909 the tunnel had been advanced steadily and was in over 12,000 feet at the end of the year but had struck no heavy flow of water. One small watercourse giving off about 100 gallons a minute had been encountered. In December, 1910, the tunnel, which was then 15,640 feet long, had entered the breccia and had passed 1,190 feet beyond the El Paso shaft, the first to be reached on the line of the tunnel. The contractors were then authorized to drive 200 feet more, unless a flow of 2,000 gallons a minute should be encountered within that distance. The cost of the tunnel was $28 a foot, and the total amount subscribed for it was $550,000. Early in 1911 a 540-foot crosscut was driven from the tunnel to cut the C. K. & N. vein, and later subscriptions were raised to drive the tunnel 1,200 feet farther into Beacon Hill. Although the tunnel had been driven far beyond the distance originally planned, the flow of water through it and the subsidence of the water in the mines, when work was stopped in April, 1911, were so slow that it was not possible to begin development work at lower depth.

Work on the tunnel was resumed in October, 1911, and early in 1912 the El Paso mine was dry to tunnel level; the Gold Dollar and Mary McKinney mines were practically drained to the lowest workings; the Portland mine was able to open its 1,500-foot level and the Independence its 1,000-foot level; the water in the Elkton, Strong, and Granite mines was subsiding steadily; and the Cresson shaft was lowered 100 feet. The Vindicator and Golden Cycle mines did not benefit from the increased drainage. Work was continued until March, 1912, when the tunnel was 16,857 feet long.

No more work was then done on the tunnel in 1912 or in 1913, but the drainage was not satisfactory. The recession in 1913 at the Elkton was 84 feet and that at the Portland 102 feet, and no results were obtained at the Golden Cycle-Vindicator. New subscriptions were accordingly made in 1913 to resume work, and drilling was started again on August 3, 1914. Work progressed until November 4, 1914, when the El Paso plant was completely destroyed by fire and work in the tunnel had to be stopped, for the waste had been hoisted through the El Paso shaft. During 1914 the recession of the water at the Elkton was 99 feet and the recession in the Portland was 132 feet, or 30 per cent more than in 1913.

On March 2, 1915, work was resumed on the tunnel and progressed steadily. At this time the tunnel was 17,127 feet long. It was proposed to drive it to the Golden Cycle-Vindicator mines, a distance of 10,250 feet farther. In 10 months of 1915 the tunnel was driven 1,920 feet. The total recession of water that year was 149 feet. At the end of 1915 the Elkton shaft was 1,460 feet deep, and in February, 1916, the recession at the Elkton shaft had enabled the company to sink the main shaft 125 feet below the sixteenth level and the tunnel had reached a heading under the Elkton about 150 feet below the bottom of the shaft. During 1915 also the Portland No. 2 shaft was sunk 154 feet and the eighteenth level (1,870 feet below the surface) was reached.

In April, 1915, two large electric pumps, which had been in use on the sixteenth level of the Golden Cycle mine, together with an additional unit of the same size, were installed on the eighteenth level, where the pumps were to remain and pump water from the eighteenth level to the level of the La Bella tunnel, 300 feet below the surface. Three new centrifugal pumps were ordered at the end of 1915 to raise the water from the twentieth level into the pumps on the eighteenth. This installation indicated that the Golden Cycle mine was not materially benefited by the drainage and that the tunnel would not be driven to that mine.

During 1916 the tunnel was driven 2,311 feet, and its total length from portal to breast at the end of the year was 22,100 feet. In addition considerable work was done in the vicinity of the Elkton main shaft. The Portland No. 2 shaft was completed to the 1,900-foot level, 1,972.73 feet vertically below the surface. In 1917 the tunnel was driven 1,746 feet, making its total length nearly 24,000 feet. The discharge of water at the portal was 8,500 gallons a minute on January 1, 1917, and it steadily diminished to 4,000 gallons at the end of 1917. The tunnel had by this time lowered the general underground water level of the district 700 feet. In 1918 the tunnel was completed. The end of the tunnel is about at the middle of the northern part of the Portland Gold Mining Co.'s property, 24,255 feet from its portal. On January 1, 1919, a branch tunnel was completed from the main tunnel to the Portland No. 2 shaft at a depth of 2,131 feet from the surface. In 1918, at the Vindicator-Golden Cycle, after work had been discontinued on the bottom levels of the mine, the nineteenth and twentieth levels were cleared up, all pipe, track, and other materials were removed, and the work of dismantling and removing the pumps from the bottom level was commenced. This work was completed and pumping was stopped on June 27, 1918. The lower workings of the mine filled rapidly, but it was hoped that when the water rose to the level of the tunnel the volume of water to be handled would decrease and that pumping would be unnecessary. This hope proved futile, however, and on July 27, 1918, the Vindicator Co. had to resume pumping from the eighteenth level of the mine.

No work has been done on the Roosevelt tunnel since January 1, 1919.

MILLING IN THE CRIPPLE CREEK DISTRICT

Lindgren and Ransome[46] trace the history of reduction processes up to 1904 as follows:

The history of the development of the processes of reduction for the Cripple Creek gold ores is in many respects of great interest. Stamp milling, long the recognized mode of treatment of gold quartz, was first tried. During 1892 and 1893 ten stamp mills of the Gilpin County type with slow drop and light stamps were erected, aggregating 270 stamps, the largest being the Rosebud and the Gold and Globe mills, having, respectively, 60 and 40 stamps, both situated along Cripple Creek below the town. A short trial sufficed to demonstrate their inefficiency to deal with the free gold, on account of difficulties of amalgamation due to a tarnish supposed to be tellurite of iron. Percussion tables and blankets were introduced to improve the gold saving, but even then the extraction was lamentably low. The matter was made worse by the appearance of unoxidized tellurides, and in a few years this process was entirely abandoned.

Smelting was early recognized as a proper method of treatment for rich ores, and an increasing amount of such material soon found its way to the smelting works at Denver and Pueblo.

The first chlorination plant was erected by Edward Holden in 1893, and by January, 1895, the first well-designed mill, of 50 tons daily capacity, was completed at Gillett, a few miles northeast of Cripple Creek. The process employed was the barrel chlorination used in South Carolina and the Black Hills.

About the same time experiments were made with the cyanide process, the first mill being erected at Brodie in 1892. In 1895 the Metallic Extraction Co.'s mill was built near Florence and gradually enlarged to a capacity of 170 tons per day. At that time began the struggle for supremacy between the chlorination and cyanide processes, from which the former [emerged] victorious.

Another change soon began to be apparent. With the advent of improved railroad facilities the lower valleys were found to be better adapted for the location of great reduction works, Colorado Springs and Florence being the most favorable points selected. In 1899 there were still four plants in operation at Cripple Creek, but in 1903 only one mill was active, aside from two smaller plants, for direct cyanide work.

In 1904 the different plants were located as follows:

Location and capacity, in tons, of reduction plants in Cripple Creek district in 1904

Cripple Creek:
    Economic mill (roast; barrel chlorination)_____ 300
    Homestake mill (direct cyanide)_____ 200
    Sioux Falls mill (direct cyanide)_____ 100
Colorado Springs:
    Portland mill (roast; barrel chlorination)_____ 300
    Telluride mill, General Metals Co. (roast; barrel
        chlorination)_____ 300
    Standard mill, United States Reduction & Refining
        Co. (roast; barrel chlorination)_____ 450
Florence:
    Dorcas mill (roast; cyanide)_____ 150
    United States Reduction & Refining Co. (roast; barrel
        chlorination)_____ 400

[46] Op. cit., pp. 138, 140, 142.

In 1905 the small nonroasting cyanide plants for oxidized ore in the Cripple Creek district treated more ore than ever before. Of these plants the Wild Horse, Dexter, Anaconda-Homestake, Los Angeles, and Sioux Falls were operated, and small plants were erected on the Santa Rita and Home Run mines. About two-thirds of the ore was sent to the chlorination plants of the Portland Co. and the United States Reduction & Refining Co., at Colorado City, and the Economic mill, at Victor. In 1905 construction was started on the roast-cyanide Golden Cycle mill at Colorado Springs. This plant was set in motion in February, 1907, and was destroyed by fire in August, 1907. Reconstruction was begun at once, and the plant was started again in December, 1907, and from that time on it has handled most of the ore mined in the Cripple Creek district as a roast-cyanide mill employing subsidiary amalgamation. In 1906 the Portland Co. erected a 300-ton cyanide plant at Colorado City to treat the tailings from its chlorination plant, and the United States Reduction & Refining Co. built a cyanide annex to their Standard plant at Colorado City for the treatment of tailings. In March, 1906, the Dorcas mill, at Florence, was destroyed by fire. The Economic chlorination mill was operated during part of 1906 but was destroyed by fire early in 1907. In 1906 the Jo Dandy purchased the cyanide mill of the Cripple Creek Cyanide Co. at Gillett, dismantled and removed it to Raven Hill and remodeled it. Of the nonroast cyanide plants the Home Run, Santa Rita, and El Paso (which was built on the dump of the El Paso mine) were operated at intervals; the Dexter and Sioux Falls were idle; the Los Angeles was destroyed by fire; the Homestake-Ironclad mill was run for a while; but the 200-ton Homestake mill was closed. The Little Giant cyanide, Wishbone roast cyanide, and Blue Flag cyanide mills were built in 1906 but were not operated until 1907. Tests on the ore of the Independence dump, at Victor, led to the decision to build a large wet concentration, nonroast cyanidation plant at this mine. This mill started operations in April, 1909, and its capacity was gradually increased until it reached 10,000 tons a month. It was operated successfully until July 1, 1915, when the Independence mine and mill were purchased by the Portland Co.

During 1907 the Isabella, small Ironclad, large Ironclad, Jo Dandy, Blue Flag, and Wishbone mills were operated.

During 1908 the Golden Cycle roast-cyanide mill and the Portland and Standard roast-chlorination mills, which employed cyanidation for the tailings, divided the bulk of the ore. In the Cripple Creek district, the Isabella, Blue Flag, Jerry Johnson, Ironclad, Jo Dandy (which was destroyed by fire in November), W. P. H., and Wild Horse plants were operated periodically. The 10-stamp experimental plant of the Portland Co., at Victor, demonstrated that the raw ores of the Portland mine district could be successfully treated without roasting. In 1909 the Golden Cycle roast-cyanide plant handled 60 per cent of the milling ore and the Portland, Standard, and Union plants, by a combined process of chlorination and cyanidation, handled the rest. The small non-roast cyanidation mills in the district did little that year.

The chlorination of Cripple Creek ores ceased in 1911. At the end of that year both the chlorination and the cyanide plants of the United States Reduction & Refining Co. at Colorado City were stopped, never to be revived. During that year the Portland chlorination plant at Colorado City was gradually transformed into a roast-cyanide plant, a new 500-ton roast-cyanide plant was built at Florence to treat the tailings of the old Union chlorination plant there, the Stratton's Independence wet concentration, non-roast cyanide mill increased its capacity, the Portland-Victor 300-ton concentration cyanide mill (completed June, 1910) was operating successfully, the Ajax-Clancy cyanide mill (later changed to a concentration cyanide mill) was built and put into operation, and several of the small cyanide mills in the Cripple Creek district were operated. The conditions remained much the same until 1915, except that the Florence cyanidation mill closed in 1913.

In 1915 experiments led to the installation of flotation at a new mill of the Vindicator Co., at Victor. In that year the Portland Co. bought the Stratton's Independence cyanide mill, operated it for a while as a cyanide plant, later experimented with the flotation process, and finally enlarged it in 1918 to a capacity of 1,500 tons a day as a cyanidation concentration plant. Ores from other dumps and mines also were purchased for treatment. The Victor mill was closed July 30, 1918, and was afterward dismantled. About 2,000,000 tons of low-grade ore were then on the Vindicator dump and about 15,000,000 tons of low-grade ore in the mines and dump of the Portland properties, all of it amenable to treatment by cyanidation or flotation. The small cyanide mills were idle in 1918 or were operated for only a short time, and some were dismantled.

During 1918 the Golden Cycle mill, at Colorado Springs, was operated steadily, despite the high cost of material and labor and the difficulty of getting labor. The Portland 400-ton roast-cyanide concentration mill at Colorado Springs stopped treating ore March 31, 1918. The production of high-grade ores in the Cripple Creek district had then fallen off so much by reason of the World War that it was impossible to operate at a profit both of the reduction plants at Colorado Springs. As the Portland was the smaller of the two it was decided to discontinue its operation and to have the Portland high-grade ores treated at the Golden Cycle plant. From 1919 to 1924 the Golden Cycle mill and the Portland Independence mill served the district.

Cripple Creek metallurgy is unique in its picking and washing houses, for most of the gold content is in the small fragments and the dust that clings to rock as hoisted from the mine. By picking and washing and preserving the "fines" from 50 to 80 per cent of the rock as hoisted is discarded, obviating the paying of freight and treatment on worthless or very low-grade rock. One of the most elaborate washing plants was installed at the Vindicator mine [47] in 1914.

THE SULPHURIC ACID INDUSTRY

The chemical industry of Colorado, particularly the manufacture of sulphuric acid, is closely linked to the mining of sulphide ores. According to Hosker,[48] the first attempt at manufacturing sulphuric acid in this State was made by William West, who erected a small plant at Blackhawk in 1875 and produced about 2,000 pounds daily. He continued the business about three years, but the demand for his product was so slight and the expense of manufacturing it on a small scale so great that he was compelled to suspend business. In 1879 Richard Pearce and others built a small plant at Golden, which proved to be a failure, running only about a year. The next attempt was made by the Denver Smelting & Chemical Co., which was organized in 1881, with William West as manager. This company purchased ground near Valverde and erected one chamber and other necessary apparatus for the manufacture of sulphuric acid. This venture also failed.

In 1885 the Western Chemical Works Co. was organized and commenced to manufacture various chemicals on the Valverde site. The history of this company is a record of many and varied struggles, but after 1892 the industry became one of the most successful in Denver. The company was reorganized in 1902 under the name of the Western Chemical & Manufacturing Co. The products manufactured embrace commercial sulphuric, muriatic, and nitric acids; liquid and anhydrous ammonia; chemically pure acids and ammonia; liquid carbonic acid gas; copperas; and other chemicals.

In consequence of the decreasing demand for sulphuric acid because of the change from chlorination to cyanidation at some of the mills treating Cripple Creek ores, the officers of the company decided to erect a mill and to engage in the zinc milling industry. At the same time they expected to produce an ore suitable for roasting in the Herreshoff furnaces and subsequently to separate the zinc and lead contents as by-products. Their plant, which was

[47] For description see U. S. Geol. Survey Mineral Resources, 1914, pt. 1, pp. 310-311, 1916.

[48] Hosker, R. B., History of chemical industry in Colorado: Mines and Mining, Jan. 3, 1908, pp. 5-8.

put into operation in 1907, comprised a sampling or coarse-crushing mill, driers, fine-crushing mill, roasters, magnetic separating mill, and wet-concentrating mill.

Up to 1906 the Western Chemical Works Co. and its successor, the Western Chemical & Manufacturing Co., bought most of its sulphur-bearing material in the form of pyrite, galena, and chalcopyrite ores and concentrates from Clear and Gilpin counties. The residues went to the Globe and Pueblo smelters. Beginning in 1906, the plant purchased most of its sulphur-bearing material in the form of pyritic gold and silver bearing ores from Leadville, also similar ores from Kokomo and Red Cliff. The company also competed for lead-zinc gold and silver bearing sulphides from Leadville (with small lots from Kokomo and Rico) which were given a slight roast in preparation for the removal of the pyrite by magnetic separation. The separated pyrite was then roasted for its sulphur content, and some of the residue contained sufficient gold, silver, copper, lead, and iron to be sold as a flux. The lead and zinc sulphides were separated in the wet concentrating mill, and the lead concentrate (galena) and zinc concentrate (sphalerite) were sold (without any attempt to remove the sulphur) to zinc retort furnace smelters and lead-bullion smelters. During 1919 electrolytic zinc was manufactured and some zinc oxide. In 1920 the plant was sold to the General Chemical Co., and Texas sulphur replaced Colorado pyrite in the roasters. The milling of lead-zinc ores ceased in 1919. The products of this plant, which were first used in refining oil at Florence, Colo., were after 1910 used also by the several Wyoming oil refiners.

In 1917 and 1918 the E. I. Du Pont de Nemours powder plant at Louviers used much pyritic ore from Leadville and Red Cliff.

RAILROAD BUILDING IN COLORADO

A notable stimulus was given to the development of mining by the completion of the Union Pacific Railroad as far as Cheyenne, Wyo., in 1867, and particularly by the completion, in the summer of 1870, of the Denver Pacific (now the Union Pacific) between Denver and Cheyenne. In August, 1870, the Kansas Pacific (now the Union Pacific) reached Denver, making two rail connections with the East. In 1870 a narrow-gage railroad was completed from Denver to Golden; in 1872 it was extended to Blackhawk, in 1873 to Floyd Hill, and in 1877 to Georgetown. It is now operated by the Colorado & Southern Railway.

The building of the Denver & Rio Grande Western Railroad and many of its branches is intimately associated with the development of mining in Colorado. The following table, compiled from the official records of the Denver & Rio Grande Western Railroad Co., shows the dates of completion of the various narrow-gage, three-rail, and standard gage lines as they existed in 1918.

Dates of completion of lines of the Denver & Rio Grande Western Railroad

Denver, Colo., to Salt Lake City, Utah, by way of Tennessee Pass

From—	To—	Narrow-gage	Three-rail	Standard-gage	Branch
Denver	Pueblo	1872	1881	1902	
Pueblo	Florence	1872	1887		
Do.	Canon City	1874	1888		
Canon City	Westcliffe by way of Grape Creek.	1881			Westcliffe.*
Do.	Salida	1880	1890		Do.
Texas Creek Junction.	Westcliffe			1900	Do.
Salida	Malta	1880	1890		Leadville.
Malta	Leadville	1880	1888–1890		Do.
Leadville	Leadville Junction.	1887		1890	Do.
Malta	Eilers		1890		Do.
Eilers	Leadville		1888		Do.
Oro Junction	Iron Silver Co.'s mines.	1880	1890	1902	Do.
Leadville	Ibex	1898			Blue River.b
Do.	Wheeler (by way of Kokomo).	1881			
Wheeler	Dillon	1882			Do.b
Malta	Rock Creek	1881		1890	
Rock Creek	Minturn	1887		1890	
Minturn	Glenwood Springs.	1887		1890	
Glenwood Springs	Aspen	1887		1890	Aspen.
Minturn	Newcastle	1889		1890	
Newcastle	Rifle	1890			
Do.	Grand Junction.			1890	

Denver, Colo., to Salt Lake City, Utah, by way of Marshall Pass

From—	To—	Narrow-gage	Three-rail	Standard-gage	Branch
Salida	Gunnison	1881			
Gunnison	Sapinero	1882			
Sapinero	Montrose	1882			
Montrose	Grand Junction.	1882	1906	1906	
Poncha Junction	Maysville	1881			Monarch.
Maysville	Monarch	1883			Do.
Gunnison	Crested Butte.	1881			Crested Butte.
Crested Butte	Anthracite	1882			Do.
Crested Butte by way of Irwin.	Floresta	1893			Do.
Mears Junction	Villagrove	1881			Villagrove.
Villagrove	Orient	1881			Do.
Do.	Alamosa	1890			Do.
Moffat	Crestone to Cottonwood.	1902			Do.
Sapinero	Lake City	1889			Lake City.
Montrose	Ouray by way of Ridgway.	1887			Ouray.
Ridgway	Vance Junction	1890			Ouray to Telluride, Rico, and Durango.
Vance Junction	Telluride	1890			Do.
Telluride	Pandora	1891			Do.
Vance Junction	Rico	1891			Do.
Rico	Durango	1891			Do.

Pueblo, Alamosa, and Durango, Colo., and Santa Fe, N. Mex.

From—	To—	Narrow-gage	Three-rail	Standard-gage	Branch
Pueblo	La Veta	1876			
Do.	Minnequa		1881		
Minnequa	Walsenburg		1888		
Do.	La Veta			1890	
La Veta c	Wagon Creek Junction by way of La Veta Pass.	1877			
Wagon Creek Junction.	Alamosa	1878			
La Veta	...do			1899	
Alamosa	South Fork	1881			Creede.
South Fork	Wagonwheel Gap.	1883			Do.
Wagonwheel Gap.	Creede	1891			Do.
Alamosa	Del Norte		1901	1902	Do.
Del Norte	Creede			1902	Do.
Alamosa	Chama	1880			Durango.
Do.	Antonito		1901		Do.
Chama	Durango	1881			Do.
Durango	Silverton	1882			Silverton.
Antonito	Espanola	1880			
Espanola	Santa Fe, N. Mex.	1882–1886			

* Abandoned, 1888; rails removed, 1890.
b Abandoned.
c Abandoned, 1899.

MINING DISTRICTS IN COLORADO

The Gregory lode (Gilpin County) was discovered in May, 1859. It was taken in claims 100 feet long, and Gregory, the discoverer, was allowed two claims. On June 8, 1859, the pattern after which the mining districts and miners' laws were fashioned was roughly outlined and adopted at a miners meeting in Gregory diggings. Some of the later arrivals demanded a redistribution of the claims on the Gregory, Hunter, Bates, Bobtail, Gregory Second, and other lodes and asked that the claims be cut down to 25 feet in length. The committee formed to consider this question brought in the following report:

1. *Resolved*, That this mining district shall be bounded as follows: Commencing at the mouth of the North Fork of Clear Creek, following the divide between said stream and Ralston Creek, running 7 miles up the last named stream to a point known as "Miners' Camp"; thence southwest to the divide between the North Fork of Clear Creek and the South Fork of the same to the place of beginning.

2. *Resolved*, That no miner shall hold more than one claim except by purchase or discovery; and in any case of purchase, the same shall be attested by at least two disinterested witnesses and shall be recorded by the secretary, who shall receive in compensation a fee of one dollar.

3. *Resolved*, That no claim which has or may be made shall be good and valid unless it be staked off with the owner's name, giving the direction, length and breadth, also the date when said claim was made; and when held by a company the name of each member shall appear plainly.

4. *Resolved*, That each miner shall be entitled to hold one mountain claim, one gulch claim, and one creek claim for the purpose of washing: the first to be 100 feet long and 50 wide; the second 100 feet up and down the river or gulch, extending from bank to bank.

5. *Resolved*, That mountain claims shall be worked within ten days from the time they are staked off, otherwise forfeited.

6. *Resolved*, That when members of a company, constituted of two or more, shall be at work on one claim of the company, the rest shall be considered as worked by putting a notice of the same on the claim.

7. *Resolved*, That each discovery claim shall be marked as such and shall be safely held, whether worked or not.

8. *Resolved*, That in all cases priority of claim, when honestly carried out, shall be respected.

9. *Resolved*, That when two parties wish to use water on the same stream or ravine for quartz-washing, it shall be equally divided between them.

10. *Resolved*, That when disputes shall arise between parties in regard to claims, the party aggrieved shall call upon the secretary, who shall designate nine miners, being disinterested persons, from which number the parties shall alternately strike out one until the names of but three remain, who shall at once proceed to hear and try the case; and should any miner refuse to obey their decision, the secretary shall call a meeting of the miners, and if their decision is the same, the party refusing to obey shall not be entitled to hold another claim in this district; and the party against whom the decision is given shall pay to the secretary and referees the sum of fifteen dollars each for their services.

On July 9, 1859, another mass meeting was held at Gregory Point, at which it was

Resolved, That for the settlement of difficulties and the purpose of preventing disputes, the miners of this district hereby enact: That there shall be elected in this district, by ballot, a president, a recorder of claims, and a sheriff, for the term of one year from this date. That the president, secretary, and one assistant, to be chosen by the people, be tellers of said election, and that it take place immediately.

On February 11, 1860, a meeting of the citizens of Gregory district was held at Mountain City, and a committee was appointed to codify and amend the laws of said district and report to an adjourned meeting to be held on February 18.

The United States Mining Laws of May 10, 1872, changed the size of claims as follows:

SEC. 2320. Mining claims upon veins or lodes of quartz or other rock in place bearing gold, silver, cinnabar, lead, tin, copper, or other valuable deposits, heretofore located, shall be governed as to length along the vein or lode by the customs, regulations, and laws in force at the date of their location. A mining claim located after the tenth day of May, eighteen hundred and seventy-two, whether located by one or more persons, may equal, but shall not exceed, one thousand five hundred feet in length along the vein or lode; but no location of a mining claim shall be made until the discovery of the vein or lode within the limits of the claim located. No claim shall extend more than three hundred feet on each side of the middle of the vein at the surface, nor shall any claim be limited by any mining regulation to less than twenty-five feet on each side of the middle of the vein at the surface, except where adverse rights existing on the tenth day of May, eighteen hundred and seventy-two, render such limitation necessary. The end lines of each claim shall be parallel to each other.

Many new districts were formed in 1859 and 1860 out of Gregory district, the country adjoining, and throughout the mountains, wherever deposits were discovered, all of which copied their laws and customs from those of the parent district, though often modifying them in important particulars. The mining districts and miners' courts lost their importance when the Territory was organized into counties. The claims were then recorded by the county recorder, but many of the district names were and are still used in the descriptions of locations and patents. Many of the names used represent nothing more than the guess or whim of the locator, and many of the commonly used names of local districts are carried miles away and across county lines. Mining districts have no defined boundaries.

All names that have ever been used to designate mining districts in Colorado up to June 1, 1924, as shown from the records of the United States Surveyor General at Denver, are given below:

Mining districts in Colorado

Name	Other names used	County	Approximate location
Alhambra	Freshwater, Guffey, Red Ruth.	Fremont	Tps. 14 and 15 S., R. 73 W. of the sixth principal meridian.
Alicante	Arkansas	Lake	Secs. 13, 14, 15, 16, 21, 22, 23, 24, 25, 26, 27, and 28, T. 8 S., R. 79 W.; secs. 18, 19, 30, T. 8 S., R. 78 W.
Allens Park		Boulder	Secs. 23, 24, 25, 26, T. 3 N., R. 73 W.
Alpine	Chalk Creek (overlaps Monarch).	Chaffee	T. 51 N., Rs. 80 and 81 W.
Alsace		Eagle	Secs. 3, 4, and 5, T. 6 S., R. 82 W.
Animas	Putnam subdistrict within Animas.	San Juan	Tps. 40, 41, and 42 N., Rs. 6, 7, and 8 W.
Argentine (Clear Creek County).	Queens, East Argentine, and West Argentine; overlaps Griffith.	Clear Creek	Tps. 4 and 5 S., R. 75 W.
Argentine (Summit County).	Peru (Montezuma).	Summit	T. 5 S., R. 75 W.
Arkansas	Alicante	Lake	
Atlantic	Dailey	Clear Creek	T. 3 S., R. 75 W.
Avalanche	Carpenter, Bald Mountain, Lincoln, Swan River.	Summit	Tps. 6 and 7 S., R. 77 W.
Badger Creek		Fremont	T. 50 N. R. 76 W.
Bald Mountain	Avalanche	Summit	
Banded Peak		Archuleta	T. 34 N., Rs. 2 and 3 E.
Banner	Subdistrict Idaho Springs.	Clear Creek	Sec. 38, T. 3 S., R. 73 W.
Bare Hills	South Cripple Creek.	Fremont	T. 16 S., Rs. 70 and 71 W.
Barnard Creek	North Cripple Creek.	Teller	T. 15 S., R. 70 W.
Bath		Park	T. 13 S., R. 76 W.
Battle Mountain	Red Cliff (Belden)	Eagle	T. 6 S., Rs. 80 and 81 W.
Bay State		Gilpin at corner of Jefferson and Clear Creek.	Sec. 35, T. 3 S., R. 72 W.
Bear Creek	Jefferson	Jefferson	T. 5 S., R. 71 W.
Beaver Creek	Fair Play	Park	Tps. 8 and 9 S., R. 77 W.
Beaver Creek	Midland	Gunnison and Saguache.	T. 47 N., R. 1 W
Beaver Dam	Snake River (Montezuma).	Summit	T. 5 S., Rs. 76 and 77 W.
Beulah		Pueblo	T. 22 S., R. 68 W.
Bevan	Utah, Miners, Minnesota, McBarnes (Breckenridge).	Summit	Tps. 6 and 7 S., R. 77 W.
Big Creek		Jackson	Sec. 6, T. 11 N., R. 81 W.
Big Horn		do	T. 11 N., R. 82 W.
Birdseye	Overlaps Alicante	Lake	Secs. 25, 26, 27, 34, 35, T. 8 S., R. 79 W.
Black Mountain		Park	T. 15 S., Rs. 73 and 74 W.
Do		Fremont	T. 16 S., R. 75 W.
Blake		Saguache	Tps. 44, 45, 46 and 47 N., Rs. 10 and 11 E.
Blanca	Sierra Blanca	Huerfano	T. 28 S., R. 7 W.
Blue Ridge		Grand	T. 1 S., R. 79 W.
Bonanza	Kerber Creek	Saguache	Tps. 47 and 48 N. Rs. 7 and 8 E
Boulder		Boulder	T. 1 N., R. 71 W.
Bowman	Needle Mountain	La Plata	T. 38 N., R. 8 W.
Box Canon	Tomichi	Gunnison	Tps. 49 and 50 N., Rs. 4 and 5 E.
Box Creek		Lake	Secs. 4, 5, and 6, T. 11 S., R. 80 W.
Buckhorn		Larimer	T. 7 N., R. 71 W.
Buckskin	Buckskin Joe		
Buckskin Joe		Park	Tps. 8 and 9 S., R. 78 W.
Buffalo Mountain		Jackson and Grand.	T. 4 N., R. 78 W.
Buffalo Peaks		Park	T. 12 S., R. 78 W.
California	California Gulch, Iron Hill (local); overlaps Alicante, Birdseye, Little Evans Gulch, Iowa Gulch, Big Evans Gulch.	Lake	Tps. 8 and 9 S., Rs. 79 and 80 W.
Do	Junction Creek, La Plata.	La Plata and Montezuma.	Tps. 36 and 37 N., Rs. 10 and 11 W.
Calumet		Chaffee	T. 51 N., R. 77 W.
Cameron. See Cleora.	Cameron Mountain.		
Camp San Diego	Sierra Blanca	Huerfano	T. 27 S., R. 72 W.
Campbell		Grand	T. 5 N., R. 76 W.
Canon		Clear Creek	Secs. 2 and 3, T. 4 S., R. 72 W.
Carbonate	Overlaps Black Mountain.	Fremont	T. 16 S., R. 74 W.
Do		Garfield	Secs. 34, 35, 36, T. 3 S., R. 89 W.
Carpenter	Overlaps Lincoln and Georgia Pass.	Summit	Secs. 5 and 6, T. 7 S., R. 76 W.; secs. 1 and 12, T. 7 S., R. 77 W.

Mining districts in Colorado—Continued

Name	Other names used	County	Approximate location
Carrizo		Baca	T. 34 S., R. 50 W.
Carson		Hinsdale	Tps. 41 and 42 N., Rs. 4 and 5 W.
Carter		Larimer	Tps. 6 and 7 N., R. 70 W.
Cascade	Democrat, Ottawa	Clear Creek	T. 4 S., Rs. 73 and 74 W.
Castle Creek	Columbia, Roaring Fork.	Pitkin	Tps. 10 and 12 S., R. 85 W.
Cebolla	White Earth	Gunnison	T. 46 N., R. 2 W. of the New Mexico principal meridian.
Central	Utillah	Boulder	Tps. 1 and 2 N., Rs. 71 and 72 W.
Central City		Gilpin	T. 3 S., R. 73 W.
Chalk Creek	Alpine	Chaffee	T. 15 S., Rs. 80 and 81 W. of the sixth principal meridian; T. 51 N., R. 6 E. of the New Mexico principal meridian.
Cheyenne	Pikes Peak, Cheyenne Mountain.	El Paso	T. 14 S., R. 67 W.
Chipeta		Montrose	Tps. 47 and 49 N., R. 16 W.; T. 46 N., R. 15 W
Chloride		Las Animas	T. 30 S., R. 64 W.
Cimarron		Hinsdale	T. 45 N., R. 5 W.
Cleora. (See Cameron).	Overlaps Turret Mountain.	Chaffee	T. 15 S., Rs. 76 and 77 W. of the sixth principal meridian; T. 51 N., Rs. 9 and 10 E., T. 50 N., Rs. 8 and 9 E.; and T. 49 N., R. 9 E. of the New Mexico principal meridian.
Cleora	Cameron	Fremont	T. 16 S., R. 76 W.
Cochetopa		Gunnison	T. 49 N., R. 2 E. of the New Mexico principal meridian.
Do	Green Mountain	Saguache	Tps. 48 and 49 N., R. 2 E.
Columbia	Buckskin Joe, Consolidated Montgomery, and Beaver Creek	Park	
Columbia	Roaring Fork, West Castle, Castle Creek.	Pitkin	Tps. 11 and 12 S., Rs. 84 and 85 W.
Conejos	Ute	Conejos	T. 36 N., R. 4 E.
Congers	Included in Breckenridge.	Summit	Secs. 17 and 18, T 7 S., R. 77 W.
Consolidated Montgomery.		Park	T. 8 S., Rs. 77 and 78 W.
Consolidated Tenmile.	Tenmile Consolidated (includes Wilkinson), Kokomo-Robinson-Frisco.	Summit	T. 5 S., R. 78 W.; Tps. 7 and 8 S., Rs. 78 and 79 W.
Consolidated Union	Union Consolidated.	do	T. 6 S., Rs. 77 and 78 W.
Coon Park	Cleora	Chaffee	T. 15 S., R. 77 W.
Copper Creek		Mesa	T. 12 S., R. 101 W.
Copperdale		Jefferson	Secs. 8, 9, 10, 15, 16, and 17, T. 2 S., R. 71 W.
Copper King	Red Gorge	Grand	T 1 S., R. 82 W.
Copper Ridge		Jackson	T. 11 N., R. 78 W.
Coral	Cascade	Clear Creek	Secs. 32, 33, T. 3 S., R. 73 W.; secs. 4 and 5, T. 4 S., R. 73 W.
Cotopaxi		Fremont	T. 48 N., R. 11 E.
Cottonwood		Chaffee	T. 13 S., R. 79 W.; Tps. 14 and 15 S., Rs. 79 and 80 W.
Crater		Mineral	T. 37 N., R. 2 E. of the New Mexico principal meridian.
Crestone	Eldorado	Saguache	Tps. 43 and 44 N., R. 12 E.
Creswell		Jefferson	Sec. 27, T. 4 S., R. 71 W.
Cripple Creek	Womack	Teller	Tps. 14, 15, and 16 S., Rs. 68, 69, and 70 W.
Cross Mountain		Gunnison	T. 15 S., R. 82 W.
Crosson	Princeton	Jefferson and Park.	Secs. 30 and 31, T. 7 S., R. 71 W.; secs. 25 and 26, T. 7 S., R. 72 W.
Crystal Hill		Saguache	T. 43 N., R. 6 E.
Crystal River	Rock Creek	Gunnison	T. 12 S., R. 87 W.
Current Creek		Fremont	T. 16 S., R. 72 W.
Dailey	Atlantic	Clear Creek	T. 3 S., Rs. 75 and 76 W.
Decatur	Summit	Rio Grande and Conejos.	T. 37 N., Rs. 4 and 5 E.; T. 36 N., R. 3 E. of the New Mexico principal meridian.
Deer Creek		Jefferson	T. 6 S., R. 70 W.
Defiance		Garfield	T. 5 S., R. 88 W.
Democrat	Democratic, Cascade, Ottawa.	Clear Creek	Secs 11, 12, 13, 14, 23, and 24, T. 4 S., R. 74 W.
De Sobe		Douglas	Secs. 29, 30, 31, and 32, T. 7 S., R. 69 W.
Dewey	Hope, Granite	Lake	Secs. 29 and 30, T. 11 S., R. 79 W.

Mining districts in Colorado—Continued

Name	Other names used	County	Approximate location
Difficult	Difficult Creek	Pitkin	Secs. 25, 26, 27, 34, 35, and 36, T. 11 S., R. 84 W.
Domingo	Vulcan, Willow Creek	Gunnison	T. 47 N., Rs. 1 and 2 W. of the New Mexico principal meridian.
Douglas Mountain	Escalante Hills	Moffat	T. 7 N., Rs. 100, 101, and 102 W.
Downieville	Morris, Lawson	Clear Creek	T. 3 S., Rs. 73 and 74 W.
Dry Pine	Frying Pan	Pitkin	T. 9 S., R. 84 W.
Eagle River	Eagle; Headwaters of Eagle River. Includes Fairview Hill.	Eagle	Tps. 7 and 8 S., Rs. 79 and 80 W.
East Argentine	Argentine	Clear Creek	Secs. 35 and 36, T. 4 S., R. 75 W.
East Beaver Creek	East Beaver, East Cripple Creek	Teller and Fremont.	T. 16 S., R. 68 W.
Eightmile	McCourt	Fremont	T. 17 S., R. 69 W.
El Dorado	Crestone	Saguache	Tps. 43 and 44 N., R. 12 E. of the New Mexico principal meridian.
Elk Mountain		Gunnison	Tps. 12 and 13 S., R. 85 W.; Tps. 11 and 12 S., Rs. 86 and 87 W.
El Paso	Turkey Creek	El Paso	T. 16 S., R. 67 W.
Empire	Upper Union	Clear Creek	T. 3 S., Rs. 74 and 75 W.
Do	California	Lake	T. 10 S., R. 79 W.
Enterprise	Central	Boulder	Secs. 10, 11, 14, 15, T. 2 N., R. 72 W.
Do		Gilpin	Secs. 7, 8, 9, T. 3 S., R. 72 W.
Eureka		do	Secs. 4 and 9, T. 3 S., R. 73 W.
Do	Includes part of Red Mountain.	San Juan	Tps. 41, 42, and 43 N., Rs. 6 and 7 W. of the New Mexico principal meridian.
Fairmount		Park	Secs. 13, 14, 23, and 24, T. 10 S., R. 78 W.
Fairfield	Fairview	Gilpin	Secs. 26, 27, 34, and 35, T. 2 S., R. 73 W.
Fairview	Fairfield		
Fairview Hill	Eagle River	Eagle	Secs. 25, 26, 35, and 36, T. 7 S., R. 80 W.
Fall City		Larimer	Secs. 11, 12, 13, 14, T. 8 N., R. 75 W.
Fall River	Lower Fall River	Larimer	
Farnsworth		Larimer	Secs. 26, 27, 34, and 35, T. 7 N., R. 71 W.
Ferrari		Park	Secs. 13, 14, 23, and 24, T. 5 S., R. 72 W.
Findley Gulch		Saguache	Secs. 34 and 35, T. 47 N., R. 7 E. of the New Mexico principal meridian.
Fish Creek	Florissant	Park	Secs. 8, 9, 16, 17, T. 13 S., R. 71 W.
Florida	Needle Mountain	La Plata	T. 38 N., R. 7 W.
Florissant		Park	Secs. 16, 17, 20, 21, T. 13 S., R. 71 W.
Ford Creek		Saguache	Secs. 15, 16, 17, 20, 21, and 22, T. 46 N., R. 7 E. of the New Mexico principal meridian.
Forest Hill	Taylor River	Gunnison	Secs. 10, 11, 14, and 15, T. 13 S., R. 83 W.
Fountain Creek	Pikes Peak	Teller	Secs. 13, 18, 19, and 24, T. 13 S., R. 69 W.
Fourmile		Chaffee	Tps. 13 and 14 S., R. 78 W.
Do		Park	T. 15 S., R 72 W.
Do		Moffat	T. 12 N., Rs. 91 and 92 W.
Fraser		Grand	Tps. 1, 2, and 3 S., Rs. 75, 76, and 77 W.
Free Gold		Chaffee	Secs. 15, 16, 21, and 22, T. 14 S., R. 78 W.
Fremont	Cotopaxi	Fremont	Secs. 19 and 30, T. 48 N., R. 12 E.; secs. 24 and 25, T. 48 N., R. 11 E.
Fremont County	Fremont		
Freshwater	Red Ruth, Guffey, Alhambra.	Park	Tps. 14 and 15 S., Rs. 72 and 73 W.
Frisco	Part of Tenmile	Summit	Secs. 16, 17, 20, and 21, T. 6 S., R. 78 W.
Front Range		El Paso	T. 12 S., R. 67 W.
Frying Pan		Eagle	T. 7 S., R. 83 W.
Do		Lake	Secs. 29, 30, 31, and 32, T. 9 S., R. 80 W.
Do		Pitkin	Tps. 8 and 9 S., Rs. 83 and 84 W.
Fulford	Overlaps Mount Egley.	Eagle	Tps. 6 and 7 S., Rs. 82 and 83 W.
Galena		Hinsdale	Tps. 43 and 44 N., Rs. 4, 5, and 6 E. of the New Mexico principal meridian.
Gem	Texas Creek	Fremont	Secs. 28, 29, 32, 33, T. 20 S., R. 73 W

Mining districts in Colorado—Continued

Name	Other names used	County	Approximate location
Georgia Pass		Summit	Secs. 7, 12, and 13, T. 7 S., R. 76 W.
Geneva	Snake River, Montezuma.	Summit and Clear Creek.	Secs. 29, 30, 31, and 32, T. 5 S., R. 75 W.; secs. 25, 26, 35, 36, T. 5 S., R. 76 W.
Glenwood		Garfield	Secs. 2, 3, 10, and 11, T. 6 S., R. 89 W.
Gold Basin	Adjoins Green Mountain.	Gunnison	Secs. 5, 6, 7, and 8, T. 48 N., R. 1 E.
Gold Blossom		Routt	T. 11 N., R. 87 W.
Gold Brick		Gunnison	Tps. 50 and 51 N., R. 83 W. of the New Mexico principal meridian.
Gold Dirt		Clear Creek	Secs. 10, 12, 13, and 14, T. 4 S., R. 73 W.
Gold Hill	Overlaps Sunshine, Ward, and Sugar Loaf.	Boulder	T. 1 N., Rs. 71, 72, and 73 W.
Golden City		Jefferson	Secs. 3, 4, 9, and 10 T. 4 S., R. 70 W.
Goose Creek	White Earth	Gunnison	Corner of Tps. 47 and 48 N., Rs. 2 and 3 W. of the New Mexico principal meridian.
Gordon		Fremont	Secs. 22 and 23, T. 16 S., R. 69 W.
Gothic		Gunnison	Secs. 13, 14, 23, and 24, T. 14 S., R. 85 W.
Grand Island		Boulder	T. 1 S., Rs. 72, 73, and 74 W.
Grand Lake		Grand and Boulder line.	Secs. 21, 22, 27, and 28, T. 1 N., R. 74 W.
Grand River		Mesa	Secs. 17, 18, 19, and 20, T. 10 S., R. 103 W.
Granite	Overlaps Hope	Chaffee	Secs. 31, 32, 33, 34, and 35, T. 11 S., R. 79 W.; Secs. 2, 3, 4, 5, and 6, T. 12 S., R. 79 W.
Do	do	Lake	Secs. 19, 20, 21, 22, 28, 29, and 30, T. 11 S., R. 79 W.
Grape Creek	Overlaps Greenhorn.	Fremont	Secs. 9, 10, 11, 14, 15, and 16, T. 19 S., R. 71 W.
Grass Valley	Idaho Springs	Clear Creek	Secs. 35 and 36, T. 3 S., R. 73 W.; secs. 1 and 2, T. 4 S., R. 73 W.
Greenhorn	Overlaps Grape Creek	Fremont	Secs. 7, 8, 9, 16, 17, 18, 19, 20, and 21, T. 19 S., R. 71 W.
Green Mountain	Cochetopa	Gunnison and Saguache.	T. 48 N., Rs. 1 and 2 E. of the New Mexico principal meridian.
Green Mountain Falls.		El Paso	T. 13 S., R. 68 W.
Gregory	Overlaps Lake Gulch.	Gilpin	Sec. 18, T. 3 S., R. 72 W.; sec. 13, T. 3 S., R. 73 W.
Griffith	Queens	Clear Creek	Secs. 9, 10, 11, 12, 13, 14, 15, and 16, T. 4 S., R. 75 W.; secs. 7, 8, 9, 10, 15, 16, 17, 18, 19, 20, and 21, T. 4 S., R. 74 W.
Guffey	Freshwater, Red Ruth.	Park	T. 15 S., R. 73 W.
Gunnison		Gunnison	T. 49 N., R. 1 E.; T. 48 N., R. 1 W. of the New Mexico principal meridian.
Hahns Peak		Routt	Tps. 9, 10, and 11 N., R. 85 W.
Half Moon	Lackawanna	Lake	Secs. 26, 27, 28, 29, 30, 31, 32, 33, 34, and 35, T. 10 S., R. 81 W.; secs. 25, 26, 27, 33, 34, 35, and 36, T. 10 S., R. 82 W.
Hall Gulch	Hall Valley		
Hall Valley	Overlaps Montezuma, Snake River, Platte, Middle Swan.	Park and Clear Creek.	T. 6 S., Rs. 75 and 76 W.
Hardscrabble		Custer	Tps. 21 and 22 S., Rs. 70, 71, and 72 W.
Do		Fremont	T. 20 S., R. 71 W.
Harmon		Grand	Sec. 6, T. 2 N., R W.
Hartsel		Park	T. 12 S., R. 75 W.; Tps. 13 and 14 S., R. 74 W.
Hawkeye		Gilpin	Sec. 23, T. 2 S., R. 73 W.
Hayden Creek		Fremont	Secs. 3, 4, 5, 6, 7, 8, 9, and 10, T. 47 N., R. 11 E. of the New Mexico principal meridian.
Highland		Pitkin	Secs. 35 and 36, T. 10 S., R. 85 W.; secs. 1 and 2, T. 11 S., R 85 W.

Mining districts in Colorado—Continued

Name	Other names used	County	Approximate location
High Lonesome		Grand	Secs. 1, 2, 11, and 12, T. 1 N., R. 75 W.
High Park	West Cripple Creek.	Teller	T. 15 S., R. 71 W.
Holy Cross		Eagle	T. 7 S., Rs. 81 and 82 W.
Homestake		Lake	Secs. 14, 15, 22, 23, 26, and 27, T. 8 S., R. 81 W.
Hoosier Gulch	Overlaps Pollock	Summit	Secs. 6 and 7, T. 8 S., R. 77 W.; secs. 1 and 12, T. 8 S., R. 78 W.
Hope	Dewey, Granite	Lake and Chaffee.	Sec. 30, T. 11 S., R. 79 W.; secs. 25, 26, 27, 31, 32, 33, 34, 35, 36, T. 11 S., R. 80 W.
Horn Silver	Red Cliff	Eagle	Secs. 31, 32, and 33, T. 6 S., R. 80 W.; secs. 4, 5, and 6, T. 7 S., R. 80 W.
Horseshoe		Park	Secs. 26, 27, 28, 30, 32, and 33, T. 9 S., R. 78 W.
Hotchkiss	White Earth	Gunnison	Secs. 4, 5, 8, and 9, T. 46 N., R. 2 W. of the New Mexico principal meridian.
Hot Springs	Glenwood Hot Springs.		
Howard		Fremont	T. 49 N., R. 11 E. of the New Mexico principal meridian.
Howes Gulch		Larimer	Secs. 9, 10, 11, and 12, T. 7 N., R. 70 W.
Huerfano		Huerfano	Secs. 29, 30, 31, and 32, T. 27 S., R. 70 W.
Do	Spanish Peaks, West Spanish Peaks.	Huerfano and Las Animas.	Secs. 33, 34, 35, and 36, T. 30 S., R. 68 W.; secs. 1, 2, 3, 4, T. 31 S., R. 68 W.
Hull		Saguache	T. 48 N., R. 9 E.
Hunter Creek	Woody	Pitkin	Secs. 4, 5, 6, 7, 8, T. 10 S., R. 84 W.
Hydraulic		Mesa	Secs. 7, 8, 17, 18, 19, 20, 21, 22, 27, 28, T. 48 N., R. 17 W. of the New Mexico principal meridian.
Ice Lake		San Juan	Secs. 18, 19, 30, T. 41 N., R. 8 W.; secs. 13, 23, 24, 25, 26, T. 41 N., R. 9 W. of the New Mexico principal meridian.
Idaho or Virginia	Included Idaho Springs.	Clear Creek	Secs. 25 and 36, T. 3 S., R. 73 W.; secs. 30 and 31, T. 3 S., R. 72 W.
Illinois Central	Illinois	Gilpin	Secs. 9, 10, 11, 12, 13, 14, 15, 16, and 17, T. 2 S., R. 73 W.
Independent	Overlaps Illinois, Pine, Union.	do	Secs. 8, 9, 10, and 11, T. 2 S., R. 73 W.
Independence	Independent; overlaps St. Kevin.	Lake	Secs. 7 to 23, inclusive, T. 9 S., R. 81 W.; secs. 5, 6, 7, and 8, T. 9 S., R. 80 W.
Indian Creek	Included in Kerber Creek	Saguache	T. 47 N., R. 7 E., and T. 48 N., R. 6 E., of the New Mexico principal meridian.
Iowa	Included Idaho Springs.	Clear Creek	Sec. 28, T. 5 S., R. 73 W.
Iowa Gulch	California	Lake	T. 9 S., R. 79 W.
Iron		Eagle	Sec. 1, T. 5 S., R. 86 W.
Iron Hill		Park	Secs. 29, 30, 31, and 32, T. 10 S., R. 78 W.
Iron Hill (local). *See* California.			
Iron Mountain		Saguache	Sec. 32, T. 45 N., R. 8 E. of the New Mexico principal meridian.
Iron Springs		San Miguel	Secs. 25 to 36, inclusive, T. 42 N., R. 9 W.; secs. 1, 2, 3, 4, 9, 10, 11, and 12, T. 41 N., R. 9 W. of the New Mexico principal meridian.
Jackson	Idaho, Virginia, Ohio.	Clear Creek	Secs. 2, 3, 10, and 11, T. 4 S., R. 73 W.
James Peak		Grand	Secs. 20 and 29, T. 2 S., R. 74 W.
Jefferson	Bear Creek	Jefferson	Sec. 18, T. 4 S., R. 71 W.; sec. 10, T. 5 S., R. 71 W.
Jo Davis	Near Delaware Flats.	Summit	Secs. 16, 17, 20, and 21, T. 6 S., R. 77 W.
Junction Creek	California	La Plata	Secs. 24, 25, and 26, T. 37 N., R. 11 W.; secs. 19, 20, 29, and 30, T. 37 N., R. 10 W. of the New Mexico principal meridian.
Kansas		Gilpin	Secs. 35 and 36, T. 1 S., R. 74 W.

Mining districts in Colorado—Continued

Name	Other names used	County	Approximate location
Kerber Creek	Bonanza	Saguache	Tps. 46 and 47 N., Rs. 7 and 8 E. of the New Mexico principal meridian.
Kezar Basin		Gunnison	Secs. 4, 5, 8, and 9, T. 48 N., R. 2 W. of the New Mexico principal meridian.
King Solomon	Sunnyside	Mineral	T. 42 N., R. 1 E. and R. 1 W. of the New Mexico principal meridian, and northward to county line.
Lackawanna		Lake	Secs. 1, 2, 3, 4, 9, 10, 11, and 12, T. 11 S., R. 82 W.; secs. 29, 30, 31, and 32, T. 10 S., R. 81 W.
Lake	Lake Gulch; overlaps Gregory.	Gilpin	Sec. 18, T. 3 S., R. 72 W.; sec. 13, T. 3 S., R. 73 W.
Do		Hinsdale	Tps. 43 and 44 N., R. 4 W. of the New Mexico principal meridian.
Do	Lake Creek, Twin Lakes.	Lake	T. 11 S., Rs. 80 and 81 W.; part of T. 12 S., R. 81 W.
Lake George	Springer; includes Pulver.	Park	Tps. 12 and 13 S., R. 72 W.
La Plata		Grand	Secs. 27, 28, 33, and 34, T. 3 S., R. 76 W.
Do		Chaffee	T. 12 S., Rs. 80 and 81 W.; north half of T. 13 S., R. 81 W.
Larimer County		Larimer	T. 11 N., R. 71 W.
La Sal	Paradox Valley	Montrose	Tps. 47 and 48 N., R. 19 W. of the New Mexico principal meridian.
Lead Mountain		Grand	Tps. 5 and 6 N., R. 76 W.; T. 5 N., R. 75 W.
Lincoln	Upper Fall River	Clear Creek	Secs. 26, 27, 28, 29, 32, 33, 34, 35, and 36, T. 2 S., R. 74 W.; secs. 1, 2, 3, 4, 5, 9, 10, 11, 12, T. 3 S., R. 74 W.; secs. 5, 6, 7, and 8, T. 3 S., R. 73 W.
Do	Lincoln Gulch	Pitkin	Secs. 1, 2, 3, 10, 11, 12, 13, and 14, T. 12 S., R. 83 W.
Do	Overlaps South Swan River, Avalanche, Minnesota.	Summit	Secs. 1 and 2, T. 7 S., R. 77 W.
Little Deer Creek	North Bear Gulch	Jefferson	Secs. 5 and 6, T. 7 S., R. 69 W.
Little Evans	California	Lake	Secs. 17, 18, 19, and 20, T. 9 S., R. 79 W.
Lone Cone		Dolores	Secs. 32 and 33, T. 41 N., R. 11 W.; secs. 4 and 5, T. 40 N., R. 11 W.
Long Gulch		Larimer	T. 8 N., R. 70 W.
Las Animas. *See* Animas.			
Lower Fall River	Spanish Bar	Clear Creek	Secs. 15, 16, 17, 18, 21, and 22, T. 3 S., R. 73 W.
Lower San Miguel		San Miguel	T. 44 N., R. 11 W. and T. 43 N., R. 10 W., of the New Mexico principal meridian.
Lower Tarryall		Park	T. 12 S., R. 72 W.
Lower Uncompahgre		Ouray	T. 46 N., R. 8 W. of the New Mexico principal meridian.
Lower Union		Clear Creek	Secs. 31, 32, 33, and 34, T. 5 S., R. 72 W.
Magnolia	Overlaps Sugar Loaf.	Boulder	Secs. 25, 26, 35, and 36, T. 1 N., R. 72 W.; secs. 29, 30, 31, and 32, T. 1 N., R. 71 W.; secs. 1, 12, 13, 24, and 25, T. 1 S., R. 72 W.; secs. 7, 18, 19, 30, and 31, T. 1 S., R. 71 W.
Mancos		Montezuma	T. 37 N., Rs. 11 and 12 W. of the New Mexico principal meridian.
Manhattan		Larimer	Secs. 24 and 25, T. 9 N., R. 80 W.; secs. 19 and 30, T. 9 N., R 79 W.
Manitou	Pikes Peak	El Paso	T. 13 S., R. 67 W.; T. 14 S., R. 68 W.
Do	Overlaps Kerber Creek	Saguache	Secs. 28, 29, 30, 31, 32, and 33, T. 48 N., R. 8 E., and secs. 4, 5, 6, 7, 8, and 9, T. 47 N., R. 8 E. of the New Mexico principal meridian.

Mining districts in Colorado—Continued

Name	Other names used	County	Approximate location
Marion	Pagosa Springs	Archuleta	Secs. 15 and 24, T. 35 N., R. 2 W., and secs. 18 and 19, T. 35 N., R. 1 E. of the New Mexico principal meridian.
Maroon		Pitkin	T. 11 S., Rs. 85 and 86 W.; T. 12 S., R. 85 W.
Maysville		Larimer	Secs. 4, 5, 6, 7, 8, and 9, T. 8 N., R. 73 W.
McBarnes	Miners (Breckenridge)	Summit	Secs. 8, 9, 17, and 16, T. 7 S., R. 77 W.
McCourt	Eightmile	Fremont	Secs. 21, 22, 27, and 28, T. 17 S., R. 69 W.
McDonough	Hotchkiss	Gunnison	Secs. 33 and 34, T. 47 N., R. 2 W., and secs 2 and 3, T. 46 N., R. 2 W. of the New Mexico principal meridian.
McKay	Union	Summit	Secs. 18, 19, T. 6 S., R. 77 W.
Middle Swan	Rexford, Missouri	do	Secs. 20, 21, 28, and 29, T. 6 S., R. 76 W.
Midland	Beaver Creek	Gunnison	Secs. 1, 2, 3, 10, 11, and 12, T. 47 N., R. 1 W., of the New Mexico principal meridian.
Mill Creek		Clear Creek	Secs. 10, 11, 14, and 15, T. 3 S., R. 74 W.
Miners	McBarnes	Summit	
Minnesota	Utah, Bevan	do	
Missouri		Lake	Secs. 13, 14, T. 8 S., R. 80 W.
Do	Middle Swan	Summit	
Monarch		Chaffee	T. 49 N., Rs. 6, 7, and 8 E. of the New Mexico principal meridian; T. 51 N., Rs. 79 and 80 W., and T. 50 N., R. 80 W. of the sixth principal meridian.
Montana	Overlaps Morris	Clear Creek	Secs. 25, 26, 34, 35, and 36, T. 3 S., R. 74 W.; secs. 30 and 31, T. 3 S., R. 73 W.
Montezuma	Snake River	Summit	T. 5 S., R. 76 W.; secs. 1, 2, 3, 4, 5, and 6, T. 6 S., R. 76 W.
Montgomery. See Consolidated Montgomery			
Morris	Montana	Clear Creek	
Mosquito			Secs. 1, 12, 13, and 24, T. 9 S., R. 79 W.; secs. 6, 7, 18, and 19, T. 9 S., R. 78 W.
Mountaindale	Adjoins Tarryall Springs	Park	Secs. 20, 21, and 22, T. 11 S., R. 72 W.
Mountain House		Gilpin	Secs. 19, 20, 29, and 30, T. 2 S., R. 72 W.
Mount Egley		Eagle	Secs. 15, 16, 21, and 22, T. 6 S., R. 82 W.
Mount Rosa	Cripple Creek	Teller	Secs. 1, 2, 11, 12, 13, 14, 23, and 24, T. 15 S., R. 68 W.
Mount Rose. See Mount Rosa.			
Mount Sneffels	Sneffels	Ouray	T. 43 N., R. 8 W., and secs. 1 and 2, T. 42 N., R. 8 W. of the New Mexico principal meridian.
Mount Wilson		San Miguel	T. 42 N., R. 10 W. of the New Mexico principal meridian.
Mount Zion		Lake	Sec. 10, T. 9 S., R. 80 W.
Muddy		Delta	Secs. 36, T. 10 S., R. 91 W.; sec. 1, T. 11 S., R. 91 W.
Music		Saguache	Secs. 17, 18, 19, and 20, T. 25 S., R. 73 W.
Myers Creek		Saguache	Secs. 10, 11, 12, 13, 14, and 15, T. 41 N., R. 4 E. of the New Mexico principal meridian.
Needle Mountain	Needle Mountains	La Plata	Tps. 38 and 39 N., R. 7 W., and T. 37 N., R. 8 W. of the New Mexico principal meridian.
Nevada		Gilpin	Secs. 10, 11, and 12, T. 3 S., R. 73 W.
Night Hawk		Douglas	Sec. 18, T. 8 S., R. 69 W.; sec. 13, T. 8 S., R. 70 W.
North Bear Gulch	Little Deer Creek	Jefferson	
North Cheyenne		El Paso	Secs. 27, 28, 29, 32, 33, and 34, T. 14 S., R. 67 W.

Mining districts in Colorado—Continued

Name	Other names used	County	Approximate location
North Cottonwood		Chaffee	Secs. 29, 30, 31, and 32, T. 13 S., R. 79 W.; secs. 25, 26, 27, 34, 35, and 36, T. 13 S., R. 80 W.; secs. 4, 5, 6, 7, 8, and 9, T. 14 S., R. 79 W.
North Cripple Creek		Teller	Secs. 1, 2, and 3, T. 11 S., R. 69 W.
North Park		Jackson	Secs. 13, 14, 23, and 24, T. 11 N., R. 82 W.
North Swan	Snake River	Summit	Secs. 7, 8, 17, and 18, T. 6 S., R. 76 W.
Ohio	Jackson, Idaho, Virginia	Clear Creek	
Oro		Montrose	T. 46 N., R. 14 W. of the New Mexico principal meridian.
Oro Fino		Jefferson	Secs. 1, 2, T. 10 S., R. 71 W.
Ottawa	Cascade, Democrat	Clear Creek	
Paquin	Part of Uncompahgre	Ouray	Secs. 16, 17, 18, 19, 20, and 21, T. 44 N., R. 7 W. of the New Mexico principal meridian.
Parachute		Garfield	Secs. 15, 16, 17, 20, 21, and 22, T. 6 S., R. 91 W.
Paradox Valley		Montrose	T. 47 N., R. 19 W. of the New Mexico principal meridian.
Park		Hinsdale	Secs. 19, 20, 21, 22, 27, 28, 29, 30, 31, and 32, T. 43 N., R. 5 W.; secs. 34, 35, and 36, T. 43 N., R. 6 W.; and secs. 1, 2, and 3, T. 42 N., R. 6 W.
Paynes Bar	Part of Idaho Springs	Clear Creek	Secs. 26, 34, and 35, T. 3 S., R. 73 W.
Pearl Pass		Gunnison	Secs. 28, 29, 30, 31, 32, and 33, T. 12 S., R. 84 W.
Pennsylvania		Park	Secs. 15, 16, 21, and 22, T. 9 S., R. 78 W.
Peru	Includes Argentine and Snake River	Clear Creek, Park, and Summit	Tps. 5 and 6 S., R. 75 W.
Phoenix	Overlaps Pine	Gilpin	Secs. 31, 32, and 33, T. 1 S., R. 73 W.
Pikes Peak	Fountain Creek	Teller	Secs. 12, 13, and 24, T. 13 S., R. 69 W., secs. 7, 18, and 19, T. 13 S., R. 68 W.
Pine	Pine Creek	Gilpin	Secs. 33 and 34, T. 1 S., R. 73 W.
Pine Creek	Part of La Plata	Chaffee	Secs. 5, 6, 7, 8, T. 13 S., R. 80 W.
Pioneer	Overlaps Lone Cone	Dolores	Tps. 40 and 41 N., Rs. 10 and 11 W. of the New Mexico principal meridian. T. 36 N., R. 4 E.
Platoro	Ute	Conejos	Secs. 14 and 15, T. 6 S., R. 76 W.
Platte	Included in Hall Valley	Summit and Park	
Pleasant Park	Overlaps Union	Summit	Secs. 7, 8, 9, 16, and 17, T. 6 S., R. 77 W.
Pleasant Valley	Overlaps Wellsville	Fremont	Sec. 19, T. 49 N., R. 10 E.; sec. 24, T. 49 N., R. 9 E.
Do	Overlaps Russell, Gulch and Lake Gulch	Gilpin	Secs. 17, 19, and 20, T. 3 S., R. 72 W.; sec. 24, T. 3 S., R. 73 W.
Pollock	Overlaps Hoosier Gulch and Tenmile Consolidated	Summit	Secs. 21 to 29 and 32 to 36, inclusive, T. 7 S., R. 78 W.; secs. 19, 30, 31, T. 7 S., R. 77 W.; secs. 1 and 2, T. 8 S., R. 78 W.
Princeton	Adjoins Crosson	Park and Jefferson	Secs. 25, 26, 35, and 36, T. 7 S., R. 72 W.
Puma		Park	Secs. 34 and 35, T. 10 S., R. 74 W.
Pulver. See Lake George.			
Putnam	Included in Animas	San Juan	Secs. 27, 28, 33, 34, and 35, T. 41 N., R. 8 W. of the New Mexico principal meridian.
Quartz Creek	Quartz; overlaps Gold Brick and Tin Cup	Gunnison	T. 51 N., Rs. 81 and 82 W.
Quartz Mountain	Overlaps Union and Spaulding	Summit	Secs. 23, 24, 25, and 26, T. 6 S., R. 78 W.
Quartz Valley	Overlaps Pine and Independent	Gilpin	Secs. 3, 4, and 5, T. 2 S., R. 73 W.
Queens	Overlaps Griffith and Argentine	Clear Creek	Secs. 10 to 16 and 22 to 24, inclusive, T. 4 S., R. 75 W.

Mining districts in Colorado—Continued

Name	Other names used	County	Approximate location
Red Cliff	Includes Horn Silver.	Eagle	Secs. 33 and 34, T. 6 S., R. 80 W.; secs. 3, 4, and 5, T. 7 S., R. 80 W.
Red Gorge	Copper King	Grand	Secs. 29, 30, 31, and 32, T. 1 S., R. 82 W.
Red Gulch		Fremont	Secs. 7, 8, 17, and 18, T. 49 N., R. 12 E. of the New Mexico principal meridian.
Red Mountain		Chaffee	T. 12 S., R. 82 W.; secs. 32 to 36, inclusive, T. 11 S., R. 82 W.
Do		Lake	Secs. 13 to 17 and 20 to 29, inclusive, T. 11 S., R. 82 W.
Do	Overlaps Mount Sneffels and Uncompahgre.	Ouray	Secs. 19 to 22 and 28 to 33, inclusive, T. 43 N., R. 7 W., and secs. 7, 8, 9, 17, and 18, T. 42 N., R. 7 W. of the New Mexico principal meridian.
Do	Overlaps Eureka	San Juan	Secs. 14, 15, 22 to 28, and 32 to 36, inclusive, T. 42 N., R. 8 W.; secs. 16 to 21 and 28 to 32, inclusive, T. 42 N., R. 7 W.
Red Ruth. *See* Freshwater.			
Rexford. *See* Middle Swan.		Summit	
Reynolds Switch	Overlaps Lake George.	Park	Secs. 11 to 15, inclusive, T. 13 S., R. 72 W.
Roaring Fork	Overlaps Castle Creek and Hunter Creek and by some is considered to overlap Columbia.	Pitkin	Secs. 11, 12, 13, 14, 23, and 24, T. 10 S., R. 85 W.; secs. 7, 8, 17, 18, 19, 20, 29, and 30, T. 10 S., R. 84 W.
Rock Creek		Montrose	Secs. 1 and 12, T. 48 N.; R. 19 W.; secs. 6 and 7, T. 48 N.; R. 18 W. of the New Mexico principal meridian.
Do	Overlaps Elk Mountain.	Gunnison	Secs. 25, 26, 35, and 36, T. 11 S., R. 88 W.; secs. 31, 32, and 33, T. 11 S., R. 87 W.; secs. 1, 2, 3, 4, 5, and 6, T. 12 S., R. 87 W.
Do		Pitkin	Secs. 32 and 33, T. 9 S., R. 87 W.
Rocky. *See* Lake George.			
Round Mountain		Teller	Secs. 14, 15, 22, and 2, T. 11 S., R. 70 W.
Royal Arch		Mineral	Secs. 21, 22, 23, 26, 27, and 28, T. 40 N., R. 1 W.
Ruby		Gunnison	T. 13 S., R. 87 W.; area extends northward into T. 12 S. and southward into T. 14 S.
Russell. *See* Russell Gulch.			
Russell Gulch	Overlaps Nevada and Pleasant Valley.	Gilpin	Secs. 14, 15, 23, and 24, T. 3 S., R. 73 W.
Sacramento		Park	Secs. 29, 30, 31, and 32, T. 9 S., R. 78 W.; secs. 2, 3, 4, 8, 10, 11, 14, 15, and 16, T. 10 S., R. 78 W.
Saguache Big Park		Saguache	Secs. 4 to 9, inclusive, T. 43 N., R. 3 E.
San Miguel		San Miguel	Secs. 28 to 33, inclusive, T. 42 N., R. 10 W.
Sentinel	Overlaps Mount Sneffels.	Ouray	Secs. 10, 11, 12, T. 43 N., R. 9 W.
Sheep Mountain	Overlaps Rock Creek and Elk Mountains.	Gunnison	Secs. 19, 20, 21, 28, 29, and 30, T. 11 S., R. 87 W.
Sherman	Overlaps Park and Carson.	Hinsdale	T. 42 N., Rs. 5 and 6 W.
Sierra Blanca	Includes West Blanca and Blanca.	Costilla and Huerfano.	Tps. 27 and 28 S., Rs. 72 and 73 W.
Silver Creek	Overlaps Bonanza.	Saguache	Secs. 2, 3, and 4, T. 47 N., R. 7 E. of the New Mexico principal meridian.
Silver Horn. *See* Horn Silver.			
Silver Lake	Lower Fall River	Gilpin	Secs. 20, 21, and 22, T. 2 S., R. 73 W.
Silverside		Gunnison	Secs. 21, 22, 27, and 28, T. 13 S., R. 81 W.
Sinbad		Mesa	Secs. 5, 6, 7, 8, 17, and 18, T. 49 N.; R. 19 W. of the New Mexico principal meridian.

Mining districts in Colorado—Continued

Name	Other names used	County	Approximate location
Slide Mountain	Overlaps Tenmile and Wilkinson.	Summit	Secs. 9 and 10, T. 6 S., R. 78 W.
Smiths Gulch	Overlaps Blake	Saguache	Secs. 23, 24, 25, 26, 35, and 36, T. 46 N., R. 10 E. of the New Mexico principal meridian.
Snake River	Includes Montezuma.	Summit	T. 5 S., R. 76 W.; secs. 1, 2, 3, 4, 5, and 6, T. 6 S., R. 76 W.
Snowmass		Pitkin	Secs. 4, 5, 8, and , T. 10 S., R. 86 W.
Snowy Range	Overlaps Grand Island.	Boulder	Secs. 16 to 21, inclusive, T. 1 N., R. 73 W.
South Boulder		Gilpin	Secs. 35 and 36, T. 2 S., R. 73 W.
South Cottonwood		Chaffee	Secs. 1, 2, 11, 12, 13, and 14, T. 15 S., R. 81 W.; secs. 1 to 18, inclusive, T. 15 S., R. 80 W.
South Cripple	Overlaps Bare Hills.	Fremont	Secs. 12 and 13, T. 16 S., R. 71 W.; secs. 7 and 18, T. 16 S., R. 70 W.
South Independence		Pitkin	Secs. 35 and 36, T. 11 S., R. 83 W., sec. 31, T. 11 S., R. 82 W.; secs. 1 and 2, T. 12 S., R. 83 W.
South Swan River		Summit	Secs. 31, 32, and 33, T. 6 S., R. 76 W.
Spalding	Stillson Patch	do	Sec. 36, T. 6 S., R. 78 W.; sec. 31, T. 6 S., R. 77 W.
Spanish Bar	Lower Fall River	Clear Creek	
Spanish Peaks		Las Animas	Secs. 1, 2, and 3, T. 31 S., R. 68 W.
Spring Butte		Pitkin	Secs. 34, 35, and 36, T. 9 S., R. 88 W.
Spring Creek		Gunnison	Secs. 23, 24, 25, and 26, T. 14 S., R. 84 W.
Springer. *See* Lake George.			
Spruce Creek		Custer	Secs. 17 to 20, inclusive, T. 46 N., R. 12 E. of the New Mexico principal meridian.
St. Kevin	Overlaps Independence.	Lake	Secs. 1, 2, 3, 10, 11, and 12, T. 8 S., R. 81 W.
Stillson Patch. *See* Spalding.			
Stowes Gulch	Stowe; overlaps Kerber Creek.	Saguache	Sec. 20, T. 46 N., R. 8 E. of the New Mexico principal meridian.
Sugar Loaf	Overlaps Ward, Grand Island, Magnolia.	Boulder	Secs. 13 to 16 and 18 to 30, inclusive, T. 1 N., R. 72 W.; secs. 15 to 23 and 26 to 30, inclusive, T. 1 N., R. 71 W.; secs. 2, 3, and 4, T. 1 S., R. 72 W
Summer Coon		Rio Grande	Secs. 17, 18, 19, and 20, T. 41 N., R. 6 E.
Summit		do	Secs. 25 to 29 and 32 to 36, inclusive, T. 37 N., R. 3 E.; secs. 19 to 21 and 28 to 33, inclusive, T. 37 N., R. 4 E.
Sunshine		Boulder	Secs. 3 and 4, T. 1 N., R. 71 W.
Sunnyside	King Solomon	Mineral	T. 42 N., R. 1 W., and southwestward into T. 41 N., R. 2 W.
Swan River		Summit	Secs. 13 to 15 and 22 to 24, inclusive, T. 6 S., R. 77 W.
Tarryall		Park	Secs. 23 to 27 and 34 to 36, inclusive, T. 7 S., R. 77 W.; secs. 1 to 4 and 9 to 17, inclusive, 21, and 22, T. 8 S., R. 77 W.
Tarryall Springs		do	T. 11 S., R. 72 W.
Taylor River	Taylor Park, Tin Cup, Forest Hill.	Gunnison	T. 12 S., Rs. 83 and 84 W.; T. 13 S., R. 82 W.; T. 14 S., Rs. 82, 83, and 84 W.
Telluride	Frying Pan	Eagle	Secs. 23, 24, 25, and 26, T. 7 S., R. 82 W.
Tenmile. *See* Consolidated Tenmile.			
Texas Creek		Fremont	Secs. 26 to 29 and 32 to 34, inclusive, T. 20 S., R. 73 W.
Three Forks		Routt	Secs. 23 and 24, T. 12 N., R. 86 W.
Tin Cup	Taylor River	Gunnison	T. 12 S., Rs. 83 and 84 W.; T. 13 S., Rs. 82 and 83 W.; T. 14 S., Rs. 81, 82, and 83 W.; T. 15 S., Rs. 81, 82, and 83 W.; and into T. 51 N., R. 72 W.

Mining districts in Colorado—Continued

Name	Other names used	County	Approximate location
Tomichi		Gunnison	Secs. 13 to 16, 21 to 29, and 32 to 36, inclusive, T. 50 N., R. 4 E., and southward into T. 49 N., Rs. 4 and 5 E. of the New Mexico principal meridian.
Trail Run	Trail Creek	Clear Creek	Secs. 31 and 32, T. 3 S., R. 73 W.; secs. 5, 6, and 7, T. 4 S., R. 73 W.
Trout Creek		Chaffee	T. 13 S., R. 77 W.; T. 14 S., R. 78 W.
Trout Lake			Secs. 4 to 9, inclusive, T. 41 N., R. 9 W. of the New Mexico principal meridian.
Tungsten	Wilbur	Fremont	Secs. 3, 4, 9, and 10, T. 17 S., R. 69 W.
Turkey Creek		El Paso	Secs. 19, 20, 29, and 30, T. 16 S., R. 67 W.
Turret Mountain	Turret	Chaffee	Tps. 16 and 17 S., R. 9 E.
Tuttle Creek	Included in Kerber Creek.	Saguache	Secs. 4 to 9, inclusive, T. 47 N., R. 7 E.
Twelvemile		Park	Secs. 13 and 24, T. 10 S., R. 79 W.; secs. 17 to 20, T. 10 S., R. 78 W.
Twin Lakes	Lake Creek	Lake	Secs. 7 to 30, inclusive, T. 11 S., R. 81 W.
Two Bit		do	Secs. 1, 2, 11, and 12, T. 11 S., R. 80 W.
Tyler	West Creek	Douglas	Secs. 23 to 27, inclusive, T. 10 S., R. 70 W.
Unaweep		Mesa	Secs. 7, 8, 9, 16, 17, and 18, T. 14 S., R. 100 W.
Uncompahgre	Includes Paquin	Ouray	T. 44 N., Rs. 7 and 8 W., and T. 43 N., R. 7 W. of the New Mexico principal meridian.
Union		Gilpin	Sec. 6, T. 2 S., R. 72 W.; sec. 1, T. 2 S., R. 73 W.; sec. 31, T. 1 S., R. 72 W.
Union. *See* Consolidated Union.		Summit	
Union. *See* Upper Union		Clear Creek	
Union Consolidated. *See* Consolidated Union.			
Union Gulch	Union	Lake	Secs. 19, 21, 22, 23, T. 10 S., R. 79 W.
Upper Fall River	Lincoln	Clear Creek	Secs. 26, 27, 28, 29, 31, 32, 33, and 34 T. 2 S., R. 74 W.; secs. 2, 3, 4, 5, and 6, T. 3 S., R. 74 W.
Upper San Miguel		San Miguel	T. 43 N., R. 9 W.; secs. 29, 30, 31, 32, and 33, T. 43 N., R. 8 W.; secs. 1 to 18, inclusive, T. 42 N., R. 9 W.; secs. 3 to 10 and 16 to 21, inclusive, 29 and 30, T. 42 N., R. 8 W. of the New Mexico principal meridian.
Upper Union	Union	Clear Creek	Secs. 12 to 16, 21 to 28, and 33 to 36, inclusive, T. 3 S., R. 75 W.; secs. 7, 8, 9, 16 to 21, and 28 to 32, inclusive, T. 3 S., R. 74 W.; area extends into T. 4 S., R. 74 W.
Ute	Platoro	Conejos	Secs. 13 to 29, inclusive, T. 36 N., R. 4 E. of the New Mexico principal meridian.
Utah	Bevan	Summit	
Utillah. *See* Central.			
Verde		Custer	Secs. 18 and 19, T. 22 S., R. 73 W.; T. 45 N., R. 12 E.
Vermillion	Overlaps Pine	Gilpin	Secs. 29, 30, 31, and 32, T. 2 S., R. 73 W.
Virginia. *See* Idaho.			
Vixen		Montrose	Secs. 22, 23, and 24, T. 49 N., R. 18 W.
Ward	Overlaps Gold Hill	Boulder	Secs. 1, 2, 3, and 4, T. 1 N., R. 73 W.; secs. 2 to 11, inclusive, T. 1 N., R. 72 W.
Wellsville		Fremont	Secs. 13, 14, 23, and 24, T. 49 N., R. 9 E. of the New Mexico principal meridian.
West Argentine. *See* Argentine.			
West Blanca. *See* Sierra Blanca.			

Mining districts in Colorado—Continued

Name	Other names used	County	Approximate location
West Castle Creek	West Castle	Pitkin	Secs. 15, 16, 17, 20, 21, 22, 28, and 29, T. 11 S., R. 85 W.
West Creek		Douglas	Secs. 22 to 27, inclusive, T. 10 S., R. 70 W.; secs. 19 to 22 and 27 to 34, inclusive, T. 10 S., R. 69 W.
Do		Teller	Secs. 1 to 12, inclusive, T. 11 S., R. 70 W.; secs. 3 to 10, inclusive, T. 11 S., R. 69 W.
West Maroon		Pitkin	Secs. 25 to 28 and 33 to 36, inclusive, T. 11 S., R. 86 W.
West Spanish Peaks		Huerfano	Secs. 33 to 36, inclusive, T. 30 S., R. 68 W.
Weston Pass			Secs. 35 and 36, T. 10 S., R. 79 W.; secs. 1, 2, and 3, T. 11 S., R. 79 W.
White Earth	Includes Cebolla, Hotchkiss, McDonough, Goose Creek	Gunnison	Tps. 46 and 47 N., R. W. of the New Mexico principal meridian.
Wilbur		Fremont	Secs. 22 to 26, inclusive, T. 16 S., R. 70 W.; secs. 27 to 34, inclusive, T. 16 S., R 69 W.
Wilkinson		Summit and Eagle.	Tps. 4 and 5 S., R. 79 W., T. 5 S., R. 78 W.; area extends southward into T. 6 S., R. 78 W.
Willow Creek		Grand	Secs. 13 to 16 and 21 to 24, inclusive, T. 4 N., R. 77 W.
Do		Gunnison	Secs. 11, 12, 13, and 1 T. 47 N., R. 2 W.
Wilson Creek	Cripple Creek	Teller	Secs. 3 to 10, inclusive, T. 16 S., R. 69 W.
Wisconsin		Gilpin	Secs. 1 to 5, 8 to 17, and 21 to 25, inclusive, T. 2 S., R. 74 W.; secs. 7, 18, and 19, T. 2 S., R. 73 W.
Womack. *See* Cripple Creek.			
Woodland Park		Teller	Secs. 3 to 10 and 15 to 22, inclusive, T. 13 S., R. 69 W.
Woody		Pitkin	Secs. 19 to 30, T. 9 S., R. 84 W., and southward into T. 10 S., R. 84 W.
York. *See* Virginia.			

PRODUCTION OF THE STATE

ACCURACY OF THE FIGURES GIVEN

The figures given in the accompanying tables are derived from the published reports indicated.

The accuracy of the State totals given by Raymond and by the Director of the Mint is considered in a treatise on this subject by Preston.[49]

The reports of the Colorado State Bureau of Mines appear to be very accurate, though they may contain some errors as to the origin of ore or metal by counties, due to the fact that the shipping point of a mine was not invariably in the same county as the mine itself.

The accuracy of the mine reports is considered in a statement by McCaskey[50] on Mineral Resources for 1914, in which he compares the results of the mint and mine returns for 1905 to 1914, inclusive.

In determining the totals for the counties the figures for the State have been so prorated among the several producing counties as to make their sum equal to the

[49] Preston, R. E., Report of the Director of the Mint upon the production of precious metals in the United States in 1896, pp. 18–43, 1897.

[50] McCaskey, H. D., Method of collecting statistics: U. S. Geol. Survey Mineral Resources, 1914, pt. 1, pp. 835–836, 1916.

TABLE 1.—*Production of gold, silver, copper, lead, and zinc in Colorado, 1858–1923, by years, in terms of recovered metals* [a]

Year	Ore sold or treated (short tons)	Gold			Silver			Copper			Lead			Zinc			Total value
		Placer	Lode	Total value	Fine ounces	Average value per ounce	Value	Pounds	Average value per pound	Value	Pounds	Average value per pound	Value	Pounds	Average value per pound	Value	
1858–1867		$14,923,918	$10,097,866	$25,021,784	302,829		$406,139										$25,427,923
1868		320,000	1,690,000	2,010,000	200,716	$1.326	266,150	50,000	$0.23	$11,500							2,287,650
1869		380,000	2,800,000	3,180,000	475,472	1.325	630,000	102,000	.2425	24,735	150,000	$0.06	$9,000				3,843,735
1870		695,000	2,320,000	3,015,000	496,988	1.328	660,000	182,500	.2118	38,654	250,000	.06	15,000				3,728,654
1871		190,000	3,443,951	3,633,951	776,648	1.325	1,029,059	183,000	.2412	44,140	555,000	.06	33,300				4,740,450
1872		271,500	2,374,963	2,646,463	1,524,206	1.322	2,015,000	204,000	.3556	72,542	1,150,000	.064	73,600				4,807,605
1873		285,000	1,733,931	2,018,931	1,543,047	1.297	2,001,331	379,493	.280	106,258	1,236,400	.06	74,184				4,200,704
1874		258,997	1,893,490	2,152,487	2,348,174	1.278	3,000,966	475,541	.220	104,619	1,277,933	.06	76,676				5,334,748
1875		263,260	1,961,308	2,224,568	2,330,291	1.24	2,889,560	280,815	.227	63,745	1,636,000	.058	94,888		$0.07		5,272,761
1876		285,000	2,441,311	2,726,311	2,564,403	1.16	2,974,707	333,333	.21	70,000	1,334,020	.061	81,375		.072		5,852,393
1877		265,000	2,883,708	3,148,708	2,882,121	1.20	3,458,546	493,664	.19	93,796	4,286,364	.055	235,750		.06		6,936,800
1878		275,774	2,964,574	3,240,348	4,672,961	1.15	5,373,904	536,145	.166	89,000	13,722,222	.036	494,000		.049		9,197,252
1879		187,000	3,006,500	3,193,500	11,899,335	1.12	13,327,257	704,301	.186	131,000	47,348,000	.041	1,941,268		.05		18,593,025
1880		179,000	3,073,514	3,252,514	14,397,539	1.15	16,557,170	859,000	.214	183,826	71,348,000	.05	3,567,400		.055		23,560,910
1881		175,500	3,124,500	3,300,000	13,272,188	1.13	14,997,572	884,000	.182	160,888	81,094,000	.048	3,892,512		.052		22,350,972
1882		192,500	3,167,500	3,360,000	12,761,719	1.14	14,548,359	1,494,000	.191	285,354	110,000,000	.049	5,390,000		.053		23,583,713
1883		132,000	3,968,000	4,100,000	13,434,610	1.11	14,912,417	1,152,652	.165	190,188	141,114,000	.043	6,067,902		.045		25,270,507
1884		123,551	4,176,449	4,300,000	12,375,000	1.11	13,736,251	2,013,125	.13	261,706	126,330,000	.037	4,674,209		.044		22,972,166
1885		124,035	4,079,390	4,203,425	12,220,982	1.07	13,076,451	1,146,460	.108	123,818	106,692,000	.039	4,160,989	100,000	.043	$4,300	21,568,983
1886		163,328	4,286,672	4,450,000	12,375,000	.99	12,251,250	1,146,460	.111	127,257	118,000,000	.046	5,428,000	100,000	.044	4,400	22,260,907
1887		280,933	3,719,067	4,000,000	11,601,563	.98	11,369,534	2,012,027	.138	277,660	126,000,000	.045	5,670,000	100,000	.046	4,600	21,321,794
1888		104,500	3,653,599	3,758,099	14,695,313	.94	13,813,596	1,621,100	.168	272,345	128,404,000	.044	5,649,777	300,000	.049	14,700	23,508,517
1889		135,870	3,747,989	3,883,859	18,375,136	.84	17,272,629	1,170,053	.135	157,956	133,940,000	.039	5,223,660	300,000	.05	15,000	26,553,104
1890		126,380	4,024,752	4,151,132	18,800,000	1.05	19,740,000	3,585,691	.156	559,368	109,192,000	.045	4,913,639	300,000	.055	16,500	29,380,639
1891		125,663	4,474,337	4,600,000	21,160,000	.99	20,948,401	6,336,878	.128	811,121	126,256,000	.043	5,429,009	300,000	.05	15,000	31,803,531
1892		71,900	5,228,100	5,300,000	24,000,000	.87	20,880,000	7,593,674	.116	880,866	120,000,000	.04	4,800,001	1,125,000	.046	51,750	31,912,617
1893		98,216	7,428,784	7,527,000	25,838,600	.78	20,154,107	7,695,826	.108	831,149	110,000,000	.037	4,070,000	1,650,000	.04	66,000	32,648,256
1894		108,747	9,382,767	9,491,514	23,281,398	.63	14,667,281	6,481,413	.095	615,734	101,226,000	.033	3,340,458	1,500,000	.035	52,500	28,167,487
1895		95,499	13,209,601	13,305,100	23,398,500	.65	15,209,024	6,079,243	.107	650,479	93,968,000	.032	3,006,976	1,671,000	.036	60,156	32,231,735
1896		90,419	14,820,581	14,911,000	22,573,000	.68	15,349,642	6,022,176	.108	650,395	89,606,000	.03	2,688,178	1,292,000	.039	50,388	33,649,603
1897		130,646	19,448,787	19,579,433	21,278,202	.60	12,766,919	9,149,967	.12	1,097,995	80,794,286	.036	2,908,592	2,683,989	.041	110,044	36,462,983
1898		83,428	23,451,104	23,534,532	23,502,601	.59	13,866,532	10,870,701	.124	1,347,965	113,416,138	.038	4,309,813	3,900,656	.046	179,430	43,238,272
1899		73,589	26,435,086	26,508,675	23,114,688	.60	13,868,811	7,356,970	.171	1,258,041	138,048,446	.045	6,212,178	11,300,656	.058	655,438	48,503,143
1900		77,966	28,684,070	28,762,036	20,336,512	.62	12,608,637	7,826,815	.166	1,299,251	164,274,762	.044	7,228,090	16,282,055	.044	716,410	50,614,424
1901		87,324	27,592,119	27,679,443	18,492,563	.60	11,095,538	7,872,529	.167	1,314,712	148,111,020	.043	6,368,772	26,843,731	.041	1,100,593	47,559,058
1902		118,774	28,398,140	28,516,914	15,941,523	.53	8,449,008	8,463,938	.122	1,132,601	106,296,827	.041	4,358,169	52,582,510	.048	2,523,963	44,980,655
1903		129,049	21,476,308	21,605,357	13,245,438	.54	7,152,536	7,809,920	.137	1,069,958	101,513,414	.042	4,263,566	80,616,000	.054	4,353,263	38,444,680
1904	2,333,881	193,068	24,049,417	24,242,485	12,960,792	.58	7,517,260	9,412,707	.128	1,204,828	107,498,854	.043	4,622,453	66,771,590	.051	3,405,353	40,992,379
1905	2,504,087	99,984	25,195,238	25,295,222	12,339,435	.61	7,527,056	9,661,546	.156	1,507,201	115,746,777	.047	5,440,098	83,561,396	.059	4,930,123	44,699,700
1906	2,648,923	106,019	22,799,652	22,905,671	12,339,052	.68	8,390,553	6,618,332	.193	1,277,338	106,646,506	.057	6,078,850	86,012,903	.061	5,246,787	43,899,199
1907	2,383,128	97,219	20,210,429	20,307,648	11,599,514	.66	7,655,679	8,826,254	.20	1,765,251	89,065,232	.053	4,720,457	85,048,564	.059	5,017,865	39,466,900
1908	2,242,969	184,457	22,411,114	22,595,571	9,002,316	.53	4,771,227	10,201,123	.132	1,346,547	61,645,671	.042	2,589,118	30,130,002	.047	1,416,110	32,718,573
1909	2,219,644	457,085	21,526,923	21,984,008	8,904,701	.52	4,630,444	10,916,191	.13	1,419,105	72,162,326	.043	3,102,980	51,210,260	.054	2,765,354	33,901,891
1910	2,434,664	389,828	20,115,786	20,505,614	8,508,942	.54	4,594,829	8,359,307	.127	1,061,632	76,058,775	.044	3,346,586	77,089,648	.054	4,162,841	33,671,502
1911	2,377,936	319,038	18,682,937	19,001,975	7,330,168	.53	3,884,989	8,024,488	.125	1,003,061	69,679,289	.045	3,135,568	94,607,456	.057	5,392,625	32,418,218
1912	2,576,626	423,865	18,164,697	18,588,562	8,212,070	.615	5,050,423	7,107,303	.165	1,172,705	75,242,267	.045	3,385,902	132,222,812	.069	9,123,374	37,320,966
1913	2,734,866	408,007	17,738,909	18,146,916	9,325,255	.604	5,632,454	7,227,826	.155	1,120,313	87,897,773	.044	3,867,502	119,346,429	.056	6,683,400	35,450,585
1914	2,677,526	642,360	19,240,745	19,883,105	8,796,065	.553	4,864,224	6,639,173	.133	883,010	74,211,898	.039	2,894,264	96,774,960	.051	4,935,523	33,460,126
1915	2,737,020	693,310	21,721,634	22,414,944	7,027,972	.507	3,563,182	7,112,537	.175	1,244,694	68,810,597	.047	3,234,098	104,594,994	.124	12,969,779	43,426,697
1916	2,697,333	712,924	18,440,897	19,153,821	7,656,544	.658	5,038,006	8,624,081	.246	2,121,524	70,914,087	.069	4,893,072	134,285,463	.134	17,994,252	49,200,675
1917	2,688,706	661,028	15,068,196	15,729,224	7,304,353	.824	6,018,787	8,122,004	.273	2,217,307	67,990,012	.086	5,847,141	120,315,775	.102	12,272,209	42,084,668
1918	2,314,890	526,202	12,225,516	12,751,718	7,063,554	1.00	7,063,554	6,277,332	.247	1,550,501	65,960,760	.071	4,683,214	89,133,901	.091	8,111,185	34,160,172
1919	1,919,768	550,562	9,336,065	9,886,627	5,758,010	1.12	6,448,971	3,560,207	.186	662,198	37,070,241	.053	1,964,722	37,220,493	.073	2,717,096	21,679,614
1920	1,571,293	514,588	7,061,731	7,576,319	5,409,335	1.09	5,896,175	4,043,734	.184	744,047	46,629,788	.08	3,730,383	48,790,742	.081	3,952,050	21,898,974
1921	1,281,381	344,640	6,490,688	6,835,328	5,631,657	1.00	5,631,657	4,153,442	.129	535,794	19,660,466	.045	884,721	2,360,000	.05	118,000	14,005,500
1922	1,412,100	356,403	6,017,016	6,373,419	5,855,911	1.00	5,855,911	3,373,454	.135	455,416	23,477,200	.055	1,291,246	23,258,000	.057	1,325,706	15,301,698
1923	1,569,100	364,429	6,227,200	6,591,629	5,334,488	.82	4,374,280	4,248,109	.147	624,472	45,698,185	.070	3,198,873	54,152,000	.068	3,682,336	18,471,590
		29,674,282	643,733,608	673,061,890	628,849,400		501,733,945	263,078,559		40,327,566	4,200,637,536		189,662,179	1,739,834,985		126,216,403	1,531,001,983

[a] From 1858 to 1895 the figures for gold and silver represent chiefly United States Mint estimates of recovered metals; figures for copper, lead, and zinc, as far as possible, represent refined metals (some of them estimates from assay content of ores treated, with allowance made for losses). The figures for 1896 to 1905 differ only in that they are based on actual receipts at the mints and smelters. The figures for 1906 to 1923 also represent actual receipts at mints and smelters, supplemented by reports from mining companies. For ore and concentrates to smelters, the figures represent assay content of gold and silver but allow for losses of the base metals in treatment.

total for the State.   Anyone who should attempt to correct these tables by adding together other published county totals to obtain the total for the State would have a sum that would much more than equal the recorded total for the State, and if he should attempt to determine the total for the United States from such county totals for Colorado and other States he would find that his total for the United States would be anywhere from one and one-half to two times the recorded total.

It should be remembered that from time to time new counties have been created, some of them by dividing old ones, and that the figures here given for any county in Colorado represent the output of the area now included in that county.   It is impossible to calculate district totals.   Fortunately the yearly output for some mines from the time they were first worked to their abandonment, or to 1923, has been obtainable.

The quantity and value of the gold, silver, copper, lead, and zinc produced in Colorado from 1858 to 1922 are shown in Table 1.   The value of the gold, silver, copper, and lead produced in the State from 1868 to 1875, according to Raymond, is shown in Table 2.

SOURCES OF FIGURES BY METALS AND YEARS

The principal sources of the figures given in Table 1 showing the production of gold, silver, copper, lead, and zinc in Colorado by years from 1858 to 1923, inclusive, are stated below.

Gold and silver.—For 1868 to 1875 Raymond's reports on the statistics of mines and mining in the States and Territories west of the Rocky Mountains; for 1876 to 1879 the reports of the Director of the Mint, which for these years show for Colorado only the State total; for 1879 to 1896 the reports of the Director of the Mint (final prorated figures); for 1897 to 1904 the reports of the Colorado State Bureau of Mines, which check very well with the reports of the Director of the Mint and are in better form by counties, particularly for copper and lead; for 1905 to 1922 the volumes entitled Mineral Resources of the United States (mines reports), published by the United States Geological Survey, hereinafter called simply Mineral Resources.

Copper.—For the early years Raymond's reports and some estimates made by others showing copper not heretofore credited to Colorado; for 1874 to 1896 Kirchhoff's table showing the value of copper produced in Colorado from the beginning of mining to 1882, given in the general report on copper in Mineral Resources for 1882 (p. 228), and B. S. Butler's general report on copper, showing smelter production in Colorado from 1874 to 1910, in Mineral Resources

for 1910 (pt. 1, pp. 171–173); for 1897 to 1904 the receipts of smelter ore as shown by the Colorado State Bureau of Mines; for 1905 to 1923 Mineral Resources (mines report).

Lead.—For the early years Raymond's reports mainly, but as these do not show invariably the total content of lead in ore shipped or produced, they are supplemented by estimates made from Raymond's data showing the quantity and grade of ore shipped, thus crediting, for the first time, to certain counties and districts, and especially to Georgetown, Clear Creek County, large quantities of lead in ore shipped to smelters in the eastern United States and in Wales and Germany; for 1873 to 1882 a table given by Kirchhoff in a general report on the lead industry of the United States in Mineral Resources for 188 (p. 310), though for the earlier years this table appears to show only the output of lead from lead smelters in Colorado and not to include the lead in the ore shipped; from 1882 to 1896 Kirchhoff's annual general reports on lead in Mineral Resources; for 1897 to 1904 the reports of the Colorado State Bureau of Mines; and from 1905 to 1923 Mineral Resources (mines report).

Zinc.—For 1885 to 1891 estimates based on an oral statement by F. L. Bartlett of quantities recovered by experiments made in one of the Eastern States; for 1891 to 1901 the annual reviews in the Denver Republican and the Leadville Herald-Democrat and the annual volumes of Mineral Industry; for 1902 to 1907 reports of the Colorado State Bureau of Mines; for 1908 to 1923 Mineral Resources (mines report).

The production of zinc from ore mined in Colorado began in 1885, when low-grade zinc ore was treated by Bartlett in a plant in Portland, Maine.   In 1891 Bartlett erected a zinc oxide plant at Canon City, and from 1891 to 1897 this was the only plant that recovered zinc from Colorado ores, but in 1898 a small quantity of Colorado ore was marketed in the United States, and in 1899 ore was shipped to Belgium, and soon thereafter zinc plants in the United States began to handle the ferruginous zinc blende ores of Colorado.

SOURCES OF FIGURES BY YEARS AND COUNTIES

GOLD AND SILVER

For 1858–1867 preference is given to county figures, which include the output of the producing areas of the State that are now represented by the present Boulder, Chaffee, Clear Creek, Gilpin, Lake, Park, and Summit counties, and the State total is made the sum of the county totals.   The county figures are the totals for the areas now within the counties named, and the data for each county are derived from many sources. Raymond's reports for States and counties and for some districts do not begin until 1868. G. C. Munson,[51] assayer in charge of the United States Mint, Denver, Colo., gives the coinage value of the gold and silver produced in Colorado from 1859 to 1886. His combined figures for the years from 1859 to 1869, inclusive, are, for gold, $27,213,081; for silver, $330,000, a total of $27,543,081. The sum of the county figures here given for the years 1859–1869, inclusive, shows a total for gold of $30,211,784 and for silver (commercial value) of $1,302,289. Hollister[52] gives the following table showing the value of the bullion, chiefly gold, obtained from mines in Colorado and deposited at the United States Mint in Philadelphia and its branches and at the United States Assay Office at New York City during the fiscal years from June 30, 1859, to June 30, 1866:

1859	$4,172	1864	$2,136,685
1860	599,846	1865	1,622,249
1861	2,091,197	1866	1,018,472
1862	2,035,416		
1863	2,893,337		12,401,374

On the theory that "by as near an approximation as practicable, the amount of deposits at the mint and branches of United States embraces only one-third of the total product of the mines," Hollister estimates the bullion yield of Colorado to June 30, 1866, at $37,204,123. Preston[53] gives the yearly production of the precious metals from 1868 to 1875, according to Raymond, for Colorado and Wyoming combined (in part for Colorado alone) as follows:

figures given by Burchard[55] for the placer gold output of Lake County in 1860–1869, namely, $5,812,000, are used, the output of Chaffee and Lake counties being separated by means of data taken from Hollister. For the production of placer and lode gold from 1859 to 1872 in Park County, Raymond[56] gives $2,750,000. For Boulder County use is made of the Boulder County Metal Mining Association's pamphlet issued in 1910, the figures in which are supplemented by some estimates.

For 1868, for Boulder, Chaffee, Clear Creek, Gilpin, Lake, Park, and Summit counties, use is made of Raymond's report,[57] particularly of his statement showing bullion produced in 1868, of Munson's figures for counties,[58] and of figures given by Hollister. The total is made to conform as nearly as possible to Raymond's total of gold and silver for these years, the division into gold and silver being estimated from data given in later reports of Raymond.

For 1869, for the same counties, particular use is made of Raymond's report,[59] in which he gives the quantity of bullion sent out of Colorado during the year ended June 30, 1869, according to Wells, Fargo & Co.'s books, and of Raymond's estimate for the six months between July 1 and December 31, 1869.

For 1870, for the same counties, use is made particularly of Raymond's reports[60] and of Munson's figures.[61]

For 1871, for the same counties, use is made of Raymond's reports[62] and of Munson's figures.[63]

TABLE 2.—*Value of gold, silver, copper, and lead produced in Colorado, 1868–1875, according to Raymond*

| Year | Preston (from Raymond) [a] | Raymond | | | | | |
	Gold and silver	Gold	Silver	Gold and silver	Copper	Lead	Total
1868	$3,250,000	$1,909,491	$197,744	$2,107,235			$2,107,235
1869	4,000,000	(?)	(?)	3,800,000			3,800,000
1870	3,775,000	(?)	(?)	3,675,000			3,675,000
1871	4,763,000	(?)	(?)	4,663,000			4,663,000
1872	4,761,465	(?)	(?)	4,661,465			4,661,465
1873	4,070,263	1,404,000(?)	2,616,263(?)	4,020,263		$28,000	4,048,263
1874	5,188,510	2,102,487(?)	3,086,023(?)	5,188,510	$100,197	173,873	5,362,383
1875	5,302,810	2,224,568	3,012,902	5,237,470	64,650	689	5,302,810

[a] Credited to Colorado and Wyoming combined, but some of the figures at least are for Colorado alone, as is shown for 1874 and 1875. For 1870, 1871, and 1872 Raymond has given $100,000 to Wyoming and for 1873 $50,000.

For 1858 to 1867 for Gilpin, Summit, and Clear Creek counties, the figures given by G. C. Munson,[54] assayer in charge of the Denver mint, are used, but for the production of Lake and Chaffee counties the

For 1872, for the same counties, use is made of Raymond's report.[64]

[51] Kimball, J. P., Report of the Director of the Mint upon the production of the precious metals in the United States during the calendar year 1886, p. 178, 1887.

[52] Hollister, O. J., The mines of Colorado, p. 434, 1867.

[53] Preston, R. E., History of the methods followed in the collection of the statistics of the production of the precious metals in the United States: Report of the Director of the Mint upon the production of the precious metals in the United States during the calendar year 1896, p. 35, 1897.

[54] Munson, G. C., in Kimball, J P., op. cit. for 1887, pp. 151, 153, 1888.

[55] Burchard, H. C., Report of the Director of the Mint upon the production of the precious metals in the United States during the calendar year 1882, p. 505, 1883.

[56] Raymond, R. W., Statistics of mines and mining in the States and Territories west of the Rocky Mountains for 1872, pp. 298–299, 1873.

[57] Raymond, R. W., op. cit. for 1869, p. 339, 1870.

[58] Munson, G. C., in Kimball, J. P., op. cit. for 1887, pp. 151, 153, 1888.

[59] Raymond, R. W., op. cit. for 1869, p. 351, 1870.

[60] Raymond, R. W., op. cit. for 1870, p. 290, 1872; op. cit. for 1872, p. 265, 1873.

[61] Munson, G. C., in Kimball, J. P., op. cit. for 1887, p. 178, 1888.

[62] Raymond, R. W., op. cit. for 1871, p. 340, 1873; op. cit. for 1872, p. 265, 1873.

[63] Munson, G. C., in Kimball, J. P., op. cit. for 1887, pp. 151, 153, 1888.

[64] Raymond. R. W., op cit. for 1872, p. 266, 1873.

For 1873, for Boulder, Chaffee, Clear Creek, Gilpin, Lake, Park, Summit, Custer, Gunnison, Rio Grande, Routt, Moffat, and San Juan counties, use is made of Raymond's report,[65] of Munson's figures,[66] and of a paper by S. F. Emmons.[67]

For 1874, for Boulder, Chaffee, Clear Creek, Gilpin, Lake, Park, Summit, Custer, Rio Grande, and San Juan counties, use is made of Raymond's report,[68] of Munson's figures,[69] and of Emmons's paper.[70]

For 1875, for Boulder, Chaffee, Clear Creek, Gilpin, Lake, Park, Summit, Custer, Gunnison, Rio Grande, Routt, Moffat, Hinsdale, and San Miguel counties, use was made of Raymond's report.[71]

For 1876 to 1878, inclusive, the figures given are based on the reports of Munson[72] and of Burchard.[73]

For 1879 the figures are based on those given in Burchard's report.[74]

For 1880 the figures are based on those given in the table in Burchard's report,[75] except for gold in Lake County, for which that report elsewhere[76] gives $34,014 lode gold and $70,000 placer gold, which seems more reasonable than the $58,000 of this table. For the years since 1880 in particular, but in general throughout this report, the coinage value of silver given in the reports of the Director of the Mint is reduced to fine ounces and the value of the silver recalculated at the average yearly market price. The factor used for transforming coinage value into fine ounces is the reciprocal of 1.2929292929 +, or 0.7734375000.

For 1881 to 1884 the figures are based on Burchard's reports.[77]

For 1885 Kimball[78] gives no detailed figures for counties but gives gold as $4,200,000 and silver as $15,824,557 (coinage value) for the State. From data derived from different sources but chiefly by interpolation, the production of gold and silver in each county has been estimated for 1885 to make the total given in the table. For several small counties added after the $4,200,000 was prorated to the larger counties, the amount of the deposits from these small counties at the Denver Mint, $3,425, was added, making the State total for gold $4,203,425.

For 1886 to 1896, inclusive, the reports of the Director of the Mint give two sets of figures for gold and silver for all States. One of these sets is the report of the agents of the Mint, which was used by the Director of the Mint (after computing the production of the entire United States) to determine the figures showing the estimated production by States. The Director's figures are here taken as the more accurate, and those of the Colorado agent's report are prorated by counties to make the total agree with the Director's figures for gold and silver.

For 1897 to 1904 the figures of the Colorado State Bureau of Mines, showing the smelter and mint receipts are taken, with small readjustments as to origin by counties.

For 1904 the figures for gold and silver from Jefferson and Mineral counties are those of the United States Geological Survey Mineral Resources, and the figures for all the rest of the State are those of the State Bureau of Mines.

For 1905 to 1908, inclusive, the figures used are chiefly those of the United States Geological Survey Mineral Resources, which are taken from reports received directly from the mines, checked against the mint receipts, and balanced against the figures of the State Bureau of Mines. For some counties the figures of the Bureau of Mines are used instead of those of the United States Geological Survey Mineral Resources. For 1905, for instance, the figures for gold and silver from Mineral County and Ouray County are taken from the State Bureau of Mines and those for other counties from the United States Geological Survey Mineral Resources. For 1906, 1907, and 1908 the figures for Mineral County are taken from the report of the State Bureau of Mines. For 1909 to 1923, inclusive, the figures are taken from Mineral Resources (mines reports).

COPPER

For 1868 to 1881, inclusive, the figures are based on data in United States Geological Survey Mineral Resources[79] and in Raymond's report.[80] Raymond states that his figures for 1868 to 1870 represent estimates, which he himself thinks high. Raymond's figures for 1874 and 1875 are said to be obtained from a statement of the Blackhawk smelter, but the small plant of 1868–1870 could not have produced more than the enlarged plant of 1874–1875. For 1868 to 1870 Raymond estimates that the matte from the Blackhawk smelter contained an average of 40 per cent of copper. For 1875 Egleston[81] gives more authentic data and estimates the average content of the matte at 25 to 30 per cent of copper.

[65] Raymond, R. W., op. cit. for 1873, pp. 284 and following, 1874.
[66] Munson, G. C., in Kimball, J. P., op. cit. for 1887, pp. 151, 153, 1888.
[67] Emmons, S. F., The mines of Custer County, Colo.: U. S. Geol. Survey Seventeenth Ann. Rept., p. 420, 1896.
[68] Raymond, R. W., op. cit. for 1874, p. 358, 1875.
[69] Munson, G. C., in Kimball, J. P., op. cit. for 1887, pp. 151, 153, 1888.
[70] Emmons, S. F., op. cit.
[71] Raymond, R. W., op cit. for 1875, p. 282, 1877.
[72] Munson, G. C., in Kimball, J. P., op. cit. for 1886, p. 178, 1887; op. cit. for 1877 pp. 151, 153, 1888.
[73] Burchard, H. C., Report of the Director of the Mint upon the production of the precious metals in the United States during the calendar year 1880, pp.156–157, 1881; op. cit. for 1881, p. 354, 1882; op. cit. for 1882, pp. 394–395, 1883.
[74] Burchard, H. C., op. cit. for 1880, p. 156, 1881.
[75] Burchard, H. C., op. cit. for 1880, p. 157, 1881.
[76] Burchard, H. C., op. cit., for 1880, p. 135.
[77] Burchard, H. C., op. cit for 1881, p. 354, 1882; op. cit. for 1882, pp. 394–395, 1883; op. cit. for 1883, p. 240, 1884; op. cit. for 1884, p. 177, 1885.
[78] Kimball, J. P., op. cit. for 1885, p. 137, 1886.
[79] U. S. Geol. Survey Mineral Resources, 1882, p. 228, 1883. Also subsequent volumes.
[80] Raymond, R. W., op. cit. for 1870, p. 372, 1872.
[81] Egleston, Thomas, The Boston & Colorado smelting works: Am. Inst. Min. Eng. Trans., vol. 4, pp. 276–298, 1876.

For 1882–1885, inclusive, the figures are those given by Butler.[82]

For 1886 the figures are obtained by estimates made on the theory that in Colorado, where copper is a by-product, the rate of its production in 1882–1885 was at least maintained, in view of the fact that the rate of production of the other metals increased.

For 1887 to 1896, inclusive, the figures are those of Butler.

For 1896 to 1904, inclusive, the figures are those of the Colorado State Bureau of Mines, representing smelter receipts.

For 1905 to 1908, inclusive, the figures are those of the Colorado State Bureau of Mines and the United States Geological Survey Mineral Resources (mines reports) combined.

For 1909 to 1923, inclusive, the figures are those of the United States Geological Survey Mineral Resources (mines reports).

LEAD

For 1869 the figures are those for Clear Creek and Summit counties only. The lead for Clear Creek County is the estimated quantity and value of lead in ores shipped to the Eastern States, to England, and to Germany, none of which has heretofore been credited to Colorado. For Summit County the estimate is made from the quantity of lead-silver ore shipped.

For 1870 the figures represent ore shipped out of the State from Georgetown, Clear Creek County, and from Summit County.

For 1871 the figures represent ore shipped out of the State from Georgetown, Clear Creek County, and ore shipped to Swansea, England, from Park County.

For 1872 the figures represent shipments of lead ore from Georgetown and from Park and Summit counties.

For 1873 the figures represent shipments from Clear Creek, Gilpin, Park, and Summit counties. Raymond[83] gives the value of lead shipped in the form of pig lead as $28,000, and to this is added the value of the estimated content of lead in ore shipped out of the State from Georgetown.

For 1874 the figures represent ore shipped from Clear Creek and 216 tons of work lead (not pure lead) produced at Breckenridge, Summit County.

For 1875–1884 the figures are taken from Kirchhoff.[84]

For 1875 the figures show the estimated lead content of shipments out of Georgetown, Clear Creek County, and the estimated production of other counties as given by Raymond,[85] the total being made to equal the production of lead in Colorado as given in Mineral Resources.[86]

For 1876 and 1877 the total is that given by Kirchhoff in Mineral Resources.[87]

For 1878 the figures are the sum of county figures, nearly those given by Kirchhoff.[88]

For 1879–1884, inclusive, the figures are taken from Kirchhoff's reports.[89]

For 1885–1892 the figures are those given by Kirchhoff[90] in Mineral Resources, those for 1885 being correctly reduced 15 per cent.

For 1893 the figures are estimates by Henderson.

For 1894 to 1896 the figures are those given by Siebenthal.[91]

For 1886 to 1896 the figures represent the output of lead given by counties in the annual reports of agents of the Director of the Mint in reports of the Director of the Mint, prorated to correspond to total production of lead.

For 1897 to 1904 the figures represent the smelter receipts as given by the Colorado State Bureau of Mines.

For 1905 to 1908 the figures are those in the United States Geological Survey Mineral Resources (mines reports), combined with those of the Colorado Bureau of Mines.

For 1909 to 1915 the figures are those in the United States Geological Survey Mineral Resources (mines reports).

ZINC

For 1885–1891, inclusive, the figures represent estimates from statements of F. L. Bartlett, formerly manager of the American Zinc-Lead Co., at Canon City, who reports that between 1885 and 1891 he treated in Portland, Maine, about 3,000 tons of zinc-lead ores from Colorado.

For 1892 the figures represent the production of the American Zinc-Lead Co. at Canon City.

For 1893 and 1894 the figures are estimates by Henderson of the production of the American Zinc-Lead Co.

For 1895 to 1898, inclusive, the figures are taken from the report of the American Zinc-Lead Co. in the annual review number of the Denver Republican,[92] showing the beginning of zinc shipments to zinc plants in Kansas.

For 1899 the figures are those given in the Herald-Democrat, of Leadville, Colo., in its annual review number, January 1, 1900, showing the production of the American Zinc-Lead Co. and the shipments to Belgium and elsewhere.

[82] Butler, B. S., Copper: U. S. Geol. Survey Mineral Resources, 1910, pt. 1, pp. 172–173, 1911.

[83] Raymond, R. W., op. cit. for 1873, p. 284, 1874.

[84] Kirchhoff, Charles, Copper: U. S. Geol. Survey Mineral Resources, 1882, p. 310, 1883; op. cit. for 1883–84, pp. 416, 419, 1885; op. cit. for 1885, p. 250, 1886.

[85] Raymond, R. W., op. cit. for 1875, p. 282, 1877.

[86] U. S. Geol. Survey Mineral Resources, 1882, p. 310, 1883.

[87] U. S. Geol. Survey Mineral Resources, 1882, p. 310, 1883.

[88] Idem.

[89] U. S. Geol. Survey Mineral Resources, 1882, p. 310, 1883; op. cit. for 1883–84, pp. 416, 419, 1885; op. cit. for 1885, p. 250, 1886.

[90] U. S. Geol. Survey Mineral Resources, 1885, pp. 251, 257, 1886; op. cit. for 1886, pp. 144–146, 1887; op. cit. for 1887, pp. 105–107, 1888; op. cit. for 1888, pp. 87–88, 1890; op. cit. for 1892, p. 124, 1893.

[91] Siebenthal, C. E., U. S. Geol. Survey Mineral Resources, 1911, pt. 1, p. 319, 1912

[92] Also in Mineral Industry for 1898, pp. 724, 727, 1899.

For 1900 the figures represent the production of the American Zinc-Lead Co., as published in the annual review number of the Denver Republican, January 1, 1901, in the Leadville Herald-Democrat, and other publications. Conflicting statements are given as to the quantity of zinc-lead ore shipped from Leadville. Mineral Industry reports that 14,000 tons of zinc ore was shipped from Leadville in 1900. The zinc exports in 1900 from Galveston and New Orleans, including both the Joplin-Galena district and Colorado, amounted to 11,425 tons. H. A. Lee,[93] commissioner of mines for the State of Colorado, says that the production of zinc in Colorado in 1900 amounted to 77,984 tons of material, averaging 42 per cent of metallic zinc, equivalent to 65,506,650 pounds of zinc. The Leadville Herald-Democrat gives for Leadville only 45,270,920 pounds, 41,948,200 pounds of which was shipped to outside smelters. The figures of the commissioner seem high and have not been incorporated in the reports of the Colorado State Bureau of Mines. Later information would seem to show that a great part of the 77,984 tons represented ore concentrated, not concentrates shipped. Further, the production in 1901 (which was probably larger than in 1900), as given by both the Leadville Herald-Democrat and Lee, was only 13,427 tons of zinc, which appears not unreasonable in view of the exports of zinc ore amounting to 13,294 tons (of possibly 42 per cent grade) from Gulf ports in 1901 and the increased shipments to domestic smelters. Lee says that 70 per cent of the output in 1901 was exported and 30 per cent was treated by domestic smelters, this percentage including ore used for making zinc oxide at Canon City and possibly some ore shipped to Mineral Point, Wis., for making zinc white. The figures given for 1900 in this report are 3,682,055 pounds produced from Colorado ores by the American Zinc-Lead smelter at Canon City, and 12,600,000 pounds in ore shipped, the latter item represented by 15,000 tons of ore, averaging 42 per cent of zinc.

For 1901 the commissioner of mines for Colorado gives the production of zinc from the ores of the State as 13,427 short tons, or 26,854,000 pounds, determined from the metallic content of the ores. The sum of the detailed figures by counties is 26,843,731 pounds. The Leadville Herald-Democrat gives for 1901 a production of 21,476,000 pounds of zinc for Leadville alone. In 1901 the American Zinc-Lead Smelting Co. produced 5,712,323 pounds of zinc.

For 1902 to 1903 the figures are those given by the Colorado State Bureau of Mines, representing smelter receipts, from which loss in smelting has been deducted.

For 1904 to 1907, inclusive, the figures are those of the Colorado State Bureau of Mines and the United States Geological Survey Mineral Resources (mines reports) combined, both representing smelter receipts

or mine shipments, from which loss in smelting has been deducted.

For 1908 to 1923, inclusive, the figures are those of the United States Geological Survey Mineral Resources (mines reports).

CONTENT OF ORE AND CONCENTRATES

The following tables show the classification and content of gold, silver, copper, lead, and zinc in the ore sold or treated in Colorado from 1909 to 1923 in terms of recovered metals.[93a]

Ore treated at gold and silver mills and concentration mills and quantity of gold and silver contained in bullion produced in Colorado from 1909 to 1923, by years

Year	Ore to gold and silver mills			Ore to concentrating mills
	Ore	Gold in bullion	Silver in bullion	
	Short tons	*Fine ounces*	*Fine ounces*	*Short tons*
1909	1,320,111	679,882.55	289,782	357,658
1910	1,455,983	a 627,879.83	396,381	464,673
1911	1,373,879	b 598,820.48	b 401,603	465,283
1912	1,435,837	c 544,750.93	c 360,490	523,063
1913	1,610,335	551,722.35	443,444	459,533
1914	1,630,640	612,143.58	676,231	378,743
1915	1,611,335	662,304.68	608,072	557,267
1916	1,442,139	581,177.76	536,755	702,200
1917	1,487,304	553,553.59	400,109	699,094
1918	1,345,451	455,276.18	345,280	555,262
1919	1,195,986	346,035.56	242,269	439,373
1920	626,900	223,362.80	138,307	677,113
1921	884,763	218,732.83	15,571	206,270
1922	849,261	217,548.64	41,405	302,232
1923	817,328	221,022.18	44,257	549,522

a In addition 13,051.46 ounces of gold in Teller County from old tailings re-treated.
b In addition 6,116.80 ounces of gold and 3,275 ounces of silver in Teller County from old tailings re-treated.
c In addition 7,123.30 ounces of gold and 5,594 ounces of silver in Teller County from old tailings re-treated.

Mine production of gold, silver, copper, lead, and zinc from concentrates produced in Colorado from 1909 to 1923, by years

Year	Concentrates	Gold	Silver	Copper	Lead	Zinc
	Short tons	*Fine ounces*	*Fine ounces*	*Pounds*	*Pounds*	*Pounds*
1909	154,091	107,355.64	2,453,274	2,529,378	36,851,766	27,036,073
1910	237,342	122,162.99	2,192,315	2,419,045	46,560,748	57,683,333
1911	201,010	109,817.43	1,967,545	2,611,779	44,863,551	41,262,830
1912	224,722	133,828.61	2,415,514	2,657,882	46,092,955	46,053,954
1913	218,291	123,177.64	2,320,781	2,366,458	47,027,901	41,692,027
1914	188,770	116,263,06	2,058,557	1,914,797	35,583,330	31,419,916
1915	233,965	128,015.39	2,214,689	2,270,790	43,565,115	53,954,312
1916	288,211	106,596.20	2,409,006	2,348,220	43,939,994	78,731,912
1917	280,563	87,650.41	2,600,743	2,686,546	41,122,473	76,381,990
1918	226,292	81,395.40	2,330,637	2,514,564	34,010,088	65,914,868
1919	100,775	69,258.60	2,123,721	1,626,272	19,014,621	20,704,184
1920	119,233	72,306.27	2,406,336	2,316,572	29,043,127	28,422,325
1921	49,140	67,432.37	2,568,500	972,552	10,424,411	217,000
1922	60,290	42,667.42	3,136,236	767,083	12,227,439	1,100,000
1923	150,566	62,780.92	3,222,670	2,911,624	32,777,209	41,164,400

Mine production of gold, silver, copper, lead, and zinc from crude ore produced in Colorado and shipped to smelters from 1909 to 1923, by years

Year	Crude ore	Gold	Silver	Copper	Lead	Zinc
	Short tons	*Fine ounces*	*Fine ounces*	*Pounds*	*Pounds*	*Pounds*
1909	541,972	254,316.39	6,153,777	8,392,668	35,316,024	24,174,187
1910	514,008	209,734.19	5,915,292	5,940,262	29,498,027	19,406,315
1911	538,774	188,997.49	4,954,076	5,412,709	24,815,738	53,344,626
1912	617,726	193,013.43	5,425,366	4,449,421	29,149,312	86,168,858
1913	664,998	180,193.95	6,556,467	4,861,368	40,869,872	77,654,402
1914	668,143	202,364.40	6,053,956	4,724,376	38,628,568	65,355,044
1915	568,418	260,463.98	4,196,542	4,841,747	25,245,482	50,640,682
1916	552,904	204,304.42	4,701,696	6,275,861	26,974,093	55,553,551
1917	502,308	87,719.98	4,295,199	5,435,458	26,867,539	43,933,785
1918	414,177	54,737.76	4,380,987	3,762,768	31,950,672	23,219,033
1919	284,409	45,337.98	3,384,877	1,933,935	18,055,620	16,516,309
1920	267,280	45,942.16	2,857,976	1,727,162	17,586,661	20,368,417
1921	190,348	27,821.84	3,006,144	3,180,890	9,236,055	2,143,000
1922	260,607	30,857,09	2,674,042	2,606,371	11,249,761	22,158,000
1923	202,250	17,437.70	2,063,439	1,336,485	12,920,976	12,987,600

93 Eng. and Min. Jour., vol. 71, p. 490, 1901.

93a For explanation see footnote to Table 1, p. 69.

Content of ore sold or treated in Colorado, 1910–1923, in terms of recovered metals

Dry and siliceous ore

Year	Ore	Gold	Silver	Copper	Lead	Zinc
	Short tons	Fine ounces	Fine ounces	Pounds	Pounds	Pounds
1910	1,957,379	943,220.53	6,117,719	6,044,125	25,898,094	4,496,219
1911	1,874,103	866,252.76	5,096,026	5,614,076	18,724,095	4,994,219
1912	1,966,300	841,295.84	5,397,439	4,196,096	20,162,526	4,554,875
1913	2,161,458	817,065.19	6,270,758	4,110,475	24,056,425	6,686,586
1914	2,157,762	893,819.64	6,584,493	4,395,772	19,150,122	2,948,759
1915	2,183,431	1,012,897.34	5,195,593	4,370,439	21,811,059	6,219,566
1916	2,057,452	847,876.76	5,562,217	4,943,797	23,335,444	7,399,120
1917	1,996,590	692,605.92	4,867,628	3,630,445	18,673,295	3,098,598
1918	1,760,222	566,274.64	4,753,627	3,590,901	16,598,612	2,249,506
1919:						
Dry gold	1,111,640	396,354.31	724,260	905,068	4,113,337	....
Dry gold and silver	111,961	21,984,11	527,072	266,154	3,181,276	....
Dry silver	443,475	10,958.46	2,979,721	510,108	7,593,286	....
1920:						
Dry gold	505,885	246,061.02	187,181	358,034	715,357	77,635
Dry gold and silver	366,297	62,173.38	952,257	1,055,118	8,981,473	....
Dry silver	351,899	6,893.30	2,497,880	831,728	7,002,932	....
1921:						
Dry gold	495,853	221,292.06	81,157	54,359	119,140	....
Dry gold and silver	269,087	53,991.30	589,284	891,660	3,409,026	....
Dry silver	465,232	31,733.60	3,991,193	1,882,519	10,561,401	....
1922:						
Dry gold	449,658	217,606.90	73,191	57,478	176,202	....
Dry gold and silver	397,191	58,192.31	1,756,858	1,053,975	7,223,125	....
Dry silver	449,652	10,421.10	3,418,709	1,917,436	7,162,729	....
1923:						
Dry gold	410,468	209,867.04	56,837	30,152	113,264	....
Dry gold and silver	480,340	67,260.95	1,567,091	1,375,938	10,202,564	....
Dry silver	275,662	6,247.87	2,330,396	1,386,215	8,382,956	....
	24,698,997	9,102,346.33	71,578,587	53,472,068	267,347,740	42,725,083

Lead ore

Year	Ore	Gold	Silver	Copper	Lead	Zinc
1910	133,323	11,719.25	1,050,611	434,630	23,914,921	....
1911	158,984	17,765.42	1,265,594	492,422	26,466,369	....
1912	182,745	19,766.81	1,470,930	539,216	29,640,633	....
1913	205,774	26,845.88	1,645,186	566,401	39,397,770	....
1914	193,087	30,525.24	1,532,943	746,212	41,239,726	....
1915	142,969	26,068.24	964,224	338,041	27,482,225	....
1916	159,247	22,687.40	1,159,615	356,620	27,427,529	....
1917	103,081	19,220.68	1,167,167	784,684	20,688,207	....
1918	183,274	17,036.60	1,456,023	600,038	33,155,889	....
1919	74,918	10,205.10	1,068,360	427,618	15,085,420	....
1920	61,254	11,459.90	841,783	235,177	11,892,212	....
1921	18,289	5,089.73	524,023	81,260	4,851,872	....
1922	48,381	3,906.13	493,697	141,082	7,062,288	....
1923	90,410	3,339.22	625,974	143,791	10,675,407	....
	1,755,736	225,635.60	15,266,130	5,887,192	318,980,468	....

Copper-lead ore

Year	Ore	Gold	Silver	Copper	Lead	Zinc
1910	2,219	181.96	188,146	148,049	774,433	....
1911	1,398	882.05	90,311	93,025	326,521	....
1912	6,810	1,936.59	389,449	531,227	1,260,482	....
1913	6,417	1,709.99	491,058	640,795	812,071	....
1914	1,192	288.91	82,119	97,672	371,161	....
1915	2,674	2,098.50	98,210	212,678	641,982	....
1916	3,129	1,724.04	110,531	299,239	449,961	....
1917	2,976	1,239.89	133,128	204,563	428,115	....
1918	1,760	198.17	35,260	138,132	309,041	....
1919	1,690	339.40	34,749	193,197	342,847	....
1920	453	173.47	7,566	30,865	184,011	....
1921	820	24.60	40,949	78,448	184,221	....
1922	627	84.47	37,418	60,299	162,646	....
1923	380	122.57	16,680	23,502	76,492	....
	32,545	11,004.61	1,755,574	2,751,691	6,323,984	....

Copper ore

Year	Ore	Gold	Silver	Copper	Lead	Zinc
1910	15,078	8,961.27	436,358	1,489,664	144,796	....
1911	18,221	8,472.08	342,550	1,652,247	67,700	....
1912	13,718	8,759.86	224,327	1,622,605	102,321	....
1913	16,555	7,615.92	223,108	1,701,666	173,746	....
1914	12,196	3,243.88	173,845	1,330,056	96,930	....
1915	23,573	6,204.17	204,317	2,169,383	244,667	....
1916	34,429	15,220.87	199,824	2,977,285	718,569	....
1917	33,233	7,455.00	182,791	3,133,437	455,698	....
1918	18,088	1,968.68	166,296	1,668,087	169,957	....
1919	9,417	4,992.49	104,381	774,435	58,466	....
1920	3,835	2,150.54	100,643	383,427	41,937	....
1921	22,195	1,811.24	382,521	1,164,248	10,522	....
1922	863	556.47	31,463	105,864	18,575	....
1923	4,655	2,139.88	110,552	302,685	71,194	....
	226,056	79,552.35	2,882,976	20,475,089	2,375,078	....

Content of ore sold or treated in Colorado, 1910–1923, in terms of recovered metals—Continued

Zinc ore

Year	Ore	Gold	Silver	Copper	Lead	Zinc
	Short tons	*Fine ounces*	*Fine ounces*	*Pounds*	*Pounds*	*Pounds*
1910	82,251	591.30	32,310	21,710	1,733,700	26,987,488
1911	110,845	164.54	55,969	11,883	407,007	51,388,865
1912	177,946	532.51	130,392	67,012	1,235,019	84,989,652
1913	141,295	73.92	21,950	1,556	140,426	66,305,374
1914	145,656	2.24	1,046	435	12,491	56,294,706
1915	100,222	28.01	24,802	412	279,154	39,168,198
1916	151,497	330.62	58,722	6,349	589,991	52,873,099
1917	140,455	522.63	19,525	20,496	1,000,413	47,582,031
1918	60,350	1,098.26	75,568	11,490	1,723,508	22,166,061
1919	50,547	80.40	28,583	....	857,754	16,141,424
1920	96,232	554.91	88,837	40,004	1,705,979	33,456,257
1921	5,727	13.91	3,237	....	67,564	1,877,000
1922	43,615	1.20	291	....	11,500	15,731,000
1923	93,000	296.65	51,217	18,983	657,247	33,130,600
	1,399,638	4,291.10	592,449	200,330	10,421,753	548,091,755

Lead-zinc ore

Year	Ore	Gold	Silver	Copper	Lead	Zinc
1910	244,414	8,224.00	679,500	221,357	23,605,329	45,605,941
1911	214,385	10,215.35	476,049	160,835	23,687,597	38,224,372
1912	229,107	6,424.64	594,427	151,147	22,841,286	42,678,285
1913	203,367	4,783.04	668,632	206,933	23,317,335	46,354,469
1914	167,633	2,891.13	414,298	69,026	13,341,468	37,531,495
1915	284,151	3,487.79	532,157	21,584	18,351,510	59,207,230
1916	291,489	4,238.70	556,446	40,791	18,392,593	74,013,244
1917	412,371	7,879.86	925,812	348,379	26,744,284	69,635,146
1918	291,196	4,832.99	570,130	268,684	14,003,753	64,718,334
1919	116,120	6,717.87	283,741	483,627	5,837,855	21,079,069
1920	185,438	12,144.71	726,472	1,109,381	16,105,887	15,256,850
1921	4,178	30.60	14,851	948	456,720	483,000
1922	22,113	304.57	40,056	37,320	1,660,135	7,527,000
1923	214,185	11,966.62	571,619	966,843	15,519,061	21,021,400
	2,880,147	84,141.87	7,054,190	4,086,855	223,864,813	543,335,835

SUMMARY OF FEATURES OF PRODUCTION SHOWN BY CURVES

GOLD

[Figures 2 and 3]

1859–1863. Rapid rise in the production of gold due to placer mining and the treatment of material from the oxidized and decomposed portions of veins by amalgamation. Production principally in Clear Creek, Gilpin, Park, Lake, and Summit counties.

1863–1867. Decline in production due to the exhaustion of sluicing ground and of the oxidized or free-milling parts of veins.

1859–1867. The only period in the history of mining in Colorado when the value of placer gold produced exceeded that of the lode gold.

1869–1871. Rise in production due to increase in lode mining in Gilpin County.

1872–1876. Decline in the production of Gilpin County.

1869–1888. The part of the curve for this period represents practically the history of lode mining in Gilpin County.

1889. San Juan region began to produce gold on a large scale.

1890. Gold discovered at Cripple Creek.

1892–1900. Remarkable rise in the production of gold in Colorado due to the rapid growth of mining in Cripple Creek.

1900. Peak of production in the Cripple Creek district and in the State.

1902. Rise in production due to the increased output from San Juan region.

1903. Decrease in production of the Cripple Creek district due to labor strike.

1903–1919. This part of the curve represents the history of mining in the Cripple Creek district.

1915. Rise in production due to the discovery of high-grade ore in the Cresson mine, at Cripple Creek.

1902–1922. General downward trend of the production of gold in all districts in Colorado.

1923. Slight rise in production due to recovery at Cripple Creek.

SILVER

[Figures 4 and 5]

1859–1867. Small output of silver, recovered in connection with gold-placer mining in Clear Creek and Gilpin counties.

1868–1871. Production of silver by lode mining in Clear Creek, Gilpin, and Boulder counties.

1871. Park County began to produce silver in considerable quantity.

1875. Production of silver in Summit County began to increase.

1877. Leadville, Lake County, started to produce silver in large quantities.

1877–1880. Rise due to the rapid development of silver mining in Lake County.

1880–1887. Decline in production of silver in Lake County.

1887–1892. Rapid development of silver mining at Aspen, Pitkin County, which reached the peak of its production in 1892.

1888–1893. Increase in production from the San Juan region. Creede began producing in 1891 and reached its peak in 1893.

1893. Peak of production for the State, due principally to the peak of production in the San Juan region. Influenced largely by the output of 5,000,000 ounces at Creede. Peak of silver production at Leadville was not reached until 1895.

1894. Decline in production in Pitkin County and the San Juan region.

1895. Increase in production at Leadville.

1895–1897. Decrease in production in Lake, Pitkin, Boulder, Clear Creek, and Gilpin counties. Increase in production in San Juan region.

1898. Increase due to production in Lake County and the San Juan region.

1898–1923. General decline in the production of silver in all districts in Colorado.

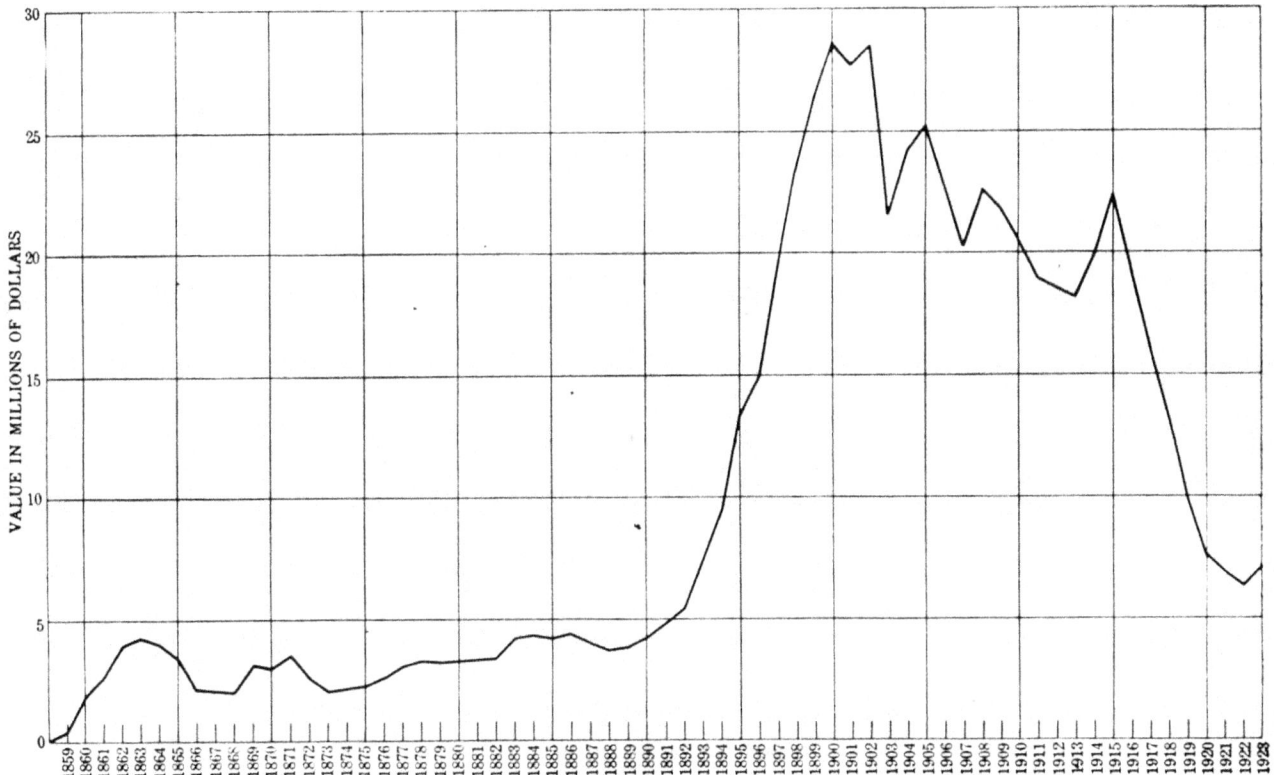

FIGURE 2.—Value in dollars of gold produced in Colorado from 1859 to 1923, by years

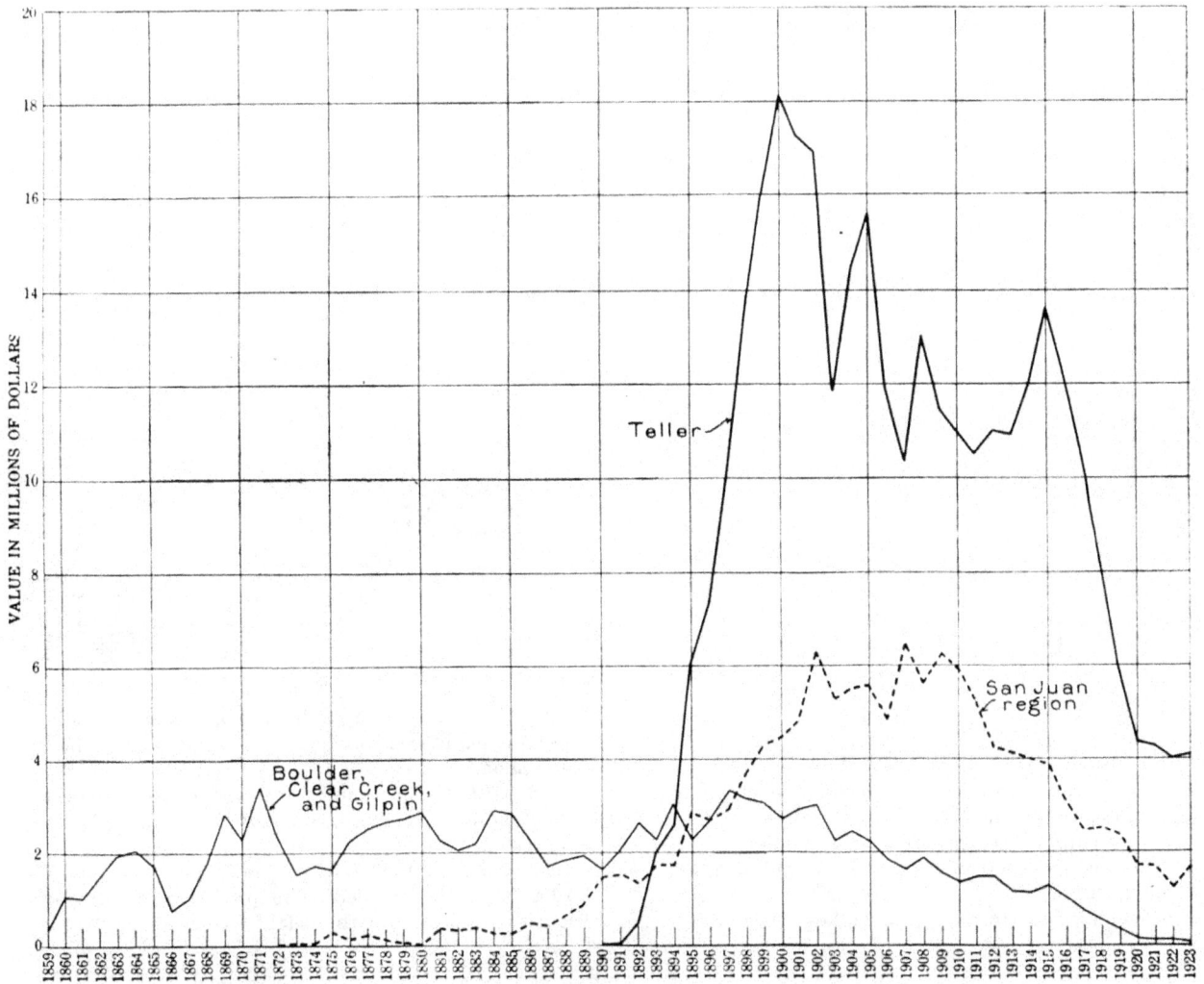

FIGURE 3.—Value of gold produced by principal counties or regions in Colorado from 1859 to 1923, by years

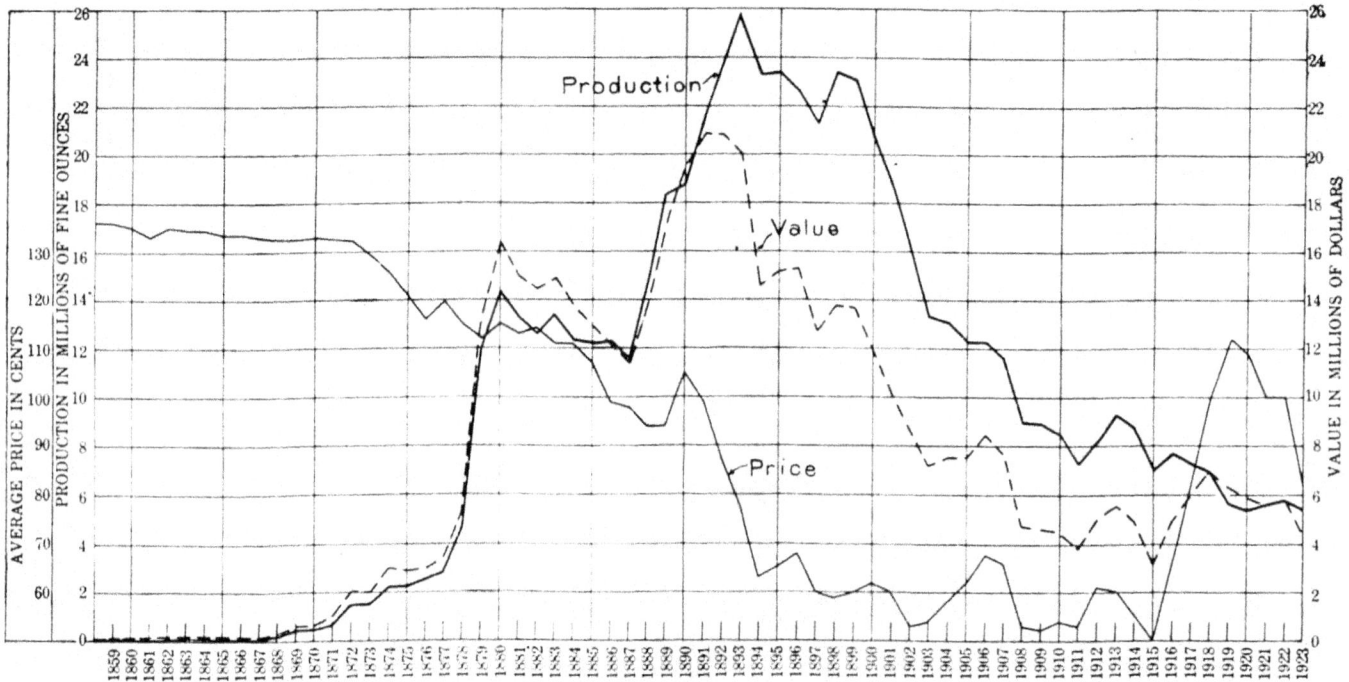

FIGURE 4.—Quantity and value of silver produced in Colorado from 1859 to 1923, by years, and average price for each year

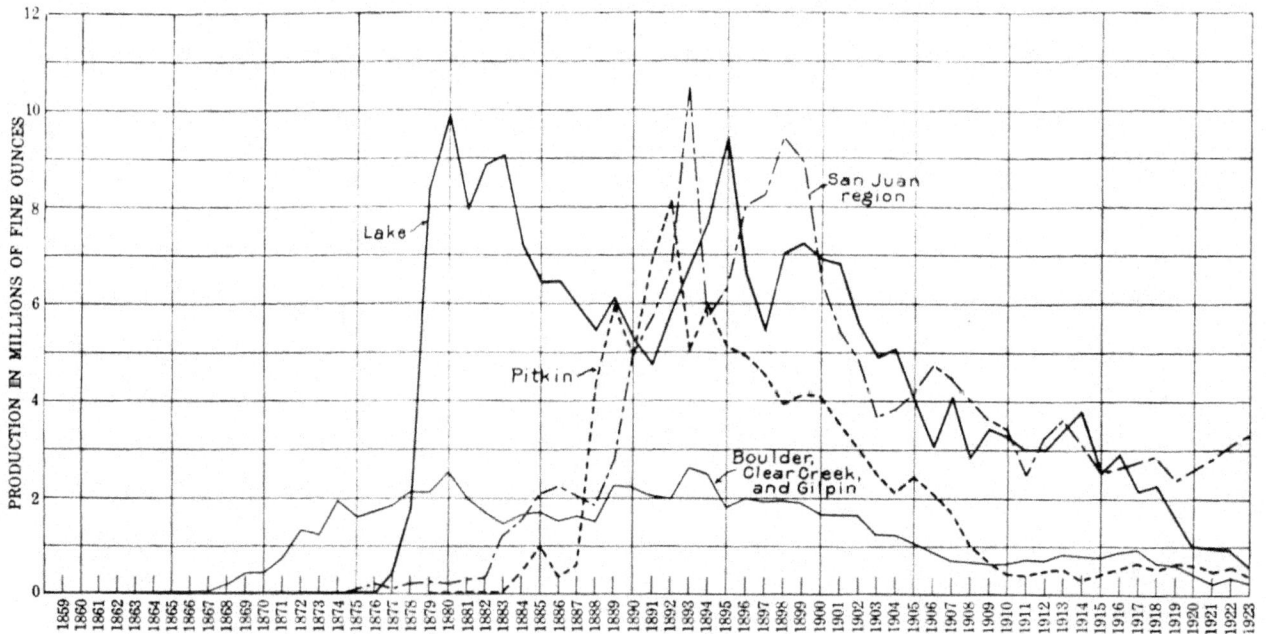

FIGURE 5.—Production of silver by principal counties or regions in Colorado from 1859 to 1923, by years

COPPER

[Figures 6, 7]

1868–1873. Copper recovered as a by-product from the complex ores of Gilpin and Clear Creek counties.

1873. First output of copper from Park County.

1876. Beginning of production of copper from the ores of the San Juan region.

1889–1892. Rise due to production at Leadville.

1892–1895. Decline in production of copper at Leadville.

1896–1898. Rise in production at Leadville and in the San Juan region.

1898–1919. Fluctuations in the production of the State due to fluctuations in the production at Leadville and in the San Juan region.

1909. Peak of production for the State.

1909–1923. General decline in production of copper in all districts in Colorado, with a slight recovery in 1923.

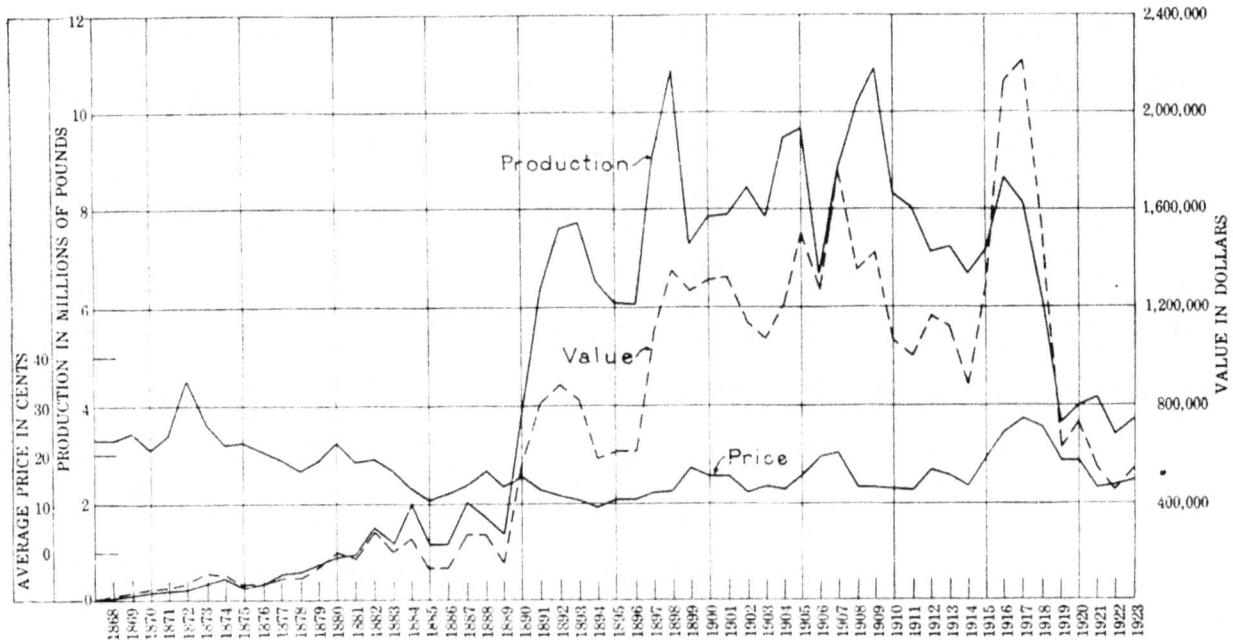

FIGURE 6 — Quantity and value of copper produced in Colorado from 1868 to 1923, by years, and average price for each year

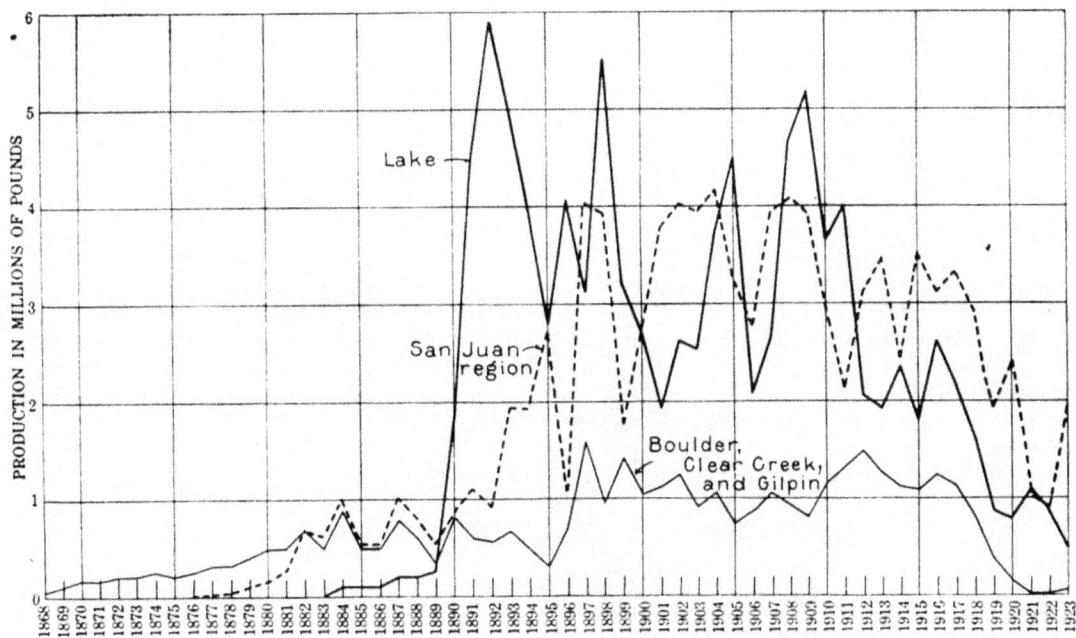

FIGURE 7.—Production of copper by principal counties or regions in Colorado from 1868 to 1923, by years

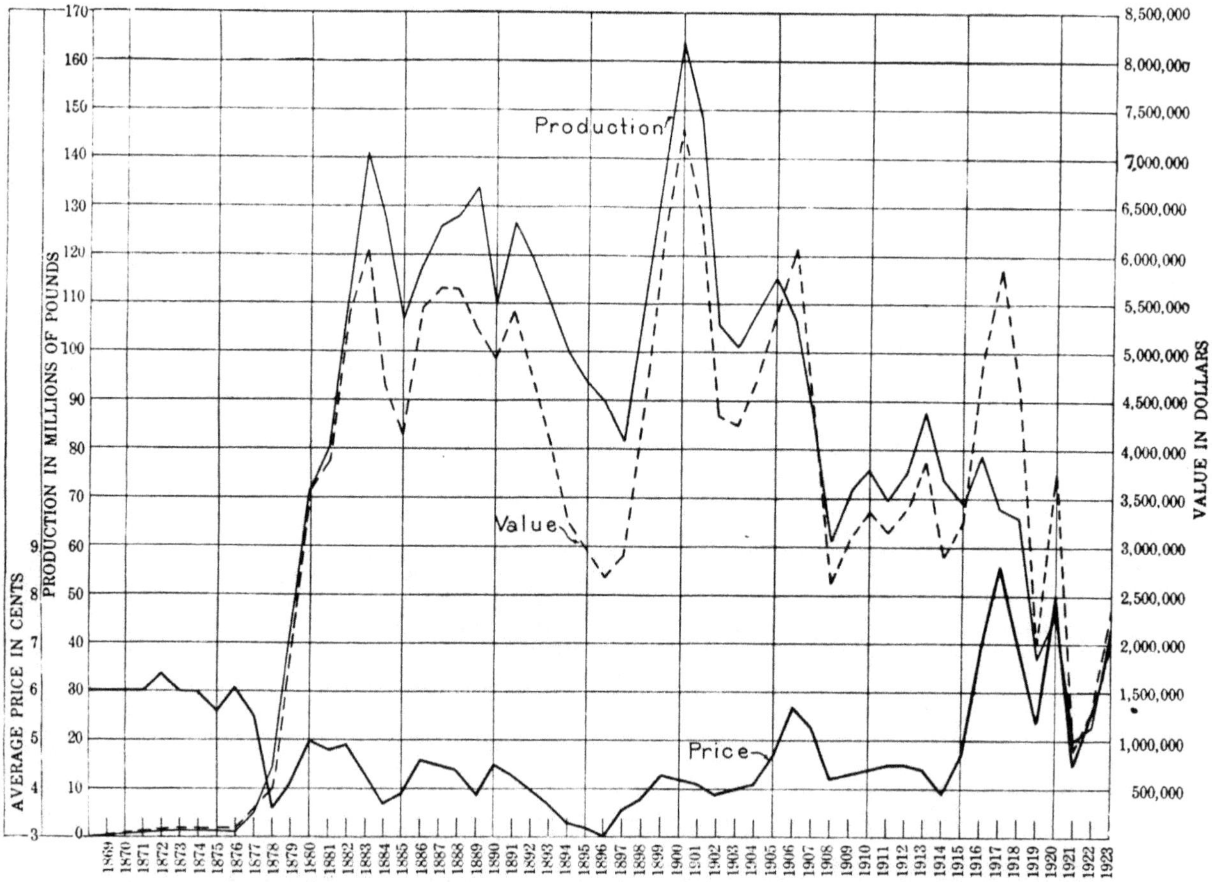

FIGURE 8.—Quantity and value of lead produced in Colorado from 1869 to 1923, by years, and average price for each year

LEAD

[Figures 8, 9]

1869–1871. Production in Clear Creek and Summit counties.

1872. Park County began to produce lead.

1873. Gilpin County began to produce lead.

1877–1883. Rapid rise due to the production of lead carbonates at Leadville.

1882. The San Juan region began to produce lead in quantity.

1883. Peak of production at Leadville.

1888. Production of lead at Aspen began in quantity.

1889–1897. Decline in production at Leadville.

1897–1900. Increase in production at Leadville and Aspen and in the San Juan region.

1900–1903. Decline in production at Leadville and in the San Juan region.

1903–1905. Increase in production at Leadville and in the San Juan region.

1905–1922. General decline in the production of lead in all districts in Colorado.

1923. Appreciable rise in production of lead, chiefly from the San Juan region.

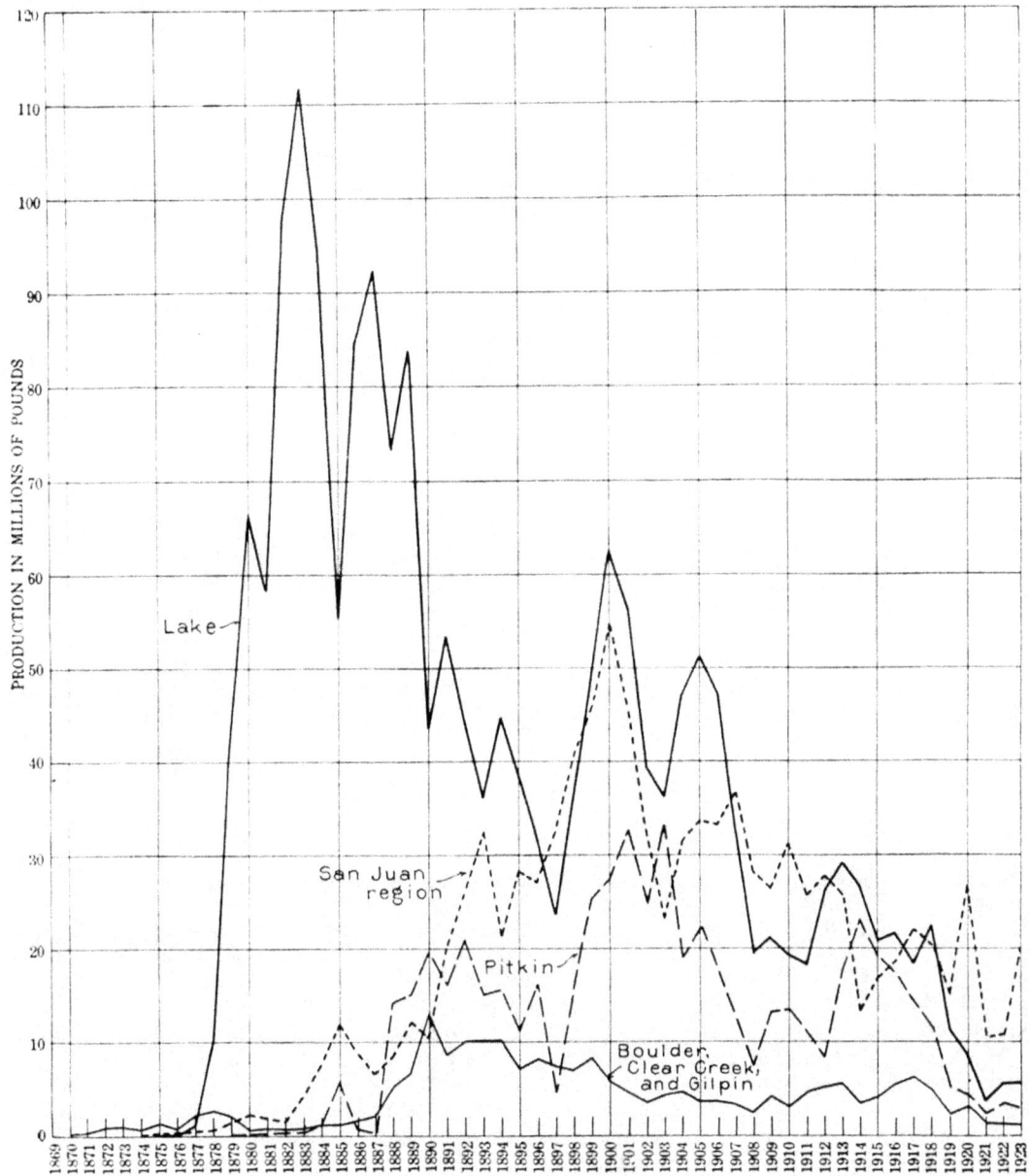

FIGURE 9.—Production of lead by principal counties or regions in Colorado from 1869 to 1923, by years

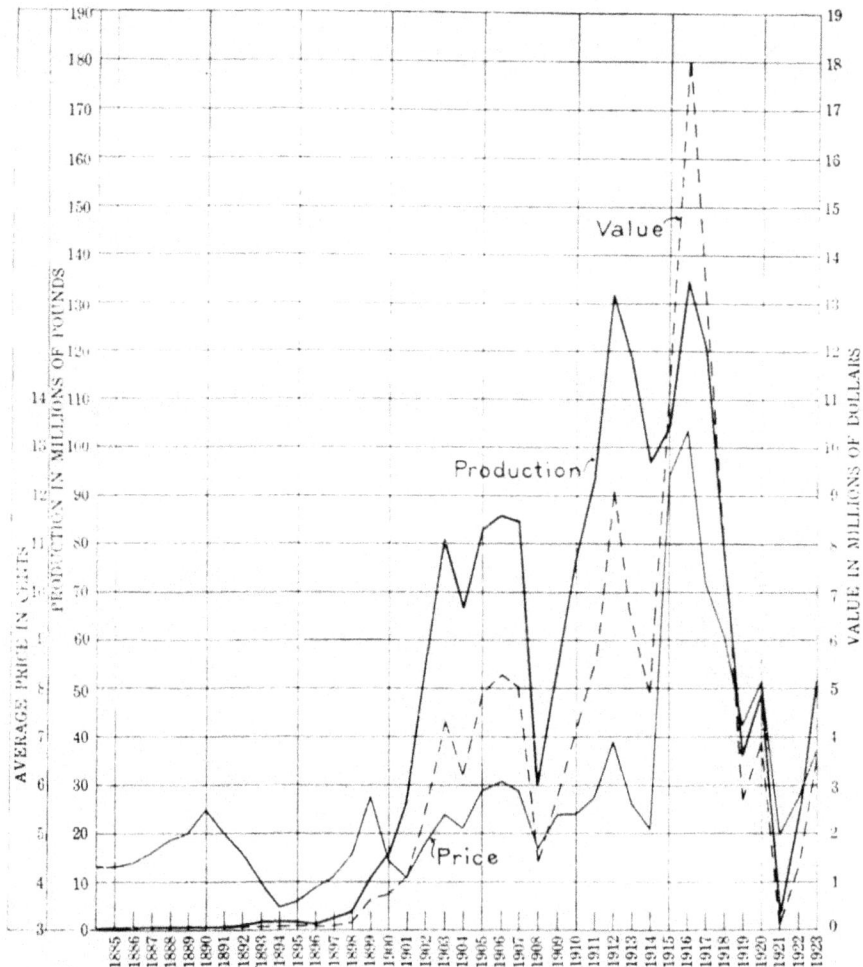

FIGURE 10.—Quantity and value of zinc produced in Colorado from 1885 to 1923, by years and average price for each year

ZINC

[Figures 10, 11]

1885–1891. Small production of zinc from Clear Creek, Lake, and Summit counties.

1890–1891. Erection of a zinc oxide plant at Canon City caused a slight increase in the production of zinc at Leadville.

1891–1907. Steady increase in production of zinc due to the development and exploitation of zinc sulphide ores at Leadville.

1892. Chaffee County began to produce zinc.

1904. Eagle County began to produce zinc.

1908. Decline in production due to the low price of zinc.

1908–1910. Increase in production at Leadville and in Summit and Eagle counties.

1910. Discovery of zinc carbonate ores at Leadville.

1910–1912. Rapid strides made in the production of zinc carbonates at Leadville brought production of the State up to the peak of 1912.

1912–1914. Decline in production at Leadville and in Summit County due to low price of zinc.

1912–1919. Steady decline in production at Leadville.

1914–1916. Rise in production in Eagle and Summit counties.

1916–1919. Decline in production in Eagle and Summit counties.

1920. Increase in Eagle and San Juan counties.

1921. Almost complete cessation of zinc mining.

1922–1923. Resumption of zinc mining late in 1922, carried through 1923, in Eagle, Lake, Summit, Gunnison, and San Juan counties.

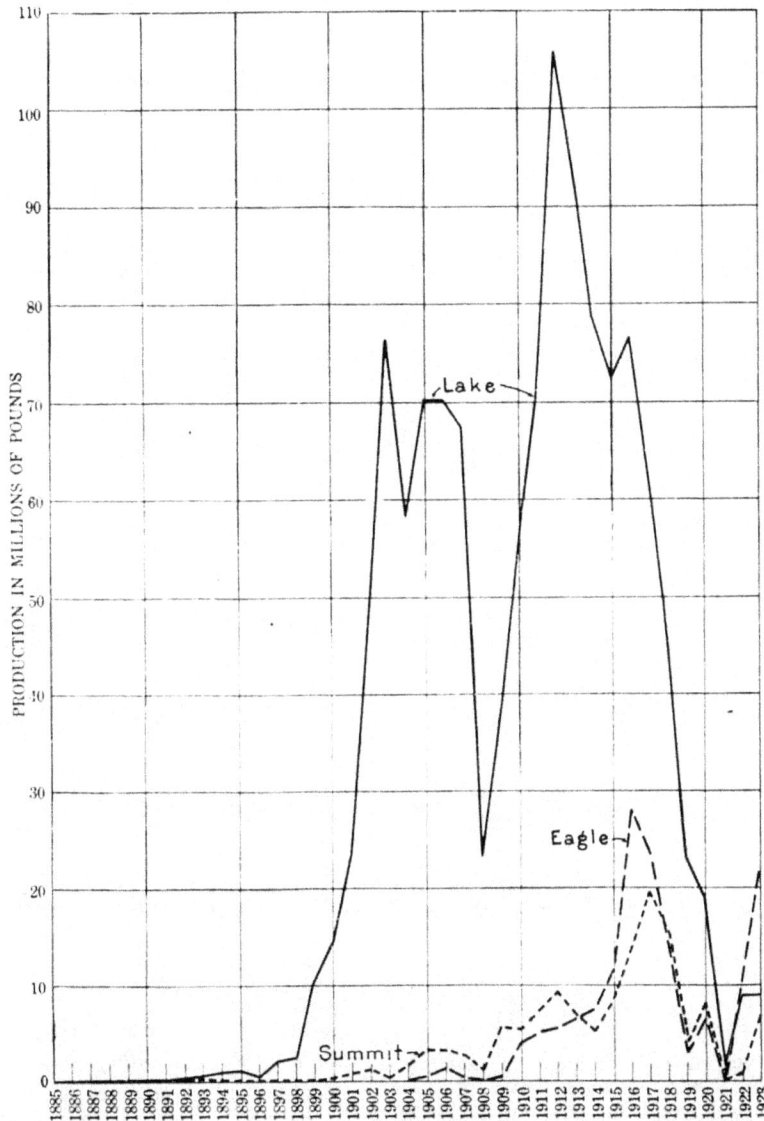

FIGURE 11.—Production of zinc by principal counties or regions in Colorado from 1885 to 1923, by years

TOTAL VALUES OF GOLD, SILVER, COPPER, LEAD, AND ZINC

[Figures 12, 13]

1859-1868. Production of gold.
1868-1876. Production of gold and silver.
1876-1880. Rise in production of silver and lead.
1880-1888. Increase in production of lead; decrease in production of silver.

1888-1893. Increase in production of gold and silver; decrease in production of lead.
1897. Increase in production of silver.
1897-1900. Increase in production of gold, lead, and zinc.
1900. Peak of production of lead and gold.
1900-1921. General decline in production of all metals except zinc; production of zinc fluctuating.
1922-1923. Slight increase owing to increases in lead and zinc.

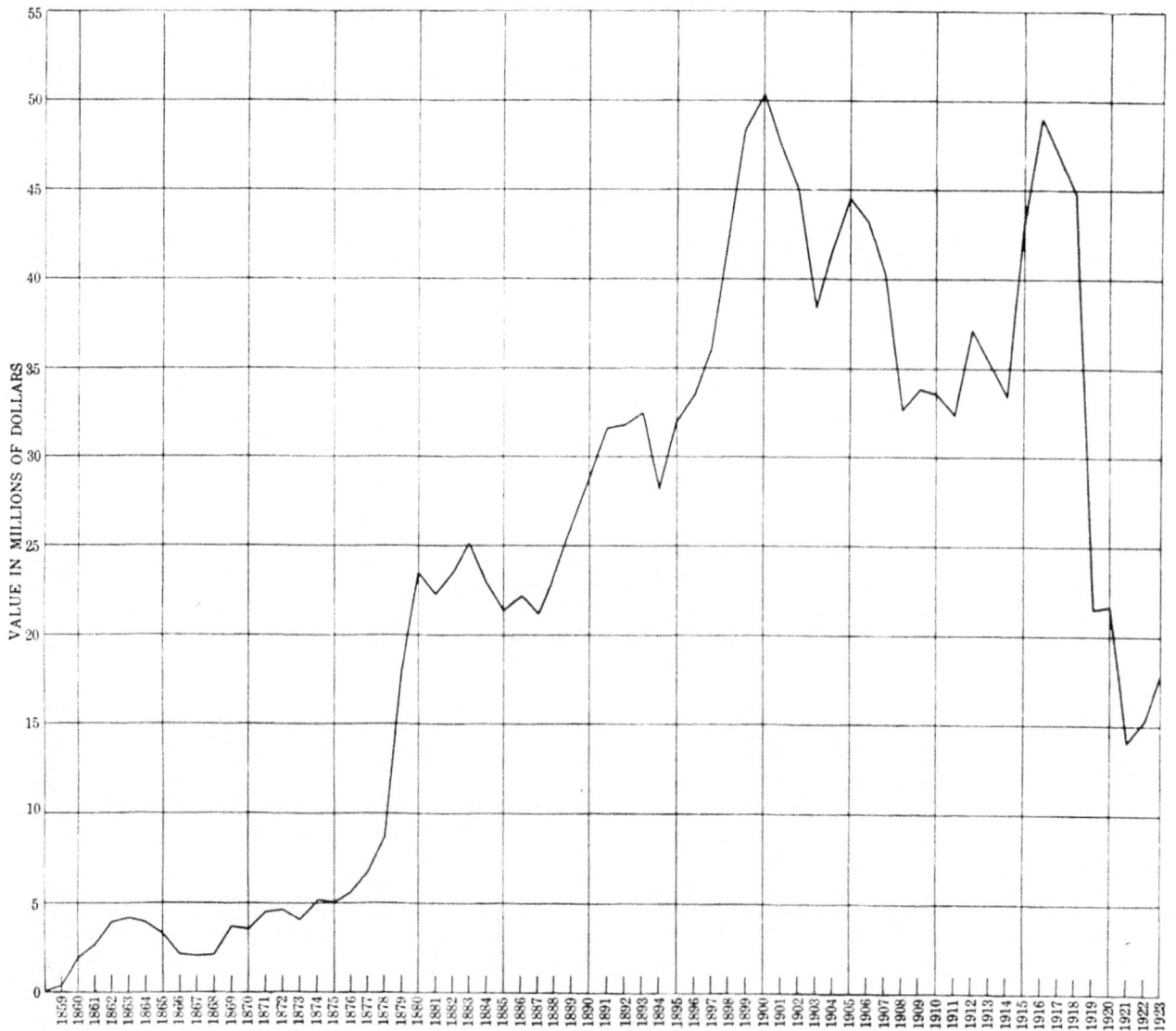

FIGURE 12.—Total value of gold, silver, copper, lead, and zinc produced in Colorado from 1859 to 1923, by years

CURVES OF PRODUCTION OF GOLD, SILVER, COPPER, LEAD, AND ZINC

[Figure 14]

A comparison of the curves shows, in general, a lack of parallelism which indicates the lack of dependence of the production of one metal upon the production of any other metal.

These curves also illustrate the relative insignificance of the production of copper in Colorado.

PRODUCTION BY YEARS AND COUNTIES

[Figures 15-20]

The value of the gold, silver, copper, lead, and zinc produced in Colorado from 1858 to 1922, inclusive, except that of the small unknown output made during the last few months of 1858, is $1,512,530,393, of which $666,470,261, or 44 per cent, was gold; $497,359,666, or 33 per cent, was silver; $39,703,094, or 3 per cent, was copper; $186,463,306, or 12 per cent, was lead; and $122,534,066, or 8 per cent, was zinc. The production in 1923, which is shown in the table

on page 103, made no appreciable change in the percentages.

In 66 years, therefore, the average production per year was $22,917,127, which according to the percentages stated above would amount annually to $10,083,536 gold; $7,562,652 silver; $687,514 copper; $2,750,055 lead; and $1,833,370 zinc. Zinc, however, was produced only in small quantities prior to 1892 and not in considerable quantity until 1897. Its production as spelter began in 1899.

In 1900 the value of the metallic output of Colorado reached its maximum—$50,614,424—of which gold amounted to $28,762,036; silver, $12,608,637; copper, $1,299,251; lead, $7,228,090; and zinc, $716,410. The output of gold in that year was the greatest ever made. The output of silver having the greatest value was made in 1891, when 21,160,000 ounces was produced, worth 99 cents an ounce, having a total value of $20,948,401. In 1893 the quantity of silver produced was the largest—25,838,600 ounces, worth

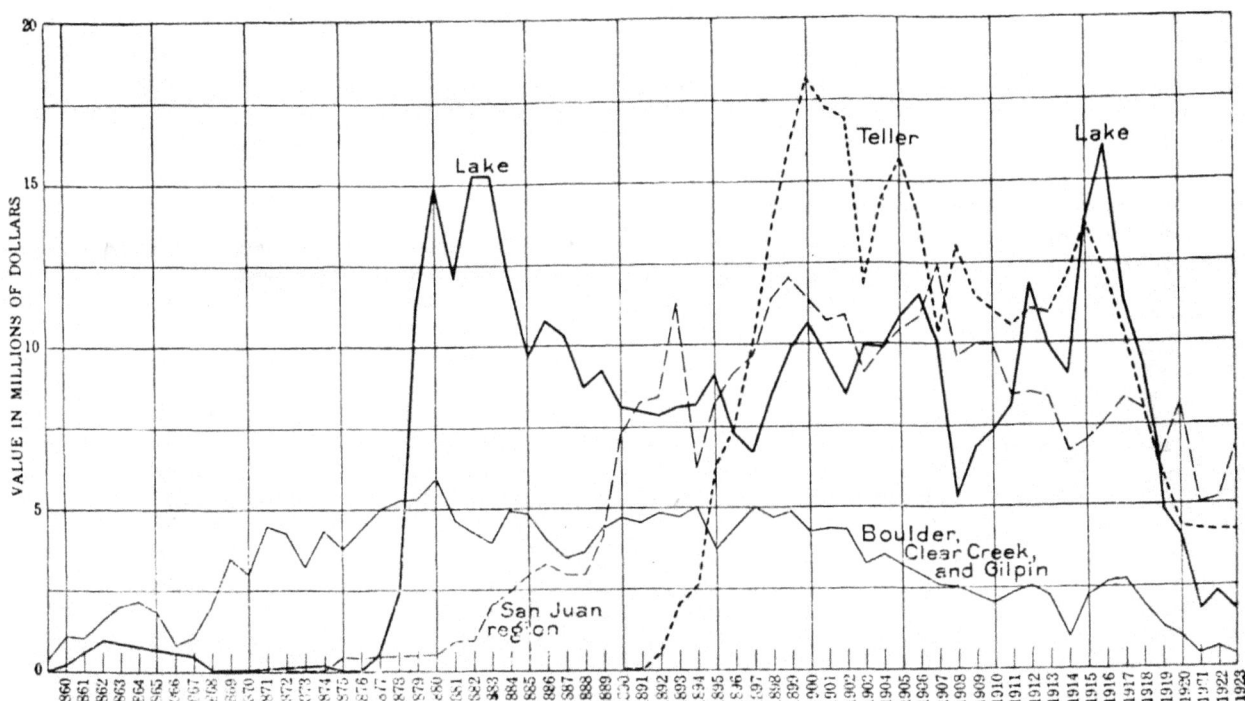

FIGURE 13.—Total value of gold, silver, copper, lead, and zinc produced by principal counties or regions in Colorado from 1859 to 1923, by years

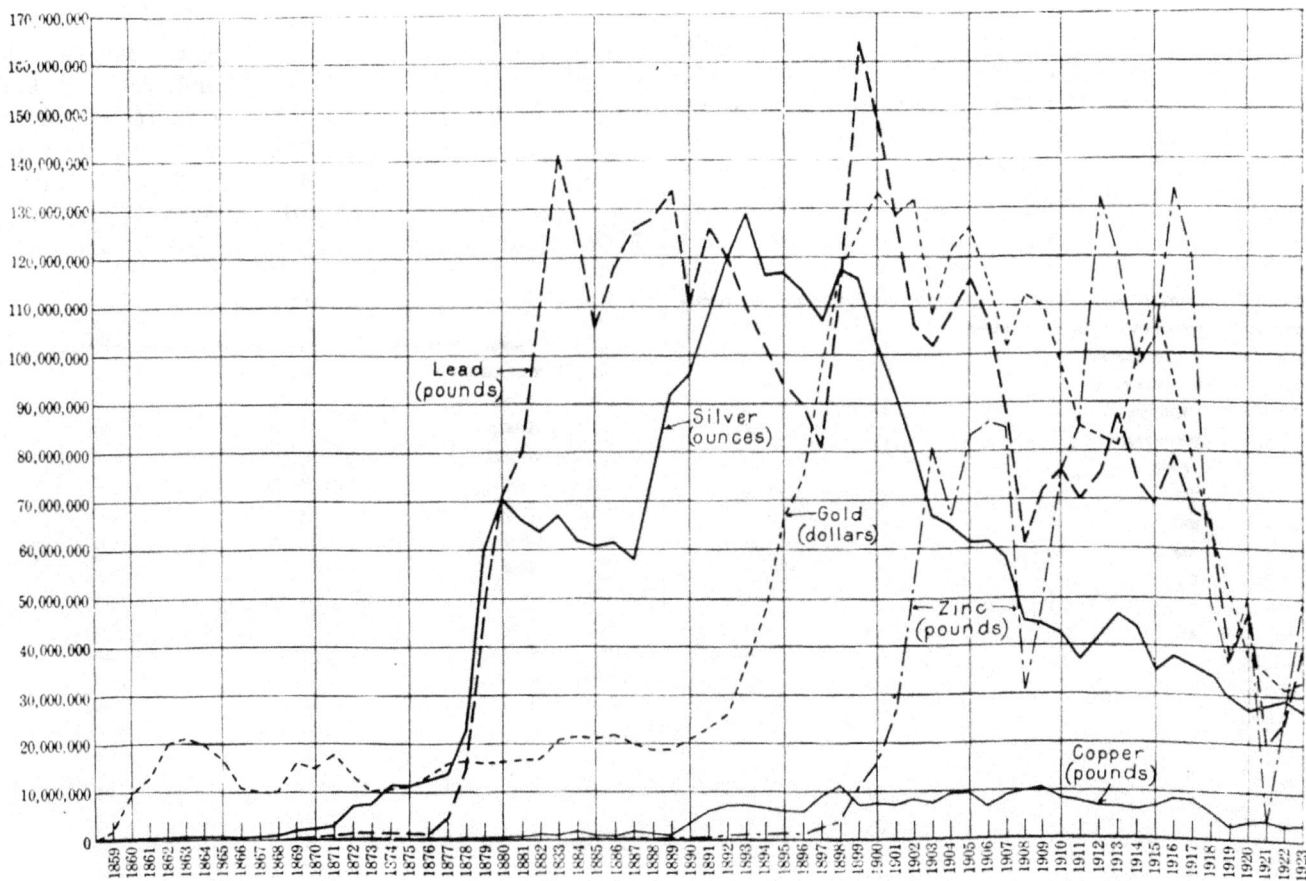

FIGURE 14.—Production of gold, in dollars, of silver, in ounces, and of copper, lead, and zinc, in pounds, in Colorado from 1859 to 1923, by years

78 cents an ounce, having a total value of $20,154,107. In 1917 the output of copper reached its greatest value—$2,217,307. The average price was then 27.3 cents a pound, and the output was 8,122,004

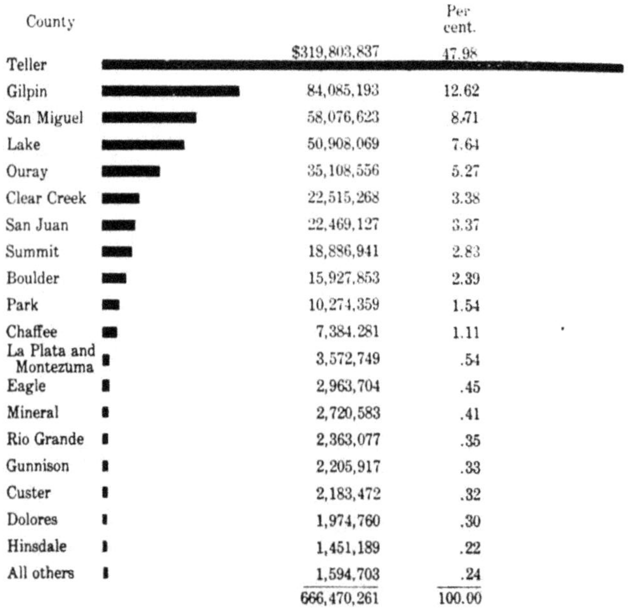

County		Per cent.
Teller	$319,803,837	47.98
Gilpin	84,085,193	12.62
San Miguel	58,076,623	8.71
Lake	50,908,069	7.64
Ouray	35,108,556	5.27
Clear Creek	22,515,268	3.38
San Juan	22,469,127	3.37
Summit	18,886,941	2.83
Boulder	15,927,853	2.39
Park	10,274,359	1.54
Chaffee	7,384,281	1.11
La Plata and Montezuma	3,572,749	.54
Eagle	2,963,704	.45
Mineral	2,720,583	.41
Rio Grande	2,363,077	.35
Gunnison	2,205,917	.33
Custer	2,183,472	.32
Dolores	1,974,760	.30
Hinsdale	1,451,189	.22
All others	1,594,703	.24
	666,470,261	100.00

FIGURE 15.—Production of gold in Colorado, 1859 to 1922, by counties

pounds. In 1909 the largest annual output of copper was made—10,916,191 pounds. The maximum for lead in both quantity and value was reached in 1900, when the output was 164,274,762 pounds, which at the price then paid, 4.4 cents a pound, was worth $7,228,090. Since 1900 the trend of the production

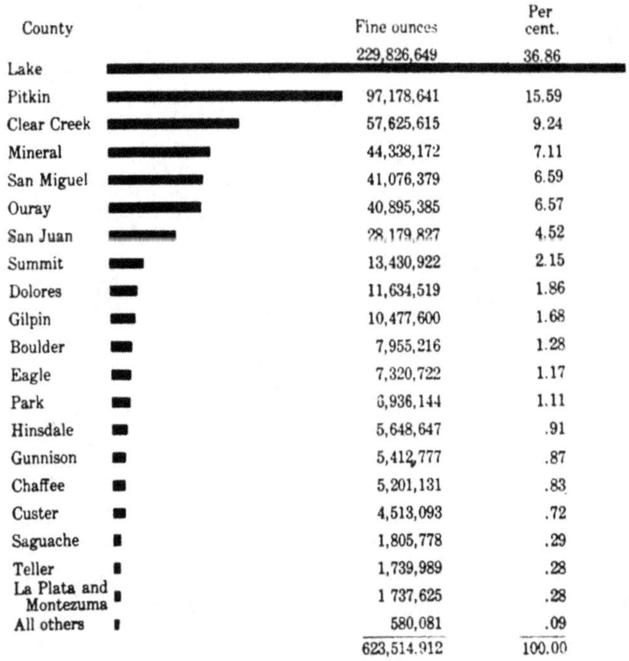

County	Fine ounces	Per cent.
Lake	229,826,649	36.86
Pitkin	97,178,641	15.59
Clear Creek	57,625,615	9.24
Mineral	44,338,172	7.11
San Miguel	41,076,379	6.59
Ouray	40,895,385	6.57
San Juan	28,179,827	4.52
Summit	13,430,922	2.15
Dolores	11,634,519	1.86
Gilpin	10,477,600	1.68
Boulder	7,955,216	1.28
Eagle	7,320,722	1.17
Park	6,936,144	1.11
Hinsdale	5,648,647	.91
Gunnison	5,412,777	.87
Chaffee	5,201,131	.83
Custer	4,513,093	.72
Saguache	1,805,778	.29
Teller	1,739,989	.28
La Plata and Montezuma	1,737,625	.28
All others	580,081	.09
	623,514,912	100.00

FIGURE 16.—Production of silver in Colorado, 1859 to 1922, by counties

of lead in Colorado has been downward, that for 1921 being the lowest recorded from 1879 to 1923. The maximum for zinc in quantity and value was reached in 1916, when the output was 134,285,463 pounds,

which, at the price then paid, 13.4 cents a pound, was worth $17,994,252.

Lake County (Leadville) holds the record for total value, its production being valued at $423,860,061 for 63 years; Teller County (Cripple Creek) is

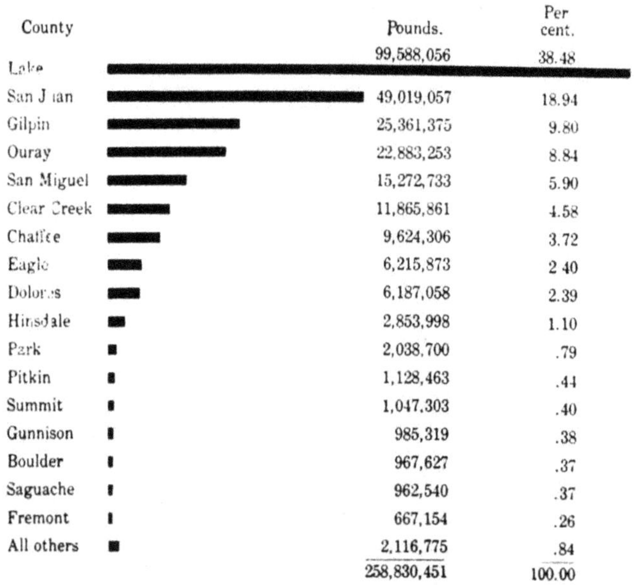

County	Pounds.	Per cent.
Lake	99,588,056	38.48
San Juan	49,019,057	18.94
Gilpin	25,361,375	9.80
Ouray	22,883,253	8.84
San Miguel	15,272,733	5.90
Clear Creek	11,865,861	4.58
Chaffee	9,624,306	3.72
Eagle	6,215,873	2.40
Dolores	6,187,058	2.39
Hinsdale	2,853,998	1.10
Park	2,038,700	.79
Pitkin	1,128,463	.44
Summit	1,047,303	.40
Gunnison	985,319	.38
Boulder	967,627	.37
Saguache	962,540	.37
Fremont	667,154	.26
All others	2,116,775	.84
	258,830,451	100.00

FIGURE 17.—Production of copper in Colorado, 1868 to 1922, by counties

second, with $320,932,424 for 31 years; San Miguel County is third, with $100,511,284 for 47 years; Pitkin County fourth, with $100,365,748 in 42 years; Gilpin County fifth, with $98,325,802 in 63 years; Clear Creek County sixth, with $86,894,724 in 63 years; Ouray County seventh, with $77,070,903 in 48 years; San Juan County eighth, with $68,204,824 in 49 years;

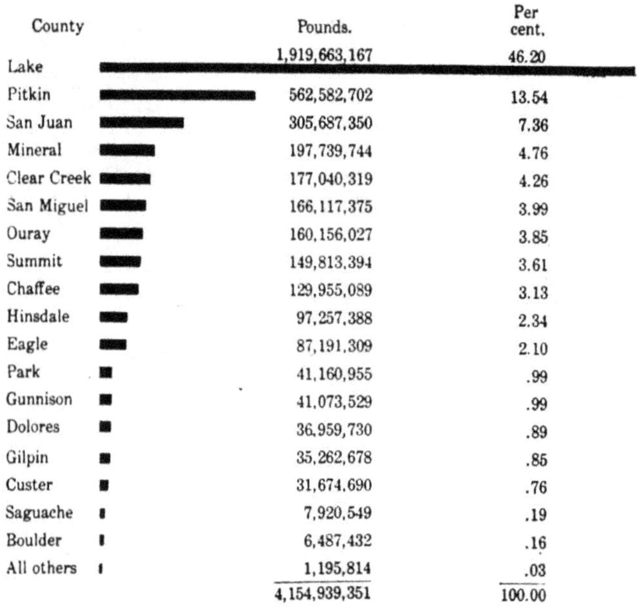

County	Pounds.	Per cent.
Lake	1,919,663,167	46.20
Pitkin	562,582,702	13.54
San Juan	305,687,350	7.36
Mineral	197,739,744	4.76
Clear Creek	177,040,319	4.26
San Miguel	166,117,375	3.99
Ouray	160,156,027	3.85
Summit	149,813,394	3.61
Chaffee	129,955,089	3.13
Hinsdale	97,257,388	2.34
Eagle	87,191,309	2.10
Park	41,160,955	.99
Gunnison	41,073,529	.99
Dolores	36,959,730	.89
Gilpin	35,262,678	.85
Custer	31,674,690	.76
Saguache	7,920,549	.19
Boulder	6,487,432	.16
All others	1,195,814	.03
	4,154,939,351	100.00

FIGURE 18.—Production of lead in Colorado, 1869 to 1922, by counties

Summit County ninth, with $47,894,768 in 63 years; Mineral County tenth, with $42,019,417 in 33 years.

The three principal counties of the San Juan region—Ouray, San Juan, and San Miguel—have produced $245,787,012 in 49 years, and the entire San Juan region, which includes Dolores, Hinsdale, La Plata, Mineral, Montezuma, Ouray, Rio Grande, San Juan, and San Miguel counties, has produced $320,259,686 in 52 years.

Teller County outranks all others in its production of gold, which amounted to $319,803,837, and Gilpin County comes next with $84,085,193, and is followed by San Miguel County with $58,076,623, Lake County with $50,908,069, Ouray County with $35,108,556, Clear Creek County with $22,515,268, San Juan County with $22,469,127, Summit County with $18,886,941, and Boulder County with $15,927,853. The three principal counties of the San Juan region— Ouray, San Juan, and San Miguel—have produced

County	Pounds	Per cent.
Lake	1,225,503,034	72.70
Eagle	131,493,129	7.80
Summit	129,810,560	7.70
San Juan	42,814,684	2.54
Clear Creek	30,399,821	1.80
Chaffee	28,449,505	1.69
Mineral	27,572,407	1.64
San Miguel	18,141,182	1.08
Pitkin	16,377,002	.97
Gunnison	16,124,550	.96
Dolores	10,648,316	.63
All others	8,348,795	.49
	1,685,682,985	100.00

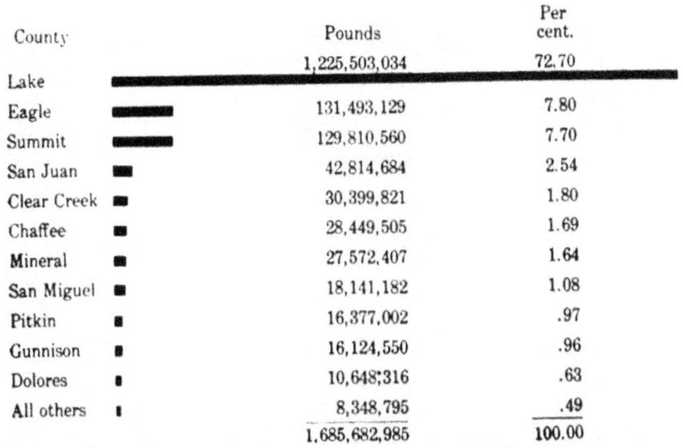

FIGURE 19.—Production of zinc in Colorado, 1865 to 1922, by counties

$115,654,306 in gold. The entire San Juan region has produced $126,285,475 in gold. Boulder, Clear Creek, and Gilpin counties combined have produced $122,528,314 in gold.

Lake County outranks all others in the production of silver, which amounts to 229,826,649 ounces, valued at $188,872,146; Pitkin County is second, with 97,178,641 ounces, valued at $72,988,357; Clear Creek County is third, with 57,625,615 ounces, valued at $52,222,478; in quantity Mineral County is fourth with 44,338,172 ounces, San Miguel is fifth with 41,076,379 ounces, and Ouray County is sixth with 40,895,385 ounces. In value, Ouray is fourth, San Miguel is fifth, and Mineral is sixth. Then follows San Juan County, 28,179,827 ounces; Summit County, 13,430,922 ounces; Dolores County, 11,634,519 ounces; and Gilpin County, 10,477,600 ounces.

Lake County holds first rank in the production of copper, its local output being 99,588,056 pounds, valued at $14,254,235. It is followed by San Juan, 49,019,057 pounds, valued at $7,578,249; Gilpin 25,361,375 pounds, valued at $4,161,058; Ouray, 22,883,253 pounds, valued at $3,300,733; San Miguel, 15,272,733 pounds, valued at $2,535,263; and Clear Creek, 11,865,860 pounds, valued at $1,919,926.

Lake County holds first rank in the production of lead, 1,919,663,167 pounds, valued at $85,061,553; Pitkin County comes next, with 562,582,702 pounds, valued at $25,573,729; and is followed by San Juan County, 305,687,350 pounds, valued at $14,621,297; Mineral County, 197,739,744 pounds, valued at $8,738,960; Clear Creek County, 177,040,319 pounds, valued at $8,026,391; San Miguel County, 166,117,375 pounds, valued at $8,313,259, this value exceeding that of Clear Creek; Ouray County, 160,156,027 pounds, valued at $7,003,622; Summit County, 149,813,394 pounds, valued at $6,541,048; Chaffee County, 129,955,089 pounds, valued at $5,710,328; Hinsdale County, 97,257,388 pounds, valued at $3,993,171; Eagle County, 87,191,309 pounds, valued at $3,871,605.

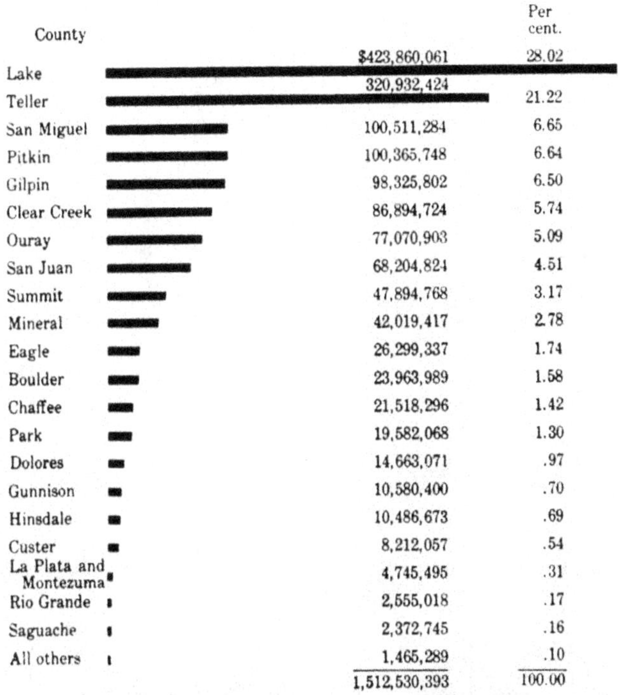

County		Per cent.
Lake	$423,860,061	28.02
Teller	320,932,424	21.22
San Miguel	100,511,284	6.65
Pitkin	100,365,748	6.64
Gilpin	98,325,802	6.50
Clear Creek	86,894,724	5.74
Ouray	77,070,903	5.09
San Juan	68,204,824	4.51
Summit	47,894,768	3.17
Mineral	42,019,417	2.78
Eagle	26,299,337	1.74
Boulder	23,963,989	1.58
Chaffee	21,518,296	1.42
Park	19,582,068	1.30
Dolores	14,663,071	.97
Gunnison	10,580,400	.70
Hinsdale	10,486,673	.69
Custer	8,212,057	.54
La Plata and Montezuma	4,745,495	.31
Rio Grande	2,555,018	.17
Saguache	2,372,745	.16
All others	1,465,289	.10
	1,512,530,393	100.00

FIGURE 20.—Total value of gold, silver, copper, lead, and zinc produced in Colorado from 1859 to 1922, by counties

Lake County also ranks first in the production of zinc—1,225,503,034 pounds, valued at $84,770,058; Eagle County is next, with 131,493,129 pounds, valued at $12,225,773; followed by Summit County, 129,810,560 pounds, valued at $10,724,763; San Juan County, 42,814,684 pounds, valued at $3,490,764; Clear Creek County, 30,399,821 pounds, valued at $2,210,661; Chaffee County, 28,449,505 pounds, valued at $2,482,051, this value exceeding that of Clear Creek; Mineral County, 27,572,407 pounds, valued at $1,511,944; San Miguel County, 18,141,182 pounds, valued at $1,323,787; Pitkin County, 16,377,002 pounds, valued at $1,028,289; and Gunnison County, 16,124,550 pounds, valued at $1,477,204, this value exceeding that of Pitkin.

Gold, silver, copper, lead, and zinc produced in Colorado, 1858–1923, by counties, in terms of recovered metals [a]

County	Ore sold or treated (short tons)	Gold			Silver		Copper		Lead		Zinc		Total value
		Placer	Lode	Total	Fine ounces	Value	Pounds	Value	Pounds	Value	Pounds	Value	
1858-1867													
Boulder			$195,000	$195,000									$195,000
Chaffee		$380,000		380,000									380,000
Clear Creek		2,100,000		2,100,000	30,349	$40,501							2,140,501
Gilpin		241,918	9,192,866	9,434,784	234,880	315,216							9,750,000
Lake		5,272,000		5,272,000	37,600	50,422							5,322,422
Park		1,780,000	710,000	2,490,000									2,490,000
Summit		5,150,000		5,150,000									5,150,000
		14,923,918	10,097,866	25,021,784	302,829	406,139							25,427,923
1868													
Boulder			50,000	50,000									50,000
Chaffee		10,000		10,000									10,000
Clear Creek		50,000		50,000	106,953	141,820							191,820
Gilpin			1,640,000	1,640,000	93,311	123,730	50,000	$11,500					1,775,230
Lake		60,000		60,000	452	600							60,600
Park		50,000		50,000									50,000
Summit		150,000		150,000									150,000
		320,000	1,690,000	2,010,000	200,716	266,150	50,000	11,500					2,287,650
1869													
Boulder			100,000	100,000	3,547	4,700							104,700
Chaffee		10,000		10,000									10,000
Clear Creek		50,000		50,000	377,359	500,000	2,000	485	100,000	$6,000			556,485
Gilpin			2,690,000	2,690,000	86,340	114,400	100,000	24,250					2,828,650
Lake		80,000	10,000	90,000	679	900							90,900
Park		40,000		40,000									40,000
Summit		200,000		200,000	7,547	10,000			50,000	3,000			213,000
		380,000	2,800,000	3,180,000	475,472	630,000	102,000	24,735	150,000	9,000			3,843,735
1870													
Boulder			100,000	100,000	60,241	80,000							180,000
Chaffee		10,000	60,000	70,000									70,000
Clear Creek		80,000		80,000	362,465	481,354	2,500	530	200,000	12,000			573,884
Gilpin			2,120,000	2,120,000	65,910	87,528	180,000	38,124					2,245,652
Lake		65,000		65,000	465	618							65,618
Park		40,000	40,000	80,000									80,000
Summit		500,000		500,000	7,907	10,500			50,000	3,000			513,500
		695,000	2,320,000	3,015,000	496,988	660,000	182,500	38,654	250,000	15,000			3,728,654
1871													
Boulder			156,605	156,605	60,377	80,000							236,605
Chaffee		10,000		10,000									10,000
Clear Creek		20,000		20,000	640,790	849,047	3,000	724	550,000	33,000			902,771
Gilpin			3,237,346	3,237,346	59,229	78,478	180,000	43,416					3,359,240
Lake		50,000	50,000	100,000	1,158	1,534							101,534
Park		40,000		40,000	15,094	20,000			5,000	300			60,300
Summit		70,000		70,000									70,000
		190,000	3,443,951	3,633,951	776,648	1,029,059	183,000	44,140	555,000	33,300			4,740,450
1872													
Boulder			221,852	224,852	199,414	263,625							488,477
Chaffee		10,000		10,000									10,000
Clear Creek		25,000		25,000	1,118,299	1,478,391	4,000	1,422	1,000,000	64,000			1,568,813
Custer					6,051	8,000							8,000
Gilpin			2,083,611	2,083,611	52,911	69,948	200,000	71,120					2,224,679
Lake		66,500	66,500	133,000	1,540	2,036							135,036
Park		50,000		50,000	142,209	188,000			50,000	3,200			241,200
Summit		120,000		120,000	3,782	5,000			100,000	6,400			131,400
		271,500	2,374,963	2,646,463	1,524,206	2,015,000	204,000	72,542	1,156,000	73,600			4,807,605
1873													
Boulder			155,000	155,000	282,326	366,177							521,177
Chaffee		10,000		10,000									10,000
Clear Creek		34,000		34,000	902,668	1,170,760	10,000	2,800	1,000,000	60,000			1,267,560
Custer					7,721	10,014							10,014
Gilpin			1,393,931	1,393,931	35,907	46,571	200,000	56,000	25,000	1,500			1,498,002
Gunnison		5,000		5,000									5,000
Lake		75,000	150,000	225,000	2,937	3,809							228,809
Park		60,000	20,000	80,000	307,633	399,000	169,493	47,458	111,400	6,684			533,142
Rio Grande			2,000	2,000									2,000
Routt and Moffat		26,000		26,000									26,000
San Juan			13,000	13,000									13,000
Summit		75,000		75,000	3,855	5,000			100,000	6,000			86,000
		285,000	1,733,931	2,018,931	1,543,047	2,001,331	379,493	106,258	1,236,400	74,184			4,200,704
1874													
Boulder			160,000	160,000	293,806	375,484							535,484
Chaffee		10,000		10,000									10,000
Clear Creek		42,500		42,500	1,634,434	2,088,807	15,000	3,300	803,983	48,239			2,182,846
Custer					17,005	21,732	5,100	1,122					22,854
Gilpin			1,525,447	1,525,447	39,418	50,376	252,050	55,451	50,000	3,000			1,634,274
Lake		70,000	143,503	213,503	2,797	3,575							217,078
Park		66,497	50,000	116,497	333,764	426,550	203,391	44,746					587,793
Rio Grande			5,000	5,000									5,000
San Juan			9,540	9,540	3,166	4,046							13,586
Summit		70,000		70,000	23,784	30,396			423,950	25,437			125,833
		258,997	1,893,490	2,152,487	2,348,174	3,000,966	475,541	104,619	1,277,933	76,676			5,334,748

[a] For explanation see footnote to Table 1, p. 69.

Gold, silver, copper, lead, and zinc produced in Colorado, 1858–1923, by counties, in terms of recovered metals—Continued

County	Ore sold or treated (short tons)	Gold Placer	Gold Lode	Gold Total	Silver Fine ounces	Silver Value	Copper Pounds	Copper Value	Lead Pounds	Lead Value	Zinc Pounds	Zinc Value	Total value
1875													
Boulder			$218,086	$218,086	203,344	$252,147							$470,233
Chaffee		$21,551		21,551									21,551
Clear Creek	12,000	68,960	3,448	72,408	1,343,610	1,666,076	15,000	$3,405	1,300,000	$75,400			1,817,289
Custer					156,142	193,616							193,616
Gilpin			1,395,566	1,395,566	62,670	77,711	193,665	43,962	50,000	2,900			1,520,139
Hinsdale			12,000	12,000	47,953	59,462							71,462
Lake		17,237	25,862	43,099	16,668	20,668							63,767
Park		80,000	24,302	104,302	412,022	510,907	72,150	16,378	25,000	1,450			633,037
Rio Grande			272,044	272,044	7,734	9,590							281,634
Routt and Moffat		3,500		3,500									3,500
San Juan			10,000	10,000	68,547	84,998			120,000	6,960			101,958
San Miguel					3,867	4,795							4,795
Summit		72,012		72,012	7,734	9,590			141,000	8,178			89,780
		263,260	1,961,308	2,224,568	2,330,291	2,889,560	280,815	63,745	1,636,000	94,888			5,272,761
1876													
Boulder			200,000	200,000	232,031	269,156							469,156
Chaffee		25,000		25,000	3,867	4,486							29,486
Clear Creek		40,000	55,161	95,161	1,421,104	1,648,481	15,000	3,150	819,672	50,000			1,796,792
Custer					38,672	44,860							44,860
Gilpin			1,990,002	1,990,002	89,365	103,663	250,000	52,500	50,000	3,050			2,149,215
Hinsdale			20,000	20,000	154,688	179,438			50,000	3,050			202,488
Lake		30,000	30,000	60,000	23,203	26,915			15,000	915			87,830
Park		40,000	20,000	60,000	386,719	448,594	68,333	14,350	50,000	3,050			525,994
Rio Grande			121,148	121,148	7,734	8,971							130,119
San Juan			5,000	5,000	48,465	56,219			249,348	15,210			76,429
San Miguel					3,867	4,486							4,486
Summit		150,000		150,000	154,688	179,438			100,000	6,100			335,538
		285,000	2,441,311	2,726,311	2,564,403	2,974,707	333,333	70,000	1,334,020	81,375			5,852,393
1877													
Boulder			400,000	400,000	232,031	278,437							678,437
Chaffee		25,000		25,000	7,734	9,281			50,000	2,750			37,031
Clear Creek		20,000	76,500	96,500	1,534,560	1,841,472	15,000	2,850	2,236,364	123,000			2,063,822
Custer		50,000		50,000	77,344	92,813							142,813
Gilpin			2,086,871	2,086,871	93,714	112,457	300,000	57,000	50,000	2,750			2,259,078
Hinsdale		25,000		25,000	92,814	111,377			100,000	5,500			141,877
Lake		30,000	25,000	55,000	458,000	549,600			1,200,000	66,000			670,600
Park		40,000	20,000	60,000	309,375	371,250	170,000	32,300	150,000	8,250			471,800
Rio Grande			195,337	195,337	7,734	9,281							204,618
San Juan			5,000	5,000	34,010	40,812	8,664	1,646	400,000	22,000			69,458
San Miguel					3,867	4,640							4,640
Summit		150,000		150,000	30,938	37,126			100,000	5,500			192,626
		265,000	2,883,708	3,148,708	2,882,121	3,458,546	493,664	93,796	4,286,364	235,750			6,936,800
1878													
Boulder			400,000	400,000	270,703	311,308							711,308
Chaffee		25,000		25,000	7,734	8,894			50,000	1,800			35,694
Clear Creek		10,000	124,000	134,000	1,759,652	2,023,600	25,000	4,150	2,722,222	98,000			2,259,750
Custer		100,000		100,000	77,344	88,946							188,946
Gilpin			2,155,708	2,155,708	96,806	111,327	300,000	49,800	50,000	1,800			2,318,635
Hinsdale		20,000		20,000	154,688	177,891			200,000	7,200			205,091
Lake		30,000	30,000	60,000	1,800,000	2,070,000			10,000,000	360,000			2,490,000
La Plata and Montezuma		1,000		1,000	1,934	2,224							3,224
Ouray		5,000		5,000	38,672	44,473							49,473
Park		40,000	20,000	60,000	309,375	355,781	175,000	29,050	150,000	5,400			450,231
Rio Grande			102,866	102,866	7,734	8,894							111,760
San Juan		6,000		6,000	24,569	28,254	36,145	6,000	400,000	14,400			54,654
San Miguel		5,000		5,000	3,867	4,447			50,000	1,800			11,247
Summit		165,774		165,774	119,883	137,865			100,000	3,600			307,239
		275,774	2,964,574	3,240,348	4,672,961	5,373,904	536,145	89,000	13,722,222	494,000			9,197,252
1879													
Boulder			400,000	400,000	348,047	389,813							789,813
Chaffee		25,000	3,500	28,500	30,938	34,651			50,000	2,050			65,201
Clear Creek		10,000	110,000	120,000	1,546,875	1,732,500	100,000	18,600	1,951,219	80,000			1,951,100
Custer			100,000	100,000	541,406	606,375							706,375
Dolores			1,500	1,500	7,734	8,662	4,301	800	10,000	410			11,372
Gilpin			2,260,000	2,260,000	232,031	259,875	300,000	55,800	100,000	4,100			2,579,775
Gunnison					19,336	21,656							21,656
Hinsdale			6,000	6,000	193,359	216,562			500,000	20,500			243,062
Lake		30,000	60,000	90,000	8,411,132	9,420,468			43,288,000	1,774,808			11,285,276
La Plata and Montezuma			2,500	2,500	3,867	4,331							6,831
Ouray			8,500	8,500	38,672	43,313			198,781	8,150			59,963
Park		40,000	20,000	60,000	324,844	363,825	200,000	37,200	300,000	12,300			473,325
Rio Grande			28,500	28,500	7,734	8,662							37,162
San Juan			6,000	6,000	30,938	34,651	100,000	18,600	500,000	20,500			79,751
San Miguel		7,000		7,000	7,734	8,662			350,000	14,350			30,012
Summit		75,000		75,000	154,688	173,251			100,000	4,100			252,351
		187,000	3,006,500	3,193,500	11,899,335	13,327,257	704,301	131,000	47,348,000	1,941,268			18,593,025
1880													
Boulder			300,000	300,000	425,391	489,200							789,200
Chaffee		25,000	6,500	31,500	61,875	71,156							107,656
Clear Creek		5,000	191,000	196,000	1,902,656	2,188,054	200,000	42,800	517,500	25,875			2,452,729
Custer			100,000	100,000	665,156	764,929							864,929
Dolores			3,500	3,500	30,938	35,579	29,000	6,206	100,000	5,000			50,285
Eagle	1,000		2,000	2,000	38,672	44,473			800,000	40,000			86,473
Gilpin			2,380,000	2,380,000	232,031	266,836	300,000	64,200	100,000	5,000			2,716,036
Gunnison					222,031	255,336			100,000	5,000			260,336
Hinsdale			6,000	6,000	116,016	133,418	30,000	6,420	50,000	50,000			195,838
Lake	140,623	70,000	34,014	104,014	9,977,344	11,473,946			66,658,000	3,332,900			14,910,860
La Plata and Montezuma			5,000	5,000	7,734	8,894							13,894
Ouray			8,500	8,500	69,610	80,052			200,000	10,000			98,552
Park		30,000	20,000	50,000	293,906	337,992	200,000	42,800	300,000	15,000			445,792

Gold, silver, copper, lead, and zinc produced in Colorado, 1858–1923, by counties, in terms of recovered metals—Continued

County	Ore sold or treated (short tons)	Gold Placer	Gold Lode	Gold Total	Silver Fine ounces	Silver Value	Copper Pounds	Copper Value	Lead Pounds	Lead Value	Zinc Pounds	Zinc Value	Total value
1880—Continued													
Pitkin					10,000	$11,500			60,000	$3,000			$14,500
Rio Grande			$6,000	$6,000									6,000
Saguache					7,734	8,891							8,894
San Juan			6,000	6,000	11,602	13,342	100,000	$21,400	430,000	21,500			62,242
San Miguel		$5,000		5,000	7,734	8,894			482,500	24,125			38,019
Summit		44,000	5,000	49,000	317,109	364,675			500,000	25,000			438,675
		179,000	3,073,514	3,252,514	14,397,539	16,557,170	859,000	183,826	71,348,000	3,567,400			23,560,910
1881													
Boulder			200,000	200,000	270,703	305,894							505,894
Chaffee	25,000		25,000	50,000	127,617	144,207			500,000	24,000			218,207
Clear Creek	5,000		195,000	200,000	1,546,875	1,747,969	200,000	36,400	815,000	39,120			2,023,489
Custer			100,000	100,000	541,406	611,789							711,789
Dolores			5,000	5,000	69,610	78,659	44,000	8,008	200,000	9,600			101,267
Eagle	2,000		4,000	4,000	79,344	89,659			1,600,000	76,800			170,459
Fremont					11,602	13,110							13,110
Gilpin			1,850,000	1,850,000	201,094	227,236	300,000	54,600	100,000	4,800			2,136,636
Gunnison			10,000	10,000	309,375	349,594			360,000	17,280			376,874
Hinsdale			10,000	10,000	123,750	139,838	40,000	7,280	1,200,000	57,600			214,718
Lake	69,000		231,000	300,000	7,966,406	9,002,039			58,464,000	2,806,272			12,108,311
La Plata and Montezuma			5,000	5,000	7,734	8,739							13,739
Ouray			55,000	55,000	85,078	96,138	100,000	18,200	230,000	11,040			180,378
Park	25,000		25,000	50,000	270,703	305,894	100,000	18,200	312,000	14,976			389,070
Pitkin			100,000	100,000	23,203	26,219			200,000	9,600			135,819
Rio Grande			290,000	290,000	7,734	8,739							298,739
Routt and Moffat	20,000			20,000									20,000
Saguache					30,938	34,960							34,960
San Juan			5,000	5,000	19,336	21,850	100,000	18,200	140,000	6,720			51,770
San Miguel	5,500	9,500		15,000	19,336	21,850			200,000	9,600			46,450
Summit	26,000	5,000		31,000	1,560,344	1,763,189			16,773,000	805,104			2,599,293
		175,500	3,124,500	3,300,000	13,272,188	14,997,572	884,000	160,888	81,094,000	3,892,512			22,350,972
1882													
Boulder			260,000	260,000	239,766	273,333							533,333
Chaffee	25,000	20,000	45,000		77,344	88,172			1,000,000	49,000			182,172
Clear Creek	6,000	214,000	220,000		1,299,375	1,481,288	300,000	57,300	815,000	39,935			1,798,523
Custer		200,000	200,000		232,031	264,515							464,515
Dolores		10,000	10,000		85,078	96,989	54,000	10,314	200,000	9,800			127,103
Eagle		5,000	5,000		98,680	112,495			2,000,000	98,000			215,495
Fremont					15,469	17,635							17,635
Gilpin		1,600,000	1,600,000		201,094	229,247	400,000	76,400	100,000	4,900			1,910,547
Gunnison		100,000	100,000		386,719	440,860			360,000	17,640			558,500
Hinsdale		20,000	20,000		61,875	70,538	40,000	7,640	600,000	29,400			127,578
Lake	63,500	256,500	320,000		8,894,531	10,139,765			97,890,000	4,796,610			15,256,375
La Plata and Montezuma		10,000	10,000		23,203	26,451							36,451
Ouray		70,000	70,000		77,344	88,172	500,000	95,500	230,000	11,270			264,912
Park	25,000	75,000	100,000		193,359	220,429	100,000	19,100	312,000	15,288			354,817
Pitkin		90,000	90,000		23,203	26,451			200,000	9,800			126,251
Rio Grande		210,000	210,000		15,469	17,635							227,635
Routt and Moffat	15,000		15,000										15,000
Saguache		10,000	10,000		77,344	88,172							98,172
San Juan		10,000	10,000		46,406	52,903	100,000	19,100	320,000	15,680			97,683
San Miguel	8,000	2,000	10,000		38,672	44,086			200,000	9,800			63,886
Summit	50,000	5,000	55,000		674,757	769,223			5,773,000	282,877			1,107,100
		192,500	3,167,500	3,360,000	12,761,719	14,548,359	1,494,000	285,354	110,000,000	5,390,000			23,583,713
1883													
Boulder			300,000	300,000	123,750	137,363							437,363m
Chaffee	25,000	25,000	50,000		204,961	227,507			4,300,000	184,900			462,407
Clear Creek	10,000	240,000	250,000		1,222,031	1,356,454	300,000	49,500	815,000	35,045			1,690,999
Custer		620,000	620,000		154,688	171,704							791,704
Dolores		5,000	5,000		193,360	214,630	100,000	16,500	200,000	8,600			244,730
Eagle		70,000	70,000		232,031	257,554			16,000,000	688,000			1,015,554
Fremont					15,469	17,171							17,171
Gilpin		1,650,000	1,650,000		154,688	171,704	200,000	33,000	100,000	4,300			1,859,004
Gunnison		100,000	100,000		502,734	558,035	30,000	4,950	500,000	21,500			684,485
Hinsdale		20,000	20,000		193,359	214,628	22,652	3,738	1,000,000	43,000			281,366
Lake	25,000	375,000	400,000		9,049,219	10,044,633			111,575,000	4,797,725			15,242,358
La Plata and Montezuma		13,000	13,000		3,867	4,292							17,292
Ouray		20,000	20,000		386,719	429,258	400,000	66,000	1,170,000	50,310			565,568
Park	20,000	180,000	200,000		135,352	150,241			312,000	13,416			363,657
Pitkin		2,000	2,000		42,539	47,218			450,000	19,350			68,568
Rio Grande		180,000	180,000		7,734	8,585							188,585
Routt and Moffat	40,000		40,000										40,000
Saguache		5,000	5,000		77,344	85,852							90,852
San Juan	7,400	35,000	35,000		270,703	300,480	100,000	16,500	1,137,000	48,891			400,871
San Miguel	2,000	123,000	125,000		193,359	214,628			782,000	33,626			373,254
Summit	10,000	5,000	15,000		270,703	300,480			2,773,000	119,239			434,719
		132,000	3,968,000	4,100,000	13,434,610	14,912,417	1,152,652	190,188	141,114,000	6,067,902			25,270,507
1884													
Boulder			350,000	350,000	100,547	111,607							461,607
Chaffee	25,000	55,000	80,000		146,953	163,118			12,000,000	444,000			687,118
Clear Creek	12,551	587,449	600,000		1,314,844	1,459,477	300,000	39,000	1,038,273	38,416			2,136,893
Custer		350,000	350,000		185,625	206,044			500,000	18,500			574,544
Dolores		1,500	1,500		54,141	60,097			152,000	5,624			67,221
Eagle	10,000	30,000	30,000		154,687	171,703			6,600,000	244,200			445,903
Gilpin		1,950,000	1,950,000		278,438	309,066	600,000	78,000	128,411	4,751			2,341,817
Gunnison		60,000	60,000		386,719	429,258			2,000,000	74,000			563,258
Hinsdale	2,184	2,500	2,500		154,687	171,703	350,000	45,500	1,000,000	37,000			256,703
Lake	202,002	30,000	470,000	500,000	7,270,313	8,070,047	100,000	13,000	93,628,000	3,464,236			12,047,283
La Plata and Montezuma		500	500		4,641	5,152							5,652
Ouray		10,500	10,500		572,344	635,302	363,125	47,206	3,000,000	111,000			804,008
Park	30,000	30,000	60,000		193,359	214,628			398,066	14,728			289,356
Pitkin		1,000	1,000		464,062	515,109			1,200,000	44,400			560,509
Rio Grande		130,000	30,000		10,828	12,019							142,019

Gold, silver, copper, lead, and zinc produced in Colorado, 1858–1923, by counties, in terms of recovered metals—Continued

County	Ore sold or treated (short tons)	Gold			Silver		Copper		Lead		Zinc		Total value
		Placer	Lode	Total	Fine ounces	Value	Pounds	Value	Pounds	Value	Pounds	Value	
1884—Continued													
Routt and Moffat		$13,000		$13,000									$13,000
Saguache			$1,000	1,000	77,344	$85,852							86,852
San Juan	8,000		40,000	40,000	464,062	515,109	300,000	$39,000	3,400,000	$125,800			719,909
San Miguel		3,000	97,000	100,000	309,375	343,406			300,050	11,100			454,506
Summit		10,000	10,000	20,000	232,031	257,554			985,250	36,454			314,008
		123,551	4,176,449	4,300,000	12,375,000	13,736,251	2,013,125	261,706	126,330,000	4,674,209			22,972,166
1885													
Arapahoe	271			271									271
Boulder			300,000	300,000	84,691	90,619							390,619
Chaffee	25,000	75,000	100,000	200,000	214,000			18,700,000	729,300			1,043,300
Clear Creek	10,000	490,000	500,000	1,356,000	1,450,920	200,000	21,600	1,038,273	40,493	25,000	$1,075	2,014,088
Conejos			277	277									277
Costilla	216			216									216
Custer			30,000	30,000	61,295	65,586			5,440,000	212,160			307,746
Dolores		4,000	4,000	70,000	74,900			100,000	3,900			82,800
Douglas	1,420			1,420	70	75							1,495
Eagle			33,000	33,000	170,156	182,067			5,950,000	232,050			447,117
Garfield			113	113	45	48							161
Gilpin		2,051,000	2,051,000	300,000	321,000	300,000	32,400	128,411	5,008			2,409,408
Gunnison			40,000	40,000	144,323	154,426			2,380,000	92,820			287,246
Hinsdale			2,000	2,000	16,320	17,462	46,460	5,018	100,000	3,900			28,380
Jefferson	697			697	5	5							702
Lake	15,000	555,000	570,000	6,441,693	6,892,612	100,000	10,800	55,522,000	2,165,358	50,000	2,150	9,640,920
La Plata and Montezuma		5,000	5,000	5,000	5,350								10,350
Mesa	431			431	3	3							434
Ouray		10,000	10,000	900,000	963,000	400,000	43,200	4,400,000	171,600			1,187,800
Park	30,000	30,000	60,000	71,310	76,302			398,066	15,525			151,827
Pitkin		1,000	1,000	1,000,000	1,070,000			5,950,000	232,050			1,303,050
Rio Grande		130,000	130,000	9,800	10,486							140,486
Routt and Moffat	23,000			23,000									23,000
Saguache			1,000	1,000	55,920	59,834							60,834
San Juan			40,000	40,000	700,000	749,000	100,000	10,800	5,300,000	206,700			1,006,500
San Miguel	3,000	97,000	100,000	400,000	428,000			300,000	11,700			539,700
Summit	15,000	185,000	200,000	234,351	250,756			985,250	38,425	25,000	1,075	490,256
		124,035	4,079,390	4,203,425	12,220,982	13,076,451	1,146,460	123,818	106,692,000	4,160,989	100,000	4,300	21,568,983
1886													
Arapahoe	293			293									293
Boulder		382,185	382,185	84,691	83,844							466,029
Chaffee	80,000	233,917	313,917	332,965	329,635			13,000,000	598,000			1,241,552
Clear Creek	10,000	599,070	609,070	1,396,364	1,382,400	200,000	22,200	1,630,000	74,980	25,000	1,100	2,089,750
Custer		21,600	21,600	61,295	60,682			4,500,000	207,000			289,282
Dolores		8,561	8,561	75,836	75,078			792,000	36,432			120,071
Eagle		423,517	423,517	569,637	563,941			2,000,000	92,000			1,079,458
Gilpin		1,337,061	1,337,061	101,784	100,766	300,000	33,300	200,000	9,200			1,480,327
Gunnison		18,226	18,226	144,323	142,880			500,000	23,000			184,106
Hinsdale		2,060	2,060	16,320	16,157	46,460	5,157	100,000	4,600			27,974
Huerfano		116	116									116
Jefferson	2,804		2,804	43	42							2,846
Lake	5,000	428,691	433,691	6,486,017	6,421,187	100,000	11,100	84,400,000	3,882,400	50,000	2,200	10,750,578
La Plata and Montezuma		10,225	10,225	4,671	4,624			100,000	4,600			19,449
Mesa	110			110									110
Montrose	281			281	3	3							284
Ouray		26,241	26,241	993,867	983,928	400,000	44,400	3,208,000	147,568			1,202,137
Park	30,000	118,284	148,284	71,310	70,597			621,000	28,704			247,585
Pitkin	5,590	17,125	17,125	399,094	395,103			800,000	36,800			449,028
Rio Grande		119,266	149,266	8,817	8,729							157,995
Routt and Moffat	16,840		16,840	387	383							17,223
Saguache		3,936	3,936	55,920	55,361							59,297
San Juan		142,799	142,799	718,523	711,338	100,000	11,100	4,300,000	197,800			1,063,037
San Miguel	3,000	214,570	217,570	430,805	426,197			300,000	13,800			657,867
Summit	15,000	149,222	164,222	422,298	418,075			1,546,000	71,116	25,000	1,100	654,513
		163,328	4,286,672	4,450,000	12,375,000	12,251,250	1,146,460	127,257	118,000,000	5,428,000	100,000	4,400	22,260,907
1887													
Arapahoe	177			177									177
Boulder		253,546	253,546	70,691	68,689			593	27			322,262
Chaffee	45,000	364,050	409,050	423,738	415,263			14,954,155	672,937			1,497,250
Clear Creek	15,000	302,214	317,214	1,284,083	1,258,401	200,000	$27,600	1,847,930	83,157	25,000	1,150	1,687,522
Custer		507	507	117,970	115,611			5,367,459	241,536			357,654
Dolores		9,743	9,743	118,262	115,897	34,000	4,692	1,000,000	45,000			175,332
Eagle		219,594	219,594	254,078	248,996			1,112,905	50,081			518,671
Fremont		186	186	474	465			5,930	267			918
Gilpin		1,134,476	1,134,476	266,281	260,955	600,000	82,800	228,622	10,288			1,488,519
Gunnison		50,506	50,506	172,616	169,164			451,351	20,311			239,981
Hinsdale		4,308	4,308	90,355	88,548	12,027	1,660	547,503	24,638			119,154
Jefferson	942		942	5	5							947
Lake		243,694	243,694	5,904,324	5,874,438	200,000	27,600	92,359,103	4,156,160	50,000	2,300	10,304,192
La Plata and Montezuma		12,473	12,473	7,126	6,983			42,210	1,899			21,355
Las Animas		1,122	1,122	8	8							1,130
Montrose	500		500	9	9							509
Ouray		22,853	22,853	952,255	933,210	666,000	91,908	2,668,135	120,066			1,168,037
Park	190,000	458,462	648,462	107,513	105,363			708,713	31,892			785,717
Pitkin		9,336	9,336	612,368	600,121			361,388	16,262			625,719
Rio Grande	5,000	117,380	122,380	7,992	7,832							130,212
Routt and Moffat	6,714		6,714	214	210							6,924
Saguache		756	756	7,196	7,052			12,582	566			8,374
San Juan		121,245	121,245	401,760	393,725	300,000	41,400	2,040,145	91,806			648,176
San Miguel	2,600	167,096	169,696	492,725	482,871			537,144	24,171			676,738
Summit	15,000	225,520	240,520	220,120	215,718			1,754,132	78,936	25,000	1,150	536,324
		280,933	3,719,067	4,000,000	11,601,563	11,369,534	2,012,027	277,660	126,000,000	5,670,000	100,000	4,600	21,321,794

Gold, silver, copper, lead, and zinc produced in Colorado, 1858–1923, by counties, in terms of recovered metals—Continued

County	Ore sold or treated (short tons)	Gold			Silver		Copper		Lead		Zinc		Total value
		Placer	Lode	Total	Fine ounces	Value	Pounds	Value	Pounds	Value	Pounds	Value	
1888													
Boulder			$189,241	$189,241	230,205	$216,393			246,282	$10,836			$416,470
Chaffee	$25,000		368,457	393,457	292,349	274,808			8,743,053	384,694			1,052,959
Clear Creek	20,000		399,821	419,821	1,148,190	1,079,299	200,000	$33,600	3,761,246	165,495	75,000	$3,675	1,701,890
Custer			120	120	3,463	3,255			4,821,143	212,130			215,505
Dolores			17,470	17,470	123,852	116,421			1,000,000	44,000			177,891
Eagle			142,002	142,002	193,489	181,880			2,370,090	104,284			428,166
Gilpin			1,250,756	1,250,756	174,364	163,902	400,000	67,200	1,288,825	56,708			1,538,566
Gunnison			18,642	18,642	60,166	56,556			1,011,792	44,519			119,717
Hinsdale			2,667	2,667	86,248	81,073	2,000	336	1,205,973	53,063			137,139
Lake			310,891	310,891	5,486,064	5,156,900	200,000	33,600	73,378,149	3,228,639	150,000	7,350	8,737,380
La Plata and Montezuma			3,574	3,574	2,294	2,156							5,730
Montrose		12,000		12,000									12,000
Ouray			24,289	24,289	789,396	742,032	579,100	97,289	3,259,904	143,436			1,007,046
Park	23,000		10,945	33,945	450,457	423,430			7,641,720	336,236			793,611
Pitkin			12,716	12,716	4,333,787	4,073,760			14,349,792	631,391			4,717,867
Rio Grande	2,000		14,260	16,260	2,923	2,748							19,008
Saguache			4,220	4,220	36,101	33,935			180,272	7,932			46,087
San Juan			190,328	190,328	223,339	209,939	240,000	40,320	2,382,358	104,824			545,411
San Miguel	7,500		417,206	424,706	663,354	623,553			636,514	28,007			1,076,266
Summit	10,000		272,209	282,209	394,058	370,415			2,126,887	93,583	75,000	3,675	749,882
Miscellaneous	5,000		3,785	8,785	1,214	1,141							9,926
	104,500		3,653,599	3,758,099	14,695,313	13,813,596	1,621,100	272,345	128,404,000	5,649,777	300,000	14,700	23,508,517
1889													
Boulder			344,503	344,503	174,471	164,003	2,748	371	51,215	1,997			510,874
Chaffee	37,000		262,853	299,853	137,759	129,493			5,000,000	195,000			624,346
Clear Creek	25,000		496,909	521,909	1,770,875	1,664,623	91,731	12,384	5,357,906	208,958	75,000	3,750	2,411,624
Custer			1,281	1,281	72,576	68,221			63,086	2,460			71,962
Dolores			77,825	77,825	618,615	581,498			2,000,000	78,000			737,323
Eagle			92,220	92,220	170,551	160,318			2,112,280	82,379			334,917
Fremont			10,841	10,841	21,683	20,382	5,317	718	466,538	18,195			50,136
Gilpin			1,054,065	1,054,065	313,071	294,287	250,110	33,765	1,411,926	55,065			1,437,182
Gunnison			39,710	39,710	48,106	45,220	556	75	485,355	18,929			103,934
Hinsdale			1,794	1,794	16,665	15,665	17,359	2,343	240,812	9,392			29,194
Lake			189,397	189,397	6,150,839	5,781,789	266,489	35,976	83,785,918	3,267,651	150,000	7,500	9,282,313
La Plata and Montezuma			4,465	4,465	1,118	1,051							5,516
Ouray			26,436	26,436	913,254	858,459	397,804	53,704	4,704,261	183,466			1,122,065
Park	42,000		82,745	124,745	224,743	211,258	855	115	4,640,682	180,987			517,105
Pitkin			35,760	35,780	5,982,238	5,623,304			15,100,807	588,931			6,212,235
Rio Grande			35,760	35,760	3,757	3,532							39,292
Routt and Moffat	8,870			8,870	189	178							9,048
Saguache									200,000	7,800			7,800
San Juan			394,873	394,873	508,328	477,828	135,018	18,227	4,096,887	159,779	•		1,050,707
San Miguel	7,000		425,588	432,588	726,456	682,869			1,166,346	45,488			1,160,945
Summit	16,000		206,724	222,724	519,842	488,651	2,066	278	3,055,981	119,183	75,000	3,750	834,586
	135,870		3,747,989	3,883,859	18,375,136	17,272,629	1,170,053	157,956	133,940,000	5,223,660	300,000	15,000	26,553,104
1890													
Boulder			380,059	380,059	118,898	124,843	90,691	14,148	45,894	2,065			521,115
Chaffee	45,300		208,950	254,250	145,674	152,958			2,400,000	108,000			515,208
Clear Creek	24,336		418,032	442,368	1,819,682	1,910,666	124,102	19,360	12,029,217	541,315	75,000	4,125	2,917,834
Custer			114,212	114,212	119,684	125,668			1,708,729	76,893			316,773
Dolores			156,297	156,297	848,785	891,224			2,000,000	90,000			1,137,521
Eagle			68,862	68,862	75,265	79,028			1,000,000	45,000			192,890
Gilpin			805,236	805,236	292,495	307,120	620,927	96,865	1,130,453	50,870			1,260,091
Gunnison			28,784	28,784	354,393	372,113	105,954	16,529	6,945,972	312,569			729,995
Hinsdale			3,697	3,697	57,387	60,256	60,584	9,451	660,708	29,732			103,136
Lake	342,163		295,063	295,063	5,313,930	5,579,627	1,766,035	275,501	43,623,477	1,963,056	150,000	8,250	8,121,497
La Plata and Montezuma			3,729	3,729	2,011	2,112							5,841
Ouray			353,133	353,133	2,791,626	2,931,207	665,754	103,858	4,228,803	190,296			3,578,494
Park	23,611		13,670	37,281	156,975	164,824			1,886,504	84,893			286,998
Pitkin					4,944,898	5,192,143			19,703,605	886,662			6,078,805
Rio Grande			25,716	25,716	1,287	1,351							27,067
Routt and Moffat	8,133			8,133	176	185							8,318
Saguache			1,745	1,745	11,988	12,587	4,290	669	176,193	7,929			22,930
San Juan			187,357	187,357	321,340	337,407	147,354	22,987	3,462,158	155,797			703,548
San Miguel	18,000		737,380	755,380	907,148	952,505			414,522	18,653			1,726,538
Summit	7,000		222,830	229,830	516,358	542,176			7,775,765	349,909	75,000	4,125	1,126,040
	126,380		4,024,752	4,151,132	18,800,000	19,740,000	3,585,691	559,368	109,192,000	4,913,639	300,000	16,500	29,380,639
1891													
Boulder			683,941	683,941	41,690	41,273							725,214
Chaffee	37,000		242,060	279,060	64,830	64,182			1,100,000	47,300			390,542
Clear Creek	22,875		415,692	438,567	1,771,055	1,753,344	57,572	7,369	7,947,786	341,755	75,000	3,750	2,544,785
Custer			49,204	49,204	48,469	47,984			838,874	36,072			133,260
Dolores			122,631	122,631	699,888	692,889			931,326	40,047			855,567
Eagle			153,453	153,453	280,168	277,366			3,776,230	162,378			593,197
Gilpin			938,016	938,016	232,001	229,681	558,298	71,462	779,837	33,533			1,272,692
Gunnison			7,402	7,402	489,268	484,375			10,340,332	444,634			936,411
Hinsdale			19,869	19,869	186,850	184,982	8,248	1,056	8,308,048	357,246			563,153
Lake	403,135	2,894	345,525	348,419	4,793,015	4,745,085	4,544,202	581,658	53,444,973	2,298,134	150,000	7,500	7,980,796
La Plata and Montezuma			23,054	23,054	3,207	3,175							26,229
Mineral			10,055	10,055	378,899	374,382			354,854	15,259			399,696
Ouray			478,750	478,750	2,273,054	2,250,323	865,044	110,726	4,168,887	179,262			3,019,061
Park	34,333		16,000	50,333	185,200	183,348			19,656	845			234,526
Pitkin			13,507	13,507	6,979,263	6,909,470			16,396,580	705,053			7,628,030
Rio Grande			38,592	38,592	7,752	7,674							46,266
Routt and Moffat	13,561			13,561									13,561
Saguache			1,422	1,422	21,285	21,072	68,047	8,710	260,577	11,205			42,409
San Juan			192,109	192,109	769,545	761,850	235,467	30,140	6,857,544	294,874			1,278,973
San Miguel	5,000		641,993	646,993	1,410,903	1,397,525			139,344	5,992			2,050,510
Summit	10,000		79,132	89,132	523,658	518,421			10,591,152	455,420	75,000	3,750	1,066,723
Teller			1,930	1,930									1,930
	125,663		4,474,337	4,600,000	21,160,000	20,948,401	6,336,878	811,121	126,256,000	5,429,009	300,000	15,000	31,803,531

Gold, silver, copper, lead, and zinc produced in Colorado, 1858-1923, by counties, in terms of recovered metals—Continued

County	Ore sold or treated (short tons)	Gold			Silver		Copper		Lead		Zinc		Total value
		Placer	Lode	Total	Fine ounces	Value	Pounds	Value	Pounds	Value	Pounds	Value	
1892													
Boulder			$982,988	$982,988	182,156	$158,476			9,697	$388			$1,141,852
Chaffee		$32,000	115,203	147,203	85,632	74,500			6,324,319	252,973	100,000	$4,000	479,276
Clear Creek	5,340		308,701	314,041	1,691,579	1,471,674	40,424	$4,689	7,916,672	316,667	250,000	11,500	2,118,571
Custer			325	325	9,635	8,382			4,963	199			8,906
Dolores			235,669	235,669	1,285,179	1,118,106	13,043	1,513	3,083,168	123,327			1,478,615
Eagle			139,299	139,299	347,954	302,720			5,259,280	210,371			652,390
Gilpin			1,358,157	1,358,157	134,462	116,982	538,988	62,523	2,232,158	89,286			1,626,948
Gunnison			6,004	6,004	146,891	127,795			525,574	21,023			154,822
Hinsdale			22,514	22,514	411,758	358,229	29,914	3,470	4,753,783	190,151			574,364
Lake	323,187	9,000	242,296	251,296	5,898,020	5,131,277	5,928,863	687,748	44,009,114	1,760,365	592,500	25,875	7,856,561
La Plata and Montezuma			34,881	34,881	3,335	2,901							37,782
Mineral			87,219	87,219	2,391,514	2,080,617			3,000,000	120,000			2,287,836
Ouray			138,688	138,688	754,114	656,079	638,875	74,109	8,012,729	320,509			1,189,385
Park		10,000	29,687	39,687	43,792	38,099			25,698	1,028			78,814
Pitkin					8,138,549	7,080,538			20,998,701	839,948			7,920,486
Rio Grande			14,487	14,487	12,526	10,898							25,385
Routt and Moffat		560		560									560
Saguache									250,000	10,000			10,000
San Juan			148,908	148,908	397,589	345,903	136,768	15,865	6,406,665	256,267			766,943
San Miguel		5,000	689,177	694,177	1,501,898	1,306,651	100,000	11,600	815,842	32,634			2,045,062
Summit		10,000	116,046	126,046	563,417	490,173	166,799	19,349	6,371,637	254,865	212,500	9,775	900,208
Teller			557,851	557,851									557,851
		71,900	5,228,100	5,300,000	24,000,000	20,880,000	7,593,674	880,866	120,000,000	4,800,001	1,125,000	51,750	31,912,617
1893													
Boulder			479,665	479,665	257,462	200,820	50,000	5,400	10,000	370			686,255
Chaffee		42,000	112,164	154,164	92,448	72,109	50,000	5,400	4,000,000	148,000	100,000	4,000	383,673
Clear Creek	5,000		579,187	584,187	2,218,377	1,730,334	40,000	4,320	8,000,000	296,000	400,000	16,000	2,630,841
Custer			4,021	4,021	32,204	25,119			150,000	5,550			34,690
Dolores			442,105	442,105	2,675,238	2,086,686	10,000	1,080	4,500,000	166,500			2,696,371
Eagle			168,867	168,867	187,658	146,373			5,000,000	185,000			500,240
Gilpin			1,218,626	1,218,626	135,850	105,963	600,000	64,800	2,000,000	74,000			1,463,389
Gunnison			7,728	7,728	144,577	112,770			500,000	18,500			138,998
Hinsdale			88,750	88,750	385,653	300,809	10,000	1,080	3,808,111	140,900			531,539
Lake	351,794		902,244	902,244	6,795,454	5,300,455	5,000,000	540,000	36,274,889	1,342,171	735,000	29,400	8,114,270
La Plata and Montezuma			37,872	37,872	4,928	3,844							41,716
Mineral			53,252	53,252	4,897,684	3,820,194			7,500,000	277,500			4,130,946
Ouray			188,854	188,854	1,221,155	952,501	600,000	64,800	8,000,000	296,000			1,502,155
Park		30,000	79,845	109,845	62,350	48,633	10,000	1,080	30,000	1,110			160,668
Pitkin					5,039,799	3,931,043			15,000,000	555,000			4,486,043
Rio Grande					796	621							621
Routt and Moffat		6,216		6,216									6,216
Saguache									250,000	9,250			9,250
San Juan			260,668	260,668	327,153	255,179	1,125,826	121,589	8,000,000	296,000			933,436
San Miguel		5,000	677,680	682,680	932,568	727,403	200,000	21,600	700,000	25,900			1,457,583
Summit		10,000	106,168	116,168	421,566	328,821			6,277,000	232,249	415,000	16,600	693,838
Teller			2,021,088	2,021,088	5,680	4,430							2,025,518
		98,216	7,428,784	7,527,000	25,838,600	20,154,107	7,695,826	831,149	110,000,000	4,070,000	1,650,000	66,000	32,648,256
1894													
Arapahoe	86		86	86									86
Boulder			489,592	489,592	75,730	47,710	50,000	4,750	10,000	330			542,382
Chaffee	35,000		85,565	120,565	25,527	16,082	50,000	4,750	1,100,000	36,300	100,000	3,500	181,197
Clear Creek	5,000		657,649	662,649	2,228,846	1,404,173	40,000	3,800	8,000,000	264,000	200,000	7,000	2,341,622
Conejos			171	171	1	1							172
Custer			118	148	1,137	716			150,000	4,950			5,814
Delta			172	172	3	2							174
Dolores			192,626	192,626	1,153,325	726,595	30,000	2,850	2,000,000	66,000			988,071
Eagle			55,521	55,521	62,543	39,402			2,000,000	66,000			160,923
Fremont			76	76	323	203							279
Garfield			63	63									63
Gilpin			1,915,863	1,915,863	228,927	144,224	400,000	38,000	2,200,000	72,600			2,170,687
Gunnison			8,052	8,052	104,938	66,111			400,000	13,200			87,363
Hinsdale			85,196	85,196	395,899	249,416	10,000	950	3,322,170	109,632			445,194
Huerfano			304	304	1	1							305
Jefferson	2,197			2,197	10	6							2,203
Lake			1,499,314	1,499,314	7,695,108	4,847,918	4,000,000	380,000	44,733,000	1,476,189	1,600,000	35,000	8,238,421
La Plata and Montezuma			114,264	114,264	417,465	263,003							377,267
Mesa	318		318	318	1	1							319
Mineral			40,336	40,336	1,866,927	1,176,164			6,500,000	214,500			1,431,000
Montrose	2,202		2,202	2,202	16	10							2,212
Ouray			178,138	178,138	995,153	626,946	600,000	57,000	4,422,000	145,926			1,008,010
Park	40,000		57,358	97,358	43,817	27,605	10,000	950	30,000	990			126,903
Pitkin			5,312	5,312	5,996,851	3,778,016			15,750,000	519,750			4,303,078
Pueblo			296	296	3	2							298
Rio Grande			16,816	16,816	1,260	794							17,610
Routt and Moffat	8,944			8,944	97	61							9,005
Saguache			17,515	17,515	608,224	383,181			250,000	8,250			408,946
San Juan			340,023	340,023	351,114	221,202	1,118,222	106,231	4,000,000	132,000			799,456
San Miguel	5,000		789,218	794,218	570,023	359,115	173,191	16,453	858,830	28,341			1,198,127
Summit	10,000		214,791	224,791	432,794	272,660			5,500,000	181,500	200,000	7,000	2,634,349
Teller			2,618,388	2,618,388	25,335	15,961							2,634,349
		108,747	9,382,767	9,491,514	23,281,398	14,667,281	6,481,413	615,734	101,226,000	3,340,458	1,500,000	52,500	28,167,487
1895													
Arapahoe		1,081		1,081	59	38							1,119
Boulder			401,926	401,926	40,685	26,445	57,864	6,191	11,439	366			434,928
Chaffee	35,000		118,629	153,629	29,630	19,260	76,070	8,140	285,056	9,122	120,000	4,320	194,471
Clear Creek	5,000		669,210	674,210	1,585,483	1,030,564	44,168	4,726	6,415,936	205,310	200,000	7,200	1,922,010
Costilla		126		126									126
Custer			68	68	88,632	57,611	4,099	439	139,768	4,473			62,591
Delta			77	77	1	1							78
Dolores			52,552	52,552	399,283	259,534	64,151	6,864	313,824	10,042			328,992
Eagle			30,900	30,900	53,421	34,724			1,770,215	56,647			122,271
Fremont			18	18									18
Garfield			153	153	1	1							154
Gilpin			1,196,319	1,196,319	190,256	123,666	209,414	22,407	844,037	27,009			1,369,401
Gunnison			36,734	36,734	114,218	74,242			201,898	6,461			117,437
Hinsdale			243,195	243,195	483,565	314,317	10,000	1,070	5,251,014	168,032			726,614

Gold, silver, copper, lead, and zinc produced in Colorado, 1858–1923, by counties, in terms of recovered metals—Continued

County	Ore sold or treated (short tons)	Gold Placer	Gold Lode	Gold Total	Silver Fine ounces	Silver Value	Copper Pounds	Copper Value	Lead Pounds	Lead Value	Zinc Pounds	Zinc Value	Total value
1895—Continued													
Huerfano			$87	$87									$87
Jefferson		$1,861	731	2,592	15	$10							2,602
Lake			1,386,359	1,386,359	9,435,413	6,133,018	2,803,550	$299,980	38,922,572	$1,245,522	1,265,000	$45,540	9,110,419
La Plata and Montezuma			3,682	3,682	99	64							3,746
Larimer and Jackson		320		320	1	1							321
Mineral			114,482	114,482	1,423,038	924,975			8,220,870	263,068			1,302,525
Montrose		1,181		1,181	11	7							1,188
Ouray			172,697	172,697	1,515,693	985,200	600,000	64,200	5,747,003	183,904			1,406,001
Park		30,000	101,761	131,761	46,658	30,328	2,938	314	98,791	3,161			165,564
Pitkin			1,387	1,387	5,131,792	3,335,665	616	66	11,163,685	357,238	21,000	756	3,695,112
Rio Grande			15,795	15,795	3,359	2,183							17,978
Routt and Moffat		5,930		5,930	86	56							5,986
Saguache			534	534	3,939	2,560			249,166	7,973			11,067
San Juan			849,411	849,411	1,894,453	1,231,394	2,057,588	220,162	8,698,800	259,162			2,560,129
San Miguel		5,000	1,421,159	1,426,159	602,039	391,325	147,727	15,807	756,809	24,218			1,857,509
Summit		10,000	225,591	235,591	288,242	187,357	1,058	113	5,477,117	175,268	65,000	2,340	600,669
Teller			6,166,144	6,166,144	68,428	44,478							6,210,622
		95,499	13,209,601	13,305,100	23,398,500	15,209,024	6,079,243	650,479	93,968,000	3,006,976	1,671,000	60,156	32,231,735
1896													
Arapahoe		1,894		1,894	19	13							1,907
Boulder			385,653	385,653	79,047	53,752	63,252	6,831	4,216	126			446,362
Chaffee		35,000	158,465	193,465	151,738	103,182	559	60	1,047,310	31,419	120,000	4,680	332,866
Clear Creek		5,000	787,631	792,631	1,626,828	1,106,243	204,519	22,088	6,438,672	193,160	400,000	15,600	2,129,722
Conejos			639	639	17	12							651
Costilla		139		139									139
Custer			42	42	60,122	40,883	1,169	120	82,105	2,463			43,508
Delta			339	339	1	1							340
Dolores			10,659	10,659	240,393	163,467			1,100,000	33,000	30,000	1,170	208,296
Eagle			16,472	16,472	65,824	44,760	2,044	221	210,717	6,322			67,775
Fremont			915	915	15	10							925
Gilpin			1,534,358	1,534,358	295,182	200,724	435,838	47,071	1,948,756	58,462			1,840,615
Grand			200	200									200
Gunnison			26,757	26,757	93,273	63,426	8,515	920	164,370	4,931			96,034
Hinsdale			212,794	212,794	510,883	347,400	13,202	1,426	5,468,856	164,066			725,686
Huerfano			169	169									169
Jefferson		1,963	16,523	18,486	4,590	3,121							21,607
Lake			1,453,458	1,453,458	6,623,764	4,504,160	4,071,761	439,750	31,993,777	959,813	642,000	25,038	7,382,219
La Plata and Montezuma			10,741	10,741	41	28							10,769
Larimer and Jackson		13		13	3	2							15
Mineral			52,238	52,238	1,560,865	1,061,388			6,021,109	180,633			1,294,259
Montrose		1,720	225	1,945	17	12							1,957
Ouray			141,046	141,046	2,371,912	1,612,900	217,310	23,469	6,599,143	197,974			1,975,389
Park		25,000	112,109	137,109	117,095	79,625	28,593	3,088	297,714	8,931			228,753
Pitkin			1,523	1,523	4,922,360	3,347,205	52,991	5,723	16,272,411	488,172			3,842,623
Pueblo			84	84	26	17							101
Rio Grande			1,870	1,870	1,353	920	1,369	148	451	14			2,952
Routt and Moffat		4,690	169	4,859	2,214	1,506			22,111	663			7,028
Saguache			331	331	2,447	1,664	241	26	65,465	1,964			3,985
San Juan			908,707	908,707	2,228,031	1,515,061	845,094	91,270	5,634,586	169,038			2,684,076
San Miguel		5,000	1,372,829	1,377,829	1,109,875	754,715	21,698	2,343	2,284,191	68,526			2,203,413
Summit		10,000	200,202	210,202	441,448	300,185	54,081	5,841	3,950,040	118,501	100,000	3,900	638,629
Teller			7,413,493	7,413,493	63,617	43,260							7,456,753
		90,419	14,820,581	14,911,000	22,573,000	15,349,642	6,022,176	650,395	89,606,000	2,688,178	1,292,000	50,388	33,649,603
1897													
Arapahoe		2,108		2,108	14	8							2,116
Archuleta			703	703	348	209							912
Boulder			512,657	512,657	138,715	83,229	58,474	7,017	309,115	11,128			614,031
Chaffee		35,000	191,936	226,936	53,859	32,315	172,891	20,747	1,686,391	60,710	100,000	4,100	344,808
Clear Creek		50,000	732,649	782,649	1,442,583	865,550	516,034	61,924	5,263,116	189,472	300,000	12,300	1,911,895
Conejos			1,054	1,054	98	59							1,113
Costilla		650	4,766	5,416	482	289	502	60	50,048	1,802			7,567
Custer			2,129	2,129	26,842	16,105	874	105	2,101,041	75,637			93,976
Delta			289	289									289
Dolores			43,469	43,469	179,901	107,941	39,654	4,758	1,093,840	39,378			195,546
Douglas		475		475	10	6							481
Eagle			34,767	34,767	46,046	27,628	2,200	264	1,144,013	41,184			103,843
Fremont			12,877	12,877	1,525	915							13,792
Garfield			310	310	42	25							335
Gilpin			2,086,471	2,086,471	374,417	224,670	1,018,595	122,231	2,007,698	72,277			2,505,629
Grand			1,943	1,943	85	51							1,994
Gunnison			40,761	40,761	103,941	62,365	2,770	332	1,013,114	36,472			139,930
Hinsdale			168,171	168,171	243,437	146,062	8,085	970	5,550,058	199,802			515,605
Huerfano			723	723	167	100	92	11	1,067	38			872
Jefferson		586	7,661	8,247	1,614	968	1,602	192	10,093	363			9,770
Lake	413,552		2,063,858	2,063,858	5,451,317	3,270,790	3,146,802	377,616	23,700,908	853,233	2,201,500	90,262	6,655,759
La Plata and Montezuma			36,944	36,944	1,514	908			857	31			37,933
Larimer and Jackson		805	2,171	2,976	97	58							3,034
Las Animas			641	641	9	5							646
Mineral			61,328	61,328	3,070,576	1,842,346	1,500	180	6,080,673	218,904			2,122,738
Montrose		1,571	4,981	6,552	851	511							7,063
Ouray			552,840	552,840	2,776,394	1,665,836	2,185,084	262,210	7,784,212	280,232			2,761,118
Park		19,000	134,619	153,619	199,945	119,967	58,002	6,962	4,517,614	162,634			443,182
Pitkin			164,430	164,430	4,599,946	2,759,968	8,360	1,003	4,456,478	160,433			3,085,834
Rio Grande			22,592	22,592	8,168	4,901	627	75	12,006	432			28,000
Routt and Moffat		5,451	4,326	9,777	7,805	4,683	958	115	88,736	3,194			17,769
Saguache			13,746	13,746	2,482	1,489	2,975	357	9,266	334			15,926
San Juan			694,326	694,326	1,101,907	661,144	1,435,203	172,224	8,021,414	288,771			1,816,465
San Miguel		5,000	1,453,144	1,458,144	869,079	521,447	354,781	42,574	4,143,767	149,176			2,171,341
Summit		10,000	263,650	273,650	514,107	308,464	133,482	16,018	1,748,761	62,955	82,489	3,382	664,469
Teller			10,131,855	10,131,855	59,879	35,927							10,167,782
	130,646	19,448,787	19,579,433	21,278,202	12,766,919	9,149,967	1,097,995	80,794,286	2,908,592	2,083,989	110,044	36,462,983	

Gold, silver, copper, lead, and zinc produced in Colorado, 1858–1923, by counties, in terms of recovered metals—Continued

County	Ore sold or treated (short tons)	Gold			Silver		Copper		Lead		Zinc		Total value
		Placer	Lode	Total	Fine ounces	Value	Pounds	Value	Pounds	Value	Pounds	Value	
1898													
Arapahoe		$703		$703	7	$4							$707
Archuleta			$145	145	40	24							169
Boulder			581,302	581,302	91,432	53,945	22,452	$2,784	8,967	$341			638,372
Chaffee		35,000	192,535	227,535	85,273	50,311	114,202	14,161	2,522,554	95,857	100,000	$4,600	392,464
Clear Creek		5,000	600,528	605,528	1,569,012	925,717	317,423	39,360	5,843,767	222,063	300,000	13,800	1,806,468
Conejos			18,355	18,355	29,777	17,568							35,923
Costilla		1,000	4,519	5,519	993	586	983	122					6,227
Custer			723	723	24,319	14,348	1,475	183	996,877	37,881			53,135
Delta			579	579	16	9							588
Dolores			88,282	88,282	463,346	273,374	149,647	18,556	686,597	26,091	400,000	18,400	424,703
Douglas		124		124									124
Eagle			30,571	30,571	70,783	41,762	71,049	8,810	1,851,512	70,357			151,500
Fremont			8,702	8,702	1,270	749			2,101	80			9,531
Gilpin			1,983,514	1,983,514	305,687	180,355	633,707	78,580	1,216,338	46,221			2,288,670
Grand			806	806	11	6							812
Gunnison			81,006	81,006	152,800	90,152	119,072	14,765	1,996,560	75,869			261,792
Hinsdale			51,282	51,282	186,456	110,009	104,038	12,901	9,828,482	373,482			547,674
Huerfano			145	145	40	24							169
Jefferson		117	1,723	1,840	102	60							1,900
Lake	517,992		2,073,036	2,073,036	7,068,727	4,170,550	5,543,954	687,450	35,945,006	1,365,910	2,673,500	122,981	8,419,927
La Plata and Montezuma			38,653	38,653	5,219	3,079	2,568	318	8,407	319			42,369
Larimer and Jackson		10,456	706	11,162	60	35	24,484	3,036					14,233
Las Animas			124	124									124
Mesa			165	165	20	12							177
Mineral			46,383	46,383	4,177,184	2,464,539	14,729	1,826	5,453,104	207,218	200,000	9,200	2,729,166
Montrose		300	2,408	2,708	6,290	3,711	34,664	4,298					10,717
Ouray			852,555	852,555	1,420,330	837,995	1,035,562	128,410	2,799,936	106,398			1,925,358
Park		7,000	152,490	159,490	198,711	117,239	20,957	2,599	1,953,001	74,214			353,542
Pitkin	165,000		71,001	71,001	3,977,270	2,346,589	4,553	565	15,903,682	604,340			3,022,495
Rio Grande			3,720	3,720	1,568	925	9,794	1,214	2,393	91			5,950
Routt and Moffat		11,728	1,025	12,753	2,173	1,282	600	74	15,477	588			14,697
Saguache			19,678	19,678	2,618	1,545	21,711	2,692	132,462	5,034			28,949
San Juan			1,132,592	1,132,592	1,048,499	618,614	2,252,421	279,300	14,659,999	557,080			2,587,586
San Miguel		2,000	1,570,677	1,572,677	2,129,082	1,256,158	360,831	44,743	6,699,712	254,589			3,128,167
Summit		10,000	333,825	343,825	415,687	245,255	9,825	1,218	4,889,204	185,790	227,156	10,449	786,537
Teller			13,507,349	13,507,349	67,799	40,001							13,547,350
	83,428		23,451,104	23,534,532	23,502,601	13,866,532	10,870,701	1,317,916	113,416,135	4,309,813	3,900,656	179,430	43,238,272
1899													
Arapahoe		269		269	2	1							270
Archuleta			103	103	43	26							129
Boulder			547,858	547,858	76,371	45,823	78,816	13,478	28,043	1,262			608,421
Chaffee		25,000	191,663	216,663	147,339	88,403	696,736	119,142	1,193,074	53,688	100,000	5,800	483,696
Clear Creek		5,000	541,825	546,825	1,502,900	901,740	292,966	50,097	7,216,260	324,732	300,000	17,400	1,840,794
Conejos			6,263	6,263	22,987	13,792							20,055
Costilla		300	506	806	126	76							882
Custer			1,054	1,054	6,004	3,602	923	158	836,894	37,660			42,474
Delta			207	207	10	6							213
Dolores			66,847	66,847	257,052	154,231	44,509	7,611	2,046,232	92,080	100,000	5,800	326,569
Douglas		83		83	24	14							97
Eagle			46,094	46,094	44,393	26,636	5,876	1,005	1,187,930	53,457			127,192
Fremont			9,405	9,405	3,974	2,384	6,698	1,145	11,443	515			13,449
Garfield			723	723	17	10							733
Gilpin			1,996,061	1,996,061	340,652	204,391	1,037,421	177,399	1,312,312	59,054			2,436,905
Grand			124	124	13	8							132
Gunnison			70,112	70,112	132,983	79,790	46,186	7,898	1,399,336	62,970			220,770
Hinsdale			38,343	38,343	155,902	93,541	49,676	8,495	10,572,353	475,756			616,135
Huerfano			124	124	5	3							127
Jefferson		542	822	1,364	351	211			770	35			1,653
Lake	525,728		2,196,498	2,196,498	7,230,118	4,338,071	3,202,828	547,684	48,598,720	2,186,942	10,575,240	613,364	9,882,559
La Plata and Montezuma			41,092	41,092	3,389	2,033	211	36	3,176	143			43,304
Larimer and Jackson		1,599	468	2,067	135	81	2,474	423					2,571
Las Animas			207	207	3	2							209
Mesa			124	124	4,120	2,472	4,650	795					3,391
Mineral		103	91,671	91,671	3,796,899	2,278,139	20,223	3,458	5,677,162	255,472	100,000	5,800	2,634,540
Montrose			620	723	46,119	27,671	75,006	12,826					41,220
Ouray			1,694,940	1,694,940	2,346,194	1,407,716	305,177	52,185	7,556,386	340,037			3,494,878
Park		20,000	133,041	153,041	72,137	43,282	7,903	1,351	540,849	24,338			222,012
Pitkin			52,233	52,233	4,158,708	2,495,225	19,351	3,309	25,458,380	1,145,627			3,696,394
Rio Grande			19,202	19,202	2,718	1,631	336	57	1,635	74			20,964
Routt and Moffat		10,693	862	11,555	1,271	763			3,405	153			12,471
Saguache			3,886	3,886	14,306	8,584	35,319	6,040	441,095	19,849			38,359
San Juan			996,273	996,273	1,191,857	715,114	1,197,661	204,800	16,011,677	720,525			2,636,712
San Miguel			1,376,705	1,376,705	1,208,395	725,037	160,239	27,401	3,918,883	176,350			2,305,493
Summit		10,000	250,566	260,566	264,872	158,923	65,531	11,205	4,032,431	181,459	125,416	7,274	619,427
Teller			16,058,564	16,058,564	82,299	49,379							16,107,943
	73,589		26,435,086	26,508,675	23,114,688	13,868,811	7,356,970	1,258,041	138,048,446	6,212,178	11,300,656	655,438	48,503,143
1900													
Arapahoe		248		248									248
Archuleta			145	145	30	18							163
Baca			103	103	102	63	8,900	1,477					1,643
Boulder			607,016	607,016	90,327	56,003	20,371	3,382	76,076	3,347			609,748
Chaffee		25,000	147,677	172,677	125,330	77,705	753,667	125,110	833,462	36,672	100,000	4,400	416,564
Clear Creek		5,000	460,447	465,447	1,358,143	842,049	244,092	40,519	4,994,263	219,748	300,000	13,200	1,580,963
Conejos			2,832	2,832	1,014	629	4,527	752	2,200	97			4,310
Costilla		200	1,867	2,067	314	195	107	18					2,280
Custer			20,835	20,835	82,605	51,215	2,301	382	709,349	31,211	20,000	880	104,523
Delta			971	971	97	60							1,031
Dolores			50,125	50,125	159,318	98,777	36,009	5,978	210,380	9,257	220,000	9,680	173,817
Douglas		62		62	24	15							77
Eagle			103,598	103,598	234,674	145,498	359,054	59,603	3,679,828	161,912	20,000	880	471,491
Fremont			8,309	8,309	2,199	1,363	6,725	1,116	8,282	365			11,153
Gilpin			1,655,502	1,655,502	236,400	146,568	799,478	132,713	735,773	32,374			1,967,157
Garfield			517	517	13	8							525
Grand			3,762	3,762	21	13							3,775
Gunnison			83,858	83,858	146,746	90,982	42,790	7,103	1,583,320	69,666	100,000	4,400	256,009
Hinsdale			56,470	56,470	155,485	96,400	29,180	4,844	9,377,062	412,591	100,000	4,400	574,706
Huerfano			124	124	20	12							136
Jefferson		78	625	703	51	32							735

Gold, silver, copper, lead, and zinc produced in Colorado, 1858–1923, by counties, in terms of recovered metals—Continued

County	Ore sold or treated (short tons)	Gold Placer	Gold Lode	Gold Total	Silver Fine ounces	Silver Value	Copper Pounds	Copper Value	Lead Pounds	Lead Value	Zinc Pounds	Zinc Value	Total value
1900—Continued													
Lake	808,071		$2,529,512	$2,529,512	6,967,279	$4,319,713	2,728,553	$452,940	62,599,654	$2,754,385	14,441,000	$635,404	$10,691,954
La Plata and Montezuma			24,927	24,927	7,187	4,456	350	58	14,500	638			30,079
Larimer and Jackson		$1,078	555	1,633	126	78	13,806	2,292					4,003
Mesa			124	124	311	193	2,150	357					674
Mineral			209,387	209,387	2,280,038	1,413,623	2,614	434	14,951,956	657,886	450,000	19,800	2,301,130
Montrose		300	1,333	1,633	19,652	12,184	32,026	5,316					19,133
Ouray			1,437,909	1,437,909	1,985,267	1,230,866	352,368	58,493	9,478,657	417,061	20,000	880	3,145,209
Park		18,000	98,558	116,558	43,138	26,746	15,000	2,490	682,107	30,013			175,807
Pitkin			13,456	13,456	4,119,116	2,553,852	6,082	1,010	27,452,260	1,207,899	20,000	880	3,777,097
Pueblo			248	248	9	5							253
Rio Grande			107,629	107,629	3,075	1,906	8,599	1,427	26,200	1,155			112,117
Routt and Moffat		3,000	287	3,287	477	296	5,765	957					4,540
Saguache			7,979	7,979	15,793	9,792	16,129	2,677	316,061	13,907			34,355
San Juan			757,204	757,204	681,317	422,416	1,972,087	327,367	17,579,177	773,484			2,280,471
San Miguel			1,827,352	1,827,352	1,136,692	704,749	311,045	51,633	3,353,425	147,551			2,731,285
Summit		25,000	313,182	338,182	403,330	250,065	53,030	8,803	5,610,710	246,871	491,055	21,606	865,527
Teller			18,149,645	18,149,645	80,792	50,091							18,199,736
		77,966	28,684,070	28,762,036	20,336,512	12,608,637	7,826,815	1,299,251	164,274,762	7,228,090	16,282,055	716,410	50,614,424
1901													
Arapahoe		331		331									331
Archuleta			124	124	18	11							135
Baca			83	83	80	48	590	99					230
Boulder			774,298	774,298	113,782	68,269	22,186	3,705	191,987	8,255			854,527
Chaffee		25,000	133,684	158,684	76,286	45,772	576,251	96,234	209,768	9,020	100,000	4,100	313,810
Clear Creek		5,000	535,975	540,975	1,271,227	762,736	374,534	62,547	3,890,216	167,279	300,000	12,300	1,545,837
Conejos			1,178	1,178	102	61	210	35	1,200	52			1,326
Costilla		200	771	971	153	92	235	39					1,102
Custer			11,120	11,120	50,394	30,236	40,528	6,768	400,481	17,221			65,345
Delta			517	517	10	6							523
Dolores			22,303	22,303	111,632	66,979	13,106	2,189	367,057	15,783	250,000	10,250	117,504
Douglas		103		103	10	6							109
Eagle			97,376	97,376	175,181	105,109	157,914	26,372	2,775,291	119,338			348,195
Fremont			6,449	6,449	933	560	15,907	2,656	33,945	1,460			11,125
Garfield			351	351	13	8							359
Gilpin			1,638,966	1,638,966	271,638	162,983	731,194	122,109	670,018	28,811			1,952,869
Grand			1,034	1,034	30	18							1,052
Gunnison			83,445	83,445	93,243	55,946	53,396	8,917	656,631	28,235	100,000	4,100	180,643
Hinsdale			76,148	76,148	152,122	91,273	12,532	2,093	7,588,675	326,313	126,591	5,190	501,017
Huerfano			83	83	10	6							89
Jefferson			310	310	20	12							322
Lake	793,014		1,776,132	1,776,132	6,830,084	4,098,050	1,930,556	322,403	56,359,708	2,423,467	23,167,140	949,853	9,569,905
La Plata and Montezuma			30,819	30,819	5,588	3,353	132	22	6,197	266			34,460
Larimer and Jackson		522	408	930	73	44	18,140	3,029					4,003
Mesa		1,940	106	2,046	155	93	7,795	1,302					3,441
Mineral			102,813	102,813	1,816,023	1,089,614	1,007	168	10,519,895	452,355	1,800,000	73,800	1,718,750
Montrose		301	1,249	1,550	101,359	60,815	55,944	9,343					71,708
Ouray			1,546,323	1,546,323	1,633,725	980,235	652,937	109,040	7,904,724	339,903			2,975,501
Park		18,000	78,322	96,322	69,175	41,505	9,657	1,613	421,955	18,144			157,584
Pitkin			4,692	4,692	3,532,863	2,119,718	50,786	8,481	32,719,511	1,408,229			3,541,120
Pueblo			165	165	52	31	210	35					231
Rio Grande			32,927	32,927	6,926	4,156	65,603	10,956	677	29			48,008
Routt and Moffat		927	3,517	4,444	239	143	500	84	2,193	94			4,765
Saguache			79,972	79,972	20,507	12,304	15,253	2,547	235,750	10,137			104,960
San Juan	242,850		962,974	962,974	784,218	470,531	2,740,042	457,587	15,473,187	665,347			2,556,439
San Miguel			2,049,472	2,049,472	916,245	519,747	308,322	51,490	3,309,517	142,309			2,793,018
Summit		35,000	303,719	338,719	368,887	221,332	17,062	2,819	4,342,437	186,725	1,000,000	41,000	790,625
Teller			17,234,294	17,234,294	89,560	53,736							17,288,030
		87,324	27,592,119	27,679,443	18,492,563	11,095,538	7,872,529	1,314,712	148,111,020	6,368,772	26,843,731	1,100,593	47,559,058
1902													
Arapahoe		227		227									227
Archuleta			83	83	10	5							88
Baca			103	103	59	31	1,929	235					369
Boulder			538,702	538,702	82,710	43,836	11,090	1,353	13,493	553			584,444
Chaffee		40,000	377,513	417,513	114,155	60,502	173,538	21,172	456,889	18,732	220,500	10,584	528,503
Clear Creek		5,000	925,481	930,481	1,279,050	677,897	473,754	57,798	3,282,270	134,573	317,705	15,250	1,815,999
Conejos			1,261	1,261	81	43	78	10					1,314
Costilla		200	978	1,178	205	109							1,287
Custer			23,708	23,708	28,189	14,940	32,945	4,019	94,662	3,881	40,500	1,944	48,492
Delta			413	413	12	6							419
Dolores			47,458	47,458	121,311	64,295	15,054	1,837	388,806	15,941	248,680	11,937	141,468
Douglas		62		62	10	6							67
Eagle			31,956	31,956	45,336	24,028	150,134	18,316	832,846	34,147			108,447
Fremont			7,379	7,379	515	273	22,300	2,721	2,836	116	22,825	1,096	11,585
Garfield			165	165	5	3							168
Gilpin			1,551,035	1,551,035	303,638	160,928	765,516	193,393	497,366	20,392			1,925,748
Grand			1,302	1,302	24	13							1,315
Gunnison			103,536	103,536	123,138	65,263	28,686	3,500	728,935	29,886	131,975	6,335	208,520
Hinsdale			98,348	98,348	117,177	62,104	8,314	1,014	6,213,763	254,764	319,000	15,312	431,542
Huerfano			847	847	260	138							985
Jefferson			517	517	3	2	2,978	363					882
Lake	748,946		1,203,924	1,203,924	5,641,857	2,990,184	2,611,167	318,562	39,450,178	1,617,457	47,637,490	2,286,600	8,416,727
La Plata and Montezuma			127,182	127,182	7,416	3,930	3,143	383	2,156	88			131,583
Larimer and Jackson		488	318	806	49	26	24,888	3,036					3,868
Mesa		84	453	537	32	17	15,000	1,830					2,384m
Mineral			112,838	112,838	1,923,917	1,019,706	2,505	306	9,291,358	380,946	2,047,555	98,283	1,611,773
Montrose		1,868	4,085	5,953	3,149	1,669	2,505	306	64	3			7,931
Ouray			2,420,726	2,420,726	789,855	418,623	526,541	64,238	4,262,063	174,745			3,078,332
Park		14,000	128,458	142,458	49,968	26,483	8,113	990	261,046	10,703			180,634
Pitkin			4,899	4,899	3,063,450	1,623,629	10,654	1,300	24,973,816	1,023,926			2,653,754
Rio Grande			14,262	14,262	3,171	1,681	1,260	154	166	7			16,104
Routt and Moffat		13,845	1,306	15,151	136	72							15,223
Saguache			5,023	5,023	10,486	5,558	13,669	1,668	454,995	18,655	267,100	12,821	43,725
San Juan	230,000		1,524,226	1,524,226	838,102	444,194	3,012,283	367,499	7,699,883	315,695			2,651,614
San Miguel			2,007,656	2,007,656	1,056,640	560,019	454,790	55,484	4,296,849	176,171			2,799,330
Summit		43,000	199,583	242,583	274,571	145,523	93,609	11,420	3,092,387	126,788	1,329,180	63,801	590,115
Teller			16,932,416	16,932,416	62,780	33,273							16,965,689
		118,774	28,398,140	28,516,914	15,941,523	8,449,008	8,463,938	1,132,601	106,296,827	4,358,169	52,582,510	2,523,963	44,980,655

Gold, silver, copper, lead, and zinc produced in Colorado, 1858-1923, by counties, in terms of recovered metals—Continued

County	Ore sold or treated (short tons)	Gold			Silver		Copper		Lead		Zinc		Total value
		Placer	Lode	Total	Fine ounces	Value	Pounds	Value	Pounds	Value	Pounds	Value	
1903													
Arapahoe		$165		$165									$165
Archuleta			$62	62	6	$3							65
Boulder			431,569	431,569	61,833	33,390	6,154	$843	115,100	$4,834			470,636
Chaffee	28,000		141,329	169,329	129,900	70,146	79,581	10,903	249,308	10,471	3,000	$162	261,011
Clear Creek	5,000		467,061	472,061	851,638	459,885	289,876	39,713	3,451,849	144,978	656,000	35,424	1,152,061
Conejos			1,220	1,220	46	25							1,245
Costilla		200	792	992	179	97							1,089
Custer			82,804	82,804	160,175	86,495	52,242	7,157	387,301	16,267			192,723
Delta			248	248	8	4							252
Dolores			43,262	43,262	103,096	55,672	147,588	20,220	143,417	6,024			125,178
Douglas		41		41	2	1							42
Eagle			16,040	16,040	27,054	14,609	32,863	4,502	677,730	28,465			63,616
Fremont			6,346	6,346	223	120	20,777	2,846	2,091	88			9,400
Garfield			103	103	3	2							105
Gilpin			1,346,113	1,346,113	375,238	202,629	611,988	83,842	945,975	39,731			1,672,315
Grand			1,426	1,426	12	6							1,432
Gunnison			48,533	48,533	65,447	35,341	15,000	2,055	127,661	5,362	55,600	3,002	94,293
Hinsdale			16,515	16,515	33,139	17,895	11,263	1,543	459,462	19,297	106,000	5,724	60,974
Jefferson			248	248	5	3							281
Lake	770,000		1,339,974	1,339,974	4,973,033	2,685,438	2,556,583	350,252	36,353,239	1,526,836	76,566,000	4,134,564	10,037,064
La Plata and Montezuma			145,331	145,331	7,716	4,167	810	111	3,017	127			149,736
Larimer and Jackson		603	1,030	1,633	10	5	56,700	7,768					9,406
Mesa		351		351	8	4							355
Mineral			178,961	178,961	1,608,788	868,746	133	18	8,600,646	361,227	2,634,000	142,236	1,551,188
Montrose		300	2,511	2,811	2,061	1,113	10,920	1,496					5,420
Ouray			2,171,508	2,171,508	417,343	225,365	380,409	52,116	3,350,569	140,724			2,589,713
Park	12,000		124,277	136,277	52,128	28,149	5,895	808	802,489	33,705			198,939
Pitkin			4,754	4,754	2,569,862	1,387,725	11,683	1,601	33,269,852	1,397,334			2,791,414
Rio Grande			12,939	12,939	3,410	1,841	5,098	698					15,478
Routt and Moffat	19,289		1,546	20,835	117	63							20,898
Saguache			2,956	2,956	22,424	12,109	67,410	9,235	376,711	15,822	44,600	2,408	42,530
San Juan			1,710,608	1,710,608	781,358	421,933	2,939,018	402,645	6,969,093	292,702			2,827,888
San Miguel	3,100		1,173,705	1,176,805	737,028	397,995	466,264	63,878	3,704,201	155,576			1,794,254
Summit	60,000		162,265	222,265	220,543	119,093	41,447	5,678	1,523,703	63,996	550,800	29,743	440,775
Teller			11,840,272	11,840,272	41,605	22,467							11,862,739
		129,049	21,476,308	21,605,357	13,245,438	7,152,536	7,809,920	1,069,958	101,513,414	4,263,566	80,616,000	4,353,263	38,444,680
1904													
Arapahoe			248	248									248
Archuleta			124	124	10	6							130
Boulder	23,905		411,581	411,581	57,424	33,306	26,115	3,343	62,111	2,671			450,901
Chaffee	12,777	15,000	49,346	64,346	69,045	40,046	263,239	33,695	652,238	28,046	294,440	15,016	181,149
Clear Creek	62,661	2,398	634,217	636,615	873,949	506,890	401,180	51,351	3,913,976	168,301	906,705	46,242	1,409,399
Conejos			827	827	52	30							857
Costilla		368	300	668	151	88							756
Custer	10,170		53,453	53,453	87,373	50,676	15,068	1,929	126,593	5,444			111,502
Delta			351	351	9	6							357
Dolores	7,727		53,783	53,783	108,301	62,815	25,392	3,250	181,229	7,793	18,196	928	128,569
Douglas		289		289	5	3							292
Eagle	1,866		30,075	30,075	27,348	15,862	32,409	4,148	375,207	16,134			66,219
Fremont			4,671	4,671	208	121	1,024	131	1,071	46			4,969
Garfield			517	517	14	8							525
Gilpin	109,557		1,403,865	1,403,865	318,406	184,675	638,945	81,785	859,293	36,950			1,707,275
Grand			641	641	13	8	1,114	143					792
Gunnison	2,067		26,024	26,024	115,153	66,789	16,233	2,078	200,462	8,620	20,010	1,021	104,532
Hinsdale	5,591		10,521	10,521	46,585	27,019	13,187	1,688	1,041,222	44,773	59,089	3,014	87,015
Jefferson			2,894	3,245	37	21	538	69					3,335
Lake	663,487		1,186,851	1,186,851	5,085,151	2,949,388	3,734,593	478,028	47,180,865	2,028,777	58,254,353	2,970,972	9,614,016
La Plata and Montezuma	3,792		130,200	130,200	31,086	18,030	1,473	189	2,177	94			148,513
Larimer and Jackson	6	141	1,037	1,178	11	6	23,028	2,948					4,132
Mesa	9	248		248	9	5							253
Mineral	124,278		222,864	222,864	1,664,633	965,487	1,337	171	13,346,436	573,897	4,402,697	224,538	1,986,957
Montrose		121	1,367	1,488	1,067	619	7,476	957					3,064
Ouray	91,244		2,174,361	2,174,361	294,028	170,536	431,048	55,174	2,044,525	87,915	5,016	256	2,488,242
Park	4,202	10,000	184,980	194,980	50,013	29,008	5,920	758	757,703	32,581			257,327
Pitkin	109,770		2,336	2,336	2,129,618	1,235,178	9,862	1,262	18,882,901	811,965	593,661	30,277	2,081,018
Rio Grande			4,010	4,010	2,281	1,323	650	83					5,416
Routt and Moffat		22,164	2,061	24,225	181	105							24,330
Saguache	499		5,519	5,519	60,506	35,093	48,722	6,236	699,312	30,070	15,585	795	77,713
San Juan	233,663		1,396,651	1,396,651	1,042,044	604,386	3,467,124	443,792	9,288,643	399,412	317,254	16,180	2,860,421
San Miguel	233,316	44,957	1,486,111	1,531,068	667,710	387,272	239,520	30,659	5,704,708	245,302			2,194,301
Summit	35,475	94,240	113,886	208,126	180,554	104,721	7,510	961	2,178,182	93,662	1,884,584	96,114	503,584
Teller	597,819		14,456,536	14,456,536	47,817	27,734							14,484,270
	2,333,881	193,068	24,049,417	24,242,485	12,960,792	7,517,260	9,412,707	1,204,828	107,498,854	4,622,453	66,771,590	3,405,353	40,992,379
1905													
Boulder	9,577		261,601	261,601	70,921	43,262	2,227	347					305,210
Chaffee	13,408	15,009	17,369	32,378	75,265	45,912	869,507	135,643	1,250,302	58,764	849,963	50,148	322,845
Clear Creek	58,775	1,881	501,817	503,698	692,437	422,387	235,669	36,764	3,270,211	153,700	1,102,301	65,036	1,181,585
Conejos	12		2,894	2,894	900	549							3,443
Custer	4,555		24,918	24,918	32,159	19,617	2,500	390					44,925
Dolores	3,826		34,766	34,766	76,526	46,681	119,821	18,692	840,319	39,495	556,266	32,820	172,454
Eagle	12,049		46,891	46,891	46,487	28,357	29,331	4,576	156,723	7,366	605,612	35,731	122,921
Gilpin	182,873		1,450,033	1,450,033	340,901	207,950	512,276	79,915	519,841	24,433	33,090	1,952	1,764,283
Grand	12		31	31	22	13	1,680	262					306
Gunnison	5,581		28,156	28,156	53,649	32,726	50,500	7,878	219,809	10,331	17,905	1,056	80,147
Hinsdale	5,041		11,991	11,991	54,419	33,196	84,485	13,180	767,681	36,081	235,178	13,876	108,324
Huerfano			269	269	617	376							645
Jefferson	15	18,088		18,088	125	76	9,000	1,404					19,568
Lake	648,464		1,180,401	1,180,401	4,033,762	2,460,595	4,486,115	699,834	51,162,040	2,404,616	70,238,634	4,144,079	10,889,525
La Plata and Montezuma	5,662		254,007	254,007	93,258	56,887	2,923	456	610	29			311,379
Mineral	91,338		216,994	216,994	1,193,442	728,000	107	17	11,880,797	558,397	2,515,628	148,422	1,651,830
Ouray	98,966		2,333,282	2,333,282	758,107	462,445	524,199	81,775	5,348,264	251,368	48,267	2,848	3,131,718
Park	6,745	2,786	318,081	320,867	49,202	30,013	12,199	1,903	543,300	25,535			378,318
Pitkin	107,927		248	248	2,469,520	1,506,407	127,094	19,827	22,386,142	1,052,149	3,854,339	227,406	2,806,037
Rio Grande			4,051	4,051	1,055	644	123	19					4,714
Routt and Moffat		6,905		6,905	30	18							6,923
Saguache	496		699	699	4,401	2,685	1,135	177	203,797	9,578	2,917	172	13,311
San Juan	204,139		1,050,971	1,050,971	750,844	458,015	2,274,109	354,761	8,045,126	378,121	163,845	9,667	2,251,535
San Miguel	291,338	21,587	1,690,266	1,711,853	1,275,079	777,798	272,513	42,512	6,970,152	327,597	17,214	1,016	2,860,776

Gold, silver, copper, lead, and zinc produced in Colorado, 1858–1923, by counties, in terms of recovered metals—Continued

County	Ore sold or treated (short tons)	Gold			Silver		Copper		Lead		Zinc		Total value
		Placer	Lode	Total	Fine ounces	Value	Pounds	Value	Pounds	Value	Pounds	Value	
1905—Continued													
Summit	36,930	$33,728	$123,748	$157,476	209,356	$127,707	44,033	$6,869	2,181,660	$102,538	3,320,237	$195,894	$590,484
Teller	716,358		15,641,754	15,641,754	56,951	34,740							15,676,494
	2,504,087	99,984	25,195,238	25,295,222	12,339,435	7,527,056	9,661,546	1,507,201	115,746,777	5,440,098	83,561,396	4,930,123	44,699,700
1906													
Boulder	5,528		188,769	188,769	21,923	14,908	3,539	683	47,491	2,707			207,067
Chaffee	14,134	31,596	27,340	58,936	54,609	37,134	349,466	67,447	1,227,019	69,940	623,955	38,061	271,518
Clear Creek	64,774	1,568	528,185	529,753	652,796	443,901	235,375	45,427	3,307,001	188,499	1,733,477	105,742	1,313,322
Conejos	85		1,474	1,474	748	509							1,983
Costilla	28		426	426									426
Custer	3,543		16,318	16,318	79,480	54,046	2,725	526	115,960	6,610	971	59	77,559
Dolores	2,242		9,398	9,398	34,290	23,317	199,379	38,480	118,229	6,739	883,533	53,896	131,830
Douglas		4		4									4
Eagle	15,986		51,561	51,561	94,912	64,540	130,233	25,135	307,755	17,542	1,426,029	86,988	245,766
Fremont	1,010		77	77	79	54					568,508	34,679	34,810
Garfield			55	55	3	2							57
Gilpin	114,662		1,115,902	1,115,902	242,478	164,885	638,002	123,134	510,791	29,115	46,000	2,806	1,435,842
Gunnison	31,103		87,505	87,505	70,798	48,143			248,737	14,178	158,198	9,650	159,476
Hinsdale	7,086		24,510	24,510	87,940	59,799	63,621	12,279	753,950	42,975	38,387	2,342	141,905
Huerfano					56	38							513
Jefferson	5						3,150	608					608
Lake	672,055	264	1,508,146	1,508,410	3,890,338	2,645,430	2,092,735	403,898	47,456,964	2,705,047	70,198,462	4,282,106	11,544,891
La Plata and Montezuma	7,757		304,633	304,633	121,721	82,770	445	86	2,228	127			387,616
Larimer and Jackson	460		904	904	1,136	772	41,331	7,977					9,653
Mesa		473		473	15	10							483
Mineral	126,164		176,150	176,150	1,254,058	852,759			14,886,356	848,522	2,892,061	176,416	2,053,847
Montrose		114		114	3	2							116
Ouray	48,468		992,179	992,179	916,256	623,054	662,111	127,787	5,721,599	326,131	10,377	633	2,069,784
Park	10,072	10,084	384,966	395,050	144,815	98,474	14,399	2,779	966,193	55,073			551,376
Pitkin	203,400		1,172	1,172	2,131,374	1,449,334	285,346	55,072	17,951,674	1,023,245	3,276,711	199,879	2,728,702
Rio Grande	70		8,580	8,580	152	103							8,683
Routt and Moffat		6,951		6,951	42	29							6,980
Saguache	999		7,628	7,628	737	501			49,141	2,801	74,302	4,532	15,462
San Juan	196,438		900,175	900,175	690,076	469,252	1,549,663	299,085	4,515,317	257,373	718,192	43,810	1,969,695
San Miguel	386,735	1,766	2,446,024	2,447,790	1,672,522	1,137,315	319,692	61,701	7,158,189	408,017			4,054,823
Summit	34,050	53,199	86,574	139,773	107,752	73,271	27,120	5,234	1,301,912	74,209	3,363,740	205,188	497,675
Teller	702,069		13,930,526	13,930,526	67,943	46,201							13,976,727
	2,648,923	106,019	22,799,652	22,905,671	12,339,052	8,390,553	6,618,332	1,277,338	106,646,506	6,078,850	86,012,903	5,246,787	43,899,199
1907													
Boulder	8,000		161,658	161,658	23,480	15,497	22,840	4,568	16,491	874			182,597
Chaffee	14,592	35,373	39,991	75,364	38,465	25,387	799,505	159,901	630,623	33,423	2,407,730	142,056	436,131
Clear Creek	79,548	11,511	511,385	522,896	518,364	342,120	171,340	34,268	2,804,172	148,621	2,771,960	163,546	1,211,451
Custer	1,601		6,845	6,845	25,995	17,157	8,420	1,684	103,585	5,490			31,176
Dolores	1,575		11,689	11,689	33,037	21,804	99,495	19,899	54,547	2,891			56,283
Douglas		49		49									49
Eagle	4,191	53,641	53,641	53,641	70,586	46,587	14,270	2,854	193,690	10,266	429,198	25,323	138,671
Fremont	162		302	302	561	370	30,330	6,066					6,738
Gilpin	87,887		938,488	938,488	209,347	138,169	874,060	174,812	611,060	32,386			1,283,855
Grand		18		18									18
Gunnison	18,078	61,569	61,569	61,569	27,277	18,003	13,690	2,738	120,226	6,372	38,224	2,255	90,937
Hinsdale	10,740		7,520	7,520	50,109	33,072	99,410	19,882	1,204,628	63,845			124,319
Huerfano	3		68	68									68
Jefferson	18				73	48	1,955	391					439
Lake	631,273	510	1,064,180	1,064,690	4,154,913	2,742,243	2,679,510	535,902	32,519,796	1,723,549	67,247,381	3,967,595	10,033,979
La Plata and Montezuma	7,812		413,034	413,034	217,579	143,602	708	142	340	18			556,796
Mesa		76		76	3	2							78
Mineral	104,977		142,803	142,803	1,246,961	822,994	12,711	2,542	12,980,288	687,955	2,691,216	158,782	1,815,076
Montrose		314		314	9	6							320
Ouray	96,662		2,415,049	2,415,049	352,519	232,663	908,675	181,735	3,606,699	191,155	30,407	1,794	3,022,396
Park	12,661	6,953	506,263	513,216	111,215	73,402			1,062,732	56,325			642,943
Pitkin	183,836		579	579	1,719,446	1,134,834	234,493	46,899	12,235,230	648,467	4,688,693	276,633	2,107,412
Routt and Moffat	3	4,908	101	5,009	429	283							5,292
Saguache	170		649	649	6,194	4,088	1,260	252	22,528	1,194			6,183
San Juan	235,639		967,732	967,732	1,175,176	775,616	2,450,280	490,056	12,483,507	661,626	1,772,764	104,593	2,999,623
San Miguel	407,491	293	2,467,223	2,467,516	1,438,299	949,270	381,437	76,287	6,499,957	344,498			3,837,578
Summit	25,127	37,214	69,376	106,590	127,847	84,379	21,865	4,373	1,915,151	101,502	2,970,991	175,288	472,132
Teller	451,082		10,370,284	10,370,284	51,630	34,076							10,404,360
	2,383,128	97,219	20,210,429	20,307,648	11,599,514	7,655,679	8,826,254	1,765,251	89,065,232	4,720,457	85,048,564	5,017,865	39,466,900
1908													
Boulder	10,296		117,234	147,234	21,498	11,394	28,955	3,822	96,503	4,053			166,503
Chaffee	8,772	16,530	32,527	49,057	35,745	18,945	337,804	44,590	1,040,238	43,690	703,706	33,074	189,356
Clear Creek	109,672	2,610	656,506	659,116	503,551	266,882	264,994	34,979	2,015,010	84,630	836,411	39,311	1,084,918
Custer	3,700		7,183	7,183	13,156	6,973	243	32	120,330	5,054			19,242
Dolores	11,024		37,238	37,238	163,563	86,688	42,495	5,609	947,962	39,814	509,184	23,932	193,281
Douglas		131		131									131
Eagle	3,009		58,131	58,131	86,715	45,959	66,141	8,731	11,204	471			113,292
Fremont	4		91	91	4	2							93
Gilpin	120,761		1,075,808	1,075,808	187,030	99,126	636,371	84,001	538,143	22,602			1,281,537
Grand	14		556	556	72	38	1,561	206	690	29			829
Gunnison	14,439		100,032	100,032	28,664	15,192	5,481	724	327,612	13,760	147,000	6,909	136,617
Hinsdale	980		2,454	2,454	29,498	15,634	188,698	24,908	280,465	11,780			54,776
Lake	408,711		1,228,449	1,228,449	2,893,496	1,533,553	4,674,502	617,034	19,646,007	825,132	23,188,080	1,089,840	5,294,008
La Plata and Montezuma	2,416		101,584	101,584	71,592	37,944	458	60	748	31			139,619
Mineral	61,131		127,549	127,549	830,951	440,404	41	5	8,238,025	345,997	1,1 0,107	51,705	965,660
Ouray	96,493		2,028,698	2,028,698	415,070	219,987	1,019,574	134,584	3,033,352	127,401			2,510,670
Park	11,372	12,066	418,742	430,808	12,047	6,385	37,106	4,898	495,985	20,831	728,000	34,216	497,138
Pitkin	133,408		538	538	1,041,700	552,101	22,474	2,967	7,568,060	317,859	722,362	33,951	907,416
Rio Grande	9		764	764									764
Routt and Moffat	4	4,858	349	5,207	1,242	658							5,865
Saguache	76		610	610	953	505	76	10	27,715	1,164			2,289
San Juan	202,643		997,824	997,824	1,004,287	532,272	2,282,738	301,321	8,402,569	352,908	10,131	476	2,184,801
San Miguel	428,231	2,892	2,314,759	2,317,651	1,543,187	817,889	562,888	74,301	7,135,863	299,706	952,872	44,785	3,554,332
Summit	14,631	145,370	41,571	186,941	66,025	34,993	28,523	3,765	1,719,190	72,206	1,232,149	57,911	355,816
Teller	601,173		13,031,917	13,031,917	52,270	27,703							13,059,620
	2,242,969	184,457	22,411,114	22,595,571	9,002,316	4,771,227	10,201,123	1,346,547	61,645,671	2,589,118	30,130,002	1,416,110	32,718,573

Gold, silver, copper, lead, and zinc produced in Colorado, 1858–1923, by counties, in terms of recovered metals—Continued

County	Ore sold or treated (short tons)	Gold			Silver		Copper		Lead		Zinc		Total value
		Placer	Lode	Total	Fine ounces	Value	Pounds	Value	Pounds	Value	Pounds	Value	
1909													
Boulder	13,188		$163,273	$163,273	48,183	$25,055	16,485	$2,143	425,605	$18,301			$208,772
Chaffee	10,214	$19,480	11,005	30,485	35,477	18,448	568,868	73,953	584,492	25,133	947,741	$51,178	199,197
Clear Creek	116,753	3,846	532,561	536,407	448,535	233,238	299,546	38,941	3,254,675	139,951	758,074	40,936	989,473
Custer	5,871		12,774	12,774	14,796	7,694	700	91	41,721	1,794	89,593	4,838	27,191
Dolores	4,787		22,266	22,266	103,646	53,896	43,538	5,660	462,373	19,882	167,574	9,049	110,753
Eagle	11,526		53,308	53,308	125,214	65,111	286,885	37,295	152,280	6,548	740,408	39,982	202,244
Fremont	5		85	85			677	88					173
Gilpin	111,118		887,311	887,311	172,010	89,445	499,146	64,889	664,581	28,577			1,070,222
Grand		1,183		1,183	9	5							1,188
Gunnison	9,071		108,493	108,493	37,423	19,460	51,815	6,736	493,070	21,202	212,093	11,453	167,344
Hinsdale	1,697		7,587	7,587	75,731	39,380	714,569	92,894	106,327	4,572			144,433
Jefferson	1		16	16									16
Lake	417,297		1,435,431	1,435,431	3,423,642	1,780,294	5,182,608	673,739	21,073,992	906,182	38,637,315	2,086,415	6,882,061
La Plata and Montezuma	4,135		127,205	127,205	74,160	38,563	484	63	2,980	128			165,959
Larimer and Jackson	48										30,722	1,659	1,659
Mineral	64,941		108,825	108,825	891,185	463,416	17,401	2,262	9,036,816	388,583	1,817,296	98,134	1,061,220
Ouray	103,864		3,044,825	3,044,825	345,815	179,824	984,269	127,955	2,813,932	120,999	19,148	1,034	3,474,637
Park	15,046	24,358	527,563	551,921	102,375	53,235	61,023	7,933	2,237,093	96,195	366,574	19,795	729,079
Pitkin	112,448		745	745	700,038	364,020	26,092	3,392	13,143,210	565,158	34,741	1,876	935,191
Routt and Moffat	24	2,418	943	3,361	3,446	1,792							5,153
Saguache	192		1,196	1,196	2,260	1,175	3,769	490	83,463	3,589			6,450
San Juan	187,041		683,267	683,267	793,637	412,691	1,653,192	214,915	9,085,068	390,658	786,518	42,472	1,744,003
San Miguel	423,609	440	2,284,611	2,285,051	1,344,152	698,959	501,285	65,167	4,941,370	212,479	804,296	43,432	3,305,088
Summit	31,098	405,360	47,406	452,766	99,763	51,877	3,839	499	3,559,278	153,049	5,798,167	313,101	971,292
Teller	575,670		11,466,227	11,466,227	63,204	32,866							11,499,093
	2,219,644	457,085	21,526,923	21,984,008	8,904,701	4,630,444	10,916,191	1,419,105	72,162,326	3,102,980	51,210,260	2,765,354	33,901,891
1910													
Boulder	14,083		139,911	139,911	46,517	25,119	16,772	2,130	53,250	2,343			169,503
Chaffee	12,496	17,010	60,142	77,152	182,003	98,282	226,772	28,800	970,523	42,703	438,539	23,681	270,618
Clear Creek	109,954	3,678	518,846	522,524	475,174	256,594	595,795	75,666	2,434,476	107,117	1,247,389	67,359	1,029,260
Costilla		2,318		2,318	9	5							2,323
Custer	7,052		9,839	9,839	7,767	4,194	3,882	493	14,796	651	6,796	367	15,544
Delta	2		110	110	139	75							185
Dolores	2,933		15,327	15,327	88,309	47,687	97,063	12,327	127,909	5,628	87,000	4,698	85,667
Douglas		83		83									83
Eagle	27,761		25,231	25,231	88,313	47,689	209,551	26,613	397,409	17,486	4,147,945	223,989	341,008
Fremont	29										18,072	976	976
Garfield	92		3,603	3,603	113	61	425	54					3,718
Gilpin	83,631		687,902	687,902	132,635	71,623	534,244	67,849	575,477	25,321			852,695
Grand	1				56	7							7
Gunnison	25,203		233,972	233,972	49,189	26,562	21,024	2,670	581,841	25,601	176,815	9,548	298,353
Hinsdale	3,468		6,320	6,320	54,422	29,388	465,472	59,115	296,182	13,032			107,855
Lake	462,033		1,213,134	1,213,134	3,322,015	1,793,888	3,645,157	462,935	19,249,503	846,978	56,367,445	3,043,842	7,360,777
La Plata and Montezuma	6,798		399,608	399,608	141,752	76,546	362	46	273	12			476,212
Mineral	62,956		121,181	121,181	773,722	417,810	29,031	3,687	8,246,000	362,824	2,421,926	130,784	1,036,286
Ouray	111,245		2,195,847	2,195,847	414,250	223,695	620,236	78,770	4,004,728	176,208	659,796	35,629	2,674,520
Park	12,329	12,846	252,701	265,547	117,037	63,200	88,748	11,271	2,041,204	89,813			465,460
Pitkin	89,037		646	646	477,813	258,019	24,843	3,155	13,408,250	589,963			851,783
Rio Grande	12		1,306	1,306	61	33	87	11	250	11			1,361
Routt and Moffat		6,689		6,689	48	26							6,715
Saguache	296		1,025	1,025	4,841	2,614	5,362	681	161,068	7,087			11,407
San Juan	206,272		710,527	710,527	782,250	422,415	1,208,496	153,479	10,688,386	470,289	3,781,259	204,188	1,960,898
San Miguel	481,000		2,494,793	2,494,793	1,144,050	617,787	544,189	69,112	7,791,841	342,841	2,193,981	118,475	3,643,008
Summit	47,040	347,204	21,562	368,766	152,250	82,215	21,740	2,761	5,015,409	220,678	5,542,685	299,305	973,725
Teller	686,941		11,002,253	11,002,253	54,263	29,302							11,031,555
	2,434,664	389,828	20,115,786	20,505,614	8,508,942	4,594,829	8,359,307	1,061,632	76,058,775	3,346,586	77,089,648	4,162,841	33,671,502
1911													
Boulder	15,816		163,174	163,174	53,753	28,489	27,752	3,469	145,955	6,568			201,700
Chaffee	7,459	5,893	59,821	65,714	92,098	48,812	88,448	11,056	1,001,651	45,074	200,509	11,429	182,085
Clear Creek	105,774	1,754	517,453	519,207	437,841	232,036	650,368	81,296	3,325,222	149,635	1,417,544	80,800	1,062,994
Costilla		21,832		21,832	96	51							21,883
Custer	3,670		5,560	5,560	13,179	6,985	1,640	205	17,511	788			13,538
Dolores	3,276		7,565	7,565	56,202	29,787	3,288	411	701,244	31,556	525,333	29,944	99,263
Douglas		166		166									166
Eagle	33,177		41,160	41,160	116,109	61,538	66,608	8,326	855,889	38,515	5,097,597	290,563	440,102
Fremont	382		178	178	1,345	713	13,976	1,747	19,904	896	140,526	8,010	11,544
Gilpin	103,038		778,774	778,774	292,659	155,109	950,240	118,780	1,239,356	55,771	23,088	1,316	1,109,750
Gunnison	11,926	1,417	143,622	145,039	32,541	17,247	9,928	1,241	631,933	28,437	557,456	31,775	223,739
Hinsdale	723		3,830	3,830	7,753	4,109	21,696	2,712	118,645	5,339	36,439	2,077	18,067
Lake	438,419		1,133,442	1,133,442	3,007,296	1,593,867	4,017,504	502,188	18,499,089	832,459	71,610,456	4,081,796	8,143,752
La Plata and Montezuma	10,059		286,953	286,953	69,444	36,805	73,911	9,239	1,511	68			333,065
Mesa		28		28									28
Mineral	65,932		179,196	179,196	545,319	289,019	33,384	4,173	7,674,556	345,355	1,258,561	71,738	889,481
Ouray	133,252		1,952,958	1,952,958	512,800	271,784	564,273	70,534	3,949,822	177,742			2,473,018
Park	5,780	24,411	34,421	58,832	69,072	36,608	24,216	3,027	923,089	41,539	407,772	23,213	163,249
Pitkin	88,823		542	542	450,772	238,909	7,408	926	11,084,334	498,795			739,172
Routt and Moffat		6,115		6,115	47	25							6,140
Saguache	184		512	512	4,664	2,472	4,984	623	74,556	3,355	46,561	2,654	9,616
San Juan	108,088		336,463	336,463	325,604	172,570	470,912	58,864	6,933,822	312,022	2,224,351	126,788	1,006,707
San Miguel	429,354		2,447,841	2,447,841	1,000,834	530,442	971,064	121,383	6,456,333	290,535	3,386,088	193,007	3,583,208
Summit	55,904	257,422	26,819	284,241	182,957	96,967	22,888	2,861	6,024,867	271,119	7,675,175	437,485	1,092,673
Teller	756,900		10,562,653	10,562,653	57,783	30,625							10,593,278
	2,377,936	319,038	18,682,937	19,001,975	7,330,168	3,884,989	8,024,488	1,003,061	69,679,289	3,135,568	94,607,456	5,392,625	32,418,218
1912													
Boulder	9,838		119,426	119,426	72,335	44,486	22,176	$3,659	305,822	13,762			181,333
Chaffee	10,287	4,619	92,870	97,489	104,686	64,382	133,570	22,039	992,578	44,666	736,392	50,811	279,387
Clear Creek	102,894	331	445,463	445,794	373,940	229,973	449,401	74,151	3,523,733	158,568	1,734,493	119,680	1,028,166
Costilla		470		470	3	2							472
Custer	4,330		16,898	16,898	25,426	15,637	2,006	331	10,444	470			33,336
Dolores	8,485		7,556	7,556	100,288	61,677	689,915	113,837	1,212,400	54,558	812,029	56,030	293,658
Douglas		75		75									75
Eagle	34,164		49,294	49,294	163,735	100,697	147,176	24,284	1,240,156	55,807	5,659,261	390,489	620,571
Fremont	1,015		253	253	3,439	2,115	35,903	5,924	55,956	2,518	447,507	30,878	41,688
Gilpin	118,652		904,505	904,505	316,205	194,466	1,025,770	169,252	1,351,600	60,822	25,377	1,751	1,330,796
Gunnison	14,046	651	124,676	125,327	29,035	17,857	8,097	1,336	306,867	13,809	483,884	33,388	191,717
Hinsdale	9,554		6,811	6,811	34,722	21,354	53,739	8,867	1,257,800	56,601	11,926	823	94,456

Gold, silver, copper, lead, and zinc produced in Colorado, 1858-1923, by counties, in terms of recovered metals—Continued

County	Ore sold or treated (short tons)	Gold			Silver		Copper		Lead		Zinc		Total value
		Placer	Lode	Total	Fine ounces	Value	Pounds	Value	Pounds	Value	Pounds	Value	
1912—Continued													
Lake	507,591		$1,103,230	$1,103,230	3,000,397	$1,845,244	2,065,800	$340,857	26,234,244	$1,180,541	105,945,783	$7,310,259	$11,780,131
La Plata and Montezuma	2,761		135,391	135,391	47,948	29,488	918	151	6,756	304			165,334
Mesa	22		9	9	257	158	5,685	938	20	1			1,106
Mineral	66,488		86,002	86,002	714,909	439,669	23,885	3,941	5,730,222	257,860	308,681	21,299	808,771
Montrose		$687		687	10	6							693
Ouray	89,975		1,049,590	1,049,590	545,177	335,284	400,552	66,091	2,989,044	134,507	140,667	9,706	1,595,178
Park	2,686	19,223	48,758	67,981	31,234	19,209	10,321	1,703	167,756	7,549	132,275	9,127	105,569
Pitkin	91,791		165	165	528,504	325,030	22,952	3,787	8,405,333	378,240	484,507	33,431	740,653
Rio Grande	133		5,549	5,549	896	551	29,673	4,896	313	14			11,010
Routt and Moffat	64	5,070		5,070	150	92	25,085	4,139					9,301
Saguache	9,459		3,805	3,805	19,309	11,875	29,479	4,864	504,845	22,718	534,928	36,910	80,172
San Juan	140,917		523,574	523,574	714,974	439,709	1,063,291	175,443	9,114,334	410,145	2,478,594	171,023	1,719,894
San Miguel	455,696		2,399,234	2,399,234	1,153,709	709,531	845,497	139,507	7,429,622	334,333	2,943,783	203,121	3,785,726
Summit	46,606	392,739	33,276	426,015	164,665	101,269	16,412	2,708	4,402,422	198,109	9,342,725	644,648	1,372,749
Teller	849,172		11,008,362	11,008,362	66,117	40,662							11,049,024
	2,576,626	423,865	18,164,697	18,588,562	8,212,070	5,050,423	7,107,303	1,172,705	75,242,267	3,385,902	132,222,812	9,123,374	37,320,966
1913													
Boulder	5,719		69,274	69,274	162,384	98,080	25,535	3,958	409,500	18,018			189,330
Chaffee	49,135	1,266	311,626	312,892	168,985	102,067	315,011	48,827	3,196,545	140,648	2,121,947	118,829	723,263
Clear Creek	104,892		432,489	432,489	408,527	246,750	426,393	66,091	3,999,614	175,983	1,489,518	83,413	1,004,726
Costilla		95		95	2	1							96
Custer	4,662		14,684	14,684	11,313	6,833	4,052	628	5,273	232			22,377
Dolores	17,802		12,432	12,432	178,816	108,005	801,819	124,282	3,079,341	135,491	2,596,232	145,389	525,599
Douglas		547		547	2	1							548
Eagle	47,488		41,220	41,220	301,380	182,034	41,368	6,412	1,351,205	59,453	6,683,643	374,284	663,403
El Paso	298						10,632	1,648					1,648
Fremont	53		92	92	78	47	4,677	725	4,591				1,467
Garfield	25		890	890	35	21	200	31					942
Gilpin	94,156		687,101	687,101	273,207	165,017	837,974	129,886	1,210,341	53,255	8,589	481	1,035,740
Gunnison	4,301	601	9,588	10,189	87,488	52,843	21,864	3,389	196,728	8,656	292,875	16,401	91,478
Hinsdale	4,329		5,280	5,280	30,477	18,408	76,304	11,827	782,318	34,422	54,732	3,065	73,002
Lake	528,311		1,023,631	1,023,631	3,400,318	2,053,792	1,923,987	298,218	29,286,183	1,288,592	93,842,857	5,255,200	9,919,433
La Plata and Montezuma	7,403		312,891	312,891	121,122	73,158	113,897	17,654	4,455	196			403,899
Mineral	56,763		50,282	50,282	805,343	486,427	31,647	4,905	3,398,364	149,528	454,875	25,473	716,615
Montrose	49	935	5	940	434	262	24,058	3,729					4,931
Ouray	97,336		959,377	959,377	537,634	324,731	500,329	77,551	2,180,591	95,946	200,429	11,224	1,468,829
Park	6,598	14,758	35,283	50,041	94,293	56,953	29,161	4,520	506,046	22,266	98,623	5,523	139,303
Pitkin	114,264		29	29	562,308	339,634	48,852	7,572	17,528,386	771,249	460,161	25,769	1,144,253
Rio Grande	6		243	243	109	66	568	88					397
Routt and Moffat	12	3,609	231	3,840	1,962	1,185	161	25	1,023	45			5,095
Saguache	980		4,243	4,243	8,694	5,251	13,277	2,058	336,886	14,823	32,964	1,846	28,221
San Juan	123,343		657,612	657,612	880,409	531,767	1,221,516	189,335	9,508,979	418,395	1,664,999	93,240	1,890,349
San Miguel	509,175		2,129,371	2,129,374	1,051,096	634,862	736,374	114,138	6,967,136	306,554	2,405,750	134,722	3,319,647
Summit	40,360	386,196	76,032	462,228	167,490	101,164	18,170	2,816	3,944,268	173,548	6,931,074	388,140	1,127,896
Teller	917,406		10,905,003	10,905,003	71,349	43,095							10,948,098
	2,734,866	408,007	17,738,909	18,146,916	9,325,255	5,632,454	7,227,826	1,120,313	87,897,773	3,867,502	119,346,429	6,683,400	35,450,585
1914													
Boulder	14,591		98,710	98,710	312,217	172,656	24,316	3,234	523,821	20,429			295,029
Chaffee	61,698	626	331,604	332,230	272,242	150,550	319,496	42,493	3,690,359	143,924	2,173,177	110,832	780,029
Clear Creek	101,366		495,275	495,275	345,387	190,999	367,790	48,916	2,435,692	94,992	1,067,314	54,433	884,615
Costilla		177		177	2	1							178
Custer	870		3,365	3,365	15,975	8,834	3,481	463	9,692	378	4,470	228	13,268
Dolores	6,905		7,973	7,973	86,526	47,849	350,278	46,587	492,023	19,189	366,549	18,694	140,292
Douglas		140		140									140
Eagle	49,377		47,194	47,194	127,080	70,275	28,105	3,738	1,177,385	45,918	7,522,098	383,627	550,752
El Paso	25						2,644	352					352
Fremont	706		1,476	1,476	1,066	589	191,917	25,525	308	12			27,602
Garfield	73		2,403	2,403	80	44	128	17					2,464
Gilpin	52,839		573,553	573,553	145,237	80,316	726,579	96,635	499,718	19,489	12,980	662	770,655
Grand	10		3	3	1,747	966			1,563	61			1,030
Gunnison	6,018	4,384	8,649	13,033	59,036	32,647	11,188	1,488	317,974	12,401	525,000	26,775	86,344
Hinsdale	118		170	170	5,987	3,311	17,098	2,274	5,723	223			5,978
Lake	547,463		1,571,451	1,571,451	3,810,830	2,107,389	2,382,910	316,927	26,784,615	1,044,600	78,763,334	4,016,930	9,057,297
La Plata and Montezuma	5,083		126,498	126,498	60,244	33,315	26,038	3,463	11,410	445			163,721
Mineral	27,952		19,304	19,304	615,734	340,501	32,586	4,334	1,401,795	54,670			418,809
Montrose	66	435	11	446	517	286	32,414	4,311					5,043
Ouray	105,560		1,211,993	1,211,993	594,289	328,642	854,038	113,587	2,119,564	82,663	44,608	2,275	1,739,160
Park	1,958	23,334	44,151	67,485	20,215	11,179	8,023	1,067	168,154	6,558	57,940	2,955	89,244
Pitkin	118,000		423	423	372,886	206,206	67,737	9,009	2,323,230	906,096	145,431	7,417	1,129,151
Rio Grande	8		474	474	16	9							483
Routt and Moffat		4,697		4,697	16	9							4,706
Saguache	1,488		16,513	16,513	18,293	10,116	35,783	4,759	534,872	20,860	8,941	456	52,704
San Juan	117,988		508,477	508,477	493,917	273,136	825,180	109,749	5,199,000	202,761	971,177	49,530	1,143,653
San Miguel	495,742		2,114,916	2,114,916	1,280,461	708,095	324,105	43,106	4,039,769	157,551			3,023,668
Summit	22,199	608,567	60,043	668,610	67,009	37,056	7,339	976	1,565,231	61,044	5,111,941	260,709	1,028,395
Teller	939,423		11,996,116	11,996,116	89,056	49,248							12,045,364
	2,677,526	642,360	19,240,745	19,883,105	8,796,065	4,864,224	6,639,173	883,010	74,211,898	2,894,264	96,774,960	4,935,523	33,460,126
1915													
Baca	1				8	4	514	90					94
Boulder	39,778		160,433	160,433	271,292	137,545	86,680	15,169	890,042	41,832			354,979
Chaffee	68,240	1,463	314,683	316,146	226,996	115,087	348,046	60,908	3,630,127	170,616	4,676,355	579,868	1,242,625
Clear Creek	121,993	859	525,724	526,583	393,108	199,306	530,949	92,916	2,527,515	118,796	1,505,032	186,624	1,124,225
Custer	1,719		4,098	4,098	31,633	16,038	12,640	2,212	89,808	4,221	30,411	3,771	30,340
Dolores	14,192		11,932	11,932	127,933	64,862	1,032,480	180,684	268,447	12,617	35,936	4,456	274,551
Douglas		596		596	4	2							598
Eagle	74,197		95,426	95,426	177,550	90,018	60,086	10,515	1,394,043	65,520	11,141,750	1,381,577	1,643,056
Fremont	1,600		674	674	3,168	1,606	127,303	22,278	30,894	1,452	228,170	28,293	54,303
Garfield	123		5,309	5,309	112	57	291	51					5,417
Gilpin	54,052		562,878	562,878	125,665	63,712	476,383	83,367	591,127	27,783	11,000	1,364	739,104
Grand		153		153	2	1							154
Gunnison	6,446		60,197	60,197	24,892	12,620	9,091	1,591	190,000	8,930	1,750,944	217,117	300,455
Hinsdale	488		737	737	9,621	4,878	9,114	1,595	266,128	12,508			19,718
Lake	481,620	69,009	2,177,143	2,246,152	2,571,062	1,303,498	1,803,433	315,599	20,957,404	984,998	72,493,178	8,989,154	13,839,401
La Plata	2,952		71,530	71,530	46,369	23,509	4,114	720	23,362	1,098			96,857
Mineral	28,071		33,039	33,039	291,807	147,946	8,943	1,565	2,382,128	111,960	85,984	10,662	305,172
Moffat		2,613		2,613	6	3							2,616

Gold, silver, copper, lead, and zinc produced in Colorado, 1858-1923, by counties, in terms of recovered metals—Continued

County	Ore sold or treated (short tons)	Gold Placer	Gold Lode	Gold Total	Silver Fine ounces	Silver Value	Copper Pounds	Copper Value	Lead Pounds	Lead Value	Zinc Pounds	Zinc Value	Total value
1915 —Continued													
Montezuma	14		$494	$494	103	$52				170	$8		$554
Montrose	169	$1,259	18	1,277	1,073	544	57,320	$10,031					11,852
Ouray	103,258		1,118,016	1,118,016	576,621	292,347	863,851	151,174	1,990,681	93,562	7,282	$903	1,656,002
Park	2,820	9,792	149,547	159,339	9,227	4,678	12,303	2,153	190,830	8,969	472,992	58,651	233,790
Pitkin	108,579		29	29	448,915	227,600	19,983	3,497	19,265,213	905,165	214,952	26,654	1,163,245
Rio Grande	1,500		14,968	14,968	325	165							15,133
Routt				371	6	3							374
Saguache	692		5,273	5,273	11,266	5,712	23,360	4,088	174,447	8,199	44,250	5,487	28,759
San Juan	147,878		583,681	583,681	480,637	218,333	1,054,463	184,531	6,791,596	319,205	2,250,226	280,144	1,585,894
San Miguel	483,954		2,069,362	2,069,362	1,096,641	555,997	562,554	98,447	5,240,277	246,293	1,040,121	128,975	3,099,074
Summit	44,602	607,195	72,949	680,144	64,223	32,561	8,646	1,513	1,916,298	90,066	8,597,411	1,066,079	1,870,363
Teller	948,082		13,683,494	13,683,494	87,767	44,498							13,727,992
	2,737,020	693,310	21,721,634	22,414,944	7,027,972	3,563,182	7,112,537	1,244,694	68,810,597	3,234,098	104,594,994	12,969,779	43,426,697
1916													
Baca	5				50	33	2,772	682					715
Boulder	33,011		119,299	119,299	292,824	192,678	64,707	15,918	864,333	59,639			387,534
Chaffee	69,358	573	184,477	185,050	100,749	66,293	1,001,455	246,358	3,016,899	208,166	4,744,985	635,828	1,341,695
Clear Creek	94,220	250	428,681	428,931	462,141	304,089	621,732	152,946	4,295,725	296,405	2,572,575	344,725	1,527,096
Custer	2,245		6,309	6,309	36,971	24,327	44,004	10,825	123,536	8,524	10,970	1,470	51,455
Dolores	6,398		7,426	7,426	77,280	50,850	419,500	103,197	588,333	40,595	182,306	24,429	226,497
Eagle	105,149	186	95,850	96,036	222,126	146,159	112,610	27,702	1,517,362	104,698	28,438,052	3,810,699	4,185,294
Fremont	1,734		786	786	4,529	2,980	101,041	24,856	31,710	2,188			30,810
Gilpin	38,913		453,259	453,259	126,553	83,272	557,317	137,100	521,334	35,972			709,603
Grand	2				134	88	760	187					275
Gunnison	10,419	2,151	29,402	31,553	29,023	19,097	84,679	20,831	313,217	21,612	1,964,873	263,293	356,386
Hinsdale	377		1,346	1,346	10,030	6,600	16,248	3,997	75,638	5,219	12,575	1,685	18,847
Jackson and Larimer	61		95	95	199	131	6,751	1,661					1,887
Lake	477,240	119,169	1,601,271	1,720,440	2,931,281	1,928,783	2,621,675	644,932	21,719,392	1,498,638	76,785,567	10,289,266	16,082,059
La Plata and Montezuma	1,688		33,055	33,055	29,380	19,332	15,142	3,725	6,667	460			56,572
Mineral	38,103		31,124	31,124	373,956	246,063	13,138	3,232	2,295,087	158,361	240,575	32,237	471,017
Montrose	197		10	10	1,132	745	100,008	24,602					25,357
Ouray	111,192		491,175	491,175	803,461	528,677	444,081	109,244	2,339,029	161,393	69,015	9,248	1,299,737
Park	3,005	10,421	223,878	234,299	13,231	8,706	22,598	5,559	330,609	22,812	47,560	6,373	277,749
Pitkin	114,330		2	2	577,863	380,234	28,991	7,117	17,519,275	1,208,830	162,574	21,785	1,617,966
Routt and Moffat	542	1,124	18	1,142	278	183	41,175	10,129					11,454
Saguache	3,338		8,024	8,024	48,959	32,215	92,581	22,775	255,449	17,626			80,640
San Juan	146,126		438,628	438,628	502,342	330,541	1,615,167	397,331	7,285,304	502,686	4,014,403	537,930	2,207,116
San Miguel	428,651		2,072,393	2,072,393	812,041	534,323	581,427	143,031	6,126,551	422,732	1,098,485	147,197	3,319,676
Summit	65,117	579,050	94,841	673,891	120,207	79,096	14,581	3,587	1,688,637	116,516	13,940,948	1,868,087	2,741,177
Teller	945,820		12,119,550	12,119,550	79,804	52,511							12,172,061
	2,697,243	712,924	18,440,897	19,153,821	7,656,544	5,038,006	8,624,081	2,121,524	70,914,087	4,893,072	134,285,463	17,994,252	49,200,675
1917													
Baca	9		3	3	57	47	6,806	1,858					1,908
Boulder	16,835		66,841	66,841	294,375	242,565	29,513	8,057	575,582	49,500			366,963
Chaffee	55,652	183	133,441	133,624	146,535	120,745	807,883	220,552	2,150,523	184,945	2,181,932	222,557	882,423
Clear Creek	84,449	435	303,549	303,984	526,750	434,042	570,091	155,635	4,836,617	415,949	3,153,030	321,609	1,631,219
Custer	5,881		7,066	7,066	88,687	73,078	88,216	24,083	228,303	19,634			123,861
Dolores	14,026		5,213	5,213	88,222	72,695	519,916	141,937	1,772,221	152,411	1,701,353	173,538	545,794
Eagle	100,875	53	41,134	41,187	136,023	112,083	53,136	14,506	2,426,988	208,721	23,715,412	2,418,972	2,795,469
Fremont	429		590	590	664	547	59,857	16,341					17,478
Garfield	18		721	721	17	14							735
Gilpin	35,289		397,087	397,087	112,585	92,770	544,648	148,689	815,906	70,168	141,490	14,432	723,146
Gunnison	12,671	327	6,308	6,635	40,272	33,184	180,121	49,173	751,000	64,586	3,054,990	311,609	465,187
Hinsdale	517		1,136	1,136	7,721	6,362	6,099	1,665	209,616	18,027	4,117	420	27,610
Lake	422,428	110,325	1,064,894	1,175,219	2,184,000	1,799,616	2,302,812	595,856	18,301,802	1,573,955	60,254,333	6,145,942	11,290,588
La Plata and Montezuma	1,772		27,952	27,952	15,521	12,782	28,333	7,735	3,745	322			48,791
Larimer and Jackson	279		587	587	602	496	23,725	6,477					7,560
Mineral	32,755		10,101	10,101	361,517	297,890	19,297	5,268	1,305,744	112,294	54,971	5,607	431,160
Montrose	64	944		944	666	549	21,275	5,808					7,301
Ouray	86,523		92,831	92,831	868,097	715,312	179,553	49,018	2,031,721	174,728	532,794	54,345	1,086,234
Park	2,693	5,451	111,907	117,358	14,705	12,117	12,824	3,501	278,709	23,969			156,945
Pitkin	124,824		105	105	662,045	545,525	27,403	7,481	14,352,523	1,234,317	571,794	58,323	1,845,751
Rio Grande	16		24	24	52	43			1,930	166			233
Routt and Moffat	75	2,359	1,056	3,415	1,341	1,105	4,326	1,181					5,701
Saguache	4,224		10,350	10,350	76,016	62,637	144,978	39,579	310,686	26,719			139,285
San Juan	145,685		318,006	318,006	658,261	542,107	1,665,923	454,797	10,515,535	904,336	3,270,500	333,591	2,553,137
San Miguel	389,293		2,009,961	2,009,961	779,364	642,196	920,425	251,276	6,205,326	533,658	1,810,245	184,645	3,621,736
Summit	66,768	540,951	62,486	603,437	175,699	144,776	25,033	6,834	915,535	78,736	19,868,814	2,026,619	2,860,402
Teller	1,084,656		10,394,847	10,394,847	64,568	53,204							10,448,051
	2,688,706	661,028	15,068,196	15,729,224	7,304,353	6,018,787	8,122,004	2,217,307	67,990,012	5,847,141	120,315,775	12,272,209	42,084,668
1918													
Boulder	10,387		52,265	52,265	156,731	156,731	17,887	4,418	262,310	18,624			232,038
Chaffee	39,598		112,478	112,478	81,187	81,187	323,830	79,986	1,885,761	133,889	2,618,769	238,308	645,848
Clear Creek	57,808		231,077	231,077	370,888	370,888	343,247	84,782	3,869,352	274,724	1,812,846	164,969	1,126,440
Custer	4,326		4,341	4,341	108,456	108,456	51,292	12,669	281,070	19,956	13,516	1,230	146,652
Dolores	9,272		3,136	3,136	54,249	54,249	618,012	152,649	517,394	36,735	661,253	60,174	306,943
Eagle	89,675		35,975	35,975	241,406	241,406	353,041	87,201	2,927,099	207,824	14,845,341	1,350,926	1,923,332
Fremont	235		312	312	639	639	22,377	5,527	1,113	79			6,557
Garfield	15		928	928	15	15							943
Gilpin	23,654		281,384	281,384	124,929	124,929	456,044	112,643	774,972	55,023	28,099	2,557	576,536
Gunnison	6,344	73	10,295	10,368	12,880	12,880	43,033	10,629	300,760	21,354	2,349,538	213,808	269,039
Hinsdale	5,222		6,249	6,249	22,245	22,245	18,308	4,522	767,972	54,526			87,542
Jefferson	23				9	9	1,000	247					256
Lake	355,840	92,066	751,173	843,239	2,290,121	2,290,121	1,626,534	401,754	22,469,915	1,595,364	46,715,736	4,251,132	9,381,610
La Plata and Montezuma	300		7,738	7,378	6,415	6,415	668	165	3,000	213			14,171
Mineral	28,372		13,943	13,943	640,959	640,959	3,490	862	989,620	70,263			726,027
Ouray	79,653		107,645	107,645	801,359	801,359	153,117	37,820	2,587,915	183,742	39,297	3,576	1,134,142
Park	2,304		63,176	63,176	18,280	18,280	12,704	3,138	233,873	16,605			101,199
Pitkin	98,413		2	2	558,722	558,722	9,684	2,392	11,666,592	828,328	145,286	13,221	1,402,665
Routt and Moffat	161	3,040	698	3,738	2,671	2,671			6,591	468			6,877
Saguache	1,716		2,553	2,553	89,510	89,510	96,866	23,926	108,253	7,686			123,675
San Juan	132,927		257,011	257,011	477,322	477,322	1,120,178	276,684	9,485,775	673,490	3,410,308	310,338	1,994,845
San Miguel	374,134		2,127,634	2,127,634	836,570	836,570	992,814	245,225	6,044,085	429,130	797,648	72,586	3,711,145
Summit	58,185	431,023	36,116	467,139	117,326	117,326	13,206	3,262	777,338	55,191	15,696,264	1,428,360	2,071,278
Teller	936,326		8,119,747	8,119,747	50,665	50,665							8,170,412
	2,314,890	526,202	12,225,516	12,751,718	7,063,554	7,063,554	6,277,332	1,550,501	65,960,760	4,683,214	89,133,901	8,111,185	34,160,172

Gold, silver, copper, lead, and zinc produced in Colorado, 1858–1923, by counties, in terms of recovered metals—Continued

County	Pro-ducing mines	Ore sold or treated (short tons)	Gold			Silver		Copper		Lead		Zinc		Total value
			Placer	Lode	Total	Fine ounces	Value	Pounds	Value	Pounds	Value	Pounds	Value	
1919														
Boulder	46	6,113		$54,653	$54,653	225,484	$252,542	11,043	$2,054	206,605	$10,950			$320,199
Chaffee	9	19,655		58,167	58,167	40,556	45,416	70,823	13,173	803,228	42,571	965,630	$70,491	229,818
Clear Creek	66	116,355		91,127	91,127	357,438	400,332	152,925	28,444	1,517,134	80,408	603,027	44,021	644,332
Custer	8	4,621		4,771	4,771	97,151	108,818	72,979	13,574	155,134	8,222			135,385
Dolores	7	4,461		2,517	2,517	35,227	39,452	264,968	49,284	98,700	5,231	67,027	4,893	101,377
Eagle	13	22,248		19,935	19,935	72,159	80,818	123,306	22,935	378,113	20,040	3,367,548	245,831	389,559
Gilpin	42	17,018		209,683	209,683	71,700	80,304	211,538	39,346	523,621	27,752			357,085
Grand	1	3		1	1	508	569			453	24			594
Gunnison	14	8,348		31,556	31,556	18,425	20,636	5,124	953	117,454	6,225	2,456,479	179,323	238,693
Hinsdale	12	1,219		8,232	8,232	22,941	25,695	7,705	1,433	55,679	2,951			38,311
Lake	62	217,667	$81,688	544,268	625,956	1,542,324	1,727,403	888,628	165,285	11,299,076	598,851	23,165,219	1,691,061	4,808,556
La Plata	12	405		5,966	5,966	6,075	6,804	167	31	2,283	121			12,922
Mineral	13	16,718		9,083	9,083	369,575	413,924	355	66	934,113	49,508	96,274	7,028	479,609
Montrose	1	172	199		199	2	2							201
Ouray	23	64,465		92,338	92,338	627,659	702,978	112,188	20,867	1,782,868	94,492	23,343	1,704	912,379
Park	9	11,210	4,135	125,422	129,557	70,949	79,463	20,436	3,801	305,908	16,213			229,034
Pitkin	12	121,534				657,058	735,905			5,310,170	281,439	80,000	5,840	1,023,184
Routt	2	172		312	312	1,283	1,437							1,749
Saguache	5	509		817	817	37,767	42,299	36,344	6,760	52,515	2,783			52,659
San Juan	20	64,899		132,560	132,560	279,667	313,227	661,667	123,070	5,443,906	288,527	1,833,768	133,865	991,249
San Miguel	13	428,585		2,105,490	2,105,490	1,100,942	1,233,055	913,925	169,990	7,636,790	404,750	515,082	37,601	3,950,886
Summit	33	17,567	464,540	11,351	475,891	87,676	98,197	6,086	1,132	446,491	23,664	4,047,096	295,438	894,322
Teller	41	775,986		5,827,816	5,827,816	35,442	39,695							5,867,511
	464	1,919,768	550,562	9,336,065	9,886,627	5,758,010	6,448,971	3,560,207	662,198	37,070,241	1,964,722	37,220,493	2,717,096	21,679,614
1920														
Boulder	37	19,476		42,428	42,428	148,834	162,229	6,685	1,230	261,088	20,887			226,774
Chaffee	6	3,900		31,302	31,302	39,211	42,740	28,195	5,188	396,250	31,700	283,235	22,942	133,872
Clear Creek	65	51,494		48,540	48,540	219,900	239,691	61,978	11,404	2,457,100	196,568	372,420	30,166	526,369
Custer	8	1,500		798	798	34,256	37,339	28,033	5,158	171,562	13,725			57,020
Dolores	7	2,752		2,350	2,350	32,167	35,062	6,804	1,252	772,588	61,807	229,865	18,619	119,090
Eagle	9	32,635		25,496	25,496	279,667	304,837	517,109	95,148	282,538	22,603	6,653,235	538,912	986,996
Gilpin	29	10,820		91,469	91,469	42,000	45,780	86,603	15,935	435,012	34,801			187,985
Grand	1	3				856	933			525	42			975
Gunnison	7	8,443		24,070	24,070	20,555	22,405			958,301	76,664	1,530,691	123,986	247,125
Hinsdale	9	568		6,151	6,151	21,522	23,459	2,625	483	80,625	6,450			36,543
Lake	60	172,988	138,864	629,501	768,365	1,099,688	1,198,660	799,744	147,153	8,590,198	687,215	18,754,531	1,519,117	4,320,510
La Plata	9	717		11,020	11,020	10,578	11,530			937	75			22,625
Mineral	12	12,597		5,710	5,710	272,322	296,831	1,120	206	531,537	42,523			345,270
Moffat	2		118		118									118
Montrose	1		198		198	2	2							200
Ouray	23	40,195		33,777	33,777	465,577	507,479	86,881	15,986	1,334,575	106,766			664,008
Park	12	5,348	526	142,632	143,158	51,023	55,615	18,674	3,436	1,085,625	86,850			289,059
Pitkin	11	125,786				625,444	681,734			4,470,300	357,624	617,790	50,041	1,089,399
Routt	1	3		44	44	100	109							153
Saguache	7	9,282		5,031	5,031	94,655	103,174	88,386	16,263	150,063	12,005			136,473
San Juan	19	201,671		266,766	266,766	746,100	813,249	1,361,391	250,496	16,601,025	1,328,082	11,837,395	958,829	3,617,422
San Miguel	14	374,169		1,340,226	1,340,226	1,064,667	1,160,487	948,696	174,560	7,571,875	605,750	175,617	14,225	3,295,248
Summit	26	48,328	374,882	30,422	405,304	106,422	116,000	359	66	477,462	38,197	8,335,963	675,213	1,234,780
Teller	41	448,618		4,323,998	4,323,998	33,789	36,830	451	83	612	49			4,360,960
	416	1,571,293	514,588	7,061,731	7,576,319	5,409,33	5,896,175	4,043,734	744,047	46,629,788	3,730,383	48,790,742	3,952,050	21,898,974
1921														
Boulder	46	13,176		34,042	34,042	112,957	112,957	302	39	140,336	6,315			153,353
Chaffee	5	2,402		32,034	32,034	27,641	27,641	8,357	1,078	318,666	14,340	39,000	1,950	77,043
Clear Creek	48	32,767		38,851	38,851	131,867	131,867	21,519	2,776	1,200,931	54,042	217,000	10,850	238,386
Costilla	1			52	52									52
Custer	5	568		184	184	19,191	19,191	37,690	4,862	106,022	4,771			29,008
Dolores	5	386		1,856	1,856	14,499	14,499	744	96	18,624	838			17,289
Douglas	2		47		47									47
Eagle	7	39,785		64,723	64,723	682,550	682,550	1,833,078	236,467	12,578	566			984,306
Gilpin	36	3,553		39,610	39,610	17,963	17,963	13,186	1,701	91,644	4,124			63,398
Gunnison	6	498		18,223	18,223	10,370	10,370			51,955	2,338			30,931
Hinsdale	6	495		3,425	3,425	32,039	32,039	9,357	1,207	65,756	2,959			39,630
Lake	41	80,501	6,184	302,960	309,144	1,043,497	1,043,497	1,107,295	142,841	3,537,889	159,205	1,821,000	91,050	1,745,737
La Plata	17	1,279		45,181	45,181	20,327	20,327			3,734	168			65,676
Mineral	8	7,076		3,816	3,816	192,468	192,468	1,899	245	156,778	7,055			203,584
Ouray	11	69,232		73,229	73,229	730,970	730,970	85,039	10,970	1,208,399	54,378			869,547
Park	10	4,929	429	40,821	41,250	47,547	47,547	7,550	974	654,090	29,434			119,205
Pitkin	17	60,476				474,225	474,225	233	30	2,395,622	107,803	283,000	14,150	596,208
Saguache	9	6,412		1,856	1,856	90,871	90,871	49,512	6,387	198,686	8,941			108,055
San Juan	17	1,164		8,272	8,272	64,179	64,179	28,558	3,684	557,555	25,090			101,225
San Miguel	15	455,281		1,468,820	1,468,820	1,776,963	1,776,963	921,573	118,883	8,436,244	379,631			3,744,297
Summit	23	17,291	337,980	20,850	358,830	104,198	104,198	27,550	3,554	504,957	22,723			489,305
Teller	47	484,110		4,291,883	4,291,883	37,335	37,335							4,329,218
	382	1,281,381	344,640	6,490,688	6,835,328	5,631,657	5,631,657	4,153,442	535,794	19,660,466	884,721	2,360,000	118,000	14,005,500
1922														
Adams	1		498		498	4	4							502
Boulder	60	3,415		37,037	37,037	121,073	121,073			68,470	3,766			161,876
Chaffee	8	6,844		19,936	19,936	26,187	26,187	20,526	2,771	661,728	36,395	178,000	10,146	95,435
Clear Creek	54	69,425		36,199	36,199	196,207	196,207	7,874	1,063	1,042,491	57,337	800,000	45,600	336,406
Custer	4	17,212		167	167	14,520	14,520	32,141	4,339	660,618	36,334			55,360
Dolores	5	678		1,953	1,953	30,267	30,267	24,089	3,252	87,200	4,796			40,268
Douglas	1		12		12									12
Eagle	7	71,892		72,111	72,111	583,737	583,737	1,330,296	179,590	322,818	17,755	11,000,000	627,000	1,480,193
Fremont	1	7		21	21	174	174	348	47	4,273	235			477
Gilpin	43	13,707	378	51,342	51,720	43,910	43,910	24,860	3,356	246,945	13,582			112,568
Gunnison	9	221		9,180	9,180	3,803	3,803	526	71	13,382	736			13,790
Hinsdale	10	1,550		1,298	1,298	50,074	50,074	14,269	1,926	114,200	6,281			59,579
Lake	54	112,547	315	412,743	413,058	952,048	952,048	871,370	117,635	5,521,818	303,700	9,003,000	513,171	2,299,612
La Plata	16	791		32,261	32,261	10,656	10,656							42,917
Mineral	9	3,978		1,654	1,654	106,903	106,903	3,422	462	153,455	8,440			117,459
Moffat	2		114		114									114
Montrose	2	251	322		322	17,968	17,968	61,119	8,251					26,544

Gold, silver, copper, lead, and zinc produced in Colorado, 1858–1923, by counties, in terms of recovered metals—Continued

County	Producing mines	Ore sold or treated (short tons)	Gold			Silver		Copper		Lead		Zinc		Total value
			Placer	Lode	Total	Fine ounces	Value	Pounds	Value	Pounds	Value	Pounds	Value	
1922—Continued														
Ouray	18	123,096		$125,960	$125,960	1,226,670	$1,226,670	58,149	$7,850	1,484,526	$81,649			$1,442,129
Park	11	1,120	$99,466	42,654	142,120	15,528	15,528	4,215	569	155,982	8,579			166,796
Pitkin	16	119,023				525,169	525,169			3,555,309	195,542			720,711
Routt	1	1				82	82							82
Saguache	4	9,671		4,849	4,849	63,542	63,542	41,622	5,619	111,782	6,148			80,158
San Juan	26	8,808		25,759	25,759	77,864	77,864	110,348	14,897	1,651,982	90,859	1,300,000	$74,100	283,479
San Miguel	26	397,840		1,077,846	1,077,846	1,645,459	1,645,459	673,867	90,972	7,060,891	388,349			3,202,626
Summit	37	17,894	255,298	26,464	281,762	119,604	119,604	94,413	12,746	559,330	30,763	977,000	55,669	500,564
Teller	55	432,129		4,037,582	4,037,582	24,462	24,462							4,062,044
	480	1,412,100	356,403	6,017,016	6,373,419	5,855,911	5,855,911	3,373,454	455,416	23,477,200	1,291,246	23,258,000	1,325,706	15,301,698
1923														
Adams	1		341		341	3	2							343
Boulder	44	1,960		27,146	27,146	39,556	32,436	7,089	1,042	26,729	1,871			61,453
Chaffee	7	2,173	707	16,366	17,073	20,762	17,025	7,140	1,091	557,429	39,020	132,000	8,976	83,136
Clear Creek	51	23,464		30,576	30,576	183,874	150,777	32,218	4,736	1,016,729	71,171	577,000	39,236	296,496
Custer	8	51,200		2,536	2,536	28,484	23,357	11,436	1,681	2,890,328	202,323			229,897
Dolores	10	1,393		2,890	2,890	39,408	32,315	56,823	8,353	162,414	11,369	138,000	9,384	64,311
Eagle	8	98,427		41,734	41,734	322,143	264,157	632,565	92,987	460,171	32,212	23,600,000	1,604,800	2,035,890
Fremont	1	44		27	27	184	151			1,999	140	20,000	1,360	1,678
Gilpin	38	5,631	133	29,063	29,196	44,942	36,852	22,884	3,364	230,157	16,111			85,523
Grand	1					323	265			314	22			287
Gunnison	4	11,019		23,854	23,854	24,939	20,450	1,788	263	1,690,430	118,330	2,889,000	196,452	359,349
Hinsdale	5	684		732	732	30,046	24,638	10,075	1,481	19,971	1,398			28,249
Lake	69	115,975	15,224	256,280	271,504	537,787	511,776	511,776	75,231	5,624,958	393,747	9,415,000	640,220	1,918,489
La Plata	8	838		15,905	15,905	17,138	14,053	816	120	1,800	126			30,204
Mineral	8	6,462		2,394	2,394	228,867	187,671	1,088	160	237,557	16,629	41,000	2,788	209,642
Montrose	2	101	177		177	10,523	8,629	17,857	2,625					11,431
Ouray	11	87,260		59,207	59,207	840,044	688,836	44,197	6,497	1,538,027	107,662			862,202
Park	11	471	144,468	16,974	161,442	18,701	15,335	5,558	817	19,401	1,358			178,952
Pitkin	17	58,641				429,581	352,256			2,972,614	208,083	465,000	31,620	591,959
Rio Grande	2	17		1,662	1,662	161	132	218	32	929	65			1,891
Saguache	5	34,456		4,229	4,229	155,723	127,693	459,477	67,543	2,919,200	204,344			403,809
San Juan	16	153,114		241,986	241,986	471,750	386,835	1,005,441	147,938	751,726	9,540,000	648,720	2,177,067
San Miguel	15	484,064		1,373,968	1,373,968	1,606,344	1,317,202	1,408,980	207,120	10,695,814	748,707			3,646,997
Summit	27	48,965	203,379	32,663	236,042	142,548	116,889	17,823	2,620	3,892,271	272,459	7,335,000	498,780	1,126,790
Teller	48	382,739		4,047,008	4,047,008	22,606	18,537							4,065,545
	417	1,569,100	364,429	6,227,200	6,591,629	5,334,488	4,374,280	4,248,109	624,472	45,698,185	3,198,873	54,152,000	3,682,336	18,471,590

Total gold, silver, copper, lead, and zinc produced in Colorado, 1858–1922, by counties and periods, in terms of recovered metals [a]

[The total for each county to the end of 1923 is given below, under "Production by counties"]

County	Period	Gold			Silver		Copper		Lead		Zinc		Total value
		Placer	Lode	Total	Fine ounces	Value	Pounds	Value	Pounds	Value	Pounds	Value	
Adams	1922	$498		$498	4	$1							$502
Arapahoe	1885–1904	8,101		8,101	101	64							8,165
Archuleta	1897–1904		$1,489	1,489	505	302							1,791
Baca	1900–1917		292	292	356	226	21,511	$4,441	6,487,432	$347,464			4,959
Boulder	1859–1922	15,927,853	15,927,853	7,955,216	7,540,178	148,494	148,494	129,955,089	5,710,328	28,449,505	$2,482,051		23,963,989
Chaffee	1859–1922	1,374,472	6,009,809	7,384,281	5,201,131	4,217,359	9,624,306	1,724,277	129,955,089	5,710,328	28,449,505	2,482,051	21,518,296
Clear Creek	1859–1922	2,852,683	19,662,585	22,515,268	57,625,615	52,222,478	11,865,860	1,919,926	177,040,319	8,026,391	30,399,821	2,210,661	86,894,724
Conejos	1861–1906		38,445	38,445	55,823	33,278	4,815	797	3,400	149			72,669
Costilla	1875–1921	28,491	14,977	43,468	2,715	1,592	1,827	239	50,048	1,802			47,101
Custer	1872–1922		2,183,472	2,183,472	4,513,003	4,522,409	553,308	104,947	31,674,690	1,386,442	217,227	14,787	8,212,057
Delta	1894–1910		4,273	4,273	306	176							4,449
Dolores	1879–1922		1,974,760	1,974,760	11,634,519	9,170,322	6,187,058	1,141,361	36,959,730	1,657,902	10,648,316	718,726	14,663,071
Douglas	1858–1922	4,509		4,509	161	128							4,637
Eagle	1879–1922	239	2,963,465	2,963,704	7,320,722	6,292,986	6,215,873	945,269	87,191,309	3,871,605	131,493,129	12,225,773	26,299,337
El Paso	1913–1914						13,276	2,000					2,000
Fremont	1881–1922		81,111	81,111	91,628	85,297	667,154	120,457	682,986	28,714	1,432,769	104,333	419,912
Garfield	1885–1918		16,924	16,924	528	327	1,044	153					17,404
Gilpin	1859–1922	242,296	83,842,897	84,085,193	10,477,600	8,510,564	25,361,375	4,161,058	35,262,678	1,541,666	329,713	27,321	98,325,802
Grand	1896–1920	1,354	11,829	13,183	3,559	2,736	5,171	805	3,231	156			16,880
Gunnison	1861–1922	14,604	2,191,313	2,205,917	5,412,777	4,886,120	985,319	180,570	41,073,529	1,830,589	16,124,550	1,477,204	10,580,400
Hinsdale	1875–1922		1,451,189	1,451,189	5,648,647	4,582,476	2,853,996	401,909	97,257,388	3,993,171	1,104,034	57,928	10,486,673
Huerfano	1875–1907		3,474	3,474	1,176	698	92	11	1,067	38			4,221
Jefferson	1858–1918	32,769	29,527	62,296	7,058	4,631	20,695	3,347	10,863	398			70,672
Lake	1859–1922	6,783,525	44,124,544	50,908,069	229,826,649	188,872,146	99,588,056	14,254,235	1,919,663,167	85,061,553	1,225,503,034	84,770,058	423,866,061
La Plata and Montezuma	1878–1922		3,572,749	3,572,749	1,737,625	1,115,815	277,675	44,903	257,906	12,028			4,745,495
Larimer and Jackson	1895–1917	16,025	8,279	24,304	2,502	1,735	235,328	38,647			30,722	1,659	66,345
Las Animas	1887–1899		2,094	2,094	20	15							2,109
Mesa	1885–1912	4,059	981	5,040	4,934	2,970	35,280	5,222	20	1			13,233
Mineral	1891–1922		2,720,583	2,720,583	44,338,172	29,003,903	274,000	44,027	197,739,744	8,738,960	27,572,407	1,511,944	42,019,417
Montrose	1886–1922	28,155	18,823	46,978	202,420	128,706	514,735	91,274	64	3			266,961
Ouray	1878–1922		35,108,556	35,108,556	40,895,385	31,557,566	22,883,253	3,300,733	160,156,027	7,003,622	1,190,650	100,426	77,070,903
Park	1859–1922	3,403,480	6,870,879	10,274,359	6,936,144	6,895,424	2,038,700	41,160,955	1,829,791	2,971,532	195,512	19,582,068
Pitkin	1880–1922		577,930	577,930	97,178,641	72,988,357	1,128,463	197,443	562,582,702	25,573,729	16,377,002	1,028,289	100,365,748
Pueblo	1894–1901		793	793	90	55	210	35					883
Rio Grande	1870–1922	7,000	2,356,077	2,363,077	176,040	170,122	123,787	19,826	46,081	1,993			2,555,018
Routt and Moffat	1866–1922	370,014	18,851	388,865	28,941	19,696	78,570	16,704	139,536	5,205			430,470
Saguache	1880–1922		261,851	261,851	1,805,778	1,498,692	962,540	180,168	7,920,659	363,953	1,072,148	68,081	2,372,745
San Juan	1873–1922		22,469,127	22,469,127	28,179,827	20,045,387	49,019,057	7,578,249	305,687,350	14,621,297	42,814,684	3,490,764	68,204,824
San Miguel	1875–1922	188,635	57,887,988	58,076,623	41,076,379	30,262,352	15,272,733	2,535,263	166,117,375	8,313,259	18,141,182	1,323,787	100,511,284
Summit	1859–1922	13,770,944	5,115,997	18,886,941	13,430,922	11,592,727	1,047,303	149,289	149,813,394	6,541,048	129,810,560	10,724,763	47,894,768
Teller	1891–1922		319,803,837	319,803,837	1,739,989	1,128,455	451	83	612	49			320,932,424
Miscellaneous	1888	5,000	3,785	8,785	1,214	1,141							9,926
		29,136,853	637,333,408	666,470,261	623,514,912	497,359,665	258,830,449	39,703,094	4,154,939,351	186,463,306	1,685,682,985	122,534,067	1,512,530,393

[a] For explanation see footnote to Table 1, p. 60

PRODUCTION BY COUNTIES

ADAMS COUNTY

Placer gold and silver produced in Adams County, 1922–1923

Year	Producing mines	Gold	Silver		Total value
			Fine ounces	Value	
1922	1	$498	4	$4	$502
1923	1	341	3	2	343
		839	7	6	845

ARAPAHOE COUNTY

Arapahoe County was organized in 1861 and named from the principal Indian tribe in the State at that time. It includes Cherry Creek from the point where the creek flows out of Douglas County to the point where it enters Denver County, which was separated from Arapahoe County in 1902. It includes Dry Creek, which rises on the ridge between Cherry Creek and the Platte and flows northwestward through Englewood and enters the South Platte just north of Petersburg. On Dry Creek and on the Platte between its mouth and the mouth of another dry creek, which enters the Platte at Alameda Avenue, Denver, placers were worked in 1858–59 and later. The county is credited with small quantities of placer gold from the earliest days of mining in Colorado. To the discovery of small quantities of gold within its boundaries in 1858 is due the beginning of the immigration that resulted in the discoveries of gold in the districts around Denver.

The figures given for 1885 in the following table are taken from the report of the Director of the Mint.[94]

The figures given for the years from 1886 to 1895, inclusive, are taken from reports of the agents of the mint in annual reports of the Director of the Mint, the gold and silver being prorated to correspond with the figures of the total production of the State as corrected by the Director of the Mint, the lead being prorated to correspond with the total production of lead in the State as given in Mineral Resources, and any "unknown production" in the State being distributed proportionately to the several counties. As with lead so with copper, but as the figures for copper given in Mineral Resources include copper from matte and ores treated in Colorado but mined in other States, the figures for copper are subject to revision.

The figures for 1891 are taken from the report of the agent of the Director of the Mint in the report on the production of the precious metals in the United States in 1891. This report shows no production for Arapahoe County in the table of counties, but the table showing gold and silver deposited at the United States Mint at Denver shows $212 in gold.

The figures for 1896 to 1904, inclusive, which are those of the Colorado State Bureau of Mines, represent smelter and mint receipts.

The reports of the Colorado State Bureau of Mines show that Arapahoe County produced $455 gold in 1905 and $248 gold in 1906, but the figures given in Mineral Resources for these years credit no production to Arapahoe County.

Gold and silver produced in Arapahoe County, 1885–1904 [a]

Year	Placer gold	Silver			Total value
		Fine ounces	Average price per ounce	Value	
1885	$271				$271
1886	293				293
1887	177				177
1894	86				86
1895	1,081	59	$0.65	$38	1,119
1896	1,894	19	.68	13	1,907
1897	2,108	14	.60	8	2,116
1898	703	7	.59	4	707
1899	269	2	.60	1	270
1900	248				248
1901	331				331
1902	227				227
1903	165				165
1904	248				248
	8,101	101		64	8,165

[a] Production began in 1858, but no records are available to show quantities or values before 1885 or between 1887 and 1893, inclusive.

ARCHULETA COUNTY

The report of the Colorado State Bureau of Mines for 1897 contains the following notes in regard to Archuleta County: [95]

This is one of the south-central border counties, with an area of about 1,100 square miles. It is bounded on the east by Conejos County, from which it was segregated by an act of legislature, approved April 14, 1885, on the south by New Mexico, on the west by La Plata, and on the north by Hinsdale and Mineral counties. * * *

Mining for the precious metals has been advanced little beyond the prospecting stage, and mere prospecting has been indulged in in desultory manner. Good values have been discovered in various sections, but the production has been practically nothing. The east and southeast section of the county is covered by one of the old Spanish land grants known as the Tierra Amarilla. Until comparatively recently the question of title has done much to retard prospecting. Sixty thousand acres of this grant have passed into the hands of a domestic corporation. This company has promulgated a set of mining rules, practically in accord with the State regulations and thrown their territory open to prospectors, and, under their regulations, guarantees title. As a result, the past year has found a number of prospectors in the field, and, if reports can be credited, their finds are worthy of careful investigation. There were engaged in mining and prospecting an average of 76 men during the past year.

For 1898 to 1904 the figures in the table, which represent smelter and mint receipts, are taken from the reports of the Colorado State Bureau of Mines.

---

[94] Wilson, P. S., agent for Colorado, in Kimball, J. P., Report of the Director of the Mint upon the production of the precious metals in the United States during the calendar year 1885, p. 136, 1886 (Gold and silver deposited at the United States Mint at Denver).

[95] Colorado State Bur. Mines Rept. for 1897, pp. 9–10, 1898.

The Colorado State Bureau of Mines gives the output of Archuleta County as $83 in gold and 15 ounces of silver for 1905 and $103 gold and 10 ounces silver for 1906. The mines report of the United States Geological Survey gives no record of any output in Archuleta County for these years.

Gold and silver produced in Archuleta County, 1897–1904

| Year | Lode gold | Silver | | | Total value |
		Fine ounces	Average price per ounce	Value	
1897	$703	348	$0.60	$209	$912
1898	145	40	.59	24	169
1899	103	43	.60	26	129
1900	145	30	.62	18	163
1901	124	18	.60	11	135
1902	83	10	.53	5	88
1903	62	6	.54	3	65
1904	124	10	.58	6	130
	1,489	505		302	1,791

BACA COUNTY

The figures given in the table for Baca County for 1900 to 1902, inclusive, are those given by the Colorado Bureau of Mines and represent smelter and mint receipts. The report of the agent of the Director of the Mint upon the production of the precious metals in the United States during the calendar year 1900 gives $93 in gold as the output of Baca County.

The production for 1915 to 1917 is taken from Mineral Resources (mines reports). The total output is credited to the Carrizo Creek district, southwest of Springfield. The ore shipped in 1915–1917 was chalcocite, partly altered to malachite and azurite. The deposits, which are in white sandstone, lie principally in T. 34 N. (New Mexico base line), R. 50 W., in secs. 10, 15, 16, 21, 22, 23, 26, 28, and 35, although there are also indications of deposits to the east, west, and south, and in northwestern Oklahoma and northeastern New Mexico.

Gold, silver, and copper produced in Baca County, 1900–1902 and 1915–1917

| Year | Ore (short tons) | Lode gold (value) | Silver | | | | Copper | | | Total value |
			Fine ounces	Average price per ounce	Value	Pounds	Average price per pound	Value		
1900		$103	102	$0.62	$63	8,900	$0.166	$1,477		$1,643
1901		83	80	.60	48	590	.167	99		230
1902		103	59	.53	31	1,929	.122	235		369
1915			8	.507	4	514	.175	90		94
1916	5		50	.658	33	2,772	.246	682		715
1917	9	3	57	.824	47	6,806	.273	1,858		1,908
		292	356		226	21,511		4,441		4,959

BOULDER COUNTY

A pamphlet entitled "Mining in Boulder County," published by the Boulder County Mining Association at Boulder, Colo., in 1910, gives the following notes on prospecting and discoveries in this area:

Charles Clouser and party pushed back from the foothills up what is now known as Sunshine Canyon, following the main ridge between Fourmile Creek and Lefthand Canyon, until they camped * * * on the west slope of Gold Hill, December, 1858. Here they found good values in the head of the gulch they afterward named Gold Run. The word that gold had been found in paying quantities spread very quickly, and claim after claim was located upon the gulch. A ditch, 6 or 7 miles long, was built to bring water from Lefthand Creek to the head of Gold Run, and during the next three or four years about $100,000 gold was produced. Many veins were discovered, and as the district laws of these days allowed only 100 feet for a claim and an extra 100 feet for the discoverer, large numbers of claims were located. The Horsfal property yielded such free-gold surface ore that it was hauled down to Gold Run and washed in sluice boxes. The Twins, Alamakee, and other mines were soon supplying surface ores to the stamp mills erected on Lefthand Creek. Fair recovery was made by sluice, arrastre, and stamp mill until oxidized ores were gone. Enterprise after enterprise failed until the Blackhawk smelter (Boston & Colorado, or Hill) was built, in 1868.

From 1860 to 1863 some placer mining seems to have been done above the town of Boulder at the mouth of Boulder Creek and above that point to the mouth of Fourmile Creek. One of these placers was at the site of the present railway bridge near the old tollhouse, above the town. Systematic placer mining was done from the mouth of Fourmile Creek upstream. In the town of Boulder and below it prospecting was done and attempts at placer mining were made, but the gravel was so heavy as to prevent systematic work of any kind.

Raymond[96] describes the operations in the Caribou district in 1869 and 1870 as follows:

This district was discovered, I believe, in 1869 or even earlier, but it was not until June, 1870, that the extraordinary value of its principal lode, the Caribou, caused it to become the object of special attention and public excitement. * * * The first work was done on this mine in 1869, when about 26 tons, containing by assay $3,217 in silver, were sold to Professor Hill, at Blackhawk. During 1870 about 425 tons of shipping ore were extracted, worth $73,772, or about $173 per ton.

In 1871, according to Raymond,[97] mining at Ward was active.

For 1872 the Georgetown Mining Review gives Boulder County $346,000 in gold and silver, but Raymond[98] says these figures are too low. He makes the following statement under the heading "Gold Hill district":[99]

The Red Cloud lode was discovered in May, 1872. * * * The mine contains both gold and silver and is remarkable on account of containing these metals in combination with tellurium, the mineral found being petzite.

[96] Raymond, R. W., Statistics of mines and mining in the States and Territories west of the Rocky Mountains for 1870, pp. 324–328, 1872.

[97] Idem for 1871, p. 360, 1873.

[98] Idem for 1872, p. 266, 1873.

[99] Idem, p. 293.

For 1873 the Georgetown Mining Review credits Boulder County with $390,000 in gold and silver, but Raymond's totals for the State that year increase the estimate.[1]

For 1874 Raymond[2] reports as follows:

Sunshine discovered. Red Cloud and Cold Spring mines at Gold Hill have produced since discovery $600,000 (400 tons at $1,500 per ton).

Other information shows that the Logan and Yellow Pine deposits at Crisman were discovered in 1874. Salina was located. Raymond says that the output of gold and silver for 1874 amounted to $539,870 and for 1875 to $480,996 (coin).[3]

For 1879 to 1884 the figures in the table are taken from reports of the Director of the Mint.[4]

[1] Idem for 1873, p. 284, 1874.
[2] Idem for 1874, p. 371, 1875.
[3] Idem for 1874, p. 358, 1875; idem for 1875, p. 282, 1877.
[4] Burchard, H. C., Report upon the production of the precious metals in the United States during the calendar year 1880, pp. 156, 157, 1881. (See the reports of this serie : or later years.)

For 1886 to 1896 the figures given in the table are derived from reports of the agents of the mint in the annual reports of the Director of the Mint, the gold and silver being prorated to correspond with the figures showing the total production of the State as corrected by the Director of the Mint, the lead being prorated to correspond with the total production of the State as reported in the annual volumes of Mineral Resources, and any unknown production in the State being distributed proportionately to the counties. As with lead so with copper, but as figures for copper given in Mineral Resources include copper obtained from matte and ores treated in Colorado, although mined in other States, the figures for copper are subject to revision.

For 1897 to 1904 the figures, which represent smelter and mint receipts, are taken from the reports of the Colorado State Bureau of Mines.

For 1905 to 1923 the figures are taken from Mineral Resources (mines reports).

Gold, silver, copper, and lead produced in Boulder County, 1859–1923

Year	Ore (short tons)	Gold	Silver			Copper			Lead			Total value
			Fine ounces	Average price per ounce	Value	Pounds	Average price per pound	Value	Pounds	Average price per pound	Value	
1859–1862		$100,000										$100,000
1863		25,000										25,000
1864		25,000										25,000
1865		20,000										20,000
1866		15,000										15,000
1867		10,000										10,000
1868		50,000										50,000
1869		100,000	3,547	$1.325	$4,700							104,700
1870		100,000	60,241	1.328	80,000							180,000
1871		156,605	a 60,377	1.325	80,000							236,605
1872		224,852	199,414	1.322	263,625							488,477
1873		155,000	282,326	1.297	366,177							521,177
1874		160,000	293,806	1.278	375,484							535,484
1875		218,086	203,344	1.24	252,147							470,233
1876		200,000	a 232,031	1.16	269,156							469,156
1877		400,000	a 232,031	1.20	278,437							678,437
1878		400,000	a 270,703	1.15	311,308							711,308
1879		400,000	348,047	1.12	389,813							789,813
1880		300,000	425,391	1.15	489,200							789,200
1881		200,000	270,703	1.13	305,894							505,894
1882		260,000	239,766	1.14	273,333							533,333
1883		300,000	123,750	1.11	137,363							437,363
1884		350,000	100,547	1.11	111,607							461,607
1885		b 300,000	b 84,691	1.07	90,619							390,619
1886		382,185	84,691	.99	83,844							466,029
1887		253,546	70,091	.98	68,689				593	$0.045	$27	322,262
1888		189,241	230,205	.94	216,393				246,282	.044	10,836	416,470
1889		344,503	174,471	.94	164,003	2,748	$0.135	$371	51,215	.039	1,997	510,874
1890		380,059	118,898	1.05	124,843	90,691	.156	14,148	45,894	.045	2,065	521,115
1891		683,941	41,690	.99	41,273							725,214
1892		982,988	182,156	.87	158,476				9,697	.04	388	1,141,852
1893		479,665	257,462	.78	200,820	50,000	.108	5,400	a 10,000	.037	370	686,255
1894		489,592	75,730	.63	47,710	50,000	.095	4,750	a 10,000	.033	330	542,382
1895		401,926	40,685	.65	26,445	57,864	.107	6,191	11,439	.032	366	434,928
1896		385,653	79,047	.68	53,752	63,252	.108	6,831	4,216	.03	126	446,362
1897		512,657	138,715	.60	83,229	58,474	.12	7,017	309,115	.036	11,128	614,031
1898		581,302	91,432	.59	53,945	22,452	.124	2,784	8,967	.038	341	638,372
1899		547,558	76,371	.60	45,823	78,816	.171	13,478	28,043	.045	1,262	608,421
1900		607,016	90,327	.62	56,003	20,371	.166	3,382	76,076	.044	3,347	669,748
1901		774,298	113,782	.60	68,269	22,186	.167	3,705	191,987	.043	8,255	854,527
1902		538,702	82,710	.53	43,836	11,090	.122	1,353	13,493	.041	553	584,444
1903		431,569	61,833	.54	33,390	6,154	.137	843	115,100	.042	4,834	470,636
1904	23,905	411,581	57,424	.58	33,306	26,115	.128	3,343	62,111	.043	2,671	450,901
1905	9,577	261,601	70,921	.61	43,262	2,227	.156	347				305,210
1906	5,528	188,769	21,923	.68	14,908	3,539	.193	683	47,491	.057	2,707	207,067
1907	8,000	161,658	23,480	.66	15,497	22,840	.20	4,568	16,491	.053	874	182,597
1908	10,296	147,234	21,498	.53	11,394	28,955	.132	3,822	96,503	.042	4,053	166,503
1909	13,188	163,273	48,183	.52	25,055	16,485	.13	2,143	425,605	.043	18,301	208,772
1910	14,083	139,911	46,517	.54	25,119	16,772	.127	2,130	53,250	.044	2,343	169,503
1911	15,816	163,174	53,753	.53	28,489	27,752	.125	3,469	145,955	.045	6,568	201,700
1912	9,838	119,426	72,335	.615	44,486	22,176	.165	3,659	305,822	.045	13,762	181,333
1913	5,719	69,274	162,384	.604	98,080	25,535	.155	3,958	409,500	.044	18,018	189,330
1914	14,591	98,710	312,217	.553	172,656	24,316	.133	3,234	523,821	.039	20,429	295,029
1915	39,778	160,433	271,292	.507	137,545	86,680	.175	15,169	890,042	.047	41,832	354,979
1916	33,011	119,299	292,824	.658	192,678	64,707	.246	15,918	864,333	.069	59,639	387,534
1917	16,835	66,841	294,375	.824	242,565	29,513	.273	8,057	575,582	.086	49,500	366,963
1918	10,387	52,265	156,731	1.00	156,731	17,887	.247	4,418	262,310	.071	18,624	232,038
1919	6,143	54,653	225,484	1.12	252,542	11,043	.186	2,054	206,605	.053	10,950	320,199
1920	19,476	42,428	148,834	1.09	162,229	6,685	.184	1,230	261,088	.08	20,887	226,774
1921	13,176	34,042	112,957	1.00	112,957	302	.129	39	140,336	.045	6,315	153,353
1922	3,415	37,037	121,073	1.00	121,073				68,470	.055	3,766	161,876
1923	1,960	27,146	39,556	.82	32,436				26,729	.070	1,871	61,453
		15,954,999	7,994,772		7,572,614	967,627		148,494	6,514,161		349,335	24,025,442

a Estimated by C. W. Henderson.　　　　b Interpolations by C. W. Henderson to agree with total for the State.

CHAFFEE COUNTY

Burchard [5] summarizes the early developments in Chaffee County as follows:

Chaffee County was organized from the southern portion of Lake County in 1879. In 1860 [more probably 1859.—C. W. H.] gold seekers first appeared along the Arkansas River, the result of which was the prosperous camp at Granite. The streams yielded considerable gold dust, especially Cache and Colorado creeks. [Colorado Creek as described by Hollister is the creek emptying into the Arkansas opposite the mouth of California Gulch, so it is therefore now in Lake County.— C. W. H.] In 1867-68 there was some lode mining, but subsequent to that time until 1880 the gold production was small. * * * Monarch mine, at Garfield, located in 1878.

Burchard [6] gives the placer gold production in Lake County from 1860 to 1869 as $5,812,000. In those years the production of Lake County included the production of placers in California Gulch, Colorado Gulch and its tributary, Little Fryingpan Gulch, Georgia Bar, Iowa Gulch, and Lake Creek (all now in Lake County), and Arkansas River, Cache Creek, Lost Canyon Gulch, Chalk Creek, Clear Creek, Cottonwood Creek, Pine Creek, Bertschey's Gulch, Kelly's Bar, Gold Run Gulch, Gilson Gulch, Oregon Gulch, Ritchey's Patch (probably all now in Chaffee County). From data presented by Hollister,[7] the area now comprised in Chaffee County is arbitrarily credited with a production of placer gold amounting to $400,000 during the years 1859-1869.

Raymond [8] gives the placer yield of Lake County in 1870 as a little more than $60,000. The lode mines included (Yankee Blade and others) are in Chaffee County. The production of the Yankee Blade for the year ending June 1, 1870, was about $60,000. Burchard [9] gives the placer and lode production of Lake County for 1870 as $125,000, quoting from the Rocky Mountain Review, of Georgetown.

[5] Burchard, H. C., Report of the Director of the Mint upon the production of the precious metals in the United States during the calendar year 1883, p. 249, 1884.
[6] Burchard, H. C., op. cit. for 1882, p. 505, 1883.

[7] Hollister, O. J., The mines of Colorado, pp. 308-320, Springfield, Mass., 1867.
[8] Raymond, R. W., Statistics of mines and mining in the States and Territories west of the Rocky Mountains for 1870, p. 332, 1872.
[9] Burchard, H. C., op. cit. for 1882, p. 505, 1883.

Gold, silver, copper, lead, and zinc produced in Chaffee County, 1859-1923

Year	Ore (short tons)	Gold Placer	Gold Lode	Gold Total	Silver Fine ounces	Silver Average price per ounce	Silver Value	Copper Pounds	Copper Average price per pound	Copper Value	Lead Pounds	Lead Average price per pound	Lead Value	Zinc Pounds	Zinc Average price per pound	Zinc Value	Total value
1859-1867		$380,000		$380,000													$380,000
1868		10,000		10,000													10,000
1869		10,000		10,000													10,000
1870		a 10,000	$60,000	70,000													70,000
1871		a 10,000		10,000													10,000
1872		a 10,000		10,000													10,000
1873		a 10,000		10,000													10,000
1874		a 10,000		10,000													10,000
1875		a 21,551	(?)	21,551													21,551
1876		a 25,000		25,000	a 3,867	$1.16	$4,486				a 50,000	$0.055	$2,750				29,486
1877		a 25,000		25,000	a 7,734	1.20	9,281				a 50,000	.036	1,800				37,031
1878		a 25,000		25,000	a 7,734	1.15	8,894				a 50,000	.041	2,050				35,694
1879		a 25,000	3,500	28,500	30,938	1.12	34,651				a 100,000	.05	5,000				65,201
1880		25,000	6,500	31,500	61,875	1.15	71,156				a 500,000	.048	24,000				218,207
1881		a 25,000	25,000	50,000	127,617	1.13	144,207				a 1,000,000	.049	49,000				182,172
1882		a 25,000	20,000	45,000	77,344	1.14	88,172				a 4,300,000	.043	184,900				462,407
1883		a 25,000	25,000	50,000	204,961	1.11	227,507				a 12,000,000	.037	444,000				687,118
1884		a 25,000	55,000	80,000	146,953	1.11	163,118				a 18,700,000	.039	729,300				1,043,300
1885		a 25,000	75,000	100,000	a 200,000	1.07	214,000				a 13,000,000	.046	598,000				1,241,552
1886		a 80,000	233,917	313,917	332,965	.99	329,635				a 14,954,155	.045	672,937				1,497,250
1887		a 45,000	364,050	409,050	423,738	.98	415,263				a 8,743,053	.044	384,694				1,052,959
1888		a 25,000	368,457	393,457	292,349	.94	274,808				a 5,000,000	.039	195,000				624,346
1889		37,000	262,853	299,853	137,759	.94	129,493				a 2,400,000	.045	108,000				515,208
1890		45,300	208,950	254,250	145,674	1.05	152,958				a 1,100,000	.043	47,300				390,542
1891		37,000	242,060	279,060	64,830	.99	64,182				6,324,319	.04	252,973	a 100,000	$0.046	$4,600	479,276
1892		a 32,000	115,203	147,203	85,632	.87	74,500				a 4,000,000	.037	148,000	a 100,000	.04	4,000	383,673
1893		a 42,000	112,164	154,164	92,448	.78	72,109	a 50,000	$0.108	$5,400	a 1,100,000	.033	36,300	a 100,000	.035	3,500	181,197
1894		a 35,000	85,565	120,565	25,527	.63	16,082	a 50,000	.095	4,750	285,056	.032	9,122	120,000	.036	4,320	194,474
1895		a 35,000	118,629	153,629	29,630	.65	19,260	76,070	.107	8,140	1,047,310	.03	31,419	a 120,000	.039	4,680	332,806
1896		a 35,000	158,465	193,465	151,738	.68	103,182	559	.108	60	1,686,391	.036	60,710	a 100,000	.041	4,100	344,808
1897		a 35,000	191,936	226,936	53,859	.60	32,315	172,891	.12	20,747	2,522,554	.038	95,857	a 100,000	.046	4,600	392,464
1898		a 35,000	192,535	227,535	82,273	.59	50,311	114,202	.124	14,161	1,193,074	.045	53,688	a 100,000	.058	5,800	483,696
1899		a 25,000	191,663	216,663	147,339	.60	88,403	696,736	.171	119,142	833,462	.044	36,672	a 100,000	.044	4,400	416,564
1900		a 25,000	147,677	172,677	125,330	.62	77,705	753,677	.166	125,110	209,768	.043	9,020	a 100,000	.041	4,100	313,810
1901		a 25,000	133,684	158,684	76,286	.60	45,772	576,251	.167	96,234	456,889	.041	18,732	220,500	.048	10,584	528,503
1902		a 40,000	377,513	417,513	114,155	.53	60,502	173,538	.122	21,172	249,308	.042	10,471	3,000	.054	162	261,011
1903		a 28,000	141,329	169,329	129,900	.54	70,146	79,581	.137	10,903	249,308	.042	10,471	3,000	.054	162	181,149
1904	12,777	a 15,000	49,346	64,346	69,045	.58	40,046	263,239	.128	33,695	652,238	.043	28,046	294,440	.051	15,016	181,149
1905	13,408	15,009	17,369	32,378	75,265	.61	45,912	869,507	.156	135,643	1,250,302	.047	58,764	849,963	.059	50,148	322,845
1906	14,134	31,596	27,340	58,936	54,609	.68	37,134	349,466	.193	67,447	1,227,019	.057	69,940	623,955	.061	38,061	271,518
1907	14,592	35,373	39,991	75,364	38,468	.66	25,387	799,505	.20	159,901	630,623	.053	33,423	2,407,730	.059	142,056	436,131
1908	8,772	16,530	32,527	49,057	35,745	.53	18,945	337,804	.132	44,590	1,040,238	.042	43,690	703,706	.047	33,074	189,356
1909	10,214	19,480	11,005	30,485	35,477	.52	18,448	568,868	.13	73,953	584,492	.043	25,133	947,741	.054	51,178	199,197
1910	12,496	17,010	60,142	77,152	182,003	.54	98,282	226,772	.127	28,800	970,523	.044	42,703	438,539	.054	23,681	270,618
1911	7,459	5,893	59,821	65,714	92,098	.53	48,812	88,448	.125	11,056	1,001,651	.045	45,074	200,509	.057	11,429	182,085
1912	10,287	4,619	92,870	97,489	104,686	.615	64,382	133,570	.155	22,039	992,578	.045	44,666	736,392	.069	50,811	279,387
1913	49,135	1,266	311,626	312,892	168,985	.604	102,067	315,011	.155	48,827	3,196,545	.044	140,648	2,121,947	.056	118,829	723,263
1914	61,698	626	331,604	332,230	272,242	.553	150,550	319,496	.133	42,493	3,690,359	.039	143,924	2,173,177	.051	110,832	780,029
1915	68,242	1,463	314,683	316,146	226,906	.507	115,087	348,416	.175	60,908	3,630,127	.047	170,616	4,676,355	.124	579,868	1,242,625
1916	69,358	573	184,477	185,050	100,749	.658	66,293	1,001,455	.246	246,358	3,016,899	.069	208,166	4,744,985	.134	635,828	1,341,695
1917	55,652	183	133,441	133,624	146,535	.824	120,745	807,883	.273	220,552	2,150,523	.086	184,945	2,181,932	.102	222,557	882,423
1918	39,598		112,478	112,478	81,187	1.00	81,187	323,830	.247	79,986	1,885,761	.071	133,889	2,618,767	.091	238,308	645,848
1919	19,655		58,167	58,167	40,550	1.12	45,416	70,823	.186	13,173	803,228	.053	42,571	965,630	.079	70,491	229,818
1920	3,900		31,302	31,302	39,211	1.09	42,740	28,195	.184	5,188	396,250	.08	31,700	283,235	.081	22,942	133,872
1921	2,402		32,034	32,034	27,641	1.00	27,641	8,357	.129	1,078	318,666	.045	14,340	39,000	.05	1,950	77,043
1922	6,844		19,936	19,936	26,187	1.00	26,187	20,526	.135	2,771	661,728	.055	36,395	178,000	.057	10,146	95,435
1923	2,173	707	16,366	17,073	20,762	.82	17,025	7,089	.147	1,042	557,429	.07	39,020	132,000	.068	8,976	83,136
		1,548,179	5,853,175	7,401,354	5,221,893		4,234,384	9,631,395		1,725,319	130,512,518		5,749,348	28,581,505		2,491,027	21,601,432

a Estimated by C. W. Henderson, with advice of Ben Stanley Revett, former general manager of the Twin Lakes Hydraulic Gold Mining Syndicate.

The mills in 1870 were the Yankee Blade 20-stamp mill, in which the ore was treated by battery amalgamation, blanket sluices. and pans for tailings, and the Treasury Mining Co.'s 15-stamp water-power mill. Hayden & Son had a 9-stamp water-power mill.

On the authority of T. F. Van Wagenen, of the Rocky Mountain Review, Raymond [10] credits $100,000 in currency ($86,200 in coin) to Lake County for 1875 but says that three-fourths of this was taken from California Gulch. The amount may be divided into $64,650 for Leadville and $21.550 for Chaffee County. Raymond adds:

In Chalk Creek, Chapman & Riggins are building works of 10 tons capacity to treat ore from their own mines, * * * [claim names] Riggins, Naomi, Tecumseh, Black Hawk, Anna, Mary Murphy, and Mount Yale. The plant of this establishment comprises two roasting furnaces and one blast furnace, not expected to be in operation before the summer of 1876. At the close of 1875 several hundred tons of ore was on the dumps.

The figures for 1879 to 1884 are taken chiefly from the reports of the Director of the Mint.[11] In the report for 1880 Burchard [12] says:

The gulch or placer mines along the Arkansas have yielded the usual amount—about $25,000 for the season. * * * There are five small and apparently inefficient smelters in Chaffee County—one at Garfield, two at Maysville, one at Poncha Springs, and one at Forest City, on Chalk Creek. None of them has done much, most of the higher grades of ore being shipped to Pueblo and Argo.

The figures for the production of the Madonna mine, in the Monarch district, from 1883 to 1911 are given by the Colorado Geological Survey.[13]

The figures for 1885 to 1896 given in the table are derived from reports of the agents of the mint, the annual reports of the Director of the Mint, the gold and silver being prorated to correspond with the figures for the total production of the State as corrected by the Director of the Mint, the lead being prorated to correspond with the total production of lead in the State as given in annual volumes of Mineral Resources, and any unknown production in the State being distributed proportionately among the counties. As with lead, so with copper, but as the figures given for copper in Mineral Resources include copper from matte and ores treated in Colorado, though produced in other States, these figures are subject to revision.

The production of the Twin Lakes placer in 1890 is taken from the Denver News for January 1, 1891.

The figure for the production of zinc in 1895 represents ore receipts of the American Zinc-Lead Co., of Canon City, as published in the Denver Republican.

The figures for 1897–1904, which represent mint and smelter receipts, are taken from the reports of the Colorado State Bureau of Mines.

The figures given in the table for 1905 to 1923 are taken from Mineral Resources (mines reports).

The output of gold for 1904 includes a little silver from placer bullion; for 1905 it includes 131 ounces of silver from placer bullion; for 1906, 279 ounces; for 1907, 238 ounces; for 1908, 158 ounces; for 1909, 156 ounces; for 1910, 157 ounces; for 1911, 47 ounces; for 1912, 39 ounces; and for 1913, 15 ounces. In 1912 the fineness of the gold in the Granite placer gold bullion, was 0.795 to 0.850 and of the silver 0.129 to 0.138. In 1913 the average fineness of the gold was 0.851 and of the silver 0.144.

For 1906. the placer production is corrected by transferring a small quantity from Lake County.

CLEAR CREEK COUNTY

From 1859 to 1867 the output of gold from the area now included in Clear Creek County was obtained from placers on South Clear Creek, chiefly in the vicinity of Idaho Springs, and from surface workings on lode deposits at Idaho Springs and Empire. The figures for gold and silver from 1859 to 1878 as given by Munson [14] have been prorated to accord with figures compiled later or with figures given in Raymond's reports for 1870 to 1875.[15]

The figures given in the table for lead from 1869 to 1876 represent lead in ores shipped to Eastern States, to England, and to Germany.

For 1878 to 1884 the figures are taken from the reports of the Director of the Mint.

The figures for 1886 to 1896 are taken from reports of the agents of the mint in annual reports of the Director of the Mint, the gold and silver being prorated to correspond with the figures showing the total production of the State as corrected by the Director of the Mint, those for lead being prorated to correspond with the total production of lead in the State as given in Mineral Resources, and any unknown production in the State being distributed proportionately among the counties. As with lead, so with copper, but as the figures given for copper in Mineral Resources include copper obtained from matte and ores treated in Colorado though produced in other States, the figures for copper are subject to revision.

The value of the output of lead from 1877 to 1879 is taken from Spurr and Garrey.[16]

The figures showing the output of zinc in 1895 represent the receipts of ore by the American Zinc-Lead Co. at Canon City.

The figures for 1897–1904, which represent smelter and mint receipts, are taken from the reports of the Colorado State Bureau of Mines.

10 Raymond, R. W., op. cit. for 1875, pp. 282, 314, 316, 1877.

11 Burchard, H. C., op. cit. for 1880, pp. 156, 157, 1881.

2 Idem, p. 153.

13 Crawford, R. D., Geology and ore deposits of the Monarch and Tomichi districts, Colo.: Colorado Geol. Survey Bull. 4, p. 239, 1913.

14 Munson, G. C., agent for Colorado, in Kimball, J. P., Report upon the production of the precious metals in the United States during the calendar year 1887, p. 153, 1888.

15 Raymond, R. W., op. cit. for 1870, p. 316, 1872; idem for 1872, pp. 277–278, 1873; idem for 1873, p. 287, 1874; idem for 1874, p. 365, 1875; idem for 1875, p. 295, 1877.

16 Spurr, J. E., and Garrey, G. H., Economic geology of the Georgetown quadrangle, Colo.: U. S. Geol. Survey Prof. Paper 63, p. 175, 1908.

The figures for 1905–1923 are taken from Mineral Resources (mine reports).

The figures for placer gold in 1907 include $7,628 in "unknown gold," probably stamp gold, not placer.

The prices of silver, copper, lead and zinc from 1850 to 1923, are those given by Loughlin.[17]

one of the original 17 counties organized by an act of Territorial legislature in November, 1861. On account of confusion liable to arise, the name of Guadalupe was soon changed to Conejos. The history of the hardships of the pioneers of this section is filled with thrilling episodes, the settlement of the whites being bitterly resented by the Indians. After a considerable season of doubt as to who should gain supremacy the

Gold, silver, copper, lead, and zinc produced in Clear Creek County, 1859–1923

Year	Ore (short tons)	Gold			Silver			Copper			Lead			Zinc			Total value
		Placer	Lode	Total	Fine ounces	Average price per ounce	Value	Pounds	Average price per pound	Value	Pounds	Average price per pound	Value	Pounds	Average price per pound	Value	
1859–1865		$2,000,000		$2,000,000													$2,000,000
1866		a50,000		a50,000	15,123	$1.339	$20,250										70,250
1867		a50,000		a50,000	15,228	1.33	20,251										70,251
1868		a50,000		a50,000	106,953	1.326	141,820										191,820
1869		a50,000		a50,000	377,359	1.325	a500,000	a2,000	$0.2425	$485	a100,000	$0.06	$6,000				556,485
1870	2,000	a80,000		a80,000	362,465	1.328	481,354	a2,500	.2118	530	a200,000	.06	12,000				573,884
1871		20,000		20,000	640,790	1.325	849,047	a3,000	.2412	724	a550,000	.06	33,000				902,771
1872	5,888	25,000		25,000	1,118,299	1.322	1,478,391	a4,000	.3556	1,422	a1,000,000	.064	64,000				1,568,813
1873	5,421	34,000		34,000	902,668	1.297	1,170,760	a10,000	.28	2,800	a1,000,000	.06	60,000				1,267,560
1874	9,490	42,500		42,500	1,634,434	1.278	2,088,807	a15,000	.22	3,300	a803,983	.06	48,239				2,182,846
1875	12,000	68,960	$3,448	72,408	1,343,610	1.24	1,666,076	a15,000	.227	3,405	a1,300,000	.058	75,400				1,817,289
1876		40,000	55,161	95,161	1,421,104	1.16	1,648,481	a15,000	.21	3,150	a819,672	.061	50,000				1,796,792
1877		20,000	76,500	96,500	1,534,560	1.20	1,841,472	a15,000	.19	2,850	2,236,364	.055	123,000				2,063,822
1878		10,000	124,000	134,000	1,759,652	1.15	2,023,600	a25,000	.166	4,150	2,722,222	.036	98,000				2,259,750
1879		a10,000	110,000	120,000	1,546,875	1.12	1,732,500	a100,000	.186	18,600	1,951,219	.041	80,000				1,951,100
1880		a5,000	191,000	196,000	1,902,656	1.15	2,188,054	a200,000	.214	42,800	a517,500	.05	25,875				2,452,729
1881		5,000	195,000	200,000	1,546,875	1.13	1,747,969	a200,000	.182	36,400	a815,000	.048	39,120				2,023,489
1882		6,000	214,000	220,000	1,299,375	1.14	1,481,288	a300,000	.191	57,300	a815,000	.049	39,935				1,798,523
1883		10,000	240,000	250,000	1,222,031	1.11	1,356,454	a300,000	.165	49,500	a815,000	.043	35,045				1,690,999
1884		12,551	587,449	600,000	1,314,844	1.11	1,459,477	a300,000	.13	39,000	a1,038,273	.037	38,416				2,136,893
1885		a10,000	490,000	b500,000	b1,356,000	1.07	1,450,920	a200,000	.108	21,600	1,038,273	.039	40,493	a25,000	$0.043	$1,075	2,014,088
1886		a10,000	599,070	609,070	1,396,364	.99	1,382,400	a200,000	.111	22,200	1,630,000	.046	74,980	a25,000	.044	1,100	2,089,750
1887		a15,000	302,214	317,214	1,284,083	.98	1,258,401	a200,000	.138	27,600	1,847,930	.045	83,157	a25,000	.046	1,150	1,687,522
1888		a20,000	399,821	419,821	1,148,190	.94	1,079,299	a200,000	.168	33,600	3,761,246	.044	165,495	a75,000	.049	3,675	1,701,890
1889		25,000	496,909	521,909	1,770,875	.94	1,664,623	91,731	.135	12,384	5,357,906	.039	208,958	a75,000	.05	3,750	2,411,624
1890		24,336	418,032	442,368	1,819,682	1.05	1,910,666	124,102	.156	19,360	12,029,217	.045	541,315	a75,000	.055	4,125	2,917,834
1891		a22,875	415,692	438,567	1,771,055	.99	1,753,344	57,572	.128	7,369	7,947,786	.043	341,755	a75,000	.05	3,750	2,544,785
1892		5,340	308,701	314,041	1,691,579	.87	1,471,674	40,424	.116	4,689	7,916,672	.04	316,667	a250,000	.046	11,500	2,118,571
1893		a5,000	579,187	584,187	2,218,377	.78	1,730,334	a40,000	.108	4,320	a8,000,000	.037	296,000	a400,000	.04	16,000	2,630,841
1894		a5,000	657,649	662,649	2,228,846	.63	1,404,173	a40,000	.095	3,800	a8,000,000	.033	264,000	a200,000	.035	7,000	2,341,622
1895		5,000	669,210	674,210	1,585,483	.65	1,030,564	44,168	.107	4,726	6,415,936	.032	205,310	a200,000	.036	7,200	1,922,010
1896		5,000	787,631	792,631	1,626,828	.68	1,106,243	204,519	.108	22,088	6,438,672	.03	193,160	a400,000	.039	15,600	2,129,722
1897		50,000	732,649	782,649	1,442,583	.60	865,550	516,034	.12	61,924	5,263,116	.036	189,472	a300,000	.041	12,300	1,911,895
1898		a5,000	600,528	605,528	1,569,012	.59	925,717	317,423	.124	39,360	5,843,767	.038	222,063	a300,000	.046	13,800	1,806,468
1899		a5,000	541,825	546,825	1,502,900	.60	901,740	292,966	.171	50,097	7,216,260	.045	324,732	a300,000	.058	17,400	1,840,794
1900		a5,000	460,447	465,447	1,358,143	.62	842,049	244,092	.166	40,519	4,994,263	.044	219,748	a300,000	.044	13,200	1,580,963
1901		a5,000	535,975	540,975	1,271,227	.60	762,736	374,534	.167	62,547	3,890,216	.043	167,279	a300,000	.041	12,300	1,545,837
1902		a5,000	925,481	930,481	1,279,050	.53	677,897	473,754	.122	57,798	3,282,270	.041	134,573	317,705	.048	15,250	1,815,999
1903		a5,000	467,061	472,061	851,638	.54	459,885	289,876	.137	39,713	3,451,849	.042	144,978	656,000	.054	35,424	1,152,061
1904	62,661	2,398	634,217	636,615	873,949	.58	506,890	401,180	.128	51,351	3,913,976	.043	168,301	906,705	.051	46,242	1,409,399
1905	58,775	1,881	501,817	503,698	692,437	.61	422,387	235,669	.156	36,764	3,270,211	.047	153,700	1,102,301	.059	65,036	1,181,585
1906	64,774	1,568	528,185	529,753	652,796	.68	423,901	235,375	.193	45,427	3,307,001	.057	188,499	1,733,477	.061	105,742	1,313,322
1907	79,548	11,511	511,385	522,896	518,364	.66	342,120	171,340	.20	34,268	2,804,172	.053	148,621	2,771,960	.059	163,546	1,211,451
1908	109,672	2,610	656,506	659,116	503,551	.53	266,882	264,994	.132	34,979	2,015,010	.042	84,630	836,411	.047	39,311	1,084,918
1909	116,753	3,846	532,561	536,407	448,535	.52	233,238	299,546	.13	38,941	3,254,675	.043	139,951	758,074	.054	40,936	989,473
1910	109,954	3,678	518,846	522,524	475,174	.54	256,594	595,795	.127	75,666	2,434,476	.044	107,117	1,247,389	.054	67,359	1,029,260
1911	105,774	1,754	517,453	519,207	437,841	.53	232,056	650,368	.125	81,296	3,325,222	.045	149,635	1,417,544	.057	80,800	1,062,994
1912	102,894	331	445,463	445,794	373,940	.615	229,973	449,401	.165	74,151	3,523,733	.045	158,568	1,734,493	.069	119,680	1,028,166
1913	104,892		432,489	432,489	408,527	.604	246,750	426,393	.155	66,091	3,999,614	.044	175,983	1,489,518	.056	83,413	1,004,726
1914	101,366		495,275	495,275	345,387	.553	190,999	367,790	.133	48,916	2,435,692	.039	94,992	1,067,314	.051	54,433	884,615
1915	121,993	859	525,724	526,583	393,108	.507	199,306	530,949	.175	92,916	2,527,575	.047	118,796	1,505,032	.124	186,624	1,124,225
1916	94,220	250	428,681	428,931	462,141	.658	304,089	621,732	.246	152,946	4,295,725	.069	296,405	2,572,575	.134	344,725	1,527,096
1917	84,449	435	303,549	303,984	526,750	.824	434,042	570,091	.273	155,635	4,836,617	.086	415,949	3,153,030	.102	321,609	1,631,219
1918	57,808		231,077	231,077	370,888	1.00	370,888	343,247	.247	84,782	3,869,352	.071	274,724	1,812,846	.091	164,969	1,126,440
1919	116,355		91,127	91,127	357,439	1.12	400,332	152,925	.186	28,444	1,517,134	.053	80,408	603,027	.073	44,021	644,332
1920	51,494		48,540	48,540	219,900	1.09	239,691	61,978	.184	11,404	2,457,100	.08	196,568	372,420	.081	30,166	526,369
1921	32,767		38,851	38,851	131,867	1.00	131,867	21,519	.129	2,776	1,200,931	.045	54,042	217,000	.05	10,850	238,386
1922	69,425		36,199	36,199	196,207	1.00	196,207	7,874	.135	1,063	1,042,491	.055	57,337	800,000	.057	45,600	336,406
1923	23,464		30,576	30,576	183,874	.82	150,777	32,218	.147	4,736	1,016,729	.07	71,171	577,000	.068	39,236	296,496
		2,852,683	19,693,161	22,545,844	57,809,489		52,373,255	11,898,079		1,924,662	178,057,048		8,097,562	30,976,821		2,249,897	87,191,220

a Estimated by C. W. Henderson.   b Interpolated by C. W. Henderson to agree with total for the State.

CONEJOS COUNTY

The early history of Conejos County is summarized by the Colorado State Bureau of Mines [18] as follows:

What is now known as Conejos County was originally organized under the name of Guadalupe, in honor of the patron saint of Mexico. As originally constituted, it embraced nearly all the territory in the southern portion of the State. It was

Indians were compelled to fall back before the advance of civilization. With later development Conejos County has been reduced to an area of 1,200 square miles, with the seat of government at Conejos. * * *

The western portion of the county includes some mineral, from Platoro in the north to the Banded Peaks and Antonito districts in the south. While prospected to a limited extent, it can properly be classed as one of the undeveloped reserves of the State. * * *

The county records show 1,094 lode claims recorded, 68 being patented, 3 placer claims, and 3 tunnel sites. Beyond the annual

17 U. S. Geol. Survey Mineral Resources, 1923, pt. 1, 1924.
18 Colorado State Bur. Mines Rept. for 1897, pp 23–24, 1898.

assessment work, little systematic exploring was prosecuted during the past year. An average of 46 men [was] employed during the year.

Patton [19] gives notes on the mining in this county from 1870 to 1917. (See also Rio Grande County, p. 201.)

For 1885 the deposits at the Denver Mint indicate the production of Conejos County.[20]

For 1894 to 1896 the figures given are based on reports of the agents of the mint in annual reports of the Director of the Mint, the gold and silver being prorated to correspond with the figures showing the total production of the State as corrected by the Director of the Mint, the lead being prorated to correspond with the total production of lead in the State as given in Mineral Resources, and any unknown production in the State being distributed proportionately among the counties. As with lead, so with copper, but as the figures for copper given in Mineral Resources include copper derived from matte and ores treated in Colorado, though produced in other States, the figures for copper are subject to revision.

For 1897 to 1904 the figures, which represent smelter and mint receipts, are taken from the reports of the Colorado State Bureau of Mines.

For 1905 and 1906 the figures are taken from Mineral Resources (mines reports).

did not invite prospecting. Comparatively recently a code of rules has been formulated in accordance with and in some respects more liberal than the State mining laws. Under these provisions, prospecting is permitted and titles guaranteed.

In what is known as the El Plomo district [on Rito Seco northeast of San Luis] the existence of * * * ore deposits has been known for some years. Spasmodic efforts of development have been made during the past nine years. During 1896 and 1897, a systematic effort has demonstrated an ore body that bids fair to make mining the leading industry of the county. The ore is low grade, with principal values in gold, and has been developed by a series of cuts and shallow shafts to demonstrate its extent. The main development is a 400-foot drift or tunnel driven into the mass and a 75-foot winze sunk at end of same. The values have been determined by a series of carefully made tests. During last year a 10-stamp amalgamation mill, with concentrating tables, was constructed. The report from a three-month's run shows an average value of $4 per ton, 70 per cent being amalgamated, and a pay product being obtained on the tables by concentrating 25 tons into 1. The ore body is quartzite, interlaced with quartz seams and charged with auriferous iron pyrite. The manager, after nine years' exploiting, states as his belief the demonstration of an ore body 1,800 feet wide and 2,600 feet long and 100 feet thick, which may be quarried to suit mill capacity. Further, that he has demonstrated that * * * the body carries an average gold value of $2.50 to $4 per ton. * * *

At Placer, a camp near Veta Pass, on the headwaters of Sangre de Cristo Creek, many improvements have been made during the past year. The placer bars that have been the seat of several excitements were again worked and yielded fair returns from the sluice box. A group of these claims has been

Gold, silver, copper, and lead produced in Conejos County, 1861–1906

Year	Ore (short tons)	Lode gold	Silver			Copper			Lead			Total value
			Fine ounces	Average price per ounce	Value	Pounds	Average price per pound	Value	Pounds	Average price per pound	Value	
861–1884		(a)										(a)
1885		$277										$277
1894		171	1	$0.63	$1							172
1896		639	17	.68	12							651
1897		1,054	98	.60	59							1,113
1898		18,355	29,777	.59	17,568							35,923
1899		6,263	22,987	.60	13,792							20,055
1900		2,832	1,014	.62	629	4,527	$0.166	$752	2,200	$0.044	$97	4,310
1901		1,178	102	.60	61	210	.167	35	1,200	.043	52	1,326
1902		1,261	81	.53	43	78	.122	10				1,314
1903		1,220	46	.54	25							1,245
1904		827	52	.58	30							857
1905	12	2,894	900	.61	549							3,443
1906	85	1,474	748	.68	509							1,983
		38,445	55,823		33,278	4,815		797	3,400		149	72,669

a Production unrecorded.

COSTILLA COUNTY

The history of Costilla County is summarized by the Colorado State Bureau of Mines as follows: [21]

The history of the county reveals several mining excitements. None of these until within the past years resulted in any systematic exploration of these mineral deposits. This may be largely due to the fact that the main portion of the county is held under a Spanish land grant. Until within a few years the owners of this land, known as the Sangre de Cristo grant,

purchased by eastern capital, which has expended a large amount in the erection of a steam excavation plant and anticipates * * * returns during this year.

On Mount Blanca a small number of properties have been operated steadily and produced small shipments of ore. During 1897 the entire mountain section has been better prospected than for many years. * * *

The county and grant records show 215 lode claims and 40 placer claims recorded. During the summer months 200 men were employed and 11 properties working.

The following details of the development of Costilla County are given in a report by Patton and others: [22]

[19] Patton, H. B., Geology and ore deposits of the Platoro-Summitville mining district, Colo.: Colorado Geol. Survey Bull. 13, 1917.

[20] Wilson, P. S., agent for Colorado, in Kimball, J. P., Report of the Director of the Mint upon the production of the precious metals in the United States during the calendar year 1885, p. 136, 1886.

[21] Colorado State Bur. Mines Rept. for 1897, pp. 24-26, 1898.

[22] Patton, H. B., and others, Geology of the Grayback mining district, Costilla County, Colo.: Colorado Geol. Survey Bull. 2, pp. 83-96, 1910.

The earliest knowledge of the presence of metals in this region was several years prior to 1875, when the Hayden expedition traversed the southern part of Colorado. The report of that survey mentions the recovery of gold from alluvials in Placer and Grayback creeks. In those days of Indian outbreaks a Government fort was maintained at what is now the town of Fort Garland. The soldiers there stationed, it would appear, had knowledge of the values in the gravels mentioned and, for a time, a prominent bar or bench of this ground lying at the confluence of the two creeks was handled, in a very small way, for the recovery of gold. This piece of land is still known as Officers' Bar. Later on the same ground was attacked by Chinamen, it is said, with very satisfying results.

Squatters were not, however, permitted to continue their operations very long, for in 1877 the Trinchera estate authorities stopped all kinds of mining operations within its holdings. Up to this time the owners of the estate had received no income from operations. Feeling that the ground was worthy of mineral development, it was deemed proper to devise methods whereby prospectors could locate and secure mining claims Under certain restrictions, therefore, the domain was opened to location.

Under the rules which now govern, the mining lands in the Trinchera estate are open to location upon very much the same plan as prevails with similar lands upon the public domain. Lode claims are laid out 300 feet wide by 1,500 feet long. * * *

The washing of alluvials for gold has been, as already stated, one of the features of mining in this region for years and promises to be a very prominent industry in the future of this country.

In 1898 the Badger State Placer Mining Co. built a large steam shovel and gold-saving structure. This machine, pretentious in its day, was intended to handle 1,000 to 2,000 cubic yards of gravel per day. There was little knowledge derivable from experience in such matters at that time, so it is little wonder that unforeseen defects in mechanical construction and gold-saving apparatus prevented continuous and successful operation of this machine.

The steam shovel, of dipper type, discharged each load into a revolving screen. Here the gravel was disintegrated and washed by jets of water, the undersize being conducted thence through a riffled sluice 3 feet wide and 30 feet long. As the shovel brought up about 1 cubic yard at a time, the apparatus would one monemt be congested with dirt, while a few moments later the screen and sluice would be running empty. Even under such conditions it is reported that this device treated about 2,700 cubic yards, from which was obtained an average yield of 24.6 cents per yard. The dirt averaged about 18 feet in depth. Naturally, in face of the inexperience of the operators and the shortcomings of the machine, the costs of operation were in excess of the recovery and the closing down followed.

The most extensive mining operations ever conducted in this region were those of the early eighties, which were carried on by the Colorado Coal & Iron Co., the present Colorado Fuel & Iron Co. This company developed what was then called the Placer [iron] mine, in some of its records, because of its location near the town of Placer, now changed to the town of Russell.

In 1910 the Colorado Gold Dredging Co. built and operated on Placer Creek, near Russell, for a short period its Hammond dredge, of the close-connected type, of 54 buckets of 4 cubic feet capacity each. The total capacity of the dredge was 2,000 yards a day. It was operated for about six months in 1911 and was then overturned. It has not been righted nor operated since.

For 1885 the figures given in the table below represent gold and silver deposited in the Denver Mint.[23]

For 1895 and 1896 the figures given in the table are based on reports of the agents of the mint in annual reports of the Director of the Mint, the gold and silver being prorated to correspond with the figures for the total production of the State as corrected by the Director of the Mint, the lead being prorated to correspond with the total production of lead in the State as given in Mineral Resources, and any unknown production in the State being distributed proportionately among the counties. As with lead, so with copper, but as the figures for copper given in Mineral Resources include copper from matte and ores treated in Colorado, though produced in other States, the figures for copper are subject to revision.

For 1897 to 1904 the figures, which represent smelter and mint receipts, are taken from the reports of the Colorado State Bureau of Mines.

For 1906 to 1921 the figures are taken from Mineral Resources (mines reports).

[23] Wilson, P. S., agent for Colorado, in Kimball, J. P., Report of the Director of the Mint upon the production of the precious metals in the United States during the calendar year 1885, p. 136, 1886.

Gold (placer and lode), silver, copper, and lead produced in Costilla County, 1875–1921

Year	Gold			Silver			Copper			Lead			Total value
	Placer	Lode	Total	Fine ounces	Average price per ounce	Value	Pounds	Average price per pound	Value	Pounds	Average price per pound	Value	
1875–84	(a)												(a)
1885	$216		$216										$216
1895	b 126		126										126
1896	b 139		139										139
1897	b 650	$4,766	5,416	482	$0.60	$289	502	$0.12	$60	50,048	$0.036	$1,802	7,567
1898	b 1,000	b 4,519	5,519	993	.59	586	983	.124	122				6,227
1899	b 300	b 506	806	126	.60	76							882
1900	b 200	b 1,867	2,067	314	.62	195	107	.166	18				2,280
1901	b 200	b 771	971	153	.60	92	235	.167	39				1,102
1902	b 200	b 978	1,178	205	.53	109							1,287
1903	b 200	b 792	992	179	.54	97							1,089
1904	b 368	b 300	668	151	.58	88							756
1906		426	426										426
1910	2,318		2,318	9	.54	5							2,323
1911	21,832		21,832	96	.53	51							21,883
1912	470		470	3	.615	2							472
1913	95		95	2	.604	1							96
1914	177		177	2	.553	1							178
1921		52	52										52
	28,491	14,977	43,468	2,715		1,592	1,827		239	50,048		1,802	47,101

a Production unrecorded.   b Estimated by C. W. Henderson.

CUSTER COUNTY

Emmons [24] reviews the history of the mines of Custer County as follows:

The mine first discovered in the region was the Senator (since rechristened the Maverick), which was located in the autumn of 1872 on an outcrop of cavernous quartz, with galena and silver glance in the cavities, which was extremely rich. After it had been opened to a depth of 50 to 60 feet, however, the ore seemed to pinch out, and the mine was for the time abandoned.

In April, 1874, a thin seam containing carbonates of copper, with native silver, was discovered on the southern slopes of the hills back of Rosita. This was the outcrop of the since famous Humboldt-Pocahontas vein, which runs northwest and southeast. On this were located first the Humboldt, then the Pocahontas, and later the Virginia and East Leviathan claims, as extensions to the southeast and the northwest, respectively. The Humboldt-Pocahontas mine has been the most permanent and steady producer of any in the region, and was worked more or less continuously for nearly 15 years, producing, according to Census reports, over $900,000 worth of ore.

The hills around were rapidly honeycombed by the excavations of prospectors, and a picturesque mining town, known as Rosita, soon sprang up and reached the height of its prosperity, with a population of 1,200 to 1,500 inhabitants, in the years 1875–1877. Many small bodies of extremely rich ore were discovered in the neighboring hills, but few were developed to a sufficient depth to be worthy of the names of mines.

The next great mine was discovered about 2 miles north of Rosita, in 1877, by E. G. Bassick, a sailor turned prospector. * * * It proved so very rich that he found no difficulty in raising capital for its development, and around the mine soon sprang up the little town of Querida. It is said that ore worth $500,000 in gold and silver was taken out of the mine in the first year and a half and that in 1879 he sold it to eastern capitalists, receiving half a million dollars in cash and a tenth interest in the stock of the new company. This company, with a nominal capital of $10,000,000, is said to have produced over $1,500,000 in gold and silver, of which $425,000 was paid in dividends to the stockholders. There seems, nevertheless, to have been something the matter with the management, and it was finally closed down in 1885, when its shaft had reached a depth of 1,400 feet, although, according to miners working in the mine at the time, the ore body appeared as large and as rich as ever. Since that time it has been sold several times at sheriff's sale to satisfy judgments obtained against it, and there has been much litigation in regard to its ownership, which is apparently not yet settled, since it has not been reopened.

In 1901–2 a new shaft, which had been sunk some time before to a depth of 650 feet, was deepened to 1,500 feet. Mining in 1901 ceased in March, but the work on this shaft was continued until May, 1902. The old shaft, which began at a large chamber excavated 400 feet from the mouth of a tunnel, was 1,300 feet below the level of the tunnel. In 1904 ore was shipped by the Bassick company from the Maine and Lookout claims. Ore was also shipped in 1905, 1906, and 1907, and the dump was treated by cyanidation from 1908 to 1915, and again in 1923.

Emmons continues:

When the Rosita Hills had been pretty thoroughly prospected attention was directed to a lower outlying group of hills which rise out of the gently sloping plains on the immediate border of the bottom lands of the Wet Mountain Valley. In 1878 three lumbermen—Edwards, Hafford, and Powell—who were engaged in hauling timber from the Sangre de Cristo Range to Rosita, had their curiosity aroused by the black cliffs that stood out so prominently on the southern end of the White Hills, which they passed in going to and from their work. One day they climbed over the cliffs and gathered several pieces of a dark, greasy-looking mineral (horn silver), which when heated in the stove, melted into a metal resembling silver. When some of these pieces had been sent to an assayer and it was learned that the mineral, when pure, contained 75 per cent of silver, another "boom" or mining excitement ensued. The discoverers promptly located the Racine Boy, Silver Cliff, and other claims; the surface of the ground in this portion of the valley was soon pitted with prospect holes; and a third town grew rapidly up, named from the mine, Silver Cliff.

In the Blue Mountains, an isolated group of hills about 2 miles north of this town, the discovery was soon after made of the remarkable deposit which later became widely known as the Bull-Domingo mine, from the two claims of which it was the consolidation.

This review of the history is followed by a list of the reduction mills.

Munson [25] gives $40,000 as the value of the output of Custer County up to and including 1874. He divides this amount as follows: For 1872, $8,000; for 1873, $10,014; and for 1874, $21,986 (17,005 ounces at $1.29+).

Raymond [26] gives the total shipments of silver ore for 1874 from Custer County as 129 tons, which yielded by mill returns $21,986. This amount represents the total output of silver from this county. He also credits the county with 8½ tons of copper ore, yielding about 30 per cent of copper.

The following table shows the value of the output of the mines of Custer County up to and including 1880, according to the Silver Cliff Mining Gazette. The figures for gold and silver have been segregated by Emmons. [27]

Output of mines in Custer County for 1880 and previous years

	Gold	Silver (coining value)	Total
Bull-Domingo		$290,000	$290,000
Silver Cliff plateau		566,956	566,956
Rosita and vicinity	$350,000	1,016,025	1,366,025
	350,000	1,872,981	2,222,981

Of other metals produced, lead is the only one of importance, and the Bull-Domingo is the only mine within the area mapped that has produced any considerable amount of this. Outside this area, but still within the county, the Terrible mine, at Ilse, on Oak Creek, has produced $759,717 worth of lead, according to mint reports, which has all gone to a single smelter.

[24] Emmons., S. F., The mines of Custer County, Colo.: U. S. Geol. Survey Seventeenth Ann. Rept., pt. 2, pp. 412–419, 1896.

[25] Munson, G. C., agent for Colorado, in Kimball, J. P., Report of the Director of the Mint upon the production of the precious metals in the United States during the calendar year 1887, p. 153, 1888.

[26] Raymond, R. W., Statistics of mines and mining in the States and Territories west of the Rocky Mountains for 1874, p. 386, 1875.

[27] Emmons, S. F., op. cit., p. 420.

Hunter[28] has described the geology of the Terrible mine.

For 1879 to 1884 the figures given in the table below are taken from the reports of the Director of the Mint.[29]

The figures for lead in 1885, representing the output of the Terrible and other mines, are taken from Mineral Resources.[30]

The figures given for 1886 to 1896 are based on reports of the agents of the mint in annual reports of the Directors of the Mint, the gold and silver being prorated to correspond with the figures showing the total production of the State as corrected by the Director of the Mint, the lead being prorated to correspond with the total lead produced in the State as given in Mineral Resources, and any unknown production in the State being distributed proportionately among the counties. As with lead, so with copper, but as the figures for copper given in Mineral Resources include copper derived from matte and ores treated in Colorado smelters, though produced in other States, the figures are subject to revision.

The figures for 1897 to 1904, which represent smelter and mint receipts, are taken from reports of the Colorado State Bureau of Mines.

The figures for 1905 to 1926 are taken from Mineral Resources (mines reports).

[28] Hunter, J. F., Some cerusite deposits in Custer County, Colo.: U. S. Geol. Survey Bull. 580, pp. 25-37, 1914.

[29] Burchard, H. C., Report of the Director of the Mint upon the production of the precious metals in the United States during the calendar year 1880, pp. 156-157, 1881. (See the reports for later years in this series.)

[30] U. S. Geol. Survey Mineral Resources, 1885, p. 257, 1886.

Gold, silver, copper, lead, and zinc produced in Custer County, 1872-1923

Year	Ore (short tons)	Gold	Silver Fine ounces	Silver Average price per ounce	Silver Value	Copper Pounds	Copper Average price per pound	Copper Value	Lead Pounds	Lead Average price per pound	Lead Value	Zinc Pounds	Zinc Average price per pound	Zinc Value	Total value
1872			6,051	$1.322	$8,000										$8,000
1873			7,721	1.297	10,014										10,014
1874			17,005	1.278	21,732	5,100	$0.22	$1,122							22,854
1875			156,142	1.24	193,616										193,616
1876			38,672	1.16	44,860										44,860
1877		$50,000	77,344	1.20	92,813										142,813
1878		100,000	77,344	1.15	88,946										188,946
1879		100,000	541,406	1.12	606,375										706,375
1880		100,000	665,156	1.15	764,929										864,929
1881		100,000	541,406	1.13	611,789										711,789
1882		200,000	232,031	1.14	264,515										464,515
1883		620,000	154,688	1.11	171,704										791,704
1884		350,000	185,625	1.11	206,044				a 500,000	$0.037	$18,500				574,544
1885		b 30,000	b 61,295	1.07	65,586				5,440,000	.039	212,160				307,746
1886		21,600	61,295	.99	60,682				b 4,500,000	.046	207,000				289,282
1887		507	117,970	.98	115,611				5,367,459	.045	241,536				357,654
1888		120	3,463	.94	3,255				4,821,143	.044	212,130				215,505
1889		1,281	72,576	.94	68,221				63,086	.039	2,460				71,962
1890		114,212	119,684	1.05	125,668				1,708,729	.045	76,893				316,773
1891		49,204	48,469	.99	47,984				838,874	.043	36,072				133,260
1892		325	9,635	.87	8,382				4,963	.04	199				8,906
1893		4,021	32,204	.78	25,119				b 150,000	.037	5,550				34,690
1894		148	1,137	.63	716				b 150,000	.033	4,950				5,814
1895		68	88,632	.65	57,611	4,099	.107	439	139,768	.032	4,473				62,591
1896		42	60,122	.68	40,883	1,109	.108	120	82,105	.03	2,463				43,508
1897		2,129	26,842	.60	16,105	874	.12	105	2,101,041	.036	75,637				93,976
1898		723	24,319	.59	14,348	1,475	.124	183	996,877	.038	37,881				53,135
1899		1,054	6,004	.60	3,602	923	.171	158	836,894	.045	37,660				42,474
1900		20,835	82,605	.62	51,215	2,301	.166	382	709,349	.044	31,211	a 20,000	$0.044	$880	104,523
1901		11,120	50,394	.60	30,236	40,528	.167	6,768	400,481	.043	17,221				65,345
1902		23,708	28,189	.53	14,940	32,945	.122	4,019	94,662	.041	3,881	40,500	.048	1,944	48,492
1903		82,804	160,175	.54	86,495	52,242	.137	7,157	387,301	.042	16,267				192,723
1904	10,170	53,453	87,373	.58	50,676	15,068	.128	1,929	126,593	.043	5,444				111,502
1905	4,567	24,918	32,159	.61	19,617	2,500	.156	390							44,925
1906	3,543	16,318	79,480	.68	54,046	2,725	.193	526	115,960	.057	6,610	971	.061	59	77,559
1907	1,601	6,845	25,995	.66	17,157	8,420	.20	1,684	103,585	.053	5,490				31,176
1908	3,700	7,183	13,156	.53	6,973	243	.132	32	120,330	.042	5,054				19,242
1909	5,871	12,774	14,796	.52	7,694	700	.13	91	41,721	.043	1,794	89,593	.054	4,838	27,191
1910	7,052	9,839	7,767	.54	4,194	3,882	.127	493	14,796	.044	651	6,796	.054	367	15,544
1911	3,670	5,560	13,179	.53	6,985	1,640	.125	205	17,511	.045	788				13,538
1912	4,330	16,898	25,426	.615	15,637	2,006	.165	331	10,444	.045	470				33,336
1913	4,662	14,684	11,313	.604	6,833	4,052	.155	628	5,273	.044	232				22,377
1914	870	3,365	15,975	.553	8,834	3,481	.133	463	9,692	.039	378	4,470	.051	228	13,268
1915	1,719	4,098	31,633	.507	16,038	12,640	.175	2,212	89,808	.047	4,221	30,411	.124	3,771	30,340
1916	2,245	6,309	36,971	.658	24,327	44,004	.246	10,825	123,536	.069	8,524	10,970	.134	1,470	51,455
1917	5,881	7,066	88,687	.824	73,078	88,216	.273	24,083	228,303	.086	19,634				123,861
1918	4,326	4,341	108,456	1.00	108,456	51,292	.247	12,669	281,070	.071	19,956	13,516	.091	1,230	146,652
1919	4,621	4,771	97,159	1.12	108,818	72,979	.186	13,574	155,134	.053	8,222				135,385
1920	1,500	798	34,256	1.09	37,339	28,033	.184	5,158	171,562	.08	13,725				57,020
1921	568	184	19,191	1.00	19,191	37,690	.129	4,862	106,022	.045	4,771				29,008
1922	17,212	167	14,520	1.00	14,520	32,141	.135	4,339	660,618	.055	36,334				55,360
1923	51,200	2,536	28,484	.82	23,357	11,436	.147	1,681	2,890,328	.07	202,323				229,897
		2,186,008	4,541,577		4,545,766	564,744		106,628	34,565,018		1,588,765	217,227		14,787	8,441,954

a Estimated by C. W. Henderson.   b Interpolated by C. W. Henderson to correspond with total production of the State.

DELTA COUNTY

The Colorado State Bureau of Mines says of Delta County: [31]

Delta is one of the west-central counties of the western slope and has an area of about 1,150 square miles. It was segregated from Gunnison County in 1883 by an act of the general assembly. The adjoining counties are Gunnison on the east, Montrose on the south, and Mesa on the north and west. * * *

Metalliferous mines occur in the eastern portion of the county but remain undeveloped and practically not prospected.

For 1894 to 1896 the figures given are based on reports of the agents of the mint in annual reports of the Director of the Mint, the gold and silver being prorated to correspond with the figures of the total production of the State as corrected by the Director of the Mint, the lead being prorated to correspond with the total production of lead in the State as given in Mineral Resources, and any unknown production in the State being distributed proportionately among the counties. As with lead so with copper, but as the figures for copper given in Mineral Resources include copper from matte and ores treated in Colorado, although produced in other States, the figures for copper are subject to revision.

For 1897 to 1904 the figures, which represent smelter and mint receipts, are taken from reports of the Colorado State Bureau of Mines.

For 1910 the figures are taken from Mineral Resources (mine reports).

Gold and silver produced in Delta County, 1894–1910

Year	Lode gold	Silver			Total value
		Fine ounces	Average price per ounce	Value	
1894	$172	3	$0.63	$2	$174
1895	77	1	.65	1	78
1896	339	1	.68	1	340
1897	289				289
1898	579	16	.59	9	588
1899	207	10	.60	6	213
1900	971	97	.62	60	1,031
1901	517	10	.60	6	523
1902	413	12	.53	6	419
1903	248	8	.54	4	252
1904	351	9	.58	6	357
1910	110	139	.54	75	185
	4,273	306		176	4,449

DOLORES COUNTY

Ransome [32] gives the following notes on the history of the Pioneer district (Rico), most of the information being obtained from an article entitled "The early trail blazers," published in the Rico News of June, 1892.

It is possible that some of the early Spanish explorers found their way up to the valley of the Dolores River, but the first party of white men known to have penetrated this region con-

sisted of about 60 trappers from St. Louis, under the leadership of William G. Walton.

1833. This party set out from the trading post of Taos, N. Mex., and spent the summer of 1833 along the Dolores River and in camp near Trout Lake, about 13 miles northeast of the present site of Rico.

1861. In 1861 Lieutenant Howard and other members of John Baker's expedition into the San Juan region made their way over the mountains from the east and prospected the Dolores River, afterward rejoining the main party at Bakers Park, where the town of Silverton now stands.

1866. Five years later a party from Arizona, under Col. Nash, following the Santa Fe and Salt Lake trail, reached the Big Bend of the Dolores (where the town of Dolores now stands), and explored the river to its source. Thence they crossed the divide to Trout Lake and proceeded down the San Miguel River.

1869. Sheldon Shafer and Joseph Fearheiler reached the site of Rico on their way from Santa Fe to Montana. They were well provided with tools and provisions and, struck by the indications of mineral wealth which the region afforded, decided to thoroughly prospect the district. They built a cabin on Silver Creek, near the spot where the South Park mine was afterward opened, and located, in July, 1869, a claim which they called the Pioneer, a name that afterward became the official designation of the mining district. It covered portions of what are now the Shamrock, Smuggler, and Riverside claims. They also made a location which they named the Nigger Baby, on account of the abundant black oxide of manganese found in the vein. Although this claim afterward became part of the Phoenix mine, the name was perpetuated as Nigger Baby Hill. In the autumn of 1869 they built a more substantial cabin near where the Rico State Bank now stands and worked on the Pioneer claim through the winter.

1870. R. C. Darling, engaged in surveying the boundaries of the Ute Indian Reservation, passed up the Dolores on his way to Mount Sneffels. He found Fearheiler and Shafer at work, located some claims near them, and proceeded on his way up stream. His name survives in Darling Ridge, one of the spurs of Expectation Mountain. During the same year Gus Begole, John Echols, Dempsey Reese, and Pony Whittemore came into the district from New Mexico and discovered the Aztec and other lodes. On the approach of winter all of the prospectors relinquished their work and left •the district. Fearheiler never returned, being killed by Indians on his way to Fort Defiance.

1872. Apparently none of the adventurous prospectors came back to their claims in the following summer, but in 1872 Darling, who had succeeded in interesting some Army officers and capitalists from Washington, D. C., in the resources of the region, led a large party into the Pioneer district from Santa Fe. They carried with them a few lengths of board from which they constructed molds for adobe bricks, and of these they erected a Mexican smelting furnace. Ore was extracted from what are now the Atlantic Cable, Aztec, Phoenix, and Yellow Jacket claims, and three small bars of bullion were produced in this furnace. The adobe, however, was not sufficiently refractory, and the furnace soon became useless. Discouraged by their failure and by the low grade of the bullion, the claims were abandoned on the approach of winter and the party returned to Santa Fe.

1875. Two years later, members of the Hayden survey mapped the region and gave many of the existing names to the more prominent topographical features.

1877–78. Prospecting was again resumed in the Pioneer district in 1877, and in 1878 became active through the energy of John Glasgow, Sandy Campbell, David Swickhimer, and others. The Atlantic Cable, Blackhawk, Hope, Cross, Grand

[31] Colorado State Bur. Mines Rept. for 1897, p. 29, 1898.
[32] Ransome, F. L., The ore deposits of the Rico Mountains, Colo.: U. S. Geol. Survey Twenty-second Ann. Rept., pt. 2, pp. 238–242, 1901.

View, Major, Phoenix, Yellow Jacket, Pelican, Aztec, and Columbia claims were all located in 1878, but, as usual, work was abandoned when the winter snows whitened the surrounding peaks.

1879. In the spring of 1879 rich oxidized silver ore was discovered on Nigger Baby Hill, and a rush to the district from the neighboring camps followed. Several claims on Nigger Baby Hill were sold to the Grand View Mining Co., in which Senator Jones and John W. Mackay, well known for their operations on the Comstock, were prominent stockholders. Ore was also found in the Chestnut vein, on Newman Hill, and a small shipment was made to Swansea. The beginnings of a settlement sprang up. The town site was surveyed and divided into lots, and E. A. Robinson became justice of the peace. The first newspaper, the Dolores News, appeared on August 21, the first seven numbers being printed in Silverton. A post office was opened, and the name Rico was given to the growing town.

1880. The Grand View smelter was begun in 1880, the machinery coming from the railway terminus at Alamosa by wagons to Mancos, and thence over the now abandoned road which reached the Dolores River by the dreaded Bear Creek Hill, 12 miles south of Rico. The freight from Alamosa was about $300 per ton. Late in the autumn the smelter began producing bullion. This same year saw the discovery of the Johnny Bull ore body.

1881. The year 1881 is notable for a punitive expedition against a party of Utes, who were overtaken near the La Sal Mountains and defeated with considerable loss of life on both sides.

1882. The following spring the Rico Mining & Smelting Co. began the erection of a second smelter in the southern end of town, and the Newman group of mines was sold to the Marrs Consolidated Mining Co. for $175,000.

1883. In 1883 the finding of ore in the South Park mine, on Silver Creek, led to active prospecting along this stream.

1884. In 1884 the Rico smelter was purchased and repaired by the Pasadena Co. and was operated as a custom plant for nearly two years.

As early as 1881 David Swickhimer, Patrick Cain, and John Gault began a shaft on the Enterprise claim on Newman Hill but subsequently sold their property for a few hundred dollars' worth of lumber. But the success of Larned and Hackett in following the veins in the Chestnut and Swansea claims led Swickhimer to repurchase a controlling interest in the Enterprise, and in October, 1887, he struck ore at a depth of 262 feet. This was the first discovery of the so-called "contact" or blanket ore, and the shaft had fortunately cut the edge of the largest and richest ore body ever found on Newman Hill.

The result of Swickhimer's discovery was to infuse new life into the district. Large bodies of ore were found in the Blackhawk, Logan, and Rico-Aspen mines, and the future of Rico looked brighter than ever before

The Enterprise group was sold in 1890 to a Pittsburgh company, and the same year saw the advent of the Rio Grande Southern Railroad. Litigation sprang up between the Enterprise and Rico-Aspen companies, but production went on, and when the suit was finally won by the Enterprise, the ore in the disputed territory had been extracted, largely by the Rico-Aspen Co.

Since 1895 the output of the Pioneer district has decreased. The large bodies of rich "contact" ore have been mined out, and many of the veins have been worked down to a depth at which the ore no longer pays for shipment. Masses of ore often proved to be curiously limited, owing to various conditions that are characteristic of the region. * * * The declining price of silver has had a depressing effect on this, as on other districts, where this metal forms a large part of the output. But nearly all the important ore bodies formerly exploited were sufficiently rich to be workable to-day had they not been exhausted.

In the year 1900 the only ore being shipped from the district was an occasional carload taken out by leasers working small areas of unexplored ground in the larger mines.

In 1897, according to the reports of the Director of the Mint, two 20-stamp mills were in operation at Rico and one smelter and one mill were in course of erection at Dunton. In 1898 two concentrating mills were projected at Rico, and a 10-stamp mill was in operation at Dunton. In 1900 three concentrating mills, having a capacity of 100 tons a day, were in operation at Rico, but they were not worked to their full capacity. In the report of the Director of the Mint for 1901 the shipments of zinc from Rico in prior years are said to have been made to Belgium and to Mineral Point, Wis. In 1901 the Rico Mining & Milling Co. built a zinc concentration mill containing 20 stamps, 6 Wilfley tables, 1 roadster, and a line of the inventor's magnetic separators, which were intended to part the lead from the zinc and iron and the iron and copper from the zinc.

The Wilfleys furnished two grades of lead concentrates, and the zinc-iron residue after being treated by the roaster, went to the magnetic separators.

In 1901 many of the mines were consolidated under the name of the United Rico Mining Co., including the Rico-Aspen Enterprise, Rico Townsite & Milling Co., Swansea, Atlantic Cable, Rico Mine Co. (which had a smelter of 150 tons and a mill of 100 tons daily capacity), Grand View, New Year's, Hope and Cross, the Group, Lexington, Onomo, and Syndicate tunnels, and the Grand View Coal Co., whose holdings comprise a half section of coal lands and coke ovens.

The Pro Patria Co. drove a crosscut tunnel 2,600 feet on the western slope of Dolores Mountain. This tunnel was designed to connect with a 100-ton concentrating plant by means of a 3,800-foot tramway having a fall of 15 per cent. Lead concentrate was to form the jig product, and the iron and zinc tailings were to be reground and the iron magnetized by roasting. The magnetic separator would then part the iron and zinc. At the Emma mine, at Dunton, a Krupp mill of 80-stamp capacity was added to the 20-stamp mill.

In 1902 Rico was very quiet, only small lots of ore being shipped by lessees and small lots from an experimental plant owned by the United Rico Mines Co. and from the new mill owned by the Pro Patria Co. The process used at the test mill, which treated ore from the Atlantic mine, was as follows:

The ore was first crushed, then sent to the 20 stamps, and then to the Frue vanners, after being sized by passing over pointed boxes. The lead, iron, and zinc were made into two products, the lead kept by itself.

The iron and zinc product was then passed through a contrivance known as the "National magnetic mineral separator." This separator removed the magnetic iron pyrites, of which the mixture carried about 40 per cent, and the remaining 60 per cent was put through a light roasting furnace and returned to the magnetic separator, where nearly all the rest of the iron was extracted. This separator consisted of three canvas belts revolving in the fields of three different electromagnets, each of which was of a different magnetic strength. The roasted or mixed ore passed over the belt affected by the weakest magnet first and afterward over the belts affected by the stronger ones. During the test runs the following products were obtained and shipped:

Content of concentrates made by experimental plant of United Rico Mines Co. at Rico, Colo., in 1902

Concentrates	Tons	Percentage of—		
		Zinc	Iron	Lead
Zinc	87.60	51.17		
	252.70	53.17		
	361.89	46.30		
Iron	132.35		59.11	
Lead	203.00			44.84

The zinc concentrates did not carry, at the highest, more than 3 per cent of iron, and the iron concentrates showed only about 4 or 5 per cent of zinc. The lead concentrates showed, in general, a low percentage of zinc, although several lots gave more than the limit of 10 per cent allowed by the smelter.

The Pro Patria Co. erected a 100-ton water-power jig concentrating mill. A tramway 3,600 feet long connected the mine and mill. Originally the company planned to work the ore entirely on jigs, but afterward they decided to put in tables and treat the middlings from the jigs after regrinding. This mill had a separator of the magnetic variety and also roasted its concentrates before treating. The mill was finished near the end of the year, and the clean zinc concentrates obtained amounted to very nearly 150 tons and averaged 50 per cent of metallic zinc.

At Dunton the Emma group of mines was in active operation with a concentration mill of 125 tons daily capacity.

According to the mine reports in Mineral Resources the milling results were not satisfactory, and the production of zinc fell off until 1905 and 1906. The year 1907 was a dull one for the district, but in 1908 the Pro Patria mill was again set in motion and produced zinc and lead concentrates. This mill was destroyed by fire in October, 1908, and no milling was done again until 1913, when the mill was remodeled into a straight wet-concentration plant and operated during that year only. In 1912 the district took on new life through shipments of copper and lead-zinc smelting ore, and in 1913 the output had a value for silver, copper, lead, and zinc larger than that of the output in 1894. In 1914 and 1915 the lead and zinc shipments decreased, and in 1914 the shipments of copper also decreased, but in 1915 the output of copper was the largest ever made by the camp. From 1916 through 1918 copper ores and lead-zinc ores were shipped in considerable quantities. From 1919 to 1923 mining was not very active in this county.

For 1879 to 1884 the figures given in the table are obtained from the reports of the Director of the Mint.[33] For 1879 and 1880 Dolores, San Juan, San Miguel, Ouray, and Rio Grande are estimated separately.

For 1879 to 1896 the figures given are based on reports of the agents of the mint in annual reports of the Director of the Mint, the gold and silver being prorated to correspond with the figures of the total production of the State as corrected by the Director of the Mint, the lead being prorated to correspond with the total production of lead for the State as given in Mineral Resources, and any unknown production in the State being distributed proportionately among the counties. As with lead so with copper, but as the figures for copper given in Mineral Resources include copper from matte and ores treated in Colorado, though produced in other States, the figures for copper are subject to revision.

The production of lead in 1887 was 1,000,000 pounds, according to an estimate in Mineral Resources.[34]

For 1897 to 1904 the figures, which represent smelter and mint receipts, are taken from the reports of the Colorado State Bureau of Mines.

For 1905 to 1923 the figures are obtained from Mineral Resources (mine reports).

[33] Burchard, H. C., Report of the Director of the Mint upon the production of the precious metals in the United States during the calendar year 1880, pp. 156-157, 1881. (See the later reports of this series.)
[34] U. S. Geol. Survey Mineral Resources, 1886, p. 145, 1887.

Gold, silver, copper, lead, and zinc produced in Dolores County, 1879–1923

Year	Ore (short tons)	Lode gold	Silver			Copper			Lead			Zinc			Total value
			Fine ounces	Average price per ounce	Value	Pounds	Average price per pound	Value	Pounds	Average price per pound	Value	Pounds	Average price per pound	Value	
1879		a $1,500	a 7,734	$1.12	$8,662	b 4,301	$0.186	$800	b 10,000	$0.041	$410				$11,372
1880		a 3,500	a 30,938	1.15	35,579	b 29,000	.214	6,206	b 100,000	.05	5,000				50,285
1881		5,000	69,610	1.13	78,659	b 44,000	.182	8,008	b 200,000	.048	9,600				101,267
1882		10,000	85,078	1.14	96,989	b 54,000	.191	10,314	b 200,000	.049	9,800				127,103
1883		5,000	193,360	1.11	214,630	b 100,000	.165	16,500	b 200,000	.043	8,600				244,730
1884		1,500	54,141	1.11	60,097				b 152,000	.037	5,624				67,221
1885		b 4,000	b 70,000	1.07	74,900				b 100,000	.039	3,900				82,800
1886		8,561	75,836	.99	75,078				792,000	.046	36,432				120,071
1887		9,743	118,262	.98	115,897	34,000	.138	4,692	1,000,000	.045	45,000				175,332
1888		17,470	123,852	.94	116,421				1,000,000	.044	44,000				177,891
1889		77,825	618,615	.94	581,498				2,000,000	.039	78,000				737,323
1890		156,297	848,785	1.05	891,224				2,000,000	.045	90,000				1,137,521
1891		122,631	699,888	.99	692,889				931,326	.043	40,047				855,567
1892		235,669	1,285,179	.87	1,118,106	13,043	.116	1,513	3,083,168	.04	123,327				1,478,615
1893		442,105	2,675,238	.78	2,086,686	b 10,000	.108	1,080	4,500,000	.037	166,500				2,696,371
1894		192,626	1,153,325	.63	726,595	b 30,000	.095	2,850	2,000,000	.033	66,000				988,071
1895		52,552	399,283	.65	259,534	64,151	.107	6,864	313,824	.032	10,042				328,992
1896		10,659	240,393	.68	163,467		.108		1,100,000	.03	33,000	a 30,000	$0.039	$1,170	208,296
1897		43,469	179,901	.60	107,941	39,654	.12	4,758	1,093,840	.036	39,378				195,546
1898		88,282	463,346	.59	273,374	149,647	.124	18,556	686,597	.038	26,091	a 400,000	.046	18,400	424,703
1899		66,847	257,052	.60	154,231	44,509	.171	7,611	2,046,232	.045	92,080	a 100,000	.058	5,800	326,569
1900		50,125	159,318	.62	98,777	36,009	.166	5,978	210,380	.044	9,257	a 220,000	.044	9,680	173,817
1901		22,303	111,632	.60	66,979	13,106	.167	2,189	367,057	.043	15,783	a 250,000	.041	10,250	117,504
1902		47,458	121,311	.53	64,295	15,054	.122	1,837	388,806	.041	15,941	248,680	.048	11,937	141,468
1903		43,262	103,096	.54	55,672	147,588	.137	20,220	143,417	.042	6,024				125,178
1904	7,727	53,783	108,301	.58	62,815	25,392	.128	3,250	181,229	.043	7,793	18,196	.051	928	128,569
1905	3,826	34,766	76,526	.61	46,681	119,821	.156	18,692	840,319	.047	39,495	556,266	.059	32,820	172,454
1906	2,242	9,398	34,290	.68	23,317	199,379	.193	38,480	118,229	.057	6,739	883,533	.061	53,896	131,830
1907	1,575	11,689	33,037	.66	21,804	99,495	.20	19,899	54,547	.053	2,891				56,283
1908	11,024	37,238	163,563	.53	86,688	42,495	.132	5,609	947,962	.042	39,814	509,184	.047	23,932	193,281
1909	4,787	22,266	103,646	.52	53,896	43,538	.13	5,660	462,373	.043	19,882	167,574	.054	9,049	110,753
1910	2,933	15,327	88,309	.54	47,687	97,063	.127	12,327	127,909	.044	5,628	87,000	.054	4,698	85,667
1911	3,276	7,565	56,202	.53	29,787	3,288	.125	411	701,244	.045	31,556	525,333	.057	29,944	99,263
1912	8,485	7,556	100,288	.615	61,677	689,915	.165	113,837	1,212,400	.045	54,558	812,029	.069	56,030	293,658
1913	17,802	12,432	178,816	.604	108,005	801,819	.155	124,282	3,079,341	.044	135,491	2,596,232	.056	145,389	525,599
1914	6,905	7,973	86,526	.553	47,849	350,278	.133	46,587	492,023	.039	19,189	366,549	.051	18,694	140,292
1915	14,192	11,932	127,933	.507	64,862	1,032,480	.175	180,684	268,447	.047	12,617	35,936	.124	4,456	274,551
1916	6,398	7,426	77,250	.658	50,850	419,500	.246	103,197	588,333	.069	40,595	182,306	.134	24,429	226,497
1917	14,026	5,213	88,222	.824	72,695	519,916	.273	141,937	1,772,221	.086	152,411	1,701,353	.102	173,538	545,794
1918	9,272	3,136	54,249	1.00	54,249	618,012	.247	152,649	517,394	.071	36,735	661,253	.091	60,174	306,943
1919	4,461	2,517	35,225	1.12	39,452	264,968	.186	49,284	98,700	.053	5,231	67,027	.073	4,893	101,377
1920	2,752	2,350	32,167	1.09	35,062	6,804	.184	1,252	772,588	.08	61,807	229,865	.081	18,619	119,090
1921	386	1,856	14,499	1.00	14,499	744	.129	96	18,624	.045	838				17,289
1922	678	1,953	30,267	1.00	30,267	24,089	.135	3,252	87,200	.055	4,796				40,268
1923	1,393	2,890	39,408	.82	32,315	56,823	.147	8,353	162,414	.07	11,369	138,000	.068	9,384	64,311
		1,977,650	11,673,927		9,202,637	6,243,881		1,149,714	37,022,144		1,669,271	10,786,316		728,110	14,727,382

a Estimated by C. W. Henderson.   b Estimated by C. W. Henderson by interpolations to correspond to total production of the State.

DOUGLAS COUNTY

The early history of Douglas County is summarized by the Colorado State Bureau of Mines as follows: [35]

While the county is usually classed among the plains counties its west and southwestern portions are quite rugged and are traversed by a spur of the main mountain chain known as Rampart Range.

Plum Creek and its tributaries afford the drainage for the east slope, and the South Platte River and tributaries the drainage for the west slope of this range. Lying east of Plum Creek Valley is what is known as Cherry Creek Plateau, drained by Cherry Creek and tributaries.

The early history of the county reveals it to be one of the first settled in the State, the search for gold being the primitive cause. Some gold was found and removed and has been every year up to the present time. The beds, however, were never very lucrative. * * *

Following the rush into the Cripple Creek district, the Rampart Range [including Devils Head Peak] was the scene of considerable excitement in 1895. During 1896 towns sprang up, and the southwest portion of the county gave evidence of a permanent producing district. The results of development during 1897 have not been as successful as expected. * * * The district is in a prospective stage. Ore carrying good values exists but has not yet been demonstrated in paying quantities. The somewhat recent reports of gold value in the sandstone beds along West Creek, if reliable, will bring the district * * * a producer. During the year an average of 77 men were employed.

The leading town and the county seat of the county is Castle Rock.

For 1885 the figures in the table represent deposits of gold and silver at the Denver Mint.[36]

For 1897 to 1904 the figures, which represent smelter and mint receipts, are taken from reports of the Colorado State Bureau of Mines.

For 1906 to 1922 the figures are taken from Mineral Resources (mines reports).

35 Colorado State Bur. Mines Rept. for 1897, pp. 32–33, 1898.

36 Wilson, P. S., agent for Colorado, in Kimball, J. P., Report of the Director of the Mint upon the production of the precious metals in the United States during the calendar year 1885, p. 136, 1886.

Gold and silver produced in Douglas County, 1858-1922

Year	Placer gold	Silver			Total value
		Fine ounces	Average price per ounce	Value	
1858-1884	(a)	----	----	----	(a)
1885	$1,420	70	$1.07	$75	$1,495
1897	475	10	.60	6	481
1898	124	----	----	----	124
1899	83	24	.60	14	97
1900	62	24	.62	15	77
1901	103	10	.60	6	109
1902	62	10	.53	5	67
1903	41	2	.54	1	42
1904	289	5	.58	3	292
1906	4	----	----	----	4
1907	49	----	----	----	49
1908	131	----	----	----	131
1910	83	----	----	----	83
1911	166	----	----	----	156
1912	75	----	----	----	75
1913	547	2	.604	1	548
1914	140	----	----	----	140
1915	596	4	.553	2	598
1921	47	----	----	----	47
1922	12	----	----	----	12
	4,509	161	----	128	4,637

a Small production unrecorded.

EAGLE COUNTY

The report of the Director of the Mint for 1883 gives the following information concerning Eagle County:[37]

Eagle County was formerly a part of Summit and embraces what were known as Eagle River and Holy Cross mining districts. * * * The chief camp is Red Cliff, in the vicinity of which are the largest producers in the county, viz, the Eagle Bird group, Belden, Clinton, Little Chief, Crown Point, Spirit, Kingfisher, Discovery, Silver Age, and a few minor properties. These are located on Battle Mountain. * * * The Belden [was] located in 1879. Eagle Bird Consolidated Mining Co. consists of 25 claims, among which [are] Silver Wave, Eagle Bird, Indian Girl, May Queen, Cleveland, and Black Iron.

For 1880 Burchard[38] gives the production of silver in the Eagle River district, under Summit County, as $50,000 (38,672 ounces at $1.29+), and adds: "Several promising contacts, notably the Belden group, have been opened the past summer."

In his report for 1881, under Summit County, Burchard says:[39]

On Eagle River, at Red Cliff, the Belden Mining Co., and the Battle Mountain smelter have produced a large quantity of base bullion. * * * No ore is being shipped except from the Belden Co.'s mine, and from this but four carloads per day. The ore, containing as it does a large percentage of lead, forms an excellent flux for the dry ores and will aid materially in smelting.

In Gold Park, near the mount of the Holy Cross, a considerable camp has been established, and a large stamp mill erected to crush the gold ores found there. The veins are quite strong, the quartz at the surface to a depth of 12 to 15 feet much decomposed, and the yield $10 to $25 per ton under stamps. Below the decomposition the ores are mainly white iron pyrites. This section is scarcely more than a year old.

In his report for 1882, under Summit County, Burchard says: "At Red Cliff the Belden is the principal mine."[40]

For 1883 the figures in the table are taken from Burchard's report.[41]

Burchard's report for 1884 contains the following notes on mining in Eagle County:[42]

The Belden shipped during the year 1,640 tons of ore, which returned as follows:

Ores		Quantity	Price	Value
Lead	tons	568	$72.00	$40,896
Silver	ounces	16,400	1.10	18,040
Gold	do	21½	20.00	430
				59,366

The Eagle Bird shipped 2,090 tons of ore, assaying 31 per cent lead and 8½ ounces silver, giving 647 tons of lead and 17,765 ounces of silver.

Black Iron produced 378 tons of ore, assaying 41 per cent lead and 9 ounces silver, giving 154.98 tons lead and 3,402 ounces silver.

The May Queen forwarded 30 tons, which returned 39 per cent lead and 8 ounces silver, giving 11 tons lead and 240 ounces silver.

The total of these mines, comprising the Cheeseman and Clayton group, was as follows:

Ores		Quantity	Price	Value
Lead	tons	814.58	$72.00	$58,659.76
Silver	ounces	21,407.00	1.10	23,547.70
Gold	do	157.00	20.00	3,140.00
				85,347.46

The shipments from the Little Chief, Crown Point, Kingfisher, Iron Mask, Spirit, Clinton, Cleopatra, Great Western, and Potvin have amounted to some 597 carloads of mineral, averaging 10 tons to the car; 32 per cent lead and 8 ounces silver per ton would give 5,970 tons of mineral yielding as follows:

Ores		Quantity	Price	Value
Lead	tons	1,910.4	$72.00	$137,548.88
Silver	ounces	4,776	1.10	52,536.00
Gold	do	250	20.00	5,000.00
				195,084.88

On the lower end of Battle Mountain is found what is now termed the quartzite, which extends on the west side of Rock Creek and on the north side of Eagle River. The most promising of its mines are the Ground Hog group, Combined Discovery, Uncle Sam, Horn Silver, and Highland Mary. * * *

The total yield of Red Cliff Camp has been stated to be, from the carbonate mines, $339,798.26; from other mines $125,000; total, $464,798.26 (which includes the value of lead).

In the vicinity of Taylor Hill much work is being done in the way of development, and McClelland's stamp mill has been running most of the year on ore from this locality. It is claimed about $12,000 was turned out during this time.

---

37 Burchard, H. C., Report of the Director of the Mint upon the production of the precious metals in the United States during the calendar year 1883, pp. 240, 290-294, 1884.
38 Burchard, H. C., op. cit. for 1880, p. 152, 1881.
39 Burchard, H. C., op. cit. for 1881, p. 435, 1882.

40 Burchard, H. C., op. cit. for 1882, p. 559, 1883.
41 Burchard, H. C., op. cit. for 1883, pp. 240, 294, 1884.
42 Burchard, H. C., op. cit. for 1884, pp. 177, 207-209, 1885.

At Holy Cross exploitation only has been done; bodies of ore have been developed, but the actual yield has been almost nothing. The production of the camp, reported from ore treated was $6,400.

The figures for lead for 1884 in the table are taken from Mineral Resources.[43]

For 1885 the production amounted to 11,000 tons of ore containing 3,500 tons of lead gross; a deduction

For 1897 to 1907 the figures, which represent smelter and mint receipts, are taken from the reports of the Colorado State Bureau of Mines.

For 1905 to 1923 the figures are obtained from Mineral Resources (mines reports).

The geology of the ore bodies at Red Cliff has been studied by Means,[45] and the siderite at Red Cliff has been described by Argall.[46]

Gold, silver, copper, lead, and zinc produced in Eagle County, 1880-1923

Year	Ore (short tons)	Gold			Silver			Copper			Lead			Zinc			Total value
		Placer	Lode	Total	Fine ounces	Average price per ounce	Value	Pounds	Average price per pound	Value	Pounds	Average price per pound	Value	Pounds	Average price per pound	Value	
1880	1,000[a]		$2,000[a]	$2,000[a]	38,672[a]	$1.15	$44,473				800,000[a]	$0.05	$40,000				$86,473
1881	2,000[a]		4,000	4,000[a]	79,344[a]	1.13	89,659				1,600,000[a]	.048	76,800				170,459
1882	2,500[a]		5,000	5,000[a]	98,680[a]	1.14	112,495				2,000,000[a]	.049	98,000				215,495
1883	19,859		70,000	70,000	232,031	1.11	257,554				16,000,000	.043	688,000				1,015,554
1884	10,000[a]		30,000	30,000	154,687	1.11	171,703				6,600,000	.037	244,200				445,903
1885	11,000		33,000	33,000[b]	170,156[b]	1.07	182,067				5,950,000	.039	232,050				447,117
1886			423,517	423,517	569,637	.99	563,941				2,000,000	.046	92,000				1,079,458
1887			219,594	219,594	254,078	.98	248,996				1,112,905	.045	50,081				518,671
1888			142,002	142,002	193,489	.94	181,880				2,370,090	.044	104,284				428,166
1889			92,220	92,220	170,551	.94	160,318				2,112,280	.039	82,379				334,917
1890			68,862	68,862	75,265	1.05	79,028				1,000,000	.045	45,000				192,890
1891			153,453	153,453	280,168	.99	277,366				3,776,230	.043	162,378				593,197
1892			139,299	139,299	347,954	.87	302,720				5,259,280	.04	210,371				652,390
1893			168,867	168,867	187,658	.78	146,373				5,000,000	.037	185,000				500,240
1894			55,521	55,521	62,543	.63	39,402				2,000,000[a]	.033	66,000				160,923
1895			30,900	30,900	53,421	.65	34,724				1,770,215	.032	56,647				122,271
1896			16,472	16,472	65,824	.68	44,760	2,044	$0.108	$221	210,717	.03	6,322				67,775
1897			34,767	34,767	46,046	.60	27,628	2,200	.12	264	1,144,013	.036	41,184				103,843
1898			30,571	30,571	70,783	.59	41,762	71,049	.124	8,810	1,851,512	.038	70,357				151,500
1899			46,094	46,094	44,393	.60	26,636	5,876	.171	1,005	1,187,930	.045	53,457				127,192
1900			103,598	103,598	234,674	.62	145,498	359,054	.166	59,603	3,679,828	.044	161,912	20,000[a]	$0.044	$880	471,491
1901			97,376	97,376	175,181	.60	105,109	157,914	.167	26,372	2,775,291	.043	119,338				348,195
1902			31,956	31,956	45,336	.53	24,028	150,134	.122	18,316	832,846	.041	34,147				108,447
1903			16,040	16,040	27,054	.54	14,609	32,863	.137	4,502	677,730	.042	28,465				63,616
1904	1,866		30,075	30,075	27,348	.58	15,862	32,409	.128	4,148	375,207	.043	16,134				66,219
1905	12,049		46,891	46,891	46,487	.61	28,357	29,331	.156	4,576	156,723	.047	7,366	605,612	.059	35,731	122,921
1906	15,986		51,561	51,561	94,912	.68	64,540	130,233	.193	25,135	307,755	.057	17,542	1,426,029	.061	86,988	245,766
1907	4,191		53,641	53,641	70,586	.66	46,587	14,270	.20	2,854	193,690	.053	10,266	429,198	.059	25,323	138,671
1908	3,009		58,131	58,131	86,715	.53	45,959	66,141	.132	8,731	11,204	.042	471				113,292
1909	11,526		53,308	53,308	125,214	.52	65,111	286,885	.13	37,295	152,280	.043	6,548	740,408	.054	39,982	202,244
1910	27,761		25,231	25,231	88,313	.54	47,689	209,551	.127	26,613	397,409	.044	17,486	4,147,945	.054	223,989	341,008
1911	33,177		41,160	41,160	116,109	.53	61,538	66,608	.125	8,326	855,889	.045	38,515	5,097,597	.057	290,563	440,102
1912	34,164		49,294	49,294	163,735	.615	100,697	147,176	.165	24,284	1,240,156	.045	55,807	5,659,261	.069	390,489	620,571
1913	47,488		41,220	41,220	301,380	.604	182,034	41,368	.155	6,412	1,351,205	.044	59,453	6,683,643	.056	374,284	663,403
1914	49,377		47,194	47,194	127,080	.553	70,275	28,105	.133	3,738	1,177,385	.039	45,918	7,522,098	.051	383,627	550,752
1915	74,197		95,426	95,426	177,550	.507	90,018	60,086	.175	10,515	1,394,043	.047	65,520	11,141,750	.124	1,381,577	1,643,056
1916	105,149	$186	95,850	96,036	222,126	.658	146,159	112,610	.246	27,702	1,517,362	.069	104,698	28,438,052	.134	3,810,699	4,185,294
1917	100,875	53	41,134	41,187	136,023	.824	112,083	53,136	.273	14,506	2,426,988	.086	208,721	23,715,412	.102	2,418,972	2,795,469
1918	89,675		35,975	35,975	241,406	1.00	241,406	353,041	.247	87,201	2,927,099	.071	207,824	14,845,341	.091	1,350,926	1,923,332
1919	22,248		19,935	19,935	72,159	1.12	80,818	123,306	.186	22,935	378,113	.053	20,040	3,367,548	.073	245,831	389,559
1920	32,635		25,496	25,496	279,667	1.09	304,837	517,109	.184	95,148	282,538	.08	22,603	6,653,235	.081	538,912	986,996
1921	39,785		64,723	64,723	682,550	1.00	682,550	1,833,078	.129	236,467	12,578	.045	566				984,306
1922	71,892		72,111	72,111	583,737	1.00	583,737	1,330,296	.135	179,590	322,818	.055	17,755	11,000,000	.057	627,000	1,480,193
1923	98,427		41,734	41,734	322,143	.82	264,157	632,565	.147	92,987	460,171	.07	32,212	23,600,000	.068	1,604,800	2,035,890
		239	3,005,199	3,005,438	7,642,865		6,557,143	6,848,438		1,038,256	87,651,480		3,903,817	155,093,129		13,830,573	28,335,227

[a] Estimated by C. W. Henderson.   [b] Interpolated by C. W. Henderson to correspond with the total production of the State.

of 15 per cent for losses gives 2,975 tons of lead, according to estimates in Mineral Resources.[44]

For 1886 to 1896 the figures given are based on reports of the agents of the mint in annual reports of the Director of the Mint, the gold and silver being prorated to correspond with the figures of the total production of the State as corrected by the Director of the Mint, the lead being prorated to correspond with the total production of lead for the State as given in Mineral Resources, and any " unknown production " in the State being distributed proportionately among the counties. As with lead so with copper, but as the figures for copper given in Mineral Resources include copper from matte and ores treated in Colorado, though produced in other States, the figures for copper are subject to revision.

EL PASO COUNTY

A little copper was produced in the Blair Athol district in El Paso County in 1913 and 1914. The figures shown in the table below are taken from Mineral Resources.[47]

Copper produced in El Paso County, 1913-14

Year	Ore (short tons)	Copper		Total value
		Pounds	Value	
1913	298	10,632	$1,648	$1,648
1914	25	2,644	352	352
	323	13,276	2,000	2,000

[43] U. S. Geol. Survey Mineral Resources, 1883-84, p. 422, 1885.

[44] U. S. Geol. Survey Mineral Resources, 1885, p. 257, 1886.

[45] Means, A. H., Geology and ore deposits of Red Cliff, Colo.: Econ. Geology, vol. 10, pp. 1-27, 1915.

[46] Argall, Philip, Siderite and sulphides in Leadville ore deposits: Min. and Sci. Press, vol. 109, p. 52, 1914.

[47] U. S. Geol. Survey Mineral Resources, 1913, pt. 1, p. 253, 1914; idem, 1914, pt. 1, p. 282, 1916.

FREMONT COUNTY

The figures for production in Fremont County from 1881 to 1889 are taken from the reports of the Denver agents of the Director of the Mint.[48]

The report for 1882 gives a list of mines, showing location and ownership. The reports for 1883 and 1884 describe prospects. The report for 1885 contains no detailed table of production or description of mines. The report for 1886 shows no production for Fremont County. The report for 1887 contains a list of producing and nonproducing mines and gives the output of the producing mines. A small output of copper was made in this year. The report for 1889 contains a list of producing mines and shows their output.

For 1889 to 1896 the figures given are based on reports of the agents of the mint in annual reports of the Director of the Mint, the gold and silver being prorated to correspond with figures for the total production of the State as corrected by the Director of the Mint, the lead being prorated to correspond with the total production of lead in the State as given in Mineral Resources, and any unknown production in the State being distributed proportionately among the counties. As with lead so with copper, but as the figures for copper given in Mineral Resources include copper from matte and ores treated in Colorado, though produced in other States, the figures for copper are subject to revision.

For 1897 to 1904 the figures, which represent smelter and mint receipts, are taken from the reports of the Colorado State Bureau of Mines.

In the report of the Director of the Mint for 1902, Downer[49] says:

Dawson district.—Copper King Free Gold Mining Co. * * * completed a concentrating mill and treated some 500 tons of ore.

Copper Gulch.—Greenhorn Mining Co. * * * completed a lixiviation mill during the last of the year and successfully handled the oxidized ore from their properties.

Currant Creek district.—Considerable prospecting * * * bodies of zinc and silver-lead ores.

For 1905 to 1923 the figures given are taken from Mineral Resources (mines reports), published by the United States Geological Survey.

Lindgren[50] has described the Cotopaxi mine at Cotopaxi and the Red Gulch district (Copperfield and Springfield), 9 miles north of Cotopaxi.

[48] Burchard, H. C., Report of the Director of the Mint upon the production of the precious metals in the United States during the calendar year 1881, pp. 354, 440, 441, 1882; idem for 1882, pp. 394, 574-575, 1883; idem for 1883, pp. 240, 294-295, 1884; idem for 1884, pp. 177, 210, 1885. Kimball, J. P., idem for 1885, pp. 133-138, 1886; idem for 1886, pp. 176-178, 1887; idem for 1887, pp. 167, 190, 1888; idem for 1888, p. 130, 1889. Leech, E. O., idem for 1889, pp. 147, 154, 1890.

[49] Downer, F. M., agent for Colorado, in Roberts, G. E., Report of the Director of the Mint upon the production of the precious metals in the United States during the calendar year 1902, pp. 120-121, 1903.

[50] Lindgren, Waldemar, Copper in Chaffee, Fremont, and Jefferson counties, Colo.: U. S. Geol. Survey Bull. 340, pp. 157-174, 1908.

Gold, silver, copper, lead, and zinc produced in Fremont County, 1881–1923

Year	Ore (short tons)	Lode gold	Silver Fine ounces	Silver Average price per ounce	Silver Value	Copper Pounds	Copper Average price per pound	Copper Value	Lead Pounds	Lead Average price per pound	Lead Value	Zinc Pounds	Zinc Average price per pound	Zinc Value	Total value
1881			11,602	$1.13	$13,110										$13,110
1882			15,469	1.14	17,635										17,635
1883			15,469	1.11	17,171										17,171
1887		$186	474	.98	465				5,930	$0.045	$267				918
1888				.94											
1889		10,841	21,683	.94	20,382	5,317	$0.135	$718	466,538	.039	18,195				50,136
1894		76	323	.63	203										279
1895		18													18
1896		915	15	.68	10										925
1897		12,877	1,525	.60	915										13,792
1898		8,702	1,270	.59	749				2,101	.038	80				9,531
1899		9,405	3,974	.60	2,384	6,698	.171	1,145	11,443	.045	515				13,449
1900		8,309	2,199	.62	1,363	6,725	.166	1,116	8,282	.044	365				11,153
1901		6,449	933	.60	560	15,907	.167	2,656	33,945	.043	1,460				11,125
1902		7,379	515	.53	273	22,300	.122	2,721	2,836	.041	116	22,825	$0.048	$1,096	11,585
1903		6,346	223	.54	120	20,777	.137	2,846	2,091	.042	88				9,400
1904	1,010	4,671	208	.58	121	1,024	.128	131	1,071	.043	46				4,969
1906	162	77	79	.68	54							568,508	.061	34,679	34,810
1907	4	302	561	.66	370	30,330	.20	6,066							6,738
1908	5	91	4	.53	2										93
1909	29	85				677	.13	88							173
1910												18,072	.054	976	976
1911	382	178	1,345	.53	713	13,976	.125	1,747	19,904	.045	896	140,526	.057	8,010	11,544
1912	1,015	253	3,439	.615	2,115	35,903	.165	5,924	55,956	.045	2,518	447,507	.069	30,878	41,688
1913	53	92	78	.604	47	4,677	.155	725	4,591	.044	202	7,161	.056	401	1,467
1914	706	1,476	1,066	.553	589	191,917	.133	25,525	308	.039	12				27,602
1915	1,600	674	3,168	.507	1,606	127,303	.175	22,278	30,894	.047	1,452	228,170	.124	28,293	54,303
1916	1,734	786	4,529	.658	2,980	101,041	.246	24,856	31,710	.069	2,188				30,810
1917	429	590	664	.824	547	59,857	.273	16,341							17,478
1918	235	312	639	1.00	639	22,377	.247	5,527	1,113	.071	79				6,557
1922	7	21	174	1.00	174	348	.135	47	4,273	.055	235				477
1923	44	27	184	.82	151				1,999	.07	140	20,000	.068	1,360	1,678
		81,138	91,812		85,448	667,154		120,457	684,985		28,854	1,452,769		105,693	421,590

GARFIELD COUNTY

The history of mining in Garfield County prior to 1897 is summarized in the report of the Colorado State Bureau of Mines for 1897 as follows:[51]

The first mineral discoveries in this county in 1878 were later intensified by reports of 1879. At this time the territory now embraced by Garfield County boundaries formed a part of the Ute Indian Reservation, and the magnified reports were largely due to the risks incurred in attempting to appropriate a section not open to settlement. In 1880, notwithstanding conditions, prospectors were numerous. The ore discovered was an argentiferous lead, in form of sulphide and carbonate, carrying low percentage of silver. At that time ores of such low grade were valueless. From 1881 to the present a small amount of development has been made each year with little or no remunerative returns. During 1897 about fifty prospectors were engaged.

For 1885 the figures in the table represent deposits of gold and silver at the Denver Mint.[52]

For 1894 and 1895 the figures given are based on reports of the agents of the mint in annual reports of the Director of the Mint, the gold and silver being prorated to correspond with the figures for the total production of the State as corrected by the Director of the Mint, the lead being prorated to correspond with the total production of lead in the State as given in annual volumes of Mineral Resources, and any "unknown production" in the State being proportionately distributed among the counties. As with lead so with copper, but as the figures for copper given in Mineral Resources include copper from matte and ores treated in Colorado, though produced in other States, the figures for copper are subject to revision.

For 1897 to 1904 the figures, which represent smelter and mint receipts, are taken from the reports of the Colorado State Bureau of Mines.

For 1905 to 1918 the figures are taken from Mineral Resources (mines reports).

Gold, silver, and copper produced in Garfield County, 1885–1918

Year	Ore (short tons)	Lode gold	Silver			Copper			Total value
			Fine ounces	Average price per ounce	Value	Pounds	Average price per pound	Value	
1885		$113	45	$1.07	$48				$161
1894		63							63
1895		153	1	.65	1				154
1897		310	42	.60	25				335
1899		723	17	.60	10				733
1900		517	13	.62	8				525
1901		351	13	.60	8				359
1902		165	5	.63	3				168
1903		103	3	.54	2				105
1904		517	14	.58	8				525
1906		55	3	.68	2				57
1910	92	3,603	113	.54	61	425	$0.127	$54	3,718
1913	25	890	35	.604	21	200	.155	31	942
1914	73	2,403	80	.553	44	128	.133	17	2,464
1915	123	5,309	112	.507	57	291	.175	51	5,417
1917	18	721	17	.824	14				735
1918	15	928	15	1.00	15				943
		16,924	528		327	1,044		153	17,404

[51] Colorado State Bur. Mines Rept. for 1897, p. 47, 1898.
[52] Wilson, P. S., agent for Colorado, in Kimball, J. P., Report of the Director of the Mint upon the production of the precious metals in the United States during the calendar year 1885, p. 136, 1886.

GILPIN COUNTY

How much of the early production of Gilpin County should be credited to placer gold can not be estimated. Although for the first few years most of the material was taken out by placer methods it consisted chiefly of the oxidized decomposed portion of the lodes. By 1863 placer mining in Gregory Gulch had ceased. In later years no placer gold worthy of mention has been taken out, but placer deposits near Perigo and Rollinsville are worthy of exploration.

In the report of the Director of the Mint for 1887 Munson shows the value of the gold and silver produced in Gilpin County from 1859 to 1887.[53] The gold and silver have been separated in the table for the years preceding 1874 in proportion to the amounts for 1874. For 1876 to 1878 they have been separated according to the figures for 1879. The figures for 1868 to 1872 have been increased to agree with Raymond's figures for the State for these years. For 1874 and 1875 the figures for gold and silver have been derived from Raymond's reports.[54] For 1879 to 1884 the figures for gold and silver have been derived chiefly from Burchard's reports.[55]

For 1886 to 1896 the figures given in the table are based on reports of agents of the mint in annual reports of the Director of the Mint, the gold and silver being prorated to correspond with the figures for the total production of the State as corrected by the Director of the Mint, the lead being prorated to correspond with the total production of lead in the State as given in Mineral Resources, and any unknown production in the State being distributed proportionately among the counties. As with lead so with copper, but as the figures for copper given in Mineral Resources include copper from matte and ores treated at Colorado smelters, though produced in other States, the figures for copper are subject to revision.

Hollister[57] credits the J. C. Lyons smelter with 100 to 200 tons of copper matte on hand in 1867. Raymond[58] says that the Lyons smelter produced 80 to 100 tons of matte in eight months beginning in the summer of 1866 and that 100 tons more was found later when the plant was torn down. There is no record of assay of copper in this matte or record of its sale.

The figures for copper in 1868 to 1873 and for lead in 1873 to 1885 represent estimates derived from

[53] Munson, G. C., agent for Colorado, in Kimball, J. P., Report of the Director of the Mint upon the production of the precious metals in the United States during the calendar year 1887, p. 151, 1888.
[54] Raymond, R. W., Statistics of mines and mining in the States and Territories west of the Rocky Mountains for 1874, p. 360, 1875; idem for 1875, p. 289, 1877.
[55] Burchard, H. C., Report of the Director of the Mint upon the production of the precious metals in the United States for the calendar year 1880, pp. 156–157, 1881; idem for 1881, p. 354, 1882; idem for 1882, p. 394, 1883; idem for 1883, p. 240, 1884; idem for 1884, p. 177, 1885.
[57] Hollister, O. J., The mines of Colorado, p. 357, 1867.
[58] Raymond, R. W., op. cit. for 1869, p. 359, 1870.

Gold, silver, copper, lead, and zinc produced in Gilpin County, 1859–1923

Year	Ore (short tons)	Gold Placer	Gold Lode	Gold Total	Silver Fine ounces	Silver Average price per ounce	Silver Value	Copper Pounds	Copper Average price per pound	Copper Value	Lead Pounds	Lead Average price per pound	Lead Value	Zinc Pounds	Zinc Average price per pound	Zinc Value	Total value
1859		$241,918		$241,918	5,943	$1.36	$8,082										$250,000
1860		(?)	$870,903	870,903	21,553	1.35	29,097										900,000
1861		(?)	725,753	725,753	18,231	1.33	24,247										750,000
1862		(?)	1,161,204	1,161,204	28,738	1.35	38,796										1,200,000
1863		(?)	1,548,272	1,548,272	38,460	1.345	51,728										1,600,000
1864		(?)	1,741,806	1,741,806	43,267	1.345	58,194										1,800,000
1865			1,451,505	1,451,505	36,272	1.337	48,495										1,500,000
1866			725,753	725,753	18,108	1.339	24,247										750,000
1867			967,670	967,670	24,308	1.33	32,330										1,000,000
1868			1,640,000	1,640,000	93,311	1.326	123,730	50,000	$0.23	$11,500							1,775,230
1869			2,690,000	2,690,000	86,340	1.325	114,400	100,400	.2425	24,250							2,828,650
1870			2,120,000	2,120,000	65,910	1.328	87,528	180,000	.2118	38,124							2,245,652
1871			3,237,346	3,237,346	59,229	1.325	78,478	180,000	.2412	43,416							3,359,240
1872			2,083,611	2,083,611	52,911	1.322	69,948	200,000	.3356	71,120							2,224,679
1873			1,393,931	1,393,931	35,907	1.297	46,571	200,000	.28	56,000	25,000	$0.06	$1,500				1,498,002
1874			1,525,447	1,525,447	39,418	1.278	50,376	252,050	.22	55,451	50,000	.06	3,000				1,634,274
1875			1,395,566	1,395,566	62,670	1.24	77,711	193,665	.227	43,962	50,000	.058	2,960				1,520,139
1876			1,990,002	1,990,002	89,365	1.16	103,663	250,000	.21	52,500	50,000	.061	3,050				2,149,215
1877			2,086,871	2,086,871	93,714	1.20	112,457	300,000	.19	57,000	50,000	.055	2,750				2,259,078
1878			2,155,708	2,155,708	96,806	1.15	111,327	300,000	.166	49,800	50,000	.036	1,800				2,318,635
1879			2,260,000	2,260,000	232,031	1.12	259,875	300,000	.186	55,800	100,000	.041	4,100				2,579,775
1880			2,380,000	2,380,000	232,031	1.15	266,886	300,000	.214	64,200	100,000	.05	5,000				2,716,036
1881			1,850,000	1,850,000	201,094	1.13	227,236	300,000	.182	54,600	100,000	.048	4,800				2,136,636
1882			1,690,000	1,690,000	201,094	1.14	229,247	400,000	.191	76,400	100,000	.049	4,900				1,910,547
1883			1,650,000	1,650,000	154,688	1.11	171,704	200,000	.165	33,000	100,000	.043	4,300				1,859,004
1884			1,950,000	1,950,000	278,438	1.11	309,066	600,000	.13	78,000	128,411	.037	4,751				2,341,817
1885			2,051,000	2,051,000	300,000	1.07	321,000	300,000	.108	32,400	128,411	.039	5,008				2,409,408
1886			1,337,061	1,337,061	101,784	.99	100,766	300,000	.111	33,300	200,000	.016	9,200				1,480,327
1887			1,134,476	1,134,476	206,281	.98	200,955	600,000	.138	82,800	228,622	.045	10,288				1,488,519
1888			1,250,776	1,250,776	174,364	.94	163,902	460,000	.168	67,200	1,288,825	.044	76,708				1,538,566
1889			1,054,065	1,054,065	313,071	.94	294,287	270,110	.135	33,765	1,411,926	.039	55,065				1,437,182
1890			805,236	805,236	292,495	1.05	307,120	620,927	.156	96,865	1,130,453	.045	50,870				1,260,091
1891			938,016	938,016	232,001	.99	229,681	558,298	.128	71,462	779,837	.043	33,533				1,272,692
1892			1,358,157	1,358,157	134,462	.87	116,982	538,988	.116	62,523	2,232,158	.04	89,286				1,626,948
1893			1,218,626	1,218,626	135,850	.78	105,963	a 600,000	.108	64,800	2,000,000	.037	74,000				1,463,389
1894			1,915,863	1,915,863	228,927	.63	144,224	a 400,000	.095	38,000	2,200,000	.033	72,600				2,170,687
1895			1,196,319	1,196,319	190,256	.65	123,666	209,414	.107	22,407	844,037	.032	27,009				1,369,401
1896			1,534,358	1,534,358	295,182	.68	200,724	435,838	.108	47,071	1,948,756	.03	58,462				1,840,615
1897			2,086,471	2,086,471	374,417	.60	224,650	1,018,595	.12	122,231	2,007,698	.036	72,277				2,505,629
1898			1,983,514	1,983,514	305,687	.59	180,355	633,707	.124	78,580	1,216,338	.038	46,221				2,288,670
1899			1,996,061	1,996,061	340,652	.60	204,391	1,037,421	.171	177,399	1,312,312	.045	59,054				2,436,905
1900			1,655,502	1,655,502	236,400	.62	146,568	799,478	.166	132,713	735,773	.044	32,374				1,967,157
1901			1,638,966	1,638,966	271,638	.60	162,983	731,194	.167	122,109	670,018	.043	28,811				1,952,869
1902			1,551,035	1,551,035	303,638	.53	160,928	765,516	.122	93,393	497,366	.041	20,392				1,925,748
1903			1,346,113	1,346,113	375,238	.54	202,629	611,988	.137	83,842	945,975	.042	39,731				1,672,315
1904	109,557		1,403,865	1,403,865	318,106	.58	184,675	638,945	.128	81,785	859,293	.043	36,950				1,707,275
1905	182,873		1,450,033	1,450,033	340,901	.61	207,950	512,276	.156	79,915	519,841	.047	24,433	33,090	$0.059	$1,952	1,764,283
1906	114,662		1,115,902	1,115,902	242,478	.68	164,885	638,002	.193	123,134	510,791	.057	29,115	46,000	.061	2,806	1,435,842
1907	87,887		938,488	938,488	209,347	.66	138,169	874,060	.20	174,812	611,060	.053	32,386				1,283,855
1908	120,761		1,075,808	1,075,808	187,030	.53	99,126	636,371	.132	84,001	538,143	.042	22,602				1,281,537
1909	111,118		887,311	887,311	172,010	.52	89,445	499,146	.13	64,889	664,581	.043	28,577				1,070,222
1910	83,631		687,902	687,902	132,635	.54	71,623	534,244	.127	67,849	575,477	.044	25,321				852,695
1911	103,038		778,774	778,774	292,659	.53	155,109	950,240	.125	118,780	1,239,356	.045	55,771	23,088	.057	1,316	1,109,750
1912	118,652		904,505	904,505	316,205	.615	194,466	1,025,770	.165	169,252	1,351,600	.045	60,822	25,377	.069	1,751	1,330,796
1913	94,156		687,101	687,101	273,207	.604	165,017	837,974	.155	129,886	1,210,341	.044	53,255	8,589	.056	481	1,035,740
1914	52,839		573,553	573,553	145,237	.553	80,316	726,579	.133	96,635	499,718	.039	19,489	12,980	.051	662	770,655
1915	54,052		562,878	562,878	125,665	.507	63,712	476,383	.175	83,367	591,127	.047	27,783	11,000	.124	1,364	739,104
1916	38,913		453,259	453,259	126,553	.658	83,272	557,317	.246	137,100	521,334	.069	35,972				709,603
1917	35,289		397,087	397,087	112,585	.824	92,770	544,648	.273	148,689	815,906	.086	70,168	141,490	.102	14,432	723,146
1918	23,654		281,384	281,384	124,929	1.00	124,929	456,044	.247	112,643	774,972	.071	55,023	28,099	.091	2,557	576,536
1919	17,018		209,683	209,683	71,700	1.12	80,304	211,538	.186	39,346	523,621	.053	27,752				357,085
1920	10,820		91,469	91,469	42,000	1.09	45,780	86,603	.184	15,935	435,012	.08	34,801				187,985
1921	3,553		39,610	39,610	17,963	1.00	17,963	13,186	.129	1,701	91,644	.045	4,124				63,398
1922	13,707	378	51,342	51,720	43,910	1.00	43,910	24,860	.135	3,356	246,945	.055	13,582				112,568
1923	5,651	133	29,063	29,196	44,942	.82	36,852	22,884	.147	3,364	230,157	.07	16,111				85,523
	242,429	83,871,960		84,114,389	10,522,542		8,547,416	25,384,259		4,164,422	35,492,835		1,557,777	329,713		27,321	98,411,325

a Estimated by C. W. Henderson.

Mineral Resources,[59] from Raymond,[60] and from Egleston.[61] Raymond says that his figures for 1868-1869, and 1870 are estimates, which he thinks are high. From his figures for 1874 and 1875, said to be taken from a statement of the Blackhawk smelter, it is seen that the small plant of 1868–1870 could not have produced more than the enlarged plant of 1874–75. For 1868 to 1870 Raymond estimates that the matte averaged 40 per cent copper. Egleston's figures, 25 to 30 per cent copper, are more authentic.

GRAND COUNTY

The following account of Grand County has been abstracted, with slight changes from the report of the Colorado State Bureau of Mines:[62]

Grand County lies in the north-central part of the State. It was originally segregated from Summit County by an act of the Territorial legislature in 1874, but its area has since been reduced by subsequent acts. As now constituted, the county has an area of 1,866 square miles. The adjoining counties are Jackson on the north; Larimer, Boulder, Gilpin, and Clear

[59] U. S. Geol. Survey Mineral Resources, 1882. pp. 228, 310, 1883; idem for 1885, p. 257, 1886 (reduction of 15 per cent).

[60] Raymond, R. W., op. cit. for 1870, p. 372, 1872; idem for 1874, p. 360, 1875; idem for 1875, p. 294, 1877.

[61] Egleston, Thomas, The Boston & Colorado smelting works: Am. Inst. Min. Eng. Trans., vol. 4, pp. 276-298, 1876.

[62] Colorado State Bur. Mines Rept. for 1897, pp. 48-49, 1898.

Creek on the east; Clear Creek, Summit, and Eagle on the south; and Routt on the west. It is very irregular in form and is outlined almost entirely by the summits of mountain ranges. The northern and eastern boundaries are defined by the Continental Divide or Colorado Front Range. The Williams Range forms a large part of the southern and the Park Range the western boundary. The county is drained by Colorado River and its tributaries. This stream rises in the northeastern and eastern parts of Grand County and drains through several forks the area from Longs Peak on the east to Mount Richtofen in the northeast. These forks unite a short distance below Grand Lake and flow in a southwesterly course through the county and out near the southwest corner. The main tributaries from the north are Soda, Stillwater, Willow, Troublesome, Muddy, Red Dirt, and Stampede; from the south, Fraser River and tributaries and Williams Fork and tributaries. Between Fraser River and Williams Fork lie the Vasquez Mountains. In this area there are mountain peaks that range in altitude from 11,000 to more than 14,000 feet.

This region, which was a favorite spot with the Indians, was entered by the white man in 1859. It has been the scene of several mining excitements, but its inaccessibility has prevented development.

The mining operations in 1897 were confined almost exclusively to the northern slope of the range, near Clear Creek County. Some promising developments were made and small lots of ore were produced. To yield a profit over charges for transportation and treatment the ore was hand sorted and the grade raised to the utmost. In this process ore that would yield a good profit in either Clear Creek, Gilpin, or Boulder counties was thrown away. In 1897 about 48 men were mining and prospecting.

For 1896 and 1897 the figures in the table are based on reports of the agents of the mint in annual reports of the Director of the Mint, the gold and silver being prorated to correspond with the figures for the total production of the State as corrected by the Director of the Mint, the lead being prorated to correspond with the total production of lead for the State as given in Mineral Resources, and any unknown production in the State being distributed proportionately among the counties. As with lead so with copper, but as the figures for copper given in Mineral Resources include copper from matte and ores treated in Colorado, though produced in other States, the figures for copper are subject to revision.

For 1898 to 1904 the figures, which represent smelter and mint receipts, are taken from reports of the Colorado State Bureau of Mines.

For 1905 to 1923 the figures are taken from Mineral Resources (mines reports).

In 1905 Lindgren [63] writes as follows concerning Grand County:

Grand County * * * has at present only an insignificant production of gold, silver, and copper. Up to the recently begun building of the Moffat Railroad from Denver to Salt Lake this county has been very inaccessible, but it is expected that its mineral resources will be more actively exploited from now on. Little is known about the geological features of the ore deposits. In the northeastern corner of the county are the old Wolverine and other silver-lead deposits, located 12 miles northwest of Grand Lake, which, again, is 15 miles north of Granby railroad station. Twelve miles east from Granby and also near the Boulder County line are promising copper prospects, some of which are owned by the Monarch Mining Co., of Boulder. A branch railroad is projected from Granby to these mines.

A third mining district is the La Plata, which is located in the southeastern part of the county, on the headwaters of Williams Fork. The deposits are reported to contain gold, silver, and copper, and some development work is in progress.

For 1905 the figures given in the table for gold represent the output from the Copper King mine of the Monarch Co.

For 1908 the figures given in the table for gold and silver represent the output of the Grand Lake and Williams Fork districts; the figures for copper represent the output of the Grand Lake district; and the figures for lead represent the output of the Williams Fork district. For 1909 the figures represent the output of the Radium district, for 1910 the output of the Parshall district, and for 1914 the output of the Williams Fork district.

63 Lindgren, Waldemar, U. S. Geol. Survey Mineral Resources, 1905, pp. 200-201, 1906.

Gold (lode and placer), silver, copper, and lead produced in Grand County, 1896-1923

Year	Ore (short tons)	Gold			Silver			Copper			Lead			Total value
		Placer	Lode	Total	Fine ounces	Average price per ounce	Value	Pounds	Average price per pound	Value	Pounds	Average price per pound	Value	
1896			$200	$200										$200
1897			1,943	1,943	85	$0.60	$51							1,994
1898			806	806	11	.59	6							812
1899			124	124	13	.60	8							132
1900			3,762	3,762	21	.62	13							3,775
1901			1,034	1,034	30	.60	18							1,052
1902			1,302	1,302	24	.53	13							1,315
1903			1,426	1,426	12	.54	6							1,432
1904			641	641	13	.58	8	1,114	$0.128	$143				792
1905	12		31	31	22	.61	13	1,680	.156	262				306
1907		$18		18										18
1908	14		556	556	72	.53	38	1,561	.132	206	690	$0.042	$29	829
1909		1,183		1,183	9	.52	5							1,188
1910	1							56	.127	7				7
1914	10		3	3	1,747	.553	966				1,563	.039	61	1,030
1915		153		153	2	.507	1		.175			.047		154
1916	2				134	.658	88	760	.246	187		.069		275
1919	3		1	1	508	1.12	569		.186		453	.053	24	594
1920	3				856	1.09	933		.184		525	.08	42	975
1923	2				323	.82	265				314	.07	22	287
		1,354	11,829	13,183	3,882		3,001	5,171		805	3,545		178	17,167

GUNNISON COUNTY

Hollister [64] says of the area now included in Gunnison County:

Since up Lake Creek (emptying into Twin Lakes, Lake County) goes the old Ute Pass from the Arkansas to the Gunnison, we may as well note here that both quartz and placer mines are known to exist throughout the immense park which pours into one channel the waters of the Gunnison. Among the paying gulches discovered and worked are Taylors, Kents, Union, Washington, and German. Ores were found there quite recently like those which occur elsewhere in the territory. But it is out of the world, as it were, and there has been no surplus energy to spare for the improvement of this vast wilderness or even for its exploration.

The early history of Gunnison County is summarized by the Colorado State Bureau of Mines as follows: [65]

Captain Gunnison, for whom the county was named, met his death at the hands of the Indians while in charge of a United States engineering corps surveying a favorable route across the mountains to the far West.

As originally organized, Gunnison County had an area of 10,600 square miles and embraced a large portion of the central territory west of the Continental Divide. Subsequent legislative enactments, authorizing the organization of Pitkin, Delta, Mesa, and Montrose counties, have reduced original territory to an area of about 3,200 square miles. * * *

In 1861 gold was discovered on Taylor River in what has since been known as the Tin Cup district, the name arising from the character of utensil used to determine the presence of gold. Almost simultaneous with this was the discovery of gold in Washington Gulch, in the northern part of the county.

In Washington Gulch the gold was "coarse" and yielded large returns. The amount removed must have been considerable, but the value is unattainable. * * *

From 1861 to 1879 various parties entered this country with variable success.

Burchard [66] says of Gunnison County:

This is the latest of the mining regions of Colorado which has been discovered, and bids fair to be a very productive one; but owing principally to its being a new and far-away region, it is doubtful whether much valuable ore has as yet been shipped from it. Its production has been variously estimated. I have given it for the fiscal year 1880 about $300,000, all in silver. It is doubtful whether the actual production has been as high as the figures stated. [Yet on p. 156 he gives $300,000 silver to Gunnison for the calendar year 1879 and on p. 157 gives $300,000 for the calendar year 1880 and $300,000 for the fiscal year 1880. At the top of p. 157 the Denver Tribune for January 1, 1881, is quoted as having ascertained that the export of Gunnison County for 1879 was less than $25,000.—C. W. H.]

The following interesting account of the development of the Gunnison country and a review of its mines and their workings, is taken from the Rocky Mountain Mining Review:

The Gunnison country, which has created such an extended furore during the past few years, * * * was but little known to the civilized world prior to 1861, when discoveries of the precious metals were made in Washington Gulch, Union [Canon], and Taylor Park.

Until 1872 little was done, whilst in that year important discoveries of silver-bearing rocks were made in the Elk Mountains. During the next five years there was a small accession of settlers.

[The year] 1878 proved to be the hardest of all for the settlers. Leadville drew off large numbers, but still others came. But there was no business of any kind of importance. A smelter was being put up at Crested Butte, a place started, yet little was done. In the fall of that year mines were opened up in the eastern part of the county, in the Carbonate field, which led to the speedy settlement and development of the whole county. The news of the rich carbonate strikes spread far and wide, and 1879 opened with a good prospect for all interested in the welfare of Gunnison County. Hillerton, Virginia City, Pitkin, Gothic, and Irwin were all laid out and built to some extent that year, and the town of Gunnison kept pace with all and improved rapidly.

In the report cited descriptions of Cochetopa district, Ruby Camp, Ohio City, Aspen City (now Pitkin County), and Elk Mountain follow.

For 1873 Raymond says: [67]

According to the Georgetown Mining Review, the total product of Lake County in 1873, in gold and value of ores shipped, was $225,000. To this I add $5,000, the product of some small diggings on Gunnison River in the northwest part of the county [then part of Lake County]. The same item is to be deducted from the product of Summit County, in which it was included in the Review.

For 1879 Burchard [68] gives $300,000 for the coining value of the silver produced; others, among whom is Munson,[69] gives $25,000. In 1880 the production was only $300,000, including the new camp of Aspen. The production in 1881 was only $400,000, and there was "very little development of mines until 1881." Arbitrarily, therefore, Gunnison is given $25,000 for 1879 and $275,000 in silver is given to Lake County.

For 1880 Burchard [70] gives $300,000 for the value of the silver produced (232,031 ounces at the coining rate of $1.29+), but 10,000 ounces is deducted and given to Aspen.

For 1881 Burchard [71] says:

Very little development of the mines occurred, however, until 1881. * * * In the autumn of that year, the Denver & Rio Grande Railroad was completed to Gunnison, and later to Crested Butte, a few miles from Ruby Camp, the present limit of settlement in that direction.

Immediately beyond Ruby lies the vast region recently occupied by the Ute Indians, which remains to be carefully explored for mineral treasures.

The most productive camp in the county is known as Tin Cup. Its practical development began in the summer of 1880 after a long and very severe winter, in which the whole country was literally buried in snow. The season was quite short, as the snows of 1880-81 began the following October. * * * During 1881, however, practical results were achieved in the form of highly creditable yields of bullion. Two smelters have been erected, one by the Virginia City Mining & Smelting Co. and the other by the Willow Creek Reduction Co. The two have a capacity of about 50 tons per day. Both will be enlarged this year to meet the increased supply of ore from the mines. * * *

[64] Hollister, O. J., The mines of Colorado, p. 316, 1867.
[65] Colorado State Bur. Mines Rept. for 1897, p. 54, 1898.
[66] Burchard, H. C., Report of the Director of the Mint upon the production of the precious metals in the United States during the calendar year 1880, p. 149, 1881.
[67] Raymond, R. W., Statistics of mines and mining in the States and Territories west of the Rocky Mountains for 1873, p. 309, 1874.
[68] Burchard, H. C., op. cit. for 1880, p. 156, 1881.
[69] Munson, G. C., agent for Colorado, in Kimball, J. P., Report of the Director of the Mint upon the production of the precious metals in the United States during the calendar year 1887, p. 133, 1888.
[70] Burchard, H. C., op. cit. for 1880, p. 157, 1881.
[71] Burchard, H. C., op. cit. for 1881, pp. 354, 394-402, 1882.

In Quartz Creek district, in neighborhood of Pitkin, there are a number of valuable mines. The Red Jacket was one of the early discoveries in 1878, and the Fairview and Silver Islet were located shortly afterward. A number of good prospects were discovered during 1879, * * * and in the spring of 1880 a great rush to Gunnison County took place, of which Pitkin had a large share. * * *

In Gothic district there are many newly discovered mines, but the developments so far made are not important.

In Copper Creek district the chief mine is the Sylvanite. * * *

The chief smelter in this section of the county is at Gothic. It is very complete as to details and has a capacity of 15 tons per day. It has not as yet been in blast. * * *

Ruby Camp, or Irwin, is situated in the westerly part of the Elk Mountain range, on the Ruby Silver belt, which gave the district its name, and is 10,500 feet above sea level. This district is within the Ute Indian Reservation. From 1,200 to 1,500 mining locations have been made, but less than 50 are in paying condition. * * * The total shipments aggregate 225 tons, having an average of 178 ounces of silver per ton. * * *

In his report for 1882 Burchard [72] mentions the Tomichi district as first developed in 1880 but says that not much was done till 1881: in 1882 activity increased. He continues:

Tin Cup district, Virginia City smelter, located 2 miles below Tin Cup in first-class running order, turning out a carload of base bullion every four days. * * *

Gothic district, including Washington Gulch. * * *

The Eureka, on Treasury Mountain, has produced more than any other mine in the locality. About 10 carloads of ore of various grade were shipped to the Boston & Colorado works at Argo, the average returns of which were $90 for gold, silver, and copper contained. The galena and antimonial silver ores were shipped to Meyer's smelter, near Kansas City; the average returns were 20 ounces of silver per ton, and 60 per cent of lead.

In his report for 1883 Burchard [73] mentions Tomichi, Tin Cup, Quartz Creek, Ruby and Irwin, Rock Creek and Gothic, Spring Creek, and Poverty districts.

Tomichi district: * * * North star mine, one of the largest producers; * * * mineral mostly galena, averaging 110 ounces silver per ton and 50 per cent lead.

Quartz district: * * * The ore is much like that of Gilpin County. * * * The Silent Friend, largest producer; * * * general character of ore, galena and carbonates.

Descriptions of many other properties are included.

Tin Cup district: The smelter has been under very poor management, and but little ore has been sold. * * * The Gold Cup mine * * * in 1878–1880 over $450,000 worth of ore taken out. * * * Over 1,200 tons were shipped to Denver in 1883. * * * The average grade of present output is 168 ounces silver and 1½ ounces gold.

Descriptions of many mines are given.

The Ruby district has added $100,000 to the production of precious metals during the year, but most of this comes from one mine. This amount would have been much larger but the concentrator did not get running until late in the summer.

Elk Mountain: * * * little has been done, owing to backwardness of the concentrators getting to work.

Poverty Gulch district, between Ruby and Gothic, in Elk Mountain district. * * *

Eureka, on Treasury Mountain, * * * put in shape for production; * * * ore of two grades, the first runs 65 to 110 ounces silver and 8 to 12 per cent copper; the second runs 22 ounces silver and 74 per cent lead.

The Eureka * * * has produced more than any other mine. * * * Several carloads of various grades of ore were shipped to Boston & Colorado works, which averaged about $300 in gold, silver, and copper. The galena and antimonial silver ores were shipped to Kansas City and assayed 25 ounces silver and 60 per cent lead.

Rock Creek, * * * a concentrator has been erected at Scofield, ready for next summer's work.

The Cochetopa district, * * * but little developments other than discovery shafts * * * until 1883.

In his report for 1884 Burchard says: [74]

In Ruby district little was done during the year, and the production and shipments have mostly been from the Forest Queen mine, * * * [where] a concentrating mill was erected during the year.

Tomichi district; * * * the North Star * * * output has been 4 to 6 carloads of ore per week.

Quartz Creek has not enjoyed a very lively season.

Tin Cup has had a little boom in placer mining, and the veins have not received much attention.

Kirchhoff [75] says of the White Pine district in 1885:

The White Pine district, in Gunnison County, is a lead carbonate camp, with the Eureka Nest Egg as the most prominent mine, operated by the American Mining & Smelting Co., which ships the product to the Royal Gorge smelter at Canon City, owned by the same company. The majority of the other mines are mere prospects, and the ore of several of them carries an undesirable percentage of zinc carbonate.

For 1887 to 1895 the figures given in the table are based on reports of the agents of the mint in annual reports of the Director of the Mint, the gold and silver being prorated to correspond with the figures for the total production of the State as corrected by the Director of the Mint, the lead being prorated to correspond with the total production of lead for the State as given in Mineral Resources, and any unknown production in the State being distributed proportionately among the counties. As with lead so with copper, but as the figures for copper given in Mineral Resources include copper from matte and ores treated in Colorado, though produced in other States, the figures for copper are subject to revision.

For 1896 to 1904 the figures, which represent smelter and mint receipts, are taken from the reports of the Colorado State Bureau of Mines.

For 1905 to 1923 the figures are taken from Mineral Resources (mines reports).

[72] Burchard, H. C., op. cit. for 1882, pp. 394, 465–472, 1883.
[73] Burchard, H. C., op. cit. for 1883, pp. 240, 308–328, 1884.
[74] Burchard, H. C., op. cit. for 1884, pp. 177, 215–218, 1885.
[75] Kirchhoff, Charles, jr., Lead in Colorado: U. S. Geol. Survey Mineral Resources 1885, pp. 255–256, 1886.

Gold (placer and lode), silver, copper, lead, and zinc produced in Gunnison County, 1861–1923

Year	Ore (short tons)	Gold			Silver			Copper			Lead			Zinc			Total value
		Placer	Lode	Total	Quantity (fine ounces)	Average price per ounce	Value	Pounds	Average price per pound	Value	Pounds	Average price per pound	Value	Pounds	Average price per pound	Value	
1861–1872	(a)																(a)
1873		$5,000		$5,000													$5,000
1879					19,336	$1.12	$21,656										21,656
1880					222,031	1.15	255,336				b100,000	$0.050	$5,000				260,336
1881			$10,000	10,000	309,375	1.13	349,594				b360,000	.048	17,280				376,874
1882			100,000	100,000	386,719	1.14	440,860				b360,000	.049	17,640				558,500
1883			100,000	100,000	502,734	1.11	558,035	b30,000	$0.165	$4,950	b500,000	.043	21,500				684,485
1884			60,000	60,000	386,719	1.11	429,258				b2,000,000	.037	74,000				563,258
1885			c40,000	c40,000	c144,323	1.07	154,426				2,380,000	.039	92,820				287,246
1886			18,226	18,226	144,323	.99	142,880				b500,000	.046	23,000				184,106
1887			50,506	50,506	172,616	.98	169,164				451,351	.045	20,311				239,981
1888			18,642	18,642	60,166	.94	56,556				1,011,792	.044	44,519				119,717
1889			39,710	39,710	48,106	.94	45,220	556	.135	75	485,355	.039	18,929				103,934
1890			28,784	28,784	354,393	1.05	372,113	105,954	.156	16,529	6,945,972	.045	312,569				729,995
1891			7,402	7,402	489,268	.99	484,375				10,340,332	.043	444,634				936,411
1892			6,004	6,004	146,891	.87	127,795				525,574	.04	21,023				154,822
1893			7,728	7,728	144,577	.78	112,770				c500,000	.037	18,500				138,998
1894			8,052	8,052	104,938	.63	66,111				c400,000	.033	13,200				87,363
1895			36,734	36,734	114,218	.65	74,242				201,898	.032	6,461				117,437
1896			26,757	26,757	93,273	.68	63,426	8,515	.108	920	164,370	.03	4,931				96,034
1897			40,761	40,761	103,941	.60	62,365	2,770	.12	332	1,013,114	.036	36,472				139,930
1898			81,006	81,006	152,800	.59	90,152	119,072	.124	14,765	1,996,560	.038	75,869				261,792
1899			70,112	70,112	132,983	.60	79,790	46,186	.171	7,898	1,399,336	.045	62,970				220,770
1900			83,858	83,858	146,746	.62	90,982	42,790	.166	7,103	1,583,320	.044	69,666	b100,000	$0.044	$4,400	256,009
1901			83,445	83,445	93,243	.60	55,946	53,396	.167	8,917	656,631	.043	28,235	b100,000	.041	4,100	180,643
1902			103,536	103,536	123,138	.53	65,263	28,686	.122	3,500	728,935	.041	29,886	131,975	.048	6,335	208,520
1903			48,533	48,533	65,447	.54	35,341	15,000	.137	2,055	127,661	.042	5,362	55,600	.054	3,002	94,293
1904	2,067		26,024	26,024	115,153	.58	66,789	16,233	.128	2,078	200,462	.043	8,620	20,010	.051	1,021	104,532
1905	5,581		28,156	28,156	53,649	.61	32,726	50,500	.156	7,878	219,809	.047	10,331	17,905	.059	1,056	80,147
1906	31,103		87,505	87,505	70,798	.68	48,143				248,737	.057	14,178	158,198	.061	9,650	159,476
1907	18,078		61,569	61,569	27,277	.66	18,003	13,690	.20	2,738	120,226	.053	6,372	38,224	.059	2,255	90,937
1908	14,439		100,032	100,032	28,664	.53	15,192	5,481	.132	724	327,612	.042	13,760	147,000	.047	6,909	136,617
1909	9,071		108,493	108,493	37,423	.52	19,460	51,815	.13	6,736	493,070	.043	21,202	212,093	.054	11,453	167,344
1910	25,203		233,972	233,972	49,189	.54	26,562	21,024	.127	2,670	581,841	.044	25,601	176,815	.054	9,548	298,353
1911	11,926	1,417	143,622	145,039	32,541	.53	17,247	9,928	.125	1,241	631,933	.045	28,437	557,456	.057	31,775	223,739
1912	14,046	651	124,676	125,327	29,035	.615	17,857	8,097	.165	1,336	306,867	.045	13,809	483,884	.069	33,388	191,717
1913	4,301	601	9,588	10,189	87,488	.604	52,843	21,864	.155	3,389	196,728	.044	8,656	292,875	.056	16,401	91,478
1914	6,018	4,384	8,649	13,033	59,036	.553	32,647	11,188	.133	1,488	317,974	.039	12,401	525,000	.051	26,775	86,344
1915	6,446		60,197	60,197	24,892	.507	12,620	9,091	.175	1,591	190,000	.047	8,930	1,750,944	.124	217,117	300,455
1916	10,419	2,151	29,402	31,553	29,023	.658	19,097	84,679	.246	20,831	313,217	.069	21,612	1,964,873	.134	263,293	356,386
1917	12,671	327	6,308	6,635	40,272	.824	33,184	180,121	.273	49,173	751,000	.086	64,586	3,054,990	.102	311,609	465,187
1918	6,344	73	10,295	10,368	12,880	1.00	12,880	43,033	.247	10,629	300,760	.071	21,354	2,349,538	.091	213,808	269,039
1919	8,348		31,556	31,556	18,425	1.12	20,636	5,124	.186	953	117,454	.053	6,225	2,456,479	.073	179,323	238,693
1920	8,443		24,070	24,070	20,555	1.09	22,405		.184		958,301	.08	76,664	1,530,691	.081	123,986	247,125
1921	498		18,223	18,223	10,370	1.00	10,370		.129		51,955	.045	2,338		.05		30,931
1922	221		9,180	9,180	3,803	1.00	3,803	526	.135	71	13,382	.055	736		.057		13,790
1923	11,019		23,854	23,854	24,939	.82	20,450	1,788		263	1,690,430	.07	118,330	2,889,000	.068	196,452	359,349
		14,604	2,215,167	2,229,771	5,437,716		4,906,570	987,107		180,833	42,763,959		1,948,919	19,013,550		1,673,656	10,939,749

a Production unrecorded.
b Estimated by C. W. Henderson.
c Interpolated by C. W. Henderson to correspond with total production of the State.

HINSDALE COUNTY

The early history of Hinsdale County is summarized by Irving and Bancroft [76] as follows:

Precious metal was probably first discovered in the Lake City area about 1842 [1848?] by a member of the Frémont party, but no one, not even Frémont, has been able to locate the place or even the stream from which the first small amount of gold was panned. On August 27, 1871, with the discovery of the Ute and Ulay veins by Harry Henson, Jorl K. Mullin, Albert Meade, and Charles Godwin, the history of Lake City began. At that time all the land which is now the "San Juan" belonged to the Indians. The reports of mineral wealth brought many prospectors into the region, and the red men became very much irritated at the frequent encroachments upon their domain. Finally in 1874, to avert open hostilities, a treaty was drawn up and ratified by the Senate, whereby a strip of land 60 miles wide and 75 miles long was ceded to the United States Government by the Ute Indians.

In August, 1874, Hotchkiss (the leader of the expedition that built a wagon road from Saguache to Lake City) discovered the rich vein now known as the Golden Fleece and named it the "Hotchkiss." News of the strike spread rapidly, and Lake City soon became a center of activity, the county seat being removed from San Juan to Lake City, where it has remained. During the same year reduction works were erected at Lake City. * * *

Development was continued, and new discoveries were made almost daily. The first boom attained its climax in 1876, coinciding with the opening up of Ocean Wave group and the continued production of the Hotchkiss and the Ute and Ulay mines. During the spring the erection of a concentrator was begun, and ground was broken for a smelter at the falls just above the city. Soon afterward the reaction and "lull," so characteristic of the region, began.

During the next three years work was continued on the Ute and Ulay and the Ocean Wave properties, the Excelsior mine was located (April, 1878), and the Crooke and Ocean Wave smelters were completed.

The year 1880 marked the beginning of the biggest boom in the Lake City region.

Raymond [77] in his report for 1874 mentions Lake district (of the San Juan country) and the discovery of the Hotchkiss lode of tellurides of gold and silver.

In the notes on the San Juan country, in the report for 1875, Raymond [78] says:

[76] Irving, J. D., and Bancroft, Howland, Geology and ore deposits near Lake City, Colo.: U. S. Geol. Survey Bull. 478, p. 13, 1911.

[77] Raymond, R. W., Statistics of mines and mining in the States and Territories west of the Rocky Mountains for 1874, pp. 384, 386, 1875.
[78] Raymond, R. W., op. cit. for 1875, p. 324, 1877.

Lake district includes all the locations made in Hinsdale County, except the mines situated in Burrows Park at the extreme head of Lake Fork of the Gunnison, which constitute what is known as Adams or Park district. * * *

The only mines that have been worked to a considerable extent in the country are the Hotchkiss, in Lake district; the Silver Wing, in Eureka district (San Juan County); and the Highland Mary, Aspen, Prospector, and Little Giant, in the Animas district (San Juan County).

The Hotchkiss, located by Hotchkiss Finley, is the best developed mine in the San Juan country. The strike of the vein is northeast and southwest; the vein matter is 60 feet thick, and it was only in the latter part of February that what is considered the true ore zone was found. There are two tunnels 50 and 80 feet long, respectively, which give access to the vein. The ore consists of tellurides, containing in value about equal proportions of gold and silver. Specimen assays range from $17,000 to $20,000 per ton; 18 tons of ore shipped averaged $1,318.61; 75 tons remain on the dump, valued at $150 per ton.

Burchard says that Crooke & Co., at Lake City, did most of the smelting in San Juan in 1880.[79]

In 1881 Burchard says: [80]

The chief metallurgical works are those of the Crooke Mining & Smelting Co. * * * The production for 1881 was 600 tons of lead and 75,000 ounces of silver.

The Polar Star, Ute, and Ulay mines have, up to the present time, almost entirely supplied Crooke's works to their full capacity.

The Palmetto has a 15-stamp mill, capable of handling 25 tons of ore per day. * * * There have been sent to mill 400 tons of ore which yielded $28,000 worth of silver.

In the report for 1882 Burchard says of the districts in Hinsdale County: [81]

Galena district comprises, as its name indicates, veins of principally argentiferous lead ores (sulphide and sulphate of lead), generally accompanied by auriferous copper and iron pyrites, gray copper, zinc blende, and quartz.

The best representatives of veins of the above character are the Ute and Ulay mines, on Henson Creek, 3½ miles west of Lake City. * * * The main shaft on the Ulay has reached a depth of 410 feet. During last year extensive concentration works have been erected at the openings on the Ulay, which have proved a complete success. Their capacity is 150 tons of ore per day, affording also a good opportunity for treatment of similar ores from foreign mines.

In connection with these mines are the smelting works near Granite Falls, 1 mile south of Lake City, with a capacity of 35 to 40 tons per day. The property is owned by the Crooke Mining & Smelting Co. (Ltd.), London. * * * The mines have worked for the greater part of the year and have produced over 8,100 tons of ore, of which 3,750 tons have been treated at the works. * * *

In the Ocean Wave the vein is about 3½ feet wide, with 10 to 12 inches of splendid mineral, principally gray copper and galena. Up to the year 1880 the total product of this mine, treated at the Ocean Wave works, was 110,000 ounces. Since that time mine and works have been idle. * * *

On Engineer Mountain, at the headwaters of Henson Creek, most of the large fissure veins carry high-grade silver ores consisting of ruby silver, antimonial silver, gray copper, and iron and copper pyrites impregnated through vein matter. The

best developed mine is the Palmetto. * * * During the first half of last year 400 tons of ore were extracted and treated at the Palmetto works by the amalgamation process, yielding $18,480. * * *

But chief among the Engineer Mountain properties is the Frank Hough mine, which was discovered early last January. * * * The ore is a copper ore, composed of a thorough mixture of gray copper, copper pyrites, and iron pyrites. The ore occurs in solid, large, and irregular bodies, often separated and intersected by small and large talcish and chloritish fissures traversing in every direction. The average value of the ore at present exposed is from 50 to 60 ounces of silver, a trace to 1 ounce of gold, and 20 to 28 per cent of copper per ton. Sixty tons of ore were shipped late last year, with an average value of about $125 per ton. * * *

Lake City is the central point of Lake district; all the prominent mines are within a radius of 4 miles.

In the report for 1883 Burchard says of the Ute and Ulay and the Golden Fleece properties: [82]

Crooke Mining & Smelting Co.'s properties, consisting of the famous Ute and Ulay, are the most extensively developed mines in the San Juan. * * * These mines are producing 600 tons of ore per month, and it is expected that next year they will produce 1,000 tons per month. These two mines were located in 1874 by Joseph Mullen and in 1876 were purchased by the Crooke Bros. A smelter was then erected for the treatment of the ore, and during the past year concentrators of a capacity of 150 tons per day were erected. The first-grade ore goes to the smelter, the concentrates to Pueblo and Denver. * * * The Pueblo smelter purchased $100,000 in concentrates from this company during 1883. * * *

The Golden Fleece is the modern name for the claim formerly known as the Hotchkiss, which during 1874–75 produced tellurium in large quantities. The vein of this property was lost but during November, 1883, was found, and the owners now expect to work a large force of men during the coming year.

In 1884 Burchard says: [83]

Hinsdale County's output has fallen off considerably * * * with the closing of the Crooke Mining & Smelting Co., the largest producer in the county, after but three months' production. * * *

The Frank Hough mine has been one of the main factors in swelling the output of the county.

The production of the Frank Hough mine amounted to 700 tons valued at $52,500.

For 1886 to 1896 the figures given in the table below are based on reports of the agents of the mint in annual reports of the Director of the Mint, the gold and silver being prorated to correspond with the figures for the State as corrected by the Director of the Mint, the lead being prorated to correspond with the total production of lead for the State as given in Mineral Resources, and any unknown production in the State being distributed proportionately among the counties. As with lead so with copper, but as the figures for copper given in Mineral Resources include copper from matte and ores treated in Colorado, though produced in other States, the figures for copper are subject to revision.

[79] Burchard, H. C., Report of the Director of the Mint upon the production of the precious metals during the calendar year 1880, p. 157, 1881.

[80] Burchard, H. C., op. cit. for 1881, pp. 354, 402, 1882.

[81] Burchard, H. C., op. cit. for 1882, pp. 394, 473–481, 1883.

[82] Burchard, H. C., op. cit. for 1883, pp. 240, 328–332, 1884.

[83] Burchard, H. C., op. cit. for 1884, pp. 177, 218–220, 1885.

In the mint report for 1891 the district of Creede was included in Saguache County, but in the report for 1892 Creede was put in both Hinsdale and Rio Grande counties. The output of Hinsdale, Mineral, and Rio Grande counties has been separated as accurately as possible with the data obtainable.

For 1897 to 1904 the figures, which represent smelter and mint receipts, are taken from the reports of the Colorado State Bureau of Mines.

For 1905 to 1923 the figures are taken from Mineral Resources (mines reports).

the north, Las Animas on the east, Las Animas and Costilla on the south, and Costilla and Saguache on the west.

In form the county is very irregular, being outlined largely by natural topographical divisions. With the exception of the Sangre de Cristo Mountains on the west, the Wet Mountains on the north, and the Spanish Peaks in the south, composed of metamorphic and eruptive rocks, the county is made up of sedimentary rocks, ranging from the Carboniferous to the Tertiary.

The first mining for precious metals was in 1875. Since that time small amounts have been produced annually, but this resource has not yet been developed beyond the prospect stage. Prospects are numerous around the southern end of the Wet

Gold, silver, copper, lead, and zinc produced in Hinsdale County, 1875–1923

Year	Ore (short tons)	Lode gold	Silver Fine ounces	Silver Average price per ounce	Silver Value	Copper Pounds	Copper Average price per pound	Copper Value	Lead Pounds	Lead Average price per pound	Lead Value	Zinc Pounds	Zinc Average price per pound	Zinc Value	Total value
1875		$12,000	a47,953	$1.24	$59,462										$71,462
1876		a20,000	a154,688	1.16	179,438				a50,000	$0.61	$3,050				202,488
1877		a25,000	a92,814	1.20	111,377				a100,000	.055	5,500				141,877
1878		a20,000	a154,688	1.15	177,891				a200,000	.036	7,200				205,091
1879		a6,000	a193,359	1.12	216,562				a500,000	.041	20,500				243,062
1880		a6,000	a116,016	1.15	133,418	a30,000	$0.214	$6,420	a1,000,000	.05	50,000				195,838
1881		10,000	123,750	1.13	139,838	a40,000	.182	7,280	1,200,000	.048	57,600				214,718
1882		20,000	61,875	1.14	70,538	a40,000	.191	7,640	a600,000	.019	29,400				127,578
1883		20,000	193,359	1.11	214,628	a22,652	.165	3,738	a1,000,000	.043	43,000				281,366
1884	2,184	2,500	154,687	1.11	171,703	a350,000	.13	45,500	a1,000,000	.037	37,000				256,703
1885		b2,000	b16,320	1.07	17,462	a b46,460	.108	5,018	a b100,000	.039	3,900				28,380
1886		2,060	16,320	.99	16,157	a46,460	.111	5,157	a100,000	.046	4,600				27,974
1887		4,308	90,355	.98	88,548	12,027	.138	1,660	547,503	.045	24,638				119,154
1888		2,667	86,248	.94	81,073	2,000	.168	336	1,205,973	.044	53,063				137,139
1889		1,794	16,665	.94	15,665	17,359	.135	2,343	240,812	.039	9,392				29,194
1890		3,697	57,387	1.05	60,256	60,584	.156	9,451	660,708	.045	29,732				103,136
1891		19,869	186,850	.99	184,982	8,248	.128	1,056	8,308,048	.043	357,246				563,153
1892		22,514	411,758	.87	358,229	29,914	.116	3,470	4,753,783	.04	190,151				574,364
1893		88,750	385,653	.78	300,809	a10,000	.108	1,080	a3,808,111	.037	140,900				531,539
1894		85,196	395,899	.63	249,416	a10,000	.095	950	a3,322,170	.033	109,632				445,194
1895		243,195	483,565	.65	314,317	a10,000	.107	1,070	5,251,014	.032	168,032				726,614
1896		212,794	510,883	.68	347,400	13,202	.108	1,426	5,468,856	.03	164,066				725,686
1897		168,171	243,437	.60	146,062	8,085	.12	970	5,550,058	.036	199,802				515,005
1898		51,282	186,456	.59	110,009	104,038	.124	12,901	9,828,482	.038	373,482				547,674
1899		38,343	155,902	.60	93,541	49,676	.171	8,495	10,572,353	.045	475,756				616,135
1900		56,470	155,485	.62	96,401	29,180	.166	4,844	9,377,062	.044	412,591	a100,000	$0.044	$4,400	574,706
1901		76,148	152,122	.60	91,273	12,532	.167	2,093	7,588,675	.043	326,313	a126,591	.041	5,190	501,017
1902		98,348	117,177	.53	62,104	8,314	.122	1,014	6,213,763	.041	254,764	319,000	.048	15,312	431,542
1903		16,515	33,139	.54	17,895	11,263	.137	1,543	459,462	.042	19,297	106,000	.054	5,724	60,974
1904	5,591	10,521	46,585	.58	27,019	13,187	.128	1,688	1,041,222	.043	44,773	59,089	.051	3,014	87,015
1905	5,041	11,991	54,419	.61	33,196	84,485	.156	13,180	767,681	.047	36,081	235,178	.059	13,876	108,324
1906	7,086	24,510	87,940	.68	59,799	63,621	.193	12,279	753,950	.057	42,975	38,387	.061	2,342	141,905
1907	10,740	7,520	50,109	.66	33,072	99,410	.20	19,882	1,204,628	.053	63,845				124,319
1908	980	2,454	29,498	.53	15,634	188,698	.132	24,908	280,465	.042	11,780				54,776
1909	1,697	7,587	75,731	.52	39,380	714,569	.13	92,894	106,327	.043	4,572				144,433
1910	3,468	6,320	54,422	.54	29,388	465,472	.127	59,115	296,182	.044	13,032				107,855
1911	723	3,830	7,753	.53	4,109	21,696	.125	2,712	118,645	.045	5,339	36,439	.057	2,077	18,067
1912	9,554	6,811	34,722	.615	21,354	53,739	.165	8,867	1,257,800	.045	56,601	11,926	.069	823	94,456
1913	4,329	5,280	30,477	.604	18,408	76,304	.155	11,827	782,318	.044	34,422	54,732	.056	3,065	73,002
1914	118	170	5,987	.553	3,311	17,098	.133	2,274	5,723	.039	223		.051		5,978
1915	488	737	9,621	.507	4,878	9,114	.175	1,595	266,128	.047	12,508				19,718
1916	377	1,346	10,030	.658	6,600	16,248	.246	3,997	75,638	.069	5,219	12,575	.134	1,685	18,847
1917	517	1,136	7,721	.824	6,362	6,099	.273	1,665	209,616	.086	18,027	4,117	.102	420	27,610
1918	5,222	6,249	22,245	1.00	22,245	18,308	.247	4,522	767,972	.071	54,526				87,542
1919	1,219	8,232	22,942	1.12	25,695	7,705	.186	1,433	55,679	.053	2,951				38,311
1920	568	6,151	21,522	1.09	23,459	2,625	.184	483	80,625	.08	6,450				36,543
1921	495	3,425	32,039	1.00	32,039	9,357	.129	1,207	65,756	.045	2,959				39,630
1922	1,550	1,298	50,074	1.00	50,074	14,269	.135	1,926	114,200	.055	6,281				59,579
1923	684	732	30,046	.82	24,638	10,075	.147	1,481	19,971	.07	1,398				28,249
		1,451,921	5,678,693		4,607,114	2,864,073		403,390	97,277,359		3,994,569	1,104,034		57,928	10,514,922

a Estimated by C. W. Henderson.   b Interpolated by C. W. Henderson to correspond with total production of the State.

HUERFANO COUNTY

The Colorado State Bureau of Mines summarizes the early history of Huerfano County as follows: [81]

Huerfano County occupies a south-central position in the State. It was originally organized in 1861, and the boundaries then established were reduced in 1867 by legislative enactment. As now constituted it has an area of about 1,750 square miles [1,500 in 1920 according to the figures of the Bureau of the Census]. The adjoining counties are Custer and Pueblo on

Mountains, known locally as the Greenhorn Mountains, in the north, along the eastern slope of the Sangre de Cristo in the west, and around the Spanish Peaks in the southern part of the county.

For 1886 to 1897 the figures given in the table are based on reports of the agents of the mint in annual reports of the Director of the Mint, the gold and silver being prorated to correspond with the figures for the total production of the State as corrected by the Director of the Mint, the lead being prorated to cor-

[81] Colorado State Bur. Mines Rept. for 1897, pp. 61-62, 1898.

respond with the total production of lead for the State as given in Mineral Resources, and any unknown production in the State being distributed proportionately among the counties. As with lead so with copper, but as figures for copper given in Mineral Resources include copper from matte and ores treated in Colorado, though produced in other States, the figures for copper are subject to revision.

For 1897 the figures, which represent smelter and mint receipts, are taken from the report of the Colorado State Bureau of Mines.

For 1905 the Colorado State Bureau of Mines credits Huerfano County with $269 in gold and 617 ounces of silver. The figures published in Mineral Resources, however, show no output for this county. For 1906 the Colorado State Bureau of Mines credits the county with $475 in gold and 56 ounces of silver, but Mineral Resources again gives no output to this county.

For 1907 the figures are taken from Mineral Resources (mines reports). The production for this year is of doubtful source.

what limited in extent, the available territory has been quite productive and is still worked or reworked in a desultory manner with primitive appliances. Several attempts by capital have been made comparatively recently to recover the gold and concentrated losses of the mills in Clear Creek and Gilpin counties from the stream beds. Although the existence of good values has been demonstrated and some have been recovered, the physical condition of the creek bed has so far proved a barrier to the financial success of the undertaking. Several minor excitements have been occasioned by reputed finds of gold-silver-copper bearing veins in several sections but have resulted in little or no production and very slight development.

Lindgren [86] has given some notes on the copper deposits of Jefferson County.

For 1885 to 1900 the placer and lode production has been separated on the basis of the deposits at the Denver Mint.

For 1885 the figures given in the table, which represent gold and silver deposited at the Denver Mint, are taken from Wilson.[87]

For 1886 to 1896 the figures given in the table are based on reports of the agents of the mint in annual reports of the Director of the Mint, the gold and silver

Gold, silver, copper, and lead produced in Huerfano County, 1875-1907

	Lode gold	Silver			Copper			Lead			Total value
		Fine ounces	Average per ounce	Value	Pounds	Average price per pound	Value	Pounds	Average price per pound	Value	
1875-1885	(a)										(a)
1886	$116										$116
1894	304	1	$0.63	$1							305
1895	87										87
1896	109										109
1897	723	167	.60	100	92	$0.12	$11	1,067	$0.036	$38	872
1898	145	40	.59	24							169
1899	124	5	.60	3							127
1900	124	20	.62	12							136
1901	83	10	.60	6							89
1902	847	260	.53	138							985
1905	269	617	.61	376							645
1906	475	56	.68	38							513
1907	68										68
	3,474	1,176		698	92		11	1,067		38	4,221

a Production unrecorded.

JEFFERSON COUNTY

The early history of Jefferson County is summarized by the Colorado State Bureau of Mines as follows: [85]

History reveals this section, to be one of the first settled in the State. Golden, the present county seat and main commercial center, was established in June, 1859, and made phenomenal growth until 1861. This year marked a rapid decline. In 1862 the Territorial capital was removed from Colorado City and located at Golden, where it remained until 1867. In 1868 grading for the Colorado Central Railway began, and the road opened for traffic in 1870. A season of great industrial improvement followed the advent of the railroad, and until 1878 this section as an important railroad and manufacturing center rivaled Denver.

The placer bars near Golden mark the scene of the first mining in the county and among the first in the State. While some-

being prorated to correspond with the figures for the total production of the State as corrected by the Director of the Mint, the lead being prorated to correspond with the total production of the State as given in Mineral Resources, and any unknown production in the State being distributed proportionately among the counties. As with lead so with copper, but as figures for copper given in Mineral Resources include copper from matte and ores treated in Colorado, though produced in other States, the figures for copper are subject to revision.

85 Lindgren, Waldemar, Pre-Cambrian copper deposits in Jefferson County; Notes on the copper deposits in Chaffee, Fremont, and Jefferson counties, Colo.: U. S. Geol. Survey Bull. 340, pp. 157, 167-170, 1908.

87 Wilson, P. S., agent for Colorado, in Kimball, J. P., Report of the Director of the Mint upon the production of the precious metals in the United States for 1885, p. 136, 1886.

85 Colorado State Bur. Mines Rept. for 1897, pp. 62-64, 1898.

For 1897 to 1904 the figures, which represent smelter and mint receipts, are taken from the reports of the Colorado State Bureau of Mines.

For 1905 to 1923 the figures are taken from Mineral Resources (mines reports).

In 1905 there were two dredges at Golden. The Colorado State Bureau of Mines and the report of the Director of the Mint give the deposits of gold and silver at the Denver Mint during this year.

In 1907, 1909, and 1918 the production came from copper properties at Evergreen.

According to data given by Lindgren,[88] the total figures for copper in the following table are low. The owners credit $35,000 for early production of copper from the Malachite mine, situated between Golden and Evergreen.

in former times for about 5 miles in length. The settlement here is called Oro City. * * *

Across the park of the upper Arkansas, to the southwest from Mount Lincoln, Lake Creek comes down from the main range beyond, after escaping from which it spreads out into two dark sheets of water, together about 2 miles wide by 5 long, and separated by a belt of land one-fourth of a mile wide, covered with pines. They are called the Twin Lakes and constitute the most considerable body of water in Colorado and give name to the county. Dayton, the county seat, is located under the range at their head.

Up the extreme left considerable fork of the Arkansas River is one of the easiest passes [Tennessee] through the range, opening out on Piney Creek or Eagle River, which puts into the Grand below the Middle Park. This pass is said to be 3,000 feet less than Berthoud Pass. [U. S. Geol. Survey topographic maps give 11,315 feet for Berthoud Pass and 10,300 feet for Tennessee Pass.] Nearly opposite California Gulch comes in Colorado Gulch [Lake County] from the west, more recently discovered than California, and worked from year to year with

Gold (placer and lode), silver, copper, and lead produced in Jefferson County, Colo., 1858–1918

Year	Ore (short tons)	Gold			Silver			Copper			Lead			Total value
		Placer	Lode	Total	Fine ounces	Average price per ounce	Value	Pounds	Average price per pound	Value	Pounds	Average price per pound	Value	
1858–1884	(a)													(a)
1885		$697		$697	.5	$1.07	$5							$702
1886		2,804		2,804	43	.99	42							2,846
1887		942		942	.5	.98	5							947
1894		2,197		2,197	10	.63	6							2,203
1895		1,861	$731	2,592	15	.65	10							2,602
1896		1,963	16,523	18,486	4,590	.68	3,121							21,607
1897		586	7,661	8,247	1,614	.60	968	1,602	$0.12	$192	10,093	$0.036	$363	9,770
1898		117	1,723	1,840	102	.59	60							1,900
1899		542	822	1,364	351	.60	211	254	.171	43	770	.045	35	1,653
1900		78	625	703	51	.62	32							735
1901			310	310	20	.60	12							322
1902			517	517	3	.53	2	2,978	.122	363		.041		882
1903			248	248	5	.54	3	218	.137	30		.042		281
1904		2,894	351	3,245	37	.58	21	538	.128	69		.043		3,335
1905	15	18,088		18,088	125	.61	76	9,000	.156	1,404		.047		19,568
1906	5							3,150	.193	608		.057		608
1907	18				73	.66	48	1,955	.20	391		.053		439
1909	1		16	16										16
1918	23				9	1.00	9	1,000	.247	217				256
		32,769	29,527	62,296	7,058		4,631	20,695		3,347	10,863		398	70,672

a Production unrecorded.

LAKE COUNTY

Hollister, writing in 1867, says of the mining in Lake County:[89]

Mount Lincoln is the northeast cornerstone of Lake County. Thence the boundary runs west on the thirty-ninth parallel to the western limit of the Territory, 150 miles; then south to the summit of the Uncompahgre Mountains, 110 miles; thence east-northeast along the summit to the main range; following that until it curves northward, it jumps across it and the Arkansas River to the crest of the Montgomery spur, along which it proceeds to the top of Mount Lincoln, the place of beginning. The area is about 16,000 square miles. It embraces the sources of the Arkansas River and the course and tributaries of that stream for about 50 miles. It also includes the Gunnison Fork of the Rio Colorado. It is with the former section we have chiefly to do, however, as in that are the principal settlements. On the western slope of the Montgomery spur, opposite Buckskin, heads California Gulch, worked more or less

considerable success. Just below California, on the same side, is Iowa Gulch, also worked at the present time. It is 12 miles thence down to the mouth of Lake Creek [Lake County], coming from the west, immediately below which is Cache Creek and Diggings [Chaffee County]. Then the mountains seem to crowd the river, the stupendous chasm through which it makes its way becomes gorged; the dividing ridges between the tributary streams, Clear, Pine, Chalk, Cottonwood [Chaffee County], which come out of the range at intervals of 7 or 8 miles, continue down to the very brink of the river. * * * On these tributaries the valley becomes wider, and below the mouth of Cottonwood, which is 40 miles from the source of the river, down to the mouth of the South Arkansas, a distance of 20 miles, farms line the banks of the stream. * * * The road from the south to the San Luis Parks, via Trout Creek, the Arkansas, the South Arkansas, and Poncha Pass, goes through this part of the valley. * * * From the head of the river to Canon City it is an alternation of bar and canyon, the bars masses of boulders and gravel more or less overlain with a light alluvial soil. * * *

Gold is found in the stream [the Arkansas] and on its southwestern tributaries from the Cottonwood to its extreme sources. A mining district was organized on the heads of Cottonwood in 1866, called Westphalian. About 30 lodes have been discov-

[88] Lindgren, Waldemar, Notes on copper deposits in Chaffee, Fremont, and Jefferson counties, Colo.: U. S. Geol. Survey Bull. 340, p. 169, 1908.
[89] Hollister, O. J., The mines of Colorado, pp. 306-319, Springfield, Mass., Samuel Bowles & Co., 1867.

ered here, none of them more than fairly struck, the ore, coarse iron and copper pyrites and sulphurets with some galena and zinc, coming to the surface not decomposed in the least.

On Pine Creek, 12 miles above Cottonwood, is another mining district, called Pine Creek, not now worked nor inhabited and about which little is known.

A short distance above Pine we have Clear Creek [Chaffee County], upon which is a mining district called La Plata.

Eight miles up the stream about thirty lodes have been discovered. Nothing of consequence has been done on them, but it is known that they possess the same general characteristics as those west of the Range. At the mouth of the creek is Georgia Bar, formerly worked considerably in winter with hand rockers and paying from $3 to $5 per day. This bar, Kelly's Bar, and the Arkansas River for about 30 miles below Lake Creek have always been the winter's resort for food of many of the miners of Lake County. From these gravels it is estimated $75,000 has been recovered. In 1861 a company built a dam on Lake Creek and flumed the Arkansas River at Georgia Bar for 1,000 feet at a cost of $10,000. They were nearly ready to commence sluicing the bed of the stream, which is very rich, when their dam gave way and carried off their flume. The men immediately enlisted in the First Colorado Volunteers, and no attempt has since been made to flume the Arkansas, although the sand for a hundred miles from its source, taken up on a shovel and panned down, gives a fine color. This company had obtained as high as $1.50 to the pan of dirt from the bed of the river.

About 3 miles above the Clear is Cache Creek, a creek only by courtesy, or rather by virtue of a ditch 9 miles long, costing $20,000 to construct, bringing 500 inches of water during the rainy season from the precipitous southern rim of the Twin Lakes. Cache Creek is at the lower end of a park, or bar, 3 miles up and down the river in length and 2 miles from the river west to the mountain in width. A party of men who were prospecting Arkansas River cached their provisions here, and in that way gold was discovered. In this park are Cache Creek Gulch, discovered by Campbell & Shoewalter, estimated production to date $30,000; Bertschey's Gulch, discovered by G. Bertschey & Co., estimated production to date $25,000, worked out; Gold Run Gulch, discovered by Long, West & Co., has produced $22,000, now worked out; Gibson Gulch, discovered by one Gibson in 1861, has paid to date $30,000, now worked by D. Houghton & Co. by bedrock flume 500 feet long; Oregon Gulch, discovered and worked out in 1860 by Thomas & Co., paid $11,000; Ritchey's Patch, discovered by J. Ritchey in 1864, now worked by himself and others, production to date $17,000; and Lake Creek diggings, at the northeast corner of the Park, discovered in 1860, now chiefly owned and worked by H. M. Severs & Co., who are erecting a sawmill to facilitate operations, estimated yield to date $55,000.

Cache Creek Gulch is very deep and has but a slight fall, so that it could not be worked to advantage in 100-foot claims, as all the gulch was originally taken up. It soon passed into the hands of Ramage Bros. & Co., who commenced bedrock flume in 1862. The property has since been transferred to the Gaff Mining Co. * * * Their flume is 4 feet square and now some three-fourths of a mile long. They work about 30 feet in depth by 150 feet in width. Side sluices, leading into the main one, are used when necessary. * * *

There are patch diggings in the points of the hills bordering Cache Gulch from which has been taken to date $86,000. It is thought that the whole park would pay for washing, could it be mined on a large scale, but a scarcity of water is the chief obstacle.

West of this park the range rises abruptly to an altitude of 3,000 or 4,000 feet, and near its summit on the eastern slope is Lost Canyon Gulch, discovered in 1860 by a party of prospectors

returning from the park of the Gunnison River, west of the range, and who here became convinced that they were lost. Of course they prospected—a lost prospector will always find himself when he can get a fair color—and as good luck would have it they struck a rich spot. About $60,000 was taken out in 1861. Next year the pay streak was lost and, although hunted for diligently ever since, has not again been found. So that the spot is doubly lost. All this section we have last described belongs in Hope mining district.

On the opposite side of the river a very rich lode was discovered in 1866, upon which a mining district was organized and called Lake Falls. A load of quartz from the lode, which is called Hattie Jane, treated by the Bertola process, in Clear Creek County, is said to have yielded at the rate of $200 per ton. We have no figures as to the width of crevice from which the quartz was taken.

Lake Creek empties into the Arkansas at the upper end of Cache Creek Park. It is 5 miles from its mouth to the head of the Twin Lakes, which we have referred to before. * * * South of them the range rises abruptly from their very edge; on the west there is a large bottom, and here is Dayton, the county seat. Hence to the Red Mountain at the extreme sources of the creek is about 15 miles in a west-southwest direction. Red Mountain seems to be in a belt of lodes, some 3 miles in width, which here crosses the range in the true course—northeast and southwest. From the top of the Red Mountain at the head of the left fork of Lake Creek other red mountains can be seen both to the east and west. Eight miles west, in an air line, a Boston company did some work in 1866, finding similar ore to that found here. This point is about 100 miles west of Pikes Peak. * * * In the streams and where the creek escapes from the mountains numerous well-defined lodes have been discovered, not greatly different in width, lineal extent, and character of ores from those of other parts of the Territory. Like them, too, they vary in richness. Some of them are absolutely barren, and some contain $100 gold to the ton, as tests of which we were personally cognizant have indicated.[90]

It is believed the requisite capital has been secured to establish at once one or two mills in this, called Red Mountain district. There is not now a quartz mill in the county, nor a shaft more than 30 feet in depth, although the Berry tunnel near the head of California Gulch has been driven 100 feet. In this pay vein 6 feet in width is claimed. The ore assays $70 or $80 a ton in gold and silver and is very rich in copper. But nothing very definite with reference to the quartz mines of the county is known.

About 12 miles above Dayton is California Gulch, divided into three districts, Arkansas Independent, California, and Sacramento. It stands second if not first among all the gulches ever worked in Colorado for extent and yield of gold. It was discovered by Slater & Co. late in 1859, but its richness was not developed until the next April. Adventurers poured in at once, and it was soon preempted in 100-foot claims for 7 miles in length. Discovery claim produced $60,000 that season. Nos. 5 and 6, above, produced $65,000. No. 1, below, has paid $55,000. A large quartz vein runs through Discovery and No. 1, below, from which $216 was sluiced by three men in 1863 in a half day. No. 4, below, prospected from $1 to $10 to the pan on the pay streak and has produced to date $75,000. No. 5, below, has paid $55,000. Nos. 11 and 12, below, paid $26,000. From 13 to 35 the average yield was $10 a day to the hand. Nos. 26 and 27 paid $50,000. Nos. 28 to 35 paid $15, and 36 to 41 $25 per day to the man. From 42 to 45 there was no pay, but thence to 56 the average was

---

[90] Ex-Secretary Elbert, of Colorado, had 46 samples of ore from 43 of these lodes assayed by Behr & Keith, of Blackhawk. They varied from $59 to $441 per ton, averaging $138.25. One sample yielded 70 per cent of copper.

$18 a day per hand. No. 57 never paid anything, nor did any ground below that, which was but little more than a mile from Discovery. These figures are given to show the spotted character of the best gold-producing ravines.

Nos. 14 to 18, above, paid $20,000 each. Nos. 19 and 20 yielded $80,000 in three months. No. 21 paid $15,000. Nos. 22 to 28, as well as other claims above Discovery not here specified, paid from $3,000 to $5,000 each. From 28 to 36 the yield gradually fell off. No. 25 paid $15,000. No. 30 paid $6,000. Nos. 31 to 35, inclusive, did not pay wages. No. 36 paid $3,500.

From No. 22 to the head of the gulch occurs a black cement, 1 to 12 feet thick, too loose to blast to advantage, too hard to be decomposed by water, hydraulics with 100-feet head having no effect on it.

For the first three years mining was carried on by sluices, long toms, Georgia and hand rockers, and pans. Since that time the claims have been largely consolidated and mining done by ground-sluicing and hydraulics. Thus, W. H. Jones now owns and works 1, 2, and 3, below; Leahy & Co., 26 and 27, below; and White, Burroughs & Co., 14 to 30, above. The pay streak in these claims yields from $10 to $15 per day per man. Not more than half the gulch is considered worked out. Rich quartz veins traverse it in four or five different places, though these have never received any attention beyond preempting and sluicing out their dirt crevices, some of which paid extraordinarily. A tunnel called the Berry tunnel at the head of the gulch discovers, a short distance from its mouth, a strong vein of excellent ore. Although a few parties still make enough to support them through the winter by working this gulch during the summer, its future prosperity, and this is true of the entire county, must depend on the development of its quartz mines. Nothing, indeed, can be more deceiving or more ephemeral than the feverish prosperity of a placer mining country. California Gulch, which six years ago was infested by 5,000 to 6,000 people, is now almost deserted. The relics of former life and business, old boots and clothes, cooking utensils, rude house furniture, tin cans, gold pans, worn-out shovels and picks, and the remains of toms, half buried sluices and riffle boxes, dirt-roofed log cabins tumbling down, and the country turned inside out and disguised with rubbish of every description, are most disagreeably abundant and suggestive. * * *

Colorado Gulch [Lake County] puts into the Arkansas opposite the mouth of California Gulch. It is 5 miles long and worked more or less its entire length, employing during the summer from 50 to 100 hands. It is chiefly owned by McCannon & Co., De Mary & Co., Long & Co., and Breece & Co. Most of them now have bedrock flumes, 16 inches square, and from 600 to 1,500 feet in length. It is worked from 60 to 100 feet in width and is from 12 to 15 feet deep, the pay being confined to within a foot of the bedrock. Gold was first discovered in the gulch July 4, 1863, and since that time, much of the work having necessarily been of a preparatory nature, about 2,000 ounces have been taken out.

There are other less important and far less worked gulches in the valley or park of the upper Arkansas; among them Iowa (Adams district), just below and on the same side of the river as California, and the Little Fryingpan, a tributary of Colorado Gulch. From Sacramento Flats, 3 miles above the mouth of California Gulch, nearly $200,000 has been taken first and last. The Arkansas River for 70 miles from its sources, with its tributaries and bars, is gold-bearing; and when placer mining shall come to be conducted on a larger scale and with more appropriate means it may, and probably will, become, in Lake County, a business of considerable importance.

The approaches to the county are all through the South Park, and with proper care in laying out and working the roads might be made very easy. From Fairplay to Dayton, crossing the Montgomery Spur north of Buffalo Peaks, is about 40 miles.

The route goes up a tributary of the Platte and by an easy grade, across a low notch between snow banks, and follows down an affluent of the Arkansas, also by an easy grade. The true route into the county, however, is via Canon City and the Colorado Salt Works, crossing the Montgomery Spur south of Buffalo Peaks, where it has not an altitude of more than 500 feet above the South Park, striking the Arkansas at Mayol's Ranch, 20 miles below Dayton, thence up or down the river. There is besides a trail from Buckskin over the spur into California Gulch.

The figures given in the table (p. 176) are for the present county of Lake, including chiefly the Leadville district, but also the placers of California Gulch, Iowa Gulch, Colorado Gulch, Lake Creek, and of several tributaries of Arkansas River within the present confines of Lake County, and lode mines in the St. Kevin, Sugar Loaf, Twin Lakes, Lackawanna Gulch, Tennessee Pass, and Big English Gulch districts. Only for the years 1911 to 1923 is the production of these districts shown separately in the mines reports in Mineral Resources. It appears impossible to separate these districts for the other years. Sugar Loaf district has made a regular but comparatively small production since 1881 and possibly before that year.

Raymond [91] says that in 1875 gulch gold was worth $17.75 to $19 per crude ounce (say $18 average) and that mill gold was worth $15 in coin. These figures have been used to obtain the content of silver in the production from 1860 to 1874.

In obtaining the total production of Lake County other compilers have used the coinage value of silver from 1874 to 1896, this being the value given in the reports of the Director of the Mint, which furnish the only available information for the production of the county in these years. The difference between the coinage value and the commercial value is as follows:

Coinage value of 130,983,284 ounces of silver, 1874–1896, at $1.2929292929+ per ounce	$169,352,125
Commercial value of 130,983,284 ounces	127,190,121
Difference	42,162,004
Total production Lake County, 1859–1923, if the silver is taken at its commercial value from 1874 to 1896	425,784,550
Calculated total production of Lake County, 1859–1923, if the silver is taken at its coinage value from 1874 to 1896	467,946,554

This $467,946,554 is the total production obtained by using the early figures hitherto generally credited to Lake County. The difference still existing is due in part to the fact that considerable silver and lead has been subtracted from Lake County and credited to Red Cliff (Eagle County), Aspen (Pitkin County),

[91] Raymond, R. W., op. cit. for 1875, p. 316, 1877

and Summit County, which was smelted at Leadville and not heretofore credited to original source. If the silver produced from 1859 to 1923 were calculated at the coinage rate the total production would be as follows:

Total production of silver, 1859–1923, inclusive:

230,482,487 ounces at coinage value	$297,996,549
230,482,487 ounces at commercial value	189,409,933
Difference	108,586,616

Calculated total production Lake County, 1859–1923:

When silver is taken at commercial value	425,784,550
When silver is taken at coinage value	534,371,166

In his report for 1869 Raymond[92] says that gold was discovered in California Gulch in 1859 and was developed in 1860. Emmons[93] says that placer gold was discovered in California Gulch in 1860 and that the Printer Boy lode was discovered in 1868 and was worked until 1877.

Burchard[94] gives placer gold from Lake County for 1860 to 1869 as $5,812,000. As Lake County in those years included Chaffee County, these figures include the placer production of California Gulch, Colorado Gulch and its tributary Little Fryingpan Gulch, Georgia Bar, Iowa Gulch, and Lake Creek (all now in Lake County), and Arkansas River, Cache Creek, Clear Creek, Chalk Creek, Cottonwood Creek, Lost Canyon Gulch, Pine Creek, Bertschey's Gulch, Kelly's Bar, Gold Run Gulch, Gibson Gulch, Oregon Gulch, and Ritchey's Patch (all now in Chaffee County). On the basis of Hollister's figures, Chaffee County is credited with $400,000 of placer gold for the years 1859 to 1869, leaving $5,412,000 for Lake County from 1859 to 1869.[95]

For 1870 Raymond[96] gives the placer yield as a little over $60,000 and credits the remainder of the production to lode mines (Yankee Blade, etc.) in Chaffee County. Burchard[97] gives the placer and lode production of Lake County for 1870 as $125,000, following the Rocky Mountain Review, of Georgetown.

In his report for 1871 Raymond says:[98]

In Lake County the placer-mining industry has suffered from the same causes which affected Summit. In California Gulch, a tributary of the South Arkansas, the most work has

been done, and a few men were at work as late as October. Since the discovery of gold in this gulch it is estimated to have yielded over two and one-half millions of dollars. The yield this year has not been as large as usual. * * * Of veins, the Printer Boy, Pilot, Five-Twenty, American Flag, and Berry tunnel have been the main objects of attention.

The Printer Boy was discovered in June, 1868. * * * One hundred and forty-five cords of ore, * * * treated at the Five-Twenty mill, gave an average yield of 18 ounces [gold] per cord. In November the mill (one battery) was running on wall rock that yielded from 3 to 6 ounces [gold] per cord. * * * The company intend to put up a mill of their own next year, which is to be located in Iowa Gulch and driven by water power. * * *

Adjoining this on the north, Messrs. Breece & Co. are working their mine. The main shaft, 130 feet in depth, carries a crevice of pay ore 6 or 8 inches in width. In the drift running south, 18 feet from the shaft, is a crevice of pay ore from 6 to 10 inches in width. In the breast of the drift running north there was, in November, an inch of rich gold ore. In this mine the gold is found in pockets that yield from 5 up to 1,000 ounces [gold].

East of the Printer Boy, John Hoover discovered a lode last summer, which he christened the American Flag. The first ore treated gave a yield of 8 ounces [gold] per cord. In the bottom, 58 feet from the surface, the crevice has split. On the footwall the pay is 4 or 5 inches in width and about the same on the hanging wall, a horse 4 feet in width being between the pay streaks.

The Five-Twenty, Printer Boy, American Flag, and Berry tunnel lodes are in granite, as also is the western wall of the Pilot. Overlying the granite, about 50 feet from this wall, is a stratum of limestone. From here to the Mosquito Range this limestone overlies the whole country, with here and there ledges of schist and granite breaking through it.

Probably next to the Printer Boy in richness is the Berry tunnel lode owned by Capt. S. D. Breece. A tunnel 100 feet in length has been driven on the vein, the breast of which is 40 feet from the surface. Work has been suspended for several years, no attempt having been made until within the last year to introduce a process for reducing the sulphuret ores of this locality. Careful assays show that this ore contains a large percentage of gold, silver, and copper. The tunnel is now badly caved in. Within a hundred yards of this lode, to the westward, the limestone makes its appearance.

The Pilot is now opened by the main shaft and three levels, 50 feet of stoping ground being between each two of them and between the first and the surface. About 20 tons of rich gold ore have been beneficiated, and much galena is out awaiting the erection of reduction works.

For 1871 the figures for gold and silver given in the table are taken from Burchard's report.[99]

Raymond[1] sketches the developments in 1872 as follows:

The placer and lode mines of California Gulch have been worked vigorously this summer. The Printer Boy, owned by the Boston & Philadelphia Gold Mining Co., J. Marshall Paul, agent, and by Captain Breece, has a crevice from 12 to 17 feet in width. * * * I am informed that the whole of this enormous vein yields under stamps at the rate of 17 ounces per cord, or $45 per ton. The mine is, however, very pockety and sometimes incredibly rich bunches of mineral occur. I am informed by one of the proprietors that a panful of dirt has been taken from one of these pockets which yielded 132 ounces

[92] Raymond, R. W., Statistics of mines and mining in the States and Territories west of the Rocky Mountains for 1869, p. 344, 1870.
[93] Emmons, S. F., Geology and mining industry of Leadville, Colo.: U. S. Geol. Survey Mon. 12, p. 9, 1886.
[94] Burchard, H. C., op. cit. for 1882, p. 502, 1883.
[95] Hollister, O. J., The mines of Colorado, pp. 308-320, 1867.
[96] Raymond, R. W., op. cit. for 1870, p. 332, 1872.
[97] Burchard, H. C., Report of the Director of the Mint upon the production of the precious metals in the United States during the calendar year 1882, p. 503, 1883.
[98] Raymond, R. W., op. cit. for 1871, pp. 364-365, 1873.

[99] Burchard, H. C., op. cit. for 1882, p. 504, 1883.
[1] Raymond, R. W., op. cit. for 1872, pp. 299-300, 1873.

in gold. The mine has a shaft on it 140 feet in depth, with levels run at regular intervals, and a large amount of backs yet untouched.

The Berry tunnel, owned by Captain Breece, has a large crevice and assays well in gold, silver, and copper, but the ore can not be treated advantageously by the common mill process.

The Five-Twenty has a 7-foot crevice, with a pay streak 14 inches wide, which yields $20 per ton.

There are a great number more of these gold lodes, which only require development and means of reduction to make them valuable. They are all situated in quartzite [in his report for 1871 Raymond says "in granite"], dipping west.

The total yield of the placers in this gulch has been estimated at $3,000,000, and there are still some 5 square miles of gravel deposits which will pay $5 per day.

The Homestake mine, near Tennessee Pass, has been extensively worked this summer by Mr. McFadden. This vein has been traced for about 3,000 feet and has an average width of 16 inches. The mineral is principally galena, with some copper and iron pyrites. About 50 tons have been sold this summer, yielding $125 per ton.

In Iowa Gulch Breece, Paul & Co. are constructing a ditch 14 miles in length and 6 feet wide to convey water to these very rich diggings.

The gulch is 4 miles long and 40 to 150 feet wide, with an average grade of $3\frac{1}{2}$ inches to 12 feet. The depth to bedrock is about 12 feet, and the whole gulch is rated as 12-ounce diggings.

In his report for 1873 Raymond says:[2]

Although but few placer mines have been steadily worked in this county during the year, the yield is somewhat larger than in the previous year.

In the Printer Boy vein, in California Gulch, several very rich strikes have been made during the year, one of which, in August last, furnished $3,000 of leaf gold in a small pocket. The mine is opened to a depth of over 300 feet, and even at this depth the ore is still decomposed brown quartz. Horizontally the mine is explored for about 500 feet. At the Territorial fair at Denver in the fall of 1873 there was on exhibition from this mine a collection of ore carrying leaf gold, which, to my knowledge, has never been surpassed in this country except by the Cederberg mine in California. The great mass of the ore from this vein, which is about 6 feet wide, carries about $30 in gold per ton, which is easily extracted in the stamp mill. The mine has produced very regularly about $9,000 per month throughout the year.

The Gray Eagle lode, in the same gulch, is also reported to have been doing well.

The Homestake, in Tennessee Gulch, is a vein containing ores carrying from 30 to 60 per cent of lead, from 200 to 250 ounces of silver. Such ore has been shipped in fair quantity (considering that the vein is yet in the course of development) to the smelting works at Golden.

The mine is opened by a cross tunnel 75 feet in length, from which levels have been run along the vein both ways, respectively 200 and 275 feet long. A second cross tunnel is in the course of construction, 75 feet below the first. This vein is remarkable because it contains in its ore an arsenical nickel mineral, which, from the small specimens shown me, I presume to be gersdorffite ($NiS_2 + NiAs_2$). This mineral has caused the formation of considerable nickel speiss at the Golden smelting works.

In the western portion of Lake County, the country has been much overrun by prospectors during the summer and fall, and at the headwaters of the Arkansas a very large number of lead veins have been found and located.

According to the Georgetown Mining Review the total output of Lake County in 1873, in gold and in value of ores shipped, was $225,000. To this should be added $5,000, the product of some small diggings on Gunnison River, in the northwest part of the county.

In his report for 1874 Raymond says:[3]

The operations in this river [the Arkansas] and its branches have been confined to sluicing and hydraulic mining in California, Colorado, Cash,[4] Chalk, Iowa, and Clear creeks. The bed of the river is also being worked to a moderate extent. The result of the season's work has been about $80,000 worth of gold, although the supply of water was small and the summer shorter than usual. Most of the old ground is now considered as worked out, and, as a consequence, attention is given to the upper levels of the creeks mentioned and to the bars of the main stream. New ditches are being dug to carry water to the heads of the richest gulches, and large areas of ground in the Arkansas Valley proper are being taken up and prepared for work next season. The success which has attended the introduction of the automatic boom (described in my last year's report) in the Blue [River] Valley has induced several parties to try this method on the poor grounds in the Arkansas, and it is confidently expected that the result will be a large increase in the gold washed during 1875. The most important new enterprise is that of the Oro Ditch Flume & Mining Co., which has already built a ditch 9 miles long, tapping the main stream near its head and carrying sufficient water to wash the upper ground on most of the eastern tributaries of the Arkansas. * * *

"[The Upper Arkansas] district embraces about a dozen gulches among the headwaters of the Arkansas, carrying both gold and silver veins. Its production last year was small, not over $145,000, which was derived mainly from the gold mines. There is as yet no market for silver ore, and most of that which has been produced has been shipped to Golden City for treatment. The belt at the upper end of the valley is undoubtedly a reappearance of the same belt that courses through Montezuma and Breckenridge and is almost exclusively argentiferous. Lower down gold veins appear, which are most strongly developed in California and Colorado gulches. The occurrence of ores of nickel and cobalt in the ores of the upper valley is perhaps the only matter of mineralogical interest that has been shown by the year's operations. These metals are found with argentiferous galenas, but in small quantities. From a number of tons of Homestake ore, treated by Mr. West, at Golden City, about 500 pounds of nickel speiss was run out, carrying from 2 to 12 per cent of that metal. The great distance to the mines from railroads prevents this metal from affording any profit to the miner.

Burchard[5] gives $223,503 as the value of gold and silver produced in Lake County in 1874.

In his report for 1875 Raymond says:[6]

The valley of the upper Arkansas is the only mineral district in this large county yet opened by the miner, if I except the workings on the headwaters of the Gunnison and Uncompahgre, which are included in the present report under the head of the San Juan region.

The industry is mainly placer mining, though a few quartz ledges are operated. Mr. Van Wagenen gives the product of the county as follows:

[2] Raymond, R. W., op. cit. for 1873, pp. 308–309, 1874.

[3] Raymond, R. W., op. cit., pp. 374–383, 1875.
[4] The Cash Creek of Raymond is no doubt the Cache Creek of Hollister and of the U. S. Geological Survey's map of the Leadville quadrangle.
[5] Burchard, H. C., op. cit. for 1882, p. 504.
[6] Raymond, R. W., op. cit. for 1875, pp. 313–317, 1877.

	Gold coin	Currency
Quartz gold from Printer Boy	$25,862	$30,000
Quartz gold from other mines (a little from Chaffee County	2,586	3,000
	$28,448	
Gulch gold (Chalk Creek and California Gulch)	36,207	42,000
Silver ore (upper Arkansas River)	21,551	25,000
	86,206	100,000
In coin		ᵃ 86,200

ᵃ Of this total product the Leadville region produced $64,650 and Chaffee County $21,550.—C. W. H.

Mr. Maurice Hayes, the Territorial assayer at Oro City, gives the production of the county as follows:

	Gold coin	Currency
Gold from the gulch mines and Printer Boy mine	$82,707	$95,940
Silver, copper, and lead from the lodes	21,551	25,000
	104,258	120,940

The last item, Raymond says, probably includes ore mined but not treated. He continues:

* * * There is still no market for silver ores in this county, and this circumstance hinders all mining enterprises in that direction. Several attempts are making to correct this disadvantage, and it is probable that in another year the Arkansas Valley will be supplied with one or more works. In California Gulch Captain Breece is building chlorination works, under the superintendence of Mr. R. Keck, and expects them to be in operation by February. At the mouth of the gulch another company has put up smelting works, which have, however, not been in operation, and in Chalk Creek Messrs. Chapman & Riggins are building works of 10 tons' capacity to treat ores from their own mines. The plant of this establishment comprises two roasting furnaces and one blast furnace. It is not expected to be in operation before the summer of 1876.

In the mines there is but little new. Those which were worked last year are, with a very few exceptions, still in operation and attaining just the degree of success which encourages a continuation of labor. The experience of years in the gulches emptying into this valley shows that the ores are, as a rule, very low in grade and require more capital than the prospector and the miner generally possess to develop them to a point where they can steadily pay.

The Printer Boy was worked steadily during the latter part of the year. Developments have not been pushed much below the depth at which work was last stopped, but the mine has been explored laterally with fair results. The yield of the mine was about $30,000. A full account of the workings is given below.

The Berry tunnel, which runs upon a lode of the same name, is one of the important enterprises of the county. The lode has a width of 8 to 12 feet and carries copper and iron pyrites, partly rich in gold and silver. As far as opened the vein shows plenty of ore but of low grade. The chlorination works spoken of above are being built for the special purpose of treating the material from the mine, though if they prove successful they will be enlarged for custom work.

Excepting a little "gouging" done by lessees, the Homestake, which at one time was considered one of the finest mines of Colorado—the best certainly on the Arkansas—has been idle during the year. Differences among the owners and disappointment in deep developments have been the cause. The property has now, however, fallen into the hands of one of the former owners, Capt. James Archer, and it has been the expectation to reopen it during the winter. It was currently reported some time ago that the mine was exhausted, but this

is not the case. At present there is but little ore exposed in the workings (which have been quite extensive), but this is the result of poor management in handling the mine. It will be remembered by readers of former reports that it was from the ore of this mine, treated at the Golden City works, that so much nickel was taken. * * *

Of the other mines in the Arkansas Valley that have been worked more or less, may be mentioned the Yankee Blade [Chaffee County], Pilot, Five-Twenty, American Flag, Mike, Gray Bird, Hidden Treasure, and Mary Francis.

The placer mines of the Arkansas Valley have produced about $42,000, considerably less than last year. California and Cash [Cache?] creeks have been the most actively worked; but even in these localities operations have been languid and intermittent, little more being done than was required to procure gold for the immediate personal necessities of claim owners.

In regard to California Gulch, which is by far the most important mining camp of Lake County, I have received later notes, which I owe to the courtesy of Mr. Rudolph Keck, M. E., formerly Territorial assayer at Fairplay and now engineer of the beneficiating works connected with the Berry tunnel enterprise.

Near Upper Oro City California Gulch runs east and west. On the mountain side south of it several parallel lodes running north and south have been discovered in porphyry, the most noted being the Printer Boy, which has produced at least $600,000 during the few years since its location. The vein, like the parallel lodes on both sides, is filled with porphyry, which is, however, softer than the country rock, and of a different color. It contains in very irregular distribution, nests of carbonate of lead with native gold, the latter occurring in particles far smaller than those found in the placer mines of the gulch. .In the lowest workings of the mine the same gangue material has lately been reached as was found some time ago on the north side of California Gulch, in the Berry tunnel, namely, a talcose mass of auriferous iron and copper pyrites with a little galena and tennantite. According to a certificate of the Territorial assayer of the county, a selected specimen of this ore contained 122 ounces of gold per ton. Several assays by Doctor Loescher, of the Malta Smelting Works, showed from 3 to 4 ounces of gold per ton.

The vein is opened by means of three shafts and several levels and is split in two places for distances varying from 200 to 400 feet.

Most of the mining work has been done between the main shaft and the line or middle shaft, in the split highest on the hill and in the eastern branch of the vein. It was here that rich nests of carbonate of lead, filled with leaf gold, were repeatedly found. The thickness of the vein proper and its branches varies between 1 inch and 4 feet but may be called on an average 7 inches, the eastern branch averaging 6 and the western 8 inches. Besides the two splits referred to, the vein shows the peculiarity that it is, from the surface down to a depth varying between 100 and 200 feet, filled with cross seams in the porphyry mass, which are from 2 to 3 feet thick and cut off abruptly by the steep eastern wall, while on the western wall they often continue for a short distance outside of the vein. They are filled with the same auriferous ore as occurs in the vein itself, only of different color and hardness. In addition to this the gangue mass, as far as its contents of gold and the differing hardness are concerned, shows a diverging vein system within the fissure from the surface toward depth, something like the spread fingers of a hand held downward. Whenever such soft veins are joined by cross seams the richness of the ore is said to be greatest. The inconsiderable difference of outer appearance between the porphyry of the walls and that of the vein matter, which can be distinguished with still less certainty in the comparatively dark workings, renders it often very difficult to follow the real ore deposit.

The line or middle shaft has been sunk on the line between two claims, the upper one of which belongs to the Philadelphia & Boston Gold Mining Co. and the lower or northerly one to a few inhabitants of the vicinity. The latter is, however, leased to the company just mentioned, and it is here that the rich pyrites spoken of above has recently been found. It is to be regretted that for the present at least it can not be extracted, because without powerful pumps the water struck at the same time can not be overcome. The material is much desired and needed by the Malta Works as a flux.

The ore in the western branches of the two splits is decidedly softer than that in the eastern ones, but so far it has not shown any such rich pockets as the eastern upper branch between the line and the main shaft. The eastern lower branch between the line and lower shaft has so far not been developed. At the lowest depth, just before the rich pyrites was struck, the contents of the ore in gold were very small. At the same time it must be remarked that the vein above this point is by no means exhausted, and, considering the former carelessness of management, this field is very promising. A little over 100 feet deep in the main shaft a mass of boulders, with a little iron pyrites and fine gold, was found, which Mr. Keck thinks may either be taken for the bed of a former stream or for the remaining moraine of a former glacier. I do not find it necessary to adopt either hypothesis. The presence of rounded boulders, unless they are clearly of a material different from the country rock, may be the result of attrition and water between the vein walls. This phenomenon is expressly considered by Von Weissenbach, in the classification of "veins of attrition," contained in his "Theory of veins," as published in Von Cotta's "Gangstudien."

At the depth of 200 feet in the line-shaft and of 100 in the main shaft the cross seams mentioned above were no longer met with, and south of the latter the vein is not split at all. At this point, however, so little systematic work has been done that no conclusions as to increase or decrease of richness can be drawn from the altered geological conditions. Indeed, little systematic work has been done on any part of the whole vein. Former operations were principally confined to robbing the rich pockets, while good milling ore was left standing.

The gold contained in the pyrites just discovered, although it can be partly washed out, can not be directly amalgamated, behaving in this respect like that in the ores of the Berry tunnel.

Among the veins running parallel with the Printer Boy, the Five-Twenty is at present the most promising. The ore from this mine yielded in the battery alone 8 ounces gold per cord, or about 1 ounce per ton (8 tons to a cord). The mill gold of these veins is usually worth $15, coin, per crude ounce, while the wash gold of the gulch is worth from $17.75 to $19.

Of the production of Lake County during 1875, now estimated at over $120,000, currency, three-quarters are said to come from California Gulch.

The placer mining of the gulch, an industry which has now been in existence for 16 years, is really a still worse robbery of the gold deposits than that carried on so long in the veins. The gold occurs in these placers, notably in the upper part of the gulch, in two different layers. The upper one consists of gravel and conglomerate and is the deposit which alone has been washed; the lower one consists of so-called cement, a hydrated oxide of iron combined with a feldspathic mass to a very hard layer, which contains not only fine and very fine gold dust but also coarse gold. As the hardness of this material precludes washing without a preliminary crushing, this layer is to-day virtually virgin ground, a fact which is the more remarkable since assays of average samples have never yielded less than an ounce of gold per ton.

At the Berry tunnel Mr. Keck has completed his beneficiating works as far as was intended for the present. While up to the end of the year only the common ores of the Berry tunnel (talcose gangue with iron and a little copper pyrites) were subjected to the process [7] employed, the mine has been better developed, and now there is a considerable quantity of more solid and richer ore ready for extraction, similar to the pyrites described above in connection with the Printer Boy. This ore is now assorted by hand, dried, in order to stamp it without water, and subjected to the rest of the beneficiating process. It is to be regretted that for the amalgamation of the residues, arrastres only are at Mr. Keck's disposal, since the gold is in this way not extracted as perfectly as could be done in pans.

The Malta Smelting Works are built on the slope of a hill and intended for lead smelting. Besides the necessary buildings and apparatus for crushing, sampling, storage of wood, etc., they contain a long reverberatory furnace for roasting (without a hearth for slagging purposes), a shaft furnace of the Kast pattern of a capacity of at least 15 tons per day, and an English cupelling furnace. The blast is furnished by a Sturtevant blower No. 4. The establishment impresses the visitor favorably, and it is only to be hoped that in the coming summer the argentiferous lead mines of the vicinity (so far containing principally cerusite) may be more energetically attacked than has been the case heretofore, in order that the metallurgical enterprise may not be crippled (as so many in Colorado have been) by the lack of material suitable to the processes employed.

Burchard [8] gives $104,258 (gold and silver) as the output for 1875, following the Central City Register.

For 1876 Burchard [9] gives gold, silver, and lead $90,900, which may be divided into $30,000 placer gold, $30,000 lode gold, $30,000 (at $1.2929+ per ounce) silver, and 15,000 pounds of lead.

For 1877 and 1878 the figures given in the table represent estimates made on the basis of the value of the total production as given by the Director of the Mint [10] and by the table in Mineral Resources [11] showing the production of bullion at Leadville. The figures for lead, silver, and gold in the table in Mineral Resources represent the contents of the lead bullion produced at Leadville, for which no tonnage of lead bullion or original ore tonnage is given, and the figures for tonnage of ore and value of ore shipments represent only the ore shipped out of Leadville.

For 1877 to 1884 the figures for lead represent estimates made on the basis of the total production of lead in the State as shown in Mineral Resources and also the production of Leadville as given by the Leadville Herald and Leadville Democrat (later the Herald-Democrat), as well as the production of other districts in the State.

For 1879 [12] Burchard gives figures for the calendar year as distinguished from the fiscal year, to which $275,000 in silver (coining value) is added from Gunnison County.

For 1880 Burchard says: [13]

---

[7] The process followed consists in stamping, dressing on fine-grain jigs, roasting with salt, lixiviation, precipitation of copper and silver, and amalgamation of the auriferous residues.—R. W. R.

[8] Burchard, H. C., op. cit. for 1882, p. 504, 1883.

[9] Burchard, H. C., op. cit. for 1882, p. 504, 1883.

[10] Burchard, H. C., op. cit. for 1883, pp. 504, 505, 1884.

[11] U. S. Geol. Survey Mineral Resources, 1885, p. 251, 1886.

[12] Burchard, H. C., op. cit. for 1880, p. 156, 1881.

[13] Burchard, H. C., op. cit. for 1880, pp. 134-143, 1881.

Production of bullion and shipments of ore were made by firms as follows: Grant Smelting Co., La Plata Mining & Smelting Co., Billing & Eilers Smelter, Cummings & Son's Smelter, Eddy, James & Co. (ore shippers), Harrison Reduction Works, Ohio & Missouri Smelter, M. E. Smith & Co. (California), Elgin Smelting Works, American Milling & Smelting Co., Malta Smelter, Gage, Hagaman & Co., Little Chief Smelter, Leadville Smelter, Robert E. Lee [mine] (shipped to Golden), Taylor-Brunton Stamp Mill, Tabor Milling Co., Colorado Prince Stamp Mill, and Oro Stamp Mill. * * *

Detailed information on the Evening Star, Yankee Doodle, Little Giant, Morning Star, Dunkin, Iron, Fryer Hill mines, Hibernia, Colorado Prince, Highland Chief, Matchless, Robert E. Lee (with crushing works), Amie, and Climax. * * * Limerock from the Pendery-Glass and iron ore which the Amie is producing and selling for purposes of flux.

For the calendar year 1880 as distinguished from the fiscal year, on page 157, Burchard gives $58,000 in gold, but on page 35 he gives $104,014, which seems as reasonable.

In his report for 1881 Burchard says:[14]

The following smelting and reduction works and mines of Lake County have returned reports of production: American Co., Billing & Eilers, Cummings & Finn, Eddy & James, Elgin Co., Fohr & Bunsen Bros., Harrison Reduction Works, Ohio & Missouri, Tabor, Taylor-Brunton, Grant Smelting Co., La Plata, Shield's Mill & Mine, Dry Placer Amalgamating Co., Annie Consolidated, A. Y., Catalpa, Climax, Crescent, Highland Chief, Iowa Gulch, Iron Silver, Dunkin, Long & Derry, Matchless, Robert E. Lee, Small Hopes, Wolftone & Agassiz, Chrysolite, Consolidated Pigar, Denver City, Silver Cord, Carbonate Hill, Colorado Prince, Evening Star, Hennett, Hibernia Consolidated, Leadville Consolidated, Little Emma, Silver Wave, Artora, Big Pittsburgh, Henriette, Little Chief, Miner Boy, Morning Star Consolidated, and Little Pittsburgh. * * * J. B. Grant & Co. * * * at present time have only five furnaces in blast, but they are all of the largest patterns, and during the past month reduced 4,300 tons of ore. * * *

The American has * * * four furnaces in blast. * * *
The Harrison Reduction Works * * * during the preceding 90 days have been engaged in enlarging and improving the establishment, and the product given above is the yield of only one furnace, which has been in blast since September 6. In a few days, however, another large furnace will be blown in, and ere long two more furnaces will be added. During the summer months this establishment has added two 50 or 60 ton furnaces, and rebuilt its old ones, giving it four splendid furnaces. * * *

Back of the smelter, one of the most complete sampling mills in the State has been erected, supplied with crushers, rolls, sampling mills, and other modern appliances for reducing and sampling ores. * * *
A. R. Meyer & Co. * * * shipped 594¾ tons * * * to the Kansas City Smelting & Refining Co.
The Miner Boy * * * quartz mill, * * * 9 tons a day average. * * *
The Shields mill, situated in Colorado Gulch, * * * roasting cylinders.

In his report for 1882 Burchard says:[15]

The Chrysolite Consolidated Mining Co. * * * was organized in 1879 with a capitalization of $10,000,000. Previous to the property passing into the hands of the company it had produced about $1,000,000. * * * The mine has paid in dividends to its original owners and to the company over $3,100,000 in clear profits. * * * According to the annual report * * * 10,774 tons of ore sold netted $401,816.89, or $37.29 per ton. * * *

During the year the company erected an experimental mill for concentrating low-grade ores, and although the machinery is in very imperfect running order it nevertheless demonstrates that the Leadville ores can be concentrated with sufficient success and economy to insure a profitable enterprise. The percussion tables and sizing machine have not been in use so far, owing to some irregularity in the machinery, and the ore was passed from the stamps directly over the Frue vanner. The result was concentrates running 38 ounces in silver and 21 per cent in lead from iron ore that runs only 4 ounces in silver to the ton and a trace of lead. With the addition of sizing and the further use of percussion tables, the result * * * is expected to be much more satisfactory. It is understood that during 1883 a large concentrating mill constructed on practical working principles will be erected.

The Robert E. Lee mine * * * is one that has few parallels in the records of mining in the United States. One instance proving the richness of its ore, which consists of chlorides of silver, may be stated. On January 14, 1880, ore was taken out to the value of $118,500 in the space of 17 hours. During the month the production amounted to over $300,000. Since development first began the value of the ore sold has reached the total of $3,000,000. * * *

This company was the first to introduce sampling works of [its] own as an auxiliary to successful and economical mining. * * *

The annual report of the Catalpa Co., for 1882, shows that the total receipts from silver, iron, and interest accounts were $95,450 and expenditures $58,332. * * * During the year, 5,202 tons of silver ore and 2,176 tons of iron [ore] were mined. * * *

During the year considerable attention was attracted to Sugar Loaf Mountain, one of the foothills of Mount Massive, and at the head of Little Fryingpan Gulch, about 2½ miles back of Soda Springs. In Little Fryingpan Gulch the Shields, Venture, and T. L. Welsh mines have become known as producers, but it was not until last fall that the discoveries were made that have attracted so much attention to Sugar Loaf Mountain.

The developments in the Shields mine, in Little Fryingpan Gulch, were sufficient to warrant the erection of a 10-stamp mill, which, under the management of Maj. A. V. Bohn, has been operated successfully on ore from the Shields, Venture, Welsh, and other mines that have been opened in the neighborhood.

Among the most important of the discoveries that have been made on Sugar Loaf Mountain, and one in which the richest ore has been found, is the Birdie R. * * * Shipments were begun from this mine at a depth of 13 feet, the ore going to the Shields mill and running from 68 to 109 ounces in silver. * * * The Orinoco, Dinero, Gunnison, Juliet, Whittlesey, and Sawyer are promising discoveries. * * *

The Grant Smelting Co., started in 1878, built two furnaces in that year and made several additions the following year. In January, 1880, * * * a number of very important changes were made, including the construction of new furnaces and the introduction of large new engines. * * * Two hundred and twenty-five tons of ore per day were being treated until May 24, 1882, when the works were enveloped in flames; * * * nothing but the smokestacks were left. * * *

Elgin Smelter, which had been idle during the early part of the year, was leased by Grant Smelting Co. * * * From July until the furnaces were blown out on December 28, the two stacks were kept constantly going. * * *

14 Burchard, H. C., op. cit. for 1881, pp. 403-418, 1882.
15 Burchard, H. C., op. cit. for 1882, pp. 481-506, 1883.

La Plata Mining & Smelting Co., the first of the large buildings, was built in June, 1878, by Berdell & Witherell, who were then engaged in the business of sampling and crushing ores. A furnace of 30 tons capacity was started in October, 1878, and the second one built in February, 1879. On June 14, 1879, the new company purchased the entire property and took possession of the smelting works, together with 24 acres of ground and about 20 buildings located just below the city of Leadville. At the same time three claims in California Gulch, known as the La Plata mines, were transferred. The new company constructed a third furnace, which was put in operation on August 1, and the fourth was started on the 22d of December, 1879. * * *

Two new furnaces were built during the year 1881, one started in April and the sixth put in blast on the 13th of December. All the six furnaces are now running and reducing about 160 tons daily. Several large additions to the buildings have also been made, and a large roaster has been placed in service during the year. * * *

American Smelter, * * * one of the most successful smelting companies in California Gulch, in addition to the treatment of ores from the mines of Leadville, owned by outside parties, * * * handles the output of several mines which belong to it. There are four large [furnaces] and one small furnace, all of which are kept constantly running. * * *

Arkansas Valley Smelter, at the western end of California Gulch, * * * formerly known as the Utah or Billings & Eilers, and it is recognized among the most complete and successful works about Leadville. In the latter part of 1881 Mr. Gustav Billing purchased the entire property and conducted [it] as sole owner until the beginning of 1882, when the A. R. Meyer & Co. Sampling Works incorporated with Billing & Eilers, forming the Arkansas Valley Smelting Co. * * *

The Harrison Reduction Works, which was [one of] the pioneer smelting works of Leadville, were entirely rebuilt during the year. Two years ago there were two small furnaces, which were closed down in January. The company then commenced remodeling the entire works. Large sampling works were built, new engines were added, and four new furnaces put up. Two of these new furnaces were put in blast on the 7th of September, 1881, and have since been kept in blast. During the past year the works have been further improved and enlarged * * * . The management has in contemplation the erection of a 100-ton furnace. * * *

Eddy, James & Co., composed of the same members as the Grant Smelting Co., * * * does an immense business in buying ores and bullion. * * * In addition to supplying the Elgin Works, the firm has sent large quantities to the Grant Works at Denver and also to the works at Pueblo. * * *

A. R. Meyer & Co. * * * formerly conducted a business similar to Eddy, James & Co., but in March, 1882, in consequence of the incorporation of the Arkansas Valley Co., attention has been confined principally to smelting. * * * The sampling works are now used for the private accommodation of the smelting company.

Cummings & Finn * * * [works] have had large additions during the past year and are now among the largest and most successful works about Leadville. There are six large furnaces, all in successful operation. New dust chambers have been introduced, and are giving the best of satisfaction. * * *

Leadville Gold & Silver Mill Co., * * * [formerly] the amalgamating works of Taylor & Brunton. Neither [member] of the old firm is any longer interested. The works are largely automatic and contain a number of improvements in the machinery. * * *

Shields mill, located in Colorado Gulch, at the mouth of Little Fryingpan, * * * production * * * silver bars. * * *

The old Oro mill has not been worked steadily during the past year, but at intervals during the last three months the machinery has been in motion.

In his report for 1882 Burchard also gives a list of producing mines in the county.

In his report for 1883 Burchard [16] gives a list of Leadville mines, their capital stock, development, production, and other information, and adds in regard to smelting operations:

In 1879, when the majority of smelters operated had only two or three 42-inch furnaces, the expense for labor was $5.20 per each ton of ore treated, and in 1882, with four to five large furnaces, it was reduced to $2.80 per ton.

The losses of metals in reduction by the Leadville smelters average about as follows: In silver, 3 to 4 per cent; in lead, from 13 to 15 per cent. The excessive loss in lead is due largely to the low per cent of this metal in smelting charges, being sometimes as low as 8 per cent.

The smelting and milling facilities of Leadville consist of six smelting establishments containing 24 furnaces of large capacity. From 23,000 to 24,000 tons of ore are treated monthly at these works. About 5,000 tons of coke and many thousands of bushels of charcoal are consumed monthly. There are about 1,000 men employed.

The American Smelting Works contain four large furnaces and treat about 4,000 tons per month.

The Arkansas Valley Smelter contains five furnaces, 4 feet 6 inches by 6 feet 6 inches at the tuyère, and measuring 12 feet from the tuyère to the charge door.

The Fryer Hill, formerly the Cummings & Finn Smelter is running three furnaces and treats about 100 tons per day. The establishment was purchased recently by members of the Omaha & Grant Smelting & Refining Co.

The Elgin Smelting Works were leased by the J. S. D. Manville Smelting Co. in the early part of the year. Considerable ore accumulated by April 24, when the first furnace was blown in. * * * More ore is coming in than can be reduced by the two medium-sized furnaces at the works, and a third is being erected.

The Harrison Smelting Works has four large furnaces, which turned out 960 tons of base bullion in November. The gross production of the Harrison for the year was greatly reduced by a disastrous fire, by which the furnace building was destroyed. The establishment was again closed down. The Harrison Reduction Works is the only one that has provided roasting furnaces for the preparation of refractory sulphide ores that are produced by Leadville mines.

The La Plata Smelting Works has five furnaces in blast. One is new and one of the largest in the State. New dust chambers, fume stack, and other additions have been made.

In reviewing the mining operations Burchard adds:

Fryer Hill.—The Chrysolite Consolidated Mining Co. has produced, since the organization of the company in 1879, about 77,027 tons of ore, containing something like 4,302,993 ounces of silver and 29,525,050 pounds of lead. * * *

The Little Pittsburgh Mining Co., another property that was supposed to have been worked out, has during the year shipped 11,500 tons of ore to the smelter. * * *

The Amie, which is consolidated with the Deer Lodge, possesses about 15 acres adjoining the Little Pittsburgh property. During 1879 and 1880 this property produced about $700,000 worth of ore. Since 1881 the property has been leased.

16 Burchard, H. C., op. cit. for 1883, pp. 236, 237, 238, 240, 332-368, 1884.

The Dunkin Mining Co., although not one of the largest producers, has been a steady one, having paid to date $210,212 in dividends. * * * The ore is a galena. * * *

The Climax * * * lower level this season produced 4,000 tons of high-grade ore. The upper levels have been worked for three years. On the 60-foot level the grade of ore has varied from 20 ounces to 500 ounces; its average value, including 25 per cent lead, has been $35 a ton.

The Matchless * * * ore is dry, heavily charged with chlorides. * * * There are 3 feet of ore, prolific in horn silver. * * * The Matchless mine has now produced and shipped ore to smelters to the value of $1,175,000, exclusive of smelting and milling charges.

Yankee Hill.—* * * The Luzerne * * * ore consists of sand and hard carbonates and ocher, in addition to carrying an appreciable amount of lead. * * *

The Small Hopes Mining Co. consists of a consolidation of the Small Hopes, Gone Abroad, Ranchero, Result, Robert Emmet, and Forest City claims. * * * The ore body is large and will average $110 per ton in silver, very little lead being present. * * *

Carbonate Hill.—The Henriette & Maid of Erin * * * shipments in 1883 have averaged 800 tons per month. The galena shipments average 200 ounces silver per ton. * * *

In the Morning Star main shaft * * * the hard carbonate streaks * * * are found containing considerable chlorobromide of silver, in large flakes, sometimes covering the whole fracture. * * * There are two classes of ore, the one a very heavy lead sand, assaying from 65 to 70 per cent lead and 16 ounces in silver; the other lower in lead but higher in silver. The latter is more siliceous, which kind does generally exceed the pure lead sands in the contents of silver. Assays of this second class run up to 90 and 150 ounces. There is no gold in this mine to speak of.

Production Morning Star Consolidated Mining Co.

1883	Tons
Ten months, net	10,288
Two months, estimated	2,057
	12,345

Average, 20 ounces silver and 34.20 per cent lead.

1882	Tons
Twelve months	19,060

Average, 30.25 ounces silver and 38.33 per cent lead.

The Evening Star * * * has a good record, and its shipments of carbonates, from a far more limited area than the Morning Star, have been astonishing. * * *

Production of the Evening Star, 1879–1882

Year	Tons	Average per ton	
		Ounces silver	Per cent lead
1879	1,980	54.80	22.90
1880	6,200	49.30	25.00
1881	15,625	50.80	28.40
1882	27,618	43.30	21.50

The Glass-Pendery mine, which consists of two claims, formerly known as the Glass and Pendery, was organized in 1879. Very high grade chloride ores were discovered in the Pendery, and from their proceeds a dividend of 10 cents a share was paid. Since then—for the past four years—the mine has not been profitable. This year * * * the output has been about 10 or 12 tons a month of chloride ore, worth from $200 to $400 a ton.

The Crescent * * * old workings, the ore is a sand carbonate of a very good grade, in addition to large bodies of iron ore.

The Catalpa has been producing pretty regularly during the year, the output being the result, mainly, of tribute workers. * * *

The Yankee Doodle * * * product has continued steadily at about 175 to 200 tons of ore during the year. * * *

The Aetna * * * ore is an excellent fluxing ore and carries from 2 to 10 ounces silver per ton.

The Big Chief and Castle View mines, property of the Big Chief Mining Co.; * * * ore runs 50 to 60 per cent lead and 6 to 10 ounces in silver and is a fine gray sand. The iron also runs well in silver.

The Adams Mining Co. was incorporated in December, 1883, and is a consolidation of the Saint Bernard and Brookland Mining Cos. * * * The property consists of the Clontarf, Brookland, and Moyamensing claims. * * * The Clontarf has been a regular shipper of considerable quantities of excellent lead sand all the year past. The sump of the shaft is in about 8 feet of solid carbonates; below it is iron ore through which the shaft is now continued. It is a favorable deep-black iron, with some manganese. * * *

The Wolftone & Agassiz Consolidation owns the Wolftone, Agassiz, and other claims on Carbonate Hill. * * * Only the upper stratum of ore has been worked during 1883, and this on lease. * * *

The Gone Abroad: * * * present resources are confined to large bodies of iron ore, with occasional pockets of chlorides. The first-class material has an average value of $125 per ton, the second-class from $25 to $35 per ton.

Iron Hill.— * * * The production of carbonate is and will remain for several years preponderating over that of sulphide, but in the course of time the latter will gain the ascendancy.

The Iron Silver Mining Co. has been one of the largest producers in Leadville since 1878. It included nine claims, extending over a mile in length, and stretched nearly from Stray Horse Gulch on the north to California Gulch on the south. They were the Iron, Iron Hat, Porphyry, Dome, Rock, Stone, Lime, Bull's Eye, and Law. Since then the company has purchased the Luella, Tucson, half of the Moyer placer, and other properties. * * *

The Silver Cord combination, * * * consisting of the Silver Wave, Silver Cord, Cleora, Delta, St. Theresa, Holy Terror, Minnie Lee, Eagle, Ottawa, A. P. Willard, and south half of the Rubie, east half of the Bull's Eye, and some minor fractions. * * *

The Smuggler * * * ore ranges from 40 to 60 ounces silver per ton and contains from 35 to 50 per cent lead.

The Iron Hill Consolidated Mining Co. * * * are the Forfeit, Ocean Wave, Little Missouri, Norman Boardman, White Cap, and Imes. * * *

The A. Y., * * * one of the oldest in Leadville, was located in July, 1876, by A. Y. Corman and others, and passed into the hands of its present owners, Mr. Samuel Harsh and associates, in 1879. It consists of one full claim, patented. The first shipments were made in July, 1880, and in the following month regular shipments were begun; * * * the owners continued to market something like 200 tons a month in the last four months of 1880. In 1881 and 1882 the output did not vary much from 50 tons a day. In 1883 the shipments have been 9,284 tons. There are in the bins 500 tons of sulphide ore, all sorted and ready for shipment, but the amount of carbonate ores is not as large as at times it has been. * * * The ore of the A. Y. is of two kinds—carbonate ore, which for the year has averaged 21 ounces in silver and 21 per cent in lead; and sulphide ore, of which the last shipments ran from 50 to 66 ounces in silver and from 15 to 33 per cent in lead. * * *

The Collateral mine in California Gulch has been a regular shipper during the year. The ore is a fine lead carbonate, and runs about 35 ounces silver and 40 per cent lead per ton. * * *

The Ruby mine, * * * on which work was commenced in the winter of 1879, * * * so far has shipped but limited quantities of ore. * * *

Rock Hill.—The La Plata Mining & Smelting Co. owns the La Plata mine * * * the ore is a fine sand carbonate, carrying from 15 to 30 ounces per ton and about 40 per cent lead. * * * The Gilt-Edge mine, owned by the Elgin Mining & Smelting Co., is in California Gulch, and is one of the largest producers in that locality. It is operated entirely by lessees. * * * The Florence mine, consisting of two claims, the Florence and J. D. Ward, * * * has been a prolific producer of lead carbonates, running well in silver and gold, since 1878. * * *

Breece Hill.—* * * Quite a large territory, with numerous developments, but as to pay there is only the Breece Iron mine, which has been producing * * * nothing but iron, with a very high percentage of metallic iron and very little silver. There is no demand for it at present, as the Fryer Hill and Stray Horse mines yield a very good iron, with more silver than the Breece iron. In 1879 and 1880 the excitement over the large bodies of sand and hard carbonates in the Highland Chief and Highland Mary caused a ready transfer of interests and much prospecting, but the returns did not come. * * *

The Standard Mining Co. owns the St. Louis and Black Prince mines; the St. Louis is a consolidation of what was formerly known as the Colorado Prince and Miner Boy mines. The ore is principally quartz, carrying gold which is found in veinlets ranging in width from 2 inches to 2 feet. The lower levels of the mine show an extensive body of sulphide ore, containing iron, copper, lead, zinc, silver, and gold. * * * A contract has been recently let to build a large concentrator for the treatment of the low-grade ore. * * *

The Little Jonny mine shaft has a depth of about 120 feet, with two stations and levels at 82 and 112 feet, respectively, from the surface. At the first level, a drift extends to the southward for about 50 feet, where a fine body of ore is encountered, possessing a thickness of nearly 4 feet. From this point an incline follows on the dip of the ore in a southeasterly direction for 50 or 60 feet, showing from 2 to 4 feet of fine sand and hard carbonates, running 40 to 50 per cent lead and 6 to 15 ounces in silver to the ton. * * *

The Little Ellen is the property of the American Mining & Smelting Co. At present the product of the mine is regulated by the requirements of the smelter. * * *

The Cleveland * * * is being explored and developed by lessees, who occasionally strike small pockets of ore. The mineral contains silver and lead and small quantities of gold; it is low grade and very siliceous. * * *

Sugar Loaf Mountain and Tennessee Park.—The mines, though not large producers, are coming to the front. * * * Among the most prominent are the Birdie R., Dinero, Gunnison, and Sundown. * * *

The Gerald Griffin mine is a new location. The developments now consist of a shaft 65 feet deep, from which 37 tons of ore have been shipped, taken out during the sinking of the shaft.

The Gunnison mine is the largest producer in this locality. The development consists of one working shaft, 215 feet in depth, and three levels. * * * The ore is a sulphuret, first class running 2,000 ounces silver and one-fourth ounce gold. The second class runs about 400 ounces silver and about the same in gold, and the third class runs from 50 to 70 ounces silver and one-tenth ounce gold per ton. * * *

The Dinero, in Sugar Loaf Gulch * * * is the oldest shipper in this locality. * * *

In his report for 1884 Burchard says: [17]

The Leadville smelters secured during the year, as formerly, the largest portion of the product of the camp.

The Harrison, with a considerable stock of siliceous and refractory ores on hand, limited its purchases to desirable smelting ores. It increased its roasting capacity by adding a new set of kilns.

The Arkansas Valley smelter, with its sampling works, handled more ore than any other in Leadville.

The business of the La Plata Mining & Smelting Co. fell off and its furnaces were not very busy, but considerable sulphuret ore was treated at its works.

The furnaces of the American Mining & Smelting Co. have been kept supplied by its own and mines which it controls. It also made shipments to its smelter in Canon City, the Royal Gorge.

The Fryer Hill Smelting Co. had two furnaces in blast all the year, and in the latter part three.

The Manville smelter had three furnaces going the latter part of the year and has built extensive roasting works.

The Omaha & Grant sampling works have not handled the usual quantity of ore.

The Colorado & Utah sampler limited its purchase principally to iron ores for shipment to Utah.

The Oro & Antioch mills were supplied with free-milling ores by the mines on Printer Boy Hill, the increased output of which also caused the erection of a new mill by the Lilian Mining Co.

Fryer Hill, the great bonanza field of the past, still retains its productive importance. The Chrysolite, Little Pittsburg, Little Chief, Climax, Amie, and Dunkin, the big producers of the past, still continue to send forth their riches. While some have been irregular in shipments, explorations have been continued on all.

The official report of the treasurer of the Chrysolite Co. for the fiscal year ending October 8, 1884, has recently been made public. During the year 3,143 tons of ore mined by the company produced $108,974.77, and 1,123 tons mined by lessees $13,317.42, a total of $122,292.19, to which are added sundry other receipts, carrying the total up to $127,482.43. The expenses amounted to $95,994.18, of which $42,095.57 were for labor, $6,289.25 for timber, $2,119.02 for ore hauling, $16,031.48 for coal, and $26,193.90 for contracts. In addition to the above expenses $4,000 were paid as an installment on the purchase of the mill, $6,546.66 for labor and supplies for the mill, and $1,162.56 for grading for the railroad branch. The cash on hand October 8, 1884, was stated to be $187,546.27, against $192,866.14 at the same time in 1883.

The general manager reports that during the year 2,420 feet of drifts, 472 of winzes and rises, and 4,605 feet of shafts were driven, a total to date of 33,565 feet. The ore sales aggregated 4,424 tons gross weight, or 3,532 tons net weight, containing 145,300 ounces of silver and 897,143 pounds of lead, for which the company received $134,086.65. Particular attention, during the first part of the year, was given to ascertaining the value of the second and third bodies of iron. * * *

A small concentrating mill was built, which is now running and returning a satisfactory profit, and an amalgamating mill was leased to work the low-grade material on the dumps by the well-known pan-amalgamation process. For concentrating the fine material, carrying both silver and lead, which is obtained from underground and the dumps, six hutches have been built and connected to the sampling mill engine. From these have been shipped 330.5 tons of ore and received $13,569.85, at a total expense (including construction) of $6,294.89, leaving a profit of $7,275.06. At present the mill

---

17 Burchard, H. C., op. cit. for 1884, pp. 220-231, 1885.

can only be run through the six clement months of the year. No crushing is done. The hard iron is saved for amalgamation. It was found by amalgamating about 1,200 tons of dump ore, containing 10 ounces of silver to the ton, that 48 per cent of the silver could be saved. The company therefore leased, with option to purchase, the Leadville Gold & Silver mill of this place for a term of three months, beginning September 1. It is now run on the company's dump ores.

The Little Chief Mining Co. has been worked during the year and paid during that time two dividends of $20,000 each. The production was 2,674 tons of ore, which, in addition to paying two dividends, has left a surplus in their treasury.

The operations of the Little Pittsburg Mining Co. during the year were chiefly directed toward the prospecting of the large iron ore shoot in the northern part of the Little Pittsburg. A new lead started from shaft No. 6, under this ore, struck water, which will need to be drained. Leases of dumps and old workings and other parts worked by the company have produced during the past year 7,200 tons of ore, for which was received $80,000 mill returns. * * *

The Amie & Deer Lodge has also been a producer of no small dimensions, 1,100 tons being the product for 1884. * * *

The Dunkin Mining Co. continued their production. * * * Over 2,000 tons of ore were sold during the year. * * *

The Climax has been quite actively worked and with good results. About 1,500 tons of ore were extracted and sold, and about 900 feet of new development work done, making in all about 4,600 feet. The value of the ore in this property varies considerably, but a fair average would be about $32 per ton. * * *

The Matchless mine has been worked continually, and the output has been equal to that of last year. * * * The total amount of development is about 10,000 feet. During 1884 about 5,400 tons of ore were produced, mostly chloride of silver, which ran $50 and upward per ton.

Yankee Hill remains in about the same condition as at the beginning of 1884. Many claims have been worked to the extent of the annual assessment, and the producing mines of a year ago have continued shipments quite regularly, the most notable being the Small Hopes Mining Co., which paid during the year fourteen dividends, amounting to $850,000; the largest amount paid by any mining company in the State. This satisfactory record is due to the extensive developments in the Forest City. * * *

The New Pittsburg Mining Co. shipped during the year about 6,000 tons of ore and made about 7,000 feet of development. Almost the entire property is worked by lessees, who, as a rule, are doing well.

On the northern part of the Big Pittsburg ore was struck by using the Stonewall Jackson shaft and drifting thence into the Pittsburg. A new shaft about 170 feet deep, with a 75-foot drift and a 200-foot 35° incline to the east, was sunk by other lessees. The ore is a heavy lead sand and carbonate, running high in lead and low in silver. The ore has been followed down on its dip to within a short distance of the Hibernia line. The body dipping thus toward it is in places from 9 to 12 feet thick. The leases have produced largely, from 15 to 30 tons a day each.

On Carbonate Hill * * * the Morning Star Mining Co. has confined its operations during the year to development work and the extraction of ore from the McHarg shaft. The explorations of late have led to the discovery of a low-grade and dry ore, underlain in part by marketable iron, of which shipments have recently been made. The minerals occur irregularly in streaks and bunches all through the contact matter. * * * The lead and dry ores from these workings run 40 ounces in silver and upward. The lower 400-foot level of the McHarg has been extended in mineral, mostly lead sand, high in lead but

low in silver, some distance toward the north, and connection has been made at one point with the Henriette, an adjoining mine on the north. In the Henriette an incline has been driven nearly to the Maid of Erin line, mostly in the same contact and similar ore, 15 feet thick, with occasional pockets of high-grade argentiferous galena. The Morning Star Co. shipped during the year 15,477 tons of ore, of the value of $173,638, exclusively from the upper contact.

Prospecting work has been continued from the upper shaft of the Evening Star mine. The ore now being extracted consists of low-grade lead and marketable iron ore. The old workings are leased and producing but little as yet. The shipments of the Evening Star mine during the year are reported to have been 3,380 tons, valued at $43,143.

The Big Chief Mining Co. has made some extensive developments during the year. The shaft has reached a depth of about 605 feet. At 468 feet the shaft pierced the regular porphyry-limestone contact and disclosed some very fine sand ore. After working out the ore in the immediate vicinity, the shaft was sunk deeper and levels driven, opening new ore bodies to the eastward. Iron was encountered which returned 6 ounces in silver to the ton, and at 9 feet below the limestone fine galena and sand carbonates were met with. The shaft was sunk about 20 inches into the ore body, when the increased flow of water enforced a suspension of work.

The ore extracted last year by the Adams Mining Co. by way of the Clontarf, came mainly from stopes to the east of the Clontarf drift. The central part of the Clontarf-Brookland and the grounds east of the Brookland drift contain the largest ore bodies opened up. Both the north and south drifts of the Brookland have ore in the breast. The north end of the Moyamensing, also the property of the Adams Mining Co., has already yielded a large amount of valuable chloride ore of the same character as that of the Forest City. The output of the company from the Clontarf-Brookland alone was 11,000 tons, netting about $220,000, of which $82,500 were paid in dividends.

The Agassiz has shipped during the year about 3,000 tons of ore, averaging from 25 to 40 ounces in silver, a fair aggregate of 30.

Iron Hill still maintains its position as the largest ore producer of the many hills around Leadville.

The Terrible Mining Co. produced from their two claims, the Terrible and Adelaide, over 150 tons of ore. The property is worked under lease, five different parties being at work on it at present.

The claims of the Argentine Mining Co. produced last year 3,580 tons of ore, containing 94,493 ounces of silver, 363 ounces of gold, and 533 tons of lead, which cost for treatment $32,642 and yielded $91,600. The ore ran 31 ounces of silver, 0.12 ounce of gold, and 17.6 per cent of lead.

The operations of the Iron Silver Mining Co. have been directed toward the exploration of the Colonel Sellers sulphide ore shoot and the cleaning up of old stopes by lessees. Although the company has remained the foremost shipper of carbonates, it has not been successful in a financial point of view. With the exception of the work on the Moyer shaft, and a limited amount of work in the Iron Silver, nearly all the old mines of the company, such as the Bull's Eye, Codfish Balls, Lime, Kaiserin, Stone and Daisy, Dome and Rock, have been worked by lessees the larger part of the year and have yielded quite handsomely. The low price of lead, however, greatly reduced the receipts therefor.

During the past year the A. Y. mine has not been an extensive shipper, but the development of the sulphide ore shoot of the mine has continued without interruption. The silver contents of the A. Y. sulphides are above the average of the carbonate ores of this section. In the sulphide shoot are extensive bodies of ore running 25 per cent lead, 22 per cent zinc, and 14 ounces of silver. This sulphide ore shoot lies to the southeast of the mine. * * *

At the Colonel Sellers, shaft No. 1 has been sunk to a depth of 400 feet, shaft No. 2 to about 500 feet, shaft No. 3 326 feet, and shaft No. 4 215 feet deep. At shafts Nos. 1 and 2 the mineral is said to be from 40 to 45 feet thick, at No. 3 20 feet, and the far northwest workings 18 feet. In the triangle formed by shafts 1, 2, and 3 there is a large accumulation of ore. From this body the bulk of the best ore is produced. The ore in the winze connecting with the drift from shaft No. 4, in the gulch, runs about 45 per cent in lead but lower in silver than the other. In the central part of the mine, from which most of the shipments have been made, the average of all the lots shipped has been about 45 ounces silver, 27 per cent lead, and 18 per cent zinc. At this mine there are four shafts and three levels. The lowest level starts from shaft No. 2, at a depth of 400 feet, with 100 feet of dump below. * * *

The mine shipped in 1884 about 13,400 tons. Four thousand two hundred tons have been shipped to the Colorado Smelting Co., at Pueblo; the balance was shipped to Leadville smelters, chiefly to those which have made the treatment of sulphide ores a special business by the erection of roasters—the Harrison Reduction Works, the Arkansas Valley, and the Manville smelter.

On the Minnie a shaft was sunk to the depth of 275 feet. Drifts were started in various directions in ore, partly low grade. Toward the west a drift was run, mostly in lime, without discovering valuable mineral. North and east of the shaft fair ore was encountered. The best, however, was struck south of the shaft, where from a small stope, since April last year, $180,000 worth of ore is said to have been taken. The present output goes mainly to the Manville smelter. * * *

At the Ruby the ore shoot is opened a considerable distance to the westward of the level. At the intersection of the level with the shoot a drift follows on the ore body for 80 feet, showing ore of variable grade for the entire distance. The width or thickness of the ore body is still unknown, but sufficient work has been done to prove it to possess unusual strength. The average of a lot of ore in bins was shown by assay to be 24 ounces in silver, 0.15 of an ounce in gold, and 24 per cent in lead.

The Great O'Sullivan is located between the Oro City and the property of the Emmet Mining Co. The shaft is about 100 feet above the bed of the gulch and is 150 feet deep. In the present main workings the ore varies from 15 to 30 inches, with occasional larger pockets. The ore is of a uniform good grade. * * * The ore output varies from 15 to 20 tons. The production for 1884 is said to have amounted to $65,000.

The property of the La Plata Mining & Smelting Co. is reported to have produced 15,060 tons of ore in 1884.

The Emmet Mining Co. produced last year 2,478 tons of ore, for which was received $127,741, or an average of $57.55 per ton.

The Florence mine shipped 1,133 tons of ore during the year. It and several adjacent mines were consolidated and incorporated in a stock company, known as the Lilian Mining Co. of Leadville. The consolidation now embraces about 100 acres.

The Pilot mine is opened by a tunnel 600 feet in length, and has lately cut quite a streak of ore containing gold.

On Breece Hill the only properties that have been actively worked are the Little Prince and Little Jonny. The Little Jonny has been shipping regularly. The 115-foot level follows on the vein to the southward for about 187 feet, ending in ore of fair grade. The mineral carries silver, gold, and lead in quantities that leave a fair profit above the cost of mining and smelting. The 166-foot level, running in the same direction, is in 201 feet and shows ore in a number of places.

The Leadville smelter of the American Co. contains four large furnaces, possessing a capacity of 150 tons of ore per diem. The furnace charges being rather light in lead ores, the daily base-bullion product of the four furnaces averages 20 tons. The base bullion contains from 90 to 100 ounces in silver to the ton and 1½ ounces in gold.

The company is also operating the Royal Gorge smelter, at Canon City, which contains two furnaces, with an ore capacity per 24 hours of 70 tons, yielding about 15 tons of base bullion daily. The company secured the Canon City establishment exclusively for the reduction of Gunnison and San Juan ores. The requisite amount of lead ore for these works will be supplied by the Eureka, one of the company's mines situated in Gunnison County. The mineral consists of galena and sulphate and carbonate of lead, ranging in value from 70 to 80 ounces in silver to the ton and carrying 50 to 70 per cent in lead. The dry ores for the Canon City works are purchased throughout the various mining districts in southwestern Colorado.

The Little Ellen mine, the most prominent of the company's mines, situated on Little Ellen Hill, is shipping 50 to 60 tons daily, employing about 30 men. The developments of the mine along the incline now exceed 1,100 feet. Water has been encountered in the mine, and two large steam pumps were used but without success; therefore, a large compressor has been purchased for that purpose.

The output of the Little Chief mine is nearly as great, so far as tonnage is concerned, as that of the Little Ellen. * * * The ore obtained from the Little Chief mine is a most desirable smelting material, carrying a great deal of lead and a large excess of iron over silica. The ore contains on an average about 0.4 ounce in gold to the ton, while the Little Ellen ores range from 1 to 1½ ounces in gold to the ton.

The Chicago Smelting & Refining Co., which is an auxiliary to the American Mining & Smelting Co., is in successful operation. It possesses a capacity for desilverizing and refining lead of 120 tons in 24 hours. About one-half of the furnaces in the establishment are at present employed and some tons of base bullion handled daily. Of the bullion treated, 20 tons are supplied by the American smelter at Leadville, 15 tons by the Royal Gorge smelter at Canon City, and 25 tons are purchased from other smelting works.

At the head of Big Evans Gulch is the New York mine. Considerable work was done on this property during the early part of 1884, but owing to the continued deep snows during the winter months it was impossible to ship the ore as fast as extracted and between 100 and 200 tons of fine ore accumulated before shipment began. Since the roads were opened the output has been quite regular. * * *

The placer ground in Lake County possessing value is as follows:

California Gulch, length 5 miles, width 100 to 600 feet; Iowa Gulch, length 4 miles, width 50 to 200 feet; moraines west of Leadville containing patches of good pay, embracing about 25 square miles; Arkansas River Valley, length 20 miles, width 100 feet to one-half mile; West Fork of Arkansas, length 8 miles, width 25 to 500 feet; Lake Fork Creek, length 3 miles, width about 200 feet; Colorado and Little Fryingpan gulches, length 2½ miles, width 40 to 125 feet; Half-Moon Gulch and its branches, 7 miles in length and 50 to 200 feet in width; Twin Lakes Creek, its tributaries and surrounding moraines, Georgia, Thompson, Empire, Union, and a dozen other minor gulches, containing many acres of good ground.

A list of mines showing reported development to date and number of tons produced in 1884 follows.

In regard to smelting charges and development in 1885 Kirchhoff says: [18]

[18] Kirchhoff, Charles, Lead: U. S. Geol. Survey Mineral Resources, 1885, pp. 251-257, 1886.

Through the active competition of the smelters, both at Leadville and in the "Valley," Leadville has become, comparatively speaking, one of the most favorable ore markets to the miner in the world. This will be clearly apparent from the following quotations, which have been the basis of transactions during the year 1885. They represent the prices paid at Leadville, delivered at the sampling works. Classifying the ores, we have:

(1) *Carbonate lead ores.*—The principal producers of this class of ore are the Iron Mining Co., the Silver Cord, Adams Mining Co., the Little Ella, the Carbonate Hill mines, among which are prominent the Crescent, the Morning Star, the Evening Star, etc., and finally, a number of mines at Red Cliff, a tributary camp.

The highest bid for this class of ore has been: New York quotation for silver, less 5 per cent; $20 per ounce for gold; and 40 cents per unit, or each per cent, for lead, if under 40 per cent of lead, and 45 cents per unit if over 40 per cent of lead, regardless of the actual New York price for lead. The bid provided that there be no smelting charge whatever.

The following is the general price list for lead ore during 1885: Silver, New York quotation, less 5 per cent; gold, $19 per ounce, if over 0.1 ounce per ton; lead, 45 cents per unit when the New York quotation of lead is 4.25 cents, 40 cents per unit when the New York quotation for lead is under 4.25 cents.

Working charges, per ton of 2,000 pounds, dry weight

Kind of lead ore:

30 per cent	$3
25 per cent	4
20 per cent	5
15 per cent	6
10 per cent	7
Under 10 per cent	8

(2) *Dry oxidized silver ores.*—The leading producers of this class of ore are the Matchless, the Forest City, the Denver City, the Robert E. Lee, the Silver, and the May Queen mines.

Silver: New York quotation, less 5 per cent; working charges: $12 per ton of 2,000 pounds.

(3) *Argentiferous iron ore.*—The leading mines which market this class of ore are the Morning Star, the Henriette, the Denver City, the Dunkin, the Robert E. Lee, and the Matchless. The following is the basis on which this ore is sold. The contents of the ore, in percentages, of metallic iron and of metallic manganese is added, and from this is deducted the contents in per cent of silica. The figure thus arrived at is called the "base excess," and the tariff standard is based on the assumption that this "base excess" is 40 per cent. The prices paid during 1885 were:

(a) For low-grade iron ore, carrying 12 ounces of silver or less per ton: Silver, 50 cents per ounce, working charge, none; iron, 10 cents added for each per cent iron or manganese over 40 per cent "base excess"; 10 cents deducted for each per cent iron or manganese under 40 per cent "base excess."

(b) For first-grade iron ore, carrying over 12 ounces of silver per ton: Silver, New York quotation, less 5 per cent; working charge, $6 per ton; iron, 10 cents added for each per cent iron or manganese over 40 per cent "base excess," and 10 cents deducted for each per cent iron or manganese under 40 per cent "base excess."

(4) *Sulphurets carrying galena.*—The principal producers are the Colonel Sellers and the Minnie and the A. Y. mines. The lowest price paid until March, 1885, was as follows: Silver, 90 per cent of New York quotations; lead, 25 cents per unit when the New York quotation is 4 cents for lead. Five cents per unit is added or deducted for each 5 cents per 100 pounds advance or decline in the New York quotation for lead. Zinc, 12 per cent is the standard. For each unit of zinc above the standard 50 cents is deducted. Working charge, $21.50 per ton of 2,000 pounds.

The latest and highest prices paid in 1885 for this class of ore were the following: Silver, 93 per cent of New York quotations; lead, 35 cents per unit, when the ore contains 30 per cent of lead, 40 cents per unit when the ore contains over 30 per cent of lead; zinc, standard, 20 per cent; for each unit above 20 per cent 50 cents is deducted.

Working charges, per ton of 2,000 pounds

Kind of lead ore:

20 per cent	$19
20 to 25 per cent	18
25 to 30 per cent	17
Over 30 per cent	16

(5) *Sulphurets with no galena and a little zinc.*—These ores, high in iron pyrites, which are produced by the Mike & Star mine, are sold on the following basis: Silver, 95 per cent of New York quotations; copper, $1 per unit; working charges, $13 per ton, free on board cars.

(6) *Sulphurets with no galena and 10 to 15 per cent zinc.*—This class of ore is produced chiefly by the Forepaugh mine and is paid for at the following rates: Silver, 95 per cent of New York quotations; working charges, $15 per ton, free on board cars.

When it is considered that coke, containing 20 per cent of ash, costs at Leadville $13 per ton, and that the cost of smelting (placing it at a low figure) is $8 per ton, run of ore, it will be understood by an examination of the above figures that more is paid for the lead in the ores than can be recovered for it.

During the current year smelting charges have been more favorable thus far to the furnace men, without rising so much as to discourage mining.

Mr. D. Bauman, of Buena Vista, under date of January 26, has furnished an estimate of the probable output of lead ores in Colorado, based upon a careful study of the conditions existing at that time. Such an estimate is of course subject to many contingencies affecting individual producers, entire groups of mines, or the whole industry. The total may be swelled by the striking of exceptionally high-grade bodies of great magnitude or by temptingly high prices. It may be diminished by accidents, labor troubles in allied industries, or a decline in values. As it is, however, it constitutes a thorough and clear review of the actual status of the industry of more immediate interest than a historical sketch of the happenings of the past year.

In Leadville the first group of mines to be considered is that of Main Fryer hill. The record of the Chrysolite, Little Chief and Little Pittsburgh shows that whatever vitality they may possess as producers of dry and milling iron ores, their output will not be greater than 3,000 tons of lead ore, averaging 12 per cent. Adding 50 tons of 20 per cent ore from a few smaller mines, a total of 370 tons of lead is reached. The product of the Matchless, New Pittsburgh, and Hibernia, on East Fryer hill, yielding ore of higher grade, is estimated at 750 tons of metallic lead contents. The mines on Yankee Hill, of which the leading ones are the Moyamensing, May Queen, Forest City, Denver City, Lee Basin, Alleghany, Scooper and Chieftain, carry but little lead in the siliceous and ferruginous ores they turn out. It is estimated that 300 tons of lead will cover the product.

The three sections of Leadville thus far mentioned are those lowest in lead, though their importance is great in other respects as producers of iron and dry chloride and iron sulphide ores. Those acquainted with that section look forward to the discovery of lead sulphide ores both on East Fryer and Yankee hills, but it is not believed, even if this possibility is verified, that it will have any appreciable effect upon the current year's markets.

Carbonate Hill is looked forward to as one of Leadville's chief sources of lead ore supply, and it is likely that fully nine-tenths of it will come from the Maid of Erin, Henriette, Brookland-Clontarf or Adams mines, the Wolftone-Agassiz, and Morning Star. It is believed that these mines combined will produce not less than 9,000 tons of metallic lead, provided the Maid of Erin and Henriette ship 10,000 tons of ore, which it is believed that they will do, though it is not likely that they will exceed it much. The Wolftone-Agassiz are counted upon to contribute a like amount to the market. A number of other mines, among them the Catalpa, Carbonate, Leadville, Glass-Pendery, Crescent, Aetna, Modest Girl, and a few others, may yield 200 tons of metallic lead. Thus far the preparations for treating the sulphides of Carbonate Hill are not extensive, and until now the development of that class of ore in a large body has been limited to the Wolftone.

On Iron Hill the mines of the Iron Silver Co., including its California Gulch properties, may be credited with an output of 24,000 tons, including concentrates, the whole averaging about 16 per cent, or 3,840 tons of metallic lead. The Silver Cord mines, which shipped last year nearly 10,000 tons, of which a large percentage was high-grade ores, now show more sulphides, which are too low to be marketed without previous concentration, which has not been provided for as yet. They will not, therefore, in 1886 occupy the same prominent position as contributors to the lead supply. On the other hand, another lead mine, called the Benton, has been opened in the rear of Iron Hill, in what is known as Adelaide Park. It is an extension of the regular deposits of the Park mine, which a few years since made considerable shipments. During the greater part of 1885 the Benton mine sent to market from 500 to 600 tons per month of ore carrying 36 per cent lead. It may be credited with 4,000 tons of ore or 1,600 tons of lead in 1886, the policy of the management at present being rather to push development work than to extract ore. The Argentine, Terrible, Humboldt, Newton, and Silver Cord may be estimated at 1,000 tons of metallic lead. The Louisville, Colorado No. 2, and North Ruby may produce 7,500 tons of 30 per cent ore. It is possible that the grade may be lower, but in that case it is likely that the tonnage will be heavier, so that the product of metal may be put at 2,250 tons. The Smuggler, Iron Hill Consolidated, and the A. Y. and Minnie mines may be relied upon for 6,000 tons of carbonates, equivalent to 1,200 tons of lead. The great sulphide deposits of the Colonel Sellers, A. Y., Minnie, Accident, Sierra Nevada, and Moyer, which form one enormous ore shoot, believed to be continuous through the William Moyer placer, the eastern part of the Silver Cord, and the Ruby, is an uncertain element in the question. It depends upon the success of the concentrating works built already and the activity with which building of new works will progress during the year. The sales of ore of the Colonel Sellers mine and the output of its concentrating plant may be estimated at 18,000 tons, equivalent to 4,250 tons, and the yield of the A. Y. and Minnie mines may be placed at 3,000 tons of sulphides or 1,000 tons of metallic lead.

On Rock Hill the Sullivan, Only Chance, and Emmet mines will produce not less than 7,000 tons of lead ore, aggregating 2,500 tons of lead. The La Plata, Crown Point, Pinnacle, Montgomery, Gilt Edge, Sequin, Pease, Willis, and Moyer properties are believed to be good for 10,000 tons of lead ore carrying 2,500 tons of lead. The Lilian, Brian Boru, G. M. Favorite, and other mines of Printer Boy hill will in all likelihood yield about 5,000 tons of ore or 1,000 tons of lead. The Upper Iowa Gulch mines and Ball Mountain will probably not exceed 500 tons of lead ore or 200 tons of metal, and will do well if they produce that.

On Little Ellen Hill the New Year property is known to contain large ore bodies, but it will hardly come into the market this year, since the incline being driven toward them will probably not reach the ore until the close of the year. All the other mines of this section, with the exception of the Little Ellen mine proper, have been poorly worked and are partly unsafe. So far as lead ore is concerned they are pockety. The Little Ellen and the majority of the other mines are worked by lessees. It is not likely that all of them together will ship more than 6,000 tons of ore or 1,200 tons of lead, unless the New Year mine begins active extraction earlier than expected.

Mount Sheridan, Sugar Loaf, Mount Kevin, and Little Fryingpan mines, in the vicinity of Leadville, may produce 300 tons of metallic lead.

In his report for 1885 Wilson [19] gives a table which shows the production of the principal smelters from Colorado ore during the year ending December 31, 1885, including the American smelter, Harrison Reduction Works, La Plata smelter, Manville smelter, Fryer Hill smelter, and Chrysolite mill.

Mr. F. L. Bartlett began buying lots of lead-zinc ore and concentrates for experiment in the East, which later resulted in the establishment of the American Zinc Lead Co.'s zinc-oxide plant at Canon City in 1891.

The Canadian commission that investigated the zinc resources of British Columbia, in its report published in 1906, gives the following account of the production of zinc ores at Leadville: [20]

In the early years there was no market for the zinc ore as such, and as a constituent of silver-lead ore it was a detriment to the value of the latter. The object of the miner, in order to conform to the requirements of the silver-lead smelters, was consequently to keep the percentage of zinc in the ore shipped as low as possible. Large quantities of zinc ore were, therefore, removed from the ores by hand sorting or mechanical concentration and thrown away; in many cases beyond recovery, in a few cases into separate dumps where it could be held pending the development of a market, which in the United States was foreseen by a few mine owners as far back as 1885. Whenever possible, zinc was passed by in the mines.

Previous to 1899, the supply of zinc ore smelted in the United States was obtained chiefly from Missouri, Kansas, Wisconsin, New Jersey, and Virginia, with comparatively small quantities from Tennessee and Arkansas. There was no ore received from the country west of Kansas, except from a group of mines near Hanover, N. Mex., whence some shipments were made about 1893, and possibly some small, spasmodic shipments from other localities, of which no record has been preserved.

The utilization of the zinc resources of the far West was early considered. In 1885, Eugene and Alfred Cowles patented an electric furnace for the reduction of zinc ore, and I believe they had in mind the treatment of mixed ore from New Mexico. H. C. Rudge built Belgian furnaces at Denver, Colo., in 1888 and actually smelted a small quantity of ore from Leadville, but because of ignorance the venture proved a failure. Messrs. Ingalls, Argall, and Wood, who have been associated in the investigation of the zinc resources of British Columbia, formu-

[19] Wilson, P. S., agent for Colorado, in Kimball, J. P., Report of the Director of the Mint upon the production of the precious metals in the United States during the calendar year 1885, p. 138, 1886.

[20] Report of the Commission (Walter Renton Ingalls, Philip Argall, and A. C. Gardé) appointed to investigate the zinc resources of British Columbia and the conditions affecting their exploitation, Mines Branch, Department of the Interior, Ottawa, Canada, pp. 5-9, 1906.

lated extensive plans for zinc development in 1889, but these proved to be premature.

The ore from Hanover was sent to Mineral Point, Wis., and Waukegan, Ill. The freight rate to those points was $12 per 2,000 pounds. Under the market conditions of that time, it being a period of general industrial depression and low prices, there was no profit in the business, and the exploitation of the mines ceased.

In the summer of 1899 certain smelters in Kansas received small shipments of blende concentrate from Creede, Colo. The real development of the zinc industry west of the Rocky Mountains may be dated from this time. At first, the Colorado ore was regarded askance, although that received from Creede was really a superior ore by any standard save that which existed among Kansas-Missouri smelters, who based their ideas at that time upon the ore of remarkable purity which was afforded by the Joplin district. They considered an iron content of upward of 2 per cent in a zinc ore to be highly objectionable; and in fact, in view of their smelting methods and equipment at that time, it was objectionable. The attempts to smelt even the comparatively clean ore from Creede in 1899 were disastrous.

A combination of circumstances, however, caused the possibility of obtaining an ore supply from Colorado and elsewhere in the far West to be kept in mind. There was at about this time a concerted effort on the part of the miners of the Joplin district to raise the price for ore. Their mines were in fact unable to furnish the supply required except at an enhanced price. On the other hand, the price for spelter left the smelter an insufficient margin, and he was keenly looking out for supplies of cheaper raw material. European smelters were in somewhat the same position.

The great deposits of mixed sulphide ore at Leadville, Colo., had been worked since about 1885 for the lead content of the low grade of ore, the zinc and most of the iron being thrown out upon the tailings pile. The concentrating mills were of the old conventional design, crushing the ore comparatively coarse and cleaning it chiefly by jigging, but there was no great profit in the operation, and it was after a while abandoned. The success in the treatment of similar ore by fine crushing at Broken Hill, New South Wales, the invention of the Wilfley table in 1895 (affording a greatly improved means for cleaning fine ore), reductions in the cost of mining, etc., led to the erection of a type of mill more especially suited to the particular ore, and although these were designed especially for the production of galena concentrate, it was found that a fair grade of zinc concentrate could be made at the same time as a by-product. The advent of an enterprising broker, acquainted with the needs of the European zinc smelters, developed an export business, which in 1900 and in two or three years subsequent attained large proportions. The factors enabling this to be done were the low price at which the miners were willing to sell the ore, and the low freight rate which was obtained, via Galveston, to Swansea and Antwerp. At first, the ore produced assayed about 45 per cent of zinc, 12 per cent of iron, and 6 per cent of lead; the average subsequently ran down to about 38 per cent of zinc, 17 per cent of iron, and 3 per cent of lead. The buyers paid a flat price for the ore, at first only $5 per 2,000 pounds, f. o. b. cars at Leadville, and shipped it to European ports at a cost of $9.50 per 2,000 pounds. The miners were well satisfied with this price because the ore was distinctly a by-product and anything realized for it was so much gain. The increasing demand for zinc ore of any kind led, however, to competition for the Leadville ore, a gradual increase in the price for it, and the producers became firm in holding out for the best bargains.

The smelters of Kansas continued their experiments on the smelting of Colorado ores with many discouraging experiences, but after a few years they succeeded in treating them profitably and drove the European buyers out of the Colorado market.

The increased price for spelter, the continuing shortage of ore, and the severe competition for what Joplin could supply forced these smelters more and more into the country west of the Rocky Mountains for their ore supply and raised the prices for such ore, in which the smelters were aided by gradual improvements in their processes.

In order to illustrate the magnitude which the zinc industry west of the Rocky Mountains has attained, I may be permitted to quote from an article by myself * * * as follows: [21]

Statistics of the production of zinc ore in Missouri and Kansas (Joplin district) and New Jersey are available for a long series of years. Up to a few years ago these were sufficient, inasmuch as nearly the whole spelter output of the United States was derived from those sources. In 1899 zinc ore from Colorado began to appear in the market, and during the last two or three years that ore, together with ore from other States and Territories west of the Rocky Mountains, and from British Columbia and Mexico, has been figuring largely in the market. It is, therefore, important to know definitely as to the production of these sources of ore supply. Such statistics respecting them as have previously been published are incomplete and of doubtful accuracy.

Statistics of zinc ore production indicate directly the magnitude of the mining industry, by showing the tonnage of ore produced and moved. In connection with spelter production, however, it is necessary to examine them with a knowledge of what they represent.

The ore production of the Joplin district is of two classes, viz, blende and calamine. The former averages about 58 per cent of zinc; in round numbers, two tons of this ore make one ton of spelter. The calamine of the district is entirely zinc silicate. It may be assumed as averaging a little better than 40 per cent of zinc, three tons of ore being required, roughly, to make one ton of spelter. The total production of zinc ore in the Joplin district in 1905 was 252,435 tons. No attempt was made to classify this as blende and calamine, but in recent years the output of the latter class of ore has amounted to 10,000 to 16,000 tons per annum, and it may be reasonably assumed that the production in 1905 was something between those figures.

A small amount of calamine, both carbonate and silicate is produced in southeastern Missouri, especially by the Valle mines. This ore goes chiefly to St. Louis, and amounts to 3,000 to 6,000 tons per annum.

The zinc ore produced in the States and Territories west of the Rocky Mountains is both blende and calamine, the latter being chiefly zinc carbonate produced in Mexico and New Mexico. The production of Colorado, Utah, Idaho, Montana, and British Columbia is chiefly, if not entirely, blende. This ore varies generally in grade from 30 per cent to 50 per cent zinc. In a few cases, as at Creede, Colo., and the output of handsorted lump ore of one mine in British Columbia, it exceeds 50 per cent, the Creede ore (mill concentrate) in fact being almost as high in zinc as the average Joplin product, but although low in iron it is higher in lead than the Joplin ore. The average zinc content of the western ore, both blende and calamine, may be assumed at 38 per cent. From $3\frac{1}{4}$ to 3 tons of this ore are required to produce one ton of spelter. This sulphide ore is comparatively high in iron and lead; some of it is very high in those elements. It is mostly produced as a concentrate from mixed sulphides, the lead product being shipped to the silver-lead smelters. The Iron Silver Mining Co. however, ships a good deal of hand-sorted lump ore from its Moyer mine, at Leadville.

Wisconsin produces a blende concentrate, which after magnetic separation, is practically as high in zinc as the average Joplin ore, and when well prepared is comparatively low in

21 Ingalls, W. R., Spelter statistics for 1905: Eng. and Min. Jour., vol. 81, pp. 909-911, 1906; Mineral Industry, vol. 14, pp. 562-570, 1906.

iron and lead, the blende itself being only slightly ferruginous and the iron content of the marketed ore being chiefly intermixed marcasite. Wisconsin also produces carbonate ore, which is used at Mineral Point for the manufacture of zinc oxide.

The large output of zinc ore in New Jersey is entirely from the Franklin mine of the New Jersey Zinc Co. It is the mixed franklinite-willemite, averaging about 20 per cent zinc, which is separated into one product (willemite) for spelter manufacture and another product (franklinite) for the manufacture of zinc oxide and spiegeleisen.

Of the western zinc-mining districts, the most important single district is Leadville, Colo. Other important single districts are Creede, Colo., Magdalena, N. Mex.; Park City and Frisco, Utah; Monterey, and Las Plomosas (near San Sostenes, on the Kansas City, Mexico & Orient Railway), Chihuahua, Mexico; and the Slocan, British Columbia. Outside of these districts, the zinc ore production west of the Rocky Mountains comes from many scattered localities. In New Mexico, besides Magdalena, Hanover is a small producer, and there are several other promising districts. In Montana, Butte is the principal source. In Idaho, the Wood River district is the most important, although some ore was obtained in 1905 from the Coeur d'Alene. In Utah, the Daly West Mining Co., of Park City, and the Horn Silver Mining Co., of Frisco, have large zinc resources; the former did not produce in 1905 but the latter shipped 8,145 tons. In Colorado, besides Leadville and Creede, zinc ore is produced at Rico, and by many small mines in Clear Creek and Summit counties. In Mexico the Calera mine, of the State of Chihuahua, was a considerable shipper of mixed sulphide ore to Pueblo, Colo. Arizona and Nevada both figured as small producers in 1905. The ores of Magdalena, N. Mex., were shipped chiefly to Missouri, Kansas, and Wisconsin, for the manufacture of zinc oxide. Other western ores are shipped to Mineral Point, Wis., for the manufacture of zinc oxide.

The figures for zinc in 1885 to 1891, given in the table on page 176, represent estimates made on the basis of the statement of F. L. Bartlett, formerly manager of the American Zinc-Lead Co., at Canon City, that between 1885 and 1891, he treated in the East about 1,500 tons of Leadville zinc-lead ore averaging 25 per cent zinc. These figures represent gross content; no attempt has been made to estimate recovered zinc in zinc oxide produced.

In his report for 1886 Kirchhoff says: [22]

The following figures relate to Leadville (made up of two items, (1) Leadville smelters' lead bullion product with contents of lead, silver, and gold, and (2) ore shipped outside of Leadville, with value of ore shipped).

Years	Leadville smelters' bullion product			Ore shipped out of Leadville	
	Lead	Silver	Gold	Ore	Value of ore shipments
	Short tons	Ounces	Ounces	Short tons	
1877	175	376,827	3,750	3,300	$400,000
1878	2,324	450,476	897	15,840	2,360,503
1879	17,650	6,004,416	1,100	18,549	2,851,850
1880	33,551	8,999,399	1,687	12,410	1,460,363
1881	38,101	7,162,909	12,192	15,630	1,016,044
1882	39,864	8,376,802	12,615	22,416	1,872,604
1883	36,870	5,057,990	22,330	(?)	6,420,692
1884	35,296	5,720,904	22,626	(?)	(?)
1885	19,128	5,099,271	8,262	137,869	(?)
1886	25,963	4,569,013	22,504	138,335	6,135,585

[22] Kirchhoff, Charles, jr., Lead: U. S. Geol. Survey Mineral Resources, 1886, pp. 144-145, 1887.

The lead contents of the ore shipments [for 1886] are estimated at 22,526 short tons. Leadville is, however, credited with considerable ore derived from tributary camps, notably Red Cliff. Mining developments in Leadville have been, generally speaking, favorable to a continuance of a heavy lead output, the principal feature being the looming up as lead mines of the Maid of Erin and the Henrietta, which have been struggling against heavy flows of water. * * *

A good deal of progress has been made during 1886 in the direction of concentrating low-grade ores, a number of plants having been built, to which others are to be added during the current year. The sulphurets continue to be troublesome to smelters on account of their high percentage of zinc.

Roasting in stalls, as predicted by the best authorities, has proved very inadequate and has been quite generally abandoned. Even ore roasted in reverberatory furnaces, at a cost of $3.50 per ton, occasions losses and mechanical difficulties in smelting.

For 1886 to 1896 the figures given in the table on page 176 are based on reports of the agents of the mint in annual reports of the Director of the Mint, the gold and silver being prorated to correspond with the figures for the total production of the State as corrected by the Director of the Mint, the lead being prorated to correspond with the total production of lead in the State as given in annual volumes of Mineral Resources, and any "unknown production" in the State being distributed proportionately among the counties. As with lead, so with copper, but as the figures for copper given in Mineral Resources include copper from matte and ores treated in Colorado, though produced in other States, the figures for copper are subject to revision. However, the item of copper from lead desilverizers in the figures for copper in Mineral Resources may belong in large measure to Leadville, thereby balancing in a manner any error.

In his report for 1887 Kirchhoff says: [23]

Leadville continues to be overwhelmingly the heaviest producer of lead ores in the State, its shipments of base bullion during 1887 having been 30,575 tons, while the shipments of lead, siliceous, and sulphide ores to valley smelters aggregated 34,600 tons. A considerable quantity of ore treated in Leadville has, however, come from other quarters. At the close of 1887 the three railroads entering Leadville reduced the freight on bullion from Leadville to Colorado Springs, Denver, or Pueblo from $12 to $10 and lowered the cost of coke on cars to the smelters to $10 a ton, while the rate on ore from Leadville to valley smelters at Pueblo and Denver was lowered from $5 to $4.70 per ton for the old tariff for ore valued at $100 or under. The unfavorable position of the Leadville works in their competition with the valley smelters has thus been improved. The mines of the district maintained their productiveness, although the most readily smelted ores are growing scarce. Concentrating equipment has been increased, and the outlook points to a continuance of the present rate of supply of the metal for the current year.

Munson [24] describes the mills of Lake County in his report for 1887 as follows:

[23] Kirchhoff, Charles, jr., Lead in Colorado: U. S. Geol. Survey Mineral Resources, 1887, p. 105, 1888.

[24] Munson, G. C., agent for Colorado, in Kimball, J. P., Report of the Director of the Mint upon the production of the precious metals in the United States during the calendar year 1887, pp. 151-152, 173-175, 189, 193, 1886.

The subject of ore dressing, as applied to the handling of low-grade sulphide ores, received a good deal of attention during the past year at Leadville, in Lake County. The total capacity of the various concentrating mills, independent of the ores treated by hand jigs, was 570 tons per day. Four large mills were in operation, the largest of which handled 200 tons per day and the smallest 100 tons per day. The average value of the ores did not exceed $5 in gold, silver, and lead, with an average of 15 per cent of zinc sulphide. The general character of these mills is the same. The ore is first crushed in rock breakers, then passed through rolls, sized in revolving screens, and treated on the ordinary four-compartment Hartz jigs. The slimes are handled upon various kinds of tables of the endless-belt pattern. The work accomplished by these mills is claimed to be highly satisfactory, leaving in the concentrates not more than 3 or 4 per cent of zinc sulphide.

Munson gives a list of producing and nonproducing mines in Lake County during 1887, which shows the production of each producing mine. No placers are included. The list of smelters, with the production of each, includes the American, Arkansas Valley, Harrison Reduction Works, La Plata, and Manville, at Leadville.

In his report for 1888 Kirchhoff says: [25]

The Leadville smelters lost some ground during 1888, their total production being 22,490 tons of base bullion, against 30,575 in 1887. The ore production of the camp fell off some, owing chiefly to the cessation of shipments by the small slopes and the reduction of the product of some of the larger mines.

Munson [26] gives a table showing the producing and nonproducing mines in 1888 and the production of each producing mine; also the production in Sugar Loaf and St. Kevins districts, outside of Leadville. No placer production is given.

The following description of the plan of the Ovoca Zinc Ore Co. for the recovery of zinc from complex ores, prepared in 1889 by Ingalls, Argall, and Wood, shows in part the possibilities of producing zinc ore and the cost of smelting lead, of freight, and other costs. Only one copy of this plan, which has been kindly loaned by Mr. Philip Argall, of Denver, Colo., is in existence.

Introduction.—There is now in sight in many of the mines of Leadville, notably the A. Y. and Minnie, the Colonel Sellers, the Sierra Nevada, the Silver Cord, and the Moyer mine of the Iron Silver Mining Co., enormous bodies of argentiferous sulphide ore, which is of too low grade to be mined and smelted in the ordinary manner at a profit, even at this time, when the cost of both mining and smelting in Colorado has been reduced to a minimum.

Cost of smelting.—The amount of zinc in these ores is such that the lead smelters will not buy them without making a very high charge for treatment, zinc being such a very undesirable element in the lead furnace. At the present rates for smelting, the silver in these ores in question would be paid for upon a basis of 90 per cent of their contents, at current New York quotations; the lead upon an arbitrary basis of 20 cents per unit, which is equivalent to 1 cent per pound of metallic lead; a deduction of from $15 to $20 per ton is then made for cost of

smelting, and of course nothing is paid for the zinc, which is lost in all methods of lead smelting, and that is the only form of smelting applicable to these ores, as their character renders them entirely unfit for reduction to spelter by any of the ordinary zinc methods.

Now, as the average gross value of this low-grade sulphide ore is less than $15 per ton, the fact that it can not be profitably treated at the present time, when the only market for silver ores in Colorado is with the lead smelters, is very evident; and equally evident is the great opportunity, now open, for any works in which these ores can be profitably treated.

Amount of ore.—In the A. Y. and Minnie mines there is now estimated to be in sight about 500,000 short tons of this low-grade ore; in the Colonel Sellers mine 250,000 tons; in the Moyer 400,000 tons; in the Sierra Nevada and Silver Cord, about 50,000 tons each. Here, then, is a million and a quarter tons of ore, standing in these five mines, which is at the present time no better than so much waste rock. Hundreds of analyses have shown that its average grade is from 10 to 15 ounces per ton in silver; and that its contents are about 10 per cent lead and from 20 to 30 per cent zinc, the remainder being iron and sulphur, with less than 5 per cent silica.

There is, moreover, in the tailings pile below the Minnie mill 60,000 tons of low-grade blendous silver ore and in the dump of the Colonel Sellers mill 90,000 tons. The average content of these tailings is 10 ounces silver per ton; 6 per cent lead; and 30 per cent zinc.

Dressing.—This low-grade sulphide ore can not be successfully dressed, as has been proved by the works of the A. Y. and Minnie and Colonel Sellers mines, which have now been in operation for three years, because the silver in the ore is almost equally distributed between its three component minerals—galena, iron pyrites, and blende—and in washing out the latter the silver which it contains is of course lost. The experience in the two mills in question has been that a saving of about 40 per cent of the silver, and 75 per cent of the lead is all that can be effected, even with the best possible work. Moreover, the Colonel Sellers mill has already stopped running for insufficient room to store tailings, and the Minnie mill will soon be obliged to stop also, for similar reason.

Smelting this ore being thus entirely out of the question, and dressing being so very dissatisfactory, it is clear that the process by which the ore may be successfully treated must be one by which the zinc in it, as well as the silver and lead, may be recovered.

Supply of ore.—The low-grade blendous silver ore of the mines of Leadville, assaying 10 ounces per ton silver and 10 per cent lead and 25 per cent zinc, can be bought from the mines at a cost of $5 per ton, in Leadville. The ore can be mined at an average cost of about $2 per ton, so that the mines can well afford to sell it at that price. The five mines named can produce this ore at the rate of more than 250 tons per day. As the Ovoca works will at first use only 50 tons per day, and the amount of ore in sight in the mines of Leadville is so enormous, there will clearly never be any difficulty in securing an ample supply of the ore for the works. The existence of this immense amount of ore, which can not be sold at all now, will of itself prevent any combination on the part of the mines to raise the price of it. Furthermore, in other parts of Colorado, there are bodies of argentiferous galena-blende ore opened of grade, character, and composition similar to that of Leadville.

The 150,000 tons of ore in the tailings dumps before mentioned, assaying 10 ounces silver, 5 per cent lead, and 30 per cent zinc, can be bought for $2 per ton, delivered on board cars, in Leadville.

Moreover, there is a very large amount of blendous ore in Leadville, not included in the foregoing estimates at all, which contains enough silver to enable it to be mined and smelted at

[25] Kirchhoff, Charles, Jr., Lead: U. S. Geol. Survey Mineral Resources, 1888, p. 87, 1890.

[26] Munson, G. C., agent for Colorado, in Kimball, J. P., op. cit. for 1888, pp. 112-116, 132, 1889.

a profit, even when a smelting charge of $18 to $20 per ton is made. It costs the lead smelters nearly that amount to treat those ores. By the Parnell process the Ovoca Zinc Ore Co. can treat them for less than $6 per ton and consequently can enter the market with the lead smelters, compete successfully with them, and can make a great profit on the high-grade ores as well as those of low grade. The ore supply upon which the Ovoca Zinc Ore Co. may draw is thus almost inexhaustible.

Smith [27] in his report for 1889 gives a list of producing mines showing individual production. No placer production is given.

In his report for 1890 Smith says: [28]

The reserves of low-grade ores known to exist in very many of the larger properties are immense. This is especially true of the bodies of sulphide ores in the Iron-Silver mine, which shows almost inexhaustible quantities in the Moyer ore shoot; [29] also in the Silver Cord, A. Y. and Minnie, and Colonel Sellers mines; the first mentioned having little or no association with zinc, the last three requiring concentration on account of zinc.

The feature of the greatest interest in the mines of Lake County during the past year is the discovery of a large copper deposit in the Henriette and Maid of Erin properties.

Smith gives a list of producing mines with individual outputs. No placer production is shown. The Henriette and Maid of Erin shipped considerable copper.

An editorial in the Mining and Scientific Press [30] describes the addition of copper ore to charges of lead furnaces at Leadville in 1890.

Figures showing tons of ore produced from 1890 to 1900 are smelter figures and are given by J. D. Irving; [31] their source is unknown.

In his report for 1891 Smith says: [32]

The falling off in the product of Lake County was to a considerable extent due to the fact that at the commencement of the year the bins of the smelters at Leadville were filled to overflowing and contracts for the output of a few large properties were considered inadvisable. The capacity of the smelting and reduction works in this county has been enlarged during the year.

Smith gives a list of the producing mines and their individual output which shows that the Arnold & Thompson placer produced $2,894. The copper produced was obtained almost entirely from the Henriette and Maid of Erin, though some was produced by the Little Jonny.

In his report for 1892 Smith [33] gives a list of producing mines showing individual production. The copper came from the Henriette and Maid of Erin. The Arnold placer yielded $2,500, the Star placer $3,000, and the Thompson placer $4,500.

Emmons [34] gives some additional information. Weeks [35] says of the manganiferous ores:

Character of the manganiferous iron ores of Colorado.—No manganese ores are mined in Colorado. Considerable iron and manganiferous iron ores are mined in the Leadville district, being used either as a flux in the smelting of silver or at Pueblo in the manufacture of spiegeleisen. Some of the ore containing the highest percentage of manganese has been sent to the Illinois Steel Co. at Chicago. Analyses of these ores, carrying about 20 per cent and over of manganese, are as follows:

Analyses of manganiferous iron ores in Colorado

Component parts	Catalpa	Crescent No. 1	Crescent No. 2	Hull	Emmett Mining Co.	
Iron	34.90	17.80	21.15	35.00	11.00	11.45
Silica	6.90	6.30	7.00	3.83	8.06	5.02
Manganese	21.30	34.00	31.00	19.30	35.36	38.22
Alumina	4.15			2.00	2.37	
Lime	.34			.46	1.23	
Magnesia	.07			.45	1.36	
Sulphur	.06	.027			.33	
Phosphorus	.04	.056		.03	.111	.073
Copper	Trace.			.63		
Oxide of lead				1.85		
Volatile matter				9.36		
Water				2.96	18.06	

Production of manganiferous silver ores.—All the manganiferous silver ores produced in the United States in 1892 of which we have any report were from Colorado and entirely from the Leadville region. Some ores of this character were produced in Montana, but no record appears to have been kept, or at least none is available.

Colorado produces two classes of manganese-bearing ores, a manganiferous iron ore used to some extent in the production of spiegeleisen, and a manganiferous silver ore used as a flux in the smelting of silver-lead ores. The manganiferous iron ores carry, as a rule, but little silver, though some of the slags from the blast furnaces of the Colorado Coal & Iron Co., at Pueblo, where these manganiferous ores are used in the manufacture of spiegeleisen, are so high in silver as to make it profitable to rework them for the recovery of silver. Occasionally some of the manganiferous ores are sent to the Illinois Steel Co., at Chicago.

The total production of manganiferous iron ores in Colorado in 1892 was 3,100 tons, worth at the mines $15,500, or $5 a ton. These ores carried from 25 to 38 per cent of manganese. The indications are that the production of these ores in 1893 will be considerably in excess of that of 1892.

In most of the mines of the Leadville district are found considerable quantities of what have been termed and described in another portion of this report as manganiferous silver ores. It is stated that there are not more than three properties in the Leadville district where the ores do not carry a percentage of iron and manganese. A full description of these ores is given under the title "Character of the manganiferous silver ores of the United States," elsewhere in this report, and of the prices obtained and the methods of payment under the title "Price of manganese and manganiferous ores in 1892."

The total amount of manganiferous silver ores shipped in 1892 was 62,309 tons, of which 2,732 tons contained an average of 34 per cent of manganese, 14,315 tons an average of 24.9 per cent, and 45,262 an average of 12 per cent. The total value of this manganiferous silver ore was $323,794, an average of $5.20 a ton.

[27] Smith, M. E., agent for Colorado, in Leech, E. O., Report of the Director of the Mint upon the production of the precious metals in the United States during the calendar year 1889, pp. 149–150, 155, 1890.

[28] Smith, M. E., agent for Colorado, in Leech, E. O., op. cit. for 1890, pp. 126–127, 135–136, 142, 1891.

[29] Abandoned in 1916 as exhausted for company work.—C. W. H.

[30] Smelting in Colorado: Min. and Sci. Press, vol. 109, p. 941, 1914.

[31] Unpublished manuscript of J. D. Irving.

[32] Smith, M. E., agent for Colorado, in Leech, E. O., op. cit. for 1891, pp. 173–174, 181–182, 187, 1892.

[33] Smith, M. E., agent for Colorado, in Leech .E. O., op. cit. for 1892, pp. 126–127, 1893.

[34] Emmons, S. F., Progress of the precious metal industry in the United States since 1880: U. S. Geol. Survey Mineral Resources, 1892, pp. 66–67, 1893.

[35] Weeks, J. D., Manganese: U. S. Geol. Survey Mineral Resources, 1892, pp. 183–184, 191, 194–195, 1893.

The Denver Republican in its review of the year 1892, published on January 1, 1893, says:

A new industry.—The zinc smelter at Canon City was erected in 1891, though it has been operated practically in 1892. This is the first time [but see report of British Columbia Zinc Commission under year 1885— C. W. H.] that metallic zinc has been produced in Colorado, and this must now be added to the mineral product of Colorado.

Production of American Zinc-Lead Co., Canon City, Colo., 1892

Zinc-lead, white	pounds	2, 500, 000
Copper (fine)	do	360, 000
Silver	ounces	137, 000
Gold	do	120
Equivalent of lead in pigments, as metals	pounds	625, 000
Equivalent of zinc in pigments, as metals	pounds	1, 125, 000
Tons of ore treated		12, 000
Total value		$335, 000

Estimated stock on hand December 31, 1892

Gold	ounces	100
Silver	do	80, 000
Copper	pounds	300, 000
Lead	do	300, 000
Zinc	do	1, 500, 000

Mineral Industry for 1892, in the article entitled "Zinc," says: [36]

In Colorado the American Zinc-Lead Co. ran its work at Canon City steadily through the year and reduced (by the Bartlett process) about 12,000 tons of low-grade complex sulphide ore.

The figures in the table on page 176, showing the zinc produced from Lake County in 1892, represent half the total production of the American Zinc-Lead Co. at Canon City. This company began to produce zinc in 1891.

The Denver Republican, in its review for the year 1893, published December 31, 1893, says:

The American Zinc-Lead Works, at Canon City, which were first operated in 1892, show a great increase for the year. There were 1,650,000 pounds of zinc from Colorado mines treated at the works during 1893. * * * 12,000 tons received; 9,000 tons treated.

The blast furnaces were in operation eight months and six zinc furnaces in operation 10 months.

Production for the year, 2,500 tons of pigment, which was shipped to the refinery at Chicago.

Production for the year

	Colorado	Total
Gold	ounces	370	460
Silver	ounces	95, 000	175, 000
Lead, as metal	pounds	940, 000	1, 140, 000
Copper, as metal	pounds	300, 000	475, 000
Zinc, as metal	pounds	1, 650, 000	2, 080, 000

The figures for lead in 1893, 1894, and 1899, given in the table on page 176, are taken from the Leadville Herald-Democrat.

[36] Mineral Industry for 1892, vol. 1, p. 466, 1893.

Of the developments in 1894 Puckett says: [37]

The world-renowned carbonate camp of Leadville has, during the year, been astonishingly metamorphosed into a gold region of splendid results and rich promise. At an average depth of 500 feet well-known silver and lead properties have run into high-grade gold ore. * * *

The principal mines thus far opened on the gold belt which have produced or are producing are the following: The Ibex Co. (comprising Little Jonny, Uncle Sam, Little Stella, etc.), Little Vinnie, Nevada, Little Ella, Valley, Midnight, Australian, Virginius, Fanny Rawlings, St. Louis (Colorado Prince and Miner Boy), Eliza, Highland Chief, Nettie Morgan, Great Hope; and those which are being worked at present, with every promise of success, are the Resurrection (to open the northern extension of the Little Ella ore shoot), Triumph, Irene, Garbutt, Antelope, Ocean Wave, Black Prince, Curran, Chemung, and many others. The above are all in the vicinity of Idaho Park.

The oldest gold-producing mine on the hill (Breece) is the Antioch, at the head of White's Gulch.

The Denver Republican, in its review for 1894, published January 1, 1895, says:

American Zinc-Lead Works, Canon City, Colo., production in 1894, over 1,500,000 pounds of zinc, 900,000 pounds lead, and 200,000 pounds copper was Colorado product.

In his report for 1895 Puckett says: [38]

The Leadville district materially increased its gold output. * * * An extensive tunnel proposition to thoroughly drain the mines * * * is being launched.

Of the production of lead in 1895 Kirchhoff says: [39]

In Colorado, Leadville has more than held its own. According to the Herald-Democrat, the mines in 1895 produced 330,933 tons of ore, of which 70,429 tons were carbonate, 86,243 tons iron, 116,975 tons sulphide, and 57,286 tons silicate. The leading producers of carbonate were the Maid of Erin, 27,614 tons; the Starr lease, 7,497 tons; Bon Air, 4,929 tons; Bison, 4,100 tons; and Welden, 3,931 tons. The Wolftone raised 35,508 tons of sulphide; the A. Y. and Minnie, 25,765 tons; Boreel, 10,172 tons; Union Leasing Co., 13,233 tons; and the Small Hopes, 9,917 tons. From the smelters' returns it appears that they treated 394,710 tons of Leadville ores, containing 31,236 tons of lead.

Of the production of zinc in the United States in 1895 Mineral Industry says: [40]

The production remained nearly the same in 1895 as in 1894, although there was a decrease from 1893. Zinc oxide is made in nearly all the producing districts, furnaces for this purpose being found at Joplin, Mo., Mineral Point, Wis., Waukegan, Ill., Lynchburg, Va., and Florence, Pa. The largest output is from Canon City, Colo., where it is made by the Bartlett process.

The Denver Republican, in its review for 1895, published January 1, 1896, gives the following table:

[37] Puckett, W. J., agent for Colorado, in Preston, R. E., Report of the Director of the Mint upon the production of the precious metals in the United States during the calendar year 1894, pp. 70-71, 73, 1895.

[38] Puckett, W. J., agent for Colorado, in Preston, R. E., op. cit. for 1895, pp. 73, 74, 75, 76, 1896.

[39] Kirchhoff, Charles, Lead: U. S. Geol. Survey Mineral Resources, 1895, p. 152, 1896.

[40] Mineral Industry for 1895, vol. 4, p. 582, 1896.

Production of American Zinc-Lead Co.'s smelter

District	Gold	Silver	Lead	Copper	Zinc
	Fine ounces	*Fine ounces*	*Pounds*	*Pounds*	*Pounds*
Lake	149	41,020	910,800		1,265,000
Clear Creek	76	3,910		5,760	200,000
Pitkin	42	2,200	26,000	875	21,000
Chaffee		840		63,800	120,000
Summit	24	2,570		1,500	
Teller	275				
Miscellaneous	28	3,100	155,000	78,487	65,000
Total, Colorado	594	53,640	1,091,800	150,422	1,671,000
New Mexico	960	19,200	28,000	404,000	17,000
Arizona	34	12,715	73,000	100,000	27,500
Total	1,588	85,555	1,192,800	654,422	1,715,500

In his report for 1896, Puckett says:[41]

At Leadville, from June 1 (when the miners struck for higher wages) to September 20 (when a violent attack was made against the Coronado mine) not one important mine was working. After September 20 production was under the protection of the State militia, and with the aid of imported miners slowly progressed. The approximate loss in output attributable to the strike was $3,000,000. The most significant progress of the year hung upon discoveries in the Lilian, on Printer Boy Hill; the Mahala, on Carbonate Hill; and the Sedalia, on Little Ellen Hill; but exploration on Fryer Hill, Poverty Flat, in Iowa Gulch, and other sections, was conspicuous.

The Denver Republican, in its review for 1896, published January 1, 1897, says:

Production of American Zinc-Lead Co., Canon City, Colo.

Pigment, 1,800 tons at $75	$135,000.00	
Copper, 667,240 pounds, at 11 cents	73,436.40	
Silver, 148,720 ounces, at 67 cents	99,642.40	
Gold, 2,165 ounces at $20	43,300.00	
	351,378.80	

District	Gold	Silver	Lead	Copper	Zinc
	Fine ounces	*Fine ounces*	*Pounds*	*Pounds*	*Pounds*
Leadville	210	38,460	805,140		642,000
Miscellaneous	1,323	59,220	423,160	93,240	650,000
Total, Colorado	1,533	97,680	1,228,300	93,240	1,292,000
New Mexico	612	34,440		395,600	
Arizona	15	14,420	69,100	91,200	428,000
Utah	5	2,180	43,600	87,200	80,000
	2,165	148,720	1,341,000	667,240	1,800,000

Bartlett describes the methods and practices at Leadville in 1896 as follows:[42]

Separation by concentration.—This method is practiced largely at Leadville, Kokomo, Georgetown, Creede, and many other places in Colorado, as well as at Park City, Utah, and some points in Montana. * * *

Judging from the assays of many thousands of zinc tailings purchased at the American zinc-lead works, Canon City, Colo., from many different concentrators, the best work done has been to reduce the lead to 2 per cent in the tailings, while they often run 8, 10, and 12 per cent lead, with an average of about 6 per cent, and a saving of 50 to 70 per cent of the gold and silver is as much as can be expected in zincky ores by concentration. The lead may be saved in the proportion of from 60 to 80 per cent, according to the class of ore.

Smelting at the Canon City plant.—The plant was devised by the writer for the purpose of utilizing the zinc and lead, as well as the gold, silver, and copper in the ore. The process is based on the result of many years' work with complex zincky ores. The original and fundamental idea is to drive off the zinc and lead in a mixture which can be refined into a white pigment used as a white lead substitute, and without loss of silver or gold. The writer found that a fractional distillation can be made of sulphide ores, driving off the lead and most of the zinc without undue loss of silver, provided there is present an excess of sulphur, or sulphide of iron or copper, sufficient to make a small amount of matte. The obstacles to be overcome were to prevent the formation of acid compounds destructive to the collecting apparatus, and to make a merchantable pigment when using ordinary bituminous slack coal and sulphide ores, the subsequent smelting of the cinder after eliminating the zinc being an exceedingly simple matter.

The process at Canon City, Colo., is in its simplest form as follows: All ores containing 20 per cent or more of zinc are crushed to pea size, mixed with 10 to 25 per cent of pea and dust coal, and blown up in a specially constructed furnace, using a blast of 4 to 8 ounces pressure. It requires from 20 to 40 minutes only to drive off the lead and the greatest part of the zinc and sulphur. A cinder is formed containing the silver, gold, and copper, in a matte mixed with more or less slag. The cinder is smelted in an inclined low-blast furnace, mixed with other suitable ores and fluxes to produce a high-grade matte. The zinc in this charge may reach from 15 to 20 per cent and is mostly driven off as fume. The fume from the "blowing-up" furnaces and from the blast furnaces is caught in bags, mixed, and refined in a suitable furnace, whereby carbon, arsenic, sulphur, and other impurities are eliminated, and a pure-white pigment produced, suitable, after being ground in oil, to be used as a white lead substitute.

Description of the furnaces.—As the processes used at Canon City are patented, and the manipulations are peculiar to this particular branch of work, it will be uninteresting to enter into minute details. Some ideas have been worked out, and some old theories have been exploded.

The loss of silver in the fume is mentioned on page 628.

Notes on the zinc-lead pigment.—The refined pigment produced at the Canon City works consists of an intimate mixture of zinc and lead in very stable form.[43] Chemists differ in their analysis of the refined product; broadly it may be said to consist of zinc oxide and lead sulphate, containing an excess of oxygen. The metallic constituents from an ordinary sample are as follows:

Zinc, metallic	47.33
Lead, metallic	24.92
Sulphur	2.96
Oxide of iron, etc	.45
Oxygen	24.34

This process has been used in a commercial way at Canon City for 10 years[44] and the pigment made has secured a regular market demand. This may not be the best or the most economical method of handling zincky ores, but it has been successful, the zinc and lead produced in pigment bringing as much in the market as they would were they reduced to metal, and the demand is fully as large.

Modifications of the Canon City process.—Zincky ores do not always carry enough in gold and silver to ship to smelting centers. The result is that there are vast quantities on the

[41] Puckett, W. J., agent for Colorado, in Preston, R. E., op. cit. for 1896, pp. 157, 158, 159, 1897.

[42] Bartlett, F. L., The treatment of zinc-lead sulphide ores: Mineral Industry for 1896, vol. 5, pp. 619-631, 1897.

[43] U. S. Patent No. 477,488.

[44] Telephone conversation with Mr. Bartlett, Jan. 24, 1914. Mr. Bartlett says that he started a plant at Canon City in 1890. Before this he had a plant in the East.—C. W. Henderson.

dumps and in the mines which carry less than $8 to $10 value in gold and silver, and from 25 to 35 per cent of zinc and 5 to 10 per cent of lead. Owing to the distance from fuel and high costs generally, such works as those at Canon City can not be carried to the mines, and it is doubtful if such ores will ever be worked. On the other hand, when the value in silver, gold, and copper is large enough so that the zinc and lead can be thrown away and still leave a profit, then a modified process can be applied to good advantage. Especially is this the case when fuel and labor are reasonably cheap, and other ores can be had for mixtures. Even heap roasting and treatment in the Canon City blast furnaces for the production of coarse matte is often admissible and more profitable than long shipments of the raw ore to the markets. The commendable features of the Canon City process are cheapness of treatment and a fair saving of value.

The operations in 1897 are described by Hodges as follows:[45]

The Leadville strike, which terminated in March, 1897, was the single labor trouble occurring in any camp of the State during the year. * * *

The labor strike in Leadville in 1896 and 1897 caused the suspension of pumping in the large down-town mines. Up to this time the operators have been unable to effect an agreement in adjusting pumping expenses, which has caused a large number of Leadville's greatest mines to lie idle the entire year. * * * Arrangements for unwatering and reopening these mines are in a fair way to be consummated. * * *

Placer mining, once the chief source of gold production, was relatively a small factor in 1897. * * * At Granite, Chaffee County; at Twin Lakes, in Lake County; and in Park and Summit counties placer mining was conducted on a more pretentious scale and was generally remunerative. [The Twin Lakes Placer Co.'s property is in Chaffee County, so no placer is given to Lake County for this year.] * * *

The deepest shaft (at Leadville) is 1,250 feet and is pumping 1,000 gallons of water per minute. There are six shafts from 900 to 1,200 feet deep, and the Yak tunnel is now in 6,800 feet and progressing at about 225 feet per month. * * *

There are two smelters operating, having a capacity of 600 tons. Three concentrating plants are of 300 tons capacity per day. The labor trouble of 1897 decreased the output of the district about 30 per cent.

The Denver Republican, in its review for 1897, published January 1, 1898, gives the following table:

Production of American zinc-lead smelter, Canon City, Colo.

			Price	Value
Gold	ounces	588.761	$20.00	$11,775.22
Silver	do	164,524.00	.62	102,003.88
Lead	pounds	2,149,237	.03½	75,223.29
Copper	do	545,769	.11	60,034.59
Zinc	do	3,578,652	.04	143,186.08
				392,223.06

The Leadville Herald-Democrat, in its review for 1897, published January 1, 1898, gives the following statement:

Bullion produced from Leadville district ores by outside smelters.

American Zinc-Lead Co., Canon City, Colo.

4,403 tons smelted.
149.46 ounces gold.
49,012 ounces silver.
1,570,000 pounds lead produced.
2,201,500 pounds zinc produced.
(Average zinc recovered per ton smelted, 25 per cent.)

Mineral Industry for 1897 says:[46]

The production of zinc oxide in the United States in 1897 was 26,262 short tons, against 15,863 short tons in 1896. This was produced chiefly at the three works of New Jersey by the Wetherill process; but there was also a good deal of zinc-lead pigment produced in Colorado, which has been reckoned as zinc oxide on the basis of its tenor in zinc.

For 1897–1904 the figures given in the table on page 176, which represent smelter and mint receipts, are taken from the reports of the Colorado State Bureau of Mines.

In his report for 1898 Hodges says:[47]

The several pumping plants are now handling from 6,000 to 6,500 gallons per minute, fully one-half coming from what is known as the down-town mines, which have been idle some three years. By May, 1899, these mines will have been drained and mining resumed. This down-town drainage will be felt in sections of nearly 4 miles in length.

The official tonnage shipments of Leadville district in 1898 were as follows:

	Short tons
Sulphide ores	206,555
Oxidized iron ores	150,980
Carbonate ores	82,650
Siliceous ores	60,170
Manganese ores	17,637
Total for 1898	517,992
Total for 1897	413,552
Increase	104,440

The flow of water in 1898, as compared with 1897, is about the same.

Projected drainage tunnel.—A drainage tunnel is projected, starting near Malta station, on the Denver & Rio Grande Railway, some 5½ miles west of the center of the productive portion of the camp, which will reach a depth of several hundred feet below most of the mines, excepting the Mahala and Rialto claims and workings.

The Arkansas Valley smelter, located at Leadville, has a daily capacity of 1,000 tons and is the largest in Colorado. About the deepest workings reached is 1,200 feet, in Mahala and Rialto ground. * * *

Dividends paid for working Leadville mines in 1898, $1,655,000.

In his report for 1898 Kirchhoff says:[48]

An important event which has since transpired was the culmination, in 1899, of negotiations for the consolidation of the principal lead-smelting and desilverizing plants in the United States under the title of the American Smelting & Refining Co., with an issue of $27,400,000 of 7 per cent cumulative stock and

45 Hodges, J. L., agent for Colorado, in Roberts, G. E., Report of the Director of the Mint upon the production of the precious metals in the United States during the calendar year 1897, pp. 111, 112, 113, 124-125, 126-127, 1898.

46 Mineral Industry for 1897, vol. 6, p. 661, 1898.
47 Hodges, J. L., agent for Colorado, in Roberts, G. E., op. cit. for 1898, pp. 86-87, 99-100, 1899.
48 Kirchhoff, Charles, Lead: U. S. Geol. Survey Mineral Resources, 1898, pp. 221-222, 1899.

$27,400,000 of common stock outstanding. The bonded indebtedness is $1,133,000 6 per cent bonds of the Omaha & Grant Smelting Co., due March 1, 1911, and $1,000,000 6 per cent mortgage bonds of the Consolidated Kansas City Smelting & Refining Co., due May 1, 1900.

The company is the owner of all the property, rights, and assets of every kind owned by the following corporations: The United Smelting & Refining Co., Helena and Great Falls, Mont.; National Smelting Co., Chicago, Ill.; Omaha & Grant Smelting Co., Omaha, Nebr., and Denver, Colo.; San Juan Smelting & Refining Co., Durango, Colo.; Pueblo Smelting & Refining Co., Pueblo, Colo.; Colorado Smelting Co., Pueblo, Colo.; Hanauer Smelting Works, Salt Lake City, Utah; Pennsylvania Lead Co., Pennsylvania Smelting Co., Salt Lake City, Utah, and Pittsburgh, Pa.; Globe Smelting & Refining Co., Denver, Colo.; Bimetallic Smelting Co., Leadville, Colo.; Germania Lead Works, Salt Lake City, Utah; Consolidated Kansas City Smelting & Refining Co., Kansas City, Mo., and El Paso, Tex.; Chicago & Aurora Smelting & Refining Co., Chicago and Aurora, Ill., and Leadville, Colo.

This list includes all the more important lead-smelting plants in the Rocky Mountain region, with the exception of the Philadelphia Smelting & Refining Co., at Pueblo, Colo., and the Tacoma Smelting & Refining Co., at Tacoma, Wash., the Puget Sound Reduction Co., at Everett, Wash., and the Selby Smelting & Lead Co., at San Francisco, Calif.

It includes all the desilverizing and refining plants, except that of the Selby Smelting & Lead Co. and the Puget Sound Reduction Co., the latter having started a new plant in 1898, nor does it include the two tidewater plants of the Guggenheim Smelting Co., at Perth Amboy, N. J., and the Balbach Smelting & Refining Co., at Newark, N. J., both of which are almost exclusively engaged in refining foreign base bullion.

The Denver Republican, in its review for 1898, published January 1, 1899, gives the following table:

Production of the American Zinc-Lead Smelter, Canon City, Colo., 1898

Metal		Quantity	Price	Value
Gold	ounces	967.10	$20.6718	$19,996.69
Silver	do	101,483	.5825	59,113.84
Lead	pounds	1,701,098	.0363	61,749.85
Copper	do	351,166	.12	43,339.92
Zinc	do	3,900,656	.06	234,039.36
Total from Colorado ores, 1898				418,239.60
Total from Colorado ores, 1897				404,005.39

The Leadville Herald-Democrat, in its review for 1898, published January 1, 1899, gives the output of Lake County as 459,056 tons. The bullion produced from ores from the Leadville district by outside smelters during 1898 includes the following amounts:

American Zinc-Lead Co., Canon City, Colo.

4,289 tons smelted.
128.67 ounces gold.
30,412 ounces silver.
76,927 pounds copper.
833,189 pounds lead produced.
2,673,400 pounds zinc produced.
Average zinc recovered per ton smelted, 31 per cent.

Mineral Industry for 1898 says: [49]

Zinc oxide.—Aside from the New Jersey Zinc Co., which in 1898 made zinc white at Jersey City, Newark, and Bethlehem, this pigment was produced by an allied company at Mineral Point, Wis. Page & Krause, of St. Louis, Mo., also made a small amount of oxide, not more than 500 tons per annum, while the Standard Oil Co. recovers a little as a by-product at Williamsburg, Brooklyn, N. Y. The only other producer in the United States is the American Zinc-Lead Co., of Canon City, Colo., which makes a zinc-lead pigment.

Spelter.—A small amount of zinc blende concentrates was shipped in 1898 from a mine at Creede. In general, Colorado zinc ore carries too much iron to be desirable for zinc smelting.

In his report for 1899 Hodges says: [50]

The Leadville district was never more active since the famous carbonate discoveries of 1878 than in the year 1899, especially during the latter half of the year. Well-known producers were worked to the limit, and an exceptionally large amount of exploration and new work was undertaken. Many drawbacks were, however, encountered.

The severity of the weather from January to April, 1899, seriously crippled the camp's efforts. Snow slides were numerous, the drifts for months being of great depth, and blockades on the railroads, which largely defied the mammoth steam plows, were the order of these arctic days. Many railroad spurs to the mines were necessarily abandoned. Fuel was dangerously scarce and a food famine threatened.

The smelter shutdown for two months pending settlement of the eight-hour law also militated seriously against the year's production.

After the termination of the trouble in the courts a vexing car shortage ensued, and weeks elapsed ere the accumulated ore product was given car accommodation.

The stupendous task of unwatering the Leadville basin or Downtown mines was completed in May by the Home Mining Co. This company was capitalized at 50,000 shares, par value $1. It successfully drained these properties, cleared away the drifts, and speedily discovered high-grade chloride iron and lead bodies. Shipments were made on a large scale from the Penrose, Bon Air, and Starr shafts, the Penrose being especially prominent in the output of silver chlorides. * * *

The result of this pumping victory was the resumption of work by the Wolftone, Weldon, Bohn, Northern, Midas, Colonnade, and Sixth Street properties. * * *

The possibility of a tunnel from Malta, 5 miles from Leadville and at the foot of the slope which it crowns, to tap the district's large mines at great depth, is actively discussed and promises realization at an early day. * * * [Not realized to date, July 1, 1925.—C. W. H.]

The building of additional spurs by the Denver & Rio Grande and Colorado & Southern railroads to the important locations has rendered the entire district readily accessible and makes possible the shipment of low-grade silver-lead and gold ores in great quantity.

Four smelters are now local to the district, a new pyritic plant, the Boston Gold-Copper Smelting Co., having reinforced the Arkansas Valley, Union, and Bimetallic.

Bismuth and zinc ores.—Valuable bismuth ore is shipped to England by the Ballard mine, carrying rich value in gold.

The zinc ores of the district have largely found a market in Belgium.

The manganese comes in greatest quantity from Carbonate Hill, and Chicago is the depot of their shipment, its prominent producers being the Catalpa, Crescent, Seneca, Garden City, and Lost Chip.

Much leasing is incident to Fryer Hill. Iron predominates in the ore formation, but lead and silver are also carried. Fryer's main exponents are the Robert E. Lee, Augusta, Matchless, Gambetta, Niles, Chrysolite, Dunkin, and Cady.

[49] Mineral Industry for 1898, vol 7, pp. 724, 727, 1899.

[50] Hodges, J. L., agent for Colorado, in Roberts, G. E., op. cit. for 1899, pp. 100-102, 121-122, 1900.

In the Graham Park sulphide belt the Maid of Erin, Wolftone, and Adams have prepared to attain great depth.

The A. Y. and Minnie and Rubie and Moyer produced largely, and the Mike and Starr marketed fine bodies of iron sulphides.

The gold district.—On Breece Hill the Penn shafts, Ballard, Fanny Rawlings, St. Louis tunnel, Little Vinnie, and Fraction were heavy shippers.

The Ibex group employs over 400 men and boasts nearly 60 miles of workings. The tonnage is about 60 per cent sulphide and 40 oxide, gold bearing, and was very large throughout the year.

The Triumph and Modoc are active in this neighborhood.

The Resurrection ore bodies have been continuously large and of fair grade. Extensive improvements on this property promise great results for its extensive acreage.

The old Lilian, in Iowa Gulch, continues its output.

On Little Ella Hill the New Year and Little Ella produced steadily. Large bodies of sulphides were uncovered by the New Monarch.

The Penfield and Fortune were new shippers, and the Reindeer to a degree.

The Dolly B. found good ore throughout the year, and the company is sinking a large shaft on the Board of Trade group.

The Yak Mining & Milling Co. is pushing its great bore from California Gulch to the Breece Hill gold section, and it has reached a length of 8,000 feet.

On Iron Hill the Iron Silver Mining Co. was a big shipper, working through its Moyer shaft. Iron sulphides have predominated in the shipments, but Wilfley concentrating tables are now handling the lead-zinc portion of the ores.

In his report for 1899 Kirchhoff says:[51]

The high prices of zinc ore attracted attention to the mineral in Colorado, and considerable quantities were shipped during 1899 from Leadville, Creede, and Montezuma, a part going via Galveston.

Kirchhoff gives a table of domestic exports of zinc, by customs districts, during the calendar year 1899, which shows that of the 5,847 tons of ore exported from Galveston nearly all was mined in Colorado. Another table, showing the distribution of exports in 1899, includes 439 tons shipped from New Orleans.

Mineral Industry[52] for 1899 reports that 8,000 tons of zinc ore was shipped to foreign smelters by way of Galveston, and that zinc ore was shipped from Creede, Leadville, and Montezuma to Kansas, Illinois, Indiana, and Belgium.

The Leadville Herald-Democrat, in its review for 1899, published January 1, 1900, gives the following figures:

Zinc-lead ore from Leadville produced and smelted in 1899

	Ore (tons)	Gold (ounces)	Silver (ounces)	Lead (pounds)	Copper (pounds)	Zinc (pounds)	Average per cent of zinc
American Zinc-Lead Co	3,343	167.17	36,777	1,003,020	66,868	2,016,040	30
Zinc ore sent to smelters outside of Colorado	10,699					8,559,200	40

51 Kirchhoff, Charles, Zinc: U. S. Geol. Survey Mineral Resources, 1899, pp. 254, 260, 261, 1901.
52 Mineral Industry for 1899, vol 8, pp. 636, 650, 1900.

Zinc ore produced by mines at Leadville in 1899, by mines

	Tons
Iron Silver	4,125
A. Y. and Minnie	665
Maid & Henriette	5,343
Louisville	200
Boreel	366
	10,699

Classification of ore

	Tons
Carbonate	32,050
Iron oxide	123,787
Sulphide	238,514
Zinc sulphide	10,699
Siliceous	105,025
Manganese (silver) oxide	15,653
	525,728

Redick R. Moore (agent shipping zinc ore to Belgium) says in part: "Owing to the high prices for spelter prevailing during the past year it was found possible to export several thousand tons of selected ores. These ores, carefully mined and in most cases carefully hand sorted, varied from 40 to 45 per cent."

The Iron Silver.—Since early in 1899, when Mr. T. E. Schwarz succeeded Captain Robinson in the management, a large tonnage of sulphide has been maintained. The largest * * * shipments have been iron sulphides, carrying a heavy iron excess, with low gold and silver values. Lead-zinc-iron sulphides also occur, for the treatment of which the old mill on the property has recently been remodeled. Zinc ores have been and are still being shipped crude from the Moyer workings to Belgium. This property was the first to furnish crude ore carrying 40 to 50 per cent zinc for export to Antwerp and has continued the heaviest individual producer of this class of ore. The zinc product is sampled and sacked in Leadville at Norton's Sampling Works and shipped via Galveston.

The mill is equipped with two pairs of coarse rolls and two of finishing rolls and 14 Wilfley tables, the tailings from the coarse tables being reground for the final treatment on the fine tables. Its capacity is about 80 tons per 24 hours. The product is a lead-iron concentrate low in zinc. It is also adapted to making a high-grade zinc concentrate from certain zincky ores.

In 1899 the Engineering and Mining Journal says:

Zinc mining.—One zinc milling scheme just perfected is known as the Golob-Colley Milling Co., the promoters being James Golob, of the Ballard mine, and James Colley, who have made experiments and have leased the old Tabor mill on the Arnold placer in California Gulch for two years. The mill has 20 stamps, and some new concentrators are to be purchased at once. The company will get its supply from the A. Y. and Minnie and Maid of Erin mines and will ship to Indiana.[53]

Zinc output.—The A. Y. and Minnie will be the next regular zinc producer. Lessees Newton and Douglas have arranged for a 2,000-ton shipment to Germany. They have large zinciferous deposits in the Minnie workings.[54]

Zinc.—There is a general scramble for good zinc ores. Mr. R. R. Moore is arranging to make another shipment to Antwerp by September. Among the new zinc producers the Louisville and the A. Y. and Minnie will largely increase their tonnage.[55]

Foreign zinc shipment.—A solid train of 30 cars, or 25,000 sacks, has been loaded with zinc ores for the Vieille Montagne at Antwerp. The cars go to Galveston, and the shipment is made

53 Eng. and Min. Jour., vol. 67, p. 657, 1899.
54 Idem, vol. 68, p. 226, 1899.
55 Idem 346

through A. J. Davis for Jacobson & Co., of New York City. The shipment is the largest of any kind of ore ever sent out to any foreign port. It consists of Maid of Erin and Moyer ores and concentrates from the Golob-Colley mill.[56]

The Denver Republican, in its review for 1899, published January 1, 1900, gives the following table:

Production of American Zinc-Lead Smelter, at Canon City, Colo., from Colorado ores in 1899

Metal		Quantity	Price	Value
Gold	ounces..	865.23	$20.67	$17,884.39
Silver	do	98,493	.5987	58,967.76
Lead	pounds..	1,802,100	.0426	76,769.46
Copper	do	325,170	.1657	53,930.67
Zinc	do	4,100,670	.09	369,060.30
				576,612.49

Mineral Industry for 1899 says:[57]

Zinc oxide.—The production of zinc oxide in the United States in 1899 amounted to 39,663 short tons, an increase of 6,916 tons over the production of 1898. Zinc oxide has been in large demand, stimulated by the comparatively low prices at which contracts for the year were made. This branch of the industry is still on a sound basis. There was an increase in average value at the works from $68 per ton in 1898 to $84 per ton in 1899, due to the advanced cost of labor, ore, supplies, and freights. The greater part of the output was made by the New Jersey Zinc Co., at Jersey City, Newark, N. J., Bethlehem, Pa., and Mineral Point, Wis. Page & Krause, at St. Louis, make annually from 400 to 500 tons and the American Zinc-Lead Co., of Canon City, Colo., produce a pigment which is a mixture of oxidized compounds of zinc and lead.

Production of zinc oxide in the United States

Year	Short tons	Total	Per short ton
1893	25,000	$1,875,000	$75.00
1894	22,814	1,711,275	75.00
1895	22,690	1,588,300	70.00
1896	15,863	1,189,725	75.00
1897	26,262	1,686,020	64.20
1898	32,747	2,226,796	68.00
1899	39,663	3,331,692	84.00

Colorado.—Considerable ore was shipped from Creede, Leadville, and other points to the zinc smelters for experiments involving the use of large quantities in the furnaces.

While the zinc ores of Colorado are generally too impure from their iron and lead content to be desirable for zinc smelting, there have been some clean concentrates made that, considering their low price, were available for profitable treatment.

About 8,000 tons at a nominal price were shipped to foreign smelters via Galveston.

The high price for zinc ore that prevailed in the Joplin district during the first three quarters of 1899 led the smelters to look for supplies of cheaper ores elsewhere. As a result of such inquiries a considerable amount of ore was obtained from other States, especially from Colorado. Zinc ore was shipped from at least three points in Colorado, namely, Creede, Leadville, and Montezuma. These shipments went to works in Kansas, in Illinois, and in Indiana, besides a considerable quantity which was exported to Belgium, via Galveston. In general, the Colorado ore is both ferruginous and lead bearing, and can not be freed cleanly from either of the undesirable elements, on which account it is an inferior ore at best. However, some of the

Colorado ore has been dressed so cleanly as to be of very high grade and a desirable product to buy, in view of the lower cost at which it can be had.

In his report for 1900 Hodges [58] says:

With the exception of some new discoveries of oxidized ore on Carbonate Hill and the desultory extraction of small bodies left in the old workings of once prominent mines on Fryer Hill, the principal production of carbonates and oxides is from the comparatively new zone located east of the Ball Mountain fault, represented by the Resurrection group, and the workings in the western end of the field by the Home, Midas, and other companies. Iron Hill should be included in this class of production, although its tonnage of oxidized ores is a variable quantity as the years pass. A very large part of the year's production of oxidized ores is due to the work of lessees, as the lease system gains in favor each year. As the former immense bodies of high-grade carbonates gave out new conditions have presented. * * *

The advancement in mining in the Leadville district during the past year has been along the line of consolidation of old properties into new companies. This has been especially emphasized in the formation of home companies. The most striking example is that incorporation known as the Home Mining Co. This company secured a large territory within the city limits, worked through the Penrose, Star, and Bon Air shafts. This area is made up of a number of long-time leases and is entirely in the Leadville basin. During the miners' strike of 1896 a portion of this ground was worked by the owners of the Maid of Erin and Henriette properties, but owing to the continuance of the strike these gentlemen decided that as the mine was filling rapidly with water they would take out their pumps and abandon this ground. Two years ago the effort was made to get a sufficient number of home people interested to furnish the capital to unwater this territory and prospect the ground for the great ore shoots which were thought to continue from Carbonate Hill into this ground.

It is the history of a long and desperate struggle with a heavy flow of water, with insufficient pumping arrangements, and of final triumph in placing the Penrose and Bon Air shafts again in the list of producers. The year 1900 brought this enterprise to the front rank as a producer. * * *

The Midas is one of the most important of the downtown shippers. During the past year its shipments have averaged 5,000 tons a month, the ore being encountered at a depth of 515 feet, and has been developed 500 feet to the Penrose mine and 300 feet toward the Coronado.

The Penrose, of the Home Co., is producing daily 350 tons, and the Bon Air and Star shafts, of the same company, 200 tons more. * * *

The Pumping Association.—Perhaps the factor which entered most largely into the success of the Home Mining Co. and the other enterprises of a similar character was the Pumping Association, formed for the purpose of unwatering the Leadville basin. This association includes nearly all the leasing companies, as well as the owners of territory embraced in the Leadville basin. All mines operating within the association territory bear the cost of pumping in proportion to their output, based on net smelter returns, less cost of hauling. By means of counters on these pumps the amount pumped is computed in gallons and charged to the association at the rate of 10 cents per 1,000 gallons. Those mines which pump are credited with the amount of water they have raised.

Taking the entire district, investigation shows that the flow of water which must be handled is not less than 15,000,000 gallons a day. Comparing this amount of water with the

[56] Eng. and Min. Jour., vol. 68, p. 587.
[57] Mineral Industry for 1899, vol. 8, pp. 635, 636, 650, 1900.

[58] Hodges, J. L., agent for Colorado, in Roberts, G. E., op. cit. for 1900, pp. 115-124, 136, 1901.

average daily tonnage of the district for the past year, we find that 28.6 tons of water are raised for every ton of ore raised. Careful estimates of the cost of pumping have been compiled and show that it costs 4 cents to pump each ton of water to the surface. Hence the cost of pumping referred to the ore makes a charge of $1.14 per ton extracted. As the amount of ore shipped during the past year was about 803,000 tons, the cost of pumping for the year amounted to $915,000.

The total amount of water is decreasing in the area affected as the country is kept drained and other mines in that area start pumping on their own account. Thus in November, 1899, the association pumped 68,018,592 gallons, while in November, 1900, the amount had fallen to 48,622,546 gallons, a decrease of nearly 30 per cent. * * *

The mining industry in the Leadville district has shown a great advancement over the previous year, due in a large measure to recovery from the smelter strike of 1898 and the extremely severe winters of 1898 and 1899, but the increased output seems to have its real basis in the opening up of new territory to the southwest and the northeast.

The possibilities in both of these directions are very good. The outlying districts have not changed materially, except that more activity in prospecting has been noticeable in the Sugar Loaf district, although no important strikes are reported from there.

The Arkansas Valley smelting plant, of the American Smelting & Refining Co., has expended some $375,000 in enlarging its furnaces and buildings, perhaps the most elaborate system of fume-condensing chambers or flues in the country. This plant has 10 furnaces, 4 of which have a capacity of 150 tons daily, and the balance are being changed to the same size. This will give this plant a daily capacity of 1,500 tons, making it by far the largest lead-smelting plant in the country. The improvements include also the addition of a large amount of trackage for the handling of increased supplies of ore and fluxes and the using of large slag pots handled entirely by animal power, light tracks running to each of the furnaces. * * *

The American Smelting Co. owns two other smelting plants in Leadville—the Union and Bimetallic smelters. The Union has been put in thorough order.

New pyritic smelter.—In November, 1899, a pyritic smelter was started by the Boston & Colorado Smelting Co. [Boston Gold-Copper Smelting Co.], which has been running continuously during the past year, except when increasing its plant. Starting with one furnace, it now has three, each o which can handle 200 tons of charge per day. The average percentage of fuel to the charge in these furnaces is 6½. The fuel used is coke. The ore being charged as a raw sulphide, the burning sulphur replaces the excess of fuel used in the lead smelters. No roasting is required for the ores, as the product, when obtained, is in iron or copper matte. Very low-grade ores are treated in this smelter.

Bismuth ores.—As an interesting and comparatively new product from the Leadville mines may be mentioned the bismuth ores on Breece Hill. This ore seems to be a mixture of bismuth carbonate and oxide, or a bismuth ocher with the carbonate predominating. The values of the ore range from 2 to 40 per cent bismuth and from one-half ounce to 70 ounces in gold to the ton, but of the amounts shipped the average values are bismuth, 8 to 10 per cent, and gold, 1½ ounces to the ton. The producing properties are the Ballard, Big Six, and Penn groups of mines.

All the output of these properties in this class of product has been contracted for by an English metallurgical establishment for the next two years, the gold being paid for at its full value and the bismuth at the rate of $15.50 for each unit.

The shipments during 1900 from the above-mentioned mines were as follows:

	Tons
Ballard	140
Penn	70
Big Six	70
	280

The manganese production.—The manganese production of the district fell off during the year. The principal amounts of this ore have come from Fryer, Carbonate, and Rock hills and seem to have formed a selvage to the large oxidized iron bodies in those hills. As the requirements of the steel works have called for an ore with a high percentage of manganese, the reserves of this grade have gradually been exhausted. The original requirements called for an ore of not less than 30 per cent, but finally an ore of 20 per cent was accepted. The principal shippers of manganese ore—the Catalpa-Crescent properties—shut down during June, 1900. * * *

Leasing and royalties.—The great bulk of the ore extracted in the Leadville district is done through the leasing system, comparatively few owners of mines operating their own properties. The royalties paid are usually fixed on a sliding scale, according to the value of the ore, from 10 per cent to 50 per cent of the net smelter returns. Perhaps there is no other mining district in the State where lessees are willing to expend such large amounts of money in the installation of plants of machinery before any ore is extracted. A case in point is the A. M. W. lease, where the leasing company has expended over $100,000 in surface, plant, and pumps, in unwatering the mine, and enlarging the main working shaft to suit the increased hoisting plant. * * *

Lead-zinc ores.—The lead-zinc ores of the district have received a good deal of attention, the effort being to separate by concentration the zinc and iron from the lead, making a product which can be handled by the smelters, and then separating the iron sulphides as far as possible from the zinc sulphides, a product suitable to the zinc smelters. As the zinc product obtained has only found a market during the last two years, it is a comparatively new industry, as heretofore the zinc paid a penalty at the lead smelters.

The vast bodies of zinc-lead sulphides which exist in the lower workings of Carbonate and Iron hills will ultimately be worked for their zinc contents, as well as for the lead, but the present outlook is in the direction of treating this class of ores by some leaching process, which will extract the zinc and leave a residue which can be handled profitably by the lead smelters.

Value of zinc-ore shipments.—The amount of zinc ore shipped during the past year, nearly one-half going to smelting works in Europe, is estimated at about 111,000 tons, which brought about $10 per ton, or a valuation of $1,110,000.

H. A. Lee, commissioner of mines for the State of Colorado, in his report for 1900 says: [59]

The zinc production [of Colorado] for 1900 amounted to 77,984 tons, averaging 42 per cent metallic zinc. The prices on board cars at place of production ranged from $4.50 to $15 per ton. About 40 per cent of this production was exported to Europe or Canada, the remainder being marketed in the United States. Lake, Mineral, Chaffee, Eagle, Gunnison, Saguache, Clear Creek, Custer, Pitkin, Ouray, Dolores, and Summit counties produced the zinc ores.

The output mentioned, 77,984 tons of 42 per cent zinc, represents 32,753.28 tons, or 65,506,560 pounds

---

[59] Eng. and Min. Jour., vol. 71, p. 490, 1901.

of zinc, which seems out of reason in comparison with other data that follow.

The Leadville Herald-Democrat, in its review for 1900, published January 1, 1901, says:

Zinc-lead ore produced and smelted at Leadville in 1900

	Tons	Gold (ounces)	Silver (ounces)	Lead (pounds)	Copper (pounds)	Zinc (pounds)	Per cent of zinc
American Zinc-Lead Co...	5,769	173.04	43,684	1,384,980	8,463	3,322,720	28.8
Zinc ore sent to smelters outside of Colorado.....	59,926					41,948,200	35.0
						45,270,920	

Zinc-lead ore produced by mines at Leadville in 1900

	Tons
Henriette and Maid of Erin............	53,374
A. Y. and Minnie	1,200
Small Hopes..........................	147
Marion Lease.........................	1,172
Boreel...............................	222
Moyer................................	1,840
Denver City..........................	821
Habendum.............................	150
Yak..................................	1,000
	59,926

The Henriette and Maid of Erin mine produced only 5,343 tons in 1899.

A. M. W. mines.—The great zinc production comes from the Adams at the 620-foot level and has added largely to the value of the lease during the year. One important part of the A. M. W. surface equipment is the concentrating mill capable of handling 100 tons of ore daily. The lead, zinc, and iron sulphides which abound in the ground controlled by the company are, when not of sufficiently high grade to ship, susceptible of concentration. The ore is crushed in Huntington mills, the intention being to make two products, a lead-iron, which is desirable for the smelters, and a zinc concentrate, which is sufficiently free from iron to make a salable product.

Mineral Industry for 1900 says: [60]

The mines in the Leadville district, Colo., have contributed a considerable portion of zinciferous tailings and ores, material which has hitherto been regarded as waste. About 14,000 tons were shipped in 1900, of which the greater part was consigned to Europe, chiefly to the Vieille Montagne works via Galveston and New Orleans.

The production of zinc oxide in the United States in 1900 was 47,151 short tons, against 39,663 short tons in 1899, an increase of 19 per cent. The average value of this material at the works during 1900 was $80 per short ton, against $84 in the previous year. The greater part of the output was made by the New Jersey Zinc Co., operating at Jersey City and Newark, N. J., Bethlehem and Palmerton, Pa., and Mineral Point, Wis. Page & Krause, of St. Louis, Mo., produce annually between 400 and 500 tons, and the American Zinc-Lead Co., of Canon City, Colo., manufacturers of pigment called "zinc-lead," which is a mixture of oxidized compounds of zinc and lead. * * *

The main feature of the zinc mining industry during 1900 was the large increase in the shipment of zinciferous ores and concentrates. A few of the Leadville mines have always pro-

duced material of this character that involved so high a treatment charge at the smelter as to require dressing in order to remove the objectionable zincky portion. In this way large quantities of tailings have accumulated which, with the crude zinc ores formerly regarded as waste, have been recently shipped to zinc smelters, the greater portion being sent to Belgium, via Galveston and New Orleans, and the remainder to Iola, Kans.; Mineral Point, Wis.; and to Indiana. At least 14,000 tons of the material were shipped abroad in 1900, against 8,000 in 1899.

The Wilfley table has been successfully used for the separation of the lead-iron-silver product from the zinciferous portion, the former being an excellent fluxing material for the lead smelters. The purchasers of tailings and waste ores do not pay for the small amount of gold and silver contained. A shipment of zinc tailings from the A. Y. and Minnie mine was made to the works of the Nickel Copper Co. of Ontario (Ltd.), at Hamilton, for experimental purposes.

With respect to the new supply of zinc ore from Colorado, it has been known for many years that the State possessed vast resources of mixed sulphide ore, most of it argentiferous and some of it rather high in zinc but still mixed with sufficient lead and iron to make it undesirable material for the zinc smelters.

In 1899 and 1900 some of the smelters of Kansas and Missouri purchased a considerable quantity of such Colorado concentrates as were better than the average, but the chief trade in that material has been developed with smelters in Wales and Belgium, who entered the market in 1899 and in 1900 and bought largely. This trade has been made possible by the favorable freight rates which have been obtained, the ore being carried from Leadville via Galveston to Swansea or Antwerp at a cost of less than $10 per ton of 2,000 pounds. The mines receive comparatively little for the material, $5 per ton being the usual price, but inasmuch as their zinc concentrates are purely a by-product and the value of the ore is increased by their removal, the miners can very well afford to sell them even at so low a price. The fact that a market for such material has been established excited general attention in Colorado and steps were taken to increase the milling capacity. At the same time, experiments were made with the Wetherill magnetic separator and other special processes with a view to the more profitable development of the State's zinc resources.

In his report for 1900 Kirchhoff says: [61]

The New Jersey Zinc Co. has begun the erection at Canon City, Colo., of a plant for concentrating the complex zinc ores of the Leadville district.

He gives a table showing exports of zinc ore by customs districts during 1899 and 1900. The Gulf ports shipped ore from the Joplin-Galena district and from Colorado. The table of destinations of zinc ore exports shows that ore shipped from Gulf ports to Belgium, during the year 1900, amounted to 2,273 tons from Galveston and 9,150 tons from New Orleans, a total of 11,423 tons.

The Engineering and Mining Journal for the period January to June, 1900, says: [62]

Zinc production.—This output shows a steady increase. Three zinc mills are running—the Moyer on ore from the Moyer mine, the Golob-Colley on Maid of Erin ore, and the Maid mill also handles 20 tons daily from its own workings. The Maid of Erin lessees are also mining 50 to 75 tons daily of crude zinc ore, which goes to the Lanyon works at Iola, Kans. [p. 88].

[60] Mineral Industry for 1900, vol. 9, pp. 659, 660, 661, 672, 1901.

[61] Kirchhoff, Charles, Zinc: U. S. Geol. Survey Mineral Resources, 1900, pp. 216, 222, 1901.

[62] Eng. and Min. Jour., vol. 69, pp. 88, 147, 358, 417, 508, 628, 687, 718, 778, 1900.

Another important zinc shipment was made this week to the Vieille Montagne works near Antwerp, Belgium, through Jacobson & Co., of New York. It comprised 40 cars of crude ore and concentrates, making a total of 1,200 tons. It goes via Galveston. The average Leadville zinc ore nets the miner about $10 a ton [p. 147].

In order not to depend entirely on the foreign zinc market, the local zinc producers have effected arrangements to handle the Leadville zinciferous ores. The Mineral Point, Wis., and the Lanyon concern at Iola, Kans., have agreed to take 100-ton shipments of concentrates, and shipments will be begun at once. Several of the Missouri plants are also in the market for Leadville zinciferous ores [p. 358].

Probably the biggest foreign shipment ever made was this week, when 100 cars carrying 3,500 tons of zinciferous ores and concentrates were shipped from Leadville direct to the Vieille Montagne Smelting Co., at Antwerp, Jacobson & Co., of New York, brokers [p. 417].

The tonnage [of shipments] is 200 tons per day of concentrates and crude ore. Nearly all of this goes to foreign ports and will be considerably increased in May [p. 508].

Jacobson & Co. will make another zinc shipment this week of over 1,500 tons to Belgium. The stuff is now being loaded on cars for Galveston. Most of this comes from the Adams lease of the A. M. W. combination [p. 628].

The output of zinc in May was over 6,000 tons, a pretty heavy tonnage compared to no product a year ago. Another heavy shipment has just been made to Belgium [p. 687].

The zinc product from the Maid and Henriette and the two concentrating mills now averages about 300 tons per day, and several big shipments to Belgium were made this week. Nearly all of the product is sent to foreign plants [p. 718].

This [Leadville zinc production] has risen from nothing to over 5,000 tons a month, while it is intimated that the formation in Missouri of a zinc-lead combination will have a tendency to bring Leadville into that market and greatly increase the tonnage of zinciferous ores from here [p. 778].

For the period July to December, 1900, the Journal says: [63]

Big zinc shipment.—Another zinc ore shipment of 2,000 tons leaves here this week for Galveston, where it will be sent to Europe. The shipment is made through Jacobson & Co., of New York City, who are handling a great many shipments of this character from this camp [p. 18].

Jacobson arranged for 1,800 tons shipment. Most of this from Moyer of the Iron Silver Co., and the Maid [p. 78].

Jacobson & Co. make another shipment this week of 1,500 tons of zinc ore and concentrates to Europe. Most of this from Maid and Henriette [p. 137].

It is said that arrangements are being made to work Colorado zinc ores in Denver or vicinity. Some Colorado ores carrying zinc have been shipped to Indiana smelters, and their representative has recently been inquiring into the quantity of ores which can be obtained [p. 151].

Adams mill erected. Blake crusher, three 5-foot Huntington mills, 12 concentrating tables. To work on A. M. W. combination, Adams, Mab, and Wolftone.

Zinc market.—Over 10,000 tons a month are shipped. Another 4,000-ton shipment has just been made by Jacobson & Co., of New York, to Belgium [p. 257].

No foreign shipments of zinc ore will be made this month, but over 200 tons per day will be shipped to Iola, Kans., and Mineral Point, Wis., where plants have arranged to take the Leadville output at a better price than New York brokers. The new mill of the A. M. W. combination will start this week [p. 467].

The Denver Republican, in its review for 1900, published January 1, 1901, says:

Output of American Zinc-Lead Co.'s smelter, Canon City, Colo., for 1900

	Gold	Silver	Lead	Copper	Zinc
	Ounces	Ounces	Pounds	Pounds	Pounds
Colorado	374.58	68,751	1,671,886	434,999	3,682,055
Other States	40.09	12,815	137,290	26,648	179,027
	414.67	81,566	1,809,176	461,647	3,861,082

The figures for zinc in 1900 given in the table on page 176 represent estimates. The Herald-Democrat, of Leadville, gives 45,270,920 pounds and estimates that 59,926 tons of 35 per cent zinc were shipped to outside smelters, but Mineral Industry for 1900 shows that only 14,000 tons of material containing approximately 45 per cent zinc was shipped in 1900, of which the greater part went to Europe by way of Galveston.[64] The United States Customs reports of Galveston and New Orleans for 1900 show only 11,423 tons (and that including some Joplin ore), and for 1901 (with an increase in activity in zinc mining) the Colorado Bureau of Mines shows only 23,261 tons of 45 per cent zinc shipped. The Herald-Democrat for 1899 shows 10,699 tons of 40 per cent zinc shipped.

In his report for 1901 Hodges says: [65]

The output from mining operations in progress in the Leadville locality comprises sulphides, oxidized iron ores, zinciferous ores, and those carrying a commercial portion of bismuth.

The sulphides at present have a considerably limited market while oxidized iron ores find a ready market. It is believed that this production will not suffer any limitation, at least through any outside causes.

The Leadville basin is now the greatest producing locality of oxidized iron ores, and the New Home Mining Co. is believed to be the largest individual producer. This property is worked through the three shafts of the company—the Penrose, Starr, and Bon Air—all thoroughly equipped with hoisting and pumping machinery fully capable of performing all work required. The most active operations are conducted through the Penrose shaft. The quantity of ore in sight within the several excavations surpasses any heretofore revealed.

The Starr shaft is operated at the 500-foot level.

The Bon Air shaft has a good body of ore developed, from which there is a steady production. * * *

The New Monarch group, embracing the Lida, New Monarch, Little Winnie, and others within a large acreage, exhibits considerable development, outputting a large tonnage of good value. * * *

The Leadville Development & Drainage Co. has, within the year 1901, undertaken an important prospecting enterprise within an undeveloped area of their own property, northwest from Leadville. This work is carried forward by alternate diamond drill and shaft sinking. * * *

The Greenback mine is operated through a shaft, present depth 1,240 feet. * * *

It has been decided to install a new hoisting plant, with an ore capacity of 750 tons per day. This will be completed in April, 1902.

[63] Eng. and Min. Jour., vol. 70, pp. 18, 78, 137, 151, 257, 467, 1900.

[64] Mineral Industry for 1900, p. 659, 1901.
[65] Hodges, J. L., agent for Colorado, in Roberts, G. E., op. cit. for 1901, pp. 120-123, 146, 147, 1902.

Old properties leased.—The A. M. W. Co., a leasing company, is working the following-named mines: Wolftone, output 60,800 tons; Adams, output 36,700 tons; and Maid of Erin, output 4,500 tons.

The ore is iron, lead, and zinc sulphides, and in addition the Wolftone and Maid of Erin produced 995 tons of carbonates.

The Castle View Mining Co. made an output of 2,140 tons of manganiferous iron ore.

The Mab Leasing Co. shipped 1,000 tons of manganiferous iron; also 750 tons of sulphides.

Midas Mining Co. (leasing company), from the O. Z. and Dillon claims, shipped 68,000 tons of argentiferous iron ore.

Leasing and subleasing.—It is estimated that 90 per cent of all work in Leadville mines is by leasing and subleasing, at royalties that range from 5 to 40 per cent, based upon changes in ground and returns received from smelting works. The probable average of royalty paid is 10 to 15 per cent. Large leasing companies also pay taxes on property, output tax, and, in many cases, insurance; also assume cost of mining and responsibilities, and install on their leased premises hoisting and pumping plants as their operations may require.

Smelting plants.—The Arkansas Valley smelting plant, owned by the American Smelting & Refining Co., is in active operation.

It is expected that other plants owned by the American Smelting & Refining Co. will at an early date be placed in commission.

The concentrating mills are the Moyer, which has a capacity of 100 tons per day; the A. M. W., 100 tons; the California Gulch, 50 tons; and the A. Y. and Minnie, 50 tons.

The crude ore is crushed to pass a 30–40 mesh screen, then treated on table machines, making two products, the first to as near 50 per cent zinc as possible; the second is an iron and lead, containing gold and silver.

Ores containing zinc in the main part are held in reserve for the now well-assured market demand.

A new plant for treatment of zinciferous ores is in construction by the American Smelting & Refining Co. at Pueblo, and other parties are erecting a plant in Leadville for treatment of zinc-bearing ores.

Zinc concentrates.—The zinc concentrates have a market at the La Salle Smelting Works, but the larger portion is loaded at the works in bulk in railway cars, thence shipped to Galveston or New Orleans for final loading and delivery to zinc-smelting works at Antwerp.

These shipments are bought and paid for by local agent in Leadville at a flat price on a basis of 45 per cent zinc, penalties and premiums considered. The miner receives about $5 per ton; cost of mining, hoisting, and placing in loading bins is about $1 per ton.

Bismuth ores.—Ores carrying bismuth carbonates are produced by the Ballard, Bruce, and Big Six mines. Shipments have been made to St. Louis, also Liverpool, but owing to the limited demand and care in maintaining the price neither the producers nor buyers are willing to make known the status of the market. It is estimated that 1,000 tons of bismuth carbonates were shipped during 1901. * * *

Placer areas.—Near Leadville and extending through the county of Lake there are large placer areas. These have received very little attention for several years, but now the placer-mining industry has a growing recognition.

Kirchhoff,[66] in his report for 1901, gives a table showing exports of zinc ore by customs districts during 1899, 1900, and 1901. The ore sent from the Gulf ports came from the Joplin-Galena district and from Colorado. In another table he shows the destination of exports of zinc ore during 1899, 1900, and 1901. In 1901 Joplin and Colorado exported 291 tons of zinc ore from Galveston and 13,003 tons from New Orleans.

Mineral Industry for 1901 says:[67]

Spelter.—A large quantity of zinciferous lead ores from Colorado was smelted during the year and the zinc content regained. Formerly these low-grade zinciferous ores were rejected as waste material.

Zinc oxide.—The production of zinc white in the United States during 1901 was 46,500 short tons, valued at $3,720,000, as against 48,840 short tons, valued at $3,677,810, in 1900. The greater part of this output was made by the New Jersey Zinc Co., operating at Jersey City and Newark, N. J., Bethlehem and Palmerton, Pa., and Mineral Point, Wis. Page & Krause, at St. Louis, Mo., produce annually 450 short tons, and the United States Smelting Co., formerly the American Zinc-Lead Co., of Canon City, Colo., manufacture a pigment called "zinc-lead," which is a mixture of oxidized compounds of zinc and lead. The production of this special pigment during 1901 was 2,500 short tons, valued at $150,000. The Renfrew Zinc Co. put its new plant at West Plains, Mo., in operation in December, 1901, and the Ozark Zinc Oxide Co. is contemplating a new works at Joplin, Mo. The estimated annual capacity of each plant is 20,000 tons of zinc oxide.

Colorado.—According to the Hon. Harry A. Lee, commissioner of mines for Colorado, the production of zinc from the ores of that State during 1901 amounted to 13,427 short tons, of which Lake County furnished 11,573 short tons, and Mineral County 1,044 tons. These figures are based on the metallic zinc content of the ores produced. About 70 per cent of the output was exported and 30 per cent treated by domestic smelters, including ore used for the manufacture of zinc-lead pigment at Canon City, Colo., and possibly some shipped to Mineral Point, Wis., for the manufacture of zinc white. In 1901 the Leadville mills produced 23,261 tons of concentrates, averaging 45 per cent zinc, 6 per cent lead, and 10 per cent iron. Besides the Leadville product there were also shipments of zinc ore from Silver Plume, Montezuma, Creede, and other mining districts. Kansas smelters have received ore from Creede which has assayed as high as 59 per cent zinc, 3.75 per cent to 5 per cent lead, and about 2 per cent iron. The high iron content of the blende is due to the iron chemically combined, which can not be removed by any method of mechanical concentration.

The immense deposits of mixed lead and zinc sulphide at Leadville and Kokomo promise to become an important factor in the zinc industry in the near future. The ore is a silver-bearing mixture of galena, pyrite, and blende, with very little gangue. As early as 1886 it was milled for its lead content and accompanying silver, the blende being accumulated in tailings heaps. Recently by the application of improved milling methods a closer separation has been obtained, and a large quantity of this class of ore has been treated at Leadville, yielding a galena-pyrite concentrate and a ferruginous blende, the latter having a large sale, chiefly to European zinc smelters, although a small quantity has been treated at Kansas smelters. The freight from Leadville via Galveston to Swansea or Antwerp during 1901 was $9.50 per short ton, and while the miners receive $5 to $7 per ton, it is purely a by-product, and the value of the ore is increased by its removal.

The Colorado Zinc Ore Co. and the New Jersey Zinc Co. are erecting magnetic separating plants at Denver and Canon

---

66 Kirchhoff, Charles, Zinc: U. S. Geol. Survey Mineral Resources, 1901, p. 219, 1902.

67 Mineral Industry for 1901, vol. 10, pp. 650, 651, 652, 1902.

City, respectively, to produce ore containing about the same percentage of iron as that which the present concentrating mills are furnishing but several units higher in zinc and lower in lead. These ores have been smelted direct in Sadtler retorts by the Midland Smelting Co., at Bruce, Kans. The American Smelting & Refining Co. is erecting a zinc-smelting plant at Pueblo which will afford a local market for the blende concentrates at Leadville, Kokomo, Montezuma, Creede, and other mining districts in the State. The United States Smelting Co. was organized during the year to take over the holdings of the American Zinc-Lead Co., formerly operating the Canon City plant, which smelts chiefly low-grade zinciferous lead ores, largely from Utah and Leadville. The product is copper matte containing the precious metals, which is shipped or refined. The by-product consists of zinc-lead pigment used largely in the manufacture of paints. The United States Zinc Co., capitalized at $2,000,000, was organized to build a large zinc plant near Pueblo, to handle Leadville zinciferous ores, chiefly from the A. Y. and Minnie mine. [The American Smelting & Refining Co.'s zinc-smelting plant, mentioned above, and the United States Zinc Co.'s plant are the same.— C. W. H.]

The Leadville Herald-Democrat, in its review for 1901, published January 1, 1902, gives the following tables:

Zinc produced from Leadville ores, 1901, in tons

	Tons of ore	Gold	Silver	Lead	Copper	Zinc
		Ounces	Ounces	Pounds	Pounds	Pounds
American Zinc-Lead Co...	6,094	25	42,757	1,096,920	24,470	3,047,000
Zinc to smelters outside of Colorado						18,429,200
						21,476,200

Output of mines at Leadville in 1901, in tons

Mine	Zinc	Sulphide	Iron oxide	Carbonate
A. M. W.		a 102,000	b 2,140	995
A. Y. and Minnie	650			
Iron Silver	3,442			
Yak	1,562			
Small Hopes	1,014			
Boreel	275			
Jonesville	276			
Miscellaneous	1,800			
	9,019			

a Includes zinc.   b Includes manganese oxide.

Product of mines at Leadville in 1901

	Tons
Carbonate	27,483
Iron oxide	256,153
Sulphide	338,041
Zinc sulphide	23,261
Siliceous	94,021
Manganese oxide	54,055
	793,014

A. M. W. mines.—During 1901 the real work of mining commenced on the Adams, Maid of Erin, Wolftone, and Mahala.

The Denver Republican, in its review for 1901, published January 1, 1902, gives the following table:

Output of American Zinc-Lead Smelting Co., Canon City, Colo., from Colorado ores in 1901

Metal		Quantity	Price	Value
Gold	ounces..	897,097	$20.00	$17,941.94
Silver	do....	122,491	.593	72,737.16
Lead	pounds..	2,510,659	.0435	110,213.66
Copper	do....	517,874	.1625	84,144.31
Zinc	do....	5,712,323	.04	228,492.92
				513,529.99

Downer [68] gives the following account of development at Leadville in his report for 1902:

The total tonnage of the ore as shipped from Leadville to the various smelters is as follows:

	Tons
Lead carbonates	22,930
Oxidized iron	285,494
Iron sulphides	281,558
Zinc sulphides	85,699
Siliceous ores	72,215
Manganese oxides	1,050
	748,946

* * * The district has made material progress in the extension of the great Yak tunnel into the territory of the Ibex Mining Co., which will drain the greater part of Breece Hill; the purchase of the old Boston Gold-Copper smelting plant by a strong eastern syndicate, and the remodeling of the same, to be put into commission some time during the coming year; the unwatering of Fryer Hill by the Fryer Hill Mines Co., the extensive improvements made in the Arkansas Valley Smelting plant; and the erection of three large mills for treating the low-grade ores of the district, two for handling the zinc product, and one for treating the siliceous ores of Breece Hill. The last is planned to handle the ore by the cyanide treatment.

Downtown mining.—During the year a consolidation was brought about between the Morocco and the Home Extension companies, and a lease obtained from the city on the streets and alleys in the vicinity of lower Harrison Avenue.

Iron ore has been opened up on the west side of Harrison Avenue, and it is believed that it is an extension of the Bon Air shoot. * * *

The work of the Yak tunnel has been especially satisfactory, both in tonnage and development. * * *

During the year a contract was entered into between the Ibex Co. and the Yak by which the tunnel is to be driven to the No. 4 shaft of the Ibex, which will strike beneath the Ibex shaft at a depth of 200 feet. The work within the Ibex lines is done by the latter company and will drain the Breece Hill territory. The total length of the tunnel is 10,500 feet and including the main laterals it is 12,000 feet.

One lateral driven during the year makes connection with the Rubie property, this being run a distance of 1,000 feet. There is also a new lateral connecting the North Mike. * * *

A number of lessees have worked through the tunnel during the past year and shipped largely.

The Arkansas Valley Smelting plant, which is the works of the American Smelting & Refining Co. at Leadville, has made several very costly improvements. One is the enlargement of the power house.

An additional 150 feet has been built on the engine room, and two new large compound condensing engines have been

[68] Downer, F. M., agent for Colorado, in Roberts, G. E., op. cit. for 1902, pp. 93–98, 132, 1903.

installed. The plant is capable of generating 1,800 horsepower. Two new blowers have been added, making four of No. 8 and three of No. 7, of the Connersville type. The substitution of electric motors for small steam engines has been another means of economy. All pumps have been brought together in one central pump house, a brick and steel building, absolutely fireproof. Two of the smelting furnaces have been increased in size from 75 tons to 125 tons each, giving 10 large blast furnaces having a daily capacity of 1,200 tons of ore.

Roasting capacity.—Six new hand roasting furnaces were added, making 24, in addition to the Ropp and Horseshoe mechanical furnaces already in use.

The total daily roasting capacity is now 375 tons.

An important piece of construction is the new mill for crushing and sampling the sulphide ores before sending them to the roasting furnaces. Two automatic sampling devices have been installed.

During the year two reductions in treatment charges were made, one of 50 cents on the lowest grade of iron ores and the other on low-grade siliceous ores, which has already operated to move a considerable amount of this material.

Zinc ore increases.—The increase in the output of zinc ore during the year is quite remarkable and is simply the beginning of the possible output of this class of ore from the vast deposits which underlie nearly all of Carbonate Hill and portions of Breece Hill. Under the present rates paid for this class of ore by the zinc smelters it is possible for a number of the mines to hand sort their ore roughly and make a shipping product which leaves a profit. The prevailing price paid during the year was $9 a ton on the cars for 45 per cent of metallic zinc in the ore, with a dollar off for each per cent less or a dollar more for each per cent above that grade.

During the year the Moyer property shipped 44,691 tons, all hand sorted. The Small Hopes consolidation shipped 7,374 tons; the Yak tunnel, from lessees, about 7,000 tons, both of which were also hand sorted.

The Adams, Maid, and Wolftone combination, known as the A. M. W., shipped from their mill as concentrates 20,000 tons, and the A. Y. and Minnie, from their new mill, 6,000 tons, the latter mill having been in operation only during the last months of the year.

New zinc concentrators.—During the year two new zinc concentrators were erected, the Minnie mill and the mill at the Resurrection mine, on Little Ellen Hill.

The A. M. W. mill has run constantly and is very conveniently arranged. The ore is first fed into a crusher of the Blake pattern and conveyed to three 5-foot Huntington mills and crushed to pass a 30-mesh screen. This product is fed at once into two hydraulic sizers of the cone-shaped pattern, which makes two sizes, fed to six Wilfley tables. The overflow is conveyed to a series of conical settling tanks supplying six other Wilfley tables, and a separation is made, giving one product of iron and lead mixed and a second product of zinc, carrying from 10 per cent to 12 per cent of iron. The ore treated in the mill as it comes from the mine only averages about 10 per cent of silica. The capacity of the mill is 100 tons per diem and the output of zinc concentrates for the year was 20,000 tons.

The Minnie mill is built on the same lines and is of the same capacity.

The Resurrection mill will use Huntington mills for crushing and then concentrate on Wilfley tables. The concentrates will be dried over a magnetic separator to clean out a large percentage of iron so as to raise the grade of the zinc.

Leadville's first cyanide mill.—Another new departure for Leadville is the erection of a mill at the Ballard mine for treating the siliceous gold ores found in that section of Breece Hill by cyanide. This is the first cyanide mill erected in the Leadville gold belt, and experimental tests have been very satisfactory.

The outlying districts have not added very materially to the output of Lake County. In the St. Kevin district the Amity mine was the largest producer, averaging about 25 tons a week. In the Twin Lakes district the work has been principally development.

Kirchhoff says in his report for 1902: [69]

Colorado has become an important producer of zinc ore, the output for the year 1902 being placed by Mr. Harry Allen Lee, commissioner of mines, at 26,241 short tons [70] [of zinc content], valued at $2,544,993.48. Of this quantity 23,819 short tons [of zinc content] are credited to Lake County, 1,024 short tons [to] Mineral County, and 665 short tons [to] Summit County.[71] In Leadville increased attention has been paid to the handling of low-grade zinc and lead ores and dumps. The United States Zinc Co.[72] has shipped considerable quantities from the Moyer dump of the Iron Silver Co. to its Canon City plant. A mill was also successfully started by the A. Y. and Minnie Co., and a large new plant for the Resurrection mine was approaching completion at the end of 1902. Magnetic separation plants for zinc ores have been built at Denver, Colo., by the Colorado Zinc Co., and the Empire Zinc Co., controlled by the New Jersey Zinc Co., has built works at Canon City. The American Smelting & Refining Co. [United States Zinc Co.] has erected a small zinc smelting plant at Pueblo. The Colorado zinc ores have largely gone to Europe, while considerable quantities have also been shipped to the Kansas gas belt and to the zinc-white works in Wisconsin.

To the Kansas works have also gone some shipments of zinc ore from the Slocan district in British Columbia.

Mineral Industry for 1902 contains the following matter: [73]

Spelter.—An important feature of the year was the increased quantity of ore supplied by Colorado, and the plans that have been made for utilizing it on a still larger scale. Some ore was shipped to the Kansas smelters from Utah.

Zinc oxide

Year	Short tons	Value	
		Total	Per ton
1900	47,151	$3,772,000	$80.00
1901	42,266	3,720,000	80.00
1902	52,730	4,023,299	76.30

In addition to others, the United States Smelting Co. (formerly the American Zinc-Lead Co.), at Canon City, Colo., manufactures a pigment. * * * The production of this special pigment in 1902 was 4,000 tons, valued at $225,000, as compared with 2,500 tons, valued at $250,000 in 1901.

The ores of this State are rapidly becoming an important factor in the zinc industry. The Leadville district alone produced 89,669 tons of ore in 1902, and deposits were worked at Kokomo, Creede, Rico, and elsewhere. The Kansas smelters took an increased quantity of these ores, which are purchased at a liberal discount from the Joplin prices and can be mixed in considerable proportions with the higher-quality ores without materially affecting the smelting results. A large portion of the output is consumed by the United States Smelting Co.,

[69] Kirchhoff, Charles, Colorado zinc: U. S. Geol. Survey Mineral Resources, 1902, p. 221, 1904.

[70] This amount should be 26,291 tons; see Colorado Bureau of Mines Twelfth Biennial Rept., pp. 152-153, 1913.

[71] Also smaller quantities to other counties; see State report just mentioned.

[72] United States Smelting Co.; see Mineral Industry for 1902, p. 600, 1903.

[73] Mineral Industry for 1902, vol. 11, pp. 599, 600, 609, 1903.

Canon City, for the manufacture of zinc-lead pigment, and some ore is marketed at Mineral Point, Wis., and in Europe. The Colorado Zinc Co., at Denver, and the Empire Zinc Co., at Canon City, have installed magnetic separating plants.

A review of progress in the metallurgy of zinc in 1902.—Construction was begun on the works of the United States Zinc Co., affiliated with the American Smelting & Refining Co., at Pueblo, Colo., and it is expected that this will be in operation in 1903.

The Denver Republican, in its review for 1902, published January 1, 1903, gives the following table:

Zinc produced in 1902 from ore mined in Colorado

	Pounds	Value
United States Smelting Co.	3,854,400	$173,448
Shipped to Europe in ores	22,000,000	990,000
Shipped to other States in ores	12,500,000	562,500
	38,354,400	1,725,948

Mineral Industry for 1903 says: [74]

Zinc oxide.—The output of zinc oxide in the United States in 1903 was 59,562 short tons, against 52,730 short tons in 1902. The totals do not include the product of the United States Reduction & Refining Co. (operating company, United States Smelting Co.), at Canon City, Colo., which in 1903 made 4,500 short tons of zinc-lead pigment.

New spelter plants in 1903 were the United States Zinc Co. * * *

The development of a market, both abroad and at home, for such zinc ore as can be produced in the Rocky Mountains has created great interest. Leadville, Colo., has continued to be the principal point of production, but shipments have been made from Kokomo, Creede, and Rico, Colo.; from Park City and Frisco, Utah; from the Magdalena district, in New Mexico; and from the Slocan, British Columbia. * * * At present this kind of ore is shipped to spelter producers of Belgium, Kansas, and Colorado; to the U. S. Reduction & Refining Co., which manufactures zinc-white lead at Canon City, Colo.; to the Ozark Zinc Oxide Co., of Joplin, Mo., and the Mineral Point Zinc Co., of Mineral Point, Wis., both of which concerns make white lead. The works at Canon City were destroyed by fire in 1903 but have been rebuilt with increased capacity.

Magnetic separation.—The Wetherill Separating Co. reports the following number of its cross-belt separators (type E) now in use for separating zinc ore in the United States, exclusive of machines employed for experimental work:

Number of machines

New Jersey Zinc Co., Franklin Furnace, N. J.	20
Empire Zinc Co., Canon City, Colo.	5
Colorado Zinc Co., Denver, Colo.	2
Resurrection Gold Mining Co., Leadville, Colo.	3
Summit Mining & Smelting Co., Kokomo, Colo.	2
Pride of the West Milling Co., Washington, Ariz.	6
Warren Separating Co., Warren, N. H.	1
Bully Hill Mining & Smelting Co., Winthrop, Calif.	1

The Blake-Morscher separator is used at the Colorado Zinc Co., Denver; the Silver Ledge Mill, Silverton; the Harris Mill, Denver; and at Benton, Wis.

The Denver Republican, in its review for 1903, published January 1, 1904, reports that in 1903 the United States Smelting Co. purchased and treated 25,000 tons of ore mined in Colorado, which yielded metals as shown in the following table:

Quantity and value of metals produced by United States Smelting Co. from ore mined in Colorado in 1903

Metal	Quantity	Value
Gold .... ounces..	2,500	$51,675
Silver .... do....	275,000	146,850
Lead .... pounds..	4,500,000	192,600
Copper .... do....	400,000	54,360
Zinc .... do....	12,000,000	630,000
		1,075,485

Quantity and value of zinc produced from ore mined in Colorado in 1903

	Pounds	Value
United States Smelting Co., Canon City	12,000,000	$630,000
Shipped to Europe in ore	13,477,333	697,560
Shipped to smelters in other States	19,919,543	1,045,576
	45,396,876	2,373,136

In his report for 1904 Downer [75] says:

The mining industry of Lake County is in a very prosperous condition, shipments of ore for 1904 aggregating 800,000 tons. * * *

Downtown district.—The Midas Mining & Leasing Co. has sunk the Coronado, Penrose, Midas, and Sixth Street shafts, at an expense of $1,000,000, through the parting quartzite to the lower ore horizons or second contact, proving an ore body 45 feet thick, extending through 30 acres at a depth of 800 feet under the streets of the city of Leadville. The mines are wet, and pumps at the Penrose have handled as high as 1,100 gallons per minute.

The Bon Air workings have been extended through several properties. In the Wood fraction a body of silver ore was discovered and a large tonnage shipped, but the development of the ore body has only begun.

The Empire Tunnel Co., of Georgetown, controls the Cloud City, and has shipped considerable tonnage of a good grade of iron and high-grade manganese that is used as fluxing material. The company is now sinking the Cloud City shaft to catch the Midas ore shoot.

The Gold Belt.—The discovery of ore in the Sunday and the development of ore channels in the Ollie Reed and New Monarch mark important events of the year in this belt. The Ollie Reed is heavily shipping, and the New Monarch is sinking a 2,000-foot shaft.

Iron Hill.—The Moyer is the principal mine of the Iron Silver Co. and produced an average of 11,000 tons a month of low-grade sulphide, carrying a good percentage of zinc. This is the largest tonnage producer in the State, the output being 130,000 tons for the year. The ore averages $8 in value and is handled at a profit of $4 per ton.

The A. Y. and Minnie mine and mill produces and treats 120 tons of ore a day and is owned by the Western Mining Co. This corporation has another mill operated by the A. M. W. Co. that treats 160 tons per day from other mines.

The Yak tunnel.—The Yak tunnel * * * is 3 miles in length and opens up for development the entire region from California Gulch to a vertical depth of 1,300 feet under the

---

[74] Mineral Industry for 1903, vol. 12, pp. 340-362, 1904.

[75] Downer, F. M., superintendent of Denver Mint and agent for Colorado, in Roberts, G. E., op. cit. for 1904, pp. 114-116, 123-124, 1905.

Ibex. Tramming is done with an electric locomotive, hauling as many as 60 cars a trip, and 3 miles of trolley wire have been strung in the tunnel and its branches The ore shipments from the tunnel during the past year have averaged between 3,000 and 6,000 tons per month, but this output will probably be doubled. The concentrating plant at the mouth of the tunnel is being constructed at a cost of $200,000 and is nearing completion. It is the largest mill in the district and will treat low-grade sulphides and zinc ores.

Placer mining.—The Saguache Placer Mining Co., operating at Twin Lakes, will put in a new dredge and work California Gulch from Georgia Gulch to Malta this summer. Prospecting ground gave reported returns of 25 cents per cubic yard.

The Twin Lakes Placer Mining Co. works a large acreage and has obtained good returns [Chaffee County.—C. W. H.].

The Denver Republican, in its review for 1904, published January 1, 1905, gives a statement of the United States Zinc Co.'s smelter, Pueblo, Colo., showing tons of ore purchased in 1904, quantity of zinc in the ore, value of the zinc, spelter produced, value of the spelter, and quantity of pig lead, as follows:

Tons of ore	17,821
Zinc in ore, pounds	12,509,613
Average per cent zinc	35.10
Value	$593,129.43
Spelter produced, pounds	9,805,699
Value	$460,867.25
Pig lead, pounds	47,402

The recovery of zinc, according to this statement, was 78.39 per cent. The gold, silver, and lead is not extracted by the zinc company but is shipped to lead smelters.

Output of United States Smelting Co., Canon City, Colo., for 1904

Tons treated	Yield				
	Gold	Silver	Lead	Copper	Zinc
	Ounces	*Ounces*	*Pounds*	*Pounds*	*Pounds*
30,000	2,400	350,000	4,800,000	800,000	14,400,000

The gold, silver, and copper, amounting to $316,000, was sent to the Pueblo plant of the American Smelting & Refining Co., and the zinc and lead pigment aggregating $474,670 was distributed all over the United States.

The smelter obtained most of its ore near Leadville and a fractional amount from Cripple Creek; the remainder came from Arizona and New Mexico.

The Denver Republican says:

It is impossible to secure any but the Leadville details of this company's statistics.

Metals produced by United States Smelting Co. from ore mined at Leadville, Colo., in 1905

Metal		Quantity	Value
Gold	ounces	1,000	$20,000
Silver	do	250,000	140,000
Copper	pounds	75,000	6,750
Zinc-lead pigment	tons	3,400	238,000
			404,750

Leadville, therefore, contributed $404,750 of the total of $790,670, other Colorado points furnishing approximately $200,000 more, and the remainder coming from outside the State.

Lindgren [76] says of the developments in 1905:

The recent development of an important zinc-mining industry in Leadville is well known. Among the most important producers are the Iron Silver Mining Co. (Moyer mine), the Western Mining Co., the Big Chief Leasing Co., and the Boreel Co. The ore is partly shipped as sulphide concentrates, partly as crude or hand-sorted ore. After the distillation of the zinc by the zinc smelters, the silver-bearing cinders are sold to the lead smelters.

Important developments have taken place in the Downtown section. The Midas Co. has sunk the Penrose shaft to a depth of 920 feet and is draining it by means of powerful pumps. The Coronado shaft, in the same vicinity, is 790 feet deep and is now drained by a drift from the Penrose. A considerable production was maintained during the year from the Coronado shaft. These pumping operations will open a large and formerly unproductive area.

The Yak tunnel, which now is 10,800 feet in length, extends from California Gulch eastward below Iron and Breece hills. It is planned to pierce the range, and its eastern portal would be in Park County, near the London mine. The tunnel is equipped with a very complete system of electric transportation, by which a great number of mines are being served.

During the year the Yak Co. completed the Rowe mill at the mouth of the tunnel. The capacity is 250 tons per day, and it is equipped with electrostatic concentrating machine for the separation of the zinc blende.

A certain quantity of crude Leadville ores is shipped to Denver and concentrated there.

Bain [77] gives the following outline:

Character of Leadville ores; analyses of Colorado zinc ores analysis of concentrates from Adams mill, Leadville, Colo.
    Milling methods:
        Wet concentration.
        Magnetic and electrostatic separation.
            Wetherill magnetic separator.
            International or Snyder magnetic separator.
            Blake-Morscher electrostatic separator.
            Sutton-Steele dielectric separator.
            German machines at United States Zinc Co.'s plant, Pueblo, Colo.
    Smelting methods:
        American Zinc & Chemical Co. (Dewey patents), Denver, Colo. (Process consists in obtaining a solution of zinc in the form of sulphate, evaporating the sulphate to dryness, and calcining it for the production of oxide.)
        United States Zinc Co., Pueblo, Colo. (Produces spelter; description of process.)
        United States Smelting Co., Canon City, Colo. (Bartlett process, whereby a zinc-lead pigment is made, while the gold, silver, and copper values are left in the cinder on the grate, in form suitable for ordinary reduction. * * * The cinder left on the grate is mixed with low-grade copper ore and reduced in a furnace which produces a 25 per cent copper matte.)

The figures given in the table on page 176 for 1905 to 1923 are taken from Mineral Resources (mines

[76] Lindgren, Waldemar, U. S. Geol. Survey Mineral Resources, 1905, pp. 203-204, 1906.

[77] Bain, H. F., Zinc and lead ores in 1905: U. S. Geol. Survey Mineral Resources, 1905, pp. 379-392, 1906.

reports) for the calendar years, published by the United States Geological Survey.

Naramore[78] gives the following account of the developments in 1906:

The zinc industry is deserving of special mention in that it represents a gain of nearly 7,000,000 pounds and of half the county's increase in total value. The 1906 zinc product was worth more than the combined gold and silver output. Previous to 1898 zinc was a small factor in Leadville's mineral industry, inasmuch as there were only small quantities which were of shipping grade. The concentrating mills, in which the crushed ore or middlings from the concentrating tables are further separated by electromagnetic machines, have succeeded in making a marketable product by raising the zinc content from 18 and 20 per cent in the raw ore to 25 and 45 per cent in the concentrates. There were 9,000 tons of crude zinc ores of shipping grade produced and less than 1,000 tons in which no other values were saved. Thus practically the entire zinc output was derived from the concentration and separation of iron-zinc or lead-zinc ores, which occur largely as a mixture of pyrite, zinc blende, and galena, usually carrying values in silver and often a small quantity of gold.

The American Zinc Extraction Co.'s mill, located near the mouth of the Yak tunnel, is a custom plant of 250 tons capacity, in which the ores are crushed in jaw crushers, sampled, dried in an automatic drier, recrushed, and fed to International magnetic separators. The resultant products are zinc concentrates and pyrite, the latter carrying most of the precious metal values. This company buys ores with as low as 20 per cent zinc content.

The Damascus mill uses electrostatic separators and handles much of the second-grade product from the A. Y. and Minnie concentrator.

The Adams or Wolftone mill doubled its capacity during 1906 by adding three Huntington mills and thirteen tables. This makes the capacity approximately 200 tons a day. By means of the tables and a system of settling tanks, two classes of zinc and lead concentrates are produced. The A. Y. and Minnie mill, of 150 tons capacity, uses similar methods. The Boston & Arizona plant was also in operation. A considerable quantity of Leadville middlings is reconcentrated at the plant of the Colorado Zinc Co. in Denver. A large tonnage of low-grade sulphides is also handled at the concentrators of the Empire Zinc Co., near Canon [City].

The Arkansas Valley plant of the American Smelting & Refining Co., located on the southern outskirts of Leadville, treated the larger portion of the ores of the district, the zinc ores excepted.

Leadville interests have been instrumental in building up the Ohio & Colorado Co.'s smelter at Salida, and during 1906 the plant received a greatly increased tonnage from Lake County.

The placer production of Lake County in 1906 is given as 64.92 ounces of gold or $1,342, but only $264 belongs to Lake County, coming from St. Kevin district.[79] The remainder belongs partly to Chaffee County, and part of it came from surface lode workings.

In his report for 1907 Naramore says:[80]

The Leadville District Mining & Milling Co. put its mill in operation during 1907. The mill is located just north of the Arkansas Valley smelter and handles low-grade sulphide ores from the dumps of the Ibex Mining Co. By the use of jigs and concentrating tables some of the silica is eliminated and a marketable product is made from ores which are otherwise not acceptable to the smelter. The Adams or Wolftone mill handled the sulphide ores from the Wolftone mine. The A. Y. and Minnie mill was closed when the mines ceased producing.

In the report for 1907 placer gold is given as 95.88 ounces or $1,982, but only $510 was actually placer and that came from St. Kevin district. A small lot, doubtful as to source, came from either Chaffee County or Summit County.[81]

In the report for 1908 Henderson gives the following table:[82]

Classification of Leadville ores, 1908

	Short tons	Percentage
Siliceous (gold-silver) [*]	25,741	6
Sulphide:		
Zinc-iron-lead (zinc blende-pyrite, with a little galena)	95,306	23
Iron (pyrite, with a little copper, gold, silver, and lead)	151,284	37
Total sulphide	246,590	60
Oxide:		
Lead	24,616	6
Iron-manganese (silver)	111,764	28
Total oxide	136,380	34
Grand total	408,711	100

[*] All or nearly all the siliceous ore is oxide, and all is dry.

As shown by this table, the greatest tonnage from Leadville in 1908 was of sulphide ore. The chief sulphide ore was iron pyrite, carrying a little copper, gold, silver, and some lead. The bulk of this went direct to the smelter. The tonnage of this class was 151,284 short tons, or 37 per cent of the total tonnage. There were also 95,306 tons, or 23 per cent, of zinc-iron-lead-sulphide (zinc blende-pyrite with a little galena). Sulphide ore made 60 per cent of the tonnage.

The higher grade of zinc sulphide ore was shipped direct to the zinc smelters without concentration. The bulk of the zinc sulphide, however, including quite a tonnage of dump ore, was separated magnetically or concentrated at Leadville, Canon City, and Denver. The Western Chemical Co., in Denver, was the only plant to run its wet mill for the further concentration of the middlings produced at the separation and concentration mills.

Mills operating.—Adams, Leadville District, and American Zinc Extraction Co. (Rho or Yak).

---

[78] Naramore, Chester, U. S. Geol. Survey Mineral Resources, 1906, pp. 221-223, 1907.

[79] U. S. Geol. Survey Mineral Resources, 1906, p. 203, 1907.

[80] Naramore, Chester, U. S. Geol. Survey Mineral Resources, 1907, pp. 259-261, 1908.

[81] U. S. Geol. Survey Mineral Resources, 1907, p. 240, 1908.

[82] Henderson, C. W., U. S. Geol. Survey Mineral Resources, 1908, pp. 386-389, 1909.

Mineral Resources for 1909 says: [83]

Classification of Leadville ores, 1909

	Short tons	Percentage
Siliceous (gold-silver) *	28,054	6.7
Sulphide:		
Zinc-iron-lead (zinc blende-pyrite, with a little galena)	127,640	30.6
Iron (pyrite, with a little copper, gold, silver, and lead)	142,139	34.1
Total sulphide	269,779	64.7
Oxide:		
Lead	27,504	6.6
Iron-manganese (silver)	91,960	22.0
Total oxide	119,464	28.6
Grand total	417,297	100.0

* All or nearly all the siliceous ore is oxide, and all is dry.

As shown by this table, the greatest tonnage from Leadville in 1909 was of sulphide ore. The chief sulphide ore was iron pyrite, carrying a little copper, gold, silver, and some lead. The bulk of this went direct to the smelter. The tonnage of this class was 142,139 short tons, or 34.1 per cent of the total tonnage. There were also 127,640 tons, or 30.6 per cent, of zinc-iron-lead sulphide (zinc blende-pyrite with a little galena). Sulphide ore made 64.7 per cent of the tonnage.

The higher grade of zinc sulphide ore was shipped direct to the zinc smelters without concentration. The bulk of the zinc sulphide, however, including quite a tonnage of dump ore, was separated magnetically or concentrated at Leadville, Canon City, and Denver.

Mills operating.—Leadville District, Adams (until April, 1909), and Rho.

Mineral Resources for 1910 contains the following statement: [84]

The generally increased activity in Leadville mines in 1909 continued in 1910, and toward the end of the year the discovery and further verification of the existence of considerable quantities of zinc carbonate and zinc silicate in the workings and dumps of many of the mines, both idle and active, gave a material impetus to the zinc industry of Leadville. There was an increase in tonnage of ore and of old dumps sold, treated, or removed for treatment. Besides a considerable tonnage of zinc carbonate shipped, the tonnages of zinc sulphide and iron sulphide increased, but the tonnages of siliceous lead oxide and iron-manganese ores decreased.

Classification of Leadville ore, 1910

		Short tons	Percentage
Sulphide:			
Zinc-iron-lead (zinc blende; pyrite with a little galena)		163,218	35.3
Iron (pyrite with a little copper, gold, silver, and lead)	160,590		
Iron (pyrite, with a little copper, gold, silver, and over 4½ per cent lead)	7,618	168,208	36.4
Total sulphide		331,426	71.7
Oxide:			
Lead	18,581		
Copper-lead	90	18,671	4.0
Iron-manganese (silver)		82,597	17.9
Zinc carbonate and silicate		8,059	1.8
Total oxide		109,327	23.7
Siliceous (gold-silver) *		21,280	4.6
Grand total		462,033	100.0

* Nearly all the siliceous ore is also oxide, and all is dry. Pure metallic gold, in quantity 3,920.90 ounces, valued at $81,052, came from less than a ton of ore, from oxide and sulphide ore bodies.

As shown by this table the greatest (and increasing) tonnage from Leadville in 1910 was in sulphide ore. The chief sulphide ore was iron pyrite, carrying a little copper, gold, silver, and some lead. Some 7,618 tons carried lead averaging over 4½ per cent. The bulk of this iron sulphide went direct to the smelters. The tonnage of this class was 168,208 short tons, or 36.4 per cent of the total tonnage. There were also 163,218 tons, or 35.3 per cent of the total tonnage, of zinc-iron-lead sulphide (zinc blende; pyrite with a little galena). Sulphide ore made 71.7 per cent of the tonnage, as compared with 64.7 per cent in 1909.

Mineral Resources for 1911 gives a list of mills, with equipment, and contains the following table: [85]

[83] Henderson, C. W., U. S. Geol. Survey Mineral Resources, 1909, pt. 1, pp. 317-319, 1911.

[84] Henderson, C. W., U. S. Geol. Survey Mineral Resources, 1910, pt. 1, pp. 417-423, 1911.

[85] Henderson, C. W., U. S. Geol. Survey Mineral Resources, 1911, pt. 1, pp. 523, 545-551, 1912.

Classification of Leadville ore, 1911

Character		Short tons	Percentage
Sulphide:			
Zinc-iron-lead (zinc blende; pyrite with a little galena)		79,376	18
Iron (pyrite with a little copper, gold, silver, and lead)	147,535	169,167	39
Iron (pyrite with a little copper, gold, silver, and over 4½ per cent lead)	21,632		
Total sulphide		248,543	57
Oxide:			
Lead	13,123	14,338	3
Copper	1,215		
Iron-manganese (silver)		64,296	15
Zinc carbonate and silicate		83,905	19
Total oxide		162,539	37
Siliceous (gold-silver) *		24,790	6
Grand total		435,872	100

* Nearly all the siliceous ore is oxide and dry. Pure metallic gold, in quantity 2,223.13 fine ounces, valued at $45,956, came from less than a ton of ore, probably from siliceous oxide and sulphide ore bodies.

Classification of Leadville ore, 1911—Continued

Character	Quantity	Recovered content					Average recovery				
		Gold	Silver	Copper	Lead	Zinc	Gold	Silver	Copper	Lead	Zinc
	Short tons	*Fine ounces* (b)	*Fine ounces* (b)	*Pounds* (b)	*Pounds* (b)	*Pounds* (c)	*Ounces per ton*	*Ounces per ton*	*Per cent*	*Per cent*	*Per cent*
Sulphide:											
Zinc-iron-lead	79,376							4.30		d 7.34	d 23.29
Iron pyrite	147,535	32,933.55	1,970,054	3,587,982	1,849,998		0.22	13.35	1.22	.63	
Lead	21,632	2,534.95	192,085	92,941	3,481,518		.11	8.88	.21	8.05	
Oxide:											
Lead	13,123	4,317.97	85,318	33,632	1,934,868		.33	6.50	.13	7.37	
Copper	1,215	6.76	968	190,727	417		.006	.80	7.85		
Iron-manganese (silver)	64,296	1,566.16	319,574		2,353,112		.024	4.97		1.83	
Zinc	83,905					(c)					e 31.05
Siliceous	24,790	9,412.50	105,620	20,648	311,938		.38	4.26	.42	.63	
"Metallics" f		2,223.13	724								
	435,872										

b Estimated recovery as lead and iron concentrates and zinc residues.
c Zinc figures as spelter or as zinc in zinc oxide.
d Average assay crude ore, mostly milled. (Lead, wet assay.)
e Average assay crude. All to zinc smelters.
f Very rich material picked out of ore and shipped separately.

The course of treatment of the Leadville ores is shown graphically.

Burton gives the history of the discovery of zinc carbonate in Colorado and at Leadville and also summarizes the history of the industry.[86]

Mineral Resources for 1912 gives a list of mills, with equipment, and contains the following table:[87]

Classification of Leadville ore, 1912

Character		Short tons	Percentage
Sulphide:			
Zinc-iron-lead (zinc blende; pyrite with a little galena)		104,148	20
Iron (pyrite with a little copper, gold, silver, and lead)	127,575 }	140,615	28
Iron (pyrite with a little copper, gold, silver, and over 4½ per cent lead)	13,040 }		
Total sulphide		244,763	48
Oxide:			
Lead	23,754 }	24,652	5
Copper	898 }		
Iron-manganese (silver)		69,805	14
Zinc carbonate and silicate		142,782	29
Total oxide		237,239	48
Siliceous (gold-silver) a		21,368	4
Grand total		503,370	00

Character	Quantity	Recovered content					Average recovery				
		Gold	Silver	Copper	Lead	Zinc	Gold	Silver	Copper	Lead	Zinc
	Short tons	*Fine ounces* (b)	*Fine ounces* (b)	*Pounds* (b)	*Pounds* (b)	*Pounds* (c)	*Ozs. per ton*	*Ozs. per ton*	*Per cent*	*Per cent*	*Per cent*
Sulphide:											
Zinc-iron-lead	104,148						0.02	5.55		d 8.80	d 24.0
Iron pyrite	127,575	32,462.84	1,536,592	1,748,246	2,511,562		.254	12.04	0.69	.98	
Lead sulphide	13,040	2,109.02	110,092	35,582	4,479,214		.162	8.44	.14	17.17	
Oxide:											
Lead	23,754	4,470.20	308,363	1,331	4,931,803		.188	12.98		10.38	
Copper	898	96.64	1,771	101,476	2,486		.108	1.97	5.65	.14	
Iron-manganese (silver)	69,805	1,799.62	276,016		1,702,976		.026	3.95		1.22	
Zinc	142,782					(c)					e 29.2
Siliceous	21,368	8,233.38	108,485	47,617	366,923		.385	5.08	.11	.86	
"Metallics" f		1,657.66	716								
	503,370										

a All the siliceous ore is also oxide and dry. Metallic gold, in quantity, 1,657.66 fine ounces, valued at $34,267, and 716 fine ounces of silver came from less than a ton of ore, from siliceous oxide and sulphide ore bodies.
b Estimated recovery, as lead and iron concentrates and zinc residues.
c Zinc figured as spelter or as zinc in zinc oxide.
d Average assay crude ore, mostly milled. (Lead, wet assay.)
e Average assay crude. All to the smelters.
f Very rich material picked out of ore and shipped separately.

86 Burton, H. E., Leadville, Colo., zinc deposits: Mines and Minerals, vol. 31, p. 436, 1911; History of the zinc industry in Colorado: Min. Sci., vol. 64, p. 85, 1911.
87 Henderson, C. W., U. S. Geol. Survey Mineral Resources, 1912, pt. 1, pp. 657-658, 682-686, 1913.

Mineral Resources for 1913 gives the following table: [88]

Classification of Leadville ore, 1913

Character	Short tons		Percentage
Sulphide:			
Zinc-iron-lead (zinc blende; pyrite with a little galena)	152,223	97,704	19.0
Iron (pyrite with a little copper, gold, silver, and lead)	10,991	163,214	31.7
Iron (pyrite with a little copper, gold, silver, and over 4½ per cent lead)			
Total sulphide		260,918	50.7
Oxide:			
Lead	32,153	32,217	6.3
Copper	64		
Iron and iron-manganese (silver and some lead)		61,389	11.9
Zinc carbonate and silicate		135,760	26.4
Total oxide		229,366	44.6
Siliceous (gold-silver) [a]		24,316	4.7
Grand total		514,600	100.0

Character	Quantity	Recovered content					Average recovery				
		Gold	Silver	Copper	Lead	Zinc	Gold	Silver	Copper	Lead	Zinc
	Short tons	Fine ounces	Fine ounces	Pounds	Pounds		Ozs. per ton	Ozs. per ton	Per cent	Per cent	Per cent
Sulphide:		(b)	(b)	(b)	(b)	(c)	[d] 0.03	[d] 4.88		[d] 9.39	[c] 23.00
Zinc-iron-lead	97,704										
Iron pyrites	152,223	23,744.01	1,959,621	1,714,724	3,989,755		.156	12.87	0.56	1.31	
Lead sulphide	10,991	1,703.38	99,225	20,195	2,658,549		.155	9.03	.09	12.09	
Oxide:											
Lead	32,153	5,548.97	218,784	34,914	6,152,882		.173	6.80		9.57	
Copper	64	18.46	89	7,037			.288	1.39	5.58		
Iron	61,389	1,410.03	312,207	1,566	1,720,661		.023	5.09		1.40	
Zinc	135,760					(c)					[c] 27.45
Siliceous	24,316	9,859.68	127,375	50,782	470,897		.405	5.24	.10	.97	
"Metallics" [f]		1,278.56	350								
	514,600										

[a] All the siliceous ore is also oxide and dry. Metallic gold, 1,278.56 fine ounces, valued at $26,430, and 350 fine ounces of silver came from less than a ton of ore, from siliceous oxide and sulphide ore bodies.
[b] Estimated recovery, as lead and iron concentrates and zinc residues.
[c] Zinc figured as spelter or as zinc in zinc oxide.
[d] Average assay crude ore, mostly milled. (Lead, wet assay).
[e] Average assay crude. All to zinc smelters.
[f] Very rich material picked out of ore and shipped separately.

Mineral Resources for 1914 gives the following table: [89]

Classification of Leadville ore, 1914

Character	Short tons		Percentage
Sulphide:			
Zinc iron-lead (zinc blende; pyrite with a little galena)	174,610	111,947	21
Iron (pyrite with a little copper, gold, silver, and lead)	17,100	195,612	37
Iron (pyrite with a little copper, gold, silver, and over 4½ per cent dry lead)	3,902		
Iron (pyrite with gold, silver, lead, and over 2½ per cent dry copper)			
Total sulphide		307,559	58
Oxide:			
Lead	29,288	29,423	6
Copper	135		
Iron and iron-manganese (silver and some lead) flux		48,839	9
Zinc carbonate and silicate		113,881	21
Total oxide		192,143	36
Siliceous (gold-silver) [a]		33,000	6
Grand total		532,702	100

[a] All the siliceous ore is oxide and is also dry. The greater part of this siliceous oxide carries a heavy excess of silica over iron; some carries iron and silica in equal quantities; a small quantity carries iron in excess of silica but carries gold and silver in quantity to take material out of mere flux class. Metallic gold, in quantity 4,773.91 fine ounces, valued at $98,685, and 1,293 fine ounces of silver came from less than a ton of ore, from siliceous oxide and sulphide ore bodies.

[88] Henderson, C. W., U. S. Geol. Survey Mineral Resources, 1913, pt. 1, pp. 257–261, 1914.
[89] Henderson, C. W., U. S. Geol. Survey Mineral Resources, 1914, pt. 1, pp. 287–288, 1916.

Classification of Leadville ore, 1914—Continued

Character	Quantity	Recovered content					Average recovery				
		Gold	Silver	Copper	Lead	Zinc	Gold	Silver	Copper	Lead	Zinc
	Short tons	Fine ounces (b)	Fine ounces (b)	Pounds (b)	Pounds (b)	(c)	Ounces per ton	Ounces per ton	Per cent (d)	Per cent	Per cent
Sulphide:											
Zinc-iron-lead	111,947					(c)	d 0.03	d 5.37	(d)	d 8.74	d 21.2
Iron pyrites	174,610	31,770.74	2,343,943	1,720,872	4,221,091		.182	13.42	0.49	1.21	
Lead sulphide	17,100	4,989.00	290,906	84,688	5,245,333		.292	17.01	.25	15.34	
Copper sulphide	3,902	1,200.18	109,068	344,490	51,992		.308	27.95	4.41	.67	
Oxide:											
Lead	29,288	9,314.94	190,428	60,002	5,227,960		.318	6.50	.10	8.93	
Copper	135	10.70	706	14,035			.079	5.23	5.20		
Iron flux	48,839	192.45	220,831	34	1,271,781		.004	4.52		1.30	
Zinc	113,881					(c)					e 24.3
Siliceous	33,000	15,361.23	166,159	94,300	539,820		.465	5.04	.14	.82	
"Metallics" f (sulphide and oxide)		4,773.91	1,293								
	532,702										

Character	Quantity	Recovered content				
		Gold	Silver	Copper	Lead	Zinc
	Short tons	Fine ounces	Fine ounces	Pounds	Pounds	Pounds
Dry, iron, and siliceous ores g	256,449	52,098.33	2,732,226	1,815,206	6,032,692	
Copper ores h	4,037	1,210.88	109,774	359,025	51,992	
Lead ores i	46,388	14,303.94	481,334	144,690	10,473,293	
Zinc ores j	143,848	.90	398	49	11,041	55,680,382
Lead-zinc ores k	81,980	2,318.64	300,107	51,346	10,094,726	23,082,952
	532,702	69,932.69	3,623,839	2,370,316	26,663,744	78,763,334

b Estimated recovery, as lead and iron concentrates and zinc residues.
c Zinc figured as spelter or as zinc in zinc oxide.
d Average assay crude ore, mostly milled. (Lead wet assay.)
e Average assay crude. All to zinc smelters.
f Very rich material picked out of ore and shipped separately.
g Includes siliceous oxide, iron, and iron-manganese (silver), largest part of iron pyrites. Smelting ores. Also includes metallics.
h Sulphide and oxide. Smelting ores.
i Sulphide and oxide ores. Smelting ores.
j Sulphide and carbonate. To smelters.
k Sulphide. Greater part to magnetic separating mills.

Mineral Resources for 1915 gives the following table: [90]

Classification of Leadville ore, 1915

Character	Short tons	Percentage
Sulphide:		
Zinc-iron-lead (zinc blende: pyrite and galena)	136,555	29
Iron (pyrite with a little gold, silver, copper, and lead, and some zinc blende) a	161,677	} 35
Iron (pyrite with a little gold, silver, copper, and over 4½ per cent dry lead and some zinc blende) a	1,802	
Iron (pyrite with a little gold, silver, lead, and over 2½ per cent dry copper)	1,627	
Total sulphide	301,661	64
Oxide:		
Lead	16,002	3
Copper b	3,359	1
Iron flux	5,699	1
Iron-manganese (silver and some lead) flux	16,761	4
Siliceous c	44,734	9
Zinc carbonate and silicate	82,592	18
Total oxide	169,147	36
Grand total	470,808	100

a The greater part of the iron pyrite ore carries iron in excess of silica and also high silver but low gold content; a large quantity (particularly in 1915) of iron pyrite ore (from Breece Hill mines) carries silica in great excess over iron and high gold but low silver content. Some of the iron pyrite ore approaches the zinc-iron-lead classification (when zinc market gives higher returns for sale as lead-zinc sulphide than as lead sulphide to lead smelters) and vice versa. (Note only 1,802 tons of iron pyrite carrying over 4½ per cent lead in 1915, as against 17,100 tons in 1914.)
b The high price of copper during 1915 stimulated the sale of this class of ore.
c All the siliceous oxide ore is also dry. The greater part of this siliceous oxide ore carries a heavy excess of silica over iron; a small part carries iron and silica in equal quantities. Metallic gold, in quantity 4,178.75 fine ounces, valued at $86,382, and 1,538 ounces of silver, came from less than a ton of ore, from siliceous oxide and sulphide ore bodies.

[90] Henderson, C. W., U. S. Geol. Survey Mineral Resources, 1915, pp. 453–454, 1917.

Classification of Leadville ore, 1915—Continued

Character	Quantity	Recovered content				
		Gold	Silver	Copper	Lead	Zinc
	Short tons	Fine ounces	Fine ounces	Pounds	Pounds	Pounds
Sulphide:						
Zinc-iron-lead	136,555	[d] 1,595.41	[d] 404,396	[d] 6,008	[d] 12,856,824	([e])
Iron pyrites	161,677	51,716.79	1,560,191	1,092,425	3,707,810	
Lead sulphide	1,802	1,567.04	32,154	13,423	451,203	
Copper sulphide	1,627	539.74	51,707	145,783	10,131	
Oxide:						
Lead	16,002	4,497.76	87,933	38,740	2,649,869	
Copper	3,359	24.90	5,252	425,088		
Iron flux	5,699	539.55	3,992	556	514	
Iron-manganese flux	16,761	95.88	91,951		403,242	
Siliceous	44,734	32,181.21	150,257	60,212	728,814	
Zinc	82,592					([e])
"Metallics" [f] (sulphide and oxide)		4,178.75	1,538			
	470,808	96,937.03	2,389,371	1,782,235	20,808,407	72,424,873

Character	Average recovery				
	Gold	Silver	Copper	Lead	Zinc
	Ounces per ton	Ounces per ton	Per cent	Per cent	Per cent
Sulphide:					
Zinc-iron-lead	[g] 0.020	[g] 5.00	([g])	[g] 8.65	[g] 22.09
Iron pyrites	.320	9.65	0.34	1.15	
Lead sulphide	.870	17.84	.37	12.52	
Copper sulphide	.332	31.78	4.48	.31	
Oxide:					
Lead	.281	5.50	.12	8.28	
Copper	.007	1.56	6.33		
Iron flux	.095	.70			
Iron-manganese flux	.006	5.49		1.20	
Siliceous	.719	3.36	.07	.81	
Zinc					[h] 22.48
"Metallics" [f] (sulphide and oxide)					

[d] Recovered as lead and iron concentrates and in zinc residues.
[e] Zinc in terms of recovered spelter and as zinc in zinc oxide.
[f] Very rich material picked out of ore and shipped separately.
[g] Average assay crude ore, mostly milled. (Lead, wet assay.)
[h] Average assay actual content. All to zinc smelters.

Mineral Resources for 1916 gives the following table: [91]

Classification of Leadville ore, 1916

Character	Short tons	Percentage
Sulphide:		
Zinc-iron-lead (zinc blende; pyrite and galena)	147,295	32
Iron (pyrite with a little gold, silver, copper, and lead, and some zinc blende)[a]	161,096 }	35
Iron (pyrite with a little gold, silver, copper, and over 4½ per cent dry lead and some zinc blende)[a]	2,208 }	
Total sulphide	310,599	67
Oxide:		
Lead	19,716	4
Copper [b]	8,520	2
Iron flux	3,879	1
Iron-manganese flux	18,215	4
Siliceous [c]	13,876	3
Zinc	85,513	19
Total oxide	149,719	33
Grand total	460,318	100

[a] The greater part of the iron pyrite ore carries iron in excess of silica and also high silver but low gold content; a large quantity (particularly in 1916) of iron pyrite ore (from Breece Hill mines) carries silica in great excess over iron and high gold but low silver content. Some of the iron pyrite ore approaches the zinc-iron-lead classification (when zinc market gives higher returns for sale as lead-zinc sulphide than as lead sulphide to lead smelters) and vice versa.
[b] The high price of copper during 1916 stimulated the sale of this class or ore.
[c] All the siliceous oxide ore is also dry. The greater part of this siliceous oxide ore carries a heavy excess of silica over iron; a small part carries iron and silica in equal quantities. Metallic gold, in quantity 1,285.56 fine ounces, valued at $26,575, and 310 ounces of silver came from less than a ton of ore, from siliceous oxide and sulphide ore bodies.

[91] Henderson, C. W., U. S. Geol. Survey Mineral Resources, 1916, pt. 1, pp. 331-388, 1919.

Classification of Leadville ore, 1916—Continued

Character	Quantity	Recovered content				
		Gold	Silver	Copper	Lead	Zinc
	Short tons	Fine ounces	Fine ounces	Pounds	Pounds	Pounds
Sulphide:						(e)
Zinc-iron-lead	147,295	d 2,031.07	d 340,463	d 1,835	d 12,922,633	
Iron pyrites	161,096	54,574.15	2,039,723	1,688,230	3,154,198	
Lead sulphide	2,208	570.77	63,840	3,779	813,562	
Oxide:						
Lead	19,716	3,483.43	114,256	27,585	3,767,487	
Copper	8,520	4,217.54	29,573	639,672	86,085	
Iron flux	3,879	333.01	2,982			
Iron-manganese flux	18,215	28.03	110,743		245,522	
Siliceous	13,876	4,614.64	81,680	249,293	521,195	
Zinc	85,513					(e)
"Metallics" f (sulphide and oxide)		1,285.56	310			
	460,318	71,138.20	2,783,570	2,610,394	21,510,682	76,008,276

Character	Average recovery				
	Gold	Silver	Copper	Lead	Zinc
	Ounces per ton	Ounces per ton	Per cent	Per cent	Per cent
Sulphide:					
Zinc-iron-lead	g 0.028	g 5.17	(g)	g 8.37	g 20.96
Iron pyrites	.339	12.66	0.52	.98	
Lead sulphide	.259	28.91		18.42	
Oxide:					
Lead	.177	5.80	.07	9.55	
Copper	.495	3.47	3.75	.51	
Iron flux	.086	.77			
Iron-manganese flux	.002	6.08		.67	
Siliceous	.333	5.81	.90	1.88	
Zinc					h 21.52
"Metallics" (sulphide and oxide)					

d Recovered as lead and iron concentrates and in zinc residues.
e Zinc in terms of recovered spelter and as zinc in zinc oxide.
f Very rich material picked out of ore and shipped separately.
g Average assay of greater part of original ore for which data was available. (Lead, wet assay.) Average assay of all zinc sulphide was 21.69 per cent zinc.
h Average assay of original content; no deductions for loss in smelting. All to zinc plants.

Mineral Resources for 1917 gives the following table: [92]

Classification of Leadville ore, 1917

Character	Short tons	Percentage
Sulphide:		
Zinc-iron-lead (zinc blende, pyrite and galena)	148,945	36
Iron (pyrite with gold, silver, copper, and lead, and some zinc blende)a	128,533	
Iron (pyrite with gold, silver, copper, and over 4½ per cent dry lead, with some zinc blende)a	3,053	32
Iron (pyrite with gold, silver, lead, and over 2½ per cent dry copper, with some zinc blende)a	364	
Total sulphide	280,895	68
Oxide:		
Lead	16,416	4
Copper	7,524	2
Iron flux	6,050	1
Iron-manganese-silver flux	21,447	5
Siliceous b	10,233	3
Zinc	69,238	17
Total oxide	130,908	32
Grand total	411,803	100

a The greater part of the iron pyrites carries iron in excess of silica and also high silver but low gold content; a large quantity of iron pyrites carries silica in great excess over iron and high gold but low silver content. Some of the iron pyrites approaches the zinc-iron-lead classification (when zinc market gives higher returns for sale as lead-zinc sulphide than as lead sulphide to lead smelters) and vice versa.
b All the siliceous oxide ore contains lead under 4½ per cent and copper under 2½ per cent and is therefore called "dry." The greater part of the siliceous oxide ore carries a heavy excess of silica over iron; a small part carries iron and silica in equal quantities.
92 Henderson, C. W., U. S. Geol. Survey Mineral Resources, 1917, pt. 1, pp. 826-828, 1921.

Classification of Leadville ore, 1917—Continued

Character	Quantity	Recovered content				
		Gold	Silver	Copper	Lead	Zinc
	Short tons	Fine ounces	Fine ounces	Pounds	Pounds	Pounds
Sulphide:						
Zinc-iron-lead	148,945	c 3,510.02	c 343,277	c 47,715	c 10,884,497	(d)
Iron pyrites c	128,533	35,570.98	1,247,574	1,143,659	2,610,052	
Lead sulphide	3,053	975.61	31,546	14,932	586,760	
Copper sulphide	364	83.08	6,800	28,982	2,170	
Oxide:						
Lead	16,416	2,242.04	101,266	42,208	2,858,848	
Copper	7,524	2,408.00	25,465	826,970	199,478	
Iron flux	6,050	300.33	24,373	6,919	49,161	
Iron-manganese flux	21,447	291.17	114,094	2,666	515,970	
Siliceous	10,233	1,830.37	46,030	57,191	137,564	
Zinc	69,238					(d)
"Metallics" (sulphide and oxide)f		1,169.01	558			
	411,803	48,380.61	1,940,983	2,171,242	17,844,500	59,669,068

Character	Average recovery				
	Gold	Silver	Copper	Lead	Zinc
	Ounces per ton	Ounces per ton	Per cent	Per cent	Per cent
Sulphide:					
Zinc-iron-lead	g 0.033	g 4.83	(g)	g 7.79	h 20.00
Iron pyrites	.277	9.71	0.44	1.02	
Lead sulphide	.320	9.71	.24	9.61	
Copper sulphide	.228	18.68	3.98		
Oxide:					
Lead	.137	6.17	.13	8.71	
Copper	.320	3.38	5.50	1.33	
Iron flux	.050	4.03	.06	.41	
Iron-manganese flux	.014	5.32		1.20	
Siliceous	.179	4.50	.28	.67	
Zinc					h 19.84
"Metallics" (sulphide and oxide)					

c Recovered as lead and iron concentrates and in zinc residues.
d Zinc in terms of recovered spelter and recovered zinc in zinc oxide.
e Includes both ore to lead plants and that used for making sulphuric acid, the residues from which were returned to lead plants.
f Very rich material picked out of ore and shipped separately; may include some Cripple Creek and Park County metallics.
g Average assay of greater part of original ore for which data was available. (Lead, wet assay.)
h Average assay of original content; no deductions for loss in smelting. All to zinc plants.

Mineral Resources for 1918 gives the following table: [93]

Classification of Leadville ore, 1918

Character	Short tons	Percentage
Sulphide:		
Zinc-iron-lead (zinc blende, pyrite, and galena)	125,281	35
Iron (pyrite with gold, silver, copper, and lead, and some zinc blende) a	117,746	35
Iron (pyrite with gold, silver, lead, and over 2½ per cent dry copper, with some zinc blende) a	4,441	
Total sulphide	247,468	70
Oxide:		
Lead	45,508	13
Copper	3,898	1
Iron flux	2,081	1
Iron-manganese flux	26,955	7
Siliceous	6,271	2
Zinc	21,292	6
Total oxide	106,005	30
Grand total	353,473	100

Character	Quantity	Recovered content				
		Gold	Silver	Copper	Lead	Zinc
	Short tons	Fine ounces	Fine ounces	Pounds	Pounds	Pounds
Sulphide:						
Zinc-iron-lead	125,281	a 2,089.60	a 302,785	a 11,490	a 7,307,290	b 40,334,218
Iron pyrites c	91,803	5,754.37	1,112,765	558,387	1,475,416	
Lead sulphide	4,441	1,459.68	37,658	33,545	832,450	
Siliceous sulphide (carrying pyrite)	25,943	18,422.13	223,279	526,000	293,691	
Oxide:						
Lead (siliceous excess)	27,957	3,319.05	212,211	67,520	7,494,803	
Lead (iron excess)	17,551	114.10	85,284	8,005	3,948,405	
Copper	3,898	405.30	10,019	402,070	34,760	
Iron flux	2,081		18,257	3,611	14,296	
Iron-manganese flux	26,955	291.05	172,475	1,569	783,906	
Siliceous	6,271	1,741.68	57,420	7,594	187,966	
Zinc	21,292					b 6,351,683
"Metallics" (sulphide and oxide)d		780.74	260			
	353,473	34,377.70	2,232,413	1,619,781	22,373,183	46,685,901

a Recovered as lead and iron concentrates and in zinc residues.
b Zinc in terms of recovered retort zinc and recovered zinc in zinc oxide.
c Includes both ore to lead plants and that used for making sulphuric acid from which residues were returned to lead plants.
d Very rich material picked out of ore and shipped separately. May include some Cripple Creek and Park County metallics.

93 Henderson, C. W., U. S. Geol. Survey Mineral Resources, 1918, pt. 1, pp. 847–852, 1921.

Classification of Leadville ore, 1918—Continued

Character	Average recovery				
	Gold	Silver	Copper	Lead	Zinc
	Ounces per ton	Ounces per ton	Per cent	Per cent	Per cent
Sulphide:			(ª)		
Zinc-iron-lead	ª 0.030	ª 4.33		ª 6.25	ª 24.90
Iron pyrites	.063	12.12	0.30	.80	
Lead sulphide	.329	8.48	.38	9.37	
Siliceous sulphide (carrying pyrite)	.710	8.61	1.01	.57	
Oxide:					
Lead (siliceous excess)	.119	7.59	.12	13.40	
Lead (iron excess)	.007	4.86	.02	11.25	
Copper	.104	2.57	5.16	.45	
Iron flux		8.77	.09	.34	
Iron-manganese flux	.011	6.40		1.45	
Siliceous	.278	9.16	.06	1.50	
Zinc					ᶠ 18.64

ª Average assay of greater part of original ore for which data were available. (Lead, wet assay.)
ᶠ Average assay of original content; no deductions for loss in smelting. All to zinc plants.

Mineral Resources for 1919 gives the following table: [94]

Classification of Leadville ore, 1919

Character	Short tons	Percentage
Sulphide:		
Siliceous sulphide (carrying pyrite) ª	24,445	11
Zinc-iron-lead (zinc blende, pyrite, and galena)	46,967	22
Iron (pyrite with gold, silver, copper, and lead, and some zinc blende) ª	66,500	31
	137,912	64
Oxide:		
Lead-copper	653	
Lead (iron excess)	1,087	
Lead (siliceous excess)	18,916	9
Copper (siliceous excess)	139	
Iron flux	1,726	1
Iron-manganese flux	33,341	15
Siliceous	6,244	3
Zinc	16,542	8
	78,648	36
Grand total	216,560	100

Character	Quantity	Recovered content				
		Gold	Silver	Copper	Lead	Zinc
	Short tons	Fine ounces	Fine ounces	Pounds	Pounds	Pounds
Sulphide:						
Lead siliceous	907	1,064.91	18,647	11,622	216,427	
Copper siliceous	1,860	903.74	40,764	137,600	11,426	
Iron pyrites ᵇ	66,500	3,733.70	664,276	283,875	1,547,598	
Siliceous sulphide (carrying pyrite)	21,678	13,935.90	150,472	308,658	132,076	
Zinc-iron-lead	46,967	ᶜ 733.22	ᶜ 112,276	ᶜ 25,058	ᶜ 2,870,173	ᵈ 18,573,049
Oxide:						
Lead-copper	653	16.33	10,506	75,923	76,968	
Lead (iron excess)	1,087	10.50	5,624		172,702	
Lead (siliceous excess)	18,916	1,059.75	172,817	5,146	5,225,401	
Copper (siliceous excess)	139		110	18,816		
Iron flux	1,726	85.50	7,063		39,388	
Iron-manganese flux	33,341	86.41	259,865	1,985	842,365	
Siliceous	6,244	2,769.23	55,126	18,494	141,967	
Zinc	16,542					ᵈ 4,592,170
"Metallics" (sulphide and oxide) ᵉ		1,810.82	587			
	216,560	26,210.01	1,498,133	887,177	11,276,491	23,165,219

ª The greater part of the pyrite carries iron in excess of silica and also high silver but low gold content. A large quantity of pyrite carries silica in great mass over iron and of greater value than the silver. Some of the pyrite approaches the zinc-iron-lead classification (when zinc market gives higher returns for sale as lead-zinc sulphide than as lead sulphide to lead smelters) and vice versa.
ᵇ Includes both ore to lead plants and ore used for making sulphuric acid from which residues were returned to lead plants.
ᶜ Recovered as lead and iron concentrates in zinc residues.
ᵈ Zinc in terms of recovered retort zinc and recovered zinc in zinc oxide.
ᵉ Very rich material picked out of ore and shipped separately. May include some Cripple Creek and Park County metallics.

[94] Henderson, C. W., U. S. Geol. Survey Mineral Resources, 1919, pt. 1, pp. 773–774, 1922.

Classification of Leadville ore, 1919—Continued

Character	Average recovery				
	Gold	Silver	Copper	Lead	Zinc
	Ounces per ton	Ounces per ton	Per cent	Per cent	Per cent
Sulphide:					
Lead siliceous	1.17	20.56	0.64	11.93	
Copper siliceous	.49	21.92	3.70	.31	
Iron pyrites	.06	9.99	.21	1.16	
Siliceous sulphide (carrying pyrite)	.64	6.94	.71	.30	
Zinc-iron-lead	*.03	*4.11		*5.52	*28.25
Oxide:					
Lead-copper	.03	16.09	5.81	5.89	
Lead (iron excess)	.01	5.17		7.94	
Lead (siliceous excess)	.06	9.14	.01	13.81	
Copper (siliceous excess)		.79	6.73		
Iron flux	.05	4.09		1.14	
Iron-manganese flux		7.79		1.26	
Siliceous	.44	8.83	.15	1.14	
Zinc					*17.80
"Metallics" (sulphide and oxide)					

' Average assay of greater part of original ore for which data were available. (Lead, wet assay.)
' Average assay of original content; no deductions for loss in smelting. All to zinc plants.

Mineral Resources for 1920 gives the following table: [95]

Classification of Leadville ore, 1920

Character	Short tons	Percentage
Sulphide:		
Siliceous sulphide (carrying pyrite) [a]	21,702	12
Iron (pyrite with gold, silver, copper, and lead, and some zinc blende)	33,718	20
Zinc-iron-lead (zinc blende, pyrite, and galena) [a]	30,899	18
	86,319	50
Oxide:		
Lead (iron excess)	694	
Lead (silica excess)	17,826	11
Iron flux	1,990	1
Iron-manganese-silver flux	37,934	22
Siliceous	9,873	6
Zinc	16,726	10
	85,043	50
Grand total	171,362	100

Character	Short tons	Recovered content				
		Gold	Silver	Copper	Lead	Zinc
		Fine ounces	Fine ounces	Pounds	Pounds	Pounds
Sulphide:						
Copper (excess silica)	2,011	1,219.00	49,920	158,181	13,156	
Lead siliceous	411	24.80	2,798	5,109	150,769	
Siliceous (carrying pyrite) [b]	19,280	10,614.80	166,487	330,470	471,958	
Iron pyrites [b]	33,718	2,337.70	219,907	153,905	1,319,004	
Zinc-iron-lead	30,899	*303.35	*68,626	*40,004	*1,498,581	*12,923,696
Oxide:						
Lead (iron excess)	694	68.80	6,288	272	119,452	
Lead (silica excess)	17,826	1,354.60	141,055	92,361	3,666,124	
Iron flux	1,990	384.000	9,678	507	83,596	
Iron-manganese-silver flux	37,934	90.70	270,431		1,043,668	
Siliceous	9,873	12,168.70	101,282	18,088	191,092	
Zinc	16,726					5,830,835
"Metallics" (sulphide and oxide) [e]		1,690.85	628			
	171,362	30,257.30	1,037,100	798,897	8,557,400	18,754,531

[a] The greater part of the pyrite ore carries iron in excess of silica and also high silver but low gold content. A large quantity (21,702 tons) carries silica in excess (over iron) and gold of greater value than silver. Some of the pyrite ore approaches the zinc-iron-lead classification (and becomes such when zinc market gives higher returns for sale as lead-zinc sulphide than as lead sulphide to lead smelters and vice versa).
[b] Includes both ore to lead plants and ore used for making sulphuric acid from which residues were returned to lead plants.
[c] Recovered as lead and iron concentrates in zinc residues. Much lead recovered in leaded-zinc oxide.
[d] Zinc in terms of recovered retort zinc and recovered zinc in zinc oxide.
[e] Very rich material picked out of ore and shipped separately. Includes much Cripple Creek and Park County metallics.

[95] Henderson, C. W., U. S. Geol. Survey Mineral Resources, 1920, pt. 1, p. 582, 1922.

Classification of Leadville ore, 1920—Continued

Character	Average recovery				
	Gold	Silver	Copper	Lead	Zinc
	Ounces per ton	*Ounces per ton*	*Per cent*	*Per cent*	*Per cent*
Sulphide:					
Copper siliceous	0.606	24.82	3.93	0.33	
Lead siliceous	.060	6.81	.62	18.34	
Siliceous (carrying pyrite)	.551	8.64	.86	1.22	
Iron	.069	6.52	.23	1.96	
Zinc-iron-lead	*f* .025	*f* 4.00		*f* 4.44	*g* 27.13
Oxide:					
Lead (iron excess)	.099	9.06	.02	8.61	
Lead (siliceous excess)	.076	7.91	.26	10.28	
Iron flux	.193	4.86	.01	2.10	
Iron-manganese flux	.002	7.13		1.38	
Siliceous	1.233	10.26	.09	.97	
Zinc					*g* 21.79
"Metallics" (sulphide and oxide)					

f Average original assay of that part of original ore for which data were available.
g Average assay of original content; no deductions for loss in smelting. All to zinc plants.

Mineral Resources for 1921 gives the following table: [96]

Classification of Leadville ore, 1921

Character	Percentage by weight	Quantity	Recovered content				
			Gold	Silver	Copper	Lead	Zinc
		Short tons	*Fine ounces*	*Fine ounces*	*Pounds*	*Pounds*	*Pounds*
Sulphide:							
Siliceous (carrying pyrite)	9	7,446	5,132.50	101,386	222,106	153,457	
Iron pyrites *a*	51	40,337	1,965.80	682,288	829,282	1,495,480	
Zinc-iron-lead	2	1,292	*b* 11.50	*b* 2,401		*b* 62,188	*c* 610,000
Total sulphide	62	49,075	7,109.80	786,075	1,051,388	1,711,125	610,000
Oxide:							
Siliceous gold	3	2,591	4,587.10	15,829	23,182	38,575	
Siliceous silver	2	1,426	27.00	18,566	3,373	26,748	
Lead (iron excess)		435		2,263		106,464	
Lead (silica excess)	5	4,023	2,072.00	67,146		1,052,730	
Iron flux	8	6,258	82.10	18,471	11,705	234,874	
Iron-manganese flux	15	11,684	86.00	97,773	5,724	354,417	
Zinc	5	4,277					*e* 1,200,000
Total oxide	38	30,694	6,854.20	220,048	43,984	1,813,808	1,200,000
"Metallics" (sulphide and oxide) *d*			608.58	198			
Total sulphide and oxide	100	79,769	14,572.58	1,006,321	1,095,372	3,524,933	1,810,000

Character	Average recovery				
	Gold	Silver	Copper	Lead	Zinc
	Ounces per ton	*Ounces per ton*	*Per cent*	*Per cent*	*Per cent*
Sulphide:					
Siliceous (carrying pyrite)	0.689	13.62	1.49	1.03	
Iron pyrites	.049	16.91	1.03	1.85	
Zinc-iron-lead	.009	1.86		2.41	*e* 29.38
Oxide:					
Siliceous gold	1.770	6.11	.45	.74	
Siliceous silver	.019	13.02	.12	.91	
Lead (iron excess)		5.20		12.24	
Lead (silica excess)	.515	16.69		13.08	
Iron flux	.013	2.95	.09	1.88	
Iron-manganese flux	.007	8.37	.02	1.52	
Zinc					*e* 17.97

a Includes both ore to lead plants and ore used for making sulphuric acid from which residues were returned to lead plants.
b Recovered in zinc residues. Some lead recovered in leaded-zinc oxide.
c Zinc in terms of recovered zinc in zinc oxide.
d Includes some Cripple Creek and Park County metallics.
e Average assay of original content; no deductions for loss in smelting. All to zinc plants.

[96] Henderson, C. W., U. S. Geol. Survey Mineral Resources, 1921, pt. 1, p. 498, 1924.

Mineral Resources for 1922 gives the following table: [97]

Classification of Leadville ore, 1922

Character	Ore		Recovered content				
	Per cent	Short tons	Gold	Silver	Copper	Lead	Zinc
Sulphide:			*Fine ounces*	*Fine ounces*	*Pounds*	*Pounds*	*Pounds*
Siliceous (carrying pyrite) [a]	9	10,169	5,968.10	160,473	411,384	174,716	....
Iron pyrites [b]	34	37,365	1,882.00	490,905	402,357	1,464,597	....
Zinc-iron-lead	10	10,513	....	....	....	1,043,675	[c] 5,127,000
	53	58,047	7,850.10	651,378	813,741	2,682,988	5,127,000
Oxide (includes low sulphur ores):							
Siliceous gold	4	3,912	7,121.99	27,320	37,323	110,454	....
Siliceous silver	13	14,415	234.70	66,935	13,528	535,990	....
Lead (iron excess)	3	3,569	43.10	16,060	2,975	880,255	....
Lead (silica excess)	3	3,372	1,756.90	26,503	606	624,533	....
Iron-manganese flux [d]	14	15,801	30.07	92,505	1,338	649,998	....
Zinc	10	11,343	....	....	....	....	[c] 3,876,000
	47	52,412	9,196.76	229,323	55,770	2,801,230	3,876,000
"Metallics" (sulphide and oxide) [e]			2,730.77	750			
	100	110,459	19,777.63	881,451	869,511	5,484,218	[c] 9,003,000

Character	Average recovery				
	Gold	Silver	Copper	Lead	Zinc
Sulphide:	*Ounces per ton*	*Ounces per ton*	*Per cent*	*Per cent*	*Per cent*
Siliceous (carrying pyrite)	0.587	15.78	2.02	0.86	....
Iron pyrites	.050	13.14	.54	1.96	....
Zinc-iron-lead	....	....	....	4.96	[f] 30.48
Oxide:					
Siliceous gold	1.821	6.98	.48	1.41	....
Siliceous silver	.016	4.64	.05	1.86	....
Lead (iron excess)	.012	4.50	.04	12.33	....
Lead (silica excess)	.524	7.86	.01	9.26	....
Iron-manganese flux [d]	.002	5.85	....	2.06	....
Zinc					[f] 21.36

[a] Much highly siliceous, and in all silica exceeds iron content. A dry gold and silver ore.
[b] Iron content exceeds silica content. A dry silver ore.
[c] Zinc in terms of recovered zinc in zinc oxide and leaded zinc oxide.
[d] Includes 176 tons of iron flux.
[e] Probably includes some Teller County and Park County specimen gold.
[f] Average assay of original content; no deduction for loss in smelting.

[97] Henderson, C. W., U. S. Geol. Survey Mineral Resources, 1922, pt. 1, p. 540, 1924.

Mineral Resources for 1923 [97a] gives the following table:

Classification of Leadville ores, 1923

Character	Ore		Recovered content				
	Per cent	Short tons	Gold	Silver	Copper	Lead	Zinc
			Fine ounces	*Fine ounces*	*Pounds*	*Pounds*	*Pounds*
Sulphide:							
Siliceous (carrying sulphides)[a]	17	18,973	2,450.40	43,801	56,445	300,718	
Iron pyrites[b]	8	8,891	727.30	129,729	168,819	332,259	
Lead	2	2,405	441.90	73,176	26,239	395,233	
Copper (mostly excess silica over iron)	3	3,716	1,795.50	67,501	209,303	38,653	
Zinc-iron-lead	10	11,831	[c]64.73	[c]27,811		[d]471,966	[e]2,658,000
	40	45,816	5,479.83	342,018	460,806	1,538,829	2,658,000
Oxides (includes low sulphur ores):							
Siliceous gold	3	3,300	5,567.03	12,743	3,806	60,571	
Siliceous silver	17	19,021	9.60	67,428	5,005	1,080,195	
Lead (iron excess)	2	2,418	61.20	8,657		412,167	
Lead (silica excess)	10	11,345	831.60	77,212	41,274	2,079,163	
Iron-manganese-silver flux	10	12,079	18.30	89,882		436,247	
Zinc	18	20,304					[e]6,732,000
	60	68,467	6,487.73	255,922	50,085	4,068,343	6,732,000
"Metallics" (sulphide and oxide)[f]			312.58	87			
	100	114,283	12,280.14	598,027	510,891	5,607,172	9,390,000

Character	Average recovery				
	Gold	Silver	Copper	Lead	Zinc
	Ounces per ton	*Ounces per ton*	*Per cent*	*Per cent*	*Per cent*
Sulphide:					
Siliceous (carrying sulphides)	0.129	2.31	0.15	0.79	
Iron pyrites	.082	14.59	.95	1.87	
Lead	.184	30.43	.55	8.22	
Copper	.483	18.16	2.82	.52	
Zinc-iron-lead	.006	2.35		[g]2.74	[g]14.04
Oxide:					
Siliceous gold	1.687	3.86	.06	.92	
Siliceous silver		3.54	.01	2.84	
Lead (iron excess)	.025	3.58		8.52	
Lead (silica excess)	.073	6.81	.18	9.16	
Iron-manganese-silver flux	.002	7.44		1.81	
Zinc					[g]20.72

a Much highly siliceous, and in all silica exceeds iron content. Partly gold and partly silver ore.
b Iron content exceeds silica content. A dry silver ore.
c Recovered in residues.
d Recovered in residues and in leaded zinc oxide.
e Zinc in terms of recovered zinc in zinc oxide and leaded zinc oxide.
f Probably includes some Teller County specimen gold.
g Average assay of original content; no deduction for loss in smelting. Wet assay of lead.

97a Henderson, C. W., U. S. Geol. Survey Mineral Resources, 1923, pt. 1, 1925.

Gold, silver, copper, lead, and zinc produced in Lake County, 1859–1925

Year	Ore (short tons)	Gold (value) Placer	Gold (value) Lode	Gold (value) Total	Silver Fine ounces	Silver Avg. price per ounce	Silver Value	Copper Pounds	Copper Avg. price per pound	Copper Value	Lead Pounds	Lead Avg. price per pound	Lead Value	Zinc Pounds	Zinc Avg. price per pound	Zinc Value	Total value
1859–1867		$5,272,000		$5,272,000	37,600	$1.341	$50,422										$5,322,422
1868		60,000		60,000	452	1.326	600										60,600
1869		80,000	$10,000	90,000	679	1.325	900										90,900
1870		65,000		65,000	465	1.328	618										65,618
1871		50,000	50,000	100,000	1,158	1.325	1,534										101,534
1872		66,500	66,500	133,000	1,540	1.322	2,036										135,036
1873		75,000	150,000	225,000	2,937	1.297	3,809										228,809
1874		70,000	143,503	213,503	2,797	1.278	3,575										217,078
1875		17,237	25,862	43,099	16,668	1.24	20,668										63,767
1876		a30,000	30,000	60,000	23,203	1.16	26,915				15,000	$0.061	$915				87,830
1877		a30,000	25,000	55,000	458,600	1.20	549,600				1,200,000	.055	66,000				670,600
1878		30,000	30,000	60,000	1,800,000	1.15	2,070,000				10,000,000	.036	360,000				2,490,000
1879	b140,623	70,000	34,014	104,014	8,411,132	1.12	9,393,454	a100,000	$0.13	$13,000	43,288,000	.041	1,774,808				11,285,276
1880		69,000	231,000	300,000	9,977,344	1.15	11,473,946	a100,000	.108	10,800	66,658,000	.05	3,332,900				14,910,860
1881		63,500	256,500	320,000	7,966,406	1.13		a200,000	.111	11,100	58,464,000	.048	2,806,272				12,108,311
1882		25,000	375,000	400,000	8,894,531	1.14		a200,000	.138	27,600	97,880,000	.049	4,797,725				9,282,313
1883		30,000	470,000	500,000	9,049,219	1.11		265,489	.168	33,600	111,575,000	.043	4,797,610				15,256,375
1884	c232,002	a15,000	555,000	570,000	7,270,313	1.11	8,070,047	1,766,035	.135	35,976	93,628,000	.037	3,464,236				15,242,358
1885		a5,000	428,691	433,691	6,441,693	1.07	6,892,612	4,544,202	.156	275,501	55,322,000	.039	2,165,358				12,047,283
1886			243,694	243,694	6,486,047	1.07	6,421,187	5,928,823	.128	581,658	84,400,000	.046	3,882,400	50,000	$0.043	$2,150	9,640,920
1887			310,891	310,891	5,994,324	.98	5,874,438	5,000,000	.116	687,748	92,359,103	.045	4,156,160	50,000	.044	2,200	10,750,578
1888			189,397	189,397	5,486,064	.94	5,156,900	5,000,000	.108	540,000	73,378,149	.044	3,228,639	50,000	.046	2,300	10,304,192
1889			295,063	295,063	6,150,839	.94	5,781,789	4,000,000	.095	380,000	83,785,918	.039	3,267,651	150,000	.049	7,350	8,737,380
1890	342,163	2,894	345,525	348,419	5,313,930	1.05	5,579,627	2,803,550	.107	298,980	43,623,477	.045	1,963,056	150,000	.05	7,500	8,121,497
1891	403,135	9,000	242,296	251,296	4,748,015	.99	4,745,085	4,071,761	.108	439,750	53,444,202	.043	2,298,134	150,000	.055	8,250	7,980,796
1892	323,187		902,244	902,244	5,898,020	.87	5,131,277	5,598,720	.12	687,450	44,009,114	.04	1,760,365	562,500	.05	25,875	7,856,561
1893	351,794		1,499,314	1,499,314	6,705,451	.78	5,300,455	2,728,553	.124	547,546	36,274,889	.037	1,342,171	735,000	.04	29,400	8,416,727
1894	347,143		1,386,359	1,386,359	7,695,108	.63	4,843,018	2,611,167	.171	322,403	44,733,000	.033	1,476,189	a1,000,000	.035	35,000	8,238,421
1895	394,710		1,453,458	1,453,458	6,435,413	.65	6,133,018	2,556,583	.166	318,562	38,922,572	.032	1,245,522	642,000	.036	45,540	9,110,419
1896	349,333		2,053,858	2,053,858	4,623,764	.68	3,270,790	3,734,593	.122	350,252	31,983,777	.03	959,813	2,201,500	.039	25,038	7,382,219
1897	413,552		2,073,036	2,073,036	5,451,317	.60	4,170,550	4,486,115	.137	478,259	23,700,908	.036	853,233	2,673,500	.041	90,262	6,655,759
1898	525,728		2,196,498	2,196,498	7,058,727	.59	4,338,071	2,092,040	.128	403,898	35,945,006	.038	1,365,910	10,575,240	.046	122,981	8,419,927
1899	618,071		2,529,512	2,529,512	7,230,118	.60	4,089,050	2,679,510	.136	535,902	48,598,720	.045	2,186,942	14,441,140	.038	613,364	9,882,559
1900	748,946		1,776,132	1,776,132	6,967,279	.62	2,990,184	4,674,502	.156	617,034	62,599,654	.044	2,754,385	47,637,140	.044	949,853	10,691,954
1901	770,500		1,203,924	1,203,924	6,830,084	.60	2,685,438	5,182,608	.193	673,739	56,359,708	.043	2,423,467	76,566,000	.041	4,134,564	9,569,905
1902	663,487		1,339,974	1,339,974	5,641,857	.53	2,919,388	4,017,504	.20	462,935	39,450,178	.041	1,617,457	58,254,353	.048	2,970,972	10,037,064
1903	648,464		1,186,851	1,186,851	4,973,033	.54	2,460,595	2,065,980	.132	502,188	36,333,239	.042	1,526,836	70,238,634	.051	4,144,079	9,614,016
1904	631,273		1,180,401	1,180,401	5,085,151	.58	2,645,430	2,382,910	.13	340,857	47,180,965	.047	2,028,777	70,198,462	.059	4,282,106	10,889,525
1905	672,055	294	1,508,146	1,508,440	3,880,338	.61	2,742,243	2,286,183	.127	316,927	51,162,040	.057	2,494,616	67,247,381	.061	3,967,595	11,544,891
1906	408,711	510	1,064,180	1,064,690	4,154,913	.66	1,533,553	2,784,615	.125	315,599	32,519,796	.053	2,705,047	58,254,634	.059	1,089,840	10,033,979
1907	417,257		1,228,449	1,228,449	2,893,642	.53	1,780,294	2,621,675	.165	644,932	19,646,007	.012	1,723,549	70,238,381	.047	2,086,415	5,204,008
1908	472,033		1,435,431	1,435,431	3,423,015	.52	1,791,888	2,182,623	.155	595,856	21,073,992	.043	825,132	38,637,315	.054	3,043,842	6,882,061
1909	438,419		1,213,134	1,213,134	3,322,015	.54	1,593,897	1,626,534	.133	401,754	19,499,503	.042	906,182	56,367,445	.054	4,081,796	7,360,777
1910	507,591		1,133,442	1,133,442	3,007,296	.53	1,845,244	888,628	.175	165,285	18,499,089	.043	846,978	71,610,456	.057	7,310,259	8,143,752
1911	528,311		1,103,230	1,103,230	3,000,397	.615	2,053,792	799,741	.246	147,153	26,234,244	.045	832,459	105,945,783	.069	4,016,930	11,780,131
1912	547,463		1,023,631	1,023,631	3,400,318	.604	2,107,389	1,107,295	.273	142,841	29,286,183	.044	1,180,541	93,842,857	.057	4,989,154	9,919,433
1913	481,620		1,571,451	1,571,451	3,810,830	.533	1,303,498	871,370	.247	117,635	20,957,404	.039	1,288,592	78,763,334	.051	10,289,296	9,057,297
1914	477,240	69,009	2,177,143	2,246,152	2,571,002	.507	1,928,783	511,776	.186	75,231	20,957,404	.047	1,044,600	72,493,178	.124	6,145,942	13,839,401
1915	422,428	119,169	1,601,271	1,720,440	2,931,281	.507	1,799,616				21,719,392	.049	984,998	76,785,567	.134	4,251,132	16,082,059
1916	355,840	110,325	1,601,894	1,175,219	2,290,121	.658	2,290,121				18,301,802	.079	1,498,638	60,254,333	.102	1,691,061	11,290,588
1917	217,667	92,096	751,173	843,239	1,542,324	.824	1,727,403				22,469,915	.086	1,573,955	46,715,736	.073	1,519,117	9,381,610
1918	172,988	81,688	541,298	625,956	2,184,000	1.00	1,198,660				11,299,076	.071	1,595,364	23,165,219	.091	91,050	4,808,556
1919	80,501	138,864	629,501	768,365	1,099,688	1.12	1,043,497				8,590,188	.08	598,851	18,754,531	.081	513,171	4,320,510
1920	112,547	6,184	302,960	309,144	1,043,497	1.09	952,048				8,537,889	.045	687,215	1,821,000	.065	91,050	1,745,737
1921	115,975	315	412,743	413,058	952,048	1.00	952,048				5,521,818	.055	159,205	9,003,000	.057	513,171	2,299,612
1922		15,224	256,280	271,504	655,838	.82	537,787				5,624,958	.07	393,747	9,415,000	.068	640,220	1,918,489
		6,798,749	44,380,824	51,179,573	230,452,487		189,409,933	100,099,832		14,329,466	1,925,288,125		85,455,300	1,234,918,034		85,410,278	425,784,550

* Estimated by C. W. Henderson.   a Emmons, S. F., op. cit., p. 18.   b Census of 1880.

LA PLATA AND MONTEZUMA COUNTIES

Purington says of the early developments in La Plata and Montezuma counties:[98]

It was not until the year 1878 that prospecting was begun in this vicinity. In that year the mine called the Comstock was opened, and work was begun on the Cumberland or Snowstorm vein, near the head of the La Plata Valley. A stamp mill was soon erected to treat the ore of this mine, but its operation was not successful. By the end of the year 1881 many locations had been made, and the nature of the richest ores, tellurides of gold and silver, was well known. Among the mines first developed, besides those mentioned, were the Century, on Bear Creek (now Montezuma County), the Tippecanoe, the Bell Hamilton, and the Ashland.

Burchard[99] in his report for 1881 says that La Plata County (including the present San Juan County) was organized in 1874. He describes La Plata district (in which he reports the Comstock mine as having produced) and also Junction Creek, Needle Mountain, and Vallecito districts.

In the report for 1882 Burchard[1] describes Needle Mountain or Florida district, at the head of the Florida, Johnsons Fork of the Vallecito, and Needle Creek. He says that no ore had as yet been shipped and that the district was not over two years old. He describes La Plata (or California) district, in which the leading property was the Century mine on Bear Creek.

In the report for 1883 Burchard[2] describes Needle Mountain district (Las Animas River and Vallecito River watersheds); Cascade district; Florida Creek district; and California district (La Plata Mountains, including watersheds of La Plata River, Mancos River, Bear Creek, Junction Creek; California dis-

trict is therefore in both La Plata and Montezuma counties, though most of the production came from La Plata). He says:

Bear Creek district: Century mine, 46⅕ tons returned in cash $12,928.61. No shipments of consequence from Needle Mountain, Cascade, and Florida. La Plata River district: High assays from several prospects; one car from Ashland mine yielded an average value per ton of $136. The South Comstock mine, value of ore shipments so far, $10,000. The Heck mine, shipments in several ton lots. Many mines "patent applied for."

The figures given in the table for 1884 are taken from Burchard's report for that year.[3]

For 1886 to 1896 the figures given in the table are derived from reports of the agents of the mint in annual reports of the Director of the Mint, the gold and silver being prorated to correspond with the figures showing the total production of the State as corrected by the Director of the Mint, the lead being prorated to correspond with the total production of lead for the State as shown in Mineral Resources, and any unknown production in the State being distributed proportionately among the counties. As with lead so with copper, but as the figures for copper given in Mineral Resources include copper from matte and ores treated in Colorado, though produced in other States, they are subject to revision.

The production of Montezuma County is first given separately in 1894. Only a small production came from districts in this county. No statement is given as to the cause of the increase in gold and silver in 1894. The amount of silver produced seems to be erroneously stated.

The figures for 1897 to 1904 are taken from the reports of the Colorado State Bureau of Mines, which represent smelter and mint receipts.

The figures for 1905 to 1923 are taken from Mineral Resources (mines reports).

[98] Purington, C. W., in Cross, Whitman, U. S. Geol. Survey Geol. Atlas, La Plata folio (No. 60) p. 12, 1899.

[99] Burchard, H. C., Report of the Director of the Mint, upon the production of the precious metals in the United States during the calendar year 1881, pp. 354, 418, 1882.

[1] Burchard, H. C., op. cit. for 1882, pp. 394, 506-509, 1883.

[2] Burchard, H. C., op. cit. for 1883, pp. 240, 368-376, 1884.

[3] Burchard, H. C., op. cit., for 1884, p. 117, 1885.

Gold, silver, copper, and lead produced in the La Plata district, La Plata and Montezuma counties, 1878–1923

Year	Ore treated (short tons)	Lode gold	Silver			Copper			Lead			Total value
			Fine ounces	Average price per ounce	Value	Pounds	Average price per pound	Value	Pounds	Average price per pound	Value	
1878		a $1,000	a 1,934	$1.15	$2,224							$3,224
1879		a 2,500	a 3,867	1.12	4,331							6,831
1880		a 5,000	a 7,734	1.15	8,894							13,894
1881		5,000	7,734	1.13	8,739							13,739
1882		10,000	23,203	1.14	26,451							36,451
1883		13,000	3,867	1.11	4,292							17,292
1884		500	4,641	1.11	5,152							5,652
1885		b 5,000	b 5,000	1.07	5,350							10,350
1886		10,225	4,671	.99	4,625				100,000	$0.046	$4,600	19,449
1887		12,473	7,126	.98	6,983				42,210	.045	1,899	21,355
1888		3,574	2,294	.94	2,156							5,730
1889		4,465	1,118	.94	1,051							5,516
1890		3,729	2,011	1.05	2,112							5,841
1891		23,054	3,207	.99	3,175							26,229
1892		34,881	3,335	.87	2,901							37,782
1893		37,872	4,928	.78	3,844							41,716
1894		114,264	417,465	.63	263,003							377,267
1895		3,682	99	.65	64							3,746
1896		10,741	41	.68	28							10,769
1897		36,944	1,514	.60	908	420	$0.12	$50	857	.036	31	37,933
1898		38,653	5,219	.59	3,079	2,568	.124	318	8,407	.038	319	42,369
1899		41,092	3,389	.60	2,033	211	.171	36	3,176	.045	143	43,304
1900		24,927	7,187	.62	4,456	350	.166	58	14,500	.044	638	30,079
1901		30,819	5,588	.60	3,353	132	.167	22	6,197	.043	266	34,460
1902		127,182	7,416	.53	3,930	3,143	.122	383	2,156	.041	88	131,583
1903		145,331	7,716	.54	4,167	810	.137	111	3,017	.042	127	149,736
1904	3,792	130,200	31,086	.58	18,030	1,473	.128	189	2,177	.043	94	148,513
1905	5,662	254,007	93,258	.61	56,887	2,923	.156	456	610	.047	29	311,379
1906	7,757	304,633	121,721	.68	82,770	445	.193	86	2,228	.057	127	387,616
1907	7,812	413,034	217,579	.66	143,602	708	.20	142	340	.053	18	556,796
1908	2,416	101,584	71,592	.53	37,944	458	.132	60	748	.042	31	139,619
1909	4,135	127,205	74,160	.52	38,563	484	.132	63	2,980	.043	128	165,959
1910	6,798	399,608	141,752	.54	76,546	362	.127	46	273	.044	12	476,212
1911	10,059	286,953	69,444	.53	36,805	73,911	.125	9,239	1,511	.045	68	333,065
1912	2,761	135,391	47,948	.615	29,488	918	.165	151	6,756	.045	304	165,334
1913	7,403	312,891	121,122	.604	73,158	113,897	.155	17,654	4,455	.044	196	403,899
1914	5,083	126,498	60,244	.553	33,315	26,038	.133	3,463	11,410	.039	445	163,721
1915	2,966	72,024	46,472	.507	23,561	4,114	.175	720	23,532	.047	1,106	97,411
1916	1,688	33,055	29,380	.658	19,332	15,142	.246	3,725	6,667	.069	460	56,572
1917	1,772	27,952	15,512	.824	12,782	28,333	.273	7,735	3,745	.086	322	48,791
1918	300	7,378	6,415	1.00	6,415	668	.247	165	3,000	.071	213	14,171
1919	405	5,966	6,075	1.12	6,804	167	.186	31	2,283	.053	121	12,922
1920	717	11,020	10,578	1.09	11,530				937	.08	75	22,625
1921	1,279	45,181	20,327	1.00	20,327				3,734	.045	168	65,676
1922	791	32,261	10,656	1.00	10,656							42,917
1923	838	15,905	17,138	.82	14,053	816	.147	120	1,800	.07	126	30,204
		3,588,654	1,754,763		1,129,868	278,491		45,023	259,706		12,154	4,775,699

a Estimated by C. W. Henderson, by subtracting from figures for San Juan region.
b Interpolated by C. W. Henderson to correspond with total production of the State.

LARIMER AND JACKSON COUNTIES

Information concerning Larimer and Jackson counties is contained in reports by Hollister, who mentions the unsuccessful rushes to Cache la Poudre River, by the State Bureau of Mines, and by Spencer.[4]

The figures for 1895 and 1896 given in the table are based on reports of the agents of the mint in annual reports of the Director of the Mint, the gold and silver being prorated to correspond with the figures for the total production of the State as corrected by the Director of the Mint, the lead being prorated to correspond with the total production of lead in the State as given in Mineral Resources, and any unknown production in the State being distributed proportionately among the counties. As with lead, so with copper, but as the figures for copper given in Mineral Resources include copper from matte and ores treated in Colorado, though produced in other States, they are subject to revision.

The figures for 1895 to 1904, which represent smelter and mint receipts, are taken from reports of the Colorado State Bureau of Mines.

The figures for 1905 to 1922 are taken from Mineral Resources (mines reports).

In 1906 the production came from the Pearl district, Jackson County.

In 1909 the production came from Larimer County, in the district west of Fort Collins, from property on Rabbit Creek, in T. 10 N., R. 72 W., near St. Cloud. Jackson County was segregated from Larimer County, May 5, 1909.

In 1916 the production came from the Pearl district, Jackson County.

In 1917 the production came from the Pearl district, Jackson County, and the Masonville district, Larimer County.

4 Hollister, O. J., The mines of Colorado, pp. 73, 106, 1867. Colorado State Bur. Mines Rept. for 1897, pp. 73–75, 1898. Spencer, A. C., Reconnaissance examination of the copper deposits at Pearl, Colo.: U. S. Geol. Survey Bull. 213, pp. 163–169, 1903.

Gold, silver, copper, and zinc produced in Larimer and Jackson counties, 1895–1917

Year	Ore (short tons)	Gold			Silver			Copper			Zinc			Total value
		Placer	Lode	Total	Fine ounces	Average price per ounce	Value	Pounds	Average price per pound	Value	Pounds	Average price per pound	Value	
1895		[a]$320		$320	1	$0.65	$1							$321
1896		[a]13		13	3	.68	2							15
1897		[a]805	[a]$2,171	2,976	97	.60	58							3,034
1898		[a]10,456	[a]706	11,162	60	.59	35	24,484	$0.124	$3,036				14,233
1899		[a]1,599	[a]468	2,067	135	.60	81	2,474	.171	423				2,571
1900		[a]1,078	[a]555	1,633	126	.62	78	13,806	.166	2,292				4,003
1901		[a]522	[a]408	930	73	.60	44	18,140	.167	3,029				4,003
1902		[a]488	[a]318	806	49	.53	26	24,888	.122	3,036				3,868
1903		[a]603	[a]1,030	1,633	10	.54	5	56,700	.137	7,768				9,406
1904	6	[a]141	[a]1,037	1,178	11	.58	6	23,028	.128	2,948				4,132
1906	460		[a]904	904	1,136	.68	772	41,331	.193	7,977				9,653
1909	48										30,722	$0.054	$1,659	1,659
1916	61		95	95	199	.658	,131	6,752	.246	1,661				1,887
1917	279		587	587	602	.824	496	23,725	.273	6,477				7,560
		16,025	8,279	24,304	2,502		1,735	235,328		38,647	30,722		1,659	66,345

[a] Estimated by C. W. Henderson, partly on basis of deposits at Denver Mint.

LAS ANIMAS COUNTY

The Colorado State Bureau of Mines says of Las Animas County:[5]

In metalliferous mines the south slope of Spanish Peaks and east slope of Culebra Range are the only sections prospected. The lodes located are reported as well-defined fissure veins, carrying ores of too low grade for direct shipment and not enough developed to justify the erection of a proper reducing plant. Placer beds of small extent are also reported near the foot of the Culebra Range along several of the small streams but are undeveloped.

The county records show 125 lode claims, of which two are patented. The reports for 1897 show 26 men employed in mining in these sections.

The figures given in the table for 1887 are derived from reports of the agents of the mint in annual reports of the Director of the Mint, the gold and silver being prorated to correspond with figures for the total production of the State as corrected by the Director of the Mint.

The figures given for 1897 to 1899, which represent smelter and mint receipts, are taken from the reports of the Colorado State Bureau of Mines.

Gold and silver produced in Las Animas County, 1887–1899

Year	Lode gold	Silver			Total value
		Fine ounces	Average price per ounce	Value	
1887	$1,122	8	$0.98	$8	$1,130
1897	641	9	.60	5	646
1898	124				124
1899	207	3	.60	2	209
	2,094	20		15	2,109

MESA COUNTY

The Colorado State Bureau of Mines says of Mesa County:[6]

Prior to 1881 this county formed a part of the Ute Indian reservation, and after being declared open for settlement the United States troops were employed in a dual capacity, viz., keeping the anxious whites back and urging the Indians to depart for their new reservation. * * *

Of metalliferous mines nothing as yet has been developed beyond the prospect stage. The county records show 102 lode claims and 23 placer claims, 6 placers patented, duly recorded. The county is locally divided into the Elk Basin, Plateau, and Copper Creek mining districts. The Elk Basin embraces the northeast corner and the Plateau the southeast corner of the county. The Copper Creek district, better known as Unaweep, covers the south-central and southwest parts of the county. This last-named district is at present attracting considerable attention and is said to have strong veins carrying high-grade copper ores. * * * An average of 48 men were engaged in prospecting during 1897.

Butler[7] gives details of the Unaweep copper district, in Mesa County.

Wilson[8] states the gold and silver deposited at the Denver Mint during 1885, and his figures have been incorporated in the table.

For 1885 to 1904 the amounts credited to placer and lode production in the table are estimates, based partly on receipts at the Denver Mint.

The figures for 1886 to 1896 given in the table are based on reports of the agents of the mint in annual reports of the Director of the Mint, the gold and silver being prorated to correspond with the figures for the total production of the State as corrected by the Director of the Mint, the lead being prorated to correspond with the total production of lead in the State as given in Mineral Resources, and any unknown production in the State being distributed proportionately among the counties. As with lead so with copper, but as the figures for copper given in Mineral Resources include copper from matte and ores treated in Colorado, though produced in other States, they are subject to revision.

For 1897 to 1904 the figures, which represent smelter and mint receipts, are taken from the reports of the Colorado State Bureau of Mines.

For 1905 to 1912 the figures are taken from Mineral Resources (mine reports).

---

5 Colorado State Bur. Mines Rept. for 1897, p. 76, 1898.
6 Idem, pp. 77–78.

7 Butler, B. S., Notes on the Unaweep copper district, Colo.: U. S. Geol. Survey Bull. 580, pp. 19–23, 1915.
8 Wilson, P. S., agent for Colorado, in Kimball, J. P., Report of the Director of the Mint upon the production of the precious metals in the United States during the calendar year 1885, p. 136, 1886.

Gold (lode and placer), silver, copper, and lead produced in Mesa County, 1885-1912

Year	Ore (short tons)	Gold			Silver			Copper			Lead			Total value
		Placer	Lode	Total	Fine ounces	Average price per ounce	Value	Pounds	Average price per pound	Value	Pounds	Average price per pound	Value	
1885		$431		$431	3	$1.07	$3							$434
1886		110		110		.99								110
1894		318		318	1	.63	1							319
1898			$165	165	20	.59	12							177
1899			124	124	4,120	.60	2,472	4,650	$0.171	$795				3,391
1900			124	124	311	.62	193	2,150	.166	357				674
1901		1,940	106	2,046	155	.60	93	7,795	.167	1,302				3,441
1902		84	453	537	32	.53	17	15,000	.122	1,830				2,384
1903		351		351	8	.54	4							355
1904	9	248		248	9	.58	5							253
1906		473		473	15	.68	10							483
1907		76		76	3	.66	2							78
1911		28		28		.53								28
1912	22		9	9	257	.615	158	5,685	.165	938	20	$0.045	$1	1,106
		4,05 9	981	5,040	4,934		2,970	35,280		5,222	20		1	13,233

MINERAL COUNTY

W. H. Emmons[9] gives a sketch of the history of the Creede district. He says that the Holy Moses mine was located in August, 1889, and that in June, 1891, the Last Chance deposit and later the Amethyst deposit were discovered. The first train on the Denver & Rio Grande Railroad arrived at Creede December 16, 1891.

The figures for 1891 to 1896 here given are based on reports of the agents of the mint in annual reports of the Director of the Mint, the gold and silver being prorated to correspond with the figures for the total production of the State as corrected by the Director of the Mint, the lead being prorated to correspond with the total production of lead for the State as given in Mineral Resources, and any unknown production in the State being distributed proportionately among the counties. As with lead so with copper, but as the figures for copper given in Mineral Resources include copper from matte and ores treated in Colorado, though produced in other States, they are subject to revision.

Smith[10] gives the output of Mineral County in 1891 under the head of Saguache County. The output of the two counties has been separated in this report as closely as possible from the information available. He gives the output of Mineral County in 1892 under the heads of Hinsdale County and of Rio Grande County.[11] The production of the Ethel and the Holy Moses mines he says was $65,220 in gold and 59,317 ounces of silver; that of the Amethyst, Bachelor, Del Monte, and Last Chance, he gives as $25,932 in gold and 2,366,778 ounces of silver.

S. F. Emmons says:[12]

Creede credited by mint authorities with $3,500,000 (at coining rate of $1.29+) of silver in 1892, other estimates giving even a larger amount.

Mineral Industry says:[13]

Output estimated at about 5,000,000 ounces. * * * Last Chance and Amethyst mines largest producers.

Puckett[14] says that the production of lead in 1896 was 1,512,226 pounds.

For 1897 to 1908 the figures, which represent smelter and mint receipts, are taken from reports of the Colorado State Bureau of Mines.

Mineral Industry[15] says that a small amount of zinc blende concentrates was shipped in 1898 from a mine at Creede, and that in 1899 considerable zinc ore was shipped from Creede, Leadville, and other points to the zinc smelters for experiments. The quantity of zinc produced in 1901, based on metallic content of the ore, is given as 2,088,000 pounds.

For 1909 to 1923 the figures have been taken from Mineral Resources (mines reports).

[11] Leech, E. O., op. cit. for 1892, pp. 126, 129, 1893.

[12] Emmons, S. F., U. S. Geol. Survey Mineral Resources, 1892, p. 68, 1893.

[13] Mineral Industry, vol. 1, p. 177, 1892.

[14] Puckett, W. J., agent for Colorado, in Preston, R. E., Report of the Director of the Mint upon the production of the precious metals in the United States during the calendar year 1896, p. 158, 1897.

[15] Mineral Industry for 1898, p 727, 1899; idem for 1899, p. 636, 1900; idem for 1901, p. 651, 1902.

[9] Emmons, W. H., and Larsen, E. S., Geology and ore deposits of the Creede district, Colo.: U. S. Geol. Survey Bull. 718, pp. 3-5, 1923.

[10] Smith, M. E., agent for Colorado, in Leech, E. O., Report of the Director of the Mint upon the production of the precious metals in the United States during the calendar year 1891, p. 184, 1892.

Gold, silver, copper, lead, and zinc produced in Mineral County, 1891–1923

Year	Ore (short tons)	Lode gold	Silver			Copper			Lead			Zinc			Total value
			Fine ounces	Average price per ounce	Value	Pounds	Value	Average price per pound	Pounds	Average price per pound	Value	Pounds	Average price per pound	Value	
1891		$10,055	378,899	$0.99	$374,382				354,854	$0.043	$15,259				$399,696
1892		87,219	2,391,514	.87	2,080,617				3,000,000	.04	120,000				2,287,836
1893		53,252	4,897,684	.78	3,820,194				a 7,500,000	.037	277,500				4,150,946
1894		40,336	1,866,927	.63	1,176,164				a 6,500,000	.033	214,500				1,431,000
1895		114,482	1,423,038	.65	924,975				a 8,220,870	.032	263,068				1,302,525
1896		52,238	1,560,865	.68	1,061,388				a 6,021,109	.03	180,633				1,294,259
1897		61,328	3,070,576	.60	1,842,346	1,500	$0.12	$180	6,080,673	.036	218,904				2,122,758
1898		46,383	4,177,184	.59	2,464,539	14,729	.124	1,826	5,453,104	.038	207,218	a 200,000	$0.046	$9,200	2,729,166
1899		91,671	3,796,899	.60	2,278,139	20,223	.171	3,458	5,677,162	.045	255,472	a 100,000	.058	5,800	2,634,540
1900		209,387	2,280,038	.62	1,413,623	2,614	.166	434	14,951,956	.044	657,886	b 450,000	.044	19,800	2,301,130
1901		102,813	1,816,023	.60	1,089,614	1,007	.167	168	10,519,895	.043	452,355	a 1,800,000	.041	73,800	1,718,750
1902		112,838	1,923,973	.53	1,019,706				9,291,358	.041	380,946	2,047,555	.048	98,283	1,611,773
1903		178,961	1,608,788	.54	868,746	133	.137	18	8,600,646	.042	361,227	2,634,000	.054	142,236	1,551,188
1904	124,278	222,864	1,664,633	.58	965,487	1,337	.128	171	13,346,436	.043	573,897	4,402,697	.051	224,538	1,986,957
1905	91,338	216,994	1,193,442	.61	728,000	107	.156	17	11,880,797	.047	558,397	2,515,628	.059	148,422	1,651,830
1906	126,164	176,150	1,254,058	.68	852,759				14,886,356	.057	848,522	2,892,061	.061	176,416	2,053,847
1907	104,977	142,803	1,246,961	.66	822,994	12,711	.20	2,542	12,980,288	.053	687,955	2,691,216	.059	158,782	1,815,076
1908	61,131	127,549	830,951	.53	440,404	41	.132	5	8,238,025	.042	345,997	1,100,107	.047	51,705	965,660
1909	64,941	108,825	891,185	.52	463,416	17,401	.13	2,262	9,036,816	.043	388,583	1,817,296	.054	98,134	1,061,220
1910	62,956	121,181	773,722	.54	417,810	29,031	.127	3,687	8,246,000	.044	362,824	2,421,926	.054	130,784	1,036,286
1911	65,932	179,196	545,319	.53	289,019	33,384	.125	4,173	7,674,556	.045	345,355	1,258,561	.057	71,738	889,481
1912	66,488	86,002	714,909	.615	439,669	23,885	.165	3,941	5,730,222	.045	257,860	308,681	.069	21,299	808,771
1913	56,763	50,282	805,343	.604	486,427	31,647	.155	4,905	3,398,364	.044	149,528	454,875	.056	25,473	716,615
1914	27,952	19,304	615,734	.553	340,501	32,586	.133	4,334	1,401,795	.039	54,670				418,809
1915	28,071	33,039	291,807	.507	147,946	8,943	.175	1,565	2,382,128	.047	111,960	85,984	.124	10,662	305,172
1916	38,103	31,124	373,956	.658	246,063	13,138	.246	3,232	2,295,087	.069	158,361	240,575	.134	32,237	471,017
1917	32,755	10,101	361,517	.824	297,890	19,297	.273	5,268	1,305,744	.086	112,294	54,971	.102	5,607	431,160
1918	28,372	13,943	640,959	1.00	640,959	3,490	.247	862	989,620	.071	70,263				726,027
1919	16,718	9,083	369,575	1.12	413,924	355	.186	66	934,113	.053	49,508	96,274	.073	7,028	479,609
1920	12,597	5,710	272,322	1.09	296,831	1,120	.184	206	531,537	.08	42,523				345,270
1921	7,076	3,816	192,468	1.00	192,468	1,899	.129	245	156,778	.045	7,055				203,584
1922	3,978	1,654	106,903	1.00	106,903	3,422	.135	462	153,455	.055	8,440				117,459
1923	6,462	2,394	228,867	.82	187,671	1,088	.147	160	237,557	.07	16,629	41,000	.068	2,788	209,642
		2,722,977	44,567,039		29,191,574	275,088		44,187	197,977,301		8,755,589	27,613,407		1,514,732	42,229,059

a Estimated by C. W. Henderson.   b Interpolated by C. W. Henderson to correspond with the total production of the State.

MONTROSE COUNTY

The Colorado State Bureau of Mines describes Montrose County as follows:[16]

Montrose is one of the western slope border counties segregated from Gunnison by an act of legislature, approved February 11, 1883. * * * It is generally considered a valley county, noted for its agricultural and horticultural products, complete system of irrigation canals, and fine stock ranges. In addition to these, the county possesses large resources in coal, building stone, clays, and metalliferous deposits, none of which, however, are very much developed. The county records disclose 129 lode claims, 594 placer claims and 1 tunnel site, 1 patented lode claim, and 35 patented placer claims duly recorded.

The Gunnison River enters near the center of the east boundary and flows northeast through an inaccessible canyon. Cimarron Creek enters near the southeast corner and flows north, joining the Gunnison River at the mouth of the canyon. This stream is lined with alluvial deposits carrying gold and has been mined in a desultory manner for many years. Near the town of Cimarron, located 2 miles south of the Gunnison River on Cimarron Creek, a number of mineralized veins in the metamorphic granite, lying north and east, and in the trachytic capping of the Cretaceous shales, lying west and south, have been located but are little developed. * * *

Along the various stream beds in this section, placer locations are numerous. For many years hand sluicing has been spasmodically indulged in and small amounts of gold, appreciable in the aggregate, produced. During the past year a few of the placer beds along the San Miguel River have been equipped with hydraulic appliances, and larger returns are anticipated in future. In Paradox Valley a number of locations have been made upon fissure veins, cutting vertically through the sedimentary beds. The value of the ore found is principally in copper and its economic importance not yet determined. An average of 94 men were employed in mining during 1897.

Hodges [17] says of the developments in 1897:

Paradox district.—Eighty men employed; greatest depth, 100 feet; copper sulphides, carrying 10 per cent free milling gold. * * *

Chipeta district.—Thirty men employed; greatest depth, 80 feet; copper sulphides, carrying 20 per cent free milling gold. * * *

Cimarron district.—Twenty men employed; greatest depth, 120 feet; character of ore, iron and copper sulphides, carrying 60 per cent gold, 40 per cent silver. * * *

Hodges [18] describes the argentiferous copper ore shipped from La Sal district in 1898.

Emmons says of the Cashin mine:[19]

According to the books of the La Sal Copper Mining Co., the present owners, it has produced altogether 363,778 ounces of silver and 732,740 pounds of copper, not including the shipments of native copper, of which no record was available.

These figures are considerably higher than those of the Colorado State Bureau of Mines, which represent smelter receipts. For 1902 copper seems out of proportion to silver. Mr. Emmons's figures seem to

16 Colorado State Bur. Mines Rept. for 1897, pp. 80–81, 1898.

17 Hodges, J. L., agent for Colorado, in Roberts, G. E., Report of the Director of the Mint upon the production of the precious metals in the United States during the calendar year 1897, p. 119, 1898.

18 Hodges, J. L., agent for Colorado, in Roberts, G. E., op. cit. for 1898, pp. 89–90, 1899.

19 Emmons, W. H., The Cashin mine, Montrose County, Colo.: U. S. Geol. Survey Bull. 285, pp. 125–128, 1906.

have been taken from record books. Smelter receipts are sometimes in error, owing to the practice of crediting ore to county of shipping point, which for this area would be Placerville, San Miguel County.

The figures here given for 1886 to 1896 are based on reports of the agents of the mint in annual reports of the Director of the Mint, the gold and silver being prorated to correspond with the figures for the total production of the State as corrected by the Director of the Mint, the lead being prorated to correspond with the total production of lead for the State as given in Mineral Resources, and any unknown production in the State being distributed proportionately among the counties. As with lead so with copper, but as the figures for copper given in Mineral Resources include copper from matte and ores treated in Colorado, though produced in other States, they are subject to revision.

For 1886 to 1904 the placer and lode production has been separately estimated, in part on the basis of deposits at the Denver Mint.

For 1897 to 1904 the figures, which represent smelter and mint receipts, are taken from the reports of the Colorado State Bureau of Mines.

For 1905 to 1923 the figures are taken from Mineral Resources (mines reports).

For 1906 and 1907 the producing district is unknown.

For 1912 to 1922, according to Mineral Resources, the production came from La Sal district.

been had out of this vein running up to 10,000 ounces per ton. It carries silver glance in considerable quantities. Thirty thousand dollars have been offered for it and refused.

This mine is in Poughkeepsie Gulch, tributary to the Uncompahgre, but now in San Juan County. In his report for 1875 Raymond says:[22]

Mount Sneffels district includes the sections drained by the Rio San Miguel [probably all now in San Miguel County]. The Uncompahgre district includes all lands drained by the Uncompahgre River and its tributaries as far north as the Ute Reservation.

The Alaska, Saxon, and Poughkeepsie mines are in Poughkeepsie Gulch, San Juan County.

According to Burchard,[23] the Grand View mine was located in 1875 and patented in 1879.

For 1878 to 1885 the figures for the separate production of Ouray, San Juan, San Miguel, and Hinsdale counties has been estimated, the division by counties being controlled by the figures for the San Juan region prorated against the State total. Burchard[24] gives figures for the San Juan country in 1879 and 1880.

Of the developments in 1881 and 1882 Purington says:[25]

As early as 1881, however, the Virginius mine, at the head of Canyon Creek (Ouray County), was worked by three levels and two shafts, and in 1882 the amount of development amounted to 2,000 feet, with a product of $75,000.

Gold (placer and lode), silver, copper, and lead produced in Montrose County, 1886–1923

| Year | Ore (short tons) | Gold | | | Silver | | | Copper | | | Lead | | | Total value |
		Placer	Lode	Total	Fine ounces	Average price per ounce	Value	Pounds	Average price per pound	Value	Pounds	Average price per pound	Value	
1886		$281		$281	3	$0.99	$3							$284
1887		500		500	9	.98	9							509
1888		12,000		12,000		.94								12,000
1894		2,202		2,202	16	.63	10							2,212
1895		1,181		1,181	11	.65	7							1,188
1896		1,720	$225	1,945	17	.68	12							1,957
1897		1,571	4,981	6,552	851	.60	511							7,063
1898		300	2,408	2,708	6,290	.59	3,711	34,664	$0.124	$4,298				10,717
1899		103	620	723	46,119	.60	27,671	75,006	.171	12,826				41,220
1900		300	1,333	1,633	19,652	.62	12,184	32,026	.166	5,316				19,133
1901		301	1,249	1,550	101,359	.60	60,815	55,944	.167	9,343				71,708
1902		1,868	4,085	5,953	3,149	.53	1,669	2,505	.122	306	64	$0.041	$3	7,931
1903		300	2,511	2,811	2,061	.54	1,113	10,920	.137	1,496				5,420
1904		121	1,367	1,488	1,067	.58	619	7,476	.128	957				3,064
1906		114		114	3	.68	2							116
1907		314		314	9	.66	6							320
1912		687		687	10	.615	6							693
1913	49	935	5	940	434	.604	262	24,058	.155	3,729				4,931
1914	66	435	11	446	517	.553	286	32,414	.133	4,311				5,043
1915	169	1,259	18	1,277	1,073	.507	544	57,320	.175	10,031				11,852
1916	197		10	10	1,132	.658	745	100,008	.246	24,602				25,357
1917	64	944		944	666	.824	549	21,275	.273	5,808				7,301
1919		199		199	2	1.12	2							201
1920		198		198	2	1.09	2							200
1922	251	322		322	17,968	1.00	17,968	61,119	.135	8,251				26,541
1923	101	177		177	10,523	.82	8,629	17,857	.147	2,625				11,431
		28,332	18,823	47,155	212,943		137,335	532,592		93,899	64		3	278,392

OURAY COUNTY

Raymond in 1874 says of the Poughkeepsie mine:[21]

Across the Saguache Range, on the Uncompahgre, is another good belt, on which is located the Poughkeepsie. Assays have

In his report for 1881 Burchard says:[26]

[22] Raymond, R. W., op. cit. for 1875, pp. 324–325, 1877.

[23] Burchard, H. C., Report of the Director of the Mint upon the production of the precious metals in the United States during the calendar year 1883, pp. 379–380, 1884

[24] Burchard, H. C., op. cit. for 1880, pp. 156–157, 1881.

[25] Purington, C. W., Preliminary report on the mining industries of the Telluride quadrangle, Colo.: U. S. Geol. Survey Eighteenth Ann. Rept., pt. 3, p. 753, 1898.

[26] Burchard, H. C., op. cit. for 1881, pp. 354, 419–420, 1882.

[21] Raymond, R. W., Mineral resources of the States and Territories west of the Rocky Mountains for 1874, p. 385, 1875.

Three miles from the town of Ouray are the Belle of Ouray and the Union, on Bear Creek. They are being worked to their utmost capacity and dumping from 6 to 8 tons of ore per day that will average throughout from 100 to 140 ounces of silver.

In the Uncompahgre district the Silver Link and Silver Point mines, within 1 mile of Ouray, are being worked very successfully by contract. These are also paying mines. * * * Both have about 500 feet of development and are producing ore that runs from 80 to 120 ounces of silver.

About 2 miles from Ouray is the Mineral Farm, which is producing 10 to 12 tons of ore per day of a value that exceeds any ore yet extracted in this region. It is gray copper, brittle silver, and galena, averaging from 300 to 400 ounces, mill run. * * *

The principal mines of Mount Sneffels are the Virginius, Yankee Boy, Terrible, Portland, Sidney, Monongahela, El Dorado, Allied, Snow Drift, Governor, Bessie Bascom, Revenue, Ethan Allen, Young America, Flagstaff, and Ruby Trust.

In the Virginius the development consists of three levels and two shafts, from which high-grade ore is taken from a pay streak of 38 inches. * * *

Development work is being done upon all the principal mines of the district; the ore taken out is piled upon the dump or in ore cribs and will await the advent of the railroad.

In his report for 1882 Burchard says:[27]

Ouray County is situated in the southwestern part of Colorado in what is generally known as the San Juan country. It is divided into six districts, viz, Uncompahgre, Red Mountain, and Sneffels [now Ouray County], and Upper San Miguel, Lower San Miguel, and Iron Springs [now San Miguel County].

Uncompahgre district is altogether tributary to Ouray County, though it is situated partly in San Juan County. * * *

Red Mountain district, lately a portion of Uncompahgre, has recently attracted considerable attention. This portion of Ouray County was long known for its large galena lodes, but the specimens obtained hardly ever ran more than 7 or 8 ounces of silver to the ton, until last summer some parties discovered a lode which they named the Yankee Girl. This lode shows 9 feet of solid galena, running 80 ounces of silver to the ton and 65 per cent lead. Two shafts have been sunk on the vein, each over 50 feet deep, showing the same amount of ore, only sometimes changing into copper pyrites, and it is reported that some runs as high as 800 ounces of silver.

The Guston, Robinson, Genesee, Senate, Congress (San Juan County), Humboldt, Hudson, Orphan Boy, and others also have galena, but some of them have enough copper to make it advantageous to work the mine for that metal.

In Sneffels district the lodes * * * extend in many instances into the Upper San Miguel district, and * * * the lodes nearly all point to Sneffels Peak.

Then follow descriptions of mines in Sneffels district, including Revenue, Potosi, Wheel of Fortune (copy of settlement sheets of ore to Pueblo show average of 7.96 ounces gold and 176.46 ounces silver), and in Virginius Basin the Terrible, Virginius (of which Burchard says: "With about 1,500 feet of levels and 370 feet of shafts this mine has sold ore to the value of $75,000, with between 300 and 400 tons, running from 50 to 100 ounces silver to the ton"), El Dorado, Yankee Boy, and others. He continues:

All the mines mentioned, besides many others, are more or less worked and have shipped or are shipping ore, but in all cases the bulk of the ore is too low grade to be shipped by burros and

wagons and is awaiting concentrating works and transportation facilities.

Since September, 1882, the Denver & Rio Grande Railroad has been built within 30 miles, and this has increased considerably the shipments. By next summer it is expected that the railroad will be completed to Ouray.

In his report for 1883 Burchard says:[28]

By an act of the last legislature of Colorado this county was divided into Ouray and San Miguel counties. * * * The area of this county comprises all that territory drained by the headwaters of the Uncompahgre River * * * and is divided into three mining districts, Uncompahgre, Sneffels, and Red Mountain. * * *

The "Mear's system," by which roads were built during the past summer, has greatly facilitated to transportation of ore.

Red Mountain.—The Yankee Girl has shipped during the year about 5,000 tons of ore, the lowest average netting about $125, while some shipments have run from $2,000 to $4,000. The record of the National Belle is little short of this. * * *

Sneffels district continues to show well. * * *

The output of the Virginius mine during the year has been estimated at $70,000. The ore from which this was produced ran from 141 to 719 ounces of silver and 36 per cent of lead to the ton. * * *

Uncompahgre district includes the mines in the vicinity of Ouray. * * * About one-half mile below and north of the town of Ouray is the Grand View * * * located 1875, * * * sold 1877. * * * Over 150 tons of ore have been treated running from $71 to $226 in gold and from 13 to 40 ounces in silver per ton. * * * Ore was treated at Argo, Pueblo, and Lake City. The character of the ore is iron and copper pyrites, blue and green carbonates of copper in a quartz porphyry gangue.

Red Mountain district takes its name from a range of scarlet peaks part of which extend into Ouray County and part into San Juan County.

The ores are of two kinds, galena and a copper ore that looks like an antimonial copper glance; that from the Hudson mine being very dark gray color, in fact nearly black; that from the Yankee Girl being of a gray color associated with copper pyrite; and that from the Orphan Boy having the usual beautiful colors of erubescite, and all these having the appearance of solid masses of metals, without gangue in them, and not having the combed or banded texture noticeable in all the ores of our vertical fissure quartz veins.

On the 14th of August, 1881, John Robinson, while hunting in Red Mountain Park, picked up a boulder, and, being astonished at its weight, broke it in two and found it to be solid galena. He and his three partners went to work and soon discovered the enormous ore body now known as the Yankee Girl mine. As no sides nor bottom could be found they could not determine how the vein lay, and so staked off two other claims adjacent, naming them the Robinson and Orphan Boy. This comprises what is known as the Yankee Girl Mining Co. The company immediately put men to work and ore shipments began, and from that time to this no break has ever occurred in the shipments.

The ore body is about 40 feet thick, and two men have stoped all the ore that with present facilities could be shipped. * * * There are in shafts and tunnels about 1,000 feet of development, which is being pushed with all the rapidity possible. The output for the year has been 3,000 tons of ore, worth $150 per ton, with a very high percentage of lead. * * *

The National Belle Silver Mining Co. owns the National Belle. Although capable of producing 50 tons of ore per day, this company has worked its mine with a view to thorough

27 Burchard, H. C., op. cit. for 1882, pp. 394, 395, 509-515, 1883.

28 Burchard, H. C., op. cit. for 1883, pp. 240, 376-386, 1884.

development, and what ore was taken out (700 tons) was removed in the course of development, the proceeds being used to defray expenses, which it has more than done.

The ore is of the same character found in this locality, viz, gray copper and copper pyrites, with a gangue of quartz.

The Red Mountain Review, under date of January 5, 1884, gives the following estimate for the year 1883:

Product of 1,025 tons of ore shipped by agents, Pueblo Smelting & Refining Co.	$214,000
Virginius mine, 300 tons	69,928
Munn Bros. mill, 250 tons	47,531
Bell Bros. mill, 137 tons	12,930
Emma Mountain, 102 tons	9,919
National Belle, 980 tons	69,600
Hudson and Sailor Boy	25,100
Yankee Girl	400,000
Guston	57,500
Grand Exchange and 180 other mines, 1,200 tons	90,000
Congress, 2,500 tons (San Juan County)	220,000
Mineral King and 16 other mines, 1,500 tons	112,500
Shipments from Carbonate King, Galena, and others	32,750
	1,361,758

This estimate, even if it includes value of copper and lead, is more than three times greater than the estimate of the Director of the Mint.

Burchard, in his report for 1884, says of the developments in Ouray County:[29]

The output of Ouray County for 1884 shows an increase of about 30 per cent. One of the greatest drawbacks to Ouray and vicinity is the lack of economical means of transportation, as of the great bulk of mineral, after paying the costs of transporting from 35 to 50 miles in wagons, and the smelting charges, only a small profit remains for the producer.

Red Mountain district.—The output of this section has been mainly from the following mines: Yankee Girl, National Belle, Genesee, Hudson, Orphan Boy, Guston, Sailor Boy, Carbonate King, Denver, Treasure Trove, Grand Prize, Maud S., Guadaloupe, Galena, Lion, and Candice.

The two leading mines, the Yankee Girl and the National Belle, are producing some 40 tons of ore per day, and the former is reported by some of the newspapers in Ouray to have produced over $600,000 worth of bullion, but these figures are doubtless exaggerated.

In Sneffels district, the Virginius, Monongahela, and Sidney are said to have produced during the year; the Virginius, about 365 tons (some of which ran over 300 ounces in silver); the Monongahela, 315 tons; and the Sidney, 258 tons.

In Uncompahgre district considerable development work was done, but on account of the low grade of ore but few mines

shipped. The shippers are said to have been the Rose, Golden Gate, Big Pigeon, Gold Finch, and Little Pigeon.

The Dallas Placer Mining Co. was organized in 1883 to work extensive gravel deposits supposed to exist on the Uncompahgre River, at a point below the mouth of the Dallas. A hydraulic elevator capable of handling 2,000 cubic yards of gravel per day with a pressure of 300 feet has been erected, and work was commenced in October, 1883. The company are actively at work, although with what success is not known.

The Camp Bird mine, in Imogene Basin above Ouray, which was operated from 1896 to June 30, 1916, when ore ceased to be taken out to allow a low level adit 10,700 feet long to be driven to cut the vein 450 feet below the bottom of the shaft and to drain the mine, has been one of the most famous of Colorado mines, and the profit made at the mine has probably represented the highest percentage of gross value of output of any mine in the State. The profit at the mine represents 65 per cent of the money actually received at the mine for products sold, after hauling, railroad freights, and smelting charges had been deducted from the gross value of concentrates shipped, and express and freight charges had been deducted from the bullion sold. The adit was completed, but insufficient ore was found to operate the mill at a profit, so the mine has been unproductive from 1916 to 1923.

The story of the Camp Bird is given in an article in the Engineering and Mining Journal,[30] which recites the staking of the Una and Gertrude claims in 1877 and their development. The writer gives assays showing knowledge of the existence of ore carrying $12 to $20 in gold, but owing to the fact that smelters did not pay for ore carrying less than 1 ounce of gold, to the $45 pack rate of ore to Silverton, to the $35 smelting charge, and in addition to the failure of the mill and of the company organized to operate it in 1881, the claims were abandoned. In 1896 came the rediscovery of rich ore by Thomas F. Walsh, who was then operating a pyritic smelter at Ouray.

Rickard says:[31]

In September, 1896, Thomas F. Walsh examined the abandoned workings of the Gertrude claim and broke some samples, which were sent to Ouray to be assayed. They contained several ounces of gold per ton. More samples were then taken and sent to Leadville for assay. * * * He set to work to

[29] Burchard, H. C., op. cit. for 1884, pp. 177, 232-233, 1885.

[30] Denver correspondence, The true story of the Camp Bird discovery: Eng. and Min. Jour., vol. 89, p. 1266, 1910.

[31] Rickard, T. A., Two famous mines; The Camp Bird: Min. and Sci. Press, vol. 103, pp. 827-828, 1911.

acquire the ground, buying abandoned claims on tax titles, until he had consolidated a large property. He also located a claim called the Camp Bird next to the Gertrude and gave the name of this new claim to his consolidated property. In 1900 he owned 103 mining claims and 12 millsites, covering 941 acres altogether. * * *

I appraised the mine [in July, 1900] at $6,000,000. At that date the mine had yielded $2,535,512 gross and a profit of $1,650,000. * * * Messrs. Hammond and Baker resumed negotiations, but it was not until April, 1902, that they finally closed the deal. * * * In the meanwhile Walsh had extracted about $1,500,000 worth of ore from which he had made $750,000 in profit.

By the new deal Walsh was to receive $3,500,000 in cash and $500,000 in shares * * * and he was to obtain a further $2,000,000 * * * in the form of a royalty of 25 per cent on profits from ore not then considered sufficiently proved.

He received the last part of this deferred payment before his death, in March, 1910, so that he obtained a total of $6,000,000 for his mine. * * *

Production of Camp Bird mine, 1896–1916, inclusive

[Compiled by C. W. Henderson]

	Ore (short tons)	Value received of recovered metallic contents (gold, silver, lead, copper)	Average value per ton	Profit at the mines and mill, exclusive of depreciation
From 1896 to July, 1900 [a]		$2,535,512		$1,650,000
July, 1900, to April 1902 [a]		1,500,000		750,000
May 12, 1902, [b] to April 30, 1903	66,825	1,974,705	$29.55	1,293,007
Year ending April 30, 1904 [c]	70,543	1,922,261	27.24	1,217,784
Year ending April 30, 1905 [c]	74,674	2,343,553	31.38	1,656,302
Year ending April 30, 1906 [c]	66,223	1,892,203	28.57	1,231,865
Year ending April 30, 1907 [c]	38,295	1,339,864	34.99	925,313
Year ending April 30, 1908 [c]	80,087	2,071,068	[d] 25.91	1,398,559
Year ending April 30, 1909 [c]	80,157	2,269,622	28.31	1,621,484
Year ending April 30, 1910 [c]	79,714	2,645,621	33.18	1,974,212
Year ending April 30, 1911 [c]	79,186	1,812,572	22.89	1,128,468
May 1, 1911, to June 30, 1912 [c]	66,505	1,742,041	26.15	1,135,292
Year ending June 30, 1913 [c]	30,012	675,630	22.51	270,042
Year ending June 30, 1914 [c]	30,595	801,079	26.19	397,321
Year ending June 30, 1915 [c]	32,313	952,288	29.47	583,701
Year ending June 30, 1916 [c]	25,601	791,749	30.92	498,438
Total, May 12, 1902, to June 30, 1916	820,730	23,234,256		15,331,788
Total, 1896, to June 30, 1916		27,269,768		17,731,788

[a] From Rickard, T. A., op. cit.
[b] Date of acquisition by Camp Bird (Ltd.).
[c] From printed annual reports of Camp Bird (Ltd.).
[d] Average of 78,996 dry tons of new ore yielding $2,046,068. There were also treated 1,121 tons of mixed ore from stamp mill wreck of 1906.

Division of value of metallic content extracted, Camp Bird mine, 1902–1916, inclusive [a]

[Compiled by C. W. Henderson]

Period	Gold	Silver	Lead	Copper	Total
May 12, 1902, to April 30, 1903	$1,902,722	$51,219	$20,764		$1,974,705
Year ending April 30, 1904	1,857,271	50,151	14,370	$469	1,922,261
Year ending April 30, 1905	2,265,093	62,506	15,583	371	2,343,553
Year ending April 30, 1906	1,807,600	57,796	26,171	636	1,892,203
Year ending April 30, 1907	1,273,476	44,198	22,115	75	1,339,864
Year ending April 30, 1908	1,963,839	68,546	38,376	307	2,071,068
Year ending April 30, 1909	2,176,634	63,117	27,211	2,660	2,269,622
Year ending April 30, 1910	2,526,437	75,178	40,395	3,611	2,645,621
Year ending April 30, 1911	1,685,980	71,436	46,748	8,408	1,812,572
May 1, 1911, to June 30, 1912	1,548,533	130,846	37,997	24,665	1,742,041
Year ending June 30, 1913	579,291	67,018	15,775	13,546	675,630
Year ending June 30, 1914	700,880	72,206	13,378	14,615	801,079
Year ending June 30, 1915	873,717	58,015	11,191	9,365	952,288
Year ending June 30, 1916	723,421	48,441	8,988	10,889	791,749
	21,884,894	920,673	339,062	89,617	23,234,256

[a] From printed annual reports of Camp Bird (Ltd.).

Value of metallic output of Camp Bird mine, by plants, 1902–1916, inclusive [a]

[Compiled by C. W. Henderson]

Period	Stamp mill	Cyanide mill	Total	Recovery of contents of crude ore — Gold	Recovery of contents of crude ore — Gold and silver	Recovery of contents of crude ore — Total
				Per cent	Per cent	Per cent
May 12, 1902, to Apr. 30, 1903	$1,842,817	$131,888	$1,974,705			90
Year ending Apr. 30, 1904	1,786,476	135,785	1,922,261	93.5		
Year ending Apr. 30, 1905	2,203,862	139,691	2,343,553	93.4		
Year ending Apr. 30, 1906	1,731,591	160,612	1,892,203	93.4		
Year ending Apr. 30, 1907	1,235,355	104,509	1,339,864	93.75		
Year ending Apr. 30, 1908	1,880,214	190,854	2,071,068	93.84		
Year ending Apr. 30, 1909	2,074,952	194,670	2,269,622	94.08		
Year ending Apr. 30, 1910	2,448,869	196,752	2,645,621	95.00		
Year ending Apr. 30, 1911	1,688,112	124,460	1,812,572	94.87		
May 1, 1911, to June 30, 1912	1,629,136	112,905	1,742,041	94.68		
Year ending June 30, 1913	629,574	46,056	675,630	93.79		
Year ending June 30, 1914	761,331	39,748	801,079		94	
Year ending June 30, 1915	905,292	46,996	952,288	96.03	94.31	
Year ending June 30, 1916	756,240	35,510	791,750	97.13	95.19	
	21,573,821	1,660,436	23,234,257			

[a] From printed annual reports of Camp Bird (Ltd.).

Ore shipped from Calliope mine, near Ouray, Colo., and its contents in gold, silver, and lead

[Compiled by V. C. Heikes, from records of G. E. Kedzie, M. E., Ouray, Colo., Nov. 1, 1890]

Date	Weight of ore	Gold	Silver	Lead
	Pounds	Ounces	Ounces	Pounds
Prior to 1887	3,120		185	
During 1887 to July 9	13,311	2.15	1,315	
Nov. 28, 1887, to Feb. 29, 1888	356,145	57.69	44,001	66,395
Jan. 5 to Dec. 19, 1888	737,250	112.00	82,769	147,470
January to December, 1889	3,795,484	476.15	180,201	329,647
January to July, 1890	1,278,997	770.68	45,892	152,119
	6,184,307	1,418.67	354,363	695,811

The figures for 1885 in the Ouray County table have been interpolated to correspond with the total production of the State.

The figures for 1886 to 1896 have been derived from reports of the agents of the mint, in annual reports of the Director of the Mint, the gold and silver being prorated to correspond with the figures for the total production of the State as corrected by the Director of the Mint, the lead being prorated to correspond with the total production of lead for the State as given in Mineral Resources, and any unknown production in the State being distributed proportionately among the counties. As with lead so with copper, but as the figures for copper given in Mineral Resources include copper from matte and ores treated in Colorado, though produced in other States, they are subject to revision.

For 1897 to 1908 the figures in the table on page 186, which represent smelter and mint receipts, are taken from reports of the Colorado State Bureau of Mines.

For 1904 to 1923 the figures are taken from Mineral Resources (mines reports).

Gold, silver, copper, lead, and zinc produced in Ouray County, 1878–1923

Year	Ore (short tons)	Lode gold	Silver Fine ounces	Silver Average price per ounce	Silver Value	Copper Pounds	Copper Average price per pound	Copper Value	Lead Pounds	Lead Average price per pound	Lead Value	Zinc Pounds	Zinc Average price per pound	Zinc Value	Total value
1878		a $5,000	a 38,672	$1.15	$44,473										$49,473
1879		a 8,500	a 38,672	1.12	43,313										59,963
1880		a 8,500	a 69,610	1.15	80,052										98,552
1881		a 55,000	a 85,078	1.13	96,138	a 100,000	$0.182	$18,200	a 230,000	.048	11,040				180,378
1882		a 70,000	a 77,344	1.11	88,172	a 500,000	.191	95,500	a 230,000	.049	11,270				264,942
1883		20,000	386,719	1.11	429,258	a 400,000	.165	66,000	a 1,170,000	.043	50,310				565,568
1884		10,500	572,344	1.11	635,302	a 363,125	.13	47,206	a 3,000,000	.037	111,000				804,008
1885		10,000	900,000	1.07	963,000	a 400,000	.108	43,200	a 4,400,000	.039	171,600				1,187,800
1886		26,241	993,867	.99	983,928	400,000	.111	44,400	3,208,000	.046	147,568				1,202,137
1887		22,853	952,255	.98	933,210	666,000	.138	91,908	2,668,135	.045	120,066				1,168,037
1888		24,289	789,396	.94	742,032	579,100	.168	97,289	3,259,904	.044	143,436				1,007,046
1889		26,436	913,254	.94	858,459	397,804	.135	53,704	4,704,261	.039	183,466				1,122,065
1890		353,133	2,791,626	1.05	2,931,207	665,754	.156	103,858	4,228,803	.045	190,296				3,578,494
1891		478,750	2,273,054	.99	2,250,323	865,044	.128	110,726	4,168,887	.043	179,262				3,019,061
1892		138,688	754,114	.87	656,079	638,875	.116	74,109	8,012,729	.04	320,509				1,189,385
1893		188,854	1,221,155	.78	952,501	600,000	.108	64,800	8,000,000	.037	296,000				1,502,155
1894		178,138	995,153	.63	626,946	600,000	.095	57,000	4,422,000	.033	145,926				1,008,010
1895		172,697	1,515,693	.65	985,200	600,000	.107	64,200	5,747,003	.032	183,904				1,406,001
1896		141,046	2,371,912	.68	1,612,900	217,310	.108	23,469	6,599,143	.03	197,974				1,975,389
1897		552,840	2,776,394	.60	1,665,836	2,185,084	.12	262,210	7,784,212	.036	280,232				2,761,118
1898		852,555	1,420,330	.59	837,995	1,035,562	.124	128,410	2,799,936	.038	106,398				1,925,358
1899		1,694,940	2,346,194	.60	1,407,716	305,177	.171	52,185	7,556,386	.045	340,037				3,494,878
1900		1,437,969	1,985,267	.62	1,230,866	352,368	.166	58,493	9,478,657	.044	417,061	a 20,000	$0.044	$880	3,145,209
1901		1,546,323	1,633,725	.60	980,235	652,937	.167	109,040	7,904,724	.043	339,903				2,975,501
1902		2,420,726	789,855	.53	418,623	526,541	.122	64,238	4,262,063	.041	174,745				3,078,332
1903		2,171,508	417,343	.54	225,365	380,409	.137	52,116	3,350,569	.042	140,724				2,589,713
1904	91,244	2,174,361	294,028	.58	170,536	431,048	.128	55,174	2,044,525	.043	87,915	5,016	.051	256	2,488,242
1905	98,966	2,333,282	758,107	.61	462,445	524,199	.156	81,775	5,348,264	.047	251,368	48,267	.059	2,848	3,131,718
1906	48,468	992,179	916,256	.68	623,054	662,111	.193	127,787	5,721,599	.057	326,131	10,377	.061	633	2,069,794
1907	96,662	2,415,049	352,519	.66	232,663	908,675	.20	181,735	3,606,699	.053	191,155	30,407	.059	1,794	3,022,396
1908	96,493	2,028,698	415,070	.53	219,987	1,019,574	.132	134,584	3,033,352	.042	127,401				2,510,670
1909	103,864	3,044,825	345,815	.52	179,824	984,269	.13	127,955	2,813,932	.043	120,999	19,148	.054	1,034	3,474,637
1910	111,245	2,195,847	414,250	.54	223,695	62),236	.127	78,770	4,004,728	.044	176,208				2,674,520
1911	133,252	1,952,958	512,800	.53	271,784	564,273	.125	70,534	3,949,822	.045	177,742				2,473,018
1912	89,975	1,049,590	545,177	.615	335,284	400,552	.165	66,091	2,989,044	.045	134,507	140,667	.069	9,706	1,595,178
1913	97,336	959,377	537,634	.604	324,731	500,329	.155	77,551	2,180,591	.044	95,946	200,429	.056	11,224	1,468,829
1914	105,560	1,211,993	594,289	.553	328,612	854,038	.133	113,587	2,119,564	.039	82,663	44,608	.051	2,275	1,739,160
1915	103,258	1,118,016	576,621	.507	292,347	863,851	.175	151,174	1,990,681	.047	93,562	7,282	.124	903	1,656,002
1916	111,192	491,175	803,461	.658	528,677	444,081	.246	109,244	2,339,500	.069	161,393	69,015	.134	9,248	1,299,737
1917	86,523	92,831	868,097	.824	715,312	179,553	.273	49,018	2,031,721	.086	174,728	532,794	.102	54,345	1,086,234
1918	79,653	107,645	801,359	1.00	801,359	153,117	.247	37,820	2,587,915	.071	183,742	39,297	.091	3,576	1,134,142
1919	64,465	92,338	627,659	1.12	702,978	112,188	.186	20,867	1,782,868	.053	94,492	23,343	.073	1,704	912,379
1920	40,195	33,777	465,577	1.09	507,479	86,881	.184	15,986	1,334,575	.08	106,766				664,008
1921	69,232	73,229	730,970	1.00	730,970	85,039	.129	10,970	1,208,399	.045	54,378				869,547
1922	123,096	125,960	1,226,670	1.00	1,226,670	58,149	.135	7,850	1,484,525	.055	81,649				1,442,129
1923	87,260	59,207	840,044	.82	688,836	44,197	.147	6,497	1,538,027	.07	107,662				862,202
		35,167,763	41,735,429		32,246,402	22,927,450		3,307,230	161,694,054		7,111,284	1,190,650		100,426	77,933,105

a Estimated by C. W. Henderson.

PARK COUNTY

Raymond's reports give much information about the early mining developments in Park County. In his report for 1870 he says:[32]

The only quartz mining company at work in this county of which I have any information is the Pioneer, which was at work during part of the year and is reported to have produced $40,000 in four months. The placer mines of the county have yielded perhaps as much more, paying rather less than $3 per day per hand for a season of say, five months.

In his report for 1871 he describes the conditions in Park County as follows:[33]

New silver discoveries on Mounts Lincoln and Bross * * * late in July or early in August. * * *

On Mount Bross, the Moose, owned by Myers, Plummer & Dudley, is opened on the surface about 400 feet in length and in depth about 20 feet. The vein is about 2 feet wide, and the ore averages by assay $460 per ton. The company were preparing to ship 30 tons of ore to Swansea, Wales, in November; cost of shipment will not exceed $70 per ton. This company also own the Dwight, which is developed similarly to the above, and contains about the same grade of ores. Ten tons from this will be shipped, making 40 tons in all. * * *

The discoveries have been preempted as "lodes," "10-acre lots," "160 acres," and "1,500 feet square," thus showing that nobody is certain in which form the mineral bodies occur. Mr. Stevens, I am informed, started the "acre" method, and called it "placer ground."

The early history of the mines of Park County and the development in 1872 are set forth in the report for 1872 as follows:[34]

Previous condition of mining affairs in Park County.—From 1859 to the present time placer and lode mining have been conducted in Park County with greater or less success. It is impossible to form any correct estimate of the total production in gold for that time, but it will probably fall short of $2,750,000.

The placers have been mined principally by the slow and expensive method of rocking and ground-sluicing. Within the last two years more economical processes have been introduced, and gravel which a few years ago would not pay over $2 a day to the man will now yield $5 to $10 and even $20.

This increase in production and corresponding diminution in expense has been effected by substituting water power for manual labor. During the past summer several parties have introduced the system of booming. This consists in collecting water in a reservoir until a strong head is obtained and then letting it over the bank to be washed in a body. The powerful current carries down trees, boulders, and all other impediments, and moves more dirt in a single hour than could formerly be excavated in a day.

[32] Raymond, R. W., Statistics of mines and mining in the States and Territories west of the Rocky Mountains for 1870, p. 332, 1872.

[33] Raymond, R. W., op. cit. for 1871, pp. 365–366, 1873.

[34] Raymond, R. W., op. cit. for 1872, pp. 296–299, 1873.

The history of the gold lodes of this district is exactly the same as that of Central, Gold Dirt, Empire, and many other mining camps in this Territory. The top quartz, decomposed and prepared by nature for amalgamation, was easily mined and readily treated. Large companies were formed, extensive mills built, and much money expended. As soon as the sulphurets were reached, the mills were closed and operations on the mines suspended, as it was found impossible to treat the pyritous ores successfully by amalgamation. There are a large number of gold leads at Hamilton, Montgomery, and Mosquito, which only require proper treatment to render them profitable. Concentration and smelting would appear to be the most effectual and economical method of handling these ores.

Present condition of mining affairs.—Mining has received a powerful impulse during the past year from the rich silver discoveries in Lincoln and Bross mountains. Although no great amount of money has as yet been taken out, the discoveries have caused an influx of prospectors and capitalists and called attention to the wonderful riches of the whole Mosquito Range from Ute Pass (head of Michigan Creek) to Buffalo Peaks. The summer of 1872 was unusually cold and stormy, but in the face of these drawbacks, and in spite of the fact that nine men out of ten were prospecting and not mining, the South Park mines have produced about 1,500 tons of ore, which has been sold for about $150,000. I estimate the cost of mining and transporting this ore at $70,000, leaving a net profit of $80,000 for the working season of four months. This ore has been principally purchased by a branch of the Boston & Colorado Smelting Co. and by the Mount Lincoln Smelting Works. The latter works have been in successful operation since December 1 of this year, are purchasing for cash all ores offered, and smelting daily about 10 tons. They use a blast furnace 3 by 3½ feet in size and 12 feet high. The furnace has three tuyères, and the blast is furnished by a Sturtevant's No. 7 pressure blower. The products are lead riches and copper matte, all of which are shipped to Germany for further treatment. The prices paid for ore are liberal, being from $20 to $50 higher than can be realized by shipping to St. Louis or Swansea.

The silver discoveries on Lincoln and Bross this summer have been very numerous. * * *

A very large number of silver mines have been discovered on the south slope of Buckskin Mountain. * * *

Future prospect of mining affairs.— * * * Messrs. J. H. Dudley & Co. have nine men on the Moose, and next season will employ some 50 or 60 more on their other mines. Many other operators have announced their intention of mining on Lincoln and Bross and also extending their operations to Hamilton, Montgomery, Buckskin, Mosquito, and Horseshoe. It is also probable that large smelting works will be erected in the latter district, to reduce the argentiferous lead ores that are found there in great abundance. * * * It is confidently expected that a railroad will be built into the Park during the coming year.

The Phillips mine.—This mine, * * * one of the most important gold lodes of this district, * * * was discovered in Buckskin Gulch in 1860, by Joseph Higginbotham, alias "Buckskin Joe." In June, 1861, twelve persons were working on it. In September of the same year the town of Buckskin contained 1,000 inhabitants. From June 18 to October 19, Stansell, Bond, and Harris, who owned 200 feet of this lead, took out $50,000. The process they employed was very simple. The top quartz and dirt was run through sluices, and the headings were reworked in arrastres, yielding $350 per cord. The retorted gold sold for $16 per ounce, coin.

During the same season about $25,000 was taken out by other parties. The lode was worked until 1863, when sulphurets were reached, which could not be treated by ordinary mill process. The total yield of the lode has been about $250,000, although many claim it to have been much greater. * * *

The Moose.—This representative silver mine of the newly discovered limestone formation (as distinguished from the deposits in quartzite) was discovered in July, 1871, by Captain Plummer. The "Dwight," probably an extension of the same, was discovered in June, 1869, by Plummer & Myers. In 1871 this property was sold to Dudley & Co., who took out the same season from the Moose 30 tons yielding 300 ounces per ton; from the Dwight, 15 tons, yielding 275 ounces per ton. As this mineral was shipped to Swansea exactly as it was taken from the mine, without any sorting or other preparation, the yield is very remarkable. * * * The principal silver-bearing minerals are galena and various decompositions of copper pyrites; much carbonate of lead is also found. The ore requires no sorting and is easily smelted. During the summer of 1872, 300 tons of this ore were sold to the Mount Lincoln Smelting Works, yielding $350 per ton. * * * All the ore from it has been packed to timber line upon jacks. The elevation of the mine above sea level is something over 13,000 feet.

Placer mines.—There is an immense area of gravel deposits in the South Park, which, owing to the high price of labor and imperfect methods of working, has never yet been touched. From personal observation and careful compilation of the statements of our most experienced gulch miners, I estimate that there are no less than 15 square miles of placer ground which will yield $8 per day to the man by an extensive and economical method of working.

In the immediate vicinity of Fair Play since the founding of that town about $1,000,000 worth of gulch gold has been taken out, at an expense of $500,000. The gold is worth on an average $18.50 per ounce, coin.

The following analysis, made by myself, will show the nature of the alloy: Gold, 89.42 per cent; silver, 9.92 per cent; copper, trace; total, 99.34 per cent; specific gravity, 15.11.

In the neighborhood of Hamilton and Tarryall placer mining has been prosecuted since August, 1859, and has yielded about $1,000,000. There are still some 2,000 acres of gravel left, which will yield from $5 to $12 per day to the man. This gold comes principally from the lodes above Hamilton, which are numerous and large but can not be worked profitably by ordinary mill process.

The Bank mine of Messrs. Mills and Hodges, on the Platte, about 4 miles above Fair Play, has been worked for three years. In that time, 2,000 days' labor have been expended upon the mine at an expense of $3 per day. Forty-five thousand cubic yards of gravel have been washed, yielding $19,350, or 43 cents per cubic yard. Average work per day per man has been 22½ yards, producing $9.675. Their ditch is 3 miles long, 6 feet wide at the top and 4 feet at the bottom, with a quarter of an inch fall to the rod. It carries about 900 inches of water. * * * They use a hydraulic pipe with a 70-foot head and an inch and a quarter nozzle and have a 2-foot flume, 220 feet long, paved with block ripples. They intend next summer to construct a 5-foot flume and use the booming method. Their average depth of gravel is 21 feet and is increasing rapidly. The ground pays more or less from the grass roots, but the principal money is found on a stratum of hardpan 5 feet above bedrock. The gold is mostly in the shape of shot and sells for $18 per ounce.

Messrs. Pease and Freeman have been working a gulch mine on Beaver Creek, about 1 mile from Fair Play, for eight years. The first year they worked five men and took out 58 cents [per yard?]. Since then they have expended $20,000 in running a flume 2 feet wide and three-fourths of a mile long, and have opened up an inexhaustible area of half-ounce diggings. * * *

Messrs. J. W. Smith and Fred Clarke have purchased nearly all the claims on the Platte River about 2 miles above and below the town of Fair Play. They have bought out some 36 men and are running a flume 6 feet high and 6 feet wide with a grade of 2 inches to 12 feet.

The season of 1872 was spent in preliminaries, but there is no doubt that in the future the enterprise will be self-sustaining until bedrock is reached, when very large pay is expected, as the dirt has yielded as high as $41 a day to the man, by shoveling into sluices.

Burchard [35] says in his report for 1883, regarding the production prior to 1871:

Up to the time of the silver discoveries in 1871 the gold lodes and placers had produced about $2,500,000, principally obtained prior to 1866.

Raymond gives the following details in his report for 1873: [36]

The bullion product of this county shows an increase over that of last year. This is largely due to the great activity in placer mining, many hundreds of acres of new ground having been opened during the year in the South Park and elsewhere, while, at the same time, the older claims have been successfully operated as heretofore. A considerable share in the increase is, however, owing to the erection and successful work of the two smelting works which have been built in the Park. The one at Dudley, on the upper Platte, at the foot of Mounts Lincoln and Bross, was erected by Mr. Edward D. Peters, jr., mining engineer, and I am indebted to this gentleman for notes in regard to the works and the operations during the year.

Before the works were built it was the common belief that large amounts of lead ores could be secured in the vicinity, especially from the Horseshoe mine. As the rich silver ores from Lincoln and Bross contain a great deal of lime, heavy spar, and also considerable galena, Mr. Peters thought that, taking all the circumstances together, it would be best to erect a blast furnace, as ores of the above description, together with ores rich in lead, can be most advantageously treated in such works. The furnace erected was a square one, 10 feet 6 inches high from the tuyères to the charge hole, and 36 by 42 inches in section at the tuyères. Of the latter, three, of 1¾-inch nozzle diameter, were originally inserted, and a No. 7 Sturtevant blower, giving at 2,600 revolutions per minute about ¼-inch pressure of mercury, furnished the blast. Subsequently the number of tuyères was increased to six, and the results were highly beneficial.

These works were finished in November, 1872, and commenced smelting in December of that year. But as no adequate supply of lead ores could be obtained, it was decided to produce copper matte and ship it abroad for separation of the metals contained in it. This plan was adopted with much reluctance by Mr. Peters, as the large amount of sulphate of baryta, which the ores at his disposal contained, was certain to prevent the production of a very concentrated matte at the first smelting.

Three distinct varieties of ores were at Mr. Peters's disposal at the time: (1) The largest quantity came from the limestone formation of Lincoln and Bross Mountains, containing the following elements in about the following proportions: Sulphate of baryta, 55 per cent; carbonate of lime, 20 per cent; silica, 20 per cent; sulphide of lead, 5 per cent, and assaying about 140 ounces per ton in silver. (2) Copper ores from the porphyritic belt in Mosquito district, containing, approximately, sulphuret of iron, 20 per cent; copper pyrites, 20 per cent; zinc blende, 25 per cent; silica, 20 per cent, and sulphate of baryta, 15 per cent, and assaying 30 ounces in silver and 2

ounces in gold per ton. (3) Ores from the quartzite formation underlying the limestone of Buckskin district, containing silica, 70 per cent; sulphide of lead, 10 per cent; sulphide of antimony, 10 per cent; zinc blende 10 per cent, and assaying 18 ounces in silver per ton.

The only material to flux charges made out of these ores was iron pyrites from the Phillips mine in Buckskin Gulch, containing 75 per cent of sulphuret of iron and 25 per cent of gangue, consisting of silica and sulphate of baryta. This ore assayed about $6 per ton in gold and cost $10 delivered at the furnace. It was roasted in open heaps at an expense of $2 per ton. The only fuel at the disposal of the works was charcoal burnt from spruce, which was delivered at the stock bank for 15 cents per bushel of 2,650 cubic inches. It weighed only 11½ pounds per bushel, and contained 18 per cent of moisture and 2 per cent of ashes, while 30 per cent was lost by screening and handling. It was capable of reducing only 24 parts of lead oxide to lead. It is clear that this coal was, all things considered, about the poorest and most expensive fuel which could be used for such work.

The usual method of preparing and blowing in the furnace was as follows: The hearth and the forehearth were packed with a mixture of two parts of ground charcoal and three parts of clay, which was firmly tamped in with tamping irons. The crucible was cut out so that its deepest point was in front, where the tap hole was located. This was 18 inches below the level of the tuyères. From here the bottom of the crucible ascended rapidly toward the back wall to within 8 inches below the tuyères. The crucible was dried thoroughly for several days by means of charcoal brands, and, last of all, an 8-inch breast of common red brick was put in and secured by wedging. After this the furnace was gradually filled to the throat with good charcoal, great care being taken to keep it free from stones and earth. As soon as the flame came through at the throat a very light blast was put on, and 6 shovels of slag (19 pounds each) were charged, alternating with 18 scoops of charcoal (4½ pounds each). In about an hour and twenty minutes the slag sank to the tuyères, and, if it appeared perfectly liquid, the charge was gradually increased, until in about 8 hours a burden of 16 shovels of slag was obtained. If the slag then continued perfectly liquid, running over the lip in a free stream, two shovels of slag were replaced by ore, and this substitution of ore for slag was continued from time to time, until on the second day the furnace reached its regular burden of 250 pounds of ore to 80 pounds of charcoal. But this proportion could never be retained for a long time, the poor quality of the charcoal causing very frequent irregularities. The working of the furnace had, therefore, to be watched very carefully and required regulation by means of alteration in the quantity and quality of the ore charges. Usually the charge was of such a composition as to produce a slag of the following composition:

	Per cent
BaO	34
CaO	18
FeO	15
SiO₃	25
BaS, ZnS, CaS	8
	100

It is seen that in this slag the baryta is the principal base, and that from the nature of its composition it must have a characteristic earthy appearance. On account of the excess of baryta, the slag was extremely liquid, and in spite of its high specific gravity, it permitted a perfect separation of the metal.

The furnace always did its work best under the following conditions: The heat had to be kept low, the blast light, and the ore had to be charged immediately over the four back tuyères, so as to leave an unburdened column of coal in the

35 Burchard, H. C., Report of the Director of the Mint upon the production of the precious metals in the United States during the year 1883, p. 386, 1884.
36 Raymond, R. W., op. cit. for 1873, pp. 284, 304–308, 1874.

center and front of the furnace; the throat had to be kept dark, the "noses" also dark, and about 6 inches in length. Under these conditions the slag was thin and ran freely, the furnace smelted about 9 tons of charge per day, and there was no difficulty found in making blasts of 30 days' duration. The average length of the campaigns for the year was 21½ days. On the whole, there was no difficulty in using this method of beneficiation as far as the smelting proper is concerned, but a very serious drawback resulted from the presence of the large quantity of heavy spar in the ore. By far the larger part of the sulphuric acid in this mineral was reduced to sulphur, which, combining with the copper, lead, silver, and a large part of the iron, formed a very great proportion of matte, so that on an average it took only 5 tons of ore to make 1 ton of matte, which, of course, was not very rich. This matte had to be crushed and calcined in a reverberatory furnace, and then smelted a second time with the addition of 15 per cent of quartzose ores, yielding finally a regulus assaying 39 per cent of copper and 900 ounces of silver per ton. This was crushed, sacked, and shipped to Germany for separation. The slag from the concentration smelting was very basic, containing about 60 per cent of iron. It was very welcome as a valuable flux for the ore smelting.

In August, 1873, it became certain that no large quantities of lead ores would ever be available from the vicinity of the works; and as the ores containing so much baryta could not be cheaply roasted before smelting, Mr. Peters decided to erect a reverberatory smelting furnace to be used instead of the shaft furnace. This furnace has a hearth 15 feet 6 inches long and 9 feet 6 inches wide and was completed about September 10. It has been running up to the end of the year without interruption, smelting 8 tons of ore per day, with a consumption of 9 cords of mixed pine and spruce wood, and meeting fully the expectations of the metallurgist. The concentration of the matte in this furnace, where there is not a reducing but an oxidizing atmosphere, is very satisfactory considering the ores to be treated. One ton of matte is now obtained from 10 tons of ore, and the regulus assays from 1,000 to 1,500 ounces of silver per ton. Notwithstanding this high grade, the resulting slag assays only from 4 to 6 ounces of silver per ton. * * *

Toward the end of the year the Dudley works ceased shipping their regulus, owing to the high rates of freight and the great discrepancies in assays. From that time on the silver and the greater part of the copper have been separated at the works by a modification of the Patera process. About 93 per cent of the silver contents of the matte and 80 per cent of the copper are extracted, and the residue, which contains the gold, is returned to the furnace as a flux. When, in the course of a repetition of this process, the gold has so accumulated in the matte as to reach 200 ounces per ton, it is intended to extract it by means of chlorination.

The production of the smelting works at Dudley during 1873 was as follows:

	Currency
Silver	$102,756
Gold	8,549
Copper	2,359
Lead	7,520
	121,184

Estimated coin value (gold at $1.12½), $107,719.

Professor Hill has put up smelting works at Alma on the same plan which was followed at his works in Blackhawk and has been running them successfully during the greater part of the year.

The production of ore in the mining districts near the two smelting works mentioned has been, as near as can be ascertained, about 1,800 tons, for which an average of $120 per ton was paid, making a total of $216,000. The Moose mine on Bross Mountain still furnishes the most high-grade ore and

has now considerable reserves in sight, but the Dolly Varden, Hiawatha, and several others have also been very successful. Altogether, Lincoln and Bross Mountains have done far better this year than was expected in the spring. In Buckskin Gulch the production has fallen off somewhat, but from lack of capital rather than of good mines.

Mosquito district promises well for the future, rich discoveries of native silver having been made late in the fall.

In the northern part of Park County much work has been done in Hall Gulch and vicinity. This valley is located on the stage route from Denver to Fairplay, about 60 miles from the former place. The work here was mostly done by the Hall Valley Silver-Lead Mining & Smelting Co. (Ltd.). The company has expended a very large amount of money in opening its mines, building an excellent tramway about 4 miles long, and in the erection of a sawmill and dressing and smelting works. * * *

The ore consists principally of copper pyrites, galena being subordinate. Native silver is often visible in the solid ore. The gangue is principally quartz and sulphate of baryta. The ore can be easily sorted by hand into two classes—the first-class ore, being the solid mineral, assays over 350 ounces per ton; the second-class, 120 ounces. These values were obtained by sampling 5 tons of the first and 15 tons of the second class.

The Whale lode is opened by three tunnels run in along the vein and by a fourth adit, which runs across the country rock from the foot of the mountain and will strike the vein at a distance of 800 feet. The latter is intended as the principal working tunnel.

A sample of 20 tons of average ore from this mine assayed 120 ounces of silver per ton. There is a very large amount of fahlore in the ore from this mine; and by far the largest part of the gangue is heavy spar.

The Cold Spring lode, which is the principal mine of the company containing solid galena, lies further down the valley than the two mines described, but it is not so well opened.

The Hall Valley Silver-Lead Mining & Smelting Co. is very fortunate in having on its property, besides a number of valuable silver lodes, a deposit of limonite iron ore, which may serve as a flux in the smelting process. To judge from the mode of deposition observable in the ore bank, the limonite is the result of precipitation from waters saturated with sulphate of iron, which, no doubt, comes from some large vein crossing the mountain spur between the main valley and Handcart Gulch. That such a vein exists is extremely probable, from the fact that iron deposits of the same kind occur in both gulches, and nearly opposite each other in localities where very small streams of water issue from the mountain side and cause in the valleys swampy places. An analysis of this iron ore by Mr. H. Stoelting, Territorial assayer at Georgetown, gives oxide of iron, 75 per cent; quartz, 6 per cent; water, 13 per cent; and a trace of sulphur.

Charcoal is contracted for in this camp at 13 to 15 cents per bushel, delivered at the smelting works. Miners receive $4 per day. The company intends to be ready for smelting in the summer of 1874.

In his report for 1874 Raymond says: [37]

The copper production [of the State] is derived entirely from Gilpin and Park counties. * * * The only works in operation during the year were the Alma works, which produced $452,000 worth of matte. This was shipped to Blackhawk for separation. The process is an exact copy of that in use by the Boston & Colorado Smelting Co., viz, roasting and matting, and has been found to be completely successful. The Mount Lincoln Smelting works, at Dudley, * * * have lain idle during the entire year, while the Holland works [2 miles southwest of

37 Raymond, R. W., op. cit. for 1874, pp. 358–359, 373, 383, 1875.

Alma] did not prove a success, due to the lack of a sufficient quantity of galena. * * * The Hall Valley Smelting & Mining Co. * * * has several hundred tons at the dressing and smelting floors of the works. The latter are very extensive in plan, having a capacity of 40 tons per day. They are fitted up without regard to expense. Extensive separating and sizing apparatus (modeled on the German systems) receives the ore after it passes through the crushers and rolls. From these the mineral is carried on trucks to the furnace room. Here the trouble commences. Last summer three large and handsome cupolas were erected but on trial did not prove successful. Later in the year the erection of reverberatories was begun, and at its close they had not yet been put in operation. * * *

The Platte.—The production from this stream during the year has amounted to $70,000; almost the entire length of Montgomery Gulch from Hoosier Pass down to and even below Fair Play, a distance of over 12 miles, is occupied by working claims, some of which are operated extensively. Tarryall Creek also has during the past year been the scene of renewed activity, and many fine claims have been in operation, from its head down to Hamilton. The fortunes of the latter town, which has lain dead for many years, are once more on the increase. * * *

Most of the parties operating are still using the hydraulic in preference to the boom method, as the ground is generally too level for the latter. In Snowstorm Gulch, however, large reservoirs have been built, and booming is carried on with great success.

The ground of Messrs. Mills & Hodges, between Alma and Dudley, proved the most productive last year. They are operating against the left bank of the stream and have a breast from 30 to 100 feet high for many hundred feet in length. The ground is comparatively free from boulders, and is broken by the hydraulic stream with great ease. The resulting gravel is washed into narrow flumes, which empty into the main creek.

The Fair Play Gold Mining Co. owns the largest bar in Colorado and will operate on a very extensive scale. The association owns 1,100 acres of land opposite and below the town of Fair Play in a claim about 5 miles in length and 2,000 feet in width. It is supplied with two flumes, the lower of which is 4,000 feet in length, 6 feet wide, and 7 feet high; but the ground is so level that bedrock has not yet been reached, and it is estimated that over 5,000 feet more will have to be driven before it will be gained. At present only surface washings are made, and the yield is, of course, not what it will be when the bottom of the bar is reached. Water is brought from the stream above through four large ditches and gains at the upper workings a head of 140 feet, while at the lower the head will be over 220 feet. Two miles of conducting pipe are now laid for present workings. The pipe is made of sheet iron, in sections 20 feet long, which slip into each other, and taper from the reservoir to the present workings from 22 to 8 inches. This claim will be reopened in May and will employ 150 men.

In Beaver Creek Messrs. Freeman & Pease commenced to place in order their 600-acre claim last year and expect to be working on a large scale next season. * * * It was in the early days a noted creek, but owing to the high cost of working it has lain idle for many years. * * *

Mosquito district.—It is now over two years since the limestone formations of the Mosquito Range were found to contain deposits of silver and gold ores. * * *

The Mosquito Range is a spur of the Great Continental divide, breaking off from the latter about 12 miles north of Fair Play and coursing nearly south for 40 or 50 miles. From the head of this spur (Mount Lincoln) the ridge sinks slowly southward until it assumes the character of a low divide at the southwestern corner of the South Park. There it bends to the east and, gradually rising again, terminates in Pikes Peak. * * * Along its entire course it forms the divide between the South Platte and the Arkansas, and the western and southern boundary of the South Park. The floor of the latter is of sedimentary origin, consisting of sandstones and limestones, and these, abutting against the Mosquito Range, are tilted up and form a portion of its eastern slope.

All the great silver mines are found in the blue limestone belt. * * *

The ores furnished by the Lincoln and Mount Bross mines are generally sulphides. Copper, lead, iron, and antimony in a sulphureted or oxidized condition, form the mass of the material, and in these the silver is distributed as glance, native metal, and perhaps a little chloride, though the occurrence of the latter compound is very doubtful. Galena exists in very small quantities, copper to a higher percentage, while probably the largest proportion of base metals is in the various minerals of iron. All these deposits are accompanied with the gangue of heavy spar (barite), which often furnishes a clue by which to trace out hidden bodies of mineral. * * *

The Moose deposit crops out on the northeast face of Mount Bross and has been extensively developed for a distance of 500 feet along its course. The vein of ore lies nearly horizontal and is very regular in size and character. But little ore has been produced during the year. * * *

The Hiawatha is a segregation of pockets, some of which have yielded thousands of dollars. * * * As nearly as I can learn, its yield has amounted to considerably over $200,000 since first opened.

The Dolly Varden, Russia, Lincoln, Montezuma, and Elephant are all of the same class. * * * It will not be above the mark to state that the mines just mentioned have yielded during the year 1874 about $300,000 worth of ore.

Lower down on the mountain, and in the stratum of quartzite, are located the gold veins, the disintegration of which has undoubtedly furnished the float gold of the Platte placers. They are mostly old discoveries, and in the early days yielded richly from the decomposed surface quartz. When unchanged iron and copper ores were struck, they were deserted and supposed to be valueless. When, however, the Alma and the Dudley smelting works were built [Dudley lead furnace completed November, 1872, changed to copper reverberatory August, 1873, ran until January 25, 1874; Alma plant put up in 1873 and running successfully during greater part of 1873–74], a demand arose for pyrites for fluxing, and the lodes were reopened to furnish these minerals. It was then discovered that the ore was still auriferous; and since that time they have been continuously and steadily producing.

A list of this class comprises the Phillips, Orphan Boy, War Eagle, and a number of others of minor importance. The ore is mostly iron pyrites, carrying from one-half to 2 ounces of gold per ton and a small percentage of copper.

The mines of Mosquito district have produced 3,000 tons of silver ore, of an average value of $140 per ton, and $50,000 worth of auriferous pyrite.

From data above we have for Park County production for 1874:

Silver ores, 3,000 tons, at $140	$420,000
Auriferous pyrites	50,000
Platte River placer	70,000
	540,000

Raymond[38] gives the total for Park County as $596,392 in gold, silver, copper, and lead. On this basis Park County produced:

Placer gold	$66,497
Lode gold	50,000
Lode silver	431,533
	548,030
Copper	48,360
	596,390

The total copper for the State is, however, $100,197, of which Gilpin County is given $55,451, which leaves for Park County $44,746, instead of $48,360, or $49,720.[39] Raymond says:

Upper Arkansas.—This district embraces about a dozen gulches among the headwaters of the Arkansas, carrying both gold and silver veins. Its production last year was small, not over $145,000, which was derived mainly from gold mines. There is as yet no market for silver ore, and most of that which has been produced has been shipped to Golden City for treatment. The belt at the upper end of the valley is undoubtedly a reappearance of the same belt that courses through Montezuma and Breckenridge and is almost exclusively argentiferous. Lower down gold veins appear, which are most strongly developed in California and Colorado gulches. The occurrence of ores of nickel and cobalt in the ores of the upper valley is perhaps the only matter of mineralogical interest that has been shown by the year's operations. These metals are found with argentiferous galenas, but in small quantities. From a number of tons [say 15 tons averaging 200 ounces silver would yield only 3,000 ounces of silver], of Homestake ore treated by Mr. West, at Golden City, about 500 pounds of nickel speiss was run out, carrying from 2 to 12 per cent of that metal.

[Five hundred pounds of 6 per cent nickel speiss would contain 30 pounds of nickel. If all the nickel was recovered, 30 pounds of nickel in 15 tons would represent one-tenth of 1 per cent nickel in the original ore. I have credited $50,000 more gold to Colorado than the $2,102,487 given by Raymond, but his silver, $3,086,023, p. 358, carries some gold, amount unknown, in matte. As the $50,000 of auriferous pyrite went into matte at Alma, it seems not unreasonable to add this $50,000 to gold production of the State and subtract $50,000 from silver production of the State, making gold production for the State $2,152,487 and silver $3,036,023 (coinage value).—C. W. H.]

Raymond says in his report for 1875:[40]

The Moose has maintained throughout the year a steady production of about 6 tons daily, averaging in value about $150 per ton. The Dolly Varden has done well also and has yielded some exceedingly rich material. The Hiawatha has been under lease and is consequently not in as good condition as it might be. The Security has kept up a fair production. * * *

In Buckskin and Mosquito gulches a very large amount of prospecting and development has been done. * * * At the head of Buckskin is the London lode, which is attracting a great amount of attention and which undoubtedly is a rich and valuable property. The Phillips has not revived yet, though about 500 tons of ore have been taken out during the year for fluxing purposes. * * *

In Sacramento and Horseshoe gulches a small amount of ore has been produced, mostly galena. The latter has been most productive, if either can be said to have yielded. The

amount of prospecting done has also been large. * * * Still, though not lacking in mineral wealth, Park County improves but slowly. This fact may be attributed to the poor ore market which exists and to the comparative inaccessibility of the district. The latter drawback * * * will only be removed when the South Park Railroad, now connecting Denver with Morrison in the foothills, is pushed up the Platte Canyon to the Park. * * *

The placer interests (in the State) have enlarged greatly during the year, the amount of ground under improvement being fully one-third more than during 1874. The gain has been in the South Park, Bear Valley, South Clear Creek, Arkansas Valley, and at the base of Hahns Peak. * * *

The copper output remains substantially as it was last year. The only works saving this metal are at Blackhawk and Golden, and as the source of supply is mainly from the pyritous ores of Gilpin, Boulder, and Park, the production will advance slowly.

For 1879 and 1880 Burchard's figures for the total production of gold have been accepted.[41]

In his report for 1881 Burchard [42] says:

Early * * * rich placers in the vicinity of Tarryall, Fair Play, Buckskin, and Mosquito * * * are still paying fairly. * * * The Moose: * * * probably not less than 5 miles of levels, winzes, and stopes have been run on this mine, and the [total] product has been considerably over $3,000,000. The company owning the property erected a smelter at Dudley, which, after running a while and undergoing frequent changes, closed down. * * *

On Loveland Mountain is the Fanny Barrett; * * * carbonate of lead [ore]; * * * main shaft, down 140 feet, shows same class of mineral. * * * The ore runs from 25 to 600 ounces, though the bulk of it is low grade. A smelter has been erected. * * *

Some very rich free gold ore has been taken from the London. Two sacks of this ore were shipped by the manager to New York, which he is reported to have said were worth nearly $20,000. * * *

At Alma there are sampling works and two smelters, but the latter are not in operation.

In the report for 1882 Burchard [43] shows the ore receipts of the Boston & Colorado Smelting Co. at Argo, near Denver, and the Grant Smelting Works at Denver. He also describes the mines and gives a list of producing and nonproducing mines. He says:

Montgomery district was one of the earliest organized mining districts in the State, with the town of Montgomery—now long deserted—as its principal camp. The Russia, on Mount Lincoln: * * * from 75 to 100 tons were shipped to the Boston & Colorado Works, which ran from 150 ounces to 400 ounces silver per ton. * * * The Danville Consolidated: * * * from 25 to 30 tons * * * shipped that mills [meaning assays] from 50 to 100 ounces in silver per ton. The D. H. Hill: * * * 40 to 75 tons of ore during the year. The Wilson produced from 100 to 150 tons of ore that milled 75 to 150 ounces in silver per ton and contained 40 per cent lead. The Bullion * * * has produced from 25 to 40 tons of ore that mills from 100 to 225 ounces silver per ton. * * * The Nova Zembla: * * * in going down * * * over 600 tons of ore were taken out, averaging when treated $60 per ton (in gold and some silver) * * * The company has a 20-stamp

38 Raymond, R. W., op. cit. for 1874, p. 358.

39 Idem, p. 374.

40 Raymond, R. W., op. cit. for 1875, pp. 282, 285, 321-324, 1877.

41 Burchard, H. C., op. cit. for 1880, pp. 156, 157, 1881.

42 Burchard, H. C., op. cit. for 1881, pp. 354, 421-423, 1882.

43 Burchard, H. C., op. cit. for 1882, pp. 390, 391, 394, 523-533, 583-586, 1883.

amalgamation mill, which has been thoroughly refitted and furnished with late improvements and appliances for concentrating and milling ores. The Kansas * * * ore carries $10 to $150 in gold, the average being over $40 per ton. Nearly one-half of this can be saved in the stamp mill by amalgamation, and the remainder can be treated by concentration and smelting. The mine is equipped with a very complete 10-stamp mill.

Buckskin district.—* * * The Moose mine: * * * but little development was accomplished during the last year; * * * the shipments of ore amounted to from 150 to 200 tons. The ore in the Moose has run from the first to last between 60 and 900 ounces silver per ton, the average for the past two years being about 200. The Dolly Varden, south of the Moose, * * * has been steadily though not vigorously worked since its discovery. The output of mineral from 1872 to 1882 has been from 15,000 to 20,000 tons. The average of ore is not less than 150 ounces to the ton. * * * During the last year * * * 250 to 300 tons of ore produced. * * * The Silver Gem is west of the Moose; * * * the ore [is] galena and sulphurets, yielding from 60 to 250 ounces of silver per ton; * * * produced 100 tons during the year. The Wyandotte mine, in upper Buckskin Gulch; * * * lying along the footwall is the 8-inch vein of decomposed quartz, running well in gold and silver and also carrying about 12 per cent of copper and a good percentage of lead. * * * The Iron mine * * * has a horizontal vein of mineral 4 to 6 feet in width, covered by a 25-foot cap of iron and lying on the limerock. This iron cap of itself is a very valuable adjunct to the company's properties. The smelters pay $3 per ton for the iron ore on the dump at the mine, to be used for fluxing purposes. One mill run of 3½ tons from the vein proper gave 120 ounces in silver per ton. Another of 2 tons gave 260 ounces. * * * Nancy C. is in granite formation; * * * the vein matter is a porphyry gangue, carrying iron, copper, and galena, which mills about $30 per ton in gold and silver. * * * The Colorado Springs group * * * ore mills from 75 to 400 ounces in silver per ton * * * The Excelsior is one of the old gold mines, which was worked with much profit in Buckskin's early days, but after the surface ores were exhausted it remained idle until it became the property of its present owners. It is a contact vein between quartzite and lime. * * * While * * * cleaning out * * * several tons of ore were * * * treated, averaging 135 ounces of silver and 2 ounces of gold per ton. * * * The ore in the Excelsior is sulphurets of silver, galena, gray copper, and pyrites of copper, carbonate of iron and lead, with occasional intermixture of crystallized lead. * * * The Criterion, also a contact between lime and quartzite, is situated below the Excelsior. It was worked at an early day as long as the ores continued to be free-milling quartz, which paid $20 to $50 at the old stamp mills. The mine, then known as the Bates lode, began to fill with water, and the ore to run into silver and galena which could not then be treated, and the work was abandoned. * * * The Silver Wave: * * * no ore was shipped, but 75 tons are on the dump that will run from $25 to $40 to the ton in silver and gold. * * *

On Loveland Mountain, on the southwest side of Buckskin Gulch, is the Fanny Barrett; * * * all the ore which has been taken out, probably over 1,000 tons * * * [of which] 200 have been treated, the remainder, about 800 tons, remaining on the dumps for treatment in the smelter which the company have erected on their mill site at Alma. * * * The Northern Light has a record as a shipper of pay ore in 1862. * * * Like the Criterion it was worked for gold but will now be operated for * * * silver. * * *

In Mosquito district the principal mine is the London. * * * The ore body is * * * of a value ranging from 3 to 4 ounces in gold, 6 to 12 ounces in silver, and 1 to 3

per cent in copper. The chief work hitherto in the mine has been for development only. * * * The company is now building 7 miles and 1,000 feet of railway line to connect with the South Park Railroad, at the mouth of Mosquito Creek, and 200 feet east of the Platte River crossing, between Alma and Fairplay. The greater part is already completed. This road will reach to the foot of London Mountain, and connection made with the mine by wire tramway. * * * The U. P. and K. P. has been actively operated. * * * Assays have returned from 2½ to 16 ounces of gold. * * * The milling ore is treated on the ground and the smelting ore is shipped. The mill has 20 stamps and a Metcalf Concentrator for tailings.

In Horseshoe district * * * the Mudsill * * * [was] located in 1880. * * * The Last Chance has had millings of 50 to 65 ounces silver and a fair percentage of lead. * * * The general character of the ores in the Horseshoe district is a carbonate of lead, carrying from 25 to 100 ounces in silver and an average of 50 per cent lead to the ton. * * *

In Halls Gulch district, in the northwestern part of the country, * * * the Missouri * * * ore runs in bismuth, silver, gray copper, copper pyrites, and iron. The ore also carries gold, running from a trace up to 11¾ ounces. Mill runs at Argo (smelter) gave from 150 ounces to 219 per ton, with six-tenths of an ounce in gold and 10 per cent copper. * * *

The placer mines of Park County have been profitably worked for years.

The Tarryall and Fairplay mines have continued to be moderately productive.

The principal work done during the year was by the Alma Placer Co., which owns 640 acres of gravel opposite and below the town of Alma. The work is being carried on by a force of 22 men and two Little Giants having the patent deflector nozzle, one giving a 4-inch stream and another a 6-inch. The banks are 60 feet high, and giving a water pressure of 60 feet, very effective work is done. About 1,000 inches of water are used, and the giants force one-third of it against the banks. The sluices are 4 feet wide, with a 4-inch grade, and there are, with branches, over 3,000 feet of sluices, which are being added to daily. The ground shows a marked improvement over last year, and the large boulders are handled by an immense derrick, operated by a hurdy wheel, with a capacity for lifting and swinging to place a 10-ton boulder.

In his report for 1883 Burchard [44] gives the ore receipts by counties of the Grant Smelting Works, the Boston & Colorado smelter, the Miners & Smelters Works at Golden, the Golden Smelting Co., the Royal Gorge Works at Canon City, and the Matthews & Webb smelter. He shows the approximate production by counties and describes the Park County mines as follows:

Montgomery district is showing signs of life, and many properties have been developed during the year. * * * The Pioneer Mining Co. has been extensively working the Nova Zembla; * * * the ore is free gold and readily yields its precious metal under stamps. About 200 tons were treated at the company's mill. * * * The Sovereign Gold Mining Co.'s property is on the eastern slope of North Star Mountain and embraces four locations aggregating 6,000 feet on the old Harrington vein, which produced about $500,000 twenty years ago. * * * The shipments during the past season have been limited to small quantities of free milling ore. Thirty days' work on this class of material, treated at a quartz mill, yielded 360 ounces of gold, worth nearly $7,000.

---

[44] Burchard, H. C., op. cit. for 1883, pp. 236–240, 386–395, 1884.

Buckskin district.—The Fanny Barrett; * * * no stoping has been done, the ore sold coming from development only, in doing which 43 carloads were shipped. * * *

In Mosquito district * * * the London mine * * * ore is a quartz with lead, copper, gold, and silver to the average of $40 per ton, according to the assayer's returns; in seven months and a fraction since the stamp and concentrating mill commenced running at the Junction it has crushed about 6,150 tons of ore, or a monthly average of 850 tons. This ore has yielded gold bullion to the amount of $124,000. It has also yielded about 420 tons of concentrate, worth $60 a ton on an average, or a total of $25,000. About 15 cars or 210 tons of high-grade ore have been shipped out to smelters, worth at least $21,000, and it will be remembered that early in the year the London mine shipped 600 tons of ore to the Idaho Springs and Central [City] mills, which yielded an average of $18 a ton, or $10,800. This swells the total production of the mine to a trifle less than 7,000 tons, worth $181,000. We are aware that this figure will be by some considered an overestimate, but we believe it to be substantially correct. * * * The Weston: * * * there are now about 1,000 tons of ore in bins and on the dump, from which a lot of 11 tons was selected and shipped without sorting, returning $23.55 per ton in gold and silver, no account being taken of the lead and copper. * * * The company proposes to erect a concentrating mill, with capacity of 50 tons per day, and has already begun grading. * * * The Danser mine has been producing steadily; although the ore is low grade, it is of that nature that it is economically concentrated. * * *

In Horseshoe district: * * * the Mudsill mine: * * * it is the intention * * * to erect an amalgamator at the foot of Mudsill Hill, * * * the plant to be very much like the Taylor & Brunton amalgamator at Leadville. * * *

Halls Gulch district.—* * * The Whale: * * * ore argentiferous galena, with gray and yellow copper; average assay, 375 ounces silver per ton and 20 per cent lead; * * * during past season, total weight shipped 232½ tons; gross value of production, $24,515.85. * * * The Missouri has an established reputation as a producer, having shipped considerable ore up to the past season. * * * The Ypsilanti: * * * but 6 carloads of ore (62 tons) have been shipped * * * during the past season. * * * The Quincy Milling & Reduction Co. has leased the mill and buildings of the Upper Platte Silver Mining Co. and has placed therein a plant of machinery for the concentration of ores of the surrounding district. The plant consists of 2 batteries (10 stamps), 3 tables of the Embrey style, also 2 pans, and 1 large "settler," besides a retort for amalgamating purposes. Public sampling is also done at these works. Owing to the lateness of the season, when the mill was ready to run, but very little concentrates were shipped. Gross amount, 130 tons; average run of concentrates, $90 per ton. Works are shut down for the winter due to the scarcity of water and no steam in the works; machinery run entirely by water power. Farther up the gulch Messrs. McGowan & Oberkircher have erected a complete little concentrator, having in place a 5-stamp battery and two Frue Vanner tables. * * *

Placers.—The Gold Placer Mining Co. has not been worked during the year. The Alma placer has produced some gold, as also the Pease and Sidell placers. * * *

The production of lode gold in 1883 came to a large extent from the London mine.

In his report for 1884 Burchard says: [45]

The production of Park County for 1884 is not as large as that of 1883. * * *

[45] Burchard, H. C., op. cit. for 1884, pp. 177, 233-235, 1885.

The placers of 1884 have increased their yield, employing mostly Chinese labor and without modern improvements.

In Buckskin district the Moose mine, * * * after lying idle a number of years has again been worked by lessees. * * * The * * * Fanny Barrett: * * * only assessment work. The Hall & Brunk Silver Mining Co. has extracted considerable ore from the Dolly Varden mine. Developments are constantly progressing, and the shipments of ore have been regular. * * * The ore, which carries galena, gray copper, zinc blende, iron and copper pyrites, will average 90 ounces of silver per ton. Excepting at the Moose and Dolly Varden but little work has been done. * * *

In Mosquito district a small force was worked in the spring on the claims of the London Mining Co., but nothing of importance has been accomplished. The mine closed, owing to litigation. The Nestor, located about 5 miles below the London, is being worked on lease, and ore is being extracted that carries one-half to 3 ounces gold. The mill * * * is supplied with pay mineral enough to test its capacity. The predominating matter is a free milling gold quartz. The New York mine, just west of the London, has materially aided in swelling the output of the county. It is worked by lessees, who have shipped since July, on an average, two carloads of ore per week. The same might be said of the U. P. and K. P. mines, on which extensive developments have been made of late and which are now being pushed. * * * On the K. P. a tunnel 150 feet was run in addition to further developments; * * * where this tunnel intersected the vein a shaft was sunk 25 feet on a large body of lead ore, * * * the average value of which is $18 per ton. * * * The Western and Hock Hocking, two well-developed mines, have been producing some ore.

In Horseshoe district, the Last Chance * * * has executed a contract * * * for delivery to the mill now in course of construction of 50 tons of ore from the mine daily for the next twelve months, and twelve teams are now engaged in this work. * * * The Atlantic is not a very extensively developed mine but during the year has shipped considerable ore that has paid well. * * * The Copper Bonanza is but a prospect, the deepest shaft being only 18 feet, yet the owner has shipped 10 tons of ore, * * * free milling and average $50 per ton. The Peerless and Peerless Maude; * * * ore galena that carries on an average of 60 ounces per ton. Some shipments were made during 1884.

For 1885 Wilson [46] gives the deposits at the Denver Mint from Park County as $35,914.50 in gold and $805.78 (coinage value) in silver. Probably most of this is placer bullion. The table showing the production of the principal smelters does not include any smelter working in Park County.

For 1885 the figures for silver have been interpolated to correspond with the total production of the State; for lead the figures are taken from Mineral Resources. [47]

For 1886 to 1896 the figures in the table have been derived from reports of the agents of the Mint in annual reports of the Director of the Mint, the gold and silver being prorated to correspond with the figures for the total production of the State as corrected by the Director of the Mint, the lead being prorated to correspond with the total production of lead in the State as given in Mineral Resources, and any "un-

[46] Wilson, P. S., agent for Colorado, in Kimball, J. P., Report of the Director of the Mint, upon the production of the precious metals in the United States during the calendar year 1885, p. 136, 1886.

[47] U. S. Geol. Survey Mineral Resources, 1885, p. 257, 1886.

known production" of the State being distributed proportionately among the counties. As with lead so with copper, but as the figures for copper given in Mineral Resources include copper from matte and ores treated in Colorado, although mined in other States, they are subject to revision.

In 1887 Munson [48] gives the deposits at the Denver Mint from Park County as $32,894.65 in gold and $364.83 in silver (coinage rate); probably most of this was placer bullion, the individual production of Bemrose, Fairbury, and Peabody placers amounted to $29,588, the production of the Alma placer being included in confidential reports. Placer mining, he says, "has probably been more successfully carried on in Park County during the past year than in any other part of the State." He continues:

One interesting feature of the placer mining in Park County is the fact that a very large part of the gold recovered did not pass through any of the United States mints or assay offices but was shipped directly to Europe, in the original form of nuggets, for manufacture into native gold jewelry. In this manner the mines alone shipped between 7,000 and 8,000 ounces.

Munson gives the total amount confidentially reported in 1887 as $741,408 in gold and the total production of Park County as $784,864, which is reduced by prorating to $648,642, and of this amount $190,000, an exceptionally large yield, is given to placers, leaving $458,642 for lode mines. In the face of the fact that the London, New York, and other gold mines were idle, this large lode production also seems questionable. On the other hand he shows, on page 154, that men were sent into the field to collect data not reported directly to the office. The list of smelters on page 189 shows no smelter operating in Park County.

In 1888 Munson [49] gives the production of Lowe's, Fairbury, Pennsylvania, Peabody, and Roberts placers as $17,677, and the amount confidentially reported as only $3,136. The Alma placer was idle. The Hilltop mine is credited with $323,225 (coinage value) in silver and $352,000 in lead, the New York mine with $307,064 (coinage value) in silver, and the London mine is not mentioned. The value of deposits at the Denver Mint for Park County was $23,893.96 in gold and $234.51 (coinage value) in silver, probably mostly placer.

In 1889 Smith [50] gives the production of the Alma placer (which is stated as silver, an error), Lowe's, Danville, Deadwood, Fairbury, Pennsylvania, Peabody, and Tarryall placers as $41,740, and amounts confidentially reported as only $309. He gives the value of deposits at the Denver Mint for Park County as $47,529.98 in gold and $488.17 (coinage value) in

silver. The Hilltop is credited with $153,858.57 (coinage value) in silver, and $148,960 in lead; the Phillips mine with $26,610 in gold.

In 1890 Smith [51] gives the production of Lowe's, Deadwood, Fairplay, and Peabody placers as $23,611, and no gold was confidentially reported. The value of the deposits at the Denver Mint from Park County is given as $28,981.67 in gold and $326.12 (coinage value) in silver. The Hilltop mine is credited with $82,272 (coinage value) in silver, and $67,947 in lead.

In 1891 Smith [52] gives the production of the Fairplay placer as $2,500, but the amount confidentially reported is $35,212, and the value of the deposits at the Denver Mint from Park County was $41,433.49 in gold and $111.34 (coinage value) in silver.

In 1892 Smith [53] gives no detailed figures for placer production and $10,000 in gold from confidential reports. The value of the deposits at the Denver Mint was $32,696.26 in gold and $270.86 (coinage value) in silver.

In 1894 Puckett [54] gives the value of the deposits at the Denver Mint from Park County as $55,888.07 in gold and $393.31 (coinage value) in silver.

In 1895 Puckett [55] gives the value of the deposits at the Denver Mint from Park County as $46,086.05 in gold and $373.80 (coinage value) in silver.

In 1896 Puckett [56] gives the value of deposits at the Denver Mint for Park County as $25,410 gold and $211.92 (coinage value) in silver.

In 1897 Hodges [57] gives the value of deposits at the Denver Mint for Park County as $19,575.23 gold and $127.18 (coinage value) in silver.

For 1897 to 1904 the figures in the table, which represent smelter and mint receipts, are taken from reports of the Colorado State Bureau of Mines.

In 1898 Hodges [58] gives the value of deposits from Park County at the Denver Mint as $7,775.47 gold and $39.71 (coinage value) in silver.

In his report for 1899 Hodges [59] says:

During the last few years little work has been done. The outlook for all placers is brighter. The old Fairplay placers have not been worked for years, save nominally by lessees, but arrangements have been completed to work them on a very large scale the coming season with modern appliances.

[48] Munson, G. C., agent for Colorado, in Kimball, J. P., op. cit. for 1887, pp. 152, 177-179, 192, 1888.

[49] Munson, G. C., agent for Colorado, in Kimball, J. P., op. cit. for 1888, pp. 117-119, 132, 1889.

[50] Smith, M. E., agent for Colorado, in Leech, E. O., Report of the Director of the Mint upon the production of the precious metals in the United States during the calendar year 1889, pp. 150-151, 155, 1890.

[51] Smith, M. E., agent for Colorado, in Leech, E. O., op. cit. for 1890, pp. 137-138, 142, 1891.

[52] Smith, M. E., agent for Colorado, in Leech, E. O., op. cit. for 1891, pp. 183, 187, 1892.

[53] Smith, M. E., agent for Colorado, in Leech, E. O., op. cit. for 1892, pp. 119, 128, 1893.

[54] Puckett, W. J., agent for Colorado, in Preston, R. E., Report of the Director of the Mint upon the production of the precious metals during the calendar year 1894, p. 71, 1895.

[55] Puckett, W. J., agent for Colorado, in Preston, R. E., op. cit. for 1895, p. 75, 1896.

[56] Puckett, W. J., agent for Colorado, in Preston, R. E., op. cit. for 1896, p. 129, 1897.

[57] Hodges, J. L., agent for Colorado, in Roberts, G. E., Report of the Director of the Mint upon the production of the precious metals in the United States during the calendar year 1897, p. 127, 1898.

[58] Hodges, J. L., agent for Colorado, in Roberts, G. E., op. cit. for 1898, p. 99, 1899.

[59] Hodges, J. L., agent for Colorado, in Roberts, G. E., op. cit. for 1899, pp. 120-122, 1900.

The same condition applies to the Alma placers, which extend from Alma to Montgomery, now known as the Green Mountain placers, of which there has been only about 500,000 cubic yards washed out of an estimated 88,000,000 yards, and which is said to average at least 30 cents. There are a number of smaller placer propositions on the different streams that are worked as the quantity of water will permit. * * *

There are in Park County about 20 mills of different systems, of which the majority are idle through litigation involving the mine upon whose output they are dependent.

The statement of bullion operated on at the Denver Mint from Park County shows $20,545.75 in gold and $99.13 (coinage value) in silver.

In his report for 1901 Hodges says:[60]

The ore is shipped over the Colorado & Southern to the smelters at Denver. There are no public stamp mills in the county, which lack operates against the district. * * *

The London mine is shipping a sulphide ore, carrying 2 to 5 ounces gold, 3 to 10 ounces silver, and 15 per cent lead. * * * The Ling mine * * * shipped considerable ore averaging $100 per ton in gold, with but little silver. * * *

Active development is being done on the Kentucky Belle, Hock Hocking, Oliver Twist, James G. Blaine, Mascotte, and Viking, but the shipments of ore were small.

Horseshoe district.—This camp has shipped some ore from the Chance, Hill Top, and Peerless Maud. * * *

Fair Play placer district.—The placer grounds about Fair Play and Alma have for years yielded good wages to the man with the pan and also paid on the investment of capital.

The Fair Play placer extends about 2½ miles along the Platte River. Some development work was done by a new company during the year, which placed the property in good shape for the coming season.

Bedrock has never been worked, owing to inadequate machinery, but the surface gravel has proved beyond a doubt the values below.

Chinamen have for years taken out thousands of dollars on this and adjoining properties, though working with their crude and primitive methods.

Beaver Creek and Snowstorm placers.—The Beaver Creek placers have been idle for about five years, owing to litigation, which has now been settled. A consolidation of the Beaver Creek and Snowstorm placers has been accomplished.

The merging of these two properties will be beneficial in many ways to both. The Mosquito and Beaver Creek ditches and other water rights will have a carrying capacity of over 4,000 inches. The sluicing can be done with the cheapest hydraulic methods, owing to sufficient grade to allow the use of pipe line, giants, and sluice boxes. The gravel beds average about 30 feet in thickness. The Snowstorm deposit is said by old placer miners to be among the richest in the district.

Alma placer.—This property has been worked almost continuously since 1870. It is situated on the Platte River and consists of about 3,000 acres, only 30 of which have been systematically operated. A ditch about 2 miles long, taken out of the Platte, furnished 2,500 miner's inches and is adequate for the present workings. Twelve hundred cubic yards of gravel have been handled in a day, and this will be more than doubled with contemplated improvements. The gravel beds are about 50 feet in thickness and carry coarse gold. Nuggets ranging from $5 to $20 have been taken out; the average of

the present workings is 40 cents a cubic yard. Simple hydraulic methods are employed in working the placer ground.

Tarryall and Peabody placers were worked, principally by Chinamen, during the year on a small scale.

The statement of bullion from Park County operated on at the United States Mint, at Denver (p. 147) shows $18,708.89 in gold and $113.28 (coinage value) in silver.

In his report for 1902 Downer says:[61]

For a number of years the London mine has been a steady producer. * * * About 1,500 tons of ore were marketed from the old workings during 1902.

Placers.—The principal interest in Park County is centered about its placer mines. The Snowstorm placer was worked vigorously during the open months and, considering the scarcity of water, made a satisfactory showing. This ground is worked by the hydraulic method, undercutting the gravel with giants and washing into long sluice boxes. The ground is estimated to carry values of 50 cents to the cubic yard.

The Cincinnati placers, at Fair Play, have done some work during the season, but suffered from scarcity of water.

The Lowe placer, at Fair Play, had a short season, being sold early in the year to the Kansas & Colorado Hydraulic Gold Mining Co. It is expected this ground will be actively worked the coming season.

The old Alma placer had a successful season as long as the water lasted. For over 30 years this ground has been worked.

The statement of bullion from Park County operated on at the Denver Mint shows $14,613.71 in gold and $68.03 (coinage value) in silver.

In his report for 1903 Downer[62] gives a statement of the bullion operated on at the Denver Mint, which shows that Park County furnished $12,620.25 in gold and $57.73 (coinage value) in silver.

In his report for 1904 Downer says:[63]

The London mine * * * shipments averaged 350 tons per month of a $50 grade, values principally gold.

A 25-ton cyanide plant has been erected at the head of Montgomery Gulch to treat the low-grade ores of that vicinity.

The Snowstorm Placer Co. controls 4,000 acres of ground and employed a force of 50 men.

The statement of bullion from Park County operated on at the Denver Mint showed $10,991.77 in gold and $59.51 (coinage value) in silver.

For 1905 to 1922 the figures are taken from Mineral Resources (mines reports).

In 1912 the fineness of bullion from Platte River placers averaged 0.821 gold and 0.173 silver; in 1913 it averaged 0.821½ gold and 0.171 silver. In 1922 the fineness of bullion from the South Park Dredging Co.'s operations below Fair Play averaged 0.744 gold and 0.155 silver; and the Buckskin Gulch placer 0.791 gold and 0.200 silver.

---

[60] Hodges, J. L., agent for Colorado, in Roberts, G. E., op. cit. for 1901, pp. 129-130, 1902.

[61] Downer, F. M., agent for Colorado, in Roberts, G. E., op. cit. for 1902, pp. 120-132, 1903.

[62] Downer, F. M. agent for Colorado, in Roberts, G. E., op. cit. for 1903, p. 8, 1904.

[63] Downer, F. M., agent for Colorado, in Roberts, G. E., op. cit. for 1904, pp. 121-124, 136, 1905.

Gold (placer and lode), silver, copper, lead, and zinc produced in Park County, 1859–1923

Year	Ore treated (short tons)	Gold Placer	Gold Lode	Gold Total	Silver Fine ounces	Silver Avg price per ounce	Silver Value	Copper Pounds	Copper Avg price per pound	Copper Value	Lead Pounds	Lead Avg price per pound	Lead Value	Zinc Pounds	Zinc Avg price per pound	Zinc Value	Total value
1859–1867		$1,780,000	$710,000	$2,490,000													$2,490,000
1868		50,000		50,000													50,000
1869		40,000		40,000													40,000
1870		40,000	40,000	80,000													80,000
1871		40,000		40,000	15,094	$1.325	$20,000				a 5,000	$0.06	$300				60,300
1872		50,000		50,000	142,209	1.322	188,000				a 50,000	.06	3,200				241,200
1873		60,000	20,000	80,000	307,633	1.297	399,640	a 169,493	$0.28	$47,458	a 111,400	.058	6,684				533,142
1874		66,497	50,000	116,497	333,764	1.278	426,550	a 203,391	.227	44,746	a 25,000	.061	1,450				587,783
1875		80,000	24,302	104,302	412,022	1.21	510,967	a 72,150	.21	16,378	a 50,000	.055	3,050				633,037
1876		40,000	20,000	60,000	386,719	1.16	444,594	a 170,000	.19	14,350	a 150,000	.036	8,250				525,994
1877		40,000	20,000	60,000	309,375	1.20	371,250	a 175,000	.166	32,300	a 300,000	.041	8,400				471,800
1878		40,000	20,000	60,000	324,841	1.15	355,781	a 200,000	.186	29,050	a 300,000	.05	12,300				430,251
1879		40,000	20,000	60,000	298,996	1.12	363,825	a 200,000	.214	37,200	a 312,000	.048	15,000				473,325
1880		30,000	25,000	50,000	270,703	1.15	337,992	a 100,000	.182	42,800	a 312,000	.049	14,976				445,792
1881		25,000	25,000	50,000	195,359	1.13	305,894	a 100,000	.191	18,200	a 312,000	.043	15,288				389,070
1882		25,000	180,000	100,000	135,352	1.14	220,429			19,100	398,066	.037	13,416				354,817
1883		20,000	30,000	200,000	193,359	1.11	150,241				398,066	.039	14,728				363,657
1884		30,000	30,000	60,000	71,310	1.11	214,628				624,000	.046	15,525				289,356
1885		30,000	118,284	148,284	70,397	1.07	76,302				708,713	.045	28,704				151,827
1886		30,000	458,462	648,462	105,513	.99	70,397				7,641,720	.044	31,892				247,585
1887		30,000	10,945	33,945	450,457	.98	105,363				4,640,682	.039	336,226				785,717
1888		190,000	82,745	124,745	224,743	.94	423,430				1,886,504	.045	180,987				793,611
1889		23,000	13,470	37,281	186,975	1.05	211,258				19,656	.043	84,803				517,105
1890		42,745	16,000	50,333	43,792	.99	183,384	855	.135	115	25,698	.04	845				286,998
1891		23,611	29,687	39,687	62,350	.87	38,099	a 10,000	.108	1,080	a 30,000	.037	1,028				234,526
1892		34,333	79,845	109,845	43,817	.78	48,633	a 10,000	.095	950	98,791	.033	1,110				78,814
1893		10,000	57,358	97,358	46,658	.63	27,605	2,938	.107	314	297,714	.03	990				160,668
1894		30,000	101,761	131,761	117,095	.65	30,328	28,593	.108	3,088	4,517,614	.036	3,161				126,903
1895		40,000	112,109	137,109	199,945	.68	73,625	20,957	.12	2,599	1,953,001	.038	8,931				165,564
1896		30,000	134,619	153,619	72,137	.60	119,967	7,903	.124	1,351	540,849	.045	162,634				228,753
1897		25,000	152,490	159,490	43,138	.59	43,282	15,000	.171	2,490	682,107	.044	74,214				443,182
1898		19,000	133,041	153,041	69,175	.60	26,746	9,657	.166	1,613	421,935	.043	24,338				353,542
1899		20,000	98,558	116,558	49,908	.62	41,505	8,113	.167	980	261,046	.041	30,013				222,012
1900		18,000	78,322	96,322	52,128	.60	28,483	5,895	.122	758	802,489	.042	18,144				175,807
1901		18,000	128,458	142,458	50,013	.53	28,149	5,920	.137	1,903	757,703	.043	10,703				157,584
1902		14,000	124,277	136,277	49,292	.54	29,008	21,199	.128	2,779	543,303	.047	33,705				180,634
1903		12,000	184,980	194,980	144,815	.58	30,013	14,399	.156	4,898	966,193	.057	25,535				198,939
1904		10,000	318,081	320,867	111,215	.61	98,474	37,106	.193	7,933	1,062,732	.053	55,073				257,327
1905	4,202	7,000	384,966	395,050	12,047	.68	73,402	61,023	.132	11,271	495,985	.042	56,325				378,318
1906	6,745	2,786	506,263	513,216	102,375	.66	6,385	88,748	.13	3,027	2,237,093	.043	20,831				551,376
1907	10,072	10,084	418,742	430,808	117,037	.66	53,235	24,216	.127	1,703	2,041,204	.044	96,195				642,943
1908	12,661	6,953	527,563	551,921	69,072	.53	63,200	10,321	.125	4,520	923,089	.045	89,813				497,138
1909	11,372	12,066	252,701	265,547	31,234	.52	36,608	29,161	.165	1,067	167,755	.045	41,539				729,079
1910	15,044	24,358	48,758	58,832	94,293	.54	19,209	8,023	.155	2,153	506,046	.044	7,549				465,460
1911	12,329	12,846	35,283	67,981	20,215	.53	56,953	12,303	.133	5,559	168,154	.039	22,266				163,249
1912	2,686	24,411	44,151	50,041	9,227	.615	11,179	22,598	.175	3,501	190,830	.047	6,558				105,569
1913	6,598	19,223	149,547	67,485	13,231	.604	4,678	12,824	.246	3,138	330,609	.069	8,969				139,303
1914	1,958	14,758	223,878	159,339	14,705	.553	8,706	12,704	.273	3,801	278,709	.086	22,812				89,244
1915	2,820	23,334	111,967	234,299	18,280	.507	12,117	20,436	.247	3,436	233,873	.071	23,969	728,000	$0.047	$34,216	233,796
1916	3,005	9,792	63,176	117,358	70,949	.658	18,280	18,674	.186	974	305,908	.053	23,873	366,574	.054	19,795	277,749
1917	2,693	10,421	125,422	63,176	51,023	.824	79,463	7,550	.184	569	1,085,625	.08	16,605	659,796	.054	35,629	156,945
1918	2,304	5,451	142,632	159,557	47,547	1.00	55,615	4,215	.129	817	654,690	.045	16,213	407,772	.057	23,243	101,199
1919	11,210	4,135	40,821	143,158	15,528	1.12	47,547	5,558	.135		155,982	.055	86,850	132,275	.069	9,127	229,034
1920	5,348	526	42,654	41,250	18,701	1.09	15,528		.147		19,401	.07	29,434	98,623	.056	5,523	289,059
1921	4,929	429	16,974	142,120		1.00	15,335						8,579	57,940	.051	2,955	119,285
1922	1,120	99,466		161,442		.82							1,358	472,992	.124	58,651	166,796
1923	471	144,468												47,560	.134	6,373	178,652
		3,547,948	6,887,853	10,435,801	3,954,845		6,910,809	2,044,258		387,749	41,180,356		1,831,149	2,971,532		195,512	19,761,020

a Estimated by C. W. Henderson.   b Shipped to Swansea, England.

PITKIN COUNTY

Burchard in his report for 1880, under Gunnison County, from which the production of Aspen is subtracted, says: [64]

Aspen City.—The Smuggler * * * has a shaft down 45 feet. A tunnel of 150 feet is in upon the Trayner. * * * The Spar * * * has a shaft of 75 feet. The Durant * * * is another prospect. * * * The Chloride shows up well under development. The Minard makes a similar showing to the Trayner. About 250 miners are now working in the camp. Several of the mines are sending ore to Leadville for reduction, employing jack trains for its transportation across the mountains. * * * The Grand Republic is down 20 feet. * * * The ore is to be piled upon the dump, there to remain until the smelter starts up next summer.

In his report for 1881 Burchard says: [65]

Pitkin County, organized in 1880, was taken from the eastern portion of Gunnison County. The principal mining districts are Independence or Chipeta, Ashcroft, and Aspen.

In the Independence district a belt of gold-bearing veins was discovered in 1879. The ores are similar to those produced in Gilpin County and are readily reduced by the ordinary milling process. The Farwell Mining Co. own all the best mines, among which are the Independence Nos. 1, 2, 3, and 4, Last Dollar, Last Dime, Legal Tender, Bennington, Choler, Sheba, Friday, Mammoth, Dolly Varden, Gatton, Minnie, Mount Hope, and Golden Champion. The company purchased a 15-stamp mill, which has been running since January, 1881, and later have built a 30-stamp mill, which was required to treat the amount of ores extracted in the process of developing their mines. The Minnehaha, Golden Rock, Little Bobbie, and Old Grimes, composing the Minnehaha group, are all promising locations.

Ashcroft is on Castle Creek, at the base of Taylor and Elk ranges. * * * The chief group of mines is the Tam o' Shanter, on Slate Mountain. The group comprises the Tam o' Shanter, Montezuma, Borealis, Ivanhoe, Halcyon, and Last Chance. The ore from these mines is gray copper, galena, and carbonates of copper and lead. * * * On the Pearl group considerable work was done, and several tons of high-grade ore were shipped for treatment, but operations were suspended in the fall. * * * The Little Grace group comprises four locations upon which some little work has been done, sufficient to produce ore assaying rich in silver.

At Aspen, on the Roaring Fork, the Smuggler shows over 30 feet of ore between dissimilar lime walls. * * * The ore is chiefly a fine-grained galena, carrying much native silver in leaf and wire forms. There is also considerable spar ore strongly impregnated with gray copper, which is of very high grade. On the Spar the chief workings are an incline shaft of 70 feet and about 100 feet of drifts and tunnels. On the western ridge of the mountain are located the Pioneer, Trayner, and many other properties.

On the western slope of the mountains, at the head of Ophir Gulch, are the Caesar, Grand Prize, Colorado, Central, Louise, Oro, and Carrie. These are all ore producers, the general character of the mineral being similar to that found upon the other side of the mountains—galena, spar, carbonates, and chlorides.

The Morning Star and Evening Star produce free-milling ores. The Evening Star, at the depth of 40 feet from the outcrop, shows 6 feet of ore, which mill runs 50 ounces. Picked samples run as high as 6,500 ounces.

The other properties mentioned show much high-grade ore, and large bodies of mineral of lower grade, which will pay well for extraction and shipment. There are from 40 to 150 feet of work done on each of these claims, while from 10 to 50 tons of ore are lying on each of their dumps.

The Eva Bella, on Copper Hill, is from 4 to 6 feet wide and has an average value of about $100 per ton. The Oneida, on the north, is an extension of the Eva Bella and shows very fine ore, though the tunnel has not yet cut the main vein.

The Climax, on Richmond Hill, has a large body of high-grade galena and spar. The finding of this lode has stimulated work upon other claims in the neighborhood, which is already productive of promising results.

West Castle, Maroon, and Rock creeks command considerable attention from the various strikes made in those sections during the past summer.

The following mills and mines of Pitkin County have returned reports of production, viz, Farwell Consolidated, Camp Bird, Tourtellotte group.

The production during 1881 was $100,000 in gold and $30,000 in silver.

In 1881 and 1882 the gold came from the Independence district. Burchard says in the report for 1882: [66]

Independence or Sparkill district, situated in the eastern part of the county, contains a number of valuable mines, the principal of which are controlled by the Farwell Mining Co. The general formation of the district is granite and the ores are composed of auriferous copper pyrites in a compact siliceous gangue and average in value $16 per ton. In the other two districts [Aspen and Ashcroft] the ore is principally argentiferous galena, averaging 70 ounces silver and 40 per cent lead per ton. The Farwell Co. own some 12 claims, which have been extensively developed, and up to August, 1882, kept two stamp mills, of 15 and 30 stamps, respectively, in constant operation, producing in the neighborhood of $125,000 in gold. Since that time the mills have remained idle. * * * The Champion mine, near the Farwell group, * * * has already produced some very fine ore. The last lot shipped, a lot of 9 tons, gave returns of $475. Besides this the tailings, after the run of the ore through a crude stamp mill, assayed 2½ ounces of silver to a ton. The ore also contains about 15 ounces of silver to a ton, none of which was saved, other lots of ore from the mine have run much higher, one lot of 10 tons producing $1,760. * * *

Aspen district is noted for its large veins of argentiferous galena, among the principal being the Smuggler, Spar, Morning and Evening Star, Eva Bella, Camp Bird, Chloride, and Climax. * * * The mines of Aspen, taken as a whole, show but inconsiderable development, and owing to a lack of machinery the production of lead and silver has been confined to small shipments of ore to Leadville, 53 miles distant, which amounted in the aggregate to about $50,000. A 60-ton smelter is in the process of construction for the ores of this district and is expected to be in operation by the middle of January, 1883.

Ashcroft or Columbia district: * * * fissure veins of high-grade galena, gray copper, and silver glance ores * * * forms the southern boundary of Pitkin County and joins Gunnison on the north. * * * The Montezuma * * * shaft of 125 feet between the walls is turning out from 15 to 25 tons of high-grade galena and sulphuret ores daily, which, from shipments made in 1881 to outside points, indicate a character capable of milling from 200 to 300 ounces silver to the ton. * * * The veins in this district * * * in many cases carrying a very high-grade ore, which is being laid on the

[64] Burchard, H. C., Report of the Director of the Mint upon the production of the precious metals in the United States during the calendar year 1880, pp. 150–151 1881.

[65] Burchard, H. C., op. cit. for 1881, pp. 354, 423–424, 1882.

[66] Burchard, H. C., op. cit. for 1882, pp. 391, 394, 533–538, 586–587, 1883.

various dumps to await the necessary machinery, as there are no smelters in operation up to this time. Two small smelters are being built, however, one of which (the Brooks & Bethune) is expected to blow in about March 1, 1883.

In his report for 1883 Burchard says: [67]

Grant Smelting Works produced (or purchased) from Pitkin County 284,340 pounds of lead, 24,963 ounces of silver, 52 ounces of gold. Matthews & Webb purchased ore from Pitkin containing $4,856 silver and $1,800 lead.

Smuggler mine * * * averages 30 ounces silver and 20 per cent lead. Spar mine * * * [is] extracting ore, which assays from 10,000 to 19,000 ounces in silver and from 3 to 50 per cent lead. Camp Bird group * * * near Tourtelotte Park, * * * located during fall of 1879 and spring of 1880, * * * has produced something over $10,000 in mineral, the smelter returns being from 64 to 495 ounces silver per ton. The Buckhorn group * * * the mineral consists principally of spar and carbonate of lead, averaging 40 ounces silver and 50 per cent lead.

Around Ashcroft, * * * since the first of the year, the Montezuma has begun to deliver on a contract made with the smelter at Aspen. * * * The Montezuma * * * has shipped in the neighborhood of 1,000 tons of ore, averaging $60 to $65 per ton. The Montezuma and Tam o' Shanter were discovered in 1880. * * * The Montezuma is shipping its ore to Aspen for treatment, a large amount having been sent outside over Pearl Pass [to Crested Butte]. * * * The ore has milled 65 ounces of silver and 35 per cent of lead. * * *

At Independence nothing is being done. The Farwell Consolidated Mining Co., that at one time was one of the largest gold producers in Colorado, has been closed the entire year.

Burchard in 1884 says: [68]

There is a smelter at the town of Aspen of 28 tons capacity which is at present being enlarged and roasters are being added. The ores received at it show about 7 per cent zinc and 13 per cent baryta, making a very refractory combination, costing some $28 per ton to treat.

In Mineral Resources for 1883 and 1884 Kirchhoff says: [69]

One of the new camps which is beginning to attract attention and which promises to become more important, is Aspen, where the Aspen Smelting Co. started its furnace in July, 1884, and is now building a second. The ores, taken from deposits at or near the contact of dolomite and Silurian limestone, carry fair quantities of lead, considerable zinc, some copper, and heavy spar, as the chief constituent of the gangue, and are exceptionally high in silver. The district is favored by the existence of a 6-foot bed of good bituminous coal, making a good coke carrying only 10 per cent of ash; and by the exercise of metallurgical skill, a constant watching of the character of the ores and the working of the furnace, it has been possible, by running hot, to overcome the technical difficulties due to the presence of exceptionally large quantities of baryta and zinc. On the other hand, the transportation to railroad is costly, though this is counterbalanced by the exceptionally high silver value of the bullion, which carries from 500 to over 1,000 ounces of silver per ton.

Kirchhoff[70] estimates 15 per cent loss on 3,500 tons of lead in 1885. The figures given for silver for that year in the table represent rough estimates on the basis of different statements as to proportion of lead to silver, giving particular weight to Kirchhoff's figures for lead for 1885 and 1886.

For 1886 to 1896 the figures here given are based on reports of agents of the Mint in annual reports of the Director of the Mint, the gold and silver being prorated to correspond with the figures for the total production of the State as corrected by the Director of the Mint, the lead being prorated to correspond with the total production of lead in the State as given in Mineral Resources, and any "unknown production" in the State being distributed proportionately among the counties. As with lead so with copper, but as the figures for copper given in Mineral Resources include copper from matte and ores treated in Colorado, though produced in other States, they are subject to revision.

In his report for 1886 Kirchhoff says: [71]

Aspen produced very little lead bullion in 1886, and a comparatively small amount of ore. Between 5,000 and 6,000 tons were marketed. This was due to a number of causes. The bringing of suits by apex claims, and the consequent injunction placed upon claims below the apex, operated to stop production on several of the most important mines. These injunctions are still in force, and there is no prospect of a final settlement during the year 1887. Then the building of the Colorado Midland Railroad has tended to stop shipments of all ores except those of the very highest grade, because they will be much more valuable as soon as the railroad reaches Aspen in the fall. When railroad transportation has been secured it is probable that there will be a great activity in developing and working low-grade mines, and a large output of lead is looked for from Aspen in the latter part of 1887 and in 1888.

According to Spurr, [72]

The Denver & Rio Grande Railroad reached Aspen, October, 1887, while the trains of the Colorado Midland did not actually reach the town limits until February, 1888.

For 1887 and 1888 Munson[73] lists the producing and nonproducing mines and gives the individual production of some of the mines. The gold at the United States Mint came from sundry deposits.

For 1889 to 1892 Smith[74] lists the producing mines and gives individual outputs. The gold in 1891 came from the Independence mine, in the Independence district.

The figures for zinc in 1895 represent ore receipts of the American Zinc-Lead Co. of Canon City, as published in the Denver Republican in its yearly review for 1895.

[67] Burchard, H. C., op. cit. for 1883, pp. 236, 238, 240, 395–401, 1884.

[68] Burchard, H. C., op. cit. for 1884, pp. 177, 235–236, 1885.

[69] Kirchhoff, Charles, jr., Lead: U. S. Geol. Survey Mineral Resources, 1883 and 1884, p. 422, 1885.

[70] Kirchhoff, Charles, jr., Lead: U. S. Geol. Survey Mineral Resources, 1885, p. 257, 1886.

[71] Kirchhoff, Charles, jr., Lead: U. S. Geol. Survey Mineral Resources, 1886, p. 145, 1887.

[72] Spurr, J. E., Geology of the Aspen mining district, Colo.: U. S. Geol. Survey Mon. 31, p. xxi, 1898.

[73] Munson, G. C., agent for Colorado, in Kimball, J. P., Report of the Director of the Mint upon the production of the precious metals in the United States during the calendar year 1887, pp. 179–180, 1888; idem for 1888, pp. 119–120, 1889.

[74] Smith, M. E., agent for Colorado, in Leech, E. O., Report of the Director of the Mint upon the production of the precious metals in the United States during the calendar year 1889, pp. 151–152, 1890; idem for 1890, pp. 138–139, 1891; idem for 1891, p. 183, 1892; idem for 1892, pp. 128–129, 1893.

In his report for 1897 Hodges says:[75]

There are four concentrating plants in this district, having a total capacity of 500 tons per day, and a stamp and combination mill at Independence * * * of 100 tons capacity. This mill is intended to work a mine carrying gold in quartz above the water level, about 150 feet deep. Below, the ore changes to a sulphide.

For 1897 to 1903 (and from 1904 to 1908 for copper and from 1904 to 1907 for zinc) the figures, which represent smelter and mint receipts, are taken chiefly from the reports of the Colorado State Bureau of Mines. For the years 1904 to 1908, with the above exceptions, the figures are taken from Mineral Resources (mine reports).

For 1898 Hodges says:[76]

The system of concentration; * * * the Smuggler, or Hallett table, is doing remarkably good work. About 10 per cent of the * * * shipments are concentrates. Total tonnage for the camp for 1898 was 165,000 tons.

For 1900 Hodges gives the following details:[77]

Deep workings.—The deepest workings in this district are in the Gibson, which has a shaft 1,200 feet in depth but has been flooded for several years past. In order to keep the working levels dry, the Mollie Gibson has a very fine pump of the Snow pattern, triple expansion, duplex, condensing, and steam jacketed, which pumps 1,800 gallons a minute, raising the water 1,000 feet to the surface. This also drains the Argentum-Juniata.

Concentrating mills.—A number of concentrating mills have been built * * * Perhaps the best example of the work is at the two Smuggler mills, the old mill having a capacity of 150 tons in a day of 10 hours, and the new mill, * * * 125 tons. Neither of these mills work a night shift.

The Smuggler ore consists of a limestone carrying from 4 to 5 per cent galena, 10 to 15 per cent zinc carbonate, and an uncertain percentage of barium sulphate, with a small amount of silica, and about 4 to 6 ounces of silver per ton. At the old mill the ore is delivered in the upper part, where it is passed through a jaw crusher and rolls, and sized by means of revolving screens, the different sizes passing to a number of Hartz jigs, made in Germany specially for this mill. The middlings from the jigs are all conveyed to Bradley pulverizers and ground so that the coarsest material passes through a 60-mesh screen. From these pulverizers the pulp is run to a series of large V-shaped classifiers, from which the different sizes are taken to Hallett concentrating tables, which are simply modifications of the Wilfley table, and there the ore, practically in the condition of slimes, is very successfully separated into a galena and baryta sulphate as the heading, while the zinc carbonate with the iron are saved as middlings for future treatment, and the lime with some zinc and silver rejected as waste.

At the Smuggler new mill the same separations are made, but only the Hallett tables are used. The ore, being crushed to the size of grains of corn, is pulverized in Bradley mills so that the coarsest will pass through a 60-mesh screen. These Bradley mills are modifications of Chilean mills, but on the Smuggler ore, which pulverizes very readily, their capacity is 60 tons a day for each machine.

The Hallett table used in these mills is the invention of Mr. S. I. Hallett, who is the general manager of the Smuggler properties, and has been modified, until, to all intents, it is simply a Wilfley table with the riffles set diagonally on the tables instead of longitudinally. The driving mechanism is the old quick-return movement formerly used on the King-Darrah table, the patent for which is now controlled by the Hallett Co. It is claimed for this table that the values being carried up to its cleaning side, the entire length, less wash water is required, and, consequently, less dilution of the pulp is made on the table and less fine values are carried over into the tailings trough. The concentrates are carried off about midway on the end of the table instead of at the upper corner side, as in the Wilfley.

At the new mill of the Smuggler Co. two sets of tables are used, one set following the other, the second set treating the middlings from the upper or first set. The middlings from the four of the first tables are treated on one table in the second set. The barium sulphate in this ore contains so much silver that no attempt is made to keep the galena and heavy spar separate, although the smelters charge quite a heavy penalty for over 5 per cent.

The Mollie Gibson has a mill in which practically the same operations are carried on except that Frue vanners are used in place of the Hallett tables. The ores capable of concentration from the Mollie Gibson and Argentum-Juniata are worked conjointly in this mill.

Another mill, farther up the Roaring Fork, which was remodeled from a lixiviation plant into a concentration mill, is now being used as a test mill on the Della S. ore. The ore here is pulverized by stamps so as to pass a 40-mesh screen, and the pulp then sized by an upward current of water against a descending stream of pulp. The heavier sizes are conveyed to Wilfley tables and the slimes to V-shaped boxes, where the settled pulp is treated on Frue vanners. A claim is made that the combination of these two systems—the Wilfley table for the coarse sand and the Frue vanner for the slimes—effects a saving of 70 per cent of the values.

Two sampling works are kept busy in the town and handle practically all the ore mined before it is sent to the smelter. All the mills, including the sampling works, are run by electric power generated by the Castle Creek Power Co., which also furnishes the illumination for the town. A large quantity of limestone was furnished during the early part of the year to the smelters, which paid for the lime and also for the silver contents, which, while low, returned a profit. This was made possible as the limestone was shipped from the old dumps without sorting.

The only economic improvement during the past year in the camp is the extending of a railroad spur up Castle Creek to the Castle Creek tunnel and a branch up the side of Aspen Mountain for a short distance, which will reduce the cost of hauling and tramming ores.

The outlying districts have been practically idle during the year, and the Independence district, of which much was hoped as a gold-producing section, has not realized the expectations of its supporters.

For 1901 Hodges [78] gives the following details:

Tunnel outlets.—The several tunnels through which the most extensive mining operations are conducted are as follows: The Cowenhoven, 2½ miles long; Smuggler and A. J., 3,000 feet; Durant, 8,500 feet; Compromise, 7,000 feet; Newman, 5,000 feet; Robinson, 3,000 feet. * * *

Mills.—The Smuggler owns and operates two ore concentrating plants. Their combined capacity is 300 crude tons per

[75] Hodges, J. L., agent for Colorado, in Roberts, G. E., Report of the Director of the Mint upon the production of the precious metals in the United States during the calendar year 1897, pp. 125–126, 1898.

[76] Hodges, J. L., agent for Colorado, in Roberts, G. E., op. cit. for 1898, p. 92, 1899.

[77] Hodges, J. L., agent for Colorado, in Roberts, G. E., op. cit. for 1900, pp. 128–131, 1901.

[78] Hodges, J. L., agent for Colorado, in Roberts, G. E., op. cit. for 1901, pp. 130–134, 1902.

day. This company also has a sampling plant for treatment and shipment of their own ores and concentrates.

The ore concentrating mills are in construction specially arranged—one for coarse crushing and concentration, the "lead mill," and the other fine crushing and concentration, the "zinc mill." They are so called because the lead is so finely disseminated in the zinc that a final fine crushing is necessary to thorough separation.

Average value of crude ore treated was 10½ ounces silver, 8½ per cent lead, and 16½ per cent zinc per ton.

Average value of mineral concentrates saved was 48.4 ounces silver, 59.3 per cent lead, and 9.3 per cent zinc.

Settlings saved in reservoir, 9.9 ounces silver, 3 per cent lead, and 31½ per cent zinc.

Treatment averaged a reduction of 9.2 tons of ore to 1 ton of mineral concentrates.

Both of the above-named mills, being of early construction, have gradually improved as repairs were needed, and plans are now perfected for remodeling in 1902 at an expenditure of not less than $30,000.

Hunter Creek mill is located at mouth of the Cowenhoven tunnel, base of Smuggler Mountain, for the purpose of concentration treatment of ores mined and output through the said tunnel, which has penetrated Smuggler Mountain a distance of 2½ miles, close to what is known as the "contact" between the middle (shale) and lower Carboniferous (lime) rock.

In the bulk of the ore as at present delivered from the Della S. mine direct to the Hunter mill the mineral is disseminated so finely that it can seldom be directly seen in the mine, but by rubbing the ore with the pick the amount of glossy lead or silver stain determines approximately the value for concentration, which at present price of silver and lead must equal about 9 ounces of silver and 4 per cent lead.

The works were placed in commission late in October, since when they have been in active operation, yielding very satisfactory results from a very low grade silver and lead ore, and they are evidently a close approach to solving the problem of treatment of low-grade refractory ores, the output of which in the Aspen district is large.

The system of ore concentration was perfected through a series of determinative tests; likewise, all machinery and mechanical appliances employed, and these of the latest proven types and manufacture, were selected and assembled in place upon (as near as may be) intelligently predetermined methods

The installation is in duplicate, so that ore treatment may be in check and details adapted to the general varying conditions of ore, or ore from two different mines be under treatment at one time.

Inasmuch as the Hunter mill has been in operation only 60 days, its full efficiency can not be accepted so determinatively as may be after a longer campaign. The operations for December, 1901, were in the aggregate as follows: Total silver saved in concentrates, 20,117 ounces; lead, 150,851 pounds; per cent silver saved, 81.1; lead, 84.3; net saving, 83.7 per cent. Average assay of crude ore, 8.4 ounces silver; 3.18 per cent lead.

In this connection a point of interest is shown in results of January, 1902. Briefly, the mill averaged 104 tons crude ore per diem, the average saving of silver and lead being 87.5 per cent.

The system of installation at the Hunter mill was furnished by the engineer, to whom much credit is due in working out and perfecting this milling plant, Mr. Charles Anderson, manager, also manager of the Della S. mine, and from which mine all ore under treatment at present time is supplied.

The machinery consists of one 6 by 20 inch crusher, two pairs of 30 by 14 inch rolls, two 6-foot Huntington mills, twelve Wilfley concentrators, four belt machines 6 feet wide, and automatic dryers with automatic sampling apparatus through which the whole quantity of ore from the rolls, screened to $\frac{3}{16}$ inch mesh, passes into storage bins, thence to the Huntington mills, and is pulverized with pressure of water through a No. 35 mesh screen, thence through a V trough, one for each mill, with grade to slowly carry the heavier granulations forward. The bottoms of these troughs are fitted with adjustable classifiers in such a way that the horizontal surface current of the pulp is not disturbed. The thus partly classified pulp flows into two series (six in each series) of table machines. The "middlings" from each series of tables are carried by a shaking launder to an elevator, thence flow onto another table.

The slimes flow from the hydraulic classifiers into tanks, and after settling are fed automatically for each series of tables. The vanners make high-grade slime concentrates and worthless tailings.

The mineral concentrates from all machines are carried by shaking launders to automatic steam dryers, which discharge the concentrates into storage bins ready for shipment.

When the ore is broken it is loaded into the mine ore cars, five forming a train, thence by horses drawn out through the Cowenhoven tunnel, a distance of about 2½ miles, and unloaded upon a grizzly set in top floor of the Hunter mill, thence through machinery and appliances, without intervention of hand labor, until it is by one man loaded into railway cars for shipment, at a whole cost of less than 50 cents per ton.

The Hunter mill is lighted and driven by electric power transmitted from the Roaring Fork Electric Lighting & Power Co., which employs water power.

The mill is heated by steam; mineral concentrates also dried by steam.

The mining industry of Aspen is well served by the Denver & Rio Grande, also the Midland Railway system, spurs from their main tracks being laid to all the larger mines, also the concentrating and sampling works, thus affording excellent facilities of transportation of ores and concentrates without intervening cost of wagon haul.

For 1905 Lindgren says: [79]

The Smuggler mine shipped a large tonnage of lead-silver-zinc ore and slimes to Canon City for the manufacture of zinc-lead pigment.

For 1906 Naramore says: [80]

More than one-half of the crude ore mined in Pitkin County in 1906 was concentrated, and in addition, a large tonnage of zinc slimes was shipped from a pond where they had accumulated from the concentration of Smuggler ores for several years, when zinc was a detriment to the ore.

For 1909 to 1923 the figures given in the table are taken from Mineral Resources (mines reports).

[79] Lindgren, Waldemar, U. S. Geol. Survey Mineral Resources, 1905, p. 208, 1906
[80] Naramore, Chester, U. S. Geol. Survey Mineral Resources, 1906, p. 229, 1907.

Gold, silver, copper, lead, and zinc produced in Pitkin County, 1880–1923

Year	Ore treated (short tons)	Lode gold	Silver			Copper			Lead			Zinc			Total value
			Fine ounces	Average price per ounce	Value	Pounds	Average price per pound	Value	Pounds	Average price per pound	Value	Pounds	Average price per pound	Value	
1880			a 10,000	$1.15	$11,500				60,000	$0.05	$3,000				$14,500
1881		$100,000	23,203	1.13	26,219				200,000	.048	9,600				135,819
1882		90,000	23,203	1.14	26,451				200,000	.049	9,800				126,251
1883		2,000	42,539	1.11	47,218				450,000	.043	19,350				68,568
1884		1,000	464,062	1.11	515,109				1,200,000	.037	44,400				560,509
1885		1,000	1,000,000	1.07	1,070,000				5,950,000	.039	232,050				1,303,050
1886	5,500	17,125	399,094	.99	395,103				800,000	.046	36,800				449,028
1887		9,336	612,368	.98	600,121				361,388	.045	16,262				625,719
1888		12,716	4,333,787	.94	4,073,760				14,349,792	.044	631,391				4,717,867
1889			5,982,238	.94	5,623,304				15,100,807	.039	588,931				6,212,235
1890			4,944,898	1.05	5,192,143				19,703,605	.045	886,662				6,078,805
1891		13,507	6,979,263	.99	6,909,470				16,396,580	.043	705,053				7,628,030
1892			8,138,549	.87	7,080,538				20,998,701	.04	839,948				7,920,486
1893			5,039,799	.78	3,931,043				15,000,000	.037	555,000				4,486,043
1894		5,312	5,996,851	.63	3,778,016				15,750,000	.033	519,750				4,303,078
1895		1,387	5,131,792	.65	3,335,665	616	$0.107	$66	11,163,685	.032	357,238	21,000	$0.036	$756	3,695,112
1896		1,523	4,922,360	.68	3,347,205	52,991	.108	5,723	16,272,411	.03	488,172				3,842,623
1897		164,430	4,599,946	.60	2,759,968	8,360	.12	1,003	15,903,682	.038	604,340				3,085,834
1898	165,000	71,001	3,977,270	.59	2,346,589	4,553	.124	565	25,458,380	.045	1,145,627				3,022,495
1899		52,233	4,158,708	.60	2,495,225	19,351	.171	3,309	27,452,260	.044	1,207,899				3,696,394
1900		13,456	4,119,116	.62	2,553,852	6,082	.166	1,010	32,749,511	.043	1,408,229	20,000	.044	880	3,777,097
1901		4,692	3,532,863	.60	2,119,718	50,786	.167	8,481	24,973,816	.041	1,023,926				3,541,120
1902		4,899	3,063,450	.53	1,623,629	10,654	.122	1,300	33,269,852	.042	1,397,334				2,653,754
1903		4,754	2,569,862	.54	1,387,725	11,683	.137	1,683	18,882,901	.043	811,965				2,791,414
1904	109,770	2,336	2,129,618	.58	1,235,178	9,862	.128	1,262	22,386,142	.047	1,052,149	593,661	.051	30,277	2,081,018
1905	107,927	248	2,469,520	.61	1,506,407	127,094	.156	19,827	17,951,674	.057	1,023,245	3,854,339	.059	227,406	2,806,037
1906	203,400	1,172	2,131,374	.68	1,449,334	285,346	.193	55,072	12,235,230	.053	648,467	3,276,711	.061	199,879	2,728,702
1907	183,836	579	1,719,446	.66	1,134,834	234,493	.20	46,899	7,568,060	.042	317,859	4,688,693	.059	276,633	2,107,412
1908	133,408	538	1,041,700	.53	552,101	22,474	.132	2,967	13,143,210	.043	565,158	722,362	.047	33,951	907,416
1909	112,448	745	700,038	.52	364,020	26,092	.13	3,392	13,408,250	.044	589,963	34,741	.054	1,876	935,191
1910	89,037	646	477,813	.54	258,019	24,843	.127	3,155	11,084,334	.045	498,795				851,783
1911	88,823	542	450,772	.53	238,909	7,408	.125	926	8,405,333	.045	378,240	484,507	.069	33,431	739,172
1912	91,791	165	528,504	.615	325,030	22,952	.165	3,787	17,528,386	.044	771,249	460,161	.056	25,769	740,653
1913	114,264	29	562,308	.604	339,634	48,852	.155	7,572	23,233,230	.039	906,096	145,431	.051	7,417	1,144,253
1914	118,000	423	372,886	.553	206,206	67,737	.133	9,009	19,265,213	.047	905,465	214,952	.124	26,654	1,129,151
1915	108,579	29	448,915	.507	227,600	19,983	.175	3,497	17,519,275	.069	1,208,830	162,574	.134	21,785	1,163,245
1916	114,330		577,863	.658	380,234	28,931	.246	7,117	14,352,523	.086	1,234,317	571,794	.102	58,323	1,617,966
1917	124,824	105	662,045	.824	545,525	27,403	.273	7,481	11,666,592	.071	828,328	145,286	.091	13,221	1,845,751
1918	98,413	2	558,722	1.00	558,722	9,684	.247	2,392	5,310,170	.053	281,439	80,000	.073	5,840	1,402,065
1919	121,534		657,058	1.12	735,905				4,470,300	.08	357,624	617,790	.081	50,041	1,023,184
1920	125,786		625,444	1.09	681,734				2,395,622	.045	107,803	283,000	.05	14,150	1,089,399
1921	60,476		474,225	1.00	474,225	233	.129	30	3,555,309	.055	195,542				596,208
1922	119,023		525,169	1.00	525,169				2,972,614	.07	208,083				720,711
1923	58,641		429,581	.82	352,256							465,000	.068	31,620	591,959
		577,930	97,608,222		73,340,613	1,128,463		197,443	565,555,316		25,781,812	16,842,002		1,059,909	100,957,707

a Estimated by C. W. Henderson.

PUEBLO COUNTY

The figures for 1894 and 1896 given in the table below are prorated from those given by Puckett.[81] For 1900 and 1901 the figures are taken from the reports of the Colorado State Bureau of Mines, but the source of this production is in considerable doubt. In fact, it is questionable whether this yield has been properly credited.

Gold, silver, and copper produced in Pueblo County, 1894–1901

Year	Gold	Silver		Copper		Total value
		Fine ounces	Value	Pounds	Value	
1894	$296	3	$2			$298
1896	84	26	17			101
1900	248	9	5			253
1901	165	52	31	210	$35	231
	793	90	55	210	35	883

81 Puckett, W. J., agent for Colorado, in Preston, R. E., Report of the Director of the Mint upon the production of the precious metals in the United States during the calendar year 1894, p. 73, 1895; idem for calendar year 1896, p. 158, 1898.

RIO GRANDE COUNTY

Patton[82] gives a summary of the mining developments in Rio Grande County for 1870 to 1917.

In his report for 1875 Raymond[83] reviews the history of development prior to 1876 as follows:

Rio Grande County, Summit [Summitville] district.—This district is usually classed as a part of the San Juan country. * * * Summit district, which is the leading gold district of the San Juan country, is situated 27 miles southwest of Del Norte and lies just within the boundary of Rio Grande County. Its approximate latitude is 37° 30'; approximate longitude 106° 30' west from Greenwich, with a mean elevation of about 12,000 feet above the sea.

The first discovery of gold in the Summit district was made in Wightman's Gulch about the last of June, 1870, by a party consisting of James L. Wightman, E. Baker, J. Cary French, Sylvester Reese, and William Boran, Wightman getting the first "prospect." All of the party, with the exception of Wightman and Reese, left by the middle of September, the

82 Patton, H. B., Geology and ore deposits of the Platoro-Summitville district, Colo.: Colorado State Geol. Survey Bull. 13, 1917 [1918].
83 Raymond, R. W., Statistics of mines and mining in the States and Territories west of the Rocky Mountains for 1875, pp. 282, 326–334, 1877.

latter remaining, engaged in sluicing, until the 9th of November, when they left, heavily packed, and made their way out through snow waist deep, reaching the Rio Grande in three days.

In the spring of 1871 a large number of people flocked into the Summit, hundreds arriving while the snow was yet very deep and work impracticable. A general disgust soon took possession of the prospectors, and by the last of August there were but three men in the district—J. L. Wightman, P. J. Peterson, and J. P. Johnson. These then remained until about the 20th of October, Wightman and Peterson being the last to leave. They took the gold realized by sluicing to Denver and had it refined at the mint, dividing $170 between the three after paying all expenses of the season's operations; not a very encouraging yield for a hard summer's work. Several lodes had in the meantime been found, or at least lode locations made. The specimens found in the gulch indicated to the miners that they had not washed far, and they believed that parent ledges in place were close by.

In 1872 a few locations were made, and 1873 witnessed a new immigration into the district, and in that year the richest mines in the Summit [district] were located. The Esmond and Summit lodes were staked during the summer, and on September 13 F. H. Brandt and P. J. Peterson located the Little Annie, Del Norte, and Margaretta mines. * * *

During 1874 a vast number of new locations were made, and the attention of the owners was turned to the matter of getting in machinery. Dr. Richard F. Adams, after locating the Summit mine, shipped a small amount of ore to be tested, and, having become satisfied that the enterprise would pay, located a mill site and ordered a mill, which was brought in and commenced to run the following spring (1875).

During the winter of 1874–75 negotiations were opened by the mine owners with capitalists for the purpose of getting in mills. The owners of the Little Annie, Del Norte, and Margaretta, of the Golden Queen, and of the Golden Star, entered into contracts by which the parties putting in the mills were to have an interest in the mines.

The spring of 1875 marked the opening of permanent mining operations at the Summit. Dr. Adams's 5-stamp mill began work as early as the season permitted. In the latter part of May [1875] the machinery for the Little Annie and Golden Queen mills reached Del Norte from Chicago, and was drawn by mule teams on a road cut for the purpose over Del Norte Mountain, 13,000 feet high, and costing over $4,000. The machinery for the two mills weighed more than 50 tons, but was successfully transported above the lower cloud belt and placed in position before the close of September [1875].

The chief gold-producing property of Summit district, and of the territory known as the San Juan mining region, is owned by the Little Annie Mining Co. and comprises the Little Annie, Del Norte, and Margaretta gold mines, and two placer claims of 20 acres each, situated in the gulch below them, with a 10-stamp mill, business and assay office, store, bunkhouse, mess house, retort house, charcoal house, blacksmith shop, tramway, substantial mine and mill dumps, dam, flumes, sluices, etc. * * *

The ores of the Del Norte and Margaretta mines have not yet been tested in the mill. Average assay value of former, $43.37; of latter $24.20. The Del Norte has yielded exceedingly rich pan prospects, and on the 26th of August, 1876, a very rich deposit of flour gold was found on the Margaretta, the extent of which has not been determined. * * *

The Little Annie has had more development than any mine in the district, some 1,200 tons of ore having been taken out, but it is as yet only an open quarry. * * * Specimens have assayed all the way from $70 to $160,000 per ton at the Denver Mint and elsewhere. * * * The average value of the ore is best shown by the result of last autumn's [1875] mill work in which 306 tons taken from the face of the mine, without

any sorting, yielded $31,444=$102.68 per ton (currency) The gold is chiefly in the form of "flour" and for the most part invisible, although fine specimens are occasionally taken, some being of very large size. The fineness of the retorts has been about 980, as shown by certificates from the mint at Denver and from the United States assay office at New York. * * *

The Little Annie mill is a 10-stamp outfit with Blake crusher; capacity of latter, 30 tons per day of 10 hours. The mill has two batteries of 5 stamps each, 4 dolly tubs or pans, and 1 agitator. Amalgamation in batteries, on table plates, in pans, and on a second set of table plates on a floor below, over which the slimes pass before going into the discharge sluice. Fall of stamps, 11 inches. Drop rate, 60. Weight, 530 pounds. Cams, two-armed. Shoes and dies of white iron; weight of former 112 pounds, of latter 84 pounds. Length of cams from point to point, 29 inches. Battery issue, 10 inches deep. Screens, No. 1, fine, slot. Pan revolution, 65 per minute; settler, 28 per minute. Engine, 25 horsepower, burning 16 cubic feet of wood per hour. * * * Frame of mill raised 10th of August, 1875; commenced to run 28th of September following, and has been in continuous operation since, except from 8th of December, 1875, to 20th of May, 1876, during which there was no ore on hand. Capacity per day of 24 hours, 5.58 tons of ore, or 1,100 pounds per stamp. Running time in 1876, from 20th of May to 1st of September (the date of this report), gross, 103 days; stoppage of all kinds, 470 hours, 17½ minutes, net running time, 2,001 hours, 42½ minutes=83 days, 9 hours; 42½ minutes, during which time the mill has produced 1,710.125 ounces of gold, worth in coin $34,202.50.

The Little Annie mill has at present but 10 stamps, but there are 64 stamps working under contract on Annie ore at a cost to the company of $10 per ton. * * * The product in its own mill (at the average so far made of $3,000 per week) is $156,000 per annum.

Raymond gives a statement of costs in running the mill on page 330 and a balance sheet for September 1, 1876, on pages 331 and 332. He continues:

The placers have been worked only at intervals, without system and without machinery. In this mode they have yielded some very fine nuggets. * * *

The Golden Queen mine was located in the latter part of September, 1873, by Josiah Mann, O. P. Posey, John Grant, and others. It has been extensively worked, and a stamp mill has been erected 88 feet east of the Little Annie mill, which is a duplicate of the latter, except that the engine is 20 horsepower instead of 25. Assays in ore of mine have not hitherto run high, and definite information as to mill results is not at present accessible. The owners are Johnston Livingston, John J. Crooke, Adams & Posey, Arthur Burton, Peter Beeker, Joseph S. Reef, John A. McDonald, Lucius A. Winchester, J. S. Partridge, Lewis Crooke, William Beck, James L. Hill, Henry B. Clark, F. C. Day, L. C. Smith, R. C. Sheppard, and L. C. Baker. Office, Summit, Rio Grande County, Colo.

The Summit mine was located in 1873 by Dr. Richard F. Adams, and his 5-stamp mill, erected in the autumn of 1874, was the pioneer of all the machinery now in the district. The mine lies high up on the northeastern face of South Mountain, and a considerable quantity of surface ore has been taken from it. Various assays have ranged from $10 to $200. Yield in mill not ascertained. Fifteen more stamps designed for the latter are at Del Norte. The power is water, supplemented by a steam engine. The owners are R. F. Adams, Lewis Crooke, and Le Grande Dodge.

The Golden Star Gold & Silver Mining Co. owns three mines, all located by Isaac Garnett: the Golden Star No. 2, staked 19th of June, 1874; Eighth Wonder, staked 22d of June, 1874; and the Keystone, staked 3d of July, 1874. Assays have been had from the former of $20. A superior mill has been erected by the company on Wightman's Fork, with ten 650-pound stamps, and provision for ten more. The building is 32 feet by 48 feet, with addition for boiler of 15 by 30 feet, substantially constructed with double-board siding inclosing a layer of tarred paper; felt roof. Large Blake crusher; engine, double cylinder and 40 horsepower; single-armed cams. Large grinding pan, with capacity of eight tons of tailings daily. Drop rate, 90; high mortars; single discharge (as are all the mills in this district); width of issue, 12 inches; amalgamation in battery and on table plates; no blankets. Bumping table on Rittinger plan for concentration; one Wheeler & Randall pan for working concentrates, with raised patent washers; screen, No. 1 fine, slot. Machinery by Morey & Sperry, of New York. Capital stock, $1,000,000. Principal office, Chicago, Ill.; branch office, New York. Daniel Barnum, president; C. R. Brooke, secretary; J. A. Sperry, builder and agent. Majority of stock held in New York. The mill has just commenced to run; results of work not ascertained.

The San Juan Consolidated Mining Co. is a combination owning over 15,000 linear feet on South Mountain, comprising a large number of locations, of which the Ida, staked out by Colonel Gillette, in July, 1874, is at present regarded as the best. A 30-stamp mill has just been erected and has commenced running. Weight of stamps, 500 pounds; fall, 12 inches; rate, 37 per minute. Double-armed cams; issue, 13 inches wide. Engine, 45 horsepower; steam pressure, 50 pounds. Amalgamation in battery and on table plates. Blankets washed every 15 minutes. Four Bartola pans; rate of revolution 35 per minute; one settler, rate 35; screens, No. 1, fine, slot; Dodge crusher, large size. Mill building, 50 feet square. Capital stock, $3,000,000. Charles W. Tankersley, president; Thomas M. Bowen, secretary. Office of the company, Del Norte, Rio Grande County, Colo.

Cropsey's mill was erected by Col. A. J. Cropsey during the present season, to commence operations in the early part of July, 1877. The structure is 32 by 60 feet, substantially built of logs. It has four batteries of six stamps each. Weight of stamps, 500 pounds; fall, 15 inches; drop rate, 30 per minute; capacity, 20 tons daily. Water supplied to batteries and pans by a large-sized Knowles steam pump. Engine, 25 horsepower. The four silvered table plates have an aggregate surface of 160 square feet. Blanket tailings run through Bartola pans. The mill was built for custom work and has up to this time been engaged in sampling the ores of various mines.

The foregoing comprises the main points of the developments in the district. Other mines of more or less prominence are the Chicago, located in 1874; present owners, John B. Hoffy, W. W. Park, and J. W. Harris; the Dexter, located October, 1873, by Josiah Mann, Arthur Burton, and John A. McDonald; the Golden Eagle, located May, 1875, by Jos. S. Reef, Josiah Mann, and Peter Beeker; the Highland Mary, located July, 1875, by Josiah Mann, P. Beeker, J. S. Reef, and A. J. Sparks; The Missionary, located by Benjamin Burroughs, June 18, 1874; the Yellow Jacket, Rising Sun, Caribou, Little Jessie, Little Nellie, Goldie May, Mountain Queen, Wisconsin, Poorman, Des Moines, Esmond, Ellen, Odin, Centennial, Princess, Aurora, Narrow Gauge, Grey Eagle, Moltke, Tender Foot, Queen Esther, David Fulton, John J. Crooke, Captain Charley, Golden Star No. 1, Annie E. Benson, McCormick, Independence St. Louis, Amazon, St. Mary's, Washington, Columbia, and Major, from the last of which have been taken the finest specimens of free gold yet yielded by Summit district.

The population of the district is about 200. Del Norte, 27 miles distant, to which access is had by a wagon road built this year and by a trail down Pinos Creek, is its supply point. Besides the mills and their outbuildings, there are about 50 cabins in the settlement, built of logs and covered with dirt. Freight to Del Norte varies from 1½ to 10 cents a pound.

Wood is the only fuel. Cost of lumber, $30 per M; potatoes 1 to 8 cents per pound; flour, $8.75 per hundred; tea, $1.25 to $1.50 per pound; beef, 7½ cents per pound; bacon, 22½ cents per pound; sugar, 18 to 20 cents per pound; onions, 12½ cents per pound; dried apples, dried peaches, and dried currants, 20 cents per pound; rice, 20 cents per pound; crackers, 20 cents per pound; cheese, 30 cents per pound; kerosene, $1 per gallon. All supplies must be laid in before winter opens. No raising of vegetables has yet been attempted in the district, though it is possible that a very few of the hardier kinds might succeed. There have been so far only two deaths, one from a blast and one from debility—an invalid. Water boils at 182°.

The two accompanying sketches show, one the mountain and camp, as seen from the northwest, the other the network of locations covering a portion of the northern and eastern face of South Mountain. Although 2,300 claims have been recorded within its limits, the district so far has been orderly and peaceable. The sketch of the mountain and camp is by Henry Learned, associate of the Chicago Academy of Design, and the mine map by J. F. Sanders, United States deputy mineral surveyor. There are at present 89 stamps in the Summit and one arrastre. Two more mills are under contract. With the reduction appliances now on the spot or coming, the rank of the district as a gold-producing territory will soon be definitely established.

In his report for 1873 Raymond says: [84]

Conejos County: The San Juan country.—Concerning this region a great deal has been said and written in the newspapers of the Territory during 1873.

The Annie, a small mine in Summit district, west of Bakers Park, has become famous for rich specimens. During the summer there was a report of its sale at a high price, on the strength of such indications of value.

The figures for 1873 to 1880 given in the table represent estimates from the data quoted and are obtained by the elimination of other county figures from the total of the San Juan country.

In his report for 1874 Raymond says: [85]

The San Juan country.—* * * Of the mines on the eastern slope, those in Summit and adjacent districts are the most promising. What is known as the South Mountain appears to be the center of a broad belt of mines, mainly auriferous, of which the Little Annie may be taken as a type. This location is on a broad and well-defined vein of quartz carrying free gold and pyrites and has already, despite the small amount of work done on it, proved of great value. The neighboring veins appear to be in almost all respects similar, and the district on the whole seems of unusual promise.

Burchard in his report for 1881 says: [86]

Rio Grande County.—* * * There is but one mining district in the county, Summit district, which includes North and South Mountains and Lookout Peak. This district has been organized for some years, but it was only during 1881 that deposits of ores of extraordinary richness were discovered. The formation is porphyry and granite, and much of the ore is free gold bearing and readily reduced by the ordinary stamp process.

In the midst of a group of the principal mines of the district is the Little Ida, the property of the San Juan Consolidated Mining Co. At the depth of about 100 feet the vein is reported to be 17 feet wide, 10 feet of which is a decomposed quartz

[84] Raymond, R. W., op. cit. for 1873, p. 313, 1874.
[85] Raymond, R. W., op. cit. for 1874, pp. 358-385, 1875.
[86] Burchard, H. C., Report of the Director of the Mint upon the production of the precious metals in the United States during the calendar year 1881, pp. 354, 424-426, 1882.

yielding from $1,500 to $2,500 per ton. In about four months after encountering this rich deposit $250,000 was taken out. Arrangements are being made to develop the mine at a greater depth by tunnels and also put it in better shape for continuous working.

The Del Norte mine is adjoining the Little Ida on the northwest and is a continuation of the same vein. It is owned by the Little Annie Mining Co., which also owns the Little Annie and several other promising mines. In the Del Norte the vein is reported to be larger than in the Little Ida, the workings, however, being 75 feet deeper than in the latter.

The Golden Vault is a promising mine owned by the Iowa & Colorado Consolidated Mining Co., which also owns a number of other valuable claims. A tunnel is being run to intersect the Golden Vault mine, which will be cut within about 700 feet from the mouth of the tunnel. In running into the mountain, a vein of good ore was met, which will probably be exploited. The Golconda Co. are running a tunnel, now in 200 feet, to strike the Golconda lode.

At the Aztec mine only development work was done during 1881, no mill runs of the ore having been made. The mine has been put into good condition for the extraction of a large amount of ore during 1882.

Near the foot of the mountain is the Missionary mine, owned by B. Burroughs, of Quincy, Ill. The ore is refractory and requires a preparatory roasting before it can be milled.

A new location on South Mountain is the Stars and Stripes, which shows free gold-bearing ore.

John H. Bush & Co. have made a number of locations in the year past upon the belt extending south from South Mountain to Lookout Peak, and the similar formation, the same character of quartz, and other indications cause them to believe that development will show that they have the same rich ores as are found on the north slope of South Mountain.

The yield of the principal mines during the portion of the year that they were in operation was $289,000. It is probable that during the coming season a vast amount of prospecting will be done and that the development of locations already made will show that this county will be an important producer of gold in the future.

At Summitville, the chief town of the district, are located the mills of the different companies, all stamp amalgamation mills. The size of each mill is indicated by the number of stamps, as follows: The San Juan Consolidated, 30; Little Annie, 10; Morey & Sperry, 10; Golden Queen, 10; Iowa & Colorado Consolidated, 10; Missionary, 10; Aztec, 5; total, 85.

The Cropsey mill is situated in Cropsey Gulch, and has 24 stamps, making a total of 109 stamps.

In his report for 1882 Burchard says: [87]

The mining section is in the southwestern part of the county and is comprised in Summit district. * * *

The Ida mine is one of the first discoveries of the district as well as one of the largest producers. The ore is honeycombed quartz, heavily impregnated with iron oxide and carrying free gold in quantities ranging from $500 to $250,000 per ton. Owing to the reluctance of the management to furnish information it is impossible to give an extended description of this rich property. The company is running two stamp mills, one called the San Juan, owned by the company, and the other the Odin, working on ore from the Ida mine.

The Little Annie was also an early discovered mine, and for a time was considered to be the most valuable in the district. * * *

A 10-stamp mill was run continuously during the latter part of the year, and a 40-stamp mill will be erected during 1883.

Recently the Golden Queen mill has been leased and is now being put in repair for use by this company.

The Golden Queen was located shortly after the Little Annie and is one of the best known mines in the camp. * * * Owing to an entanglement of the affairs of this company, it remained idle during the year.

On the Aztec mine steady development has been made, and the ore taken out has paid for the work from the beginning. This company has shipped all the ore to the smelting works at Pueblo, and the mine has paid large dividends.

The Golconda comprises the Golconda, Game Cock, Boss, Len, Nick, and Tunnel, situated on South Mountain. When they are systematically developed they will become mines of prominence.

The Iowa & Colorado Consolidated Co.'s property is of the best character. The following description of the very complete mill of this company is from the San Juan Prospector:

The entrance to the mill is through the engine room. Here are two large boilers and two 50-horsepower engines, which furnish the power to drive the vast and complicated machinery. In the rear of the boilers is placed a No. 5 Knowles plunger pump, which furnishes the water for the boilers and batteries and for fire protection.

On the central floor are located the stamps, 50 in number, 20 of which are 650 pounds in weight, ten 700 pounds, and ten 850 pounds. In front of the batteries are 8 tables, each having 2 silver plates, and in the rear are 8 Tullup automatic feeders which supply the batteries with ore from the ore bins.

The floor below is largely fitted up as a silver mill, being the only mill of this character in this district. Its object is to save the silver, with which much of the Summit ore is impregnated.

There are here four grinding pans and two immense settlers, also three "Bee jigs" for concentrating purposes. These jigs will concentrate 15 tons of tailings into one, and the concentrations will be shipped to a smelter. Alongside of the grinding pan is a clean-up funnel, into which the amalgam is placed as it comes from the plates. Above the stamp floor is the ore bin, which has a capacity of 300 tons; and above the ore bin is the Iowa tunnel and the tramway. The tramway is 3,665 feet in length, and is supported by 36 trusses, having its upper limit at the mouth of the Iowa tunnel. It has 84 buckets, which carry 100 pounds of ore at a trip, and its capacity is 150 to 200 tons of ore per day. The ore will be crushed at the mine by a Blake crusher, which has a crushing capacity of 150 tons per day. The power to drive the crusher and also to ventilate the mine is furnished by a 25-horsepower engine. It will be seen that the ore is crushed at the mine and carried direct to the stamps by power, thus saving all handling or moving. In fact it has been the design in constructing this mill to do it in such a manner as to save by automatic power all the manual labor possible, thus saving expense. Near the mill is a retort house, 24 by 40, in which is located the assaying department of the company. Here also is located a small smelting furnace, constructed for testing the ores of the company's mines.

In his report for 1883 Burchard says: [88]

The only mine in the county in regard to which any information has been obtained is the Golconda, located on the summit between More Creek and Crooked River. Some 80 tons of ore have been produced during the year by it, averaging $50 to the ton.

In his report for 1884 Burchard gives the following details. [89]

[87] Burchard, H. C., op. cit. for 1882, pp. 391, 394, 538-539, 1883.

[88] Burchard, H. C., op. cit. for 1883, pp. 240, 401, 1884.
[89] Burchard, H. C., op. cit. for 1884, pp. 177, 236-237, 1885.

The Aztec Co. owns considerable property, the principal being the Aztec mine, a claim 50 by 1,500 feet, the oldest location in the camp and originally known as the Summit lode. The ore is a decomposed jasper, carrying black sulphurets, free gold, and tellurium. The mine has shipped ore for over two years that has averaged over $200 per ton. The Aztec mine has three shafts and two tunnels and is better developed than any mine on that portion of South Mountain.

The Iowa & Colorado Mining Co. have been idle for a long time. This company has 19 patented claims, with little work done upon them aside from that necessary to secure a patent. The main Iowa tunnel, started to cut the Parole lead, is in about 1,300 feet. This property was sold at trustee's sale and was bought in trust for the bondholders, for $67,838.15, being the amount of all their bonds, with interest to date of sale and costs and expenses of sale.

On the property of the Golconda Co. three leads have been struck in tunnel No. 1. In tunnel No. 2 fourteen blind leads have been struck, and the tunnel has not yet accomplished its design.

The report of the Golconda Co. shows the indebtedness of the company to be $106,000.

The mill-run of 1884 was far better than was anticipated. The ore was taken principally from the middle drift and stopes, mining southeasterly from the shaft, which connects the two tunnel levels, and was taken out of a width of about 6 feet. This ore ran over $18 per ton by raw amalgamation and produced gold 0.816 fine at the United States Mint at Denver, which returned to the company, in cash, from this run, $2,800.43. Besides this ore, a small lot of about 4 tons was run in one battery of 5 stamps from the vein under the ore shaft of No. 1 tunnel. This ore ran nearly, if not quite, as well as that from the middle drift, and the proceeds are included in the amount above named. Tests of ore from other parts of the mine were made, all showing up reasonably well. Returns from the Omaha and the Grant smelter, at Denver, show the concentrates to contain, per ton, 5 ounces of silver and 40 ounces of gold.

The Little Annie Consolidated Mining & Milling Co. own about 70 acres in claims and millsites.

There will be considerable work done at Summitville during the present winter, it now being the intention of the Annie Co. to run their mine and mill to the full capacity during the cold months. The mine has been cleaned out and placed in shape for working and employs a small force of miners. The ore is crushed and loaded into a bucket tramway at the mine and carried to the tramway house on top of the mill, where the ore is dumped into a series of chutes that carry it to any of the 12 batteries of 5 stamps each. Here, through self-feeders, the rock is pounded up and washed out over the battery plates, which catch a large per cent of the gold, attracted by quicksilver. The tailings are carried on to the vanner house, where 16 Frue vanners are working and form the closing process in catching the gold. The per cent of gold saved by this company is calculated at about 90. From the mill and vanner house the amalgam is taken to the retort house, where the quicksilver is extracted and a gold brick cast.

The Annie Co. is at present the only company in the camp running a mill, though considerable work is done in the mines by other parties. The Annie mill has a 125-horsepower Corliss engine. The ore of the Annie Co. is being handled at a cost of about $3 per ton. The gold will average 0.900 fine and is worth $19 [?] per ounce.

For 1885 the figures given are interpolated to correspond with the total production of the State.

For 1886 to 1896 the figures given are based on reports of the agents of the mint in annual reports of the Director of the Mint, the gold and silver being prorated to correspond with the figures for the total production of the State as corrected by the Director of the Mint, the lead being prorated to correspond with the total production of lead for the State as given in Mineral Resources, and any unknown production in the State being distributed proportionately among the counties. As with lead so with copper, but as the figures for copper given in Mineral Resources include copper from matte and ores treated in Colorado, though produced in other States, they are subject to revision.

For 1887 and 1888 Munson [90] lists the producing and nonproducing mines and gives individual reports of the producing mines.

In his report for 1889 Smith [91] lists the producing mines and gives their individual outputs. For 1890 he gives the total of the confidential reports of individual mines. For 1891 he estimates the production of the county. He also says:

The Mammoth mine at Platoro, in Rio Grande County, and the Golconda, at Summitville, are doing excellent work. The properties of the Little Annie Co., in litigation, promise to be released at an early date, and there is a probability of railway construction to Summitville camp.

For 1892 Smith [92] gives the individual production of the Ethel and Holy Moses mines, in Mineral County, and the total of the confidential reports from Rio Grande County.

For 1897 to 1904 the figures given in the table, which represent smelter and mint receipts, are taken from the reports of the Colorado State Bureau of Mines.

For 1905 to 1923 the figures are taken from Mineral Resources (mines reports).

[90] Munson, G. C., agent for Colorado, in Kimball, J. P., Report of the Director of the Mint upon the production of the precious metals in the United States during the calendar year 1887, pp. 180–181, 1888; idem for 1888, p. 120, 1889.

[91] Smith, M. E., agent for Colorado, in Leech, E. O., Report of the Director of the Mint upon the production of the precious metals in the United States during the calendar year 1889, p. 152, 1890; idem for 1890, p 139, 1891; idem for 1891, pp. 174, 183, 1892.

[92] Smith, M. E., agent for Colorado, in Leech, E. O., op. cit. for 1892, p. 129, 1893.

Gold (placer and lode), silver, copper, and lead produced in Rio Grande County, 1870–1923

Year	Ore treated (short tons)	Gold Placer	Gold Lode	Gold Total	Silver Fine ounces	Silver Average price per ounce	Silver Value	Copper Pounds	Copper Average price per pound	Copper Value	Lead Pounds	Lead Average price per pound	Lead Value	Total value
1870			(a)											(a)
1873			$2,000	$2,000										$2,000
1874			5,000	5,000										5,000
1875			272,044	272,044	7,734	$1.24	$9,590							281,634
1876			121,148	121,148	7,734	1.16	8,971							130,119
1877			195,337	195,337	7,734	1.20	9,281							204,618
1878			102,866	102,866	7,734	1.15	8,894							111,760
1879			28,500	28,500	7,734	1.12	8,662							37,162
1880			6,000	6,000										6,000
1881			290,000	290,000	7,734	1.13	8,739							298,739
1882			210,000	210,000	15,469	1.14	17,635							227,635
1883			180,000	180,000	7,734	1.11	8,585							188,585
1884			130,000	130,000	10,828	1.11	12,019							142,019
1885			130,000	130,000	9,800	1.07	10,486							140,486
1886			149,266	149,266	8,817	.99	8,729							157,995
1887		$5,000	117,380	122,380	7,992	.98	7,832							130,212
1888		2,000	14,260	16,260	2,923	.94	2,748							19,008
1889			35,760	35,760	3,757	.94	3,532							39,292
1890			25,716	25,716	1,287	1.05	1,351							27,067
1891			38,592	38,592	7,752	.99	7,674							46,266
1892			14,487	14,487	12,526	.87	10,898							25,385
1893					796	.78	621							621
1894			16,816	16,816	1,260	.63	794							17,610
1895			15,795	15,795	3,359	.65	2,183							17,978
1896			1,870	1,870	1,353	.68	920	1,369	$0.108	$148	451	$0.03	$14	2,952
1897			22,592	22,592	8,168	.60	4,901	627	.12	75	12,006	.036	432	28,000
1898			3,720	3,720	1,568	.59	925	9,794	.124	1,214	2,393	.038	91	5,950
1899			19,202	19,202	2,718	.60	1,631	336	.171	57	1,635	.045	74	20,964
1900			107,629	107,629	3,075	.62	1,906	8,599	.166	1,427	26,260	.044	1,155	112,117
1901			32,927	32,927	6,926	.60	4,156	65,603	.167	10,956	677	.043	29	48,068
1902			14,262	14,262	3,171	.53	1,681	1,260	.122	154	166	.041	7	16,104
1903			12,939	12,939	3,410	.54	1,841	5,098	.137	698				15,478
1904			4,010	4,010	2,281	.58	1,323	650	.128	83				5,416
1905			4,051	4,051	1,055	.61	644	123	.156	19				4,714
1906	70		8,580	8,580	152	.68	103							8,683
1908	9		764	764										764
1910	12		1,306	1,306	61	.54	33	87	.127	11	250	.044	11	1,361
1912	133		5,549	5,549	896	.615	551	29,673	.165	4,896	313	.045	14	11,010
1913	6		243	243	109	.604	66	568	.155	88				397
1914	8		474	474	16	.553	9							483
1915	1,500		14,968	14,968	325	.507	165							15,133
1917	16		24	24	52	.824	43				1,930	.086	166	233
1923	17		1,662	1,662	161	.82	132	218	.147	32	929	.07	65	1,891
		7,000	2,357,739	2,364,739	176,201		170,254	124,005		19,858	47,010		2,058	2,556,909

a Unrecorded.

ROUTT AND MOFFAT COUNTIES

The output of Routt and Moffat counties from 1866 to 1872 and in 1874, whatever it may have been, is possibly included in the placer production of Summit County.

George and Crawford say:[93]

The Hahns Peak placers were discovered by Joseph Hahn in 1865 and active work was begun on them in 1866. While the activity with which the placers have been worked has varied, probably not a year has passed in which there has not been some gold taken. The total production of the region is a matter of uncertainty and will never be known. The estimates are the merest guesswork, as is shown by the fact that they range from $200,000 to $15,000,000. The more conservative estimates place the amount at from $200,000 to $500,000.

For 1873 Raymond says:[94]

The Hahns Peak placer camp is situated between Elk and Snake rivers. The gold is worth $15 per ounce. About 30 men found employment there in the summer of 1873.

In his report for 1875 Raymond[95] gives the following details:

Poverty Flat.—In the districts bordering on North Park, in the gulches flowing from Hahns Peak, some very promising placers are being opened. Gold was first discovered in this camp by Sam Conger, of Caribou fame, in the summer of 1869, and, without causing much of an excitement, the district last year developed into what will without doubt prove next year a very valuable addition to our mineral lands. Two large companies, the Hahns Peak and the Purdy, have absorbed, by location and purchase, the best parts of the district, owning between them about 1,250 acres, located on Poverty Flat and Ways Gulch. Their combined capital is $6,000,000, and the total amount of improvements up to date have cost in the neighborhood of $100,000. Twenty miles of ditche have been dug, and it is proposed this year to build several new ones of a total length of 12 miles, which will bring 5,000 inches of water additional into Ways Gulch.

Poverty Flat is on Beaver Creek, one of the tributaries of the Elk River. The elevation of the Hahns Peak Co.'s claim, which is on this flat, is about 8,000 feet. The ground is from 10 to 60 feet deep, of gravel, free from large boulders, and pays well through its entire depth. Last year but one run of 25 days with one hydraulic was made, which washed out $3,500 in gold. The company is now arranging to work 12 hydraulics and has a constant water supply equal to about 2,000 inches. Five reservoirs have been built, by which means the supply can be maintained quite late in the year.

The Purdy Co., located in Ways Gulch, will put in this year a bedrock flume and probably will not take out much gold till 1877. The ground is as good as that of the Hahns Peak Co., but not so well located as to water. Hence the necessity of the 12-mile ditch from the Elk.

Several very promising gold-bearing lodes have been discovered at Hahns and Ways Peak, upon which work will be done this year. It is expected that the year's work will de-

[93] George, R. D., and Crawford, R. D., The Hahns Peak region, Routt County, Colo.: Colorado Geol. Survey First Rept., 1908, p. 221, 1909.
[94] Raymond, R. W., Statistics of mines and mining in the States and Territories West of the Rocky Mountains for 1873, pp. 300, 303, 1874.
[95] Raymond, R. W., op. cit. for 1875, pp. 318–319, 1877.

velop a splendid gold district in Grand County, which will have the effect not only to draw an agricultural population to the valleys of the Bear and Grand but to incite further explorations in the numerous mountain chains around the North and Middle parks, which are as likely to be rich in minerals as any in the Territory.

The Rabbit Ear mines [Routt, Grand, and Jackson counties].— In the range of mountains dividing Middle from North Park are the Rabbit Ear mines, about which but little can be said, except that the prospects are fair. The mines are numerous, large, and easily worked. They carry mainly silver ore, and some of them show good veins of argentiferous galena. But little development has been made, no workings having been extended deeper than 50 feet, and the majority of the claims being only sunk 10 feet. Numerous companies have been organized and several tunnel schemes proposed, and next season will demonstrate whether the discoveries are of any value. The camp is 60 miles from the nearest reducing works. It is proposed to put up a mill in the district next year, should the supply of ore warrant it.

Routt County was established by an act of the legislature approved January 29, 1877.[96]

For 1877 to 1880 no records of production have been found.

For 1881 Burchard gives the following details:[97]

Routt, the extreme northwestern county of the State, makes but a small addition to the production of precious metals in Colorado. For many years it has been the favorite hunting ground of the Ute Indians, who have jealously and successfully guarded it against the encroachments of prospectors—the advance guard of a mining population.

During the fall of 1881 silver-bearing ore was found in the Gore Range, and also in the mountains about Steamboat Springs, but no reliable reports have been received of the quantity or quality.

Placer mining has been conducted for some years, with varying results, in the vicinity of Hahns Peak, in the northeastern part of the county. Most of the placers have been worked by two companies, the International and the Hahns Peak Mining Co., the latter, under the superintendency of Messrs. McIntosh & Cody, having been quite successful during the year.

It is probable that an increased force of men will be employed in placer mining during 1882, and that, as the Indians have been removed, systematic prospecting for gold and silver bearing lodes will be conducted in the many mountain ranges traversing the county.

The only report of production received from Routt County was from the Hahns Peak mine.

The total production of the county I estimate at $20,000 in gold.

For 1882 the figures given in the table represent Burchard's estimate.[98] In his report for 1883 he says:[99]

Routt County is more devoted to agricultural pursuits than mining, but some mining has been done each year, principally on the Elk River and its forks, and considerable gold is annually produced. On Poverty Bar a claim worked by Judge McIntosh is said to have taken out some $16,000 during the past year. Carruthers & Hinman, who have a claim at the upper end of Poverty Bar, have done very well during the year. Owing to scarcity of water these claims can only be worked about two months in the spring. Mr. Hutchinson has continued work on the Purdy Ground, and Mr. Wood has some rich ground, about 300 acres, on a wash called String Ridge. I have estimated, from the reports furnished by the miners of this section, the production of the county during the calendar year 1883 to have been in gold $40,000.

For 1884 the figures are taken from Burchard's report.[1]

For 1885 the figures have been interpolated. The deposits at the Denver Mint amounted to $23,104.

For 1886 the figures are taken from Munson's report.[2]

For 1887 to 1896 the figures given are taken from reports of the agents of the mint in annual reports of the Director of the Mint, the gold and silver being prorated to correspond with the figures for the total production of the State as corrected by the Director of the Mint, the lead being prorated to correspond with the total production of the State as given in annual volumes of Mineral Resources, and any unknown production in the State being distributed proportionately among the counties. As with lead, so with copper, but as the figures for copper given in Mineral Resources include copper from matte and ores treated in Colorado, though produced in other States, they are subject to revision.

The report of the Director of the Mint for 1889 gives no production for Routt County, but $8,870 in gold and 189 ounces of silver were deposited at the Denver Mint from mines in the county.

The report of the Director of the Mint for 1890 gives no production for Routt County, but $8,133 in gold and 176 ounces of silver were deposited at the Denver Mint from mines in the county.

For 1897 to 1904 the figures, which represent smelter and mint receipts, are taken from the reports of the Colorado State Bureau of Mines. In its report for 1897 the Bureau says:[3]

The clean-up for work done on the north side of the peak [Hahns] in the winter of 1896–97 is reported to have yielded over $3,600 for four months' work of three men. * * * During the year several small shipments have been made that yielded a fair profit notwithstanding $30 freight charge with smelter charge added. The main value is in silver and lead.

For 1897 Hodges[4] gives $5,451 deposited at the Denver Mint from Routt County; for 1898, $11,728; for 1899, $10,693; for 1901, $927; for 1902 Downer gives $13,845; for 1903, $19,289 in gold and 30 ounces of silver; for 1904, for the total production of the county, he gives $22,164 in gold and 85 ounces of silver, of which $20,648 was gold deposited at the Denver Mint.

96 Colorado State Bur. Mines Report for 1897, p. 103, 1898.
97 Burchard, H. C., op. cit. for 1881, pp. 354, 425–426, 1882.
98 Burchard, H. C., op. cit. for 1882, pp. 395, 587, 1883.
99 Burchard, H. C., op. cit. for 1883, pp. 240, 401–402, 1884.

1 Burchard, H. C., op. cit. for 1884, p. 177, 1885.
2 Munson, G. C., agent for Colorado, in Kimball, J. P., Report of the Director of the Mint upon the production of the precious metals in the United States during the calendar year 1886, p. 177, 1887.
3 Colorado State Bur. Mines Rept., 1897, p. 104, 1898.
4 Hodges, J. L., agent for Colorado, 1897–1901, and Downer, F. M., agent for 1902–1904, in Roberts, G. E., Report of the Director of the Mint upon the production of the precious metals in the United States during the calendar year 1897, p. 127, 1898; idem for 1898, p. 99, 1899; idem for 1899, p. 122, 1900; idem for 1901, p. 147, 1902; idem for 1902, p. 132, 1903; idem for 1903, p. 81, 1904; idem for 1904, p. 123, 1905.

The figures for 1905–1922 are taken from Mineral Resources (mine reports), published by the United States Geological Survey. The figures for 1906 to 1910 represent the output of the Hahns Peak (Routt County) and the Lay and Fourmile (Moffat County) districts, except for 1908, when the Lay district is not included. For 1912 to 1914 Routt and Moffat counties are combined. In 1912 the output of copper came from Douglas Mountain, in Moffat County, and in 1913 the copper and lead from Hahns Peak, in Routt County.

In his report for 1881 Burchard says: [7]

Saguache County as a mining section is not more than two years old, the first location of a mineral-bearing lode being made on Kerber Creek in May, 1880. It was called the Exchequer.

He also gives descriptions of the Exchequer, Empress Josephine, Boss Mammoth, Arkansas, Revenue, Silver King, Townsend, Rawley, Rover, Whale, Superior, Lawrence, Albert, Bonanza, and other mines.

Small quantities of lead were possibly produced from 1882 to 1886.

Gold (placer and lode), silver, copper, and lead produced in Routt and Moffat counties, 1866–1922

Year	Ore treated (short tons)	Gold			Silver			Copper			Lead			Total value
		Placer	Lode	Total	Fine ounces	Average price per ounce	Value	Pounds	Average price per pound	Value	Pounds	Average price per pound	Value	
1866–1872		(a)		(a)										
1873		$26,000		$26,000										$26,000
1875		3,500		3,500										3,500
1881		20,000		20,000										20,000
1882		15,000		15,000										15,000
1883		40,000		40,000										40,000
1884		13,000		13,000										13,000
1885		23,000		23,000										23,000
1886		16,840		16,840	387	$0.99	$383							17,223
1887		6,714		6,714	214	.98	210							6,924
1889		8,870		8,870	189	.94	178							9,048
1890		8,133		8,133	176	1.05	185							8,318
1891		13,561		13,561										13,561
1892		560		560										560
1893		6,216		6,216										6,216
1894		8,944		8,944	97	.63	61							9,005
1895		5,930		5,930	86	.65	56							5,986
1896		4,690	$169	4,859	2,214	.68	1,506				22,111	$0.03	$663	7,028
1897		5,451	4,326	9,777	7,805	.60	4,683	958	$0.12	$115	88,736	.036	3,194	17,769
1898		11,728	1,025	12,753	2,173	.59	1,282	600	.124	74	15,477	.038	588	14,697
1899		10,693	862	11,555	1,271	.60	763				3,405	.045	153	12,471
1900		b 3,000	287	3,287	477	.62	296	5,765	.166	957				4,540
1901		927	3,517	4,444	239	.60	143	500	.167	84	2,193	.043	94	4,765
1902		13,845	1,306	15,151	136	.53	72							15,223
1903		19,289	1,546	20,835	117	.54	63							20,898
1904		22,164	2,061	24,225	181	.58	105							24,330
1905		6,905		6,905	30	.61	18							6,923
1906		6,951		6,951	42	.68	29							6,980
1907	3	4,908	101	5,009	429	.66	283							5,292
1908	4	4,858	349	5,207	1,242	.53	658							5,865
1909	24	2,418	943	3,361	3,446	.52	1,792							5,153
1910		6,689		6,689	48	.54	26							6,715
1911		6,115		6,115	47	.53	25							6,140
1912	64	5,070		5,070	150	.615	92	25,085	1.65	4,139				9,301
1913	12	3,609	231	3,840	1,962	.604	1,185	161	1.55	25	1,023	.044	45	5,095
1914		4,697		4,697	16	.553	9							4,706
1915		2,984		2,984	12	.507	6		.175					2,990
1916	512	1,124	18	1,142	278	.658	183	41,175	.246	10,129				11,454
1917	75	2,359	1,056	3,415	1,341	.824	1,105	4,326	.273	1,181				5,701
1918	161	3,040	698	3,738	2,671	1.00	2,671		.247		6,591	.071	468	6,877
1919	172		312	312	1,283	1.12	1,437		.186					1,749
1920	3	118	44	162	100	1.09	109		.184					271
1922	1	114		114	82	1.00	82		.135					196
		370,014	18,851	388,865	28,941		19,696	78,570		16,704	139,536		5,205	430,470

a Unrecorded. Probably included in Summit County. b Estimated by C. W. Henderson.

SAGUACHE COUNTY

The Colorado State Bureau of Mines says of the early developments in Saguache County: [5]

Saguache County * * * organized December 26, 1866. * * * The mining history practically begins with 1879–80, during the great rush to the Gunnison country. Many were attracted by the large silver-lead fissure veins on Kerber Creek, and developments seemed to justify the building of the town of Bonanza that had grown to a place of considerable importance before the fall of 1880. On the west slope of the Cochetopa mountains, the narrow veins of quartz carrying free gold likewise established the camp of Willard on Cochetopa Creek [a tributary to Gunnison River].

Patton [6] gives details of the history of the Bonanza district.

In his report for 1882 Burchard [8] gives a description of properties in Saguache County, and in his report for 1883 he says: [9]

Columbia mine * * * owns stamp mill (which is now being erected). * * * The Exchequer; * * * a great deal of the ore * * * is free milling, carrying from 50 to 90 ounces silver per ton. * * * The Rawley * * * is one of the most productive mines in the county. * * * Shipments are regular from the Rawley. * * * The Empress Josephine is probably the largest producer in the county. * * *

The Crystal Hill district (Carnevo mining camp) mines [described]. * * *

Near the town of Crestone the most promising properties are the North Andover, Lion Mountain, and Garfield.

5 Colorado State Bur. Mines Rept. for 1897, p. 106, 1898.
6 Patton, H. B., Geology and ore deposits of the Bonanza district, Saguache County: Colorado State Geol. Survey Bull. 9, 1915.
7 Burchard, H. C., Report of the Director of the Mint upon the production of the precious metals in the United States during the calendar year 1881, pp. 354, 426–427, 1882.
8 Burchard, H. C., op. cit. for 1882, pp. 395, 539–545, 1883.
9 Burchard, H. C., op. cit. for 1883, pp. 240, 402–406, 1884.

In his report for 1884 Burchard says:[10]

The old smelter at Bonanza has been idle, owing to its inability to properly treat the ores found here; but during the last few months some capitalists have remodeled the United States mill, at Parkville, into a smelter. * * *

The Paragon has shipped but a small amount of ore, which was concentrated very successfully with the hand jigs in use at the Empress Josephine.

For 1885 the figures have been interpolated.

For 1887 to 1896 the figures given have been taken from reports of the agents of the mint in annual reports of the Director of the Mint, the gold and silver being prorated to correspond with the figures for the total production of the State as corrected by the Director of the Mint, the lead being prorated to correspond with the total production of lead for the State as given in Mineral Resources, and any unknown production in the State being distributed proportionately among the counties. As with lead, so with copper; but as the figures for copper given in Mineral Resources include copper from matte and ores treated in Colorado, though produced in other States, they are subject to revision.

For 1889, 1892, and 1893 the figures, which have been interpolated, are of doubtful value. No record

is available of the location of placers that made the large output credited to lode gold in 1894.

For 1897 to 1904 the figures which represent smelter and mint receipts are taken from the report of the Colorado State Bureau of Mines.

In his report for 1900 Hodges says:[11]

This county is divided into three districts. * * *

Bonanza district.—* * * During the year two concentrating mills, each having a capacity of 50 tons per day, have been erected and successfully operated in the Bonanza district.

Crestone district.—In the southeastern part of the county is located the Crestone district. The old Baca grant, in litigation for a quarter of a century, has passed to a wealthy syndicate, which immediately prepared to develop its mineral resources. A large stamp mill was constructed and equipped with the latest appliances. Its daily capacity is 300 tons, but only 50 are now being treated. * * * Other mills are contemplated, and a railroad is now being constructed into the district.

Embargo district.—In the Embargo district, in the southwestern part of the county, considerable work has been done on old claims. The ores are heavy in iron sulphides and suitable for treatment at the matting smelter that will be erected the coming year.

For 1905 to 1923 the figures are taken from Mineral Resources (mines reports).

[11] Hodges, J. L., agent for Colorado, in Roberts, G. E., op. cit. for 1900, p. 131, 1901.

[10] Burchard, H. C., op. cit. for 1884, pp. 177, 238–239, 1885.

Gold, silver, copper, lead, and zinc produced in Saguache County, 1880–1923

	Ore treated (short tons)	Lode gold	Silver Fine ounces	Silver Average price per ounce	Silver Value	Copper Pounds	Copper Average price per pound	Copper Value	Lead Pounds	Lead Average price per pound	Lead Value	Zinc Pounds	Zinc Average price per pound	Zinc Value	Total value
1880			$7,734	$1.15	$8,894										$8,894
1881			30,938	1.13	34,960										34,960
1882		$10,000	77,344	1.14	88,172										98,172
1883		5,000	77,344	1.11	85,852										90,852
1884		1,000	77,344	1.11	85,852										86,852
1885		1,000	55,920	1.07	59,834										60,834
1886		3,936	55,920	.99	55,361										59,297
1887		756	7,196	.98	7,052				12,582	$0.045	$566				8,374
1888		4,220	36,101	.94	33,935				180,272	.044	7,932				46,087
1889									*200,000	.039	7,800				7,800
1890		1,745	11,988	1.05	12,587	4,290	$0.156	$669	176,193	.045	7,929				22,930
1891		1,422	21,285	.99	21,072	68,047	.128	8,710	260,577	.043	11,205				42,409
1892									*250,000	.04	10,000				10,000
1893									*250,000	.037	9,250				9,250
1894		17,515	608,224	.63	383,181				*250,000	.033	8,250				408,946
1895		534	3,939	.65	2,560				249,166	.032	7,973				11,067
1896		331	2,447	.68	1,664	241	.108	26	65,465	.03	1,964				3,985
1897		13,746	2,482	.60	1,489	2,975	.12	357	9,266	.036	334				15,926
1898		19,678	2,618	.59	1,545	21,711	.124	2,692	132,462	.038	5,034				28,949
1899		3,886	14,306	.60	8,584	35,319	.171	6,040	441,095	.045	19,849				38,359
1900		7,979	15,793	.62	9,792	16,129	.166	2,677	316,061	.044	13,907				34,355
1901		79,972	20,507	.60	12,304	15,253	.167	2,547	235,750	.043	10,137				104,960
1902		5,023	10,486	.53	5,558	13,669	.122	1,668	454,995	.041	18,655	267,100	$0.048	$12,821	43,725
1903		2,956	22,424	.54	12,109	67,410	.137	9,235	376,711	.042	15,822	44,600	.054	2,408	42,530
1904	499	5,519	60,566	.58	35,093	48,722	.128	6,236	699,312	.043	30,070	15,585	.051	795	77,713
1905	496	699	4,401	.61	2,685	1,135	.156	177	203,797	.047	9,578	2,917	.059	172	13,311
1906	999	7,628	737	.68	501		.193		49,141	.057	2,801	74,302	.061	4,532	15,462
1907	170	649	6,194	.66	4,088	1,260	.20	252	22,528	.053	1,194				6,183
1908	76	610	955	.53	505	76	.132	10	27,715	.042	1,164				2,289
1909	192	1,196	2,260	.52	1,175	3,769	.13	490	83,463	.043	3,589				6,450
1910	296	1,025	4,841	.54	2,614	5,362	.127	681	161,068	.044	7,087				11,407
1911	184	512	4,664	.53	2,472	4,984	.125	623	74,556	.045	3,355	46,561	.057	2,654	9,616
1912	9,459	3,805	19,309	.615	11,875	29,479	.165	4,864	504,845	.045	22,718	534,928	.069	36,910	80,172
1913	980	4,243	8,694	.604	5,251	13,277	.155	2,058	336,886	.044	14,823	32,964	.056	1,846	28,221
1914	1,488	16,513	18,293	.553	10,116	35,783	.133	4,759	534,872	.039	20,860	8,941	.051	456	52,704
1915	692	5,273	11,266	.507	5,712	23,360	.175	4,088	174,447	.047	8,199	44,250	.124	5,487	28,759
1916	3,338	8,024	48,959	.658	32,215	92,581	.246	22,775	255,419	.069	17,626				80,640
1917	4,224	10,350	76,016	.824	62,637	144,978	.273	39,579	310,686	.086	26,719				139,285
1918	1,716	2,553	89,510	1.00	89,510	96,866	.247	23,906	108,253	.071	7,686				123,675
1919	509	817	37,767	1.12	42,299	36,344	.186	6,760	52,515	.053	2,783				52,659
1920	9,282	5,031	94,655	1.09	103,174	88,386	.184	16,263	150,063	.08	12,005				136,473
1921	6,412	1,856	90,871	1.00	90,871	49,512	.129	6,387	198,686	.045	8,941				108,055
1922	9,671	4,849	63,542	1.00	63,542	41,622	.135	5,619	111,782	.055	6,148				80,158
1923	34,456	4,229	155,723	.82	127,693	459,477	.147	67,543	2,919,200	.07	204,344				403,809
		266,080	1,961,501		1,626,385	1,422,017		247,711	10,839,859		568,297	1,072,148		68,081	2,776,554

* Estimated by C. W. Henderson.

SAN JUAN COUNTY

Raymond describes the development in the San Juan country in 1873 in his report for that year as follows: [12]

Conejos County; the San Juan country.—Concerning this region a great deal has been said and written in the newspapers of the Territory during 1873.

The mines are located in Bakers Park, on the headwaters of the Las Animas, a tributary of the San Juan River. The country is accessible with some difficulty by way of Pueblo Del Norte and from there up the Rio Grande to the divide, which has to be crossed into Bakers Park. Some of the mines, such as the Little Giant, are spoken of as rich gold veins, while the majority are lead veins, rich in silver. Considerable quantities of the Little Giant ore have been worked in a small and imperfect 5-stamp mill, which is reported to have given a very satisfactory yield.

The coming summer will probably witness great activity in the San Juan country. I have no doubt smelting works will be erected there; I have knowledge of several parties who intend to put such schemes into operation immediately. The product for 1873 was about $15,000, of which about $12,000 was from the Little Giant, the only developed mine. Probably $2,000 was brought away from the district in specimens.

For 1874 the production of the San Juan country has been divided between San Juan, Rio Grande, and Hinsdale counties. Raymond gives the following details in his report for that year: [13]

The San Juan country.—The first excitement over the San Juan mines has died out, but some encouragement is given for future bona fide operators. The entire district of southwestern Colorado is now known to be very prolific of gold and silver veins, and there can hardly be a doubt that in a few years it will be proved to be a very promising mining field. So far the majority of the discoveries are of silver, and the mines bear more resemblance to those of Georgetown than to those of any other district in Colorado. * * *

The locations so far number nearly 2,000, four-fifths of which are silver lodes. A number of these are, of course, double—that is, on the same vein—so that the actual number of distinct veins discovered has probably not been over 1,200. The country is now divided into the following districts: Summit, Decatur, Alamosa, Telluric, Sangre de Cristo, Lake, Uncompahgre, Humboldt, Adams, El Dorado, Mosca, La Plata, Mancos, Animas, and Eureka.

Geologically, it is divided into two almost equal divisions by the main range of the Rocky Mountains, on the western slopes of which are the mines around and in Baker Park and those on the tributaries of the Uncompahgre and the Gunnison. East of the range are the mines clustered around the upper waters of the Alamosa and the Rio Grande del Norte.

Concerning the first the following extract from a letter received by the Georgetown Mining Review will give all that I am able to obtain of a reliable and satisfactory nature:

The mining belt of San Juan is 25 miles broad, commencing at Mineral Creek. At Mineral Creek there are a number of veins opened. These are the Bakers Park mines, and the belt extends north for a distance of about 3 miles to Hazelton Mountain, which carries the first rich belt, and although the veins are narrow (not averaging more than 3 to 5 inches of ore), a number of them are working to a good profit. The best of these are the Grey Eagle, Susquehanna, and Aspen.

The gray copper in these veins assays from 500 to 2,300 ounces per ton; the galena runs lightly, 50 to 80 ounces. A small quantity of native silver has been taken out of the Aspen.

From here to Minnie Creek the lodes are of huge dimensions, carrying galena poor in silver, but this statement has some notable exceptions. The Pride of the West, owned by two Georgetown miners, is perhaps the best lode in San Juan, the ore struck being 7 feet wide, of solid galena, interspersed with gray copper, and running well (at least in assays). The Green Mountain lode is a similar vein but has not the same true vein appearance, nor does it carry as large a body of ore as the Pride of the West. However, its gray copper assays from 800 to 1,500 ounces per ton, and the galena about 100 ounces. These lodes are located a little north of the head of Cunningham Gulch, and from here to Minnie Creek the veins are frequent, and many of them are of immense size but assay low (from 6 to 40 ounces per ton).

The mines on Minnie Creek are of large size, carrying considerable rich ore, and crossing each other at every conceivable angle. Consequently there are lots of disputes and fine chances for lawsuits. I did not take anything here, but went on to Eureka district, which, I think, will prove to be the best camp in San Juan. Here the veins have regular courses (northeast to southwest). The change is sudden and remarkable. Just south of Eureka Creek they assay light, run in all directions, and are almost numberless; while on the north side of Eureka and Niagara creeks they have a true course, are large, and about 300 feet apart.

Across the Saguache Range, on the Uncompahgre, is another good belt, on which is located the Poughkeepsie. Assays have been had out of this vein running up to 10,000 ounces per ton. It carries silver glance in considerable quantities.

Summitville district [Rio Grande County].—Of the mines on the eastern slope, those in Summit and adjacent districts are the most promising. What is known as the South Mountain appears to be the center of a broad belt of mines, mainly auriferous, of which the Little Annie may be taken as a type. This location is on a broad and well-defined vein of quartz, carrying free gold and pyrites, and has already, despite the small amount of work done on it, proved of great value. The neighboring veins appear to be in almost all respects similar, and the district on the whole seems of unusual promise.

Being located so many hundred miles away from the rest of the world, and lying under the many disadvantages caused by hard winters and almost impassable mountain barriers, the San Juan mines have as yet no market for ores, and, until this necessity is supplied, can hardly be expected to add much to the bullion product of the Territory. During last year about 25 tons of silver ore that would reach $300 to $400 per ton were shipped to various points in the East for reduction, and several minor lots have been smelted at Golden, Denver, and Blackhawk.

In Bakers Park two smelting furnaces were built in 1874, but they have not proved successful. In Summit district a stamp mill for gold quartz has been erected, and several other similar mills are contemplated.

The silver ores appear to be rather complex in composition and of that nature which the smelter denominates as "heavy." They will therefore give plenty of trouble to inexperienced metallurgists and be the cause undoubtedly of many failures. The abundance of copper, lead, and zinc in nearly all the mines ought, however, fully to compensate for the extra expense involved in a complete process, so soon as connection with the rest of the world makes it possible to send them to a market.

In his report for 1875 Raymond gives the following details: [14]

---

[12] Raymond, R. W., Statistics of mines and mining in the States and Territories west of the Rocky Mountains for 1873, p. 313, 1874.

[13] Raymond, R. W. op. cit. for 1874, pp. 358, 383–386, 1875.

[14] Raymond, R. W., op. cit. for 1875, pp. 282, 284, 285, 324–326, 1877.

The Golden Smelting Co. (Golden) under Mr. West, the Rosita Reduction Works (Rosita, Custer County), the Mount Lincoln Smelting Works (Dudley, Park County) and Greene & Co.'s Works in Bakers Park [Silverton, San Juan County] have been the additions to the smelting capacity of the Territory. * * *

The production of pig lead for the year has been small; the Lincoln City works [Summit County] have been idle for the entire year and the Golden works did not open till September. Quite a large amount of this metal has been turned out from the new works in Bakers Park, and if the run had been one of a year, instead of a few months, the total would have been a large addition to the product of the Territory. * * *

San Juan country.—This vast region of over 30,000 square miles has never been prospected or explored to any extent outside of the 5,600 square miles relinquished by the Ute Indians in their treaty of September, 1873, and now comprising La Plata, San Juan, and Hinsdale counties. The country is divided into six mining districts, all of which, except Lake and Adams districts, are in San Juan County.

The Animas district includes all locations made on the Animas River and its tributaries to a point 2 miles above Howardsville.

The Eureka district joins the Animas at this point and extends to the divide between the waters of the Animas and those of the Gunnison and Uncompahgre.

The Uncompahgre district includes all lands drained by the Uncompahgre and its tributaries as far north as the Ute Reservation.

Mount Sneffels district [now in Ouray and San Miguel counties] includes the sections drained by the Rio San Miguel.

Lake district includes all the locations made in Hinsdale County, except the mines situated in Burrows Park, at the extreme head of the Lake Fork of the Gunnison, which constitute what is known as the Adams or Park district.

Much has been accomplished during the past two years, notwithstanding the great distance from railroad communications, the inaccessibility of the country, and the want of capital for the development of a new mining camp. More than 5,000 locations have been recorded; two good wagon roads have been built into the country, one via Del Norte, up the Rio Grande, the other up the Gunnison, via Saguache. Several flourishing towns have grown up, of which Silverton, the county seat of San Juan, and Lake City, the county seat of Hinsdale County, are the principal ones, each having a population of some five hundred. The only mines that have been worked to considerable extent in the county are the Hotchkiss in Lake district; the Silver Wing, in Eureka district; and the Highland Mary, Aspen, Prospector, and Little Giant, in the Animas district. * * *

The Silver Wing mine consists of a group of ten lodes, situated on Jones Mountain, 1 mile above the town of Eureka. It is developed by one tunnel, 100 feet long; a second tunnel is under contract for 1,000 feet. This tunnel cuts all the veins from 300 to 1,000 feet below the outcrop. Assays range from $130 to $2,800; the ore contains iron, lead, and a large percentage of copper.

The Highland Mary, Rob Bruce, and Powderhouse claims are all located on one vein, all situated at the head of Cunningham Gulch, 3 miles above Howardsville. The workings consist of four tunnels, running in and along the vein, 300 feet apart, one above the other. The crevice is 15 feet wide, with an ore streak from 9 to 30 inches wide. The ore is argentiferous galena. Sample assays from first-class ore gave $2,100; second-class, $760; third-class, $170.

The Prospector lode has probably furnished more ore than any other in the San Juan country. The mine is located on Hazelton Mountain, 2 miles above Silverton. The developments consist of two shafts, 100 and 130 feet deep, with a level 100 feet long connecting the two, and 100 feet from the surface.

The Little Giant, located in Arrastre Gulch, was the first location made in the San Juan country. The mine has a pay streak of 8 inches of gold-bearing quartz. Twenty-seven tons, worked by the arrastre, produced $150 per ton. In 1872 a company was organized in Chicago, known as the Little Giant Co., which erected upon this property amalgamation works, containing a 12-horsepower engine, Dodge crusher, and ball pulverizer. The works were built 1,000 feet below the mine, with a wire tramway to bring the ore to the mill. About 100 tons of ore were milled, producing $14,500, or about 65 per cent of the assay value. The property has been involved in litigation since the spring of 1874.

Raymond then gives a list of producing mines in the San Juan country for 1875, with description of crevice in feet, pay streak in inches, character of ore, percentage of lead, number of ounces of silver per ton, coin value per ton of 2,000 pounds, and improvements. His figures for production are given below:

Total amount of ore smelted in San Juan for 1875, 172.5255 tons; average value per ton, $216.59; total coin value	$37,361.82
Total amount shipped for treatment elsewhere, 48.5765 tons; average value per ton, $805.65; total coin value	39,135.06
Total amount of ore extracted, 221.1002 tons; coin value	76,496.88
Average coin value per ton of all ore extracted and treated or shipped	345.98
Total amount of bullion produced from the 172.5255 tons was 60.25 tons, assaying $540.35 per ton; coin value	32,556.10
1,559 ounces silver refined in the country; coin value	2,015.63
Total	34,571.73

Loss in extracting, $2,790.09, equal to 7.2 per cent of the ore value.

He adds:

The coming season will witness considerable activity in both mining and smelting throughout the San Juan country. Seventeen mines are being worked during the present winter, which by June 1 will produce 500 tons in the Animas district and 300 tons in the Lake district.

The next year's product of bullion, it is estimated, will be about as follows: Greene & Co., Silverton, 1,000 tons ore, producing 400 tons bullion, value, $550 per ton. San Juan Smelting Co., [at the] forks of [the] Animas, 600 tons ore, producing 240 tons bullion, value, $400 per ton. There will be no refining done in the country next season, since the bullion can be shipped out, the lead paying the cost of transportation. Besides the two above works, which will be in operation by the first of July, the Rough & Ready Works, at Silverton, which have been lying idle for the want of capital, may be put in operation.

For 1876 to 1884 the figures given represent estimates from all available data to correspond with the total production of the State. The production of the San Juan region has been distributed among the several counties.

For 1879 and 1880 Burchard [15] gives the production of the San Juan region, and for 1880 he adds a de-

[15] Burchard, H. C., Report of the Director of the Mint upon the production of the precious metals in the United States during the calendar year 1880, pp. 155–157, 1881

scription of the Silver Wing mines of the Eureka mining district. In his report for 1881 he devotes five pages to a description of mines in San Juan County.[16]

In his report for 1882 Burchard gives the following details:[17]

The completion of the Denver & Rio Grande Railroad to Silverton, in July last [1882]. * * *

The ores of San Juan County are treated at other localities exclusively. A considerable amount has been shipped to the smelters at Durango, but the larger portion has gone to Pueblo and St. Louis, while smaller amounts were sent to Omaha and Denver. The only reduction plant in the county is at Silverton.

During the year the Martha Rose Mining & Smelting Co. erected works of 20 tons capacity and provided with excellent facilities for handling and treating the ores. After a short and unsuccessful run the works closed down. * * *

The reduction of cost of shipping ores from $35 to $40 per ton to $12, which was effected upon the completion of the railroad, has given a wonderful impetus to mining. * * *

The North Star No. 1, * * * located in 1876, has been steadily and successfully worked ever since. * * * All the ore taken out averages 70 to 80 ounces of silver per ton, and 35 per cent lead, though with the usual sorting a grade double that is readily obtained. The ore which pays is galena, but considerable gray copper of a very high grade is taken from the mine. All the ore is shipped in carload lots to St. Louis. The mine has produced over 100,000 ounces of silver and 1,500,000 pounds of lead. * * *

The Belcher * * * ore is galena, free from refractory gangue. The mine has been a constant producer. * * *

The Empire is an eastern extension of the North Star, and is a well-defined vein, carrying gray copper in paying quantities. The pay streak * * * averages 100 ounces of silver and 10 per cent of copper.

On Hazelton Mountain, about 3 miles northeast of Silverton, is the Aspen Group, which comprises the Aspen, Susquehanna, Mammoth, Legal Tender, McGregor, Matchless. * * * The veins, of 3 to 8 feet, carry galena and gray copper ore in a quartz gangue. * * * The group is the property of the San Juan & New York Mining & Smelting Co. The Gray Eagle, also on Hazelton Mountain, has produced 106 tons of ore, averaging 50 per cent lead and 60 ounces of silver. * * *

The Silver Lake shipped ore milling 50 ounces of silver and 60 per cent of lead. * * *

On Green Mountain, in Cunningham Gulch, is the Green Mountain mine. * * * The ore is galena, carrying small quantities of very high grade gray copper. The company have employed a jig with great success for dressing their fine ore. * * *

The Congress is a new discovery situated on the divide between San Juan and Ouray counties. * * * One hundred and ten tons of ore were shipped from the mine in 30 days, all taken from the shaft. The ore already shipped has netted in Silverton over $100 per ton, having given the following results per ton: 1st. 27 per cent in copper; gold 1½ ounces; silver 18 ounces; 2d. copper 33 per cent, gold 1⅓ ounces, silver 20 ounces. Occasional assays have yielded 60 per cent in copper, 3⅓ ounces in gold, and 30 ounces in silver. * * *

Three hundred feet east of the Congress lode is the Salem. This lode shows a different character of ore, running very high in copper and gold. * * * Two carloads of ore have been shipped. * * *

The Hudson is a northern extension of the Salem. It was located during the summer. * * * The mine has produced 1 ton of ore per day to the man since work was commenced in September. * * * Mill runs gave 32 per cent in copper, 29 ounces of silver, and one-half ounce gold to the ton.

The Yankee Girl group [Ouray County] comprises five or six locations, chief of which are the Yankee Girl, Orphan Boy, and Robinson. The ore is very different from the Hudson, Congress, and Salem, being a galena and copper ore running high in silver. * * * Four hundred tons of ore have been shipped.

In Eureka district there are a number of mines rapidly becoming steady producers.

On the Byron a tunnel of over 500 feet has been run on the vein, which carries remarkably fine galena with the addition of considerable zinc. Associated with the zinc, but not mixed or combined, is a streak of very high grade silver ore. * * *

Animas Forks district includes the mines tributary to Mineral Point.

Burchard then gives two pages of descriptions of mines. In his report for 1883 he gives the following details:[18]

In the northern portion of the county, in the Red Mountains, the formation and character is described in the review of this district under Ouray County. The county line passes on the summit of the divide, and for some time there existed considerable doubt regarding which county, Ouray or San Juan, quite a number of these mines were in. Regarding the location, each mine owner was requested to state whether his property was in Ouray or San Juan county, and the production reported is accordingly placed to the credit of the county, in which the same is located. * * *

The North Star No. 2 * * * [has produced] 550 tons of ore that run 100 ounces silver, one-half ounce gold, and 40 per cent lead to the ton. * * *

Belcher mine, * * * since * * * about one year ago, [has produced] * * * over 1,700 tons of ore [which] has yielded an average of 38 ounces silver and 40 per cent lead. * * *

The Silver Lake, on Round Mountain, * * * shipped to Sweet's Sampling Works 72 tons of ore that contained 2,016 ounces of silver and 7 ounces of gold, and that averaged 55 per cent lead. * * *

On Green Mountain is the property of the Green Mountain Mining Co., viz., the Green Mountain and Green Mountain Extension. * * * The production of this mine during 1883 was about 300 tons of ore, which averaged 38 ounces of silver and 45 per cent lead to the ton. * * *

The Lackawanna group of mines, on Kendal Mountain, * * * 40 tons have been sampled, and gave an average of 50 ounces silver and 20 per cent lead. * * *

The North Star, on Solomon, * * * produced quite a large amount of ore prior to the present season of an excellent grade. The past summer there have been taken out and shipped 550 tons of ore, that run 100 ounces silver, one-half ounce gold, and 40 per cent lead. * * *

The Emerald mine, on Anvil, * * * has produced considerable ore from the date of its discovery. The last year some 30 tons were shipped to Sweet's Sampling Works that averaged 75 ounces and 20 per cent lead.

On Red Mountain, near Red Mountain City, but in San Juan County, are the Congress, Senate, Salem, St. Paul, and others, and near Chattanooga numerous good prospects.

The Congress was located during the fall of 1881. Being late in the season nothing but the assessment work was done in 1881, but early the next spring the owners resumed work and

16 Burchard, H. C., op. cit. for 1881, pp. 354, 427–432, 442–443, 1882.
17 Burchard, H. C., op. cit. for 1882, pp. 390, 391, 395, 545–554, 1883.
18 Burchard, H. C., op. cit. for 1883, pp. 236, 237, 238, 240, 407–417, 1884.

sunk a 50-foot shaft all the way through solid mineral. In the summer of 1882 the Congress was bonded for $15,000, and it was sold to the present owners six days afterward for $21,000. * * * More than 100 tons of ore were shipped before the season was over, and fully as much more remains on the dump at the present time, making in all nearly 200 tons of solid mineral which has been taken out of the Congress shaft, which is now only 90 feet deep. The only drifting that has ever been done was at the 50-foot level, directly after the purchase, when they drifted 18 feet, endeavoring to find the extent of the ore body, but nothing save solid mineral was encountered. The 200 tons of ore shipped from the Congress last fall netted, after paying for mining, shipping, and smelting, $53 per ton. The crosscut is 275 feet in length and cuts the vein, intersecting the bottom of the shaft at about 85 feet depth, and runs through 15 feet of solid mineral before reaching the shaft. The ore is copper pyrites carrying gold and silver, and like the ore of the Yankee Girl it needs no sorting—is shipped just as it comes from the mine. Average ounces of gold per ton, 1½; silver, 20; and copper 30 per cent. For the year the average daily output has been about 8 tons per day. * * *

Near Chattanooga are the Jenny Lind, Windsor, Bertha and Ida, Big Four, Bonanza Boy, Dipper, Genesee, Humboldt, Providence, Silver Cup, Independence, Silver Ledge, Little Maud, and probably 20 others of as good indications. * * *

Eureka district.—* * * Very little ore has been shipped from this part of San Juan County, because the mines as a general rule are above timber line and somewhat inaccessible, and the ore has to be packed on burros quite a distance, then by wagons to Silverton for shipment to outside smelters. Concentrating works are now being erected. * * *

The Sunnyside, near the Franklin, has been working continuously, filling a contract with the Pueblo smelters for 4,000 tons of ore. The only drawback to this mine is the large per cent of zinc it contains; but for that it would be one of the most desirable dry ores found in the State. * * * The ore carries considerable gold and only about 5 to 10 per cent lead. * * * Animas Forks district includes the mines tributary to Mineral Point.

The Little Dora, in Grouse Gulch * * * shows * * * a streak of bismuth from 2 to 4 inches wide, which yields 1,128 ounces silver, 3 ounces gold per ton. * * *

The La Plata Miner publishes the following résumé of the year's output of San Juan County:

While it is impossible to give the entire output in tons and its actual value in dollars, yet we have been able to get the accurate results of over 5,955 tons of ore, which we present; and from this we assume that portion of ore that was shipped without sampling to have been of equally good grade. This will be more apparent when it is known that by far the larger portion of the unsampled ore was from the Yankee Girl [Ouray County]. The total output as shown by the Denver & Rio Grande shipping books is as follows for each month:

	Tons		Tons
January	760	August	2,130
February	480	September	1,780
March	500	October	1,770
April	510	November	1,240
May	380	December	550
June	700		
July	1,440		12,230

In addition to this there were shipped during the same time 2,110 tons of iron ore to the Durango smelter. Of this amount 5,955 were sampled in Silverton, from which we have the returns. We find that the above amount of ore contained 1,119 ounces of gold, 255,394 ounces of silver, 591,362 pounds of copper, and 3,166,860 pounds of lead. * * *

Of the ores sampled two classes would seem to cover it all—copper and lead. The average percentage of copper in the ores sampled and paid for as copper ore was from 15 to 30 per cent, while the great bulk of the ore has been lead ore and has averaged freely 30 per cent of that metal.

It will be noticed that this statement includes a large amount shipped from Silverton which was the production of the Yankee Girl. It probably also includes the shipments of the National Belle. Both of these mines are large producers, but neither is situated in San Juan but in Ouray. * * *

The Mining Record gives the output of San Juan for the same year as follows:

Since January 1, 1883, there have been shipped from Silverton over the Denver & Rio Grande Railroad over 15,000 tons of ore. * * *

The following is the number of tons shipped: Stoiber Bros. sampled and bought 3,171; sampled and shipped 793; unsampled and shipped, 927; total 4,891. E. T. Sweet has crushed, sampled, and shipped 3,760 tons, and T. B. Comstock has crushed, sampled, and shipped 2,500 tons and concentrated 800 tons. Messrs. Stoiber, of Silverton, are reported to have sampled and shipped 2,089½ tons of ore that contained 207 ounces of gold, 97,011 ounces of silver, 85,212 pounds of copper, and 977,899 pounds of lead.

In his report for 1884 Burchard says:[19]

The North Star * * * product for 1884 was 1,100 tons. * * *

Red Mountain district.—That portion of Red Mountain district lying in San Juan County has not been prolific in its output. In the vicinity of Chattanooga not much was done in development, and in the vicinity of Red Mountain City litigation has caused the closing of many properties. * * *

The shipments from the county are about 8,000 tons, showing an increase of 600 tons over last year, when the shipments from the county were about 7,400 tons. The falling off in the total shipments by way of Silverton amounts to over 1,000 tons, which can be mostly accounted for by the idleness of the Congress and Hudson and the reduction in shipments from the Yankee Girl and National Belle mines, at Red Mountain.

For 1885 the figures given in the table have been interpolated to correspond with the total production of the State.

For 1886 to 1896 the figures given have been taken from reports of the agents of the mint in annual reports of the Director of the Mint, the gold and silver being prorated to correspond with the figures for the total production for the State as corrected by the Director of the Mint, the lead being prorated to correspond with the total production of lead for the State as given in annual volumes of Mineral Resources, and any unknown production in the State being distributed proportionately among the counties. As with lead so with copper, but as the figures given in Mineral Resources include copper from matte and ores treated in Colorado, although produced in other States, they are subject to revision.

For 1887 and 1888 Munson[20] lists the producing and nonproducing mines and gives the individual output of the producing mines. The list of smelters

[19] Burchard, H. C., op. cit. for 1884, pp. 177, 239-244, 1885.
[20] Munson, G. C., agent for Colorado, in Kimball, J. P., Report of the Director of the Mint upon the production of the precious metals in the United States during the calendar year 1887, pp. 181-183, 189, 1888; idem for 1888, pp. 121-123, 1889.

operating in 1887 includes the New York & San Juan at Durango but none at Silverton.

In his report for 1889 Smith [21] gives only the estimated total production of San Juan County, but in his reports for 1890, 1891, and 1892 he lists the producing mines and gives their individual outputs.

In his report for 1897 Hodges says: [22]

Eureka district.—* * * Concentration and smelting; one 10-stamp mill.

Animas district.—* * * Amalgamation, concentration, and smelting processes; 9 mills operating 300 stamps and 2 mills using Huntington process and jigs; 3 concentrating mills projected.

The figures for 1897 to 1904, which represent receipts at smelters and the mint, are taken from the reports of the Colorado State Bureau of Mines as are the figures for zinc for 1904 to 1907.

In his report for 1898 Hodges says: [23]

The county does not have [custom] reduction works of any character, except one amalgamation mill for free gold ores. The ores are almost exclusively smelting ores. The pyritic smelter at Silverton closed down during the latter part of 1897, owing to the scarcity of iron sulphide ores, it is claimed.

Las Animas district.— This district, located north and east of the town of Silverton, has really but five large producing mines, and these properties easily furnish over one-half of the county's tonnage. Their product is almost wholly in the form of concentrates, their output of crude ore being not less than 300 tons per day. Several mills are in constant operation.

Eureka district.—This district has been a large producer for years, its heavy lead ores, in addition to their silver contents, carrying a fair amount of gold values, which, in several instances, is the principal value of the ore mined. Two mills are operated this year.

Ice Lake district.—The production of this district has been limited to the rather small output of two mines, operated during the summer season. One of these properties has a mill, but it closes down as soon as the snow begins. * * *

Red Mountain, Uncompahgre, and Cascade, the other districts in this county, have had some prospecting, but the output has been small.

In his report for 1899 Hodges says: [25]

The Red Mountain Railroad, running from Silverton to Red Mountain increased its carload shipments over 1898, the amount averaging three to four carloads per day in the fall and early winter until the deep snow put a stop to transportation over that line. * * *

Gladstone-Silverton Railroad.—A feature of the summer work was the inauguration and completion of the Gladstone, Silverton, & Northern Railroad, running a distance of 9 miles from Silverton to the Gold King group of mines, near Gladstone. * * *

The Silver Lake tramway, 14,700 feet in length, is one of the longest on the American continent. The concentrating

mill, entirely modern, handles 250 tons of crude ore each day. * * *

The Iowa group * * * has a modern concentration mill. * * *

The Sunnyside group, now equipped with a modern concentrating plant, equal to 100 tons per day. * * * In part of the group lead-zinc ore prevails, and this is also to be handled in a concentration plant after the system followed at Leadville and Creede.

The new mill of the Gold King group, recently completed, will handle 100 tons of crude ore each 24 hours.

The Empire Consolidated group, Sultan Mountain, * * * will have a new 50-ton concentrator ready for work early in the year.

San Juan's new mills.—The Red & Bonita mill in the Gladstone section was started up early in November. It has a capacity of 75 tons per day. * * * Among the modern mills are the following:

	Tons per day		Tons per day
Silver Lake	250	Boston and Silverton	75
Iowa	100	Silver Ledge	50
Gold King	100	Sunnyside	50
Terry	100	Silver Wing	40
Bonita	75		——
			840

The Iowa Gold Mining & Milling Co. shipped in 1899 9,547 tons of ore, largely concentrates, which contained 5,500 ounces of gold, 26,000 ounces of silver, and 5,500 tons of lead. * * *

The new smelter at Silverton purposes to treat particularly sulphide ore. * * *

In his report for 1900 Hodges says: [26]

There are in active operation 12 concentration mills, having a total capacity of 1,190 tons per day, besides that of the pyritic smelter.

New pyritic smelter.—One of the * * * improvements inaugurated is the Kendrick-Gelder pyritic smelter. * * * A saving has been made in the cost of treatment by installing the Billberton hot-blast matte settling attachment. * * *

The Silver Lake mines * * * average about 200 tons per day. * * * The owners * * * were the first in the State to use electric drills in their mines. * * *

The Sunnyside * * * has two mills in active operation, but owing to scarcity of water the upper one was not run continuously. Their combined capacity is 150 tons per day.

The Gold King * * * present output is about 250 tons per day.

In his report for 1901 Hodges says: [27]

The Silver Lake group, combining 175 mines, mill sites, and placer claims, the American Smelting & Refining Co. purchased for 2⅓ million dollars from its original owners, E. G. Stoiber and wife. * * *

Milling facilities.—The county is well supplied with mills, and four railroads from its different sections give ready conveyance to the ores. The mills in active operation and their daily capacity are:

	Tons
Silver Lake Mill No. 1	250
Silver Lake Mill No. 2	200
Gold King	250
Iowa Tiger	150
Sunnyside	150
Empire Hercules	100

[21] Smith, M. E., agent for Colorado, in Leech, E. O., Report of the Director of the Mint upon the production of the precious metals in the United States during the calendar year 1889, p. 152, 1890; idem for 1890, pp. 139-140, 1891; idem for 1891, p. 185, 1892; idem for 1892, p. 130, 1893.

[22] Hodges, J. L., agent for Colorado, in Roberts, G. E., Report of the Director of the Mint upon the production of the precious metals in the United States during the calendar year 1897, p. 170, 1898.

[23] Hodges, J. L., agent for Colorado, in Roberts, G. E., op. cit. for 1898, pp. 92-94, 1899.

[25] Hodges, J. L., agent for Colorado, in Roberts, G. E., op. cit. for 1899, pp. 108-110, 1900.

[26] Hodges, J. L., agent for Colorado, in Roberts, G. E., op. cit. for 1900, pp. 132-133, 1901.

[27] Hodges, J. L., agent for Colorado, in Roberts, G. E., op. cit. for 1901, pp. 134-137, 1902.

	Tons
Yukon	75
Great Mogul	75
Terry	50
Sunnyside Extension	50
Howardsville	40
San Juan Chief	40
Silver Queen	40
	1,470

Pyritic smelter.—The Kendrick-Gelder smelter, which was operated a portion of the year, proved of advantage in offering a market to such low grade as had formerly been practically negatived by freight and smelting charges. It largely expedited the development of such bodies of ore and converted into copper matte 5,500,000 pounds from the upper levels of the Henrietta.

The Gold King Consolidated mines, in the Gladstone district, shipped during the year 10,100 tons of concentrates * * *

Tungsten ores.—* * * Production 12,500 pounds evidencing from 66 to 71 per cent tungstic acid.

In his report for 1902 Downer says: [28]

The actual tonnage raised, which includes the ores sent to the concentration mills as well as the crude ore shipped direct, will not fall short of 230,000 tons. Of this product there was shipped to the Durango smelter and the Pyritic smelter, located at Silverton, nearly 40,000 tons of concentrates and about 10,000 tons of crude ore.

Gold King Consolidated.—The largest producer in the county, both in tonnage and values, is the Gold King Consolidated, its ore mined and milled during the year reaching 72,455 tons. The mill of the company consists of an amalgamation and concentration plant, with a capacity of about 200 tons a day. The 80-stamps are of the rapid-drop pattern, weighing 850 pounds each, and the screens are especially made of copper wire for these batteries and have shown a remarkable life. The gold retorts sent from the mill have reached a total value of $400,000, and the concentrates shipped amounted to 14,428 tons, with a valuation of over $700,000. * * *

Silver Lake mines.—The Silver Lake mines, which two years ago were sold to the Guggenheim Exploration Co., have made many improvements. All ore mined is treated in the large concentration mill of the company, which handles about 300 tons a day. This has been enlarged, and the coming year will see an increased capacity of 100 tons a day, making a total of 400 tons. * * *

The scheme of operation in the mill is first crushing and coarse-jigging the ore, the tailings being recrushed and sized through revolving screens and sent to Hallett concentrating tables. The middlings from these tables are reground in Chilean mills and the product passed over stationary canvas slime tables. The tonnage shipped from the Silver Lake mills during the year was 13,350, entirely lead concentrates.

The Sunnyside.—The Sunnyside mine has taken third place as a producer and during a large part of the year has kept both of its mills busy, they being of 30 tons and 100 tons daily capacity. Both are amalgamation and concentration. During the year they shipped 4,742 tons of concentrates, some 15,000 tons of ore going to both mills.

North Star.—The North Star, on Sultan Mountain, at one time the greatest producer in the county, up to the first of the year had lain idle for nearly 10 years. It is claimed that in the deep workings a large body of good grade ore was encountered. A concentration mill has been erected.

Grand Mogul.—The Grand Mogul property is located on upper Cement Creek, on the system of veins which cross from the Gold King to the Sunnyside, and, while not having been a heavy producer during the year, has shipped a number of cars of ore and kept up continuous development work. A large concentration mill will be erected.

The Esmeralda is a new producer. During the latter half of the year it shipped 100 carloads of ore carrying from an ounce to $1\frac{1}{2}$ ounces in gold and as high as 100 ounces in silver.

The Highland Mary mine, now incorporated as the Gold Tunnel & Railway Co., has become a shipper and is worked through the Gold Tunnel. This cuts the vein at a depth of 1,500 feet. A 150-ton concentrating mill was built and is working successfully.

Red Mountain district.—In the Red Mountain district a great deal of development work and prospecting has been done, but no new shippers have been recorded, the low prices of silver and copper preventing in many instances the taking out of ore which could be handled under more favorable market conditions.

Among the older mines in this section, which have shipped at intervals are the Brooklyn-Bonner, Yankee Girl, Genesee-Vanderbilt, and Silver Ledge. The last-named has just completed a 200-ton concentration mill.

Kendrick-Gelder plant.—The Kendrick-Gelder pyritic smelter, at Silverton, has only run during the summer months. As a large part of the ore treated carries considerable copper, the production of a very good grade of matte is secured. This is concentrated by being resmelted with the addition of copper ores, so that a matte is finally produced running 50 to 70 per cent in copper with high gold and silver contents. This furnace is run with a hot blast and does not use over 6 per cent of fuel. As the charge can be run with a much higher proportion of silica in the slag, and no roasting is required, the cost of smelting is materially reduced over the lead furnace. A very large part of the crude ore of the county found its way into the bins of this smelter while it was running.

In his report for 1904 Downer says: [29]

The total tonnage shipped from the various districts for 1904 amounted to 66,288, the values being in gold, silver, lead, copper, zinc, and tungsten. There has been a substantial increase over 1903, due to improved facilities for handling the output, extension of trams to the mills and reduction works, as well as economic handling of the properties.

The proportion of concentrates to crude ore is about 7 to 1. The larger producers shipped over 44,000 tons. Amalgam and bullion amounting to $571,000 has been saved by the mills. * * *

The Old Hundred has built new tramway and installed a large Rand compressor.

The Ruby Basin mill is being increased from 50 to 500 tons capacity per day.

The Gold King mill has increased its capacity 30 per cent.

The Silver Ledge has established electric magnetic separators to separate zinc from the lead and has demonstrated the commercial value by an increase in net returns of about 40 per cent, shipping a high-grade zinc concentrate that was dead loss and expense previously.

The Green Mountain Co. has almost completed a 300-ton mill. A new tramway is being pushed as rapidly as possible to replace the one burned on the Silver Lake.

For 1905 to 1923 the figures in the table have been taken from Mineral Resources (mines reports).

[28] Downer, F. M., agent for Colorado, in Roberts, G. E., op. cit. for 1902, pp. 104–106, 1903

[29] Downer, F. M., agent for Colorado, in Roberts, G. E., op. cit. for 1904, p. 119, 1905

Gold, silver, copper, lead, and zinc produced in San Juan County, 1873-1923

Year	Ore treated (short tons)	Lode gold	Silver			Copper			Lead			Zinc			Total value
			Fine ounces	Average price per ounce	Value	Pounds	Average price per pound	Value	Pounds	Average price per pound	Value	Pounds	Average price per pound	Value	
1873		$13,000													$13,000
1874		9,540	3,166	$1.278	$4,046										13,586
1875		10,000	68,547	1.24	84,998				120,000	$0.058	$6,960				101,958
1876		5,000	48,465	1.16	56,219				249,348	.061	15,210				76,429
1877		5,000	34,010	1.20	40,812	8,664	$0.19	$1,646	400,000	.055	22,000				69,458
1878		6,000	24,569	1.15	28,254	36,145	.166	6,000	400,000	.036	14,400				54,654
1879		6,000	30,938	1.12	34,651	100,000	.186	18,600	500,000	.041	20,500				79,751
1880		6,000	11,602	1.15	13,342	100,000	.214	21,400	430,000	.05	21,500				62,242
1881		5,000	19,336	1.13	21,850	100,000	.182	18,200	140,000	.048	6,720				51,770
1882		10,000	46,406	1.14	52,903	100,000	.191	19,100	320,000	.049	15,680				97,683
1883	7,400	35,000	270,703	1.11	300,480	100,000	.165	16,500	1,137,000	.043	48,891				400,871
1884	8,000	40,000	464,062	1.11	515,109	300,000	.13	39,000	3,400,000	.037	125,800				719,909
1885		40,000	700,000	1.07	749,000	100,000	.108	10,800	5,300,000	.039	206,700				1,006,500
1886		142,799	718,523	.99	711,338	100,000	.111	11,100	4,300,000	.046	197,800				1,063,037
1887		121,245	401,760	.98	393,725	300,000	.138	41,400	2,040,145	.045	91,806				648,176
1888		190,328	223,339	.94	209,939	240,000	.168	40,320	2,382,358	.044	104,824				545,411
1889		394,873	508,328	.94	477,828	135,018	.135	18,227	4,096,887	.039	159,779				1,050,707
1890		187,357	321,340	1.05	337,407	147,354	.156	22,987	3,462,158	.045	155,797				703,548
1891		192,109	769,545	.99	761,850	235,467	.128	30,140	6,857,544	.043	294,874				1,278,973
1892		148,908	397,589	.87	345,903	136,768	.116	15,865	6,406,665	.04	256,267				766,943
1893		260,668	327,153	.78	255,179	1,125,826	.108	121,589	8,000,800	.037	296,000				933,436
1894		340,023	351,114	.63	221,202	1,118,222	.095	106,231	4,000,000	.033	132,000				799,456
1895		849,411	1,894,453	.65	1,231,394	2,057,588	.107	220,162	8,098,800	.032	259,162				2,560,129
1896		908,707	2,228,031	.68	1,515,061	845,094	.108	91,270	5,634,586	.03	169,038				2,684,076
1897		694,326	1,101,907	.60	661,144	1,435,203	.12	172,224	8,021,414	.036	288,771				1,816,465
1898		1,132,592	1,048,499	.59	618,614	2,252,421	.124	279,300	14,659,999	.038	557,080				2,587,586
1899		996,273	1,191,857	.60	715,114	1,197,661	.171	204,800	16,011,677	.045	720,525				2,636,712
1900		757,204	681,317	.62	422,416	1,972,087	.166	327,367	17,579,177	.044	773,484				2,280,471
1901	242,850	962,974	784,218	.60	470,531	2,740,042	.167	457,587	15,473,187	.043	665,347				2,556,439
1902	230,000	1,524,226	838,102	.53	444,194	3,012,283	.122	367,499	7,699,883	.041	315,695				2,651,614
1903		1,710,608	781,358	.54	421,933	2,939,018	.137	402,645	6,969,093	.042	292,702				2,827,888
1904	233,663	1,396,651	1,042,044	.58	604,386	3,467,124	.128	443,792	9,288,643	.043	399,412	317,254	$0.051	$16,180	2,860,421
1905	204,139	1,050,971	750,844	.61	458,015	2,274,109	.156	354,761	8,045,126	.047	378,121	163,845	.059	9,667	2,251,535
1906	196,438	900,175	690,076	.68	469,252	1,549,663	.193	299,085	4,515,317	.057	257,373	718,192	.061	43,810	1,969,695
1907	235,639	967,732	1,175,176	.66	775,616	2,450,280	.20	490,056	12,483,507	.053	661,626	1,772,764	.059	104,593	2,999,623
1908	202,643	997,824	1,004,287	.53	532,272	2,282,738	.132	301,321	8,402,569	.042	352,908	10,131	.047	476	2,184,801
1909	187,041	683,267	793,637	.52	412,691	1,653,192	.13	214,915	9,085,068	.043	390,658	786,518	.054	42,472	1,744,003
1910	206,272	710,527	782,250	.54	422,415	1,208,496	.127	153,479	10,688,386	.044	470,289	3,781,259	.054	204,188	1,960,898
1911	108,088	336,463	325,604	.53	172,570	470,912	.125	58,864	6,933,822	.045	312,022	2,224,351	.057	126,788	1,006,707
1912	140,917	523,574	714,974	.615	439,709	1,063,291	.165	175,443	9,114,334	.045	410,145	2,478,594	.069	171,023	1,719,894
1913	123,343	657,612	880,409	.604	531,767	1,221,516	.155	189,335	9,508,979	.044	418,395	1,664,999	.056	93,240	1,890,349
1914	117,988	508,477	493,917	.553	273,136	825,180	.133	109,749	5,199,000	.039	202,761	971,177	.051	49,530	1,143,653
1915	147,878	583,681	430,637	.507	218,333	1,054,463	.175	184,531	6,791,596	.047	319,205	2,259,226	.124	280,144	1,585,894
1916	146,128	438,628	502,342	.658	330,541	1,615,167	.246	397,331	7,285,304	.069	502,686	4,014,403	.134	537,930	2,207,116
1917	145,685	318,006	658,261	.824	542,407	1,665,923	.273	454,797	10,515,535	.086	904,336	3,270,500	.102	333,591	2,553,137
1918	132,927	257,011	477,322	1.00	477,322	1,120,178	.247	276,684	9,485,775	.071	673,490	3,410,308	.091	310,338	1,994,845
1919	64,899	132,560	279,667	1.12	313,227	661,667	.186	123,070	5,443,906	.053	288,527	1,833,768	.073	133,865	991,249
1920	201,671	266,766	746,100	1.09	813,249	1,361,391	.184	250,496	16,601,025	.08	1,328,082	11,837,395	.081	958,829	3,617,422
1921	1,164	8,272	64,179	1.00	64,179	28,558	.129	3,684	25,090	.05					101,225
1922	8,808	25,759	77,864	1.00	77,864	110,348	.135	14,897	1,651,982	.055	90,859	1,300,000	.055	74,100	283,479
1923	153,114	241,986	471,750	.82	386,835	1,005,441	.147	147,800	10,738,943	.07	751,726	9,540,000	.068	648,720	2,177,067
		22,711,113	28,651,577		20,432,222	50,024,498		7,726,049	316,426,293		15,373,023	52,354,684		4,139,484	70,381,891

SAN MIGUEL COUNTY

Purington[30] gives the following details of early developments in the Telluride region in chronologic order:

1875. First (?) prospector entered region. Location made on what is now the Smuggler vein, and a ton of ore worth $2,000 was shipped to the smelter at Alamosa. After this about 60 tons of ore were shipped, of which the yield is unknown.

1876. Nevada and What Cheer claims located.

1877. Small shipments made principally to Silverton, the ores being mined for silver and running from 75 to 125 ounces to the ton.

1878. Marshall Basin produced 200 tons, and small lots were sent to the Silverton smelter from the mines about Ophir, some of the ore being exceedingly rich in silver.

1879. First attempt at milling the ores. Two arrastres were built in the vicinity of Ophir, and to one the Gold King shipped 3 tons of gold ore daily in 1879.

1881-82. The Virginius mine, at the head of Canyon Creek (Ouray County), in 1881 was worked by three levels and two shafts, and in 1882 the development amounted to 2,000 feet, with a product of $75,000 in silver.

By 1883, one hundred men were at work in Marshall Basin, and a small smelter was built at the old town of Ames. This apparently did not prove successful, as it ran only a year. At this time it is noted that a shipment of 4 tons of ore from the Smuggler vein gave a return of 800 ounces of silver and 18 ounces in gold to the ton. * * *

The total placer production to 1896 has been estimated at $100,000.

In another publication Purington gives the following note:[31]

Whether or not the early Spanish explorers of southwestern Colorado actually passed through this district is uncertain. At any rate no traces remain to indicate that exploration for the precious metals was prosecuted before the middle of the present century, and the first active search for gold and silver in the Telluride district was in 1875. In that year locations were made on the vein now called the Smuggler. At that time and in following years, down to 1882, many locations were made and small amounts of the precious metals were produced. Although the available data concerning the early developments are of the most fragmentary character, it is probable that the total product of the quadrangle previous to 1882 did not exceed $50,000. The region did not attain

[30] Purington, C. W., Preliminary report on the mining industries of the Telluride quadrangle, Colo.: U. S. Geol. Survey Eighteenth Ann. Rept., pt. 3, pp. 752-754, 1897

[31] Cross, Whitman, and Purington, C. W., U. S. Geol. Survey Geol. Atlas, Telluride folio (No. 57), 1899.

importance * * * until 1890, when the Rio Grande Southern Railroad was completed from Ridgway to the town of Telluride.

For 1875 to 1882 the figures given in the table on page 226 are estimates, corresponding to the total production of the State. The figures for placer production from 1878 to 1902 are also estimates made on this basis.

In his report for 1881 Burchard says:[32]

South of Ouray is San Miguel district, on the North Fork of the San Miguel River. The principal mines shipping ore are the Smuggler, Mendota, Cimarron, and Argentine. A number of others have been developed and contain pay ore but have not been shipping. Among them may be mentioned the Boomerang, Alta, Quail, Palmyra, Big Elephant, Silver Chief, Ausbury, Shamrock, and Hyacinth.

The Pandora & Oriental has a 40-stamp mill on the ground, and a mill on the Gold King has been completed. The mill about to be erected by the Pandora & Oriental Co. will have 16 Frue vanners, 2 to each 5-stamp battery, driven by a separate engine. The stamps weigh 825 pounds and will be driven by a 120-horsepower engine, which is of double the capacity required, it being the intention of the company to erect an additional 40 stamps, should developments on the property justify it. * * *

In the vicinity of Ophir, in Iron Springs district, 10 miles south of San Miguel, are a number of mines located on Silver and Yellow Mountains and on Wilson Creek. The ore is high grade, although the veins are narrow, the widest being that of the Osceola mine. * * *

In the valley of the San Miguel, the Keystone Co. was engaged most of last season in erecting flumes and other preparations for washing on an extensive scale. They ran just a month before closing down and cleaned up about $3,500 in gold, worth about $16 per ounce. On Bar No. 1, the St. Louis Co. ran three weeks and cleaned up about $1,500 gold. A few other groups of men working in a small way with toms, rockers, and sluices obtained lesser amounts.

At the head of the San Miguel a deposit of very rich float rock was discovered, which was crushed in mortars and washed by hand, yielding, it is said, about $10,000. The tract of country carrying this float rock is doubtless quite limited.

For 1882, Burchard gives the following details:[33]

The Upper San Miguel district, with Telluride as its central point, comprises that section of Ouray County drained by the North Fork of the San Miguel River and its tributaries and Turkey Creek, emptying into South Fork. The first discoveries were made in 1875, but owing to the inaccessibility, distance from railroads, and disadvantages arising from these sources, very little ore was shipped until 1881, when Gunnison, the nearest railroad point, was 140 miles distant, the cost of transportation then being not less than $45 per ton. * * *

Following up the river on the north side, the following tributaries empty into it: Remine, Eder, Park [Mill], Butcher, Cornet, Marshall. Remine and Eder basins have not been prospected to any extent. A few good lodes have been opened on Park Creek. * * *

In Marshall Basin the most prominent vein is the Smuggler, upon which the following locations have been made: Mendota, Sheridan, Smuggler, Union, and Cleveland. The first three of these have been working and shipping ore all the season, but I am unable to learn results.

The Pandora & Oriental is owned and operated by a French company, which has expended about $300,000 in their development and in surface improvements. The mine has upward of 1,500 feet of development; the mill is equipped with 20 stamps and 16 Frue vanners.

In the Bridal Veil Basin are prospects with from 10 to 50 feet development, carrying both gold and silver. The Lewis has 140 feet of development and 40 tons of ore on dump. This mine shipped some ore in 1882, the average value of which was $60 per ton.

On Mount Wilson there are a number of prospects carrying galena and gray copper. Among the principal ones may be mentioned the Cow Boy, Fanny Forest, American Girl, Western Wonder, Black Prince, Lone Star, and St. Julian.

In South or Gold King Basin are the Gold King and Minnie Myrtle.

The Gold King has a 4-foot pay streak, the ore from which averages by mill run $50 in gold per ton.

The Minnie Myrtle has about 250 feet of development. The ore carries both gold and silver. The owners of these mines erected a 10-stamp mill with vanners in 1880 and are increasing the capacity by 10 stamps. Their mines are located high on the mountain, and the ore is taken to the mill by a tram 2,600 feet in length.

The Alta belongs to the Silver Mountain Mining Co., which has been actively developing it since 1881. The present workings consist of 3 levels aggregating 700 feet. The ore is gray copper and sulphuret, which is found in streaks and pockets.

Placers.—The Keystone placer during the past two seasons has been operated continuously with very successful results.

The St. Louis & Lower San Miguel Placer Mining Co., after expending $30,000 and making a trial run, with the result of cleaning up $1,300 from 2,200 cubic yards of gravel, became involved in litigation and was forced to suspend operations.

The Keokuk Hydraulic Mining Co., purchased during the past 18 months the large bodies of gravel known as the Kansas City and Montana bars and proceeded to develop them by a ditch 11 miles in length, which they have recently completed.

Four miles above the mouth of Leopard Creek are situated the Wheeler bars, upon which a tunnel 8 feet high, 5 feet wide, and 96 feet long produced over $4,000. The Willow Creek bar, owned and operated by a party of Pennsylvania capitalists, has produced an average of 75 cents per cubic yard of gravel for all washings upon their claim.

Iron Springs district is situated in the southern part of the county, covering the drainage of Howards and Lake forks of the South Fork of the San Miguel River, with Ophir as the center.

He then gives descriptions of development work on the Summit, Tip Top, Lookout, Nevada, Carribeau, Butler, Vulcan, Silver Bell, Tidal Wave, Magnolia, Mohawk, San Juan, Globe, and Osceola properties at Ophir.

Burchard's report for 1883 gives the following details:[34]

San Miguel County was created by the legislature of this State during its last session, it having been a portion of Ouray County. * * *

The county is divided into three mining districts: Lower San Miguel, Upper San Miguel, and Iron Springs districts.

In Lower San Miguel placer mines are the chief source of production, but owing to the scarcity of water these have been worked but little during the past year.

Upper San Miguel includes Marshall Basin, Pandora, and the mines tributary to Telluride.

[32] Burchard, H. C., Report of the Director of the Mint upon the production of the precious metals in the United States during the calendar year 1881, p. 420, 1882.
[33] Burchard, H. C., op. cit. for 1882, pp. 515–523, 1883.
[34] Burchard, H. C., op. cit. for 1883, pp. 236, 238, 240, 417–426, 1884.

Iron Springs district embraces those mines in the vicinity of Ophir and Ames.

In Upper San Miguel district, immediately over the crest of the divide that separates Ouray and San Miguel counties, on the western slope of Virginius Pass, is Marshall Basin, and here is located the celebrated vein on which three adjoining claims, the Mendota, Sheridan, and Smuggler, * * * [have been developed].

Beginning at the Mendota, which is just over the divide from the Virginius, the claims on the vein going in a southerly direction are as follows: Mendota, Sheridan, Smuggler, Union, Cleveland, Bullion, Hidden Treasure, Ajax, and five others.

He next gives descriptions of development on the Mendota, Sheridan, and Smuggler.

The Union and the adjoining claim, the Revenue, are owned by the Dolobran Mining Co. * * * On the Cleveland, Bullion, and Hidden Treasure a small amount of work has been done and a very small amount produced. * * *

Adjoining the Hidden Treasure, and running into the Cleveland, is the Cimarron mine, one of the rare mines which has paid for its own development, and at a time when Marshall Basin was comparatively inaccessible.

The Pandora & Oriental * * * owns a 40-stamp mill. * * *

The Cincinnati * * * ore sampled by Munn Bros., Ouray, and sold to * * * Pueblo Smelting & Refining Co., 1,266 pounds, 78¼ ounces gold, 23⅛ silver. * * * The Argentine, which has been worked for several months past; * * * during the summer a large amount of ore has been reduced by a mill above Telluride, with satisfactory results. * * * Bear Creek possesses numerous prospects. * * * The Ballard has been one of the steady producers. During the past summer * * * taking out a large amount of gold-bearing quartz, which was packed down to the Golden group and Andrus mills. * * * The Nellie has shipped some ore. * * * In Turkey Creek the Gold King * * * has been continuously at work. * * * The company owns * * * a 20-stamp mill.

In Lower San Miguel district most of the placers have been idle and but few of the mines have been worked.

Iron Springs district, or Ophir district, the first name being used only in location certificates. * * * Since 1877 the town has continued to advance in prosperity.

Ore shipped during 1883 from Iron Springs district

Name of mine	Ore (short tons)	Silver (ounces)	Gold (value)	Lead (pounds)
Nevada		32,650		155,000
Silver Bell	140	16,800		67,300
Carribeau	200	10,000		200,000
Summit	140	18,480		140,000
Lookout and Tip Top	160	9,600		160,000
Mohawk	2	232		960
Globe and Suffolk	7		$910	
Grand View	8	3,600		2,400
Winnifred	3	300		2,000
Santa Cruz	35	2,800	36	21,000
Tidal Wave				
Pitkin				
Butler				
Spar	21	2,100		12,600
Deadwood				
Staatsburg				
	716	96,562	946	761,260

The ore shipped from Trout Lake mining district, and also from vicinity of Ames, San Miguel County (late part of Ouray County), Colo., during year 1883 was: From San Juan, Fox, Garibaldi, Beatty's, and Jones & Co.'s mines, Trout Lake district 2,500 ounces silver, and 6,400 pounds lead, and 5 per cent copper; from Hard Cash, Colonel Keim's, and Sunlight mines, vicinity of Ames, 3,400 ounces silver and 4,000 pounds lead.

In his report for 1884 Burchard gives the following details: [35]

The Gold King 20-stamp mill has been steadily operated and shipped over $50,000 in gold bullion.

The Golden group mill turned out $53,000 in gold from ore of the Nellie mine.

The Belle 10-stamp mill and the Bossick & Leslie mill, of 10 stamps and six Frue vanners, have been completed and will handle both gold and silver ore.

The Boomerang is erecting a tramway and has made arrangements for a 20-stamp mill and concentrator.

The 20-stamp mill of the Pandora Co. has been run on ores of that and the Oriental mine and has also been concentrating low-grade ores of the Sheridan, Cimarron, Smuggler, and Mendota.

The McFarlane 5-stamp mill has been busy during the season, as has also that of Goebel & Lane, at Ophir.

In Marshall Basin the mines that produced in 1883 have continued to improve with each foot of development. On the vein known as the Smuggler, which can be traced over 2 miles, are located the Mendota, Sheridan, Union, Cleveland, Bullion, and Hidden Treasure. The Cimarron, Revenue, and Grand Central, which cross it, are equally rich in quantity and quality.

The Sheridan, on this same vein and south of the Mendota, has been a lively shipper during 1884. * * * The season's work has been almost wholly in the line of development. * * * From this work have been taken 1,400 tons of ore, nearly 400 of which have averaged $225 to the ton. The balance of the output, averaging from $50 to $55 per ton, has been concentrated at the Pandora mill, the concentrates running up to $400 to $450 per ton. * * *

The Union and the Revenue have been doing considerable dead work. On the Cleveland, Bullion, and Hidden Treasure but a small amount of work was done during the year, and no ore shipment reported.

The Cimarron has produced a large amount of ore rich in gold, of which 320 tons were shipped to Denver and Pueblo, and over 600 tons of concentrates were worked at the Pandora mill. It has three levels, the lowest reached by a crosscut of 98 feet in length, from whence a drift has been run 176 feet on the vein, from which a winze is being raised to the second level. The drift has been continued about 200 feet from this winze and is now in ore carrying rich copper. * * *

The Flora mine, in the same basin, has been worked under contract. Work on the main shaft having been completed, the shaft being down 110 feet, another contract has been let to run a 100-foot drift on the vein, beginning at a point 150 feet from the top of the main shaft. * * *

From the levels of the Pandora & Oriental considerable ore has been stoped in 1884, which was crushed by stamps belonging to the company. Over 3,000 tons of ore have been removed, partly from the opening of the galleries and partly from the stopes. The character of the ore is a copper and iron pyrites, carrying free gold. * * *

In Bridal Veil Basin are numerous prospects. * * *

The Summit Creek mine, though possessing but a small amount of development, has shipped considerable ore that was rich in gold.

On Bear Creek is the Nellie, a mine on which dead work has been done for some years, and from which in two months' time the shipments were such that the owners were reimbursed for all expenditures, and the company now has a good working capital. * * *

[35] Burchard, H. C., op. cit. for 1884, pp. 177, 244-247, 1885.

Trout Lakes is a new district in the southern part of the county.

The San Bernardo and Honduras, two locations on the same vein in the Lake Fork Pass, and the Garibaldi, Fox, and Kent, and many others have produced some high-grade mineral.

In the vicinity of Ames is the Gold King mine. The mine was worked this season from about the first of July. The last run of the mill was very satisfactory, and although only 5 and 10 stamps were kept going the gold produced more than paid the ordinary running expenses. The mill has 20 stamps. The Hard Cash has been worked during the season, the development * * * consisting of a shaft 175 feet deep. Some rich shipments were made last fall. Seven tons, in two lots, returned 119 and 122 ounces in silver per ton.

Ophir district has now quite a number of producing mines and has added during 1884 considerable to the output of the State. The Nevada has been a steady producer in 1884. The second-class mineral runs 157 ounces; first-class, 254; while choice lots reach as high as 500 and 600. The first shipment made to Silverton this season crossed the range June 15; since then regular shipments have been made. Five tons of ore were sent to Montrose, being the first lot shipped that way.

For 1886 to 1896 the figures given are taken from reports of the agents of the mint in annual reports of the Director of the Mint, the gold and silver being prorated to correspond with the figures for the total production of the State as corrected by the Director of the Mint, the lead being prorated to correspond with the total production of lead in the State as given in Mineral Resources, and any unknown production in the State being distributed proportionately among the counties. As with lead so with copper, but as the figures for copper given in Mineral Resources include copper from matte and ores treated in Colorado, although produced in other States, they are subject to revision.

In his report for 1885 Wilson[36] gives a statement of the gold and silver of domestic production deposited at the Denver Mint during 1885 by counties of Colorado.

In Munson's report for 1887 he gives a list of producing and nonproducing mines which shows the individual output of the producing mines.[37] In this list he credits the San Miguel Hydraulic Placer Mining Co. with $2,600 in gold.

In his report for 1888 Munson[38] lists the producing and nonproducing mines and gives the individual output of the producing mines.

In Smith's report for 1889 he gives the estimated total production of San Miguel County.[39]

In his reports for 1890, 1891, and 1892 he lists the producing mines and gives their individual outputs.[40]

He credits the Keystone placers with $18,000 in gold in 1890 and the San Miguel Hydraulic Placer Co. with $93 in 1891.

In his report for 1897, Hodges says:[41]

Iron Springs district.—* * * Concentration and amalgamation; two 10-stamp mills, one 10-stamp mill in course of erection.

Iron Springs district, at Ophir.—* * * Concentration and amalgamation; three mills, operating 60 stamps; one 30-stamp mill in course of erection.

New camp of Sawpit.—* * * No mills.

Upper San Miguel district.—* * * Concentration, amalgamation, and smelting; six stamp mills, operating 280 stamps, and one Huntington plant, representing eight Huntington mills; one 10-stamp mill in course of erection.

Mount Wilson district.—* * * Amalgamation and concentration; one stamp mill.

For 1897 to 1904 (and for 1897 to 1908 for copper) the figures which represent smelter and mint receipts, are taken from the reports of the Colorado State Bureau of Mines.

In his report for 1898 Hodges gives the following details:[42]

The county of San Miguel is divided into four mining districts, known as Upper San Miguel, Lower San Miguel, Iron Springs, and Mount Wilson. Of these the Upper San Miguel district, in which the town of Telluride, the county seat, is located, contains the greatest amount of development work and the largest number of producing mines. * * *

The Upper San Miguel district has been subdivided into sections, known locally as Marshall Basin, Savage Basin, Bridal Veil Basin, Bear Creek, and Turkey Creek Basin. * * *

Marshall Basin is perhaps the best known of these different localities from the famous Smuggler-Union vein. * * * There are three mills in operation connected with the mines of this basin. Formerly the mills were used as amalgamation mills, but for several years no attempt has been made to amalgamate the ores of Marshall Basin, the mills being used entirely for concentration of the lower grades of ore. * * *

Savage Basin, lying just beyond Marshall Basin, contains several * * * gold producers, and at the same time others are producing silver-lead ores in good quantity. These mines are in several instances provided with mills for amalgamation and concentration in the case of gold ores and for concentration in the silver-lead properties.

The largest of these properties has been in active operation for about five years, and owing to its location has heretofore been worked through a tunnel from its mill, a distance of nearly 3,000 feet. During the past year a great deal of development has been accomplished, and it is estimated that the mine has several years of work for its mill in the ground already blocked out. The mill has a capacity of 160 tons every 24 hours and is in constant operation. * * *

Another of the steady producers in this basin has a very high grade gold and silver ore and concentrates its lower grades, having a 10-stamp mill operated by electricity.

The Bridal Veil district, while long known as in the gold belt of this section, has been comparatively neglected until this year.

This section can almost be considered new territory, and, although no ore has been shipped from the basin, yet a mill

[36] Wilson, P. S., agent for Colorado, in Kimball, J. P., Report of the Director of the Mint upon the production of the precious metals in the United States during the calendar year 1885, p. 136, 1886.

[37] Munson, G. C., agent for Colorado, in Kimball, J. P., op. cit. for 1887, pp. 184-185, 1888.

[38] Munson, G. C., agent for Colorado, in Leech, E. O., Report of the Director of the Mint upon the production of the precious metals in the United States during the calendar year 1888, pp. 124-125, 1889.

[39] Smith, M. E., agent for Colorado, in Leech, E. O., op. cit. for 1889, p. 152, 1890.

[40] Smith, M. E., agent for Colorado, in Leech, E. O., op. cit. for 1890, pp. 128, 140, 1891; idem for 1891, p. 184, 1892; idem for 1892, p. 129, 1893.

[41] Hodges, J. L., agent for Colorado, in Roberts, G. E., Report of the Director of the Mint upon the production of the precious metals in the United States during the calendar year 1897, p. 120, 1898.

[42] Hodges, J. L., agent for Colorado, in Roberts, G. E., op. cit. for 1898, pp. 94-98, 1899.

has been built consisting of 10 stamps, making three mills in this section during the past season, and the probabilities are that the coming season will see a very material addition to the gold output of the county from these claims.

Bear Creek lies directly west of Bridal Veil Basin. * * * One of the largest mills in the county, located here, has until recently been shut down, but is now leasing 40 of its stamps to a contiguous producing property.

Another property in this basin, which has a large vein of refractory ore, has erected a concentration mill of 40 tons daily capacity, but so arranged that its output can easily be doubled.

A 10-stamp concentrating mill of 20 tons capacity has been erected on another property, and a tramway from the mine nearly 3,000 feet long.

Turkey Creek Basin has produced gold ore for a number of years. It boasts a very strong free-milling vein, which has been worked more or less extensively for the past 18 years. The principal claim on this * * * vein is connected with its mill by a surface tramway. The mill is provided with 40 stamps and plates for amalgamating, the tailings being treated on concentrating machinery. Both at the mine and the mill the machinery is run by electricity. During the past year a new and very complete mill has been erected in this basin of 200 tons daily capacity. The crushing capacity of the mill is 300 tons per diem, and roasting furnaces of over 200 tons daily capacity are also in place. The ore passes into a 12 by 22 crusher, then through a dryer of same capacity, thence through high-speed rolls, where it is further crushed and passed through a one-eighth-inch screen. It is then sent through Bruckner roasters and again passed through rolls, being reduced to a fineness of 20 mesh, passing again through a set of finishing rolls, leaving them at a fineness of 40 mesh, when it is fed into Huntington mills and then passes over amalgamated plates.

The Iron Springs district lies in the southern part of the county of San Miguel, and includes the large number of mines and mills which have the town of Ophir as their distributing center. The mines of this section are principally gold producing, although some very rich argentiferous galena veins are worked. Some 6-stamp mills are in operation, all provided with concentrating machinery. * * *

The Mount Wilson district lies west of the Iron Springs district, on the southern line, dividing Dolores and San Miguel counties. A number of very good prospects are being developed. The most important producing property [the Silver Pick] is located on the north slope of Mount Wilson, about 10 miles due west of the town of Telluride, and is one of the highest located mines in Colorado. * * * This property has been worked continuously for 10 years and is opened by nine levels, the lowest of which is at an altitude of over 13,000 feet. * * * The mill is over a mile from the mine, connected by a wire-rope tramway.

The Lower San Miguel district is north of Telluride. Very little work was done during the year. Its principal town is a camp called Saw Pit. The mines are more of the character of deposits in the lime formation. * * * As a rule, the ores are low in grade and values chiefly in silver and lead. Some desultory placer mining has been carried on during the past season on the San Miguel River. * * *

Electric power.—Probably no section has so much machinery moved by electricity as * * * San Miguel County. The Telluride Power Transmission Co. has its generating station at Ames, utilizing the confluence of Lake and Howards forks of San Miguel River, which forms South Fork and is about 10 miles from Telluride. This company also controls the water from Trout Lake, some 2½ miles from the generating station, and obtains a fall of 1,500 feet in that distance. The fall from the waters of South Fork is about 900 feet. The * * * pressure obtained is used on two Pelton water wheels, which drive a line shaft actuating two generators. * * * Their combined capacity is estimated at 1,500 horsepower. The greatest distance to which the current is carried is to the Camp Bird mill and mine, in Ouray County, 17 miles, where 100 horsepower is used.

In Savage Basin the Tom Boy mine uses * * * 170 horsepower for its mill, from this source. The Japan mine, in the same basin, uses a 50-horsepower motor for its mill and 20 horsepower for its mine. Near by, the Columbia Menona uses two 60-horsepower motors in its mill; the Valley View mill 20 horsepower. Crossing to Bear Creek, there is the 120-stamp mill of the San Miguel Consolidated Co. using 80 horsepower; but the Nellie mine and mill are operated and lighted by their own power generator at their station on Bear Creek. The Euclid Avenue mill is furnished by power from the Transmission Company. This is 10 miles from the plant. In the Turkey Creek basin the Bessie mine and mill use 100 horsepower, and the Gold King mine and mill are lighted and worked by a 40-horsepower motor. At the Liberty Bell mill, just above Telluride, a 50-horsepower motor is used, while in the town of Telluride several small shops use the power in addition to 15 arc lights and about 1,500 incandescent lamps for illuminating the town.

Tunnels.—Several tunnel enterprises have been pushed, notably the Meldrum tunnel, which will cut through the main range from Pandora, some 2 miles from Telluride to Ironton, in Ouray County. This tunnel will cut the veins of Marshall and Savage basins in San Miguel County and the producing properties near Ironton, in Ouray County. The bore will be about 4¼ miles in length by 12 feet in height and 12 feet in breadth, and will cost between $4,000,000 and $5,000,000. It is the intention to utilize this tunnel for railroad purposes as well as for mining and drainage.

In his report for 1899 Hodges gives the following details: [43]

In 1898 the Telluride shipments reached 17,908 tons, and Ophir, 6,850 tons. Saw Pit was credited with an output of $75,000 at the mills and smelters. * * *

Shipments by months from Telluride, Colo., 1898–1899, in carload lots

Months	1898	1899	Months	1898	1899
January	94	90	August	184	131
February	87	92	September	161	240
March	94	123	October	150	124
April	103	102	November	163	125
May	134	114	December	160	130
June	141	34			
July	152	5	Total for year	1,628	1,310

At 11 tons to the car, the average figure, it will be seen that Telluride shipped 14,410 tons in 1899, compared with 17,908 in the year previous. At Ophir the Carribeau, Shoemaker, and Butterfly were closed part of the summer, hence the shipments from that camp fell from 6,850 tons in 1898 to 5,480 in 1899. * * *

As has been the case for some years past, the Smuggler Union occupies first place as shipper. The year's output from this well-developed property again approximated $1,000,000 in value [but see table, p. 224.—C. W. H.]. Early in May the Smuggler Union property was sold by its Denver and China owners to the New England Exploration Co., made up of Boston men, on an estimated basis of $2,000,000. A crushing plant was installed at the plant so that the ore is now crushed before being trammed to the reduction plant. The latter during the summer

43 Hodges, J. L., agent for Colorado, in Roberts, G. F., op. cit. for 1899, pp. 104–107, 1900.

months was doubled in size, which means a largely increased output for 1900. * * *

The second shipper on San Miguel's list is the famous Tom Boy, * * * [which was offered] in London in December, 1896, at a valuation of $2,000,000.

In June a London company, known as the Tom Boy Gold Mines, was incorporated and took the property. * * *

In his [report the manager] places the output at $13,000 to $15,000 per week, with development proceeding rapidly from the 300-foot, 500-foot, and 700-foot levels. The main shaft, now over 800 feet in depth, is being sunk still further. * * *

[The Japan-Mikado group] became prominent in 1897, when it produced $112,500 in silver, $87,500 in gold, and $50,000 in lead. In 1898 it showed a gain of 25 per cent [with a] total for the year [of] $312,500. The property includes nine patented claims on the extension of the Tom Boy vein and near it, and also the Park placers north of the Japan property. The output for 1899 was stated as $420,000. There is a concentrating plant attached. * * *

The Liberty Bell Mining Co. * * * commenced extensive development last spring and has a large area of milling ore blocked out ready for the treatment plant it proposes to erect early this year. The capacity will equal 200 to 300 tons per day, which the mine can easily supply. [The ores contain] gold, silver, and lead * * * extracted by pan amalgamation and cyanidation, the latter [method being used] for the mill tailings.

The Liberty Bell Co. owns half a dozen claims in San Miguel and as many more on the Ouray side. The company is a subdivision of the United States & British American Exploration Co.

In 1898 the Nellie and Ella group, on Bear Creek, produced gold at the rate of $30,000 per month, but litigation over the territory caused the mine to stop shipping early last spring and cut its output for the year by 50 per cent.

In Mount Wilson district the [Silver Pick mine] is prominent. It paid nothing last year because of labor and other troubles of an unexpected nature. The property * * * includes 13 claims. Concentrates produced at the company's mill are worth $200 per ton. The mine was closed during the eight-hour discussion. When this was arranged, water was met near the surface in stoping out the ore. This caused another delay, the year showing a loss of 50 per cent in output. The concentrates show 90 per cent gold and 10 per cent copper and silver. As the ore is of the arsenical-iron class, it is not easily treated. The mine is 13,000 feet above sea level.

During the second week of November last the Carribeau and Klondike properties, both in the Ophir district, changed hands. In the case of the Carribeau, the Venture Corporation (Ltd.), of London, were the purchasers. For years the Carribeau has been the chief producer of the camp, affording employment to about 100 miners. * * * There is a milling plant of 20 stamps attached. The net value of 386 samples was $20.75 per ton. The new owners will increase the concentration plant to 50 rapid-drop stamps, which will be ready for the ore by the first of May.

The following mills were in operation in San Miguel County:

	Tons per day		Tons per day
Smuggler Union	400	Allegheny	50
Power Company's (mill) Plant	250	San Bernardo	50
Liberty Bell	240	Silver Pick	45
Bessie Cyanide	200	Turkey Creek	40
Tom Boy	200	Illium	25
Japan Concentration	100	Butterfield	20
Gold King	80	Euclid Avenue	20
Suffolk	80	Carribeau	20
Hector	75	Little Mary	20
Butterfly	60		
Columbia-Menona	60	Total tons per day	2,035

This shows a gain of 575 tons capacity per day over 1898, which will be increased largely. The use of electricity as motive power and its transmission long distances by wire are features of mining and milling in San Miguel [County]. The Telluride Power Transmission Co. [also] maintains one of the largest mills in the county. The [electric] plant dates back to 1890 and has recently extended its wires to the Camp Bird group, in Ouray County. * * * The plant's location is at the junction of two forks of the San Miguel River, 12 miles from Telluride. The waters of Howards Fork and Trout Lake are aided by a flume of 4,400 feet in length, with a head of 525 feet. The Trout Lake Reservoir covers 160 acres. The power is carried 13½ miles to the Tom Boy mill, and it furnished light to the residences and business houses in Ophir and Telluride. * * *

In Middle Basin, the Montana group, with gold, silver, and lead as the values, is being exploited by the Virginius Co. The ore vein has been cut at * * * depth by the Ophir tunnel at a point where the body is 25 feet in width. The new buildings at the mine were destroyed by fire last fall, but will be speedily replaced. The ore is treated at the Cimarron mill, which will be increased to handle from 250 to 300 tons per day.

Hodges mentions the developments in 1900 in his report for that year as follows: [44]

Ophir district.—In the Ophir mining district, the Globe-Suffolk Mining Co.'s property is one of the most extensively developed and perhaps the best equipped in the district, but owing to litigation it has not produced largely. It has installed a new 40-stamp mill with the latest improved amalgamating appliances and two Frue vanners for concentration. * * * The Butterfly-Terrible group has made rapid strides during the year. It is an old property, which was worked in a desultory manner, but under new ownership and management development work has been actively carried forward. A new 30-stamp mill has been recently erected, containing Frue vanners and concentrating appliances of latest designs. * * *

Upper San Miguel district.—The Tom Boy Gold Mines Co. has accomplished about 4,000 feet of development during the year. The shaft is equipped with a double-drum hoisting plant capable of hoisting 500 tons per day. The machinery is driven by compressed air and electric power. * * * The mill, with a capacity of 200 tons per day, is situated at the mouth of the tunnel on the 600-foot level. The ore as it comes from the mine is dumped into the bins, passing thence through three crushers and two sets of rolls. From here it passes through automatic feeders into eight Huntington mills, thence to the amalgamating plates. From the plates the pulp is treated by Frue vanners, any values remaining being concentrated in this process. The ore is practically free-milling, and the mine produced the past year 73,741 tons. * * *

The Liberty Bell Gold Mining Co. has put into operation a milling process of cyaniding the ore that is unlike any other process now in operation in the State. The cyanide process is what is known as the South African method of direct treatment. As the ore leaves the mine it is fed into the crushers and then loaded into the tramway buckets automatically and conveyed to the mill, some 2 miles away. It is transported at a cost of about 20 cents per ton. * * *

At the mill the ore is discharged into bins, and from here into automatic feeders to the batteries. From the batteries the ore passes over amalgamating plates * * * and then to the Wilfley concentrating tables. From the concentrators the tailings pass into the cyaniding department. This method is designed to save the refractory contents that usually escape. The tailings from the Wilfley tables are received into tanks 33 feet in diameter and 8 feet deep, with a capacity of 300 tons.

[44] Hodges, J. L., agent for Colorado, in Roberts, G. E., op. cit. for 1900, pp. 133–136, 1901.

Western San Miguel County.—During the year there have been made a few * * * discoveries of ore deposits carrying a fair percentage of copper, in the western end of the county, close to the Utah line, and extending north into Montrose County. The long distance the ore requires to be hauled by wagons to railroad has greatly retarded development work. From tests made it * * * is believed that the lixiviation process is best adapted for treating the ores, and one plant was erected the past year.

In his report for 1901 Hodges gives the following details: [45]

In the mining districts of San Miguel County during the year 1901 * * * the tonnage [fell] off only about 10 per cent from that of the year previous, despite the bitter labor trouble in the spring of the year, which practically tied up the county's leading producer—the Smuggler Union—for more than three months, and the disastrous fire in November, which destroyed the buildings and tramway terminals of the same company at the mouth of Bullion tunnel, and caused the loss of 24 lives by suffocation, the tunnel acting as a strong flue for the ingress of heavy smoke from the burning buildings.

Happily the labor difficulty was finally adjusted, an agreement being concluded as to wages and hours of labor which is to be held inviolate for the term of three years and appears to guarantee stability of labor conditions in the Telluride district for at least that period. * * *

The tonnage loss for the year was confined to the Telluride district, the only one affected by the discouraging incidents recited. Ophir, Placerville, and Saw Pit shipping points actually advanced their records some 55 carloads over 1900.

Concentrates formed about 90 per cent of the shipments, and as high as 20 tons of crude ore were resolved into one of shipping product by the Liberty Bell, which now successfully cyanides its tailings, thereafter passing the slimes over * * * canvas tables. * * *

The holdings of the Ophir Consolidated Co. [consist] of 71 claims, covering all the acreage between the Butterfly-Terrible and Carribeau. * * * Development has been heretofore limited to the Silver Bell and Butler veins.

Fifty stamps additional to the 20 now in operation at the mill completed in midsummer will be * * * supplied, thus increasing its capacity to 250 tons. The concentrates are produced 1 ton from 5, and bring a smelter return of $350 to $500 per car, about 40 cars monthly having been shipped and four of crude ore yielding $400 per car. * * *

The Smuggler Union Co., notwithstanding the handicaps of an extended labor strike and disastrous conflagration, fell off but little in its gross product from the year previous, having milled 106,389 tons crude ore, shipped to smelter at Durango 15,246 tons crude ore and 9,546 tons of concentrates (dry weight), consigned bullion to mint 23,799 ounces, the total valuation exceeding $800,000. Development work consisted of extending drifts and sinking shafts 3,500 feet.

The Smuggler mills.—The older of its two mills at Pandora was substantially renewed and enlarged from 60 to 80 stamps, 30 of these crushing good-grade gold ore from the Contention mine, up Bear Creek, which property was purchased by the Smuggler for $100,000 and connected with its mills by a Bleichert tramway over 15,000 feet in length. * * * The completion of its modern cyanide plant of 98 by 225 feet ground dimensions, capable of treating 400 tons of tailings daily, will materially increase its showing the coming year. * * *

Tomboy gold mines.—The Tomboy Co. during 1901 shipped over 4,800 tons of concentrates and nearly 50,000 ounces of gold bullion, the total valuation of its production exceeding

$800,000. The mine levels were extended over 4,000 feet, and in the different workings 7,000 feet of development done. Much of the tonnage was from the stopes over the 300-foot level, the reserve there having been depleted about 30,000 tons. Manager Herron's estimate of available ore over the 300-foot level is 25,000 tons. Toward the year's close milling was inaugurated from the low-grade stopes over the 500-foot level, a large body having been proven in that territory. The present stopes of the shaft levels are believed to indicate 15,000 available tons.

This company has purchased the Argentine Nos. 1 and 2 lode claims, Fraction and Red Cloud claims, and Argentine Nos. 1 and 2 mill sites. Two thousand feet of development was accomplished on the Argentine and 1,000 feet on the Cincinnati, also carrying the Argentine vein.

The Tomboy Co., purposes the speedy erection of a 60-stamp mill, which will virtually double its milling capacity.

The mill of the Liberty Bell Gold-Mining Co., located just without the city limit of Telluride, treated during 1901 approximately 70,000 tons of ore, which yielded 1,020 tons of dry concentrates, 21,550 ounces gold bullion, and 35,000 ounces bullion from cyanide plant. The Liberty Bell milling plant * * * comprises 80 stamps, extensive concentrating machinery, and a cyanide adjunct, with daily capacity of 300 tons. A * * * slimes plant of 100 tons capacity is handling the residuum of the cyanide plant.

Placers.—The Keystone Hydraulic Mining Co., is prepared to begin extensive operations on San Miguel River in the spring of 1902. It has under construction a dam across East Fork of San Miguel River, a 2,500-foot flume 6 by 12, and is installing an adequate pipe line. Two 10-inch giant nozzles will be placed, capable of washing 12,000 to 20,000 cubic feet of gravel daily, which has been sampled and shown to carry from 10 cents in surface dirt per yard to $1.50 in bedrock. * * *

The Butterfly-Terrible properties, on the western extremity of Yellow Mountain during the year advanced development about 1,000 feet, laid a 1,200-foot pipe line from Wilson Creek to the mill flume, and strengthened its milling plant, which treated 14,825 tons of ore, yielding 20 per cent in concentrates and the remainder free gold.

In his report for 1902 Downer says: [46]

The output of 1902 in San Miguel County has been confined to a comparatively few mines, the Smuggler Union, Liberty Bell, and Tomboy being by far the largest producers. In addition to these, the Ophir Consolidated Mines Co., the Ella Gold Mining Co., the San Miguel Consolidated Gold Mining Co., the Alta Mines Co., and San Bernardo have all been running intermittently.

The Smuggler Union Co., had an active year in spite of the closing down of all their works during the latter part of 1902, owing to labor troubles and the assassination of the general manager, Mr. Arthur Collins.

The Contention mine, one of the properties owned by the Smuggler and situated in Bear Gulch, is connected with the mills of the company by a long aerial tramway. This, in connection with the other trams belonging to the company, all of which enter the mill from the different upper terminals, gives a total length of aerial tram of over 5½ miles.

Smuggler mills at Pandora.—The large company mill at Pandora is worked in conjunction with their smaller mill. At the upper terminals of the different tramways the ore is crushed to pass a 2½-inch ring and carried direct to the storage bins at the mills, where it is fed automatically to the stamps. These weigh 1,050 pounds each and drop 100 times a minute, discharging through a 20-mesh screen of 24 wire gage. The ore

45 Hodges, J. L., agent for Colorado in Roberts, G.E., op. cit. for 1901, pp. 137-140, 1922.

46 Downer, F. M., agent for Colorado, in Roberts, G. E., op. cit. for 1902, pp. 98-103, 1903.

then passes over copper plates to amalgamate any gold contained, and the tailings, after being * * * sized and recrushed, are sent over Wilfley and Triumph tables. The tailings from these tables are then conveyed to the cyanide plant, and slimes and sand are distributed in the large percolation tanks. These are wooden tanks 40 feet in diameter and 8 feet high, capable of holding 450 tons of material each. * * *

The overflow slimes, when filling these tanks, are conveyed in launders to the canvas-table plant, which is very extensive, each table being 12 feet square. The concentrates from the canvas tables are sized again and treated on Frue vanners, the concentrates from these concentrated slimes reaching as high as $70 a ton. All the concentrates from the mills are shipped direct to the Durango smelter, and the retorted amalgam and zinc precipitates, after refining, are melted into bars and the former sent to the United States Mint at Denver, while the latter are shipped to the plant at Omaha. * * *

The Liberty Bell mine has been working nearly all the year, but during the month of February on one day three snowslides came down over the company's property, killing 18 men and carrying away the ore bins at the upper terminals of the tramway, and with the ore houses two 10-ton crushers were carried down the hill. It required some time to rebuild the upper terminal, and, in consequence, very little work was done at the mill. During February, March, and April the output was very much restricted. For the year, however, the total amount of ore mined and treated at the mill was 73,916 dry tons. Instead of crushing at the mine, as formerly, the crushers have been moved to the mill and the crude ore crushed at that point, being delivered on to a belt conveyor. * * * After crushing to pass a 2½-inch ring the ore is fed by automatic feeders to the stamps and passed through a 20-mesh screen, as much free gold as possible is plated, and the tailings are then delivered to a double row of Wilfley tables for concentration. The mill has 80 stamps, each weighing 1,050 pounds, with 100 drops per minute. * * * The plates to the batteries are arranged in three steps, with a concave trough in front of the second step for catching any rusty gold or amalgam. The tailings from the concentrating tables are taken by launders to the cyanide plant and fed into large wooden tanks, the slimes overflowing and being conveyed to the canvas-table plant.

The method used in the cyaniding plant is similar to that in the Smuggler Union plant. As soon as the tank is filled with the tailings the required amount of lime is added, and the charge, which averages from 150 to 200 tons, is treated with a dilute solution of cyanide. After percolation the sands are dried by a pneumatic exhaust, which leaves about 15 per cent of water in the charge. When sufficiently dry, the charge is shoveled out through gates in the bottom of the tank and dropped into a corresponding tank directly below the one being discharged. Here the sands are again subjected to treatment of strong solution of cyanide. Of the total ore crushed, 30 per cent of the weight is in * * * the slimes. These are treated in the canvas plant, which consists of a building 200 feet in length and contains three banks of tables, each table being 12 feet square. The first two banks of tables face each other and have a grade of 1⅛ inches to the foot. The third bank is on a lower level and has a grade of 1¼ inches to the foot. When the slimes are received into the building they are sized by passing through four spitzkasten, which make two sizes * * * of the slimes, each of the spitzkasten receiving the same pressure in the upcast stream. The overflow from these sizers is sent to the first two banks of tables and the bottom discharge goes on the lower bank. The material caught on the tables is washed into a launder carrying the pulp to four belt machines of the Triumph pattern, but is

first sized again by settling in four spitzkasten, one over each belt machine. About $1 per ton is extracted from the slimes on the canvas tables, and the product of the belt machines is very clean iron, giving a value of about $70 per ton.

The overflow from the final spitzkasten is again treated in a canvas plant of about 12 tables, and the concentrate from this set of tables gives the remarkable value of $60 per ton. The quantity recovered, however, is comparatively small. The amalgam is retorted and run into bars, which are shipped to the mint. * * * The zinc precipitates from the cyanide plant are treated with sulphuric acid and melted into bars, weighing on an average 250 pounds, and sent to the Omaha refinery. * * *

The Tomboy mine, located in Savage Basin, at an altitude of 11,500 feet, should perhaps be classed first in this district as a dividend producer but does not rank as high in point of tonnage as either of the mines above mentioned during the past year. The ore, however, is of better grade and comes mainly from the Argentine and Cincinnati lodes, situated some half a mile from the Tomboy proper.

The vein mined averages from 8 feet to 16 feet in width and is all classed as milling ore and averages from $15 to $16 a ton, almost all of which is gold. During the past year this vein produced 35,408 tons of ore, all treated at the new mill of the company, which yielded $534,797, at an expense of $242,960. The same year the Tomboy produced 48,644 tons, yielding $329,232, giving an average of $6.77 per ton treated, at an expense of $257,958.

This company does not cyanide its tailings, the two mills being of the same character, amalgamation and concentration, the crushing being accomplished by 1,050-pound stamps and following the general practice of this district in using 20-mesh screens.

The new mill, which was completed during the early part of the season, has 60 stamps, and its record shows that it has averaged 4.32 tons to the stamp each 24 hours. The principal difficulty the mills have to contend with is a scarcity of water during the winter months. * * *

The Ella Gold-Mining Co. has been working during the last half of the year in taking out ore from the Nellie mine. Development work has been kept up during the entire year, and consists of between 6,000 and 7,000 feet of main upraises and levels. This company has in Gold King Basin a 125-stamp mill, of which 35 stamps were devoted to treating the Nellie ore. * * *

In the Ophir district nearly all the work was done by the Ophir Consolidated Mines Co., which operated the Silver Bell-Butler properties. To their 20-stamp concentrating mill they added 30 stamps, and in this part of the mill the ore is plated as well as concentrated. The capacity of the mills is 130 tons daily, and the average value of the ore treated is from $5 to $6 per ton, which is concentrated about 10 into 1. The ore is sized in Hallett sizers and in the Dimmick classifier, being handled on 4 Wilfley tables and 18 Frue vanners.

The Telluride Electric Light & Power Co.'s plant, located at Ames's station, a few miles below Telluride, furnishes the power used by nearly all the mills in operation within a radius of 20 miles.

In his report for 1904 Downer gives only a brief note in regard to conditions in San Miguel County.[47]

For 1905 to 1923 (except for copper for 1905 to 1908 from the Colorado Bureau of Mines) the figures are taken from Mineral Resources (mines reports).

47 Downer, F. M., agent for Colorado, in Roberts, G. E., op. cit. for 1904, pp. 116–117, 1905.

Production of the Tomboy Gold Mines Co. (Ltd.) from date of incorporation, June 7, 1899, to June 30, 1923

[Compiled by Chas. W. Henderson]

Period	Tons of ore crushed	Value of amalgam bullion	Combined value of amalgam bullion and concentrates	Value of concentrates from pulsating classifier	Value of concentrates	Value of cyanide bullion	Value of cyanide slag	Total value of products
June 7, 1899, to June 30, 1899	4,063		$52,983					$52,983
July 1, 1899, to June 30, 1900	52,216		577,660					577,660
July 1, 1900, to June 30, 1901	55,846		649,281					649,281
July 1, 1901, to June 30, 1902	85,726		856,065					856,065
July 1, 1902, to June 30, 1903	82,437	$600,415			$131,246			731,661
July 1, 1903, to June 30, 1904	69,580	330,278			156,324			486,602
July 1, 1904, to June 30, 1905	102,953	484,508			253,430			737,938
July 1, 1905, to June 30, 1906	104,063	574,360			373,615			947,981
July 1, 1906, to June 30, 1907	110,597	873,011			408,677			1,281,688
July 1, 1907, to June 30, 1908	104,091	657,562			341,459			999,021
July 1, 1908, to June 30, 1909	102,844	487,486			332,971			820,457
July 1, 1909, to June 30, 1910	110,560	488,304			328,061			816,365
July 1, 1910, to June 30, 1911	116,222	462,774		$31,590	318,550			812,914
July 1, 1911, to June 30, 1912	107,577	401,432		32,861	520,688			954,981
July 1, 1912, to June 30, 1913	129,618	358,732			681,625			1,040,357
July 1, 1913, to June 30, 1914	137,456	385,612			573,446			959,058
July 1, 1914, to June 30, 1915	145,857	414,114			537,494	$75,916		1,027,524
July 1, 1915, to June 30, 1916	150,488	327,519			586,093	160,476		1,074,088
July 1, 1916, to June 30, 1917	148,939	314,823			675,674	160,114		1,150,611
July 1, 1917, to June 30, 1918	151,028	306,003			661,683	166,569	$1,134	1,135,389
July 1, 1918, to June 30, 1919	155,334	265,796			437,713	158,223	2,967	864,699
July 1, 1919, to June 30, 1920	146,006	107,920			672,735	28,647	2,688	811,990
July 1, 1920, to June 30, 1921	197,557	a 83,145			a 793,211			876,356
July 1, 1921, to June 30, 1922	211,003	a 299,085			a 575,188	310		874,583
July 1, 1922, to June 30, 1923	206,146	248,599			691,615			940,214
July 1, 1923, to June 30, 1924	215,598	176,232			579,185			755,417
	3,203,805	8,647,716	2,135,989	64,451	10,630,683	750,255	6,789	22,235,883

a Includes clean-up of 60-stamp amalgamation mill, abandoned Nov. 22, 1919; replaced by new flotation plant set in full operation January, 1920.

Partial production of the Smuggler Union Mining Co., Telluride, Colo., 1882-1923

Year	Ore treated (short tons)	Gold Fine ounces	Gold Gross value	Silver Fine ounces	Silver Gross value	Lead Pounds	Lead Gross value	Copper Pounds	Copper Gross value	Total gross value	Property
1882		31.00		8,949						10,089	Mendota claim.
1883	105	94.50		18,864						20,196	Do.
1884 a										75,000	Do.
1885	266	361.00		59,805						71,946	Do.
1885-1890	1,927	15,724.00		334,334						598,161	Union claim.
1886		247.00		52,984							Mendota claim.
1887	1,026	4,104.00		256,500						335,450	Sheridan claim.
1882 b											Smuggler claim.
1883											Do.
1884											Do.
1885											Do.
1886											Do.
1887											Do.
1888											Do.
1889										412,000	Do.
1890											Do.
1891	7,928	7,088.00		161,525		4,700				c 220,360	Smuggler Union Mining Co.
1892											Do.
1893											Do.
1894											Do.
1895											Do.
1896	63,848	21,685.39		566,808						c 684,592	Do.
1897											Do.
1898											Do.
1899										470,777	Do.
1900	70,916									564,980	Do.
1901	133,938									837,418	Do.
1902	89,205			720,000						567,823	Do.
1903	114,419	21,792.00		637,121						635,530	Do.
1904	60,626	11,218.00		370,900						350,951	Do.
1905	77,588	15,509.56		463,172						297,010	Do.
1906	140,870	31,585.55		754,368						492,561	Do.
1907	161,791	33,986.74		1,007,020				1,419		588,300	Do.
1908 d	160,805	33,336.98		915,966				680		694,341	Do.
1909	186,688	37,650.75		1,210,500				651		637,976	Do.
1910 d	151,203	28,451.59	$565,493	509,232	$272,368	445,749	$20,059	1,113	$79	857,999	Do.
1911	92,183	14,635.11	291,988	355,169	186,636			481	35	478,659	Do.
1912	102,755	16,071.45	326,599	445,120	258,924	863,105	14,025			599,549	Do.
1913	140,728	18,148.00	362,494	351,649	200,356	867,505	17,005			579,855	Do.
1914	125,800	18,586.04	374,011	503,552	262,372	195,329	2,458			638,841	Do.
1915	105,493	21,625.00	437,640	363,568	172,853	1,023,100	20,718			631,212	Do.
1916	52,804	14,834.00	296,773	92,593	57,847	1,257,900	67,721			422,341	Do.
1917	50,356	11,635.62	232,824	102,526	80,727	1,026,412	59,358			372,909	Do.
1918	57,682	11,384.80	228,089	170,699	159,979	1,396,706	57,974			446,042	Do.
1919	103,826	20,127.00	339,913	487,047	521,935	2,859,257	93,592			1,015,440	Do.
1920	112,389	21,631.24	433,072	417,198	423,822	3,087,919	128,709			985,604	Do.
1921	160,633	21,842.81	435,320	1,065,708	1,006,388	3,698,357	62,029	1,852	46	1,503,783	Do.
1922	128,823	13,935.45		1,086,808		3,061,107		129,894			Do.
1923	169,185	19,024.00		811,198		2,944,522					Do.

a Record of gross returns with no year recorded; assumes year 1884.
b For seven years prior to 1889 there is no record for Smuggler claim except gross ore sales.
c Value less smelting charge and freight.
d There is a slight discrepancy between 1908 and 1910, due to overlapping of reports.

The following table shows in detail the production of the Liberty Bell Gold Mining Co. from December, 1898, to September 30, 1924:

Production of Liberty Bell Gold Mining Co., Telluride, Colo.

[Compiled by C. W. Henderson]

Period	Ore milled (dry short tons)	Conc. (dry short tons) Wilfley	Conc. Canvas	Gold Amalgam	Gold Conc. Wilfley	Gold Conc. Canvas	Gold Concentrate cyanide bullion	Gold Cyanide bullion	Gold Cyanide precip. & slag	Gold Total	Silver Amalgam	Silver Conc. Wilfley	Silver Conc. Canvas	Silver Concentrate cyanide bullion	Silver Cyanide bullion	Silver Cyanide precip. & slag	Silver Total	Gross value Total	Per ton
December, 1898, to Sept. 30, 1899	16,026	202		2,348	365					2,713	1,994	11,420				8,721	13,414	$64,263	$4.01
Year ended Sept. 30, 1900	21,500	200		3,912	379				553	4,844	2,880	10,475				6,677	22,076	114,134	5.31
1901	54,197	883		9,802	1,439			1,436	326	13,003	7,293	28,607			26,583	1,330	69,160	312,221	5.76
1902	67,439	1,200	247	15,276	2,047	450		2,425	299	20,497	10,966	47,328	12,008		43,730	230	115,362	482,427	7.15
1903	83,373	1,789	632	17,471	2,503	1,114		3,032	15	24,135	12,903	61,316	32,605		50,568		157,622	581,422	6.97
1904	45,811	938	371	13,820	1,580	656		2,044		18,100	10,197	42,560	19,927		34,708		107,392	436,246	9.52
1905	27,379	243	107	6,738	269	145		506	1,265	8,923	8,296	10,861	5,511		9,756	16,635	51,059	215,416	7.87
1906	92,900		763	21,940		272		5,106	1,848	29,166	48,273		47,408		83,446	27,385	206,512	741,128	7.98
1907	102,104	1,415		18,622	455			8,494		27,571	41,494	75,714			109,685		225,893	728,261	7.13
1908	116,353	1,758		20,244	682			12,128	107	33,161	43,514	83,691			153,956	2,308	283,469	844,226	7.26
1909	125,681	1,518		17,960	1,206			9,723		33,991	27,625	46,110			116,251	1,667	191,653	702,835	5.59
1910	133,899	2,503		23,893	3,441			11,569	141	39,044	33,976	87,231			153,078	3,858	278,143	959,874	7.17
1911	155,950	2,511		34,213	4,668			20,061	31	58,973	42,012	87,652			201,420	668	331,752	1,399,636	8.97
1912	175,340	2,892		26,147	4,402			18,444		48,993	41,979	91,784			180,082		313,845	1,203,726	6.87
1913	179,216	1,854		18,083	2,743		1,695	14,172		36,693	50,034	59,241		16,466	162,861		288,602	938,733	5.24
1914	173,840	559		10,461	412		1,612	20,665		33,150	24,734	8,593		21,176	261,416		315,919	862,662	4.96
1915	174,100	3,018		99	1,407		(a)	22,549		24,055	578	38,863		(a)	275,297		314,432	662,028	3.80
1916	165,300	3,558		b 24	4,737		(a)	25,780		30,541		35,322		(a)	229,666		256,678	791,788	4.79
1917	143,100	3,410			1,605		(a)	29,565		31,170		31,628		(a)	221,278		261,294	847,360	5.92
1918	90,700	2,349			2,349			44,512		46,861		22,430			176,161		198,591	1,153,043	12.71
1919	110,700	88			55			39,221		39,276					130,358		131,110	949,967	8.58
1920	56,100	147			664			18,275		18,939		15,529			133,047		148,576	565,536	10.08
1921	e 55,285	3,357		156	13,397			669		14,222	102	245,064			5,571		250,737	543,555	9.83
	2,366,293	38,462		261,209	50,442		3,307	313,376	4,687	633,021	408,622	1,259,630		37,642	2,757,918	69,479	4,533,291	16,100,487	6.80
1922	(d)	1,038		412	1,387					1,812	289	7,491					7,780	45,269	
1923	(d)	68		200	398			13		598	133	1,936					2,069	14,299	
1924	(d)	59		106	378					484	105	1,454					1,559	11,034	
		1,165		718	2,163					2,894	527	10,881					11,408	70,602	
		39,627		261,927	52,605		3,307	313,389	4,687	635,915	409,149	1,270,511		37,642	2,757,918	69,479	4,544,699	16,171,089	

a Not precipitated separately and included in cyanide bullion.
b Amalgamation was discontinued in 1916; the amalgam shown in 1916 was derived from the melting of old copper plates.
c Company dissolved in 1920 and since then in process of liquidation.
d Lessees' operations.

Gold, silver, copper, lead, and zinc produced in San Miguel County, 1875–1923

Year	Ore treated (short tons)	Gold Placer	Gold Lode	Gold Total	Silver Fine ounces	Silver Avg. price per ounce	Silver Value	Copper Pounds	Copper Avg. price per pound	Copper Value	Lead Pounds	Lead Avg. price per pound	Lead Value	Zinc Pounds	Zinc Avg. price per pound	Zinc Value	Total value
1875					3,867	$1.24	$4,795										$4,795
1876					3,867	1.16	4,486										4,486
1877					3,867	1.20	4,640										4,640
1878					3,867	1.15	4,447										11,247
1879		$5,000		$5,000	7,734	1.12	8,662										30,012
1880		7,000		7,000	7,734	1.15	8,894				50,000	$0.036	$1,800				38,019
1881		5,500		5,000	19,336	1.13	21,850				350,000	.041	14,350				46,450
1882		5,500	$9,500	15,000	38,672	1.14	44,086				482,500	.05	24,125				63,886
1883		8,000	123,000	125,000	193,359	1.11	214,628				200,000	.048	9,600				373,254
1884		2,000	97,000	100,000	309,375	1.11	343,406				200,000	.049	9,800				454,506
1885		3,000	97,000	100,000	400,000	1.07	428,000				782,000	.043	33,626				539,700
1886		3,000	214,570	217,570	430,805	.99	426,497				300,000	.037	11,100				657,867
1887		2,600	167,096	169,696	492,725	.98	482,871				300,000	.039	11,700				676,738
1888		2,600	417,206	424,706	663,354	.94	623,553				300,000	.046	13,800				1,076,266
1889		7,500	425,588	432,588	726,456	.94	682,869				537,144	.045	24,171				1,160,945
1890		7,000	737,380	755,380	907,148	1.05	952,505				636,514	.044	28,007				1,720,538
1891		18,000	641,993	646,993	1,410,903	.99	1,397,525				1,166,346	.039	45,488				2,050,510
1892		5,000	689,177	694,177	1,501,898	.87	1,306,651				414,522	.045	18,653				2,045,062
1893		5,000	677,680	682,680	932,568	.78	727,403	100,000	$0.116	$11,600	139,344	.043	5,992				1,457,583
1894		5,000	789,218	794,218	570,023	.63	359,115	200,000	.108	21,600	815,842	.04	32,634				1,198,127
1895		5,000	1,421,159	1,426,159	602,039	.65	391,325	173,191	.095	16,453	700,000	.037	25,900				1,857,509
1896		5,000	1,372,829	1,377,829	1,109,875	.68	754,715	147,727	.107	15,807	858,830	.033	28,341				2,203,413
1897		5,000	1,453,144	1,458,144	869,079	.60	521,447	21,698	.108	2,343	756,809	.032	24,218				2,171,341
1898		2,000	1,570,677	1,572,677	2,129,082	.59	1,256,158	354,781	.12	42,574	2,284,191	.03	68,526				3,128,167
1899			1,376,705	1,376,705	1,208,395	.60	725,037	360,831	.124	44,743	4,143,767	.036	149,176				2,305,493
1900			1,827,352	1,827,352	1,136,692	.62	704,749	160,239	.171	27,401	6,699,712	.038	254,589				2,731,285
1901	233,316		2,049,472	2,049,472	916,245	.60	549,747	311,045	.166	51,633	3,918,883	.045	176,350				2,793,018
1902	291,338		2,007,656	2,007,656	1,056,640	.53	560,019	308,322	.167	51,490	3,353,425	.044	147,551				2,709,330
1903	386,735		2,007,656	2,007,656	737,028	.54	397,995	454,790	.122	55,484	3,309,517	.043	142,309				1,794,254
1904	407,491	3,100	1,173,705	1,176,805	687,710	.58	397,272	466,264	.137	63,878	4,296,849	.042	176,171				2,194,301
1905	428,231	44,957	1,486,111	1,531,068	1,275,079	.61	777,798	239,520	.156	30,659	4,704,201	.041	155,576	17,214	$0.059	$1,016	2,860,776
1906	423,609	21,587	1,690,266	1,711,853	1,672,522	.66	1,137,315	272,513	.193	42,512	5,704,708	.043	245,302				4,054,823
1907	481,000	1,766	2,446,024	2,447,790	1,543,187	.66	949,277	319,692	.20	61,701	6,970,152	.047	327,597				3,837,578
1908	429,354	293	2,467,223	2,467,516	1,438,299	.53	817,889	381,437	.132	76,287	7,158,189	.057	408,017				3,554,332
1909	455,696	2,892	2,314,759	2,317,651	1,344,152	.52	698,959	562,888	.13	74,301	6,499,957	.053	344,498				3,305,088
1910	509,175	440	2,284,611	2,285,051	1,144,050	.54	617,787	501,285	.127	65,167	7,135,863	.042	299,706	952,872	.047	44,785	3,643,008
1911	495,742		2,494,793	2,494,793	1,000,834	.615	530,442	544,189	.125	69,112	4,941,370	.043	212,479	804,296	.054	43,432	3,583,208
1912	483,951		2,447,841	2,447,841	1,153,709	.604	709,331	971,064	.165	121,383	7,791,841	.044	342,841	2,193,981	.054	118,475	3,785,726
1913	428,651		2,399,234	2,399,234	1,051,096	.553	634,862	845,497	.155	139,507	6,456,333	.045	290,535	3,386,088	.057	193,007	3,319,647
1914	389,293		2,129,371	2,129,371	1,280,461	.507	708,095	736,374	.133	114,138	7,429,622	.045	334,333	2,943,783	.069	203,121	3,023,668
1915	428,565		2,114,916	2,114,916	1,096,641	.658	555,997	324,105	.175	43,106	6,967,136	.044	306,554	2,405,750	.056	134,722	3,099,074
1916	374,134		2,069,362	2,069,362	812,041	.824	534,323	562,554	.246	143,037	5,240,277	.039	157,551	1,040,121	.051	128,975	3,319,676
1917			2,072,393	2,072,393	779,364	1.00	642,196	581,427	.273	251,276	6,126,551	.047	246,293	1,098,485	.124	147,197	3,621,736
1918			2,009,961	2,009,961	836,570	1.12	836,570	920,425	.247	245,225	6,205,326	.069	422,732	1,810,245	.134	184,645	3,711,145
1919			2,127,634	2,127,634	1,100,942	1.09	1,233,055	992,814	.186	169,990	6,044,085	.086	533,658	797,648	.102	72,586	3,950,886
1920			2,105,490	2,105,490	1,064,667	1.00	1,160,487	913,925	.184	174,560	7,636,790	.071	429,130	515,082	.091	37,601	3,295,248
1921			1,340,226	1,340,226	1,776,963	1.00	1,776,963	948,696	.129	118,883	7,571,875	.053	404,750	175,617	.073	14,225	3,744,297
1922			1,468,820	1,468,820	1,645,459	1.00	1,645,459	921,573	.135	90,972	8,436,244	.08	605,750		.05		3,202,626
1923			1,077,846	1,077,846	1,606,344	.82	1,317,202	673,867	.147	207,120	7,060,891	.045	379,631		.057		3,646,997
			1,373,968					1,408,980			10,695,814		388,349				
Total		188,635	59,261,956	59,450,591	42,082,723		31,579,554	16,681,713		2,742,383	176,813,189		9,061,966	18,141,182		1,323,787	104,158,281

* Interpolated by C. W. Henderson to correspond with total production of the State.

SUMMIT COUNTY

Ransome [49] has given a detailed history of mining in Summit County from 1860 to 1909. According to Kimball,[50] the placers of Summit County produced $5,500,000 in gold from 1860 to 1869.

Raymond [51] gives the following note in his report for 1869:

The silver mines of Summit County have not been developed to any extent; they are mostly strong galena-bearing veins. Silver ores proper, especially brittle silver and ruby silver, are also found; a piece of the latter kind weighing 7 pounds was taken from the Anglo-Norman lode. From a couple of hundred assays made by different parties, the probable average value of the galena ores is indicated to be about $100 per ton. It must be borne in mind, however, that the value of ores can not reliably be ascertained for a whole vein by assays made of pieces. A great hindrance to the development of the lodes in this country and the beneficiation of the ores has been the enormous cost of transportation. At present, however (September, 1869), a wagon road is building from Georgetown to the Snake River mines, which will be completed within a few weeks; and doubtless the improved facilities for communication will reduce prices in every respect.

Raymond gives the following details in his report for 1870: [52]

This county is principally noted at the present time for its rich and extensive placer deposits. Commencing at the headwaters of Swan River, extending around to the head of the Blue, and down the latter stream for at least 20 miles, there is almost a continuous placer, carrying gold in profitable quantities. The ground varies in richness, paying from $3 to $30 per day per hand. Less than $5 per day per hand will probably not pay expenses where labor has to be hired. The report of the United States assistant marshal to the Census Bureau mentions but four claims in Summit County as worked during the year ending June 1, 1870, and gives less than $9,000 as the aggregate product, being about $7 per day per hand. The extreme imperfection of this return may perhaps result from the attempt to obtain information at so unfortunate a period as the 1st of June, when the season has scarcely opened and the miners can not be found. I am indebted to William P. Pollock, Esq., county clerk and recorder, for trustworthy information concerning the operations of 1870, and to Mr. R. J. Burns, of Austin, Nev., for valuable notes of a personal visit to the county.

Montezuma and Breckenridge are the principal mining towns, the former being the headquarters of quartz and the latter of placer mining. The most productive gulches near Breckenridge are Illinois, Iowa, French, Gold Run, Galena, and Georgia, and Buffalo and Delaware flats. Mr. Pollock says that Georgia Gulch alone produced about $3,000,000 from its discovery in 1859 to the close of 1862.

The placer mining season is very brief, lasting but little over five months in the year, yet several claims have each yielded $10,000 per season for several seasons past, and as high as 90 ounces, or $1,575, has been obtained from one week's run of 49 days' work. Several gold nuggets were taken out during last season; one from Georgia Gulch, weighing 9 ounces 3 penny-

weights and 9 grains; one from Galena Gulch, weighing 8½ ounces; and one from Lincoln City, weighing 9½ ounces. The amount of bullion taken out the past season, exclusive of silver, is estimated by the recorder at $350,000. The Georgetown Miner, at the close of the season, said it was nearly or quite $500,000.

The county contains 100 miles of excellently constructed ditches, many of them having several thousand inches capacity, for conveying the water to work the claims, and declarations are on file for the construction of 40 miles more next season.

Two hundred and eighty thousand feet linear measure of placer ground has been preempted since the 1st of May, 1870. The greater part of this new ground prospects very well and gives abundant indications of a large yield with proper management. The claims which have been successfully worked in past seasons, as well as those recently developed, still contain sufficient gold to occupy the miners for years, and as there is an immense quantity of ground yet unclaimed and known to contain mineral wealth in quantities which will repay active and economical working, there is no doubt that Summit County will continue to produce annually increasing amounts of gold.

There are on the county records over 4,000 lodes recorded; but very few of them have been sufficiently developed to show their real value, as the owners of most of them are working their placer mines. The majority of the lodes now under exploitation are situated at Montezuma and St. Johns, in Snake River mining district.

Montezuma is reached by stage from Denver or Idaho [Springs], or by a direct road from Georgetown across the range * * * crossing near Grays Peak. * * * In one of these parks, through which flows the South Fork of the Snake, Montezuma is situated, while Breckenridge is about 20 miles southwest. * * *

The leading mine is, at Montezuma, the Comstock, owned by the Boston Silver Mining Association, Col. W. L. Chandler, superintendent. It is situated on the southwestern face of Glacier Mountain, nearly 12,000 feet above sea level. Mr. Burns describes his visit to the mine as follows:

"Following up the toilsome trail, we reached and entered the lower tunnel of the mine. This tunnel is 150 feet long, from which a level extends 425 feet. We saw masses of ore ready for the hands of the stopers along nearly the whole length of this level. A shaft or winze 70 feet deep connects this level with the lower one, which is 200 feet long, in which we examined the same massive vein loaded with ore. Ascending to the level again, we climbed into a 'stope,' and observed a large body of ground in which the miners had been busy extracting the abundant ore. A winze of 70 feet also connects the lower and upper tunnels, and in the works of the latter the ore occurred in wide strata. The vein of the Comstock stands nearly perpendicular and varies in size. At one point it spread out to 8 feet and at another it contracted to a few feet, but it preserved a general width of 4 to 5 feet. At the point of greatest width there was a stratum of compact ore 2 feet thick upon the headwall; the same upon the footwall; while ore was disseminated through the intervening mass of feldspathic gangue. In the different works disclosing the vein the solid ore ranged from 4 inches to 2 feet thick, and I should judge it fairly averaged 18 inches. The galena was massive and formed perhaps one-third of the ore. Zinc blende and iron and copper pyrites occur also abundantly, and in the deepest works silver glance and brittle silver are not uncommon. Handsome crystals of heavy spar are of frequent occurrence. Tests of the value of the different kinds of ore ranged from $40 to $400 per ton. * * * Captain Ware informed us that it was part of his project to open the mine by a tunnel 450 feet below the present lowest tunnel, which will be 550 feet long and will cut the vein nearly 700 feet from the surface. If this plan is carried out it will open avenues to bodies of ore that will require years to extract. The amount of

9 Ransome, F. L., Geology and ore deposits of the Breckenridge district, Colo.: U. S. Geol. Survey Prof. Paper 75, pp. 16–20, 1911.

50 Kimball, J. P., Report of the Director of the Mint upon the production of the precious metals in the United States during the calendar year 1887, p. 151, 1888.

51 Raymond, R. W., Statistics of mines and mining in the States and Territories west of the Rocky Mountains for 1869, p. 378, 1870.

52 Raymond, R. W., op. cit. for 1870, pp. 328–332, 1872.

ore ready for reduction and piled at the mouth of the tunnel, in the ore house near by, and at the mill, was estimated at 3,000 tons. And if occasion should require it, Captain Ware told us he could easily set 50 men to stoping ore.

"The ore will be delivered from the mine to the mill over a tramway about 2,200 feet long. This tramway, which was building under the direction of Captain Ware, will connect with the present lowest tunnel and ultimately with the projected deep tunnel. Not only ore but the miners and all supplies for the mines will be carried over the tramway. It will be capable of discharging 100 tons of ore from the mine to the mill daily.

"* * * The reduction works of the company are of a very inferior character, * * * wholly inadequate to the treatment of the ore. They consist of a crusher and rollers, a small concentrator, and a reverberatory and a cupola furnace."

According to later reports, the mine has about 1,000 feet of stoping ground and can work 75 miners underground. Sixty tons of ore can be raised to the surface daily and delivered at the mill by the tramway at a cost of about $3.75 per ton for mining and 20 cents for transportation. The company has reduced during the summer 50 tons of ore, at a reduction cost of about $22 per ton. The average yield was $100 per ton. The imperfect apparatus was capable of treating only the galena ores—about one-fifth of the vein material; the remainder, containing from 30 to 40 ounces of silver per ton, being thrown aside. It is proposed to construct a new mill, combining amalgamation with smelting, so that the ores can be reduced.

The Chenango Co. owns the Favre, Chloride, Coley, and G. T. Clark, all highly esteemed lodes. The mine is in Glacier Mountain, about a quarter of a mile farther down the canyon than the Comstock. A tunnel, about 460 feet long in December last, had cut through two veins, assaying about 50 ounces of silver per ton. A short distance below the mine the company has a mill, which is idle, and reported to be of no value.

The Sukey lode, belonging to the Sukey Silver Mining Co., Hon. J. T. Lynch, superintendent, is opened by two tunnels, one 260 feet and the other (150 feet above) 96 feet long. One hundred and eighty feet above the upper tunnel is the discovery shaft, 40 feet deep. The vein is from 4 to 6 feet in width, with an ore streak of 20 inches to 3 feet. The ore exhibits very rich specimens, but the great bulk of it is of a low grade, the average point being between $35 and $40 per ton. The capacity of the Sukey for the production of this grade of ore is very great. The company owns a small mill, 30 by 80 feet in size, and containing five stamps, one roasting furnace, and two Blatchley pans for amalgamation. It is run by water power. Seventy tons, reduced during the summer, averaged 60 ounces of silver per ton, the cost of reduction $22 to $24 per ton. It is proposed to increase the capacity of the mill to 15 tons per day, which will reduce the cost to $15 per ton.

The St. Lawrence Silver Mining Co. owns the Silver Wing and Napoleon lodes, on the north face of Glacier Mountain, a few hundred feet above the South Fork of the Snake. The former is tunneled 30 feet, showing a vein 4 feet wide between walls with an ore streak varying from 10 to 20 inches and carrying by average assay 35 ounces of silver per ton. The Napoleon is tunneled 65 feet, with a crevice similar to the Silver Wing, and assaying about 60 ounces of silver per ton. During the past summer the company has been completing its mill, a very good one, containing a 12-stamp battery and two pans for amalgamating. Arrangements are said to have been made for the erection of a Stewart and Airey furnace for roasting and chloridizing.

The Old Settler lode, owned by Black & Milner, is tunneled 260 feet, and shows an ore streak 2 feet wide, composed of lead, zinc, gray copper, and iron sulphurets. Assays range from 20 to 100 ounces of silver per ton.

The Dysart lode, owned by George W. Packard, has a shaft 30 feet deep, showing a vein 4 feet wide between walls, and an ore streak of 18 inches. Assays give from 30 to 100 ounces of silver per ton.

The Umpire lode, owned by Sharrat & Morrow, has a shaft 20 feet deep, showing a vein 4 feet wide. Assays give from 20 to 60 ounces of silver per ton.

The North Star lode, owned by Lynch, Pratt & Co., is 4½ feet between walls, with 12 inches of ore, composed of lead, zinc, and copper sulphurets. It assays from 80 to 240 ounces of silver per ton.

Guibor's extension of the Coley lode, owned by Guibor & Co., has a shaft 60 feet deep, a vein 4 feet wide, and an ore-streak of 20 inches, assaying from 50 to 200 ounces of silver per ton.

The Tiger lode, owned by Lynch, Pratt & Co., has a shaft 20 feet deep, a vein between walls 6 feet wide, and two pay streaks, one next the north wall, 10 inches wide (heavy galena), assaying 100 ounces of silver per ton, and the other next the south wall, 6 inches wide, assaying from 1,000 to 2,500 ounces of silver per ton. The intermediate rock assays from 16 to 30 ounces of silver per ton.

The Walker lode, owned by Fix & Hewitt, is opened by a shaft and tunnel and worked by the latter, which is in 60 feet. The vein is 2 feet wide and the pay streak about 4 inches. An assay from several tons of ore reduced in the Sukey Co.'s mill gave 206 ounces of silver per ton.

The Chautauqua lode, owned by Teller & Bull, has a shaft 32 feet deep and an ore vein 6 feet wide. About 100 tons of ore are extracted, all of which contains more or less gray copper. Four samples taken from the pile—two from the inferior and two from the best quality—were assayed by Hon. J. T. Lynch, with the following result:

No. 1: 41⅙ ounces of silver; coin value _____ per ton __	$54.08
No. 2: 22¼ ounces of silver; coin value _____ per ton __	29.12
No. 3: 716⅙ ounces of silver; coin value _____ per ton __	931.84
No. 4: 672 ounces of silver; coin value _____ per ton __	873.00

Making an average of 363$\frac{2}{6}$ ounces of silver per ton.

The average of 34 assays, made by Mr. Lynch, agent of the Sukey Co., during the summer, from various lodes in this vicinity, as shown by the assay book, was $143.35 per ton.

Each of the mines above named has ore on the dump ranging from 20 to 200 tons; and there are many other lodes in the district which contain ore in paying quantities. It is believed that as soon as the late improvement made by Mr. Stetefeldt, of Nevada, for roasting and chloridizing ores, is introduced into Snake River district, which is contemplated next summer, it will be one of the most important silver-producing districts in Colorado.

There are numerous other lodes in all stages of development in other portions of the county, many of them exceedingly rich. The Bullion & Incas Mining Co., near the head of Clinton Gulch, in Tenmile district, owns some very good veins, and has run a tunnel 800 feet, passing through several lodes which are said to "prospect" very handsomely. A large number of lodes of decomposed quartz, containing free gold, have been discovered near the sources of our placer mines, and will, it is hoped, be thoroughly developed and practically worked next season. The lodes of Summit County have been neglected in the past, but the coming year will witness an era of development, both in placer mines and lodes, never before known; and it is expected that the yield of the precious metals will double that of any previous year since the settlement of the county.

Raymond's report for 1871 gives the following information: [53]

In Summit County the placer mining season of 1871 has not been as prosperous as heretofore, owing to the small amount of snow that fell during the winter and also to the scanty rain of the summer. The supply of available water has been much less than in average years, and, as a necessary consequence, the amount of bullion produced has been less than usual. Still, the yield of gold per hand per day is reported as nearly one-half ounce, and the total shipments of gold from the county are given as 3,700 ounces. Considerable new placer ground has been discovered and developed; many new ditches have been built, and some companies have made very extensive preparations for next season. Although there were not as many companies at work in French Gulch as the year before, a fair share of placer mining has been done here. * * *

French Gulch is about 5 miles long and, with Stilson Patch, has about 17 miles of ditches, 6,700 feet of flume, 5 hydraulics, and in July had a population of 165.

About the same time Gold Run Gulch was worked by several companies, working 43 men in all and averaging about $10 per day to the man. Buffalo Flats, situated at the lower end of Gold Run, were worked by 14 men. Gold Run and Buffalo Flats are covered by 2½ miles of large ditches and use seven hydraulics. Delaware Flats was being worked by nine men, and has about 6 miles of ditches and 2 hydraulics. Galena Gulch was worked by two companies—one working 10 men and running two flumes and the other, working the upper portion of the gulch, employed a number of men—both companies doing well. Galena is covered by a 5-mile ditch and uses two hydraulics. Georgia, Humbug, and American gulches were owned by six companies, and considered the richest in the county, as they yielded from 1 to 2 ounces a day to the man, with a few inches of water. One company was running a bedrock flume in the Swan, near the mouth of Georgia, for the purpose of striking the pay streak in each.

In Illinois Gulch William McFadden was working six men. He averaged $10 per hand per day—more than in previous seasons. In Salt Lick Gulch the yield was satisfactory. Toward the end of August, T. H. Fuller & Co. had finished their extensive preparations in Mayo Gulch and commenced working by the booming process, which gave them good results. They were, however, at the same time constructing a ditch from Indiana Gulch, which they hoped would give them sufficient water for groundsluicing during the next season. At the same time Greenleaf & Co. were mining extensively in Utah Gulch. They were building a ditch from the Blue River to the head of the gulch, and expected by this means to do the largest placer-mining business in the county during the next season. In Hoosier Gulch, in the extreme southeastern end of the county, Bemrose & Co. have been mining with good results, their ground, an old channel, being very rich.

Many new lodes have been discovered during the year in the county, especially in Tenmile district, but the principal work in lode-mining was done by the old companies mentioned in my last report. Prominent among these stand the Boston Silver Mining Association and the St. Lawrence Silver Mining Co.

The Comstock, the property of the Boston Silver Mining Association, was reported, in August, in shape to furnish 20 tons a day, and 1,500 tons of ore were on the dump. The company employed 100 men. A substantial tramway was constructed from the mine to the new mill, which was under construction. It is to have a capacity of 20 tons per day and will include smelting furnaces for the beneficiation of the galena ores, while the greater portion of the ore is to be roasted and amalgamated. * * * The mine was, at the time mentioned, 260 feet deep, and about 1,200 feet of stoping ground was exposed.

The St. Lawrence Silver Mining Co. has also been energetically at work. Their Silver Wing mine is about 500 feet above the works on Glacier Mountain. A tunnel was being driven in August, which was expected to be 200 feet by the 1st of September and which will give 200 feet of stoping ground. The crevice is 5 feet wide, and the vein of solid mineral about 17 inches in width, which produces about 1 ton of ore to the foot advanced in the tunnel. The ore assays from 30 to 180 ounces of silver per ton and contains brittle silver, zinc blende, antimony, gray copper, and galena. The ore has so far increased in quantity as the tunnel progressed.

About 300 feet northward is the Napoleon lode, in which a tunnel is also being driven, which will be as long and open as much stoping ground as that in the Silver Wing when the contract for running it is completed. The ore is similar to that of the Silver Wing but gives a higher assay. A track covered with sheds will connect the Napoleon with the Silver Wing. At the tunnel entrance of the latter * * * ore houses are being built for the reception of the ore from the two lodes, and from here a double-track tramway will be laid, on which the ore will be conveyed through the ore houses to the rock breaker. The ore will then pass from the rock breaker on to the drying floor, which will be heated by the escape gases from the furnace. From this it passes to the stamps and is then conveyed by two endless-chain conveyances into the weighing hopper. After weighing it is dumped into the receiving hopper at the base of the furnace. The ore is then raised by an elevator to the feeding hopper at the top of the furnace. After roasting and chloridizing it is drawn from the base of the furnace and conveyed to the cooling floor. After cooling it is passed into the concentrator, then into the amalgamation pans, after which the amalgam is retorted.

The main building is 30 by 50, contains one of Howland's 10-stamp rotary batteries, two of Wheeler & Randall's amalgamating pans (all cast iron), settler, and retorts. The furnace building will be 35 by 40 and 50 feet high, ore house 20 by 40, and the blacksmith and tool shops will be adjacent. The works will be operated by a 50-horsepower engine; their capacity will be 10 tons per day, and next spring another battery of stamps and two additional pans will be put in, which will double the capacity. The Airey furnace, conveyances, etc., will be similar to Stewart's works at Georgetown, and the furnace will be constructed by the same men who built that of Mr. Stewart. By the arrangement above described it will be seen that most of the labor will be performed by simple mechanical agencies and machinery.

The works were expected to be completed in September, but they were not ready to start at the end of the year.

Of other mines which have become well known during the year, the Chautauqua, Register, Tiger, Coley Extension, and Walker should be mentioned. They are, however, not nearly as well developed as the mines of the two companies above spoken of.

The completion of the reduction works in the early future will undoubtedly do much for the further development of the quartz interests of the county, which have so far principally suffered from want of a market for the ores.

In his report for 1872 Raymond gives the review of conditions in Summit County: [54]

This county is the extreme northwestern county in Colorado and embraces the area west of the main range and north of 39° 31' north latitude, being about 150 miles in an east and west and

[53] Raymond, R. W., op. cit. for 1871, pp. 361–364, 1873.

[54] Raymond, R. W., op. cit. for 1872, pp. 266, 282–286, 1873.

115 miles in a north and south direction and including, within these limits, about 17,000 square miles of surface.

A number of passes across the main range afford an entrance into Summit County, the principal one of which is the Tarryall Pass, from the head of Tarryall Creek (a tributary of the Platte) to Breckenridge. A wagon road from Georgetown to Montezuma also exists, but, owing to its unfortunate location and the steep grade on which it is built, it is but little used, except during the summer months, and then only as a trail for jack trains and for travel on horseback.

Aside from a partial development of a few of the mineral resources of the county, and this, too, in the small area around the headwaters of the Snake, Swan, and Blue rivers, the county is to-day nearly as wild and unimproved as it was when first entered by white men. Dense bodies of timber cover the mountains, extending from the timber line, which is on an average about 11,000 feet above sea level, to the valleys of the streams.

The principal towns of the county are Breckenridge, the county seat and headquarters of the placer-mining interest, situated on the Blue River, and Montezuma, a small mining camp on the Snake, the post office and depot of supplies for the Snake River mines.

The principal business of the county is placer mining, which has been steadily prosecuted since 1860 and which has been generally remunerative to those engaged therein.

Within the scope of country extending from the head of the Swan to the head of the Blue River, and down this latter stream for a distance of about 20 miles, are a number of tributary gulches emptying into the Swan and Blue rivers, and it is in these gulches that most of the placer mines of the county are to be found. Aside from the mineral resources of the county, the most prominent objects of interest are the Hot Springs [Glenwood], situated on the Grand River and reached by a good trail from Breckenridge. These springs are remarkable for their size and for the temperature of their waters, which is 112° F. at the point where the water issues from the earth. The water has a decided sulphurous taste and a not unpleasant sparkle. The springs are in good repute and are much frequented by Indians as well as whites.

The population of Summit County is extremely variable, being greatest in summer, when the mining is at its height, and dwindling down to a comparatively small number in the winter months, a condition of affairs peculiar to a country that relies solely upon an industry which is dependent on the seasons for its successful prosecution.

Labor, during the mining season, commands from $3 to $5 per day, averaging about $3.50 per day for the season. During the past year about 250 men have been engaged in mining and accessory work, such as ditch building, etc. The total product for the year, including the value of the ore and lead riches shipped from the Snake River mining camps, will not exceed $125,000 coin value.

During the past season the process of booming has been inaugurated in Summit County. This consists in collecting water in a proper reservoir, of large capacity, and discharging a great volume of it at once, thereby removing an amount of gravel impracticable by any other means. Notwithstanding the great volume of water used, and the amount of gravel kept in motion in the flume in a state of thick mud, the results seem to be favorable as regards the collection of gold. Large areas of ground can be worked by this method which can not be mined profitably by other ways. Ground which by the ordinary hydraulic process pays $3 per day to the hand can be made to yield $25 per day. The extensive application of booming to many of the gulches of Summit County can not fail to raise the gold yield of the county.

Commencing in the extreme southern part of the mining districts of the county, and on the head of the Blue River, he first mining camp is that of Bemrose & Co., in Hoosier Gulch, a tributary of the Blue River. * * * The ground pays about $12 per day to the hand employed.

In this immediate neighborhood is situated the Vanderbilt, a free-gold lode in quartzite. The crevice is irregular, varying from 6 to 12 inches in width. The ore averages well, and many exceedingly rich specimens have been found. The lode is not being actively worked.

On the hillside of the second gulch north of Hoosier Gulch is situated the Hunter lode, a vein of silver-bearing ore, opened by a shaft of 30 feet in depth. The ore vein varies from 3 to 8 inches in thickness. The ore consists of galena, gray copper, and blue and green carbonate of copper, and carries from 40 to 1,000 ounces of silver per ton. Several tons have been shipped to Newark, N. J., averaging about 300 ounces per ton.

On the Blue River, a short distance above, south of Breckenridge, Messrs. Crone & Fuller are sluicing on a side bar, employing three men, for an average season of four months, the ground paying about $5 per day to the man.

Below this claim Messrs. Sheppard & McNasser have ground opened preparatory to booming next season.

Immediately below Breckenridge, in what is known as "Klack's Gulch," probably an old side channel of the Blue River, Greenleaf & Co. (Springfield Mining Co.), are employing twelve men. They are using a 3-foot flume, dumping into Blue River. The ground is paying finely.

Opposite Breckenridge, in Lomax Gulch, the same firm is booming, employing three men, with fair results.

On Corkscrew patch the same firm is booming, employing two men. The ground is not yet fairly opened but prospects well.

Jones & Hunter are mining in Yuba Dam patch, below Breckenridge, employing two men in booming; pay averages $7 per day to the hand.

In Iowa Gulch Adams & Stahl are booming, employing six men. During the past season the gulch has produced about $6,000, the length of the season being about four months.

On the same hill, and south of Iowa Gulch, Hopkins & Hoopes have opened ground preparatory to booming. The ground prospects well.

In Illinois Gulch, a tributary of the Blue, and emptying into that stream just above Breckenridge, Colonel Fuller is mining. He is also booming in Mayo Gulch, a tributary of Illinois. He employs four men. The ground pays about $10 per day to the man.

Above Mayo is Pacific Gulch, also a tributary of Illinois. Messrs. Hopkins, Hoopes & Blair have here opened ground preparatory to booming. The ground prospects well.

In French Gulch, emptying into the Blue, a short distance below Breckenridge, a number of companies are mining. In the head of the gulch Messrs. Martin Day & Co. are sluicing in the gulch and booming in the hillside, employing four men; pay good. Conners & Cobb, in Rich Gulch, a tributary of French, are preparing to boom during the next season.

C. P. Clark is sluicing in French Gulch and booming in Lilian Vail patch. He employs eight men with good results. McFarland & Todd are next below, in French Gulch, employing two men; pay fair. The Badger Flume Co. comes next. The pay here is poor, the bedrock not having been reached.

In Negro Gulch, a tributary of French, T. H. Fuller is preparing to boom, with very favorable prospects.

Mower & Hangs, on Stillson's patch, in French Gulch, are sluicing with three men, using a hydraulic; pay good.

Sisler & Co., below, use sluices, employing two men, with fair results.

The Blue River Mining Co. come next in French Gulch, using sluices and a hydraulic. They employ five men; pay fair.

The U. S. Grant Mining & Smelting Co. is the lowest company in French Gulch, employing seven men; pay fair.

North of French Gulch is Gold Run, a somewhat noted gulch. In this locality D. Walker with three men, Moffat & Co. with four men, the Tiffin Gold-Mining Co. with four men, L. Peabody with six men, and G. Mumford with ten men are sluicing with excellent results.

On Delaware Flats, east of Gold Run, A. Delaine with three men, and Canfield & Johnson with two men are doing well.

On the sidehill, between Delaware Flats and Galena, the claims of Twibill & Stogsdale are located. The firm employs eight men and has the best paying ground in the county, the gross yield of which for the past season has been nearly $25,000.

In Galena Gulch the Galena Mining Co. with eight men, and Kiland & Coatney, with four men, are sluicing with good results.

On the Swan River, below the mouth of Georgia Gulch, Clegg & Young are putting in a bedrock flume, employing four men. The bedrock has not yet been reached. The prospects of the enterprise are favorable.

In Georgia Gulch George Twist is working three men; pay good.

In Humbug Gulch P. Iverson employs two men; pay fair.

In American Gulch Hitchcock & Stomes are employing three men in sluicing; pay good.

Messrs. Greenleaf & Co. are building a large ditch to carry the water of the three forks of Swan River to the head of Humbug Gulch. The ditch, when fully completed, will be about 13 miles in length; it is 4 feet wide and 4 feet high, flumed for the entire distance, and will carry 1,500 miner's inches of water. The ditch has throughout a grade of a quarter of an inch to carry 12 feet. This ditch will command the largest extent of ground left unworked in the county and will doubtless prove a remunerative enterprise.

Twelve miles below Breckenridge Salt Lick Gulch is being worked by four men. Water is brought from Tenmile Creek. The pay is good.

Comparatively little lode mining was done in the county during the past season. In Snake River district considerable prospecting was done and with favorable results. In Montezuma the St. Lawrence Co. worked the Silver Wing lode, and succeeded in developing a fair-sized vein of pay ore, specimens from which were remarkable for the amount of ruby silver contained therein. The mill of the company is a substantial structure, containing 10 stamps and an Airey furnace, which, however, has not been tested yet. The mill is not quite in running order. On Bear Creek, a tributary of the Snake, and about 1½ miles above Montezuma, is located St. John, the seat of operations of the Boston Silver Mining Association, W. L. Chandler, superintendent. This organization, although owning much other valuable property, is developing only the Comstock lode, a large vein, the ore of which carries much galena. The development of the mine has been described in former reports, since which time, except stoping ore and the commencement of another tunnel designed to cut the vein at a depth of 700 feet from the surface, but little has been done in the mine. Since the last report a large mill has been built, finished late in 1872, and a few tons of ore treated before the severity of the winter suspended operations.

The mill is built so as to receive the ore from the ore house, where it is dumped from the tramway conveying it from the mine. On its entrance into the mill it is dumped by the side of a large Blake crusher and then either fed into the stamp battery of 10 stamps or passed through the crusher and rollers and then elevated to screens preparatory to dressing. If the ore is suited for amalgamation it is passed from the crusher to the ball grinder, and from this machine to the proper bins, from which it is conveyed to the roasting furnace.

Such of the ore as consists of galena associated with baryta and quartz and carrying but little gray copper or other brittle ore of silver is passed through the sizing and dressing machinery. The latter consists of, first, two continuously working jiggers on the same floor with the stamps. These jiggers are designed for stuff of from one-fourth to 1¼ inches diameter and pass their tailings, if worthy of further treatment, directly to the stamps. From the stamps the now finely crushed stuff passes, secondly, three "spitz cutters," which size it, the coarsest going, thirdly, to two other continuously acting jiggers, designed for stuff of from one-half millimeter to 1 millimeter in size; and the finest, fourthly, to two of Rittenger's continuously working shaking tables, which treat the slimes. The ore now being prepared for smelting is dried and passed to the roasting furnaces, two in number. These furnaces have three floors and are designed each to treat 8 tons in 24 hours. Passing from one floor to the other the ore is desulphurized. * * * The charge is drawn in a liquid condition into a sand-bed on the ground floor of the furnace house, and, when cold, is broken up and taken to the shaft furnace, which is about 10 feet in height and having three common tuyères. This furnace is built of fire bricks made of material found near the works, which have stood perfectly the test imposed upon them. The furnace walls are one brick thick, and are bound with hoop and flat iron. The blast is supplied by a McKenzie blower, driven by a small steam engine. What lime is necessary for a flux is obtained in the Snake River Valley, a few miles below the works. The working results of this method of beneficiating the ores of the Comstock lode are most excellent. The slag from the shaft furnace is clear of silver or lead, and the lead produced is soft and of a good quality. At this point the operations of the company cease and the pigs of lead * * * are shipped to the East for separation.

The amalgamation ore, after being subjected to a chloridizing roasting, is further treated in two of Wheeler & Randall's pans, furnished with proper settlers, etc. The amalgam, after retorting, is smelted into bars, assayed, and stamped, and in this state shipped East.

Among the prominent discoveries of the year may be mentioned the veins found in what is known as Geneva district, a section of country embracing the main range between Peru district and one of the numerous heads of the Platte River. The district, geographically, is partly in Park and partly in Summit County, but owing to its distance from the settled portion of Park County is practically a Summit County mining camp. A number of promising veins have been located, prominent among which are the Revenue, Starr, Overland, and Loraine lodes. The Revenue has two shafts of about 20 feet in depth and carries a vein of pay ore of about 1 foot in thickness. The ore consists of gray copper and galena and is rich in silver, assays varying from 100 to 800 ounces of silver per ton. The returns from a lot of about 8 tons shipped to Georgetown and there crushed show that the first-class ore yields nearly 500 ounces per ton. The lode is remarkable for the large amount of gray copper contained in it and for the presence of beautiful specimens of native sulphate of copper. The other lodes of this district have been but little developed, but all show more or less of the same characteristics as the Revenue. The Overland carries 6 feet of mixed galena and gangue, and assays nearly 1,000 ounces to the ton.

In his report for 1873 Raymond gives the following details: [55]

Since my last report there has been but little change in mining affairs in this county. The various placer mines of the county continue to be worked, but with a slightly diminished

55 Raymond, R. W., op. cit. for 1873, pp. 284, 299–303, 1874.

force as compared with the previous year. An increased interest in lode mining is to be noticed, and a considerable amount of prospecting has been done in various parts of the county.

The erection of a smelting furnace in French Gulch has turned some attention toward the galena lodes in the vicinity of French and Illinois gulches. If the operations of the furnace prove successful, and a demand for galena ores springs up, there is no reasonable doubt that the large lodes of such ores already known to exist in that locality could be worked with profit, furnishing winter employment for the placer miners.

The season has been only a moderately good one for the county. The practice of booming has permitted the successful working of poorer ground than has been before worked in the county.

The large ditch of Fuller & Co. to carry the water of the South Swan to the Georgia Gulch country is as yet unfinished, and a large amount of ground, that will yield a fair revenue with a proper supply of water, remains unworked. The population of the county does not exceed 350, the larger part of whom are engaged in placer mining.

The production of the county for the past year has been as follows:

Placer gold, going to Denver	$75,000
Placer gold, going north to Union Pacific Railroad	26,000
Ores sold	5,000
	106,000

This [placer gold sent to the Union Pacific Railroad] is the product of new diggings at the base of Hausers Peaks [Hahns Peak, Routt County], in the extreme northeastern part of the county, near the Wyoming line. The total product of the county is given by the Georgetown Mining Review (good authority) at $111,000; but, with the concurrence of the editor, Mr. Van Wagenen, I have transferred $5,000, the product of the Gunnison River diggings, from this county to Lake County [transferred to Gunnison County by C. W. H.], where, according to Hollister's map, these diggings lie. [R. W. R. in footnote.]

The different placer claims in the county, more fully described in last year's report, will be briefly reviewed.

On the Upper Blue River and in Hoosier Gulch, Bemrose & Co. and others have mined during the short season (the altitude being over 9,500 feet above sea level and the supply of water scanty), and have realized remunerative returns.

Next below, and about 2 miles above (south of) Breckenridge, is the claim of Mr. McLeod, which has been worked to a slight extent during the summer. Below the McLeod claim is the ground owned and worked by Fuller & Crome. This claim is on a Blue River bar, the bedrock is comparatively shallow and the pay good. Messrs. Fuller & Crome work four men during the season, and during the low water of the fall months work in the bed of the river. Silver nuggets are found in this part of the Blue River, some of which have weighed as much as 1½ ounces. Immediately west of Breckenridge is the booming claim of Messrs. Jones & Greenleaf. This claim employs two men, with good results, in ground that previous to the introduction of the booming process could not be successfully worked. Below Breckenridge and on the west side of the Blue River is Iowa Gulch, owned by Adams & Twibell. This is also a booming claim and one that has yielded well. Under the old system of working the ground scarcely paid expenses. Two men are employed. Just above the mouth of Tenmile Creek, ground is being opened by Izzard & Co., but no returns will be realized until next year. Below the mouth of Tenmile, and emptying into Blue River, is Salt Lick Gulch, which employs four men during a season, which is longer than the rest of the county usually enjoys. The ground is good and the water

abundant. The yield during the past season has been about one-half ounce of gold per day per man. On the east side of the Blue, and immediately above Snake River, are Soda Gulch and its tributaries. This ground yields at least $5 a day to the hand but is sadly in need of more water. There is a very considerable area of sidehill ground around this gulch which would pay well for booming if water was brought to it.

The Springfield Mining Co., working on the east side of the Blue River, just below Breckenridge, is running two flumes and employs four men. The pay is good.

In French Gulch about the same amount of mining has been done as in former years. In the upper part of the gulch, Cobb & Co. are running 6 flumes, employing 24 men. The ground is somewhat spotted, but where pay is found it is rich.

Calvin Clark, next below, is running a flume in the gulch and also a boom on the sidehill, in both of which he has good pay. Water is somewhat scarce.

The McFarland claim, next below, has been worked during a part of the summer, and has yielded about one-half ounce per day per man. If work had been commenced during the flush of the water, in the early summer, the returns would have been much larger. The ground of the Badger Co. (a Denver * * * organization) comes next. Bedrock has been struck in this claim at a depth of 25 feet, and good pay found.

In Nigger Gulch, a tributary of French Gulch, a boom has been in running order for a short time during the season. Ten men have been employed, with fair results.

On Stilson's patch, on the west side of French Gulch, Messrs. Sissler & Mower and Mr. Hangs have been working by the hydraulic method and realizing about $6 a day to the hand. Next below is the ground of the Blue River Mining Co., employing six men, and in good pay.

Immediately below, and at the mouth of French Gulch, is the General Grant flume, employing four men and yielding about $6 a day to the man.

In Georgia Gulch, Mr. Hitchcock and Messrs. Iverson & Furth are working on good pay. Eli Young & Co. are running a bedrock flume up the Swan River, with the intention of reaching bedrock below the mouth of Georgia Gulch. This is a good enterprise and has been steadily prosecuted for several years. Bedrock has not yet been reached, although the ground now pays expenses. It is highly probable that this flume will develop some exceedingly rich ground, as it will drain the mouth of Georgia Gulch, once the richest placer fields in the county.

In Galena Gulch, Messrs. Riland and Twibell & Co. are mining with fair results. Between Delaware Flats and Galena Gulch is Strogsdale's patch, owned and worked by Strogsdale & Twibell. This is the best placer ground in the county. The yield will average fully 1 ounce per day to the man. Five men were employed during the past season.

A. Delaine, at Delaware Flats, has employed three men during the season, and realized about one-half ounce per day to the man.

In Gold Run mining has been steadily carried on during the season.

Walker & Majors have employed three men, the pay being one-half ounce a day to the man. Moffat & Canfield, three men; pay $7 a day. L. Peabody, six men; pay $7 a day. Tiffin Mining Co., five men; pay fair. J. Nolan, three men; pay fair. G. Mumford, five men; pay $7 a day.

Booming.—This method of mining poor placer ground is rapidly coming into favor in Summit County. In Mayo, Nigger, Lomax, and Iowa gulches mining is carried on in this manner with remunerative results, and it is highly probable that other localities will also adopt this process. The introduction of the "self-acting gate," whereby the opening and shutting of the gate of the reservoir is made automatic, leaves scarcely any further economy to be introduced into this method of mining. The self-acting gate now considered the best, consists of a water box suspended in guides, the rope from which passes over two pulleys, one of 12 feet and one of 5 feet, to the lower edge of the canvas gate (barred with strips of iron or 2-inch timber). When the water in the reservoir reaches the proper height, a small flume conducts it to the box, which, when full of water, has weight enough to roll up the canvas gate at the bottom of the reservoir from the bottom, allowing the water in the reservoir to issue through a gate (generally 4 by 6 feet in size). By the time the reservoir is nearly empty, the water in the weight box has discharged itself through holes made for that purpose in the bottom, and a weighted arm on the second pulley drops the gate to its place, where the pressure of the water in the reservoir keeps it in place, water tight. One man is now considered ample force to run a boom, and his duties consist mostly in clearing timber from the ground to be worked and in breaking the larger boulders into sizes small enough to go through the flume, which is usually 4 feet wide, with a grade of 1 foot in 12. The use of a boom permits the working of ground that could by no other means be made to pay. The experience of the Summit County mines goes to prove that notwithstanding the large amount of water used and the velocity with which it rushes through the flume, the gold collects readily in the upper boxes of the flume, in which mercury is generally placed. Booming may be considered as the best labor-saving invention introduced into the country of late years, and while it is not adapted to all placer claims, it permits the working of claims that would otherwise be valueless.

Lode mining.—Aside from the work done in the Snake River mines but little lode mining has been done. The Vanderbilt, a free-gold quartz lode in the quartzite in the Upper Blue, has been worked a little, but no ore has been treated. Arrastres will probably be erected during the coming year. The ore is rich and of a character easily treated, and good working results are therefore to be expected. The Hunter, east of Hoosier Gulch, has been worked during a part of the summer. The ore is good quality and will mill about 200 ounces in silver per ton.

The Cincinnati lode, on Mineral Hill, French Gulch, has been opened by Messrs. Sears & Conant. The lode has been traced for about 900 feet on the surface, and the deepest shaft is now about 40 feet. Two hundred tons of mineral are on the various dumps, and the ore vein, wherever opened, shows from 2 to 16 inches. The ore is a clean, coarse galena, free from zinc, and carries from 30 to 40 ounces of silver per ton.

A reverberatory furnace has been erected in French Gulch, and a few pigs of lead have been made. It is proposed to erect a shaft furnace for the treatment of this and ores from neighboring lodes.

Some work has been done on lodes around the head of Illinois Gulch. The ore is generally a pure galena, mixed with sulphate and carbonate of lead, assaying from 40 to 700 ounces in silver and from a trace to 6 ounces in gold per ton. The lodes around Breckenridge almost all carry galena (and its oxidized products) with traces of copper. The size of the veins is good, and the ore varies in value from 10 to 700 ounces in silver. The erection of smelting furnaces at or near Breckenridge, where water power could be obtained, would do more for the development of the mineral resources of the Blue River country than anything else that could be devised. The supply of lumber for both timber and fuel is ample, and good clay for bricks is close at hand. The ores to be treated are of a fair average value and of a class to furnish a pig lead of purity, containing from 100 to 500 ounces of silver and considerable gold per ton. The supply of ore would be ample for at least one 10-ton furnace.

In the vicinity of Montezuma much prospecting has been done and some ore shipped to Georgetown. The mill of the St. Lawrence Co. is not yet in order, and work has ceased on the mines for the winter. On Bear Creek the Boston Silver Mining Association is still working on the cross-cut tunnel to strike the Comstock lode at a considerable depth.

At the head of Decatur Gulch, about 3 miles from Montezuma, several valuable lodes have been opened. The Revenue, Starr, and Congress carry gray copper and galena ores, and the Treasure Vaults and others yield what is supposed to be a new ore, containing bismuth, sulphur, and silver. The gray-copper lodes have received most attention and development. The Revenue has three shafts, 60, 38, and 60 feet in depth. Work will be carried on in this and Congress lode during the winter. The Revenue shows from 3 to 18 inches of solid ore, which is of good quality and mills from 100 to 450 ounces per ton, the first-class ore carrying also 10 per cent of copper, and the second-class ore over 50 per cent of lead.

Of the bismuth-bearing lodes but little can yet be said. The pure ore assays from 6,000 to 7,000 ounces in silver per ton and contains 40 per cent of bismuth. The ore, which shows a slight tendency toward crystallization, is intimately mixed with well-defined quartz crystals. Comparatively little work has been done on these lodes. They are situated immediately on top of the range and were discovered late in the summer of 1873. These, as well as the gray-copper veins, are on the main range, at an elevation of fully 12,000 feet above the sea level, and are partially in Park County, the veins crossing the range diagonally. The district [Geneva] is a new one, and bids fair to be a flourishing mining camp.

With the completion of a railroad into South Park, which would give the Summit County section of the Territory freights at reduced rates, and with the successful working of a smelting furnace at or near Breckenridge, there can be no doubt that Summit County would rapidly increase in population and wealth. At present, however, with only placer mines at work, and these only during the summer, the progress of the county is far from satisfactory. The resources of the county in the form of placer and lode mines, coal, agricultural and grazing land, are ample for the support of a large population, and the amount of land on the streams in the northern and western part of the county suitable for cultivation is much larger than is generally supposed.

In his report for 1874 Raymond says: [56]

Placer workings, Blue.—Nothing new of note is to be reported from the placer workings of this valley. The production of 1874 amounted to $70,000 and was taken from the old and standard claims in French, Illinois, Iowa, Lomax, Georgia, Swan, and Indiana gulches. Early in the year a project was set on foot to carry a ditch from the Upper Swan River into the head of Georgia Gulch, which is the richest gulch in this valley. It has not yet been completed, nor am I able to ascertain whether it was commenced. The new ground to be won would amply warrant a large expenditure of money to bring water into it. In the operations on the Blue and its tributaries the system of booming, which was introduced some two years ago, has almost completely superseded all other methods of washing where there is sufficient inclination of the ground.

Lode workings, Snake River district.—This district, one of the earliest discovered in Colorado, has developed more slowly than any other, principally by reason of its great inaccessibility and the heavy and unpromising character of the ore. In fact, it was in the Coaley lode, on Glacier Mountain, that the first discovery of silver in the Territory was made; but the quantity was small, and for many years it has only been known in the dreams of prospectors. The district includes the camps of

56 Raymond, R. W., op. cit. for 1874, pp. 358, 359, 375, 376, 1875.

Montezuma and Peru. The ores are heavy, carrying both zinc and lead, and as a rule are not rich in silver. There is no market as yet nearer than Georgetown, and transportation across the Snowy Range is only possible in summer, and then at a cost of $18 to $20 per ton. Early in 1874 an excitement sprang up concerning the mines, and the district was invaded by about 200 prospectors from Georgetown, South Park, and Denver, who continued there until driven out by the deepening snows on the range. The result of the excitement was the discovery of a number of new lodes, the reopening of many of the old ones, and a lively interest among smelters and millers. There is now a company formed, with the purpose, as soon as the weather will permit, of erecting smelting works at Peru to handle the heavy galena ores of that and neighboring camps. Last year these Peru lodes were considerably developed. They are undoubtedly of great strength and value, carrying very heavy bodies of galena, that will assay from 20 to 60 ounces of silver per ton.

The Comstock property is the only claim in this district upon which continuous work has been done for the past two years. It is owned by the Boston Silver Mining Co., which has a large and fine mill, and is now engaged in driving a tunnel for the vein, which will cut it 600 feet deep. It is now within 300 feet of the lode. It is reported that the company will open its works for custom ore next season.

Breckenridge district.—The ores of this district are of low grade (in silver) but very pure galenas. The veins are numerous, but only a few have been worked, as until lately no market has been found for the ores. In the summer of 1873 Messrs. Spears & Conant put up a small reverberatory furnace at Lincoln City (French Gulch), which, however, did not prove successful. In 1874 the mode of treatment was altered by the introduction of a blast furnace, which, not proving satisfactory alone, was supplemented by the addition of two Drummond lead furnaces. These last did well, and by September they were fired up in earnest and continued running steadily and successfully until the close of the year. The product of the four months' run was 216 tons of work lead, carrying from 60 to 80 ounces of silver per ton. The ore is derived mainly from the Cincinnati or Robley lode, which has been developed to a considerable extent and is capable of producing 10 tons daily of almost pure galena. The works having proved successful, owners of other similar veins are preparing to open them next summer, and the company will probably enlarge its facilities in order to work the increased supply. The work lead is shipped by train across Hamilton Pass to the South Park, and from there it is carried to Denver, where it takes rail to Chicago, bringing 5 cents to 6 cents per pound, after refining, exclusive of the value of silver contained in it. Breckenridge district contains also a number of gold veins, but none are being worked.

In his report for 1875 Raymond says: [57]

Bismuth has been found to a high percentage in the argentiferous galena of the Geneva and Snake River mines.

The metallic product of this county is placed at $76,000, being wholly gold dust from placers. There has been also a yield of probably $10,000 from the Snake River mines, which would belong to Summit but which is included in the statement of Clear Creek, nearly all of it being sold there. The properties worked during the year are the same as those of last season, with a few additions. The Summit County placers, being located at so high an altitude, can be worked successfully in good seasons only. The last season was but a tolerable one, and expectations were as a rule not realized. Fully as much ground as usual was opened, but the water supply gave out nearly a month earlier than usual, and all ground worked by booming had to be abandoned. This system of washing is

doing good work in the Blue Valley, and though it may be considered, in comparison with the more extensive and complicated systems of California, as rather primitive, it has peculiar merits of its own not to be despised. It has been described in former reports.

In French Gulch and its tributaries, Dry, Mayo, Humboldt, and Nigger gulches, Calvin Clark, I. H. Fuller, J. J. Cobb, and the Badger Co. have been at work as usual. The production of the gulch (which is by far the best at present in the Blue Valley) has amounted to about $40,000. Georgia Gulch and Gold Run have together yielded about $13,000; about $4,000 have come from other minor localities, such as Galena, Iowa, and Lomax gulches, and Stilson, Buffalo, and Delaware Flats; and $15,000 may be credited to claims located directly in the main valley. * * *

The Lincoln City Lead Works have been idle during the year, having fallen into legal troubles.

Snake River district has been steadily but slowly improving, and of Peru district the same may be said. The Champion, Tiger, Printer's Pool, Peruvian, Blanche, Orphan Boy, Silver Wing, Potosi, Cony, Sukey, and the Comstock tunnel mines have been worked with considerable regularity during the year, and all but the last have shipped ore.

The Sukey mill made a short run in the summer but was unsuccessful. The St. Lawrence mill has been transported to Georgetown and sold to the Pelican Co.

The operations of the Boston Silver Co., at St. John's, * * * have been steadily prosecuted during the season. The company continued to drive its long tunnel into Glacier Mountain until, on the 20th of November, it had reached a length of 1,075 feet. It is well constructed throughout, being 7 feet wide by 9 feet high, and in all places where the rock is not sufficiently hard, timbered in the most substantial manner. Eight veins have been crossed so far, Nos. 1, 2, 3, 4, and 6 showing at the points of intersection quartzose gangue and iron pyrites; Nos. 5 and 8, a similar gangue, with zinc blende and galena; and No. 7, heavy spar, galena, and blende, with little iron pyrites. All the veins, with the exception of No. 8, are very large, some of them extraordinarily so, as, for instance, No. 5, which at the point of intersection is about 20 feet, and No. 2, which is 83 feet thick. The latter carries such soft vein material (broken quartz with iron pyrites) and delivered when opened such volumes of water that it was found necessary to timber closely the whole 83 feet, sides, top, and bottom, with 12-inch timbers (sets of 10-inch timbers 3 feet apart, which had first been inserted, having been broken like reeds by the enormous pressure). Just before the driving of the tunnel was temporarily stopped, a ninth vein had been struck and entered into for 3 feet, which carried quartz and iron pyrites. The opposite wall not having been reached, the value of the vein is unknown.

At the time above mentioned positive orders from the directors of the company in Boston forbade the continuation of the tunnel for the present and ordered the development of Nos. 5 and 7 by drifts and rises. This work has been going on uninterruptedly, 30 miners having been employed on an average, and with most unexpected results. The north drift of No. 5 was 200 feet long at the end of the year, the vein material having been soft throughout that distance. When the drift had progressed so far, a very serious cave occurred about 170 feet from the entrance of the drift, which delayed further driving for a month, and required timbers 2 feet in diameter for a distance of 50 feet to overcome it. A rise was started at the end of the year a little south of the cave. While at the intersection of the tunnel with No. 5, only galena, blende, and a little iron pyrites were found, containing, when solid, 60 to 70 ounces of silver per ton; the vein carried beyond a point 75 feet from the entrance northward, besides the minerals named, rich silver ores, such as ruby silver, stephanite, polybasite, and

[57] Raymond, R. W., op. cit. for 1875, pp. 282, 285, 317-321, 1876.

tetrahedrite in considerable quantity, so that the average value of the ores was more than doubled. The south drift on No. 5 was in about 50 feet at the end of the year and has since been driven to the intersection with No. 7, 140 feet from the tunnel. In this drift only one pocket of ruby silver, stephanite, and fahlore was found. The drift on No. 7 north of the tunnel was, at the end of the year, about 60 feet long, and is throughout in very soft and dangerous ground, carrying no ore. The south drift was in about 200 feet and showed ore (zinc blende and galena in heavy spar) for almost the whole distance, the vein being on an average 4 to 5 feet wide and the ore streak varying from 1 to 3 feet. At a distance of about 180 feet from the entrance, the drift encountered a horizontal fault of 8 feet (the vein being thrown to the east), and 12 feet farther a second fault of 6 feet in the same direction. Between the two the vein carried very good galena. About 6 feet beyond the second fault the galena became very solid and contained much native silver, which continued for a distance of 30 feet. The ore in No. 7, at the intersection with the tunnel, contained in solid galena from 48 to 50 ounces of silver; at the further end of the drift, where native silver was visible, it contained from 100 to 500 ounces, and in ordinary galena from 70 to 80 ounces.

Since the end of the year two rises have been started on No. 5, 170 feet apart, the northern one being at the time of this writing 80 feet and the southern 114 feet high. In the latter north and south drifts 100 feet above drift No. 1 are being started. Besides this, stoping ground for 10 men is opened, and extraordinarily rich silver ores, carrying, however, little galena, are now being extracted.

On No. 7 one rise, now 120 feet high, has been made, and drifts will be started north and south in a few days. There are three stopes opened on No. 7, two of which produce ore.

Little work has been done during the year in the upper mine on the Comstock lode (probably No. 7 of the tunnel) beyond the extraction of about 60 tons of solid galena. The stopes there opened will probably be worked during this year, but no extensive work is intended there until the connection has been made from the tunnel below.

At the end of the year the company had about 800 tons of ore on hand, three-fourths of which is dressing ore, to be concentrated in its very systematic establishment and to be smelted into pig lead.

A dozen or more veins have been worked in Snake River district by other parties at intervals during the year, and some of the veins have produced rich lead ores in small quantities. No large developments have, however, been made. The ores mined have mostly been bought by the Boston Silver Co.

Burchard gives the production of Summit County from 1860 to 1881 in his reports for 1880 and 1881.[58] The production of the area now comprised in Eagle County has been deducted to obtain the figures for these years given in the table on page 245. In his report for 1881 Burchard says:

Within the last two years lode mining commenced on an extensive scale and during the last year assumed such proportions as to place Summit County high on the list as a bullion-producing county of the State.

Tenmile district is the chief point of production. It covers the headwaters of Tenmile Creek, in the southeastern part of the county.

The principal sections of Tenmile district are Chalk, Sheep, Chicago, Elk, and Jacque mountains.

On Sheep Mountain is the most noted mine of the county, the Robinson. Reports as to the condition of this mine are conflicting and highly unsatisfactory. In November last the drifts ran into a column of very low-grade ore, the manager resigned, the usual dividend was postponed, and a general crash occurred. * * *

The Idalia, Ballarat, Sheep Mountain, and Minnie are on the same ledge as the Robinson. They will doubtless soon become producers, as systematic development is going on.

The Gray Eagle has been continuously worked, and the ore extracted sent to the Argo and Leadville smelters.

The Wheel of Fortune mine contains some high-grade ore, a quantity of which was extracted during the past year and smelted at Kokomo.

The Monitor is on Sheep Mountain, between the Robinson and Gray Eagle mines, and consists of six claims, the Aetna, Galena, Regatta, Bonanza, Pride of the West, and Tabor. The property has not yet been worked, but there is no reason to doubt that the Monitor will be one of the leading mines of the Tenmile district.

The Crown Point, Kearsarge, Tiger, Snow Bank, and Michigan are prominent mines, and in condition to make a good future showing of production.

On Elk Mountain there are a number of well-known producing mines, the Aftermath, Milo, White Quail, Badger, Silver Wave, Raven, and Colonel Sellers. They are contact veins, and large bodies of ore have been exposed by the development work of the year.

The Thunderbolt Consolidated, Meily, Lennox, Sabbath Rest, and Mercantile companies have all been organized for the purpose of opening up promising groups of mines on this mountain. The Bledsoe, at the foot of Elk Mountain, in Searle's Gulch, is one of the earliest locations. A tunnel has been run in upon it a distance of 80 feet, which shows an 8-foot body of iron. The last work on this shows the ore turning to galena, and several good seams have been developed. Assays in gold and silver have been made, returning as high as $100 to the ton. The mine has the same characteristics as those of the other prominent mines of the district and apparently needs but little development to put it on a paying basis.

The Queen of the West, Mayflower, and Ida L., on Jacque Mountain, have been actively worked and commenced shipping ore. The Mayflower has about 500 feet of dead-work development but only 90 feet of the vein. The Ida L. carries 15 per cent copper.

On Copper Mountain there are no shipping mines, only development work having been done during the year. There are a number of strong lodes, all having the same general characteristics of the ores of this section; among them may be mentioned the Ida May, Reconstruction, Graff, and Hattie.

Of the mines of Gold Hill the same may be said. The Caledonia group has been systematically worked, and some ore is on the dump awaiting shipment. The Minnie Lee, Union Consolidated, Fisher, Grand View, and Ocean Wave are other properties which have attracted a fair share of attention. The last named carries 30 per cent of copper.

The Gilpin, Golden Cork, Homestake, Mountain Stake, Spring Well, "76", Argenta, Big Injun, Aztec, Lady Alice, Little Alice, Aetna, Vesuvius, El Dorado, Falcon, Pelican, Nova Scotia Bay, Alpine, and Matchless, on Fletcher Mountain, and the Charley Ross, Condor, and Sub-Treasury, on Pacific Mountain, are locations upon which but little development work was done until late in the season of 1881. They are mostly situated high up on the rugged sides of these lofty mountains and are

[58] Burchard, H. C., Report of the Director of the Mint upon the production of the precious metals in the United States during the calendar year 1880, pp. 156–157, 1881; idem for 1881, pp. 354, 432–437, 442–443, 1882.

inaccessible except for pack trains, the trails for which in many instances have not yet been completed. With facilities for extracting the ore and getting it to market these mines will prove of undoubted value.

Northwest and 1 mile distant from Boreas is the Warrior's Mark, a late discovery, which contains an ore of gray copper of exceeding high grade. It is worked by an open cut, and ore to the value of nearly $40,000 has been taken out and worked in a concentrating mill.

The Snow Drift is in the same vicinity and is taking out considerable rich ore, which is piled on the dump, no shipments having yet been made.

The Rochester Queen is another new mine of fair promise. It is worked by a tunnel connected by crosscut and incline with the shaft. High-grade ore has been extracted, and, during three months, the product averaged $5,000 per month from shipments.

The Ontario, on Farncomb Mountain, north of Lincoln, is reported to have struck a large body of exceedingly rich ore.

The following table from the Denver Republican shows the output of the leading producing mines of Tenmile district:

Name	Mountain	Number of tons produced in 1881	Present daily output (tons)	Capable daily output (tons)
Robinson Consolidated	Sheep	115		
Gray Eagle	do			25
Wheel of Fortune	do	2,500		30
Crown Point	do			15
Tiger	do			60
Snow Bank	do			20
Kearsarge	do			5
Aftermath	Elk		25	100
White Quail	do	9,000	10	100
Milo	do	3,520	15	100
Badger	do	4,800	35	70
Silver Wave	do			40
Eagle and Raven	do	550		150
Colonel Sellers Consolidated	do			60
Queen of the West	Jacque	180		4
May Flower	do	150		10
Ida L	do			15
Gilpin	Fletcher			10
Matchless	do	225		2

The following is an approximate statement of the weekly output of the leading producing mines of Tenmile, Colo.:

	Tons
Aftermath	480
Badger	210
Eagle and Raven	60
Fletcher Mountain	20
Gray Eagle	30
Ida L	90
Little Chicago	15
Milo	240
Robinson	800
Silver Wave	90
White Quail	150
Wheel of Fortune	180
Total weekly output	2,365

Chihuahua is one of the early camps and contains a large number of valuable mines, which were worked to a greater or less extent during the year. Some ore was shipped, and a considerable amount extracted in process of development remains on the dumps awaiting shipment.

Production of Chihuahua district, 1881

Name of mine	Number of months worked in 1881	Amount of development (feet)	Character of ore	Average value of ore per ton	Number of tons shipped to home smelters	Number of tons shipped out of county	Number of tons on dump	Capable daily output (tons)
Delaware	5	300	Galena and copper	$100.00	25		75	10
Orphan Boy	3	325	Lead	92.00	20	25	30	10
Blue Lode	3	100	Brittle silver	694.00	4	10	2	1½
Peruvian	4	700	Gray copper	45.00	100		75	10
Bullion	3	100	Galena and gray copper	290.00		2	10	1
Eliza Jane	2	75	Gray copper	500.00	2	2	1	½
Chicago Lode	4	200	Lead	65.00		10	10	1
Little Chief	1	50	do	35.00	4	2	1	½
Silver Ledge	9	500	Antimony, silver, and galena	96.00		5	40	50
Jennie B	5	200	Gray copper	204.00		3	5	½
Telephone	3	75	Galena and yellow copper	187.00		2	3	¼
Atlantic & Pacific tunnel	12	500	Lead	94.00		5	100	20
New Discovery	3	100	Gray copper	210.00	1	9	2	1
Frenchman	2	100	Lead	56.00		2	18	5

Below Tenmile is Frisco, containing some excellent mines discovered in recent times and upon which comparatively but little work has been done.

The Frisco Belle shows a large body of mineral, principally galena.

The Mountain Chief and Leetsdale show well for the amount of development.

The following is a list of the stamp mills in Summit County, with their capacity:

Name	Location	Number of stamps	Remarks
Brooks & Snyder	Breckenridge	20	
Lawrence	do	10	
Gold Park	Gold Park (Eagle County)	40	Owned by the Gold Park Mining Co.
Lincoln	Lincoln	10	
Little Chief	Red Cliff (Eagle County)	30	Anglo-American mine.

The smelters of the county, with their capacity and amount of production, are as follows:

Name of works	Location	Tons capacity	Number of days in operation	Tons of bullion	Remarks
White Quail smelter	Kokomo	30	140	950	Present owner took charge April 15. Value of bullion, $175,000.
Greer smelting works	do	40	76	240	Silver in bullion, 125 ounces.
Carbonateville smelter	Carbonateville	30			Never used.
Summit smelter (two stacks)	Robinson	80	180		Treated Robinson ore, using low-grade bullion for flux.
Wilson smelter (two stacks)	Breckenridge				Produces matte.
Boston & Colorado (two stacks)	do				No figures obtained.
Elyria (two stacks)	do				Do.
Sissapo (two stacks)	Montezuma				Do.
Lincoln (two stacks)	Lincoln				Do.
Battle Mountain (two stacks)	Red Cliff				No figures obtained [now Eagle County].

The Kokomo Times estimates that the ore output for Summit County for 1881 will run from $3,500,000 to $4,000,000. Of this immense production, Tenmile will furnish about $2,000,000, while the Battle Mountain district [now in Eagle County] will probably be the next heaviest producer. Decatur and Montezuma have done reasonably well, and Breckenridge, coming in as she did in the summer and fall, will doubtless add materially to it.

The Breckenridge Leader publishes the following statement of the yield of the mines of Summit County for the years 1860 to 1880:

Production of Summit County, Colo., 1860–1881

Gold from placers, 1860 to 1869		$5,500,000
Gold from placers, 1870	$100,000	
Gold from placers, 1871	70,000	
Gold from placers, 1872	60,000	
Gold from placers, 1873	ª101,000	
Silver and lead, 1869 to 1874	200,000	
		531,000
Gold from placers, 1874	76,408	
Silver and lead	50,000	
		126,408
Gold from placers, 1875	72,413	
Silver	50,000	
		122,413
Gold from placers, 1876	150,000	
Silver and lead	200,000	
		350,000
Gold from placers, 1877	150,000	
Silver and lead	40,000	
		190,000
Gold from placers, 1878	165,774	
Silver and lead	155,000	
		320,774
Gold from placers, 1879	100,000	
Silver and lead	375,000	
		475,000
Gold from placers, 1880	50,000	
Silver and lead, 1880	400,000	
		450,000
Silver, gold, and lead, 1881		3,250,000
Total for 21 years		11,315,595

ª Includes Hahns Peak.—C. W. H.

The following smelting works, mills, and mines have returned reports of production, viz: L. W. Aldrich, Lincoln City, Summit, White Quail, Greer, Wilson, Badger, Central Fluming, Gray Eagle Consolidated, J. L. Fuller, Matchless, Robinson Consolidated, Silver King, Silver Wave, Blue River, Bell, Gold Park [Eagle County], and Belden [Eagle County].

In his report for 1882 Burchard gives the production of Summit County, and from his figures the production of the area now comprised in Eagle County has been deducted to obtain the figures for 1882 given in the table on page 245. Burchard says:[59]

The principal district of Summit County is Tenmile district, which embraces the valley of Tenmile Creek from the summit of the Continental Divide, west to the summits of Sheep, Elk, Chalk, Jacque, Tucker, and Cooper Mountains.

The principal mining towns are Kokomo, Recen, Robinson Carbonateville, and Wheeler's. Kokomo and Recen adjoin and are practically one town. * * *

[59] Burchard, H. C., op. cit. for 1882, pp. 390, 391, 395, 554-559, 590-593, 1883.

The average grade of ore is about 30 ounces in silver and 35 per cent lead. This is the showing of the producers.

He then gives descriptions of the Robinson, Idalia, Gray Eagle, Ballarat (which was located 1878, but no work of consequence was done until 1881), Wheel of Fortune, Rambler, Sarsfield, Bay City, Crown Point, Little Chief, Lucky Boy, Michigan, Snow Bank, Aftermath ("which has produced 11,000 tons of ore, averaging 30 ounces silver and 20 per cent lead; * * * the ore is chiefly sand carbonates"), Milo ("adjoining the Aftermath, * * * so far taken $200,000; * * * the output is about 5 tons per day, part of which goes to Greer's concentrating works and part to Doncaster's"), White Quail ("commenced output just below the surface, and to the present has produced 6,750 tons, of an average of 30 ounces silver, 1 ounce gold, and 4 per cent lead to the ton; the ore is treated at the White Quail Co.'s smelter, which has a capacity of 40 tons daily"), Raven and Eagle, Badger, Little Ida, Queen of the West, Mayflower, Ida Lee, Delmonico, and Selma mines. He continues:

On Fletcher Mountain, about 6 miles from Kokomo, the principal locations are owned by the Croesus Co., Peerless Co., and Matchless Co. They are fissure veins above timber line. The ores are high-grade galena, carrying native and brittle silver. * * *

The Colorado Land & Mineral Co. owns the Bernadotte, Maximus, Maximus Extension, and Orestes, known as the Maximus mines. * * * They have taken out * * * a quantity of fine mineral from the surface by opening cuts on the veins, which have given returns from the smelter running from $150 to $200 per ton.

Union district is situated northeast of Breckenridge, and comprises Gold Run basin and the adjacent slopes. Gold Run is a tributary of the Swan River, having a northwesterly course, east of Gibson Hill and west of Silver Hill, a spur from Mineral Hill running north between Galena Gulch and Gold Run. Its length is about 4 miles from its mouth to the northeast flank of Mineral Hill. In this gulch, in the early history of gulch mining, a large amount of gold was taken out. Several properties are still being operated, the principal being the central portion of the gulch owned by L. Peabody. * * * Its area is one-half mile from north to south, and three-fifths of a mile from east to west. The water to operate this property is conveyed from the Blue River at a point 4 miles south of Breckenridge, by a ditch 9½ miles long. George A. Mumford owns the lower and northern portion, embracing Buffalo and Kentucky flats, comprising 650 acres, which is regularly worked. * * * South of Peabody's claims the upper portion of the basin is the property of the Gold Run Placer Mining Co., represented by * * * 50 acres of patented ground.

In digging the ditches and working the placers, deposits of galena were discovered, and at one point carbonates were uncovered. This was near the last-named property, on the slope of Silver Hill. This was located as a claim and called the Wilkes-Barre. Other locations were made at an early day but were subsequently abandoned. During the last year some of the old workings were reopened and new discoveries made that gave an impetus to lode mining in the district.

He gives descriptions of the Orphan Boy, John J., Hard Times, Iron, Ohio (located 1881), Wild Cat, Cooper, Reese, Highland Mary, Lafayette, Ida, Lexington.

In his report for 1883 Burchard says: [60]

Summit County lies on the Pacific slope of the Snowy Range. The area of this county is much smaller than formerly, as the legislature during its last session created from the western portion Eagle and Garfield counties. With the exception of Eagle County, this county still retains the principal mining districts, viz, Union and Tenmile.

Tenmile district embraces the valley of Tenmile Creek, within whose boundaries are Jacque, Sheep, Elk, Chalk, and Cooper mountains.

Union district lies east of Tenmile and embraces Montezuma, Decatur, Chihuahua, Breckenridge, and the Blue River. Although some of the largest producing mines of this county were not worked, and others to a limited extent, the year was successful, owing to the fact that a number of new prospects became producing mines during the year. This is apparent more especially in the vicinity of Breckenridge and Montezuma, whose season has been a progressive one, and the stability of these camps has been strengthened by the improvements and developments in the mines as well as by the advent of the railroad. But in Tenmile district there is either a lack of energy, confidence, or capital among the mine owners that has thrown the district into a dormant state, from which it will require considerable time to recuperate.

The first silver-bearing lode discovered in Colorado was in this county, on Glacier Mountain, near Montezuma, in July, 1864, and though it has never amounted to anything as a producer, it led to the great silver excitement and development of the Clear Creek County silver district.

In the early years of Colorado the placers along the tributaries of the Blue River were among the most productive in the State, Georgia, French, Humbug, and Gold Run gulches being the richest; yet large quantities of the precious metal were produced from Swan River, Tenmile Creek [?], Illinois, Iowa, and Ryan gulches. The yield of the placers in this county has always been good, and mining has been carried on there every summer since the first discoveries. Tenmile district does not show the life it possessed two years ago. Almost every branch of mining is at a standstill, with a future equally dark. The ores as a general rule are low grade in silver, and on account of the lead and iron contained they are desirable ores for smelting purposes. The formation of the veins are principally contact, averaging 4 feet thick, composed mostly of carbonate of lead and carbonate of iron.

On Sheep Mountain is the property of the Robinson Consolidated Mining Co., that during 1881 and 1882 created great excitement on account of the wonderful production that was the principal cause of swelling the Summit County production in 1881. This property is now worked under lease. According to the terms of the latest lease the lessee agrees to pay the company the following royalties: On ore under 35 ounces to the ton, $3 per ton; ore of 35 ounces and not exceeding 40, $4 per ton; ore of 40 ounces and not exceeding 45, $6 per ton; ore of 45 ounces and not exceeding 50, $9 per ton; ore of 50 ounces and not exceeding 55, $12 per ton; ore of 55 ounces and not exceeding 60, $18 per ton; on all ores over 60 ounces, in the proportion of $18 per ton per 60 ounces. Settlements are to be made with the company on the first of every month. The lessee agrees to employ a force of not less than 65 men, and extract at least 1,500 tons a month, providing the ore is contained in the mine and available. He also agrees to diligently prosecute development work on the line of the ore shoots below the thirteenth level and keep development work at least 50 feet ahead of the extraction of the ore.

The Idalia, an adjoining property to the Robinson, has been idle most of the time, and but little advancement in development has been made.

The Wheel of Fortune, on top of Sheep Mountain, has been worked and some ore shipped. Developments have been pushed, and large bodies of ore are now blocked out ready for extraction. The ore is a carbonate of lead, found in the contact, and is about 4 feet thick. The grade is only fair, averaging something like 20 ounces silver per ton.

On the Gray Eagle some exploration and development work has been done, and, as reported, the prospects for finding a large body of ore are good. This property is worked by lessees, who have spent some $6,000 or $7,000 in their researches.

The Crown Point, owned by the Graphic Mining Co., has been put in good shape and is now one of the largest producers in this district. In tunnel, drifts, and levels is an aggregate of over 1,000 feet of development. The ore is better grade than found anywhere in this vicinity, the average per shipment being about $100 per ton.

The Snow Bank is developed by a shaft 100 feet deep, with an aggregate of 100 feet in levels. The ore body is from 4 to 10 feet thick, it being lead carbonates and sulphates of iron, that run from 12 to 15 ounces silver and 20 per cent lead per ton. The ore is refractory and requires concentrating to be made marketable. No shipments were made during 1883, but over 3,000 tons of ore was extracted and now lies on the dump.

The Boston & Sheep Mountain Mining Co. own seven claims on this mountain, all adjacent to the Robinson Consolidated Mining Co.'s property. No ore has been mined from this group during 1883. The principal development, and that which promises best results, was a shaft sunk 162 feet on the Zonars lode and designed to tap the Robinson ore shoot on its dip, but the work was discontinued before the desired result was attained, the necessary depth to catch this ore shoot being 320 feet.

The Little Chicago is developed by an incline shaft on the vein 288 feet deep. A tunnel 640 feet long cuts the vein intersecting the shaft at the 240 foot level. At this point there is 700 feet of drifting on the vein, which is a contact, with an almost vertical dip, and radically different from any other mineral vein on Sheep Mountain. It has a sedimentary rock, giving a strong lime reaction, for a footwall and a yellow-colored porphyry for a hanging wall. The bulk of the vein filling is feldspar and decomposed porphyry, the ore when it occurs being a coarse cube galena intimately associated with pyrites of iron.

The Nettie B. Forrest mine is being worked by an incline 120 feet long. The drifts and upraises from this incline aggregate 450 feet. From the bottom of the shaft, which is 34 feet, there are about 250 feet of drifts on the vein. The Nettie B., as it is mostly called, is a contact vein, between a micaceous sandstone cap and a lime footwall. One of the peculiarities of this vein is that it dips with the contour of the mountain at an angle of about 20°, attaining greater depth as it descends the hill, the contour of the mountain being about 15°. The vein from the outcrops to about 90 feet from the surface is badly broken, the ore being bunchy, the bulk of it being an oxide of iron of good quality, and is in demand as a smelting flux, running 45 per cent excess in iron and about 9 ounces in silver; the ore associated with it being a carbonate of lead running about 40 ounces silver and 15 per cent lead per ton. The main objection to this ore is its excessive moisture. Below the 90-foot level the character of the ore radically changes into an iron, heavy in sulphur and void of either silver or lead.

The West side, on Fletcher Mountain, was, during the early part of the year, worked by lessees, but on account of litigation was closed down. The ore is a sulphuret in quartz, running from 50 to 80 ounces silver per ton.

[60] Burchard, H. C., op. cit. for 1883, pp. 236, 237, 238, 240, 426–433, 1884.

The Nova Scotia Boy No. 2, owned by the Croesus Mining Co., is being put in condition for production in 1884. Until recently the development consisted of a few open cuts and a tunnel 25 feet long lower down the mountain. This tunnel will be pushed until the contact is found.

But very little is being done on Jacque Mountain. The Ida L. Bledsoe, Queen of the West, and the Delmonica have been worked but very little and but a very small quantity of ore shipped.

On Copper Mountain, the Reconstruction and Storm King have been shipping some ore and pushing explorations. The ore is a low-grade carbonate of lead.

The Aftermath, the property of the American Mining & Smelting Co. on Elk Mountain, has been producing regularly under lessees. The property is worked through an incline tunnel 690 feet long, having 11 levels, or a total of 2,500 feet. The ore, which is carbonate of lead, averages 25 ounces of silver and 25 per cent lead.

The Milo mine, adjoining the Aftermath, has an outlet for its production through the tenth level of the Aftermath. The ore is a carbonate of lead and sulphuret of iron, averaging 25 ounces silver per ton. The developments consist of a shaft, or rather an incline shaft, 550 feet deep, with levels at every 50 feet, which in all aggregate about 1,500 feet.

Of the White Quail, one of the best producers in this district, no accurate information of the amount of development or of the character of the ore has been obtained.

The Matchless mine has been idle.

Union district embraces that portion of Summit County in the vicinity of and to the northeast of Breckenridge; the principal camps in addition to Breckenridge are Montezuma, Lincoln, Decatur, and Chihuahua. The placers have not been yielding as largely as formerly, many of them being idle or being put in condition for future working.

The Peabody Placer, owned by Lilon Peabody, in Gold Run, has been worked with some success. The property possesses three short flumes, 200, 300, and 400 feet long, respectively; these flumes are 2 feet wide, 20 inches high, and give a supply of about 130 miner's inches.

The property of the Gold Run Placer Mining Co. was not worked during the year, but preparations are now being made for extensive work.

The Galena & Delaware Gulch placer was worked but very little on account of litigation; the property is well supplied with sluices and a Little Giant hydraulic.

The Ryan Gulch placer has likewise been idle, litigation being the cause. This property is well supplied with hydraulics, and if sufficient water could be obtained, the property would be equal in production to any in this district.

The Fuller Placer Mining Co. consists of the Maggie placer, 122 acres; the Fuller & Crome placer, about 160 acres; the May lode claim, on which is a tunnel 130 feet, the formation being quartzite and lime; the character of the ore is galena and gray copper, carrying native silver, the mill runs averaging 150 ounces silver per ton; the Nannie Houston lode, developed by a shaft 40 feet deep, showing a 20-inch pay streak similar in character to the May lode; the Germania lode, which is developed by a tunnel 280 feet long, the pay streak being from 10 to 20 inches wide, carrying sulphurets and carbonates that average 90 ounces silver per ton. None of this property has been worked during the year beyond working the annual assessment. From reliable information obtained, I understand that machinery has been purchased, and as soon as erected the property, both placer and lode claims, will begin to produce.

The Salt Lick placer has been successfully worked. This property possesses a flume about 600 feet long, the water supply being about 60 miner's inches.

The Blue River Mining Co. owns about 400 acres of good placer ground on Blue River. The surface only has been worked, the property possessing but very little development.

On Negro Hill, east of Breckenridge, is the Bunker Hill mine. The formation is porphyry hanging wall and lime footwall. The vein, which is 3 feet wide, runs due east and west. The ore averages 6 inches to a foot, the character being mostly a lead carbonate, averaging from 40 to 60 per cent lead and from 20 to 45 ounces silver and one-fourth ounce gold per ton. The developments consist of two shafts, one being 100 feet deep, the other 75 feet. A drift 30 feet below the surface extends between the two shafts, which are about 100 feet apart. A drift also extends about 20 feet west from the 100-foot shaft. Another drift, 20 feet below the one just mentioned, extends 20 feet at present but is being rapidly pushed. About 300 tons of ore were extracted while running these developments, which, on account of the low price of lead, still remains upon the dump.

The Washington mine, adjoining the Bunker Hill, together with the Uncle Sam and Ella, adjoining properties, are owned by Joseph Watson. These mines are thoroughly developed and equipped, nearly $90,000 having been expended upon them. In addition to the hoisting plant is a concentrating mill employing 20 stamps and other auxiliaries providing the facilities necessary for dressing the low-grade ores and rendering the entire product marketable. But a very small amount of ore was shipped during 1883, but the indications are that during the coming year the production will be quite large.

The Mayo is showing signs of improvement as development progresses. The ore, which is carbonate of lead, is of higher grade than that found in the Washington. Machinery has recently been erected, and development work will be pushed the entire winter.

On Mineral Hill the Cincinnati has been a regular shipper. The developments in tunnels and shafts aggregate over 1,000 feet. The mineral being a carbonate of lead averaging 65 per cent lead and 16 ounces silver per ton.

The Pennsylvania Breckenridge Mining Co. is working quite a large force in developing.

On the Seymore developments continue, and a small quantity of ore has been shipped.

The Mineral Hill Tunnel Co.'s tunnel is now about 1,225 feet long and is the most extensive continuous workings on the hill. But a very small amount of ore is extracted and shipped, the company wishing to first get their property in shape.

The Seventy-Nine has shipped some ore, although the developments hardly aggregate 100 feet. Their ore is galena, which averages 40 per cent lead, 25 ounces silver, and some gold.

The Gibson Hill Mining Co.'s property is composed of 20 patented claims, situated on the slope of Gibson Hill, McKay mining district, about a mile from Breckenridge. They are working 12 men steadily and are doing development work principally but are taking out enough ore to pay all expenses. The ore is a galena, and the coarser part is shipped as taken out and nets from $60 to $75 per ton. The second-class ore is being concentrated by Messrs. Newcomb & Musgrove at the Elyria concentrator.

The Elkhorn, on the west slope of Negro Hill, near the town of Breckenridge, has an aggregate of 490 feet of development in shafts, tunnels, and levels. About 60 or 70 tons of ore were extracted while making these developments; the average value of that extracted being $45 silver, $13 gold, and 65 per cent lead. In the workings some 300 or 400 tons of ore of equal value are in sight that will probably be extracted and shipped as soon as the mine can be cleared of water.

The Hildebrand is developed by a tunnel 175 feet long, extending through a hard quartz with small veins of mineral crossing at intervals, carrying copper pyrites and antimonial silver. This tunnel will be extended to catch a larger vein some 60 feet farther on.

The Rose of Breckenridge is developed by a tunnel 850 feet in length running to crosscut that vein, which it is expected to do in the next 50 feet. This tunnel is the Colorado & New Mexico Consolidated Mining & Smelting Co.'s property, which in its course has cut half a dozen veins that show but small pay streaks of very rich ore, the character being mostly galena and zinc blende, although some gray copper is found.

On Argentine Hill, the Fredonia, the property of the Lake Shore Silver Mining Co., is developed by an incline on the vein 100 feet. About 150 feet below and to the northwest another incline has been driven, cutting the vein about 150 feet from the surface. From this point levels are being run and stoping commenced. The vein is about 8 feet wide and the entire vein matter will average 26 ounces silver per ton. The ore is a sulphuret and lies between porphyry and dolomite. In addition to the regular shipments, fully $30,000 worth of low-grade ore has been thrown on the dump, which will be concentrated during the coming season.

About 10 miles from Breckenridge is the Warrior's Mark, which is again coming to the front as a producer. The ore averages $100 per ton, carrying brittle silver, copper glance, and baryta. The mine is said to be in a condition to continue shipments for some time to come.

On the Swan the Rochester Queen has been idle and the I X L has come to the front as a producer. This mine is probably the best developed in this locality, possessing, in addition to a tunnel 200 feet long, about 600 feet in shafts, winzes, and levels. The I X L Mining Co. owns five locations, through which the presence and continuity of a fissure vein 91 feet in width have been established. It is one of the most remarkable leads in this section, and contains throughout its entire width a very fair average grade of material, principally of a free milling character.

On Brewery Hill is the Sac Tunnel, the property of the New York & Summit County Mining Co. This tunnel and the workings aggregate about 650 feet; contracts for additional work have been let, and arrangements are now being made for active operations.

The Rock Island placer, comprising 90 acres, has been idle. This placer is well supplied with water, and the timber is unequaled by any in the State.

The Silver Eel and Silver Eel Extension are being extensively developed by a crosscut tunnel, which, when finished, will be 450 feet long, cutting the vein at a depth of 250 feet from the surface; the ore, which is free milling, yields $40 per ton silver and gold. The width of pay streak averages 30 inches.

The railroad passing near Montezuma has awakened mine owners, and during the past season more activity and real good has been accomplished than ever before during the existence of this camp. Numerous prospects have blossomed into producing mines, the only drawback being the lack of cars to transport the ores to market.

The Coaly, on Glacier Mountain, named after the discoverer, was located in July, 1864, and was the first discovery of silver in Colorado. This mine has yielded but a very small amount of bullion and has been worked but very little, save by open cuts along the surface.

The Harrison, the property of the Gem City Mining Co., on the eastern slope of Glacier Mountain, that has been idle so long on account of litigation, has finally settled all dispute and is now one of the big producers of this section. The developments on the lode are two tunnels or adits which have been driven on the vein, and three open cuts showing the crevice,

and the discovery shaft—the aggregate being about 200 feet in all. Ore is shown in all the workings and improves with depth, the character being gray copper and galena which averages about 60 ounces silver per ton.

The Belle East has shipped about 300 tons of ore during the year, it being a gray copper and galena. This property is developed by three levels, they being 85 feet, one above the other. The level No. 1, or upper level, is 120 feet long. Level No. 2 is 200 feet long, and level No. 3 is rather a drift, being 750 feet long and having an opening on the eastern slope of the mountain. Winzes connect these levels, and the outlet is through level No. 3.

The Herman mine, owned by the Herman Mining Co., has been shipping some ore. The crevice is about 6 feet wide and all the crevice matter pays for shipment, the average grade being between 60 and 70 ounces silver per ton. This is probably the largest ore body in the district, and with development will probably be one of the largest producers. The present developments consist of a drift on the vein 70 feet. About 35 feet below this drift an incline shaft has been started which will be sunk 100 feet, when levels will be run.

The Silver King was developed considerably some years ago and has lain idle until the present management took hold of the property. The developments consist of three adit levels with connecting winzes that aggregate 1,500 feet. The upper level, or level A, is 100 feet long; level B is 500 feet long; and level C is 600 feet long. Contracts have been recently let to extend levels B and C each 50 feet more. Work has been confined to the two lower levels, the ore streak having increased both in quality and quantity, it having widened from 1 inch in the upper level to nearly 2 feet in the lower, or C level. The grade of the mineral is good, carrying ruby silver, gray copper, and galena. The body of mineral in sight has warranted the company in putting the Silver King mill in thorough repair by putting in crushers, Cornish rolls, Hartz jigs, and buddles.

On the St. Elmo the drifts aggregate 450 feet long on the vein.

The Mark Twain is now developed with capacity for working 150 men in the stopes, but on account of the railroad not supplying sufficient number of cars, only about 14 men have been employed. The developments consist of a crosscut tunnel 400 feet long, cutting the Mark Twain lode and St. Elmo veins. On the Mark Twain lode is a shaft 205 feet deep and an aggregate of 630 feet in levels. The pay streak averages 8 inches wide, of galena and gray copper, carrying about 140 ounces silver per ton. The workings are supplied with T rails, ventilators, etc. Charles Buckland is manager.

The Mark Twain extension has been worked during the year.

The Montezuma Silver Mining Co. owns the Radical, Chautauqua, Erie, General Teller, and several other lodes on Glacier and Teller mountains, all of which are more or less developed by shafts, levels, and tunnels. The Radical and Chautauqua have been worked the most during the year. On the Radical the developments aggregate 400 feet of development in drifts, and on the Chautauqua the aggregate is about 1,500 feet. The shaft is 150 feet deep and the upper drift is 150 feet long. A crosscut tunnel, which at present is about 1,000 feet, is being run to intersect this lode at a depth of 350 feet. The ore found in both of these claims is galena and gray copper, carrying on an average 90 ounces silver per ton.

The Blanche has been idle most of the year. For about two months the mine was worked, and some ore extracted and shipped. The developments consist of a shaft 75 feet in depth, and an adit on the vein 180 feet long. The ore is a galena averaging 80 ounces silver per ton.

On Lenawee Mountain the Eliza Jane is being systematically developed. At present the only work done consists of a shaft 70 feet deep and a crosscut 40 feet long; but these are being pushed to put the mine in a condition for production.

On Collier Mountain the Waterman is developed by a drift 200 feet long and an open cut on the surface 75 feet. Lower down the mountain another drift has just been started, which is being pushed as fast as possible. The pay streak averages 10 inches of galena and gray copper that runs about 90 ounces silver per ton.

On Independence Mountain the Great Republic is developed by an adit 200 feet and a crosscut 100 feet long; the vein is badly broken, but with depth the mineral is more solid, it being galena that averages 50 ounces silver per ton.

In the Horseshoe, in Peru district, near Decatur, the Bufa Mining & Milling Co. is pushing development on the Bufa lode. Water filled the shaft and upper workings, and the company is now running a crosscut tunnel, which at present is 195 feet long. This tunnel will drain this property, and the ore is high-grade sulphurets, carrying gold and silver.

The Consolidated Silver Ledge Mining Co. is not working at present. They have now between 400 and 500 tons of ore on the dump ready for shipment. The ore consists of galena, gray copper, and gold, and gives returns from mill-runs of from 60 to 150 ounces per ton. The property is splendidly located for working.

The Royal Gorge Mining Co. has a very valuable property situated in close proximity to the Silver Ledge, but the ore is very different from the Ledge, as it consists of an arsenical iron, carrying gray copper, ruby silver, brittle silver, silver glance, gold, and copper. The lode is exposed by nature for a distance of 106 feet and is 2 feet in width where it crops out. At a depth of 10 feet it is 4 feet and 2 inches of the same character of mineral. The lode is clearly traceable for a distance of 1,200 feet and is being worked by two adits. The property has paid from the first hour's work, as the ore was obtained at the surface and will average in value from 114 ounces to 325 ounces in silver value.

The Delaware and Commodore mines, near Decatur, have been extracting and shipping large quantities of ore.

The Piqua also has shipped considerable ore, it being galena and gray copper that carried on an average 115 ounces silver per ton. This mine is developed by a drift 200 feet long and a shaft 30 feet deep.

The Hunkidori has been steadily worked. The ore shipped averaged $120 per ton. Developments consist of two drifts that aggregate 390 feet.

In his report for 1884 Burchard says: [61]

While the output of mineral from some portions of this county was insignificant and the total production smaller than usual, the strikes made in the latter half of the year have stimulated mining industries, and prospecting and exploring have been extensively carried on. In the Tenmile district, especially around Robinson, explorations have disclosed large ore bodies.

Gibson Hill, near Breckenridge, is booming because of the discovery of gold lodes. Chihuahua is coming to the front with new producing mines.

The South Park Railroad is offering very low rates to shippers; old mills are being repaired and new ones built; in fact, the whole county seems to have new life enthused into it, and the managers and owners are working hard through this winter to enable them, when the snow is gone and the roads open, to take their ore to market.

A strike is reported to have been made by E. C. Moody at the head of Georgia and American gulches, near Breckenridge, Summit County.

The Boston Mining Co., at St. John's, has been producing a small amount of ore.

The Blanche mine is being successfully worked, connection having been made with the shaft at a depth of 140 feet by drift-ing on the vein a distance of 300 feet. The vein is galena, carrying 40 per cent lead and 35 ounces silver, and showing ore all the way.

Chihuahua district.—Mining is said to be improving. A number of mines carrying high-grade ore are being worked, among them the Pickwick, which is said to have produced $3,000 in sinking the shaft 50 feet. The Eliza Jane has also produced some ore. The Maid of Orleans, Rothschild, Winning Card, Chihuahua, Queen, and Edith are said to be promising properties.

The Silver Queen mine, near the summit of Argentine Pass, is working a force of 10 miners, and sinking and stoping are being vigorously prosecuted. A ton of ore a day is being produced, which mill runs $400 silver. The pay streak averages from 3 to 10 inches of solid mineral, the character of which is gray copper and sulphurets.

The Rochester Queen mine employs a few men. The I X L is thought to be a fine property and has a concentrating mill of 40 tons capacity, worked by a wheel. Owing to litigation, it has not been producing.

About 2 miles below Swan City the Ouray Placer Co. are making preparations for mining gold placers. They have a saw-mill in operation cutting lumber for flumes, etc., and have built a ditch 1½ miles in length.

The Boss mine employs a few men on a vein yielding gold. On the apex of the mountain, at an altitude of some 11,000 feet, is the Ontario, and just below, on French Gulch, the Elephant, employing a dozen men.

In Tenmile district, at Felicia Grace mine, ore was discovered at a depth of 85 feet in an ore shoot similar to the old Robinson, an adjoining property, which has induced a large amount of prospecting to be done.

On neighboring claims about 100 men are employed. Three of these properties are shipping ore. The ore is sulphide and is from 12 to 18 feet thick and mills from 40 to 90 ounces in carload lots. The Felicia Grace has shipped over 700 tons of ore and is still shipping at the rate of 10 tons per day; the Result about 400 tons and averages 8 tons per day; the Last Chance 100 tons and is shipping 3 tons per day; the shipments from Felicia Grace are to the Argo works, at Denver. These shipments averaged 55 ounces of silver, leaving a handsome margin after the payment of $20 freight and reduction charges and 10 per cent off the silver. The lessees are now sinking another shaft near the end line of the claim upon the dip of the shoot.

The value of placer gold from 1884 to 1903 given in the table on page 245 has been estimated on the basis of all the information available.

The figures for gold and silver in 1885 have been interpolated to correspond with the total production of the State and with deposits of gold at the Denver Mint.

According to Wilson,[62] the gold produced in Summit County and deposited at the United States Mint at Denver during the calendar year 1885 amounted to $198,525, and the silver to $3,700 (coining value).

The figure for lead in 1886 is derived from a preliminary estimate published in Mineral Resources.[63]

For 1886 to 1896 the figures have been taken from reports of the agents of the Mint in annual reports of the Director of the Mint, the gold and silver being

[61] Burchard, H. C., op. cit. for 1884, pp. 177, 248-249, 1885.

[62] Wilson, P. S., agent for Colorado, in Kimball, J. P., Report of the Director of the Mint upon the production of the precious metals in the United States during the calendar year 1885, p. 136, 1886.

[63] U. S. Geol. Survey Mineral Resources, 1885, p. 257, 1886.

prorated to correspond with the figures for the total production of the State as corrected by the Director of the Mint, the lead being prorated to correspond with the total production of lead in the State as given in Mineral Resources and any unknown production in the State being distributed proportionately among the counties. As with lead so with copper, but as the figures for copper given in Mineral Resources include copper from matte and ores treated in Colorado, though mined in other States, they are subject to revision.

In his reports for 1887 and 1888 Munson [64] gives lists of producing and nonproducing mines and shows the individual output of the producing mines. Most of the lead produced in 1887 came from the White Quail mine, at Kokomo. The output of the Robinson mine is included in the total of confidential reports, in which lead is not shown separately. In 1888 the Victoria mine (Tenmile district) produced $100,000 in gold; the Boss (Breckenridge), $40,000 in gold; the Jumbo (Tenmile), $53,000 in gold. The Iron Mask (Tenmile), Minnie (Breckenridge), Oro (Breckenridge), Delphos (Tenmile), and Lucky (Breckenridge) mines were large producers of lead.

In his reports for 1889, 1890, 1891, and 1892 Smith [65] gives lists of the producing mines and shows their individual outputs. In 1889 the Victoria produced $68,173 in gold, and the Delphos $60,340 in gold, and $125,724 (coining value) in silver. The Delphos, Iron Mask, and Sultana and the Boss group were large producers of lead. The report for the Robinson was confidential. In 1890 the Delphos produced $43,740 in gold, and $88,721 in silver; the Boss group $126,061 in silver; the Juniata, $10,000 in gold and $10,343 in silver; the White Quail, $57,800 in gold and $144,000 in silver; the Victoria, $21,420 in gold. The value stated for silver is the coining value. The Chautauqua, Delphos, Juniata, Oro group, Ouray, Washington, and White Quail were large producers of lead. The Robinson mine is mentioned, but its production is not shown. In 1891 the Ouray, Pennsylvania, Sts. John, and Surprise were large producers of lead. Much of the production of the county is included in confidential reports. The Wilfley Co.'s large mill at Kokomo was burned. In 1892 the Decatur Mines Syndicate produced $9,000 in gold, $319,275 in silver, $58,725 in lead (at $87 per ton), and $19,800 in copper (at $220 per ton); the Juniata, $25,107 in gold, $3,534 in silver, and $5,525 in lead; the Ouray group (Breckenridge), $29,000 in gold, $154,800 in silver and $60,900 in lead; the Sts. John, $32,493 in silver and $18,105 in lead; and the White Quail, $12,420 in gold, $60,181 in

silver, and $60,378 in lead. The value stated for silver is the coining value.

In his reports for 1894, 1895, and 1896 Puckett [66] gives the deposits at the Denver Mint from mines in Summit County as follows:

In 1894, $86,257 in gold and $857 in silver; in 1895, $70,589 in gold and $663 in silver; in 1896, $59,033 in gold and $489 in silver. The value given for the silver is its coining value.

For 1897 to 1904 (and for zinc for 1906 and 1907), the figures given in the table on page 245, which represent smelter and mint receipts, are taken from the reports of the Colorado State Bureau of Mines.

In his reports for 1897 and 1898 Hodges [67] gives the deposits at the Denver Mint from mines in Summit County. In 1897 the value of the gold was $49,485, and the coining value of the silver was $489. In 1898 the value of the gold was $30,643, and the coining value of the silver was $319. In the report for 1899 Hodges says: [68]

The large output of placer gold from Summit County in the early sixties was one of the factors in spreading the fame of Colorado as a mining country, and it is estimated that about 5 per cent of its placer deposits have been worked and in a coarse and primitive manner. The bedrock on nearly all its streams and gulches is from 20 to 70 feet in depth, and only the high bars and benches have been worked, but in the last year several large companies have acquired nearly all the accessible placer ground and after having prospected thoroughly have erected large hydraulic appliances and dredges on the Blue and Swan and are now in such condition that the coming year promises a considerable increase in the yield of precious metal.

The old-time sluice boxes and giants are still worked by those who have neither the acreage nor means to equip their ground with modern appliances. * * * Several large milling plants have been built during 1899, equipped with the latest improvements especially adapted to meet the requirements of the ore of each mine, and are almost automatic in action.

In the camps of Robinson and Kokomo the production was not as large as expected, as all the large mines were compelled to shut down during the railroad blockade for lack of fuel.

This district is in the western side of the county and about 15 miles from Leadville, to which point all ore is shipped. * * *

Frisco district.—One of the districts of Summit County is Frisco, situated a few miles from Breckenridge, in the heart of the Tenmile Range. A number of mines here are constant shippers. The ore is pyritic copper, iron, and lead, ranging from $10 to $300 per ton.

In the Montezuma, Rathbone, and Smoke River districts a number of important strikes of good grade silver-lead ore were made, and this section showed more activity than for many years. Several of the properties were worked at a fair profit.

In the Swandyke region development was large. * * *

There is but one public sampler in the county, that at Breckenridge, owned by the Denver Smelting & Refining Co., and it has proved a great convenience to the smelting-ore producers.

During the blockade of last winter it continued to purchase all ore offered.

[64] Munson, G. C., agent for Colorado in Kimball, J. P., op. cit. for 1887, pp. 185-188, 192-193, 1888; idem for 1888, pp. 125-128, 1889.

[65] Smith, M. E., agent for Colorado, in Leech, E. O., Report of the Director of the Mint upon the production of the precious metals in the United States during the calendar year 1889, pp. 152-155, 1890; idem for 1890, pp. 127, 140-141, 1891; idem for 1891, pp. 175, 184-185, 1892; idem for 1892, pp. 119, 130, 1893.

[66] Puckett, W. J., agent for Colorado, in Preston, R. E., Report of the Director of the Mint upon the production of the precious metals in the United States during the calendar year 1894, p. 72, 1895; idem for 1895, p. 75, 1896; idem for 1896, p. 159, 1897.

[67] Hodges, J. L., agent for Colorado, in Roberts, G. E., Report of the Director of the Mint upon the precious metals in the United States during the calendar year 1897, p. 127, 1898; idem for 1898, p. 99, 1899.

[68] Hodges, J. L., agent for Colorado, in Roberts, G. E., op. cit. for 1899, pp. 119-120, 122, 1900.

The deposits at the Denver Mint from Summit County were $34,812 in gold and $326 in silver (coining value).

In his report for 1901 Hodges says: [69]

Tenmile district * * * embraces the camps of Frisco Robinson, and Kokomo. Considerable high and low grade smelting ore was produced during the year, carrying values in silver and lead with some gold. The Robinson Mining & Smelting Co. has nearly finished its new smelter, which will be an important addition to the district. This plant will handle the low grades of iron and sulphide ore.

Nigger and Mineral Hills have some good producers of lead and silver, carrying fair values in gold.

Montezuma district is handicapped by the lack of railroad transportation, a long wagon haul being necessary to market the ores. The veins are large and are rich in silver with some gold.

Farncomb Hill continues to produce rich free and crystallized gold, but with a limited output. Properties worked principally by lessees. During the year many mines made small shipments and are well developed for future operations.

The Denver Smelting & Milling Co. handles about 90 per cent of the Breckenridge ores. Numerous private mills treat the ores of their own properties.

Gold Pan placers.—The Gold Pan Mining Co., at Breckenridge, has the most complete plant for placer mining in the State. The company owns about 1,700 acres with ample water rights. Water is taken out of the Blue River about 4 miles above the lower end of the placer, by means of ditches and about 2 miles of steel pipe 5 feet in diameter. At the end of this pipe line a system of Evans hydraulic elevators has been installed. These elevators work down to bedrock and carry the gravel to the sluices above. Through these elevators the water will have a pressure of 150 pounds to the square inch under a head of 350 feet. With the four Evans elevators and giants already in place, the company proposes to start the coming season with capacity to handle 6,000 to 8,000 cubic yards of gravel a day. Two steel derricks over 100 feet high are to be used in removing boulders and heavy débris. Each derrick will be equipped with guy ropes, carriers, and necessary engines. * * * During the year the gravel beds were thoroughly prospected by means of drills and sand pumps. * * * The average depth of gravel is 60 feet. * * *

The Mecca Placer Co. is in French Gulch, at Breckenridge. The company has put in 3,000 feet of 60-inch pipe and installed Evans hydraulic elevators. In the past this ground has furnished large returns in nuggets and fine gold.

The Blue River Co. owns a large acreage of placer ground north of Breckenridge. During the season steam excavators made a good opening to bedrock, and the company is in shape to make a fine showing next season.

The Oro Grande Placer Co. has put in a hydraulic elevator and a new pipe line. This property is near Dillon.

The American Gold Dredging Co. is on the Swan River. The operations of the last year were highly successful. Evans elevators, giants, and dredges were used.

The deposits of bullion at the Denver Mint from Summit County were $67,015 gold and $492 in silver (coining value).

In his report for 1902 Downer says: [70]

At Robinson, during a few months, the Robinson smelter produced some matte. * * * Altogether, some 200 tons of ore were shipped from this camp for the year.

At Kokomo considerable activity was evidenced. The bulk of the ore mined at this point is of just sufficient value to permit of mining and shipping, the principal value being in silver and iron contained.

On Elk Mountain the following properties shipped: Eagle, Delaware, Summit, Kimberly, and Wilfley. The Eagle Co. probably produced the largest tonnage, averaging about 40 tons a day, but the ore rather depreciated, running about $10 a ton.

The Summit Mining & Smelting Co. has been a constant producer, its ore carrying a large percentage of zinc. It has a mill. After being crushed the ore is handled on Wilfley tables, and the concentrates are passed over a Wetherill magnetic separator to extract the iron from the zinc sulphide. A roasting furnace will be added, giving a slight roast to the concentrates with the expectation of rendering more of the iron magnetic and taking out a larger percentage of that metal. As the iron contents decrease the zinc percentage increases; a higher price is paid for the cleaner ore. The average output of this company was about 25 tons a day.

The Kimberly mine produced about 3,000 tons during the year and the Wilfley mine about 2,500 tons.

On Sheep Mountain the following properties produced: Michigan, Washington, Snow Bank, and United States.

On Jacque Mountain, the Queen of the West, which in the past produced some high-grade silver ore, has been working intermittently.

The Wintergreen shipped about 3,000 tons of very low grade iron sulphide ore.

In the Mayflower district, on Fletcher Mountain, the Bird's Nest group shipped about 200 tons of ore, averaging as high as $2\frac{1}{2}$ ounces in gold and 20 ounces in silver to the ton.

In the district surrounding Breckenridge the principal work has been the completion and operation of the American Dredging Co.'s placer plant on Swan River. During the short season of five months in which the dredges were operated it is stated that fully $100,000 in gold was saved, the first year of operations, and as only the beginning of their ground has been touched, the future of this company seems to be very bright.

The Gold Pan Co., which owns a large territory on Blue River, just above Breckenridge, has spent much money in construction of an immense dumping ground, as all of its tailings have to be elevated. A large pit was excavated to reach bedrock and a long flume built to take the entire water of the Blue, around the pit. The season was consumed in getting down to bedrock, at this point 40 feet below the surface, and the work was impeded by the presence of boulders, many of which weighed several tons. But the work of lifting these out of the pit with a very large stone bolt is accomplished by a fine mechanical arrangement of cables and cranes, the stone bolt depositing its load at the desired point automatically. The Evans hydraulic elevator has been installed and is being used successfully in the pit.

On French Gulch some very rich placer deposits were found on the Mecca Co.'s property, the pay streak averaging from $10 to $15 to the cubic yard, and at one place running as high as $42.

In the vicinity of Breckenridge lode-mining interests were not very active.

[69] Hodges, J. L., agent for Colorado, in Roberts, G. E., op. cit. for 1901, pp. 141–142, 147, 1902.

[70] Downer, F. M., agent for Colorado, in Roberts, G. E., op. cit. for 1902, pp. 118–120, 132, 190?.

Probably the mine which produced the largest amount of value was the Cashier, in Browns Gulch. This property is equipped with a 20-stamp mill.

On Bald Mountain the Mountain Pride started work in April and shipped 503 tons of ore, which averaged about $27 to the ton.

In the Swandyke district a good deal of prospecting was done.

The only new producer in the vicinity of Breckenridge was Carbonate mine, on Nigger Hill, which sent out three carloads of ore averaging $50 to the ton in gold, silver, and lead.

On Mineral Hill the June Bug had four cars, averaging $70 to the ton.

Montezuma was fairly active during the year. The Pride shipped 1,000 tons to the smelter, and the Mines Development Co., working the California Bell tunnel, produced 350 tons of good grade.

At Rathbone the only regularly producing property was the Pennsylvania, which put out 800 tons of ore and concentrates.

Bullion from Summit County was deposited at the Denver Mint amounting to $90,702 in gold and $615 in silver (coinage value).

In his report for 1903 Downer [71] gives the bullion from Summit County deposited at the Denver Mint as $121,451 in gold and $838 in silver (coinage value).

In his report for 1904 Downer says: [72]

The Gold Pan Mining Co. owns 1,700 acres of rich territory, and washed several acres of 75 feet to bedrock during the season. The company expended a large sum in the installation of a modern plant, which is the most complete in the State.

The American Gold Dredging Co. uses both dredge boat and hydraulic giants in operating.

The Reliance Gold Dredging Co. is constructing a $75,000 gold dredge, which will be completed for next season.

The Summit Banner Mining Co. and the Mekka Placer are each installing plants for the season of 1905.

The Masonton G. M. & M. Co. has completed a 20-ton cyanide plant to treat new ore bodies.

The old Union M. & M. Co. has under construction a 100-ton zinc concentrator to handle the ores from the Union and Wellington mines.

The Colorado & Wyoming Development Co., operating the Wellington and other properties on Mineral Hill, has discovered a 10-foot vein of zinc-lead ore in a 700-foot crosscut, which has been driven on 600 feet and upraised on 350 feet. Shipments of 800 tons per month have been made for the last half of the year, returns averaging 28 per cent zinc and 12 per cent lead.

The deposits of bullion from Summit County at the Denver Mint amounted to $75,557 in gold and $619 in silver (coinage value).

The figures for 1905 to 1923 (except for zinc for 1906 and 1907 from the Colorado Bureau of Mines), given in the table, are taken from Mineral Resources (mines reports).

[71] Downer, F. M., agent for Colorado, in Roberts, G. E., op. cit. for 1903, p. 81, 1904.

[72] Downer, F. M., agent for Colorado, in Roberts, G. E., op. cit. for 1904, pp. 20-121, 124, 1905.

Gold (placer and lode), silver, copper, lead, and zinc produced in Summit County, 1859-1923

Year	Ore tonnage treated (short tons)	Gold Placer	Gold Lode	Gold Total	Silver Fine ounces	Silver Avg price per ounce	Silver Value	Copper Pounds	Copper Avg price per pound	Copper Value	Lead Pounds	Lead Avg price per pound	Lead Value	Zinc Pounds	Zinc Avg price per pound	Zinc Value	Total value
1859–67		a$5,150,000		a$5,150,000													$5,150,000
1868		a150,000		a150,000	a7,547	$1.325	$10,000										150,000
1869		a200,000		a200,000	a7,907	1.328	10,500						$3,000				213,000
1870		500,000		500,000	a3,782	1.322	a5,000				50,000	$0.06	3,000				513,500
1871		70,000		70,000	a3,855	1.297	a5,000				50,000	.06					70,000
1872		120,000		120,000	23,784	1.278	30,396				100,000	.064	6,400				131,400
1873		75,000		75,000	7,734	1.24	9,590				100,000	.06	6,000				86,000
1874		70,000		70,000	154,688	1.16	179,438				423,950	.06	25,437				125,833
1875		72,012		72,012	30,938	1.20	37,126				141,000	.058	8,178				89,780
1876		150,000		150,000	119,883	1.15	137,865				100,000	.061	6,100				335,538
1877		150,000		150,000	154,688	1.12	173,251				100,000	.055	5,500				192,626
1878		165,774		165,774	317,109	1.15	364,675				100,000	.036	3,600				307,239
1879		75,000		75,000	1,560,344	1.13	1,763,189				100,000	.041	4,100				252,351
1880		44,000	a$5,000	49,000	674,757	1.14	769,223				500,000	.05	25,000				438,675
1881		26,000	5,000	31,000	270,703	1.11	300,450				a16,773,000	.048	805,104				2,599,293
1882		50,000	5,000	55,000	232,031	1.11	257,554				a5,773,000	.049	282,877				1,107,100
1883		10,000	5,000	15,000	234,351	1.07	250,756				a2,773,000	.043	119,239				434,719
1884		10,000	10,000	20,000	422,298	.99	418,075				985,250	.037	36,454				314,008
1885		15,000	185,000	200,000	230,120	.98	215,718	2,066	$0.135	$278	985,250	.039	38,425	a25,000	$0.043	$1,075	490,256
1886		15,000	149,222	164,222	394,058	.94	370,415				1,546,000	.046	71,116	a25,000	.044	1,100	654,513
1887		15,000	225,520	240,520	519,842	.94	488,651				1,754,132	.045	78,936	a25,000	.046	1,150	536,324
1888		10,000	272,209	282,209	516,358	1.05	542,176				2,126,887	.044	93,583	a75,000	.049	3,675	749,882
1889		16,000	206,724	222,724	523,658	.99	518,421				3,055,981	.039	119,183	a75,000	.05	3,750	834,586
1890		7,000	222,830	229,830	421,566	.87	490,173	166,799	.116	19,349	7,775,765	.045	349,909	a75,000	.05	4,125	1,126,040
1891		10,000	79,132	89,132	432,794	.78	328,821				10,591,152	.043	455,420	a212,500	.046	3,750	1,066,723
1892		10,000	116,046	126,046	288,242	.63	272,660	1,058	.107	113	6,371,637	.04	254,865	a415,600	.04	9,775	900,208
1893		10,000	106,165	116,165	441,448	.65	187,357	54,081	.108	5,841	6,277,000	.037	232,249	a200,000	.035	16,600	693,838
1894		10,000	214,791	224,791	514,107	.65	300,185	133,482	.124	16,018	5,500,000	.033	181,500	65,000	.036	7,000	685,951
1895		10,000	225,591	225,591	415,687	.68	308,464	9,825	.171	1,218	5,477,117	.032	175,268	82,489	.039	2,340	600,669
1896		10,000	200,202	210,202	264,872	.60	245,255	65,531	.166	11,205	3,950,040	.03	118,501	100,000	.041	3,900	638,629
1897		10,000	263,650	273,650	403,330	.59	158,923	53,030	.167	8,803	1,748,761	.036	62,955	227,156	.046	3,382	664,469
1898		10,000	333,825	343,825	368,887	.60	250,065	17,062	.122	2,849	4,889,204	.038	185,790	1,125,416	.058	10,449	786,537
1899		10,000	250,564	260,566	274,571	.62	221,332	22,888	.137	2,861	4,032,431	.045	181,459	491,055	.044	7,274	619,427
1900		25,000	313,719	338,719	220,543	.60	145,523	41,447	.128	5,678	5,610,710	.044	246,871	1,000,000	.041	21,606	865,527
1901		35,000	303,719	338,719	180,554	.53	119,093	7,510	.137	961	4,342,437	.043	186,725	1,329,180	.048	41,000	790,625
1902		43,000	199,583	242,583	209,356	.54	104,721	44,033	.128	6,969	3,092,387	.045	126,788	550,800	.054	63,901	590,115
1903		60,000	162,265	222,265	107,752	.58	127,707	21,865	.156	5,234	1,523,703	.042	63,996	1,884,584	.051	29,743	440,775
1904	35,475	94,240	113,886	208,126	127,847	.61	73,271	28,523	.193	4,373	2,178,182	.047	93,662	3,320,237	.059	96,114	503,584
1905	36,930	33,728	123,748	157,476	66,025	.68	84,379	21,740	.20	3,765	2,181,660	.057	102,538	3,363,740	.061	195,894	590,484
1906	34,050	53,199	86,574	139,773	99,763	.66	34,993	16,412	.132	499	1,301,912	.053	74,209	2,970,991	.059	205,188	497,675
1907	25,127	37,214	69,376	106,590	152,250	.53	51,877	18,170	.13	2,761	1,915,133	.042	101,502	1,232,149	.047	175,288	472,132
1908	14,631	145,370	41,571	186,941	182,957	.52	82,215	7,339	.127	2,849	1,719,190	.043	72,206	5,798,167	.054	57,911	355,816
1909	31,698	405,360	47,406	452,766	164,665	.54	96,967	8,646	.125	2,708	5,015,409	.044	220,678	5,542,685	.054	313,101	971,292
1910	47,040	347,204	21,562	368,766	167,490	.53	101,269	14,581	.165	2,816	6,024,867	.045	271,119	7,675,175	.057	299,305	973,725
1911	55,404	257,422	26,819	284,241	167,009	.615	101,164	25,033	.155	976	4,402,422	.044	198,109	9,342,725	.069	437,485	1,092,673
1912	46,606	392,739	33,276	426,015	67,009	.604	37,056	6,359	.133	1,513	9,342,725	.044	173,548	9,342,725	.056	644,648	1,372,749
1913	40,360	386,196	76,032	462,228	64,223	.553	32,561	27,550	.175	3,387	5,111,941	.039	61,044	5,111,941	.051	388,140	1,127,896
1914	22,199	608,567	60,043	680,144	120,207	.507	79,096	94,413	.129	6,834	1,365,231	.047	90,066	13,940,948	.124	260,709	1,028,395
1915	44,602	607,195	72,949	673,891	175,699	.658	144,776	17,823	.135	1,132	1,916,298	.069	116,516	19,968,814	.102	1,066,079	1,870,363
1916	65,768	579,050	94,841	603,437	87,676	.824	117,326		.147	66	1,688,637	.086	78,736	15,696,264	.091	1,868,087	2,741,177
1917	58,185	540,951	62,486	475,891	106,422	1.00	98,197		.246	3,554	915,535	.071	55,191	8,335,963	.073	2,026,619	2,860,402
1918	17,567	431,023	36,116	467,139	104,198	1.12	116,000		.273	12,746	777,338	.053	23,664		.081	1,428,360	2,071,278
1919	48,328	464,540	11,351	405,304	119,604	1.09	104,198		.247	2,620	446,491	.08	38,197		.05	295,438	894,322
1920	17,291	374,882	30,422	358,830	142,548	1.00	119,604		.186		477,462	.071	22,723		.134	675,213	1,234,780
1921	17,894	337,980	20,850	281,762		1.00	116,889		.184		504,957	.045	30,763	977,000	.057	55,689	489,305
1922		255,298	26,464			1.00					559,330	.055		7,335,000	.068	498,780	500,564
1923	48,965	203,379	32,663	236,012		.82					3,892,271	.07	272,459				1,126,790
		b13,974,323	5,148,660	19,122,983	13,573,470		11,709,616	1,065,126		151,909	153,705,665		6,813,507	137,145,560		11,223,543	49,021,558

a Estimated by C. W. Henderson.

b With the exception of an unknown quantity of placer gold in the early sixties and a small quantity in 1921 and 1922 from McNulty Gulch, in the Consolidated Tenmile district, this figure represents placer gold from the Breckenridge district.

TELLER COUNTY

The figures given for 1891 to 1896 are taken from the reports of the agents of the Director of the Mint for counties of Colorado, adjusted and prorated to agree with the estimate for the State of Colorado made by the Director of the Mint in his annual reports upon the production of the precious metals.

For 1897 to 1909 the figures for gold and silver, which represent smelter and mint receipts and are equivalent to mine shipments to smelters and mill shipments to smelters and the mint, are taken from reports of the Colorado State Bureau of Mines.

For 1901 and 1902 the figures given by the Colorado State Bureau of Mines for the production of gold in El Paso (all from Cripple Creek district) and Teller counties have been combined. The silver for the two counties has also been combined for 1901.

For 1910 to 1923 the figures are taken from Mineral Resources. The figures showing quantity of ore treated from 1904 to 1910 are taken from mines reports, and those from 1911 to 1923 from reports of mills and smelters. The figures for gold, silver, copper, and lead from 1910 to 1923 represent smelter and mint receipts and are equivalent to mine shipments direct to smelters and mill shipments to smelters and the mint.

Total gross production of Stratton's Independence mine, from 1891 to June 30, 1915 [a]

By late W. S. Stratton	$3,985,440
By first English company operating on company account, 1898–1904	11,049,030
By the same company operating by lessees, 1904–1908	4,015,290
By the Argall company, operating 1908–1915	4,571,968
	23,621,728

Earnings of Stratton's Independence mine

Profit by Mr. Stratton	$2,402,164
Dividends by first English company, 1898–1904	4,142,739
Dividends by the same company, lessor, 1904–1908	606,250
Dividends by the Argall company, operation 1908–1915	455,625
Cash as of Sept. 30, 1915 (approximately)	510,000
	8,116,778

Ore produced and sold from Stratton's Independence mine

	Tons
By W. S. Stratton	41,694
By first English company	323,270
By first English company, lessor	104,040
By the present company	104,729
Total shipping ore	573,733
Ore milled by the Argall company, operating 1908–1915	685,130
Total production to June 30, 1915	1,258,863

[a] This mine and mill were sold June 30, 1915, to the Portland Gold Mining Co.

Recapitulation of production of ore

	Short tons	Value
Portland mine:		
April 1 to December 31, 1894	7,826	$553,976
Calendar year—		
1895	31,516	1,700,095
1896	23,598	1,116,128
1897	18,852	1,177,643
1898	27,799	1,879,682
1899	38,548	1,951,219
1900	60,787	2,351,396
1901	76,906	2,408,413
1902	89,664	2,334,024
1903	90,245	2,608,994
1904	96,521	2,598,725
1905	109,234	2,422,033
1906	103,614	1,932,083
1907	79,960	1,600,950
1908	94,311	1,834,081
1909	83,909	1,438,650
1910	67,515	1,241,168
1911	50,258	1,140,054
1912	44,562	987,416
1913	53,246	1,380,713
1914	62,998	1,467,005
1915	72,192	1,710,277
1916	96,046	2,236,842
1917	86,688	1,768,972
1918	53,887	1,120,851
1919	45,417	1,173,616
1920	31,426	867,381
1921	7,806	223,667
1922	27,336	576,328
1923	30,346	675,218
Total	1,763,014	46,477,600
Average value per ton for the period		$26.36
Victor Mills:		
July 1 to December 31, 1910	46,237	113,253
Calendar year—		
1911	120,961	424,489
1912	173,361	547,424
1913	178,162	526,187
1914	210,132	539,131
1915 ᵃ	282,192	819,433
1916	322,892	803,381
1917	471,873	966,765
1918	522,756	1,071,924
1919	327,776	645,645
1920	205,498	509,998
1921	208,051	1,005,294
1922	163,842	497,578
1923	151,082	503,363
Total	3,384,815	8,973,865
Grand total of ore production to Jan. 1, 1924	5,147,829	55,451,465
Average value per ton at the Victor Mills		$2.65
Total dividends paid to January 1, 1924		$11,692,080
Percentage of gross production paid in dividends		26

ᵃ Includes six months' operation of the Independence Mill.

Recorded production of gold from certain individual mines at Cripple Creek, Colo., to December 31, 1921

[Compiled by C. W. Henderson, with the help of the operators]

Mine	Ounces	Value
Acacia	108,638.10	$2,245,748.83
Ajax	247,917.46	5,124,919.06
Anchoria-Leland	131,983.09	2,728,332.61
Blue Bird	78,770.36	1,628,327.85
Carbonate Queen	48,292.03	998,284.57
Christmas	64,589.45	1,335,182.43
Climax	13,435.58	277,738.09
Colorado Boss	12,787.84	264,348.11
Cresson	981,272.03	20,284,693.10
Dante	61,687.07	1,275,184.91
Deadwood	90,251.32	1,865,660.36
Doctor Jackpot	344,728.41	7,126,168.67
Elkton	663,908.43	13,724,205.25
El Paso	479,592.82	9,914,063.45
Gold Dollar	199,997.36	4,134,312.35
Gold King	157,248.93	3,250,623.87
Gold Sovereign	59,662.92	1,233,342.01
Golden Cycle	1,088,931.05	22,510,202.56
Granite	765,689.51	15,828,206.91
Findley	151,777.11	3,137,511.31
Hoosier	8,572.58	177,210.96
Hull City	194,016.65	4,010,680.10
Index	44,669.25	923,395.35
Isabella-Victor	710,333.34	14,683,893.32
Jerry Johnson	242,109.66	5,004,850.85
Joe Dandy	60,429.36	1,249,185.73
Lady Campbell	166.55	3,442.89
Last Dollar-Modoc	250,452.03	5,177,302.94
Lillie	89,589.89	1,851,987.39
Lincoln	20,957.81	433,236.38
Little Nell	127.73	2,640.41
Moose	28,523.63	589,635.76
Mary McKinney	517,621.10	10,700,177.76
Midget-Bonanza	159,151.98	3,289,963.41
Pinnacle	16,589.84	342,942.43
Portland	2,594,866.94	53,640,660.19
Raaler	143,843.95	2,973,518.34
Rose Nichol	46,472.39	960,669.56
Savage-Gold King	24,924.35	515,232.04
Sheriff	8,489.96	175,503.05
Stratton's Independence	1,209,375.00	25,000,000.00
Strong	538,385.92	11,029,424.69
Sunshine-Sedan	2,206.27	45,607.65
School Section	43,629.67	901,905.32
Trail	88,460.35	1,828,637.72
Theresa	29,031.07	600,125.48
Vindicator	1,026,200.99	21,213,457.13
W. P. H.		3,000,000.00
Wild Horse	84,466.96	1,746,087.03
Bull Hill (Stratton Cripple Creek Mining & Dredging Co.)	256,184.73	5,295,808.36
Globe Hill (Stratton Cripple Creek Mining & Dredging Co.)	177,081.57	3,373,851.98
		299,628,090.52

Production of gold, silver, copper, and lead in Teller County (Cripple Creek district), 1891–1922

Year	Ore treated (short tons)	Lode gold	Silver Fine ounces	Silver Average price per ounce	Silver Value	Copper Pounds	Copper Average price per pound	Copper Value	Lead Pounds	Lead Average price per pound	Lead Value	Total value
1891		$1,930										$1,930
1892		557,851										557,851
1893		2,021,088	5,680	$0.78	$4,430							2,025,518
1894		2,618,388	25,335	.63	15,961							2,634,349
1895		6,166,144	68,428	.65	44,478							6,210,622
1896		7,413,493	63,617	.68	43,260							7,456,753
1897		10,131,855	59,879	.60	35,927							10,167,782
1898		13,507,349	67,799	.59	40,001							13,547,350
1899		16,058,564	82,299	.60	49,379							16,107,943
1900		18,149,645	80,792	.62	50,091							18,199,736
1901		17,234,294	89,560	.60	53,736							17,288,030
1902		16,932,416	62,780	.53	33,273							16,965,689
1903		11,840,272	41,605	.54	22,467							11,862,739
1904	597,819	14,456,536	47,817	.58	27,734							14,484,270
1905	716,358	15,641,754	56,951	.61	34,740							15,676,494
1906	702,069	13,930,526	67,943	.68	46,201							13,976,727
1907	451,082	10,370,284	51,630	.66	34,076							10,404,360
1908	601,173	13,031,917	52,270	.53	27,703							13,059,620
1909	575,670	11,466,227	63,204	.52	32,866							11,499,093
1910	668,941	11,002,253	54,263	.54	29,302							11,031,555
1911	756,900	10,562,653	57,783	.53	30,625							10,593,278
1912	849,172	11,008,362	66,117	.615	40,662							11,049,024
1913	917,406	10,905,003	71,349	.604	43,095							10,948,098
1914	939,423	11,996,116	89,056	.553	49,248							12,045,364
1915	948,082	13,683,494	87,767	.507	44,498							13,727,992
1916	945,820	12,119,550	79,804	.658	52,511							12,172,061
1917	1,084,656	10,394,847	64,568	.824	53,204							10,448,051
1918	936,326	8,119,747	50,665	1.00	50,665							8,170,412
1919	775,986	5,827,816	35,442	1.12	39,695							5,867,511
1920	448,618	4,323,998	33,789	1.09	36,830	451	$0.184	$83	612	$0.08	$49	4,360,960
1921	484,110	4,291,883	37,335	1.00	37,335							4,329,218
1922	432,129	4,037,582	24,462	1.00	24,462							4,062,044
1923	382,739	4,047,008	22,606	.82	18,537							4,065,545
		323,850,845	1,762,595		1,146,992	451		83	612		49	324,997,969

BIBLIOGRAPHY

Many interesting details concerning the settlement of Colorado are given in Hall's, Smiley's, and Stone's histories of Colorado. The proceedings of the Colorado Scientific Society contain much information concerning the mining industry. In the preparation of this report the works listed below have been consulted.

BAIN, H. F., Zinc and lead ores in 1905: U. S. Geol. Survey Mineral Resources, 1905, pp. 379-392, 1906.

BANCROFT, H. H., History of Nevada, Colorado, and Wyoming, San Francisco, 1890.

BASTIN, E. S., and HILL, J. M., Economic geology of Gilpin County and adjacent parts of Clear Creek and Boulder counties, Colo.: U. S. Geol. Survey Prof. Paper 94, 379 pp., 23 pls., 1917.

BUREAU OF THE MINT, Report of the Director of the Mint upon the production of gold and silver in the United States during the calendar years, 1880-1923. (Separate report for each year.)

BURTON, H. E., Leadville, Colo., zinc deposits: Mines and Minerals, vol. 31, p. 436, 1911.

—— History of the zinc industry in Colorado: Min. Sci., vol. 64, p. 85, 1911.

BUTLER, B. S., Notes on the Unaweep copper district, Colo.: U. S. Geol. Survey Bull. 580, pp. 19-23, 1915.

COLORADO STATE BUREAU OF MINES Reports and Bulletins, 1897-1923. The publications of this bureau give statistics and general information concerning the mining industry of the State.

COPPER HANDBOOK, vols. 1-11, 1900-1913, succeeded by Mines Handbook, vols. 12-15, 1915-1922.
    Presents statistics and gives details concerning mining operations in the years mentioned.

CRAWFORD, R. D., Geology and ore deposits of the Monarch and Tomichi districts, Colo.: Colorado Geol. Survey Bull. 4, 317 pp., 25 pls., 1913.

CRAWFORD, R. D., and GIBSON, RUSSELL, Geology and ore deposits of the Red Cliff district, Eagle County, Colo.: Colorado Geol. Survey Bull. 30, Boulder, 1925.

CROSS, WHITMAN, U. S. Geol. Survey Geol. Atlas, La Plata folio (No. 60), 1899.

—— and PURINGTON, C. W., U. S. Geol. Survey Geol. Atlas, Telluride folio (No. 57), 1899.

CUSHMAN, SAMUEL, and WATERMAN, J. P., The gold mines of Gilpin County, Colo., Central City, 1876.

DAVIS, C. C., Olden times in Colorado, Los Angeles, Calif., 1916.
    Contains chapter on mines with review and statistics affecting the building of the Denver & Rio Grande Railroad to Leadville.

DENVER REPUBLICAN, Reviews of mining industry for the preceding calendar year published at the beginning of each year from 1893 to 1913.

EGLESTON, THOMAS, The Boston & Colorado smelting works: Am. Inst. Min. Eng. Trans., vol. 4, pp. 276-298, 1876.

EMMONS, S. F., The mines of Custer County, Colo.: U. S. Geol. Survey Seventeenth Ann. Rept., pt. 2, pp. 405-472, pl. xxxvii, 1896.

—— Geology and mining industry of Leadville, Colo.: U. S. Geol. Survey Mon. 12, xxix, 770 pp., 45 pls., and atlas of 35 sheets folio.

—— U. S. Geol. Survey Geol. Atlas, Tenmile district folio (No. 48), 1898.

EMMONS, W. H., The Cashin mine, Montrose County, Colo.: U. S. Geol. Survey Bull. 285, pp. 125-128, 1906.

—— and LARSEN, E. S., Geology and ore deposits of the Creede district, Colo.: U. S. Geol. Survey Bull. 718, pp. 9-10, 1913.

ENDLICH, F. M., Report as geologist of San Juan division: U. S. Geol. and Geog. Survey Terr. Ann. Rept. for 1874, pp. 181-240, 1876.

ENGINEERING AND MINING JOURNAL (editorial), The revelation of a metallurgical secret, vol. 87, pp. 464, 963, 1909.

FOSSETT, FRANK, Colorado—a historical, descriptive, and statistical work on the Rocky Mountain gold and silver mining region, Denver, 1876.

FOSSETT, FRANK, Colorado—its gold and silver mines, 1879.

—— Same, 1880.

GEORGE, R. D., and CRAWFORD, R. D., The Hahns Peak region, Routt County, Colo.: Colorado Geol. Survey First Rept., 1908, pp. 189-229, 1 pl., 1909.

HALL, FRANK, History of the State of Colorado, 1889.

HENDERSON, C. W., Gold, silver, copper, lead, and zinc in Colorado (mines reports): U. S. Geol. Survey Mineral Resources, 1908-1923.
    The reviews by counties in these volumes give a detailed synopsis of mining activity during the years 1908-1923, which are covered only briefly in Professional Paper 138.

HOLLISTER, O. J., The mines of Colorado, Springfield, Mass.. 1867.

IRVING, J. D., and BANCROFT, HOWLAND, Geology and ore deposits near Lake City, Colo.: U. S. Geol. Survey Bull. 478, 128 pp., 8 pls., 1911.

JONES, O. M., Bibliography of Colorado geology and mining, with subject index, from the earliest explorations to 1912: Colorado Geol. Survey Bull. 7, 493 pp., 1914.

KIRCHHOFF, CHARLES, Lead: U. S. Geol. Survey Mineral Resources, 1882-1905, inclusive.

—— Zinc: U. S. Geol. Survey Mineral Resources, 1900, pp. 213-227, 1901.

—— Zinc: U. S. Geol. Survey Mineral Resources, 1902, pp 217-229, 1904.

LEADVILLE HERALD-DEMOCRAT, Reviews of mining industry for the preceding calendar year published at the beginning of each year from 1897 to the present.

LINDGREN, WALDEMAR, [Gold and silver] Colorado: U. S. Geol. Survey Mineral Resources, 1905, pp. 185-214, 1906.

—— Notes on copper deposits in Chaffee, Fremont, and Jefferson counties, Colo.: U. S. Geol. Survey Bull. 340, pp. 157-174, 1908.

LOUGHLIN, G. F., Prices of silver, copper, lead, and zinc, 1850-1922: U. S. Geol. Survey Mineral Resources, 1922, pt. 1, p. 127A, 1925.

MINERAL INDUSTRY, vols. 1-29, 1892-1920.
    Presents statistics and gives details in regard to mining in the years mentioned.

MINING IN BOULDER COUNTY, Boulder County Mining Association, Boulder, Colo., 1910.

NARAMORE, CHESTER, [Gold and silver] Colorado: U. S. Geol. Survey Mineral Resources, 1906, pp. 199-240, 1907.

—— [Gold, silver, copper, etc., in Western States] Colorado: U. S. Geol. Survey Mineral Resources, 1907, pp. 235-279, 1908.

PATTON, H. B., Geology and ore deposits of the Platoro-Summitville mining district, Colo.: Colorado Geol. Survey Bull. 13, 122 pp., 40 pls., 1918.

—— and others, Geology of the Grayback mining district, Costilla County, Colo.: Colorado Geol. Survey Bull. 2, 111 pp., 9 pls., 1910.

PEARCE, H. V., The Pearce gold-separation process: Am. Inst. Min. Eng. Trans., vol. 39, pp. 722-734, 1908.

PURINGTON, C. W., Preliminary report on the mining industries of the Telluride quadrangle, Colo.: U. S. Geol. Survey Eighteenth Ann. Rept., pt. 3, pp. 745-850, pls. ciii-cxviii, 1898.

Ransome, F. L., The ore deposits of the Rico Mountains, Colo.: U. S. Geol. Survey Twenty-second Ann. Rept., pt. 2, pp. 229–397, pls. 26–41, 1901.

—— A report on the economic geology of the Silverton quadrangle, Colo.: U. S. Geol. Survey Bull. 182, 265 pp., 16 pls., 1901.

—— Geology and ore deposits of the Breckenridge district, Colo.: U. S. Geol. Survey Prof. Paper 75, 187 pp., 33 pls., 1911.

Raymond, R. W., Statistics of mines and mining in the States and Territories west of the Rocky Mountains, 1869–1875.

Rickard, T. A., Across the San Juan Mountains: Eng. and Min. Jour., vol. 76, pp. 307–308, 1903.

—— Two famous mines; The Camp Bird: Min. and Sci. Press, vol. 103, pp. 827–828, 1911.

—— The development of Colorado's mining industry: Am. Inst. Min. Eng. Trans., vol. 26, pp. 834–848, 1896.

Rico News, June, 1892, The early trail blazers.

Rockafellow, B. F., History of Fremont County, in History of the Arkansas Valley, Baskin & Co., 1881.

Smiley, J. C., History of Denver, with outlines of the earlier history of the Rocky Mountain country, 1903.

Spencer, A. C., Reconnaissance examination of the copper deposits at Pearl, Colo.: U. S. Geol. Survey Bull. 213, pp. 163–169, 1903.

Spurr, J. E., and Garrey, G. H., Economic geology of the Georgetown quadrangle (together with the Empire district), Colo., with general geology by S. H. Ball: U. S. Geol. Survey Prof. Paper 63, 422 pp., 87 pls., 1908.

Spurr, J. E., Geology of the Aspen mining district, Colo.: U. S. Geol. Survey Mon. 31, xxxv, 260 pp., 43 pls., and atlas of 30 sheets folio, 1898.

Stone, W. F. (editor), History of Colorado, vol. 1, S. J. Clarke Publishing Co., Chicago, 1918.

A list of works on Colorado is given on pp. 877–890, under the heading "Colorado literature," by Eugene Parsons.

CONCLUSIONS

This history of mining in Colorado will be useless unless the facts set forth for the period 1859–1923 can point in some way to the future. That Colorado has been a large producer of metals is definitely known. That it has been chiefly a producer of gold and silver is shown by the fact that of the calculated gross value of recovered gold, silver, copper, lead, and zinc, amounting to $1,531,000,000, $673,000,000 in gold, or 44 per cent of the total, and 628,850,000 ounces of silver, with a commercial value of $501,734,000, or 33 per cent, represent the gold and silver added to the world's supply. Thus 77 per cent of the total gross value of Colorado's production of these five metals is represented by gold and silver. Most of the gold is still in existence. A great part of the silver was coined and in this form represents a value of $1.29 an ounce. The copper produced, chiefly as a by-product of gold and silver mining, amounting to 263,000,000 pounds, with a gross calculated value of $40,328,000, has not all been dissipated. The enormous quantity of lead recovered, 4,200,000,000 pounds, with a gross value of $189,662,000, and the large quantity of zinc recovered, 1,740,000,000 pounds, with a gross calculated value of $126,216,000, have probably been largely used up in paint, brass, chemicals, and auto-

mobile tires. Probably little of the lead and zinc remains for the use of humanity. It seems hardly a mere coincidence that the total gross value of these five metals—$1,531,000,000 to the end of 1923, is very close to the assessed valuation of the State of Colorado for 1923, $1,550,000,000. The fact that the curve of the assessed value from the early days of Colorado—when mining or labors dependent on mining had developed the only assessable wealth—to the present time not only parallels but actually coincides with the curve of the gross production of the five metals can not be a mere accident. Denver in particular owes its growth to mining. Colorado Springs owes a great part of its development to mining. Pueblo owes its industrial existence to mining and metallurgy. These three cities are still the first three in the State.

That the surface of Colorado "has not been scratched" is a statement not borne out by facts. The surface has been well scratched and even intensively perforated with holes ranging from 10 to 3,000 feet in depth, and with tunnels as much as 5 miles in length. Much of this "scratching" was misdirected. With the exception of the men from Georgia and California most of the early gold seekers were ignorant of minerals or mining. It is a tribute to their energy that they found nearly all the placer-gold deposits in the first two years and worked out in the first five years the more easily worked deposits, making enormous outputs, such as $6,000,000 in California, Cache, Colorado, and other gulches of Chaffee and Lake counties, $5,500,000 in the vicinity of Breckenridge, $750,000 in the vicinity of Idaho Springs, $2,500,000 in the vicinity of Fairplay and Tarryall, and $200,000 at Hahns Peak after 1865. The oxidized decomposed portions of the lode veins they treated by placer methods and other laborious mechanical appliances, making constant improvements with development of the stamp mill. From 1859 to 1865 by these methods $7,741,361 in gold was recovered in Gilpin County, $1,500,000 at Empire, Clear Creek County, about $600,000 in Park County, and about $150,000 in Boulder County. They reached sulphides in the lode mines at a depth of 40 feet at Empire, at 40 to 180 feet in Gilpin County, and at similar depths in Boulder and Park counties. Later in Clear Creek and Park counties they found the oxidized outcrops of silver ore which changed to sulphides within 5 to 50 feet. They developed smelters in 1868 to treat both gold and silver ores. The oxidized silver-lead ores discovered at Leadville in 1877 in bedded deposits were enormous, and the depth of oxidation can not be expressed in terms of average depth, but where these bedded deposits cropped out, oxidation extended down the beds as far as 1,000 feet. At Red Cliff the oxidation followed the dip of the bed down 1,000 feet. Where the surface was grassy, as at Cripple Creek, the soil extended in many places to 20 feet below the grass roots. Amalgamation of the Cripple Creek oxidized ores was carried on with indifferent success for a very short time. When the unoxidized telluride ores were reached a new problem arose.

It is well to attempt to visualize the outcrops of the deposits before ore was discovered. In Gilpin County, except in places of bold outcrop, the surface in the creeks was covered with swampy soil upon which grew aspen trees, and on the hillsides was a growth of brush and even luxuriant evergreen forests. Leadville was in general heavily forested at the time of discovery of the silver-lead ores in 1877. Nearly all that timber was soon removed to make charcoal for the furnaces or for mine timbers.

Is it to be presumed that outcrops have been overlooked? No doubt some have been overlooked, but most of the easily found outcrops have been discovered.

The new silver-lead-zinc ore bodies found in 1923 and 1924 on Mount McClellan, 3 miles west of Silver Plume, and 4 miles north of the Belmont lode, which was discovered in 1864, were found by tunnels 1,600 to 3,000 feet in length. The outcrops are either completely hidden by brush and forests or are inconspicuous and uninviting. Some of the best veins of the San Juan region did not reach the surface. The "Leadville" blue limestone, so productive at Leadville, lies under morainal wash in Park County. A combination of intensive geologic examination and the energy of the informed prospector may find new surface ore bodies. For the present the old districts should be the starting points. Prospectors should have the unfavorable localities eliminated by competent geologists.

In 1899 the tungsten deposits near Nederland, Boulder County, were recognized and opened. In 1918 the enormous deposit of molybdenum near Climax, which had been known for many years, began to be worked. The ore averages less than 1 per cent MoS_2. So far Colorado has found no "porphyry copper" deposits such as those in Utah, Arizona, Nevada, and New Mexico, where the copper averages between 1 and 2 per cent. In fact, Utah has found only one such deposit so far; New Mexico two, with one only profitable to date; Nevada one; and Arizona five. Will the granite outcrops of Colorado disclose such deposits?

For the immediate future, the sulphides of lead and zinc, practically untouched because hitherto of too low grade and too complex, in several of the early bonanza surface camps, offer the best opportunity, with the application of modern metallurgy.

These lead-zinc-iron deposits of Colorado are known to be extensive. Such men as S. F. Emmons refer to enormous bodies of sulphides, in one district alone, which this history shows to be still untouched. The development of these sulphide ore bodies in the ramification of geologically guided crosscuts and drifts should also disclose good-sized deposits of oxidized lead-silver ores. Properly directed diamond drilling will uncover deposits of gold, silver, lead, and zinc, as it has in the past.

What will geology plus the drill tell us of ore at depth? Colorado's low-grade uranium and vanadium deposits are known to be very extensive. To extract oil from oil shale demands not only the eduction of the oil but also the solution of the mining problem. Colorado's fluorspar deposits constitute the basis for a new industry. Colorado's deposits of coal, limestone, gypsum, building stone, and clay are enormous and will be more extensively developed with the increase of population.

APPENDIX

The following tables show, by counties, the mine production of gold, silver, copper, lead, and zinc from crude ore shipped to smelters during the period 1909–1923.

Mine production of metals from crude ore shipped to smelters, by counties, 1909–1923, in terms of recovered metals [a]

County	Ore (short tons)	Gold (fine ounces)	Silver (fine ounces)	Copper (pounds)	Lead (pounds)	Zinc (pounds)
Baca	22	0.14	115	10,092		
Boulder	61,742	41,877.10	1,990,118	326,196	3,989,873	
Chaffee	141,878	48,673.56	1,276,335	3,424,060	12,746,039	9,577,545
Clear Creek	160,395	111,825.96	2,582,399	3,186,429	16,497,075	1,025,597
Custer	22,760	1,656.77	517,725	389,800	1,370,620	114,079
Delta	2	5.32	139			
Dolores	81,336	4,290.36	980,297	4,917,122	7,682,441	5,071,098
Eagle	248,478	32,963.39	3,332,796	5,743,263	3,151,376	20,526,803
El Paso	239			11,239		
Fremont	6,214	217.40	15,271	558,076	146,157	854,275
Garfield	346	670.19	372	1,044		
Gilpin	123,647	155,264.36	1,043,510	4,156,453	3,724,229	
Grand	21	.20	3,568	816	2,855	
Gunnison	48,114	7,851.53	341,622	385,423	4,558,097	16,866,952
Hinsdale	9,688	2,277.53	363,605	1,363,195	1,543,939	32,167
Jackson	229	19.40	591	23,725		
Jefferson	24	.78	9	1,000		
Lake	4,317,140	674,327.53	32,188,561	31,215,853	181,538,618	497,397,186
La Plata	41,922	77,453.45	670,179	258,563	72,797	
Larimer	48					30,722
Mesa	22	.44	257	5,685	20	
Mineral	280,894	7,989.62	6,824,322	142,622	12,323,210	99,131
Moffat	104	.19	185	38,444		
Montezuma	545	578.53	686	5,962	286	
Montrose	897	2.13	32,264	314,051		
Ouray	107,621	27,457.07	2,149,938	2,765,661	7,468,226	634,285
Park	61,942	88,984.55	612,570	327,479	9,095,943	2,243,532
Pitkin	552,135	42.60	5,965,574	99,638	77,252,598	3,349,215
Rio Grande	193	424.55	1,311	30,546	3,422	
Routt	594	74.28	8,356	32,303	7,614	
Saguache	16,406	2,941.49	460,794	618,302	2,522,898	148,543
San Juan	78,712	30,267.92	3,304,797	3,913,761	18,540,167	1,475,295
San Miguel	14,810	5,469.17	680,464	379,841	6,276,203	98,030
Summit	115,098	16,801.13	1,029,150	221,779	7,843,621	14,079,354
Teller	293,953	662,791.03	240,546	451	612	
	6,788,171	2,003,199.67	66,618,426	64,868,874	378,358,936	573,623,809

[a] For explanation see footnote to Table 1, p. 69.

Mine production of metals from crude ore shipped to smelters, by counties and years, 1909–1923

Baca County

[No shipments of crude ore to smelters in 1909–1914 and 1919–1923]

Year	Ore (short tons)	Gold (fine ounces)	Silver (fine ounces)	Copper (pounds)	Lead (pounds)	Zinc (pounds)
1915	8		8	514		
1916	5		50	2,772		
1917	9	0.14	57	6,806		
	22	.14	115	10,092		

Boulder County

Year	Ore (short tons)	Gold (fine ounces)	Silver (fine ounces)	Copper (pounds)	Lead (pounds)	Zinc (pounds)
1909	3,951	6,562.92	34,521	14,292	35,348	
1910	6,212	4,842.13	44,774	15,519	53,213	
1911	7,808	5,817.61	49,926	27,468	114,541	
1912	5,558	4,635.89	71,522	22,072	305,690	
1913	4,614	2,710.05	162,021	25,535	409,328	
1914	7,329	3,475.08	291,471	21,769	475,309	
1915	5,322	3,867.54	161,291	81,105	661,687	
1916	4,097	2,364.66	216,935	62,914	785,040	
1917	4,789	2,104.51	268,307	21,794	537,909	
1918	3,182	1,522.02	154,231	16,523	262,310	
1919	4,309	1,736.49	216,700	10,218	173,555	
1920	1,507	898.53	102,941	6,685	84,069	
1921	792	301.05	66,938	302	38,096	
1922	1,939	650.75	113,405		33,769	
1923	333	387.87	35,135		20,009	
	61,742	41,877.10	1,990,118	326,196	3,989,873	

Mine production of metals from crude ore shipped to smelters, by counties and years, 1909–1923—Continued

Chaffee County

Year	Ore (short tons)	Gold (fine ounces)	Silver (fine ounces)	Copper (pounds)	Lead (pounds)	Zinc (pounds)
1909	10,214	532.36	35,321	568,868	584,492	947,741
1910	12,496	2,909.37	181,846	226,772	970,523	438,539
1911	7,308	2,861.96	91,957	88,448	999,432	200,509
1912	10,287	4,492.59	104,647	133,570	992,578	736,392
1913	13,577	10,770.44	137,092	238,590	1,616,532	627,112
1914	12,034	10,611.82	229,271	166,388	1,612,720	212,326
1915	11,764	5,743.06	162,485	120,571	977,204	1,419,732
1916	16,731	1,071.32	35,525	777,095	569,501	1,729,511
1917	19,621	1,943.26	112,984	741,104	1,103,035	979,872
1918	12,162	1,780.22	50,665	260,547	932,457	1,187,604
1919	4,365	1,140.06	26,534	40,479	476,509	465,972
1920	3,900	1,514.24	39,211	28,195	396,250	283,235
1921	2,402	1,549.65	27,641	8,357	318,666	39,000
1922	2,844	961.50	20,400	17,987	638,711	178,000
1923	2,173	791.71	20,756	7,089	557,429	132,000
	141,878	48,673.56	1,276,335	3,424,060	12,746,039	9,577,545

Clear Creek County

Year	Ore (short tons)	Gold (fine ounces)	Silver (fine ounces)	Copper (pounds)	Lead (pounds)	Zinc (pounds)
1909	14,475	12,181.51	253,924	182,508	1,249,201	80,784
1910	21,137	16,411.21	297,407	464,223	971,499	284,801
1911	23,518	15,807.77	280,756	551,445	1,664,787	263,300
1912	19,162	11,709.87	249,599	327,453	1,772,722	229,039
1913	17,723	11,843.82	252,326	335,579	2,121,195	51,301
1914	14,294	11,263.78	189,976	222,360	1,272,014	15,011
1915	13,395	11,657.40	230,245	339,816	1,293,400	55,067
1916	11,039	8,437.62	204,027	328,798	1,701,191	37,403

251

Mine production of metals from crude ore shipped to smelters, by counties and years, 1909-1923—Continued

Clear Creek County—Continued

Year	Ore (short tons)	Gold (fine ounces)	Silver (fine ounces)	Copper (pounds)	Lead (pounds)	Zinc (pounds)
1917	7,600	4,600.69	143,986	198,050	1,068,412	8,891
1918	5,513	3,056.11	125,427	99,813	1,149,165	
1919	4,085	1,683.88	87,461	59,765	492,158	
1920	3,873	1,144.48	103,182	38,036	945,424	
1921	1,763	829.75	43,804	14,361	333,457	
1922	1,090	559.54	31,877	4,305	225,402	
1923	1,728	638.53	88,402	19,917	237,048	
	160,395	111,825.96	2,582,399	3,186,429	16,497,075	1,025,597

Custer County

Year	Ore (short tons)	Gold (fine ounces)	Silver (fine ounces)	Copper (pounds)	Lead (pounds)	Zinc (pounds)
1909	276	3.61	7,258	700	41,721	89,593
1910	128	5.63	3,368	1,539	7,300	
1911	350	23.91	11,033	1,640	11,649	
1912	830	255.01	22,641	2,006	10,444	
1913	309	13.93	6,812	4,052	5,273	
1914	551	102.98	15,516	3,481	9,692	
1915	1,075	106.33	30,981	12,640	89,808	
1916	2,245	303.51	36,959	44,004	123,536	10,970
1917	5,381	325.50	85,303	86,167	217,670	
1918	4,251	210.00	107,766	51,292	281,070	13,516
1919	4,049	230.80	94,995	72,979	153,455	
1920	1,500	38.60	34,256	28,033	171,562	
1921	568	8.90	19,191	37,690	106,022	
1922	547	8.08	14,520	32,141	60,618	
1923	700	19.98	27,126	11,436	80,800	
	22,760	1,656.77	517,725	389,800	1,370,620	114,079

Delta County

[No shipments of crude ore to smelters in 1909 and 1911-1923]

Year	Ore (short tons)	Gold (fine ounces)	Silver (fine ounces)	Copper (pounds)	Lead (pounds)	Zinc (pounds)
1910	2	5.32	139			

Dolores County

Year	Ore (short tons)	Gold (fine ounces)	Silver (fine ounces)	Copper (pounds)	Lead (pounds)	Zinc (pounds)
1909	1,563	756.66	88,225	37,280	312,040	167,574
1910	973	475.05	71,621	96,988	126,856	87,000
1911	1,431	114.80	41,389	3,288	700,241	525,333
1912	6,498	159.77	82,063	689,838	1,082,521	700,773
1913	11,419	533.21	115,266	799,537	917,039	395,340
1914	6,498	367.79	85,226	350,278	395,908	218,178
1915	14,192	577.21	127,933	1,032,480	268,447	35,936
1916	6,398	353.93	77,269	419,500	588,333	182,306
1917	13,422	240.75	85,490	516,493	1,634,136	1,662,513
1918	9,272	151.70	54,249	618,012	517,394	661,253
1919	4,461	121.76	35,225	264,968	98,700	67,027
1920	2,752	113.68	32,167	6,804	772,588	229,865
1921	386	89.78	14,499	744	18,624	
1922	678	94.48	30,267	24,089	87,200	
1923	1,393	139.79	39,408	56,823	162,414	138,000
	81,336	4,290.36	980,297	4,917,122	7,682,441	5,071,098

Eagle County

Year	Ore (short tons)	Gold (fine ounces)	Silver (fine ounces)	Copper (pounds)	Lead (pounds)	Zinc (pounds)
1909	6,366	2,378.21	118,866	285,567	70,131	11,910
1910	5,148	991.30	74,158	209,551	30,327	
1911	5,585	1,896.23	108,252	66,608	314,056	
1912	5,906	2,294.55	156,417	147,048	702,365	57,652
1913	6,448	1,724.53	286,284	41,308	355,206	317,256
1914	3,701	2,077.98	112,809	26,847	71,299	95,069
1915	5,407	4,353.59	151,475	59,345	59,519	681,665
1916	13,724	3,932.92	171,330	98,348	53,493	4,588,981
1917	14,992	1,354.70	97,437	38,253	614,547	2,921,564
1918	20,864	1,702.60	202,703	352,957	408,318	746,025
1919	3,423	867.05	43,462	123,306	12,755	
1920	13,517	979.30	252,508	517,109		106,681
1921	38,785	3,130.98	682,550	1,833,078	12,578	
1922	71,892	3,488.37	583,737	1,330,296	322,818	11,000,000
1923	32,720	1,791.08	290,808	613,582	123,964	
	248,478	32,963.39	3,332,796	5,743,263	3,151,376	20,526,803

El Paso County

[No shipments of crude ore to smelters reported for 1909-1912 and 1915-1923]

Year	Ore (short tons)	Gold (fine ounces)	Silver (fine ounces)	Copper (pounds)	Lead (pounds)	Zinc (pounds)
1913	214			8,595		
1914	25			2,644		
	239			11,239		

Mine production of metals from crude ore shipped to smelters, by counties and years, 1909-1923—Continued

Fremont County

[No shipments of crude ore to smelters in 1919-1921]

Year	Ore (short tons)	Gold (fine ounces)	Silver (fine ounces)	Copper (pounds)	Lead (pounds)	Zinc (pounds)
1909	5	4.11		677		
1910	29					18,072
1911	382	8.61	1,345	13,976	19,904	140,526
1912	1,015	12.24	3,439	35,903	55,956	447,507
1913	28	4.45	63	4,677		
1914	706	71.40	1,066	191,917	308	
1915	1,600	32.61	3,168	127,303	30,894	228,170
1916	1,731	38.02	4,529	101,041	31,710	
1917	429	28.54	664	59,857		
1918	235	15.09	639	22,377	1,113	
1922	7	1.02	174	348	4,273	
1923	44	1.31	184		1,999	20,000
	6,214	217.40	15,271	558,076	146,157	854,275

Garfield County

[No shipments of crude ore in 1910, 1911, 1915, and 1918-1923]

Year	Ore (short tons)	Gold (fine ounces)	Silver (fine ounces)	Copper (pounds)	Lead (pounds)	Zinc (pounds)
1909	92	174.30	113	425		
1912	25	43.05	35	200		
1913	73	116.25	80	128		
1914	123	256.82	112	291		
1916	18	34.88	17			
1917	15	44.89	15			
	346	670.19	372	1,044		

Gilpin County

Year	Ore (short tons)	Gold (fine ounces)	Silver (fine ounces)	Copper (pounds)	Lead (pounds)	Zinc (pounds)
1909	15,777	17,904.47	101,858	342,021	411,530	
1910	16,845	21,115.43	83,905	339,184	247,346	
1911	18,885	21,066.99	119,117	460,807	116,180	
1912	15,114	21,317.25	112,861	531,488	142,354	
1913	16,094	18,704.35	104,701	442,811	247,277	
1914	9,624	14,790.69	97,103	522,812	318,018	
1915	6,538	10,720.50	71,594	340,692	360,315	
1916	6,739	9,706.85	83,988	415,372	298,113	
1917	6,553	7,270.49	68,974	87,998	430,161	
1918	2,343	2,351.10	35,273	87,998	243,142	
1919	4,263	4,886.27	48,908	135,235	334,082	
1920	2,032	2,468.67	32,738	74,648	214,318	
1921	676	1,027.68	13,138	9,831	69,080	
1922	1,133	1,320.25	29,935	18,408	111,146	
1923	1,031	613.37	39,417	16,994	181,167	
	123,647	155,264.36	1,043,510	4,156,453	3,724,229	

Grand County

[No shipments of crude ore in 1909, 1911-1913, 1915, 1917, 1921, and 1922]

Year	Ore (short tons)	Gold (fine ounces)	Silver (fine ounces)	Copper (pounds)	Lead (pounds)	Zinc (pounds)
1910	1			56		
1914	10	0.15	1,747		1,563	
1916	2		134	760		
1919	3	.05	508		453	
1920	3		856		525	
1923	2		323		314	
	21	.20	3,568	816	2,855	

Gunnison County

Year	Ore (short tons)	Gold (fine ounces)	Silver (fine ounces)	Copper (pounds)	Lead (pounds)	Zinc (pounds)
1909	878	152.53	25,979	51,815	271,996	212,093
1910	531	36.61	14,939	4,076	184,549	176,815
1911	1,119	458.33	20,611	4,690	399,486	508,561
1912	724	95.03	15,056	6,180	134,683	253,857
1913	2,460	179.31	85,249	15,131	147,335	32,328
1914	1,579	99.94	54,164	564	279,320	345,500
1915	3,816	1,269.06	17,935	8,861	177,905	1,750,944
1916	5,649	1,084.29	18,032	83,024	123,568	1,829,434
1917	8,021	201.26	19,370	164,029	267,019	2,531,712
1918	6,064	377.32	11,928	42,927	298,511	2,349,538
1919	4,958	1,443.12	15,006	1,812	99,999	2,456,479
1920	3,927	1,164.39	8,844		423,758	1,530,691
1921	498	869.69	10,362		51,955	
1922	189	379.10	3,692	526	13,382	
1923	7,701	41.55	20,455	1,788	1,684,631	2,889,000
	48,114	7,851.53	341,622	385,423	4,558,097	16,866,952

Mine production of metals from crude ore shipped to smelters, by counties and years, 1909–1923— Continued

Hinsdale County

Year	Ore (short tons)	Gold (fine ounces)	Silver (fine ounces)	Copper (pounds)	Lead (pounds)	Zinc (pounds)
1909	1,697	274.77	75,656	714,569	106,327	
1910	1,255	258.67	50,689	452,077	93,405	
1911	223	161.05	6,543	13,060	9,844	
1912	334	141.60	14,171	17,691	53,019	
1913	606	159.73	22,540	64,443	153,541	15,475
1914	118	8.22	5,987	17,098	5,723	
1915	488	35.65	9,621	9,114	266,128	
1916	377	65.11	10,030	16,248	75,638	12,575
1917	517	54.95	7,721	6,099	209,616	4,117
1918	952	202.27	10,425	13,632	286,490	
1919	719	368.78	22,001	7,654	53,259	
1920	568	297.55	21,522	2,625	80,625	
1921	495	165.68	32,039	9,357	65,756	
1922	850	53.79	47,827	10,612	66,572	
1923	489	29.71	26,833	8,916	17,996	
	9,688	2,277.53	363,605	1,363,195	1,543,939	32,167

Jackson County

[No shipments of crude ore to smelters in 1909–1916 and 1918–1923]

Year	Ore (short tons)	Gold (fine ounces)	Silver (fine ounces)	Copper (pounds)	Lead (pounds)	Zinc (pounds)
1917	229	19.40	591	23,725		

Jefferson County

[No shipments of crude ore to smelters reported for 1910–1917 and 1919–1923]

Year	Ore (short tons)	Gold (fine ounces)	Silver (fine ounces)	Copper (pounds)	Lead (pounds)	Zinc (pounds)
1909	1	0.78				
1918	23		9	1,000		
	24	.78	9	1,000		

Lake County

Year	Ore (short tons)	Gold (fine ounces)	Silver (fine ounces)	Copper (pounds)	Lead (pounds)	Zinc (pounds)
1909	329,316	65,338.64	3,200,379	5,180,000	17,790,952	18,696,394
1910	324,261	53,722.65	3,044,053	3,624,237	12,424,001	16,869,270
1911	375,829	51,335.01	2,822,522	3,926,396	12,922,845	51,138,383
1912	426,659	49,557.87	2,579,156	1,937,221	16,480,884	82,254,957
1913	453,567	44,252.76	3,009,982	1,846,675	20,283,586	75,316,137
1914	472,713	67,152.03	3,552,427	2,331,476	19,752,247	63,970,666
1915	377,888	97,972.08	2,214,599	1,796,078	11,801,775	45,563,916
1916	354,665	70,341.48	2,624,792	2,616,644	10,973,143	45,136,890
1917	298,831	47,514.22	1,871,870	2,155,057	9,199,015	30,906,551
1918	256,386	34,511.72	2,054,923	1,626,534	16,801,933	16,601,986
1919	190,145	24,398.44	1,491,990	886,072	10,147,046	13,260,881
1920	165,977	26,943.37	1,086,059	799,744	8,451,945	17,600,155
1921	80,501	14,047.11	1,043,219	1,107,295	3,537,889	1,521,000
1922	112,547	17,235.67	951,298	871,370	5,521,818	9,003,000
1923	97,855	10,004.48	641,291	511,054	5,449,539	9,257,000
	4,317,140	674,327.53	32,188,561	31,215,853	181,538,618	497,397,186

La Plata County

Year	Ore (short tons)	Gold (fine ounces)	Silver (fine ounces)	Copper (pounds)	Lead (pounds)	Zinc (pounds)
1909	3,035	6,018.58	72,528	256	2,980	
1910	6,050	18,782.52	141,349	142	273	
1911	7,058	13,651.04	65,264	73,728	1,511	
1912	2,761	6,312.07	47,863	918	6,756	
1913	7,403	14,966.67	121,096	113,897	4,455	
1914	5,083	5,999.42	60,220	26,038	11,410	
1915	2,952	3,333.88	46,363	4,114	23,362	
1916	1,602	1,495.88	29,177	12,024	6,551	
1917	1,648	1,222.97	15,324	25,795	3,745	
1918	300	356.91	6,415	668	3,000	
1919	405	288.60	6,075	167	2,283	
1920	717	533.09	10,578		937	
1921	1,279	2,180.40	20,289		3,734	
1922	791	1,542.02	10,500			
1923	838	769.40	17,138	816	1,800	
	41,922	77,453.45	670,179	258,563	72,797	

Larimer County

[No shipments of crude ore to smelters in 1910–1923]

Year	Ore (short tons)	Gold (fine ounces)	Silver (fine ounces)	Copper (pounds)	Lead (pounds)	Zinc (pounds)
1909	48					30,722

Mesa County

[No shipments of crude ore to smelters reported for 1909–1911 and 1913–1923]

Year	Ore (short tons)	Gold (fine ounces)	Silver (fine ounces)	Copper (pounds)	Lead (pounds)	Zinc (pounds)
1912	22]	0.44	257	5,685	20	

Mine production of metals from crude ore shipped to smelters, by counties and years, 1909–1923— Continued

Mineral County

Year	Ore (short tons)	Gold (fine ounces)	Silver (fine ounces)	Copper (pounds)	Lead (pounds)	Zinc (pounds)
1909	26,468	914.09	795,852	10,581	1,851,088	58,131
1910	25,791	238.45	722,142	11,357	1,057,703	
1911	20,404	1,059.12	488,054	13,735	1,137,063	
1912	29,968	837.82	677,569	13,039	1,374,856	
1913	29,594	803.87	723,591	20,800	1,421,138	
1914	27,952	933.83	615,734	32,586	1,401,795	
1915	10,864	699.26	279,785	6,143	636,022	
1916	19,677	711.43	358,402	7,569	618,324	
1917	18,420	166.23	355,266	15,438	549,379	
1918	27,151	644.59	640,373	3,490	924,686	
1919	14,492	324.29	366,994	355	571,829	
1920	12,597	276.22	272,322	1,120	531,537	
1921	7,076	184.60	192,468	1,899	156,778	
1922	3,978	80.01	106,903	3,422	153,455	
1923	6,462	115.81	228,867	1,088	237,557	41,000
	280,894	7,989.62	6,824,322	142,622	12,323,210	99,131

Moffat County

[No shipments of crude ore to smelters in 1909–1911, 1913–1915, and 1918–1923]

Year	Ore (short tons)	Gold (fine ounces)	Silver (fine ounces)	Copper (pounds)	Lead (pounds)	Zinc (pounds)
1912	64		124	25,085		
1916	25	0.19	38	9,033		
1917	15		23	4,326		
	104	.19	185	38,444		

Montezuma County

[No shipments of crude ore to smelters in 1912–1914 and 1918–1923]

Year	Ore (short tons)	Gold (fine ounces)	Silver (fine ounces)	Copper (pounds)	Lead (pounds)	Zinc (pounds)
1909		4.16	8			
1910	320	353.44	189	123		
1911	1		5	183		
1915	14	23.90	103		170	
1916	86	67.82	193	3,118	116	
1917	124	129.21	188	2,538		
	545	578.53	686	5,962	286	

Montrose County

[No shipments of crude ore to smelters in 1909–1912 and 1918–1921]

Year	Ore (short tons)	Gold (fine ounces)	Silver (fine ounces)	Copper (pounds)	Lead (pounds)	Zinc (pounds)
1913	49	0.24	427	24,058		
1914	66	.53	510	32,414		
1915	169	.88	1,057	57,320		
1916	197	.48	1,132	100,008		
1917	64		653	21,275		
1922	251		17,964	61,119		
1923	101		10,521	17,857		
	897	2.13	32,264	314,051		

Ouray County

Year	Ore (short tons)	Gold (fine ounces)	Silver (fine ounces)	Copper (pounds)	Lead (pounds)	Zinc (pounds)
1909	3,459	1,881.73	110,946	478,777	458,752	19,148
1910	4,739	1,382.76	178,388	167,568	773,323	
1911	4,194	1,832.32	178,642	49,123	976,300	
1912	3,331	1,148.13	190,360	46,868	458,873	
1913	5,101	1,392.61	150,648	111,680	233,692	14,158
1914	24,536	9,531.18	171,225	537,176	366,756	8,084
1915	35,783	7,116.54	193,663	627,140	460,237	7,282
1916	7,856	908.27	116,751	313,256	550,983	51,028
1917	7,419	382.06	190,300	134,035	798,995	522,524
1918	5,445	981.26	228,109	95,811	1,040,604	
1919	2,779	286.50	180,602	93,056	561,170	12,061
1920	1,662	405.76	127,053	58,673	414,769	
1921	513	21.38	79,535	39,807	216,862	
1922	471	52.10	36,223	8,984	94,664	
1923	333	134.47	17,493	3,707	62,246	
	107,621	27,457.07	2,149,938	2,765,661	7,468,226	634,285

Park County

Year	Ore (short tons)	Gold (fine ounces)	Silver (fine ounces)	Copper (pounds)	Lead (pounds)	Zinc (pounds)
1909	13,013	25,211.52	97,055	61,023	2,237,093	366,574
1910	12,092	12,118.65	116,878	88,748	2,041,204	659,796
1911	5,630	1,639.02	69,014	24,216	923,089	407,772
1912	2,422	2,301.01	31,026	10,321	167,756	132,275
1913	5,728	1,593.58	94,118	29,161	506,046	98,623
1914	1,758	2,102.35	20,004	8,023	168,154	57,940
1915	2,620	7,083.94	9,131	12,303	190,830	472,992
1916	2,975	10,823.05	13,062	22,598	330,609	47,560
1917	1,793	5,324.87	12,199	12,380	259,071	
1918	1,334	3,051.73	12,221	11,048	191,754	
1919	1,805	6,025.10	13,332	13,319	207,661	
1920	4,352	6,898.52	45,263	17,016	1,043,203	
1921	4,929	1,974.72	47,543	7,550	654,090	
1922	1,120	2,063.39	14,529	4,215	155,982	
1923	371	773.10	17,195	5,558	19,401	
	61,942	88,984.55	612,570	327,479	9,095,943	2,243,532

Mine production of metals from crude ore shipped to smelters, by counties and years, 1909–1923—Continued

Pitkin County

Year	Ore (short tons)	Gold (fine ounces)	Silver (fine ounces)	Copper (pounds)	Lead (pounds)	Zinc (pounds)
1909	46,558	29.90	571,402	372	6,657,255	
1910	33,181		366,685	327	7,499,775	
1911	20,738		338,188		3,001,968	
1912	36,688	7.98	457,060	934	2,448,733	484,507
1913	42,442	.38	445,180	13,173	8,128,145	460,161
1914	38,525		256,800	27,560	10,207,264	145,431
1915	24,270	1.40	284,440	8,446	5,795,368	214,952
1916	35,046		363,215	19,397	7,356,212	162,574
1917	49,425	2.84	453,036	19,512	6,887,569	290,514
1918	41,200	.10	348,071	9,684	6,821,194	145,286
1919	31,634		471,091		3,148,665	80,000
1920	34,453		416,329		2,814,716	617,790
1921	36,779		425,643	233	1,925,777	283,000
1922	40,617		401,023		2,238,182	
1923	40,579		367,411		2,261,775	465,000
	552,135	42.60	5,965,574	99,638	77,252,598	3,349,215

Rio Grande County

[No shipments of crude ore to smelters in 1909, 1911, 1916, and 1918–1922]

Year	Ore (short tons)	Gold (fine ounces)	Silver (fine ounces)	Copper (pounds)	Lead (pounds)	Zinc (pounds)
1910	12	63.18	61	87	250	
1912	133	246.51	896	29,673	313	
1913	6	9.29	109	568		
1914	8	22.93	16			
1915	1	1.08	16			
1917	16	1.16	52		1,930	
1923	17	80.40	161	218	929	
	193	424.55	1,311	30,546	3,422	

Routt County

[No shipments of crude ore to smelters in 1910–1912, 1914, 1915, 1922, 1923]

Year	Ore (short tons)	Gold (fine ounces)	Silver (fine ounces)	Copper (pounds)	Lead (pounds)	Zinc (pounds)
1909	24	45.62	3,417			
1913	12	11.18	1,954	161		1,023
1916	517	.68	234	32,142		
1917	10	11.37	1,075			
1918	25	3.30	1,272			6,591
1919	2	2.13	222			
1920	3		100			
1921	1		82			
	594	74.28	8,356	32,303		7,614

Saguache County

Year	Ore (short tons)	Gold (fine ounces)	Silver (fine ounces)	Copper (pounds)	Lead (pounds)	Zinc (pounds)
1909	171	43.79	2,258	3,769	83,463	
1910	296	49.58	4,841	5,362	161,068	
1911	184	21.11	4,664	4,984	74,556	46,561
1912	760	117.29	9,825	22,395	80,488	24,400
1913	697	192.48	8,279	13,124	318,632	24,391
1914	1,488	798.82	18,293	35,783	534,872	8,941
1915	692	255.08	11,266	23,360	174,447	44,250
1916	3,338	388.16	48,959	92,581	255,449	
1917	4,224	500.69	76,016	144,978	310,686	
1918	1,716	123.50	89,510	96,866	108,253	
1919	509	39.52	37,767	36,344	52,515	
1920	852	116.72	46,958	51,304	36,220	
1921	722	57.90	60,786	39,742	148,483	
1922	471	216.07	31,978	41,432	102,109	
1923	296	20.78	9,394	6,278	81,657	
	16,406	2,941.49	460,794	618,302	2,522,898	148,543

Mine production of metals from crude ore shipped to smelters, by counties and years, 1909–1923—Continued

San Juan County

Year	Ore (short tons)	Gold (fine ounces)	Silver (fine ounces)	Copper (pounds)	Lead (pounds)	Zinc (pounds)
1909	6,100	2,015.01	432,172	306,474	1,935,144	
1910	6,052	1,730.26	406,218	211,223	2,116,963	
1911	1,919	955.52	118,453	66,257	355,940	37,918
1912	6,621	3,259.93	421,618	419,313	1,410,085	105,273
1913	10,832	3,732.49	640,490	693,389	2,543,186	
1914	4,675	3,181.76	198,300	159,776	1,249,027	
1915	5,952	4,676.17	120,777	162,340	1,523,751	
1916	11,748	4,502.00	181,278	664,055	1,650,102	
1917	11,247	2,603.43	280,867	574,987	1,938,599	
1918	4,152	1,475.44	130,896	297,954	812,146	
1919	2,849	656.74	115,761	166,316	738,101	32,104
1920	2,016	328.98	107,142	86,226	663,950	
1921	1,164	382.07	64,178	28,558	557,555	
1922	2,758	548.11	67,187	66,890	798,858	1,300,060
1923	627	220.01	19,460	10,003	246,760	
	78,712	30,267.92	3,304,797	3,913,761	18,540,167	1,475,295

San Miguel County

Year	Ore (short tons)	Gold (fine ounces)	Silver (fine ounces)	Copper (pounds)	Lead (pounds)	Zinc (pounds)
1909	2,471	2,653.55	56,492	143,833	1,012,397	
1910	1,396	430.60	57,333	14,556	499,593	
1911	1,348	108.50	95,185	16,263	715,798	
1912	1,210	303.44	70,643	12,822	601,160	
1913	907	80.75	52,462	4,394	403,761	
1914	411	117.21	13,491	2,448	105,601	
1915	332	121.99	7,190	6,717	144,535	56,625
1916	504	286.66	9,362	18,277	96,777	34,394
1917	578	411.21	18,335	20,798	83,784	
1918	1,089	189.32	34,167	43,870	420,697	7,011
1919	747	60.79	35,415	15,804	395,778	
1920	629	82.10	28,557	10,134	263,656	
1921	1,309	223.62	97,783	28,573	731,420	
1922	651	302.25	46,172	15,814	163,472	
1923	1,228	97.18	57,877	25,538	637,774	
	14,810	5,469.17	680,464	379,841	6,276,203	98,030

Summit County

Year	Ore (short tons)	Gold (fine ounces)	Silver (fine ounces)	Copper (pounds)	Lead (pounds)	Zinc (pounds)
1909	10,535	1,025.38	32,392	3,431	198,650	3,493,523
1910	2,660	345.64	28,316	6,122	238,856	834,104
1911	1,615	406.45	21,084	6,394	296,548	113,681
1912	5,042	1,089.29	83,693	11,898	867,056	742,226
1913	6,559	1,805.55	112,550	9,770	1,353,482	302,120
1914	1,926	893.86	31,612	4,810	389,568	277,898
1915	2,234	1,172.74	22,539	5,054	309,678	109,151
1916	8,756	2,845.26	64,391	8,531	785,704	1,729,925
1917	20,680	2,296.08	122,851	23,810	752,261	4,105,527
1918	9,666	634.57	81,017	9,765	439,844	1,506,814
1919	4,357	501.34	74,806	6,086	335,647	141,785
1920	11,344	1,002.91	87,573	359	275,997	
1921	8,711	776.88	64,538	13,513	289,233	
1922	15,780	1,271.69	114,346	94,413	457,330	677,000
1923	5,233	733.49	87,442	17,823	853,767	45,600
	115,098	16,801.13	1,029,150	221,779	7,843,621	14,079,354

Teller County
[No shipments of crude ore to smelters in 1921]

Year	Ore (short tons)	Gold (fine ounces)	Silver (fine ounces)	Copper (pounds)	Lead (pounds)	Zinc (pounds)
1909	45,474	108,373.88	36,157			
1910	32,309	73,261.54	25,880			
1911	33,245	69,772.14	22,072			
1912	36,617	82,717.86	22,860			
1913	28,584	64,665.23	23,192			
1914	32,460	68,644.40	30,908			
1915	40,926	99,385.27	38,765			
1916	37,111	84,570.24	31,603			
1917	6,233	8,974.57	6,243			
1918	837	1,352.00	683			
1919	50	278.40	21			
1920	99	732.92	1,817	451	612	
1922	2	28.90	3			
1923	6	33.68	342			
	293,953	662,791.03	240,546	451	612	

The following tables show, by counties, the mine production of gold, silver, copper, lead, and zinc from gold and silver mills and from concentrating mills during the period 1909–1923.

Mine production of metals from gold and silver mills and from concentrating mills, by counties, 1909–1923, in terms of recovered metals [a]

County	Ore to gold and silver mills			Ore to concentrating mills (short tons)	Concentrates produced					
	Short tons	Gold in bullion (fine ounces)	Silver in bullion (fine ounces)		Short tons	Gold (fine ounces)	Silver (fine ounces)	Copper (pounds)	Lead (pounds)	Zinc (pounds)
Boulder	63,441	16,066.20	38,863	92,233	5,600	7,261.95	329,534	23,657	1,169,575	....
Chaffee	141,698	839.85	457	131,535	34,942	36,108.02	288,073	844,309	11,110,720	12,860,666
Clear Creek	339,557	23,192.52	23,430	793,656	130,796	91,202.91	2,725,648	1,945,397	25,239,994	18,301,665
Custer	24,848	2,846.64	24,121	68,119	2,846	14.33	5,967	4,392	3,435,198	41,677
Dolores	1	72.72	44	16,409	6,535	1,073.87	132,696	12,115	2,678,770	2,499,259
Eagle	....	28.29	7	589,898	259,213	3,198.35	306,378	51,657	11,744,658	132,085,487
El Paso				84	27	....				
Fremont				25	18		15	2,037	4,591	7,161
Gilpin	338,495	48,843.49	15,032	303,929	118,934	116,773.58	983,642	2,810,963	6,047,562	250,623
Gunnison	66,722	16,514.18	9,878	20,133	12,086	16,370.17	128,209	62,855	2,356,815	1,377,686
Hinsdale		92.25	75	21,321	3,610	499.02	51,652	79,483	2,678,641	87,622
Lake	41,404	37,800.10	11,506	979,376	714,410	23,984.33	3,025,083	415,181	77,611,450	266,177,568
La Plata	4,100	955.51	477	300	145	241.76	5,854	228	....	....
Larimer	50	9.00	11							
Mineral	200	17.83	2,060	238,070	38,828	24,677.31	358,204	78,064	32,150,562	6,681,012
Montezuma		0.77	1	128	64	70.52	26	97		
Ouray	829,815	378,225.70	282,854	468,670	149,612	204,266.38	7,457,631	3,185,092	26,887,196	442,298
Park	3,296	882.17	388	13,059	881	105.36	76,860	10,875	202,426	....
Pitkin		71.50	5,262	993,834	248,539	15.82	2,080,507	184,480	91,057,763	316,021
Rio Grande	1,499	744.96	309	....						
Routt	50	38.20	12	306	41	47.07	2,679			
Saguache	21	17.72	2	66,472	11,108	440.40	265,574	507,478	3,453,873	519,101
San Juan	600,181	63,623.63	24,507	1,207,030	186,812	181,433.28	4,369,588	11,151,962	101,061,043	47,897,203
San Miguel	5,815,309	870,601.90	3,999,888	780,408	442,915	551,787.96	13,012,937	11,466,934	96,367,721	17,073,066
Summit	17,783	9,034.51	4,313	495,043	200,275	5,766.75	761,608	76,306	27,846,173	105,120,909
Teller	10,798,777	5,626,687.62	533,449	53,248	67,740	156,368.68	52,159			
	19,087,247	7,097,207.26	4,976,946	7,333,286	2,735,977	1,421,708.34	36,420,524	32,913,562	523,104,728	611,739,024

[a] For explanation see footnote to Table 1, p. 69.

Mine production of metals from gold and silver mills and from concentrating mills, by counties and years, 1909–1923

Boulder County

Year	Ore to gold and silver mills			Ore to concentrating mills (short tons)	Concentrates produced					
	Short tons	Gold in bullion (fine ounces)	Silver in bullion (fine ounces)		Short tons	Gold (fine ounces)	Silver (fine ounces)	Copper (pounds)	Lead (pounds)	Zinc (pounds)
1909	4,149	1,212.65	808	5,088	518	122.76	12,854	2,193	390,257	....
1910	7,498	1,452.14	682	373	202	473.92	1,061	1,253	37	....
1911	7,946	1,785.07	620	62	254	290.86	3,207	284	31,414	....
1912	4,058	1,070.02	577	222	54	71.33	236	104	132	....
1913	1,065	616.21	232	40	23	24.87	131	....	172	....
1914	7,160	787.65	315	102	286	512.37	20,431	2,547	48,512	....
1915	11,285	684.37	32,619	23,171	1,499	3,209.04	77,382	5,575	228,355	....
1916	14,666	2,914.90	1,594	14,248	764	491.54	74,295	1,793	79,293	....
1917	2,671	1,010.53	746	9,375	485	118.39	25,322	7,719	37,673	....
1918	410	183.50	90	6,795	245	822.80	2,410	1,364	....	....
1919	163	784.65	43	1,671	116	122.70	8,741	825	33,050	....
1920	404	633.24	10	17,565	439	520.68	45,883	....	177,019	....
1921	578	950.08	11	11,806	609	395.65	46,008	....	102,240	....
1922	526	1,139.03	373	950	48	1.88	7,295	....	34,701	....
1923	862	842.16	143	765	58	83.16	4,278	....	6,720	....
	63,441	16,066.20	38,863	92,233	5,600	7,261.95	329,534	23,657	1,169,575	....

Chaffee County

[No production in 1909, 1910, 1912, 1920, 1921, and 1923]

Year	Short tons	Gold in bullion (fine ounces)	Silver in bullion (fine ounces)	Ore to concentrating mills (short tons)	Short tons	Gold (fine ounces)	Silver (fine ounces)	Copper (pounds)	Lead (pounds)	Zinc (pounds)
1911	....	31.88	10	151	4	....	84	....	2,219	....
1913	35,558	268.91	185	....	4,077	4,035.55	31,693	76,421	1,580,013	1,494,835
1914	49,664	298.44	173	....	5,970	5,131.08	42,791	153,108	2,077,639	1,960,851
1915	56,476	73.96	35	....	8,581	9,405.77	64,466	227,475	2,652,923	3,256,623
1916	....	99.56	26	52,627	7,743	7,753.20	65,192	224,360	2,447,398	3,015,474
1917	....	67.10	28	36,031	3,669	4,444.85	33,522	66,779	1,047,488	1,202,060
1918	....	....	....	27,436	3,628	3,660.90	30,522	63,283	953,304	1,431,165
1919	....	....	....	15,290	1,170	1,673.77	14,016	30,344	326,719	499,658
1922	....	....	....	....	100	2.90	5,787	2,539	23,017	....
	141,698	839.85	457	131,535	34,942	36,108.02	288,073	844,309	11,110,720	12,860,666

Mine production of metals from gold and silver mills and from concentrating mills, by counties and years, 1909-1923 —Continued

Clear Creek County

Year	Ore to gold and silver mills			Ore to concentrating mills (short tons)	Concentrates produced					
	Short tons	Gold in bullion (fine ounces)	Silver in bullion (fine ounces)		Short tons	Gold (fine ounces)	Silver (fine ounces)	Copper (pounds)	Lead (pounds)	Zinc (pounds)
1909	52,702	2,745.18	1,160	49,576	11,221	10,835.95	193,422	117,038	2,005,474	677,290
1910	47,248	2,833.83	1,293	41,569	8,115	5,854.13	176,444	131,572	1,462,977	962,588
1911	40,489	3,498.01	2,362	41,767	9,088	5,726.01	154,697	98,923	1,660,435	1,154,244
1912	40,798	3,158.07	4,473	42,934	9,669	6,681.33	119,868	121,948	1,751,011	1,505,454
1913	47,541	2,373.59	2,542	39,628	9,999	6,704.25	153,659	90,814	1,876,419	1,438,217
1914	42,593	3,106.10	5,854	44,479	11,336	9,589.05	149,557	145,430	1,163,678	1,052,303
1915	31,665	2,538.57	4,329	76,933	12,170	11,235.93	158,528	191,133	1,234,175	1,449,965
1916	21,369	1,460.44	835	61,812	15,123	10,839.38	257,274	292,934	2,594,534	2,535,172
1917	7,115	372.41	210	69,734	16,372	9,711.08	382,549	372,041	3,768,205	3,144,139
1918	1,396	281.04	155	50,899	9,846	7,841.20	245,306	243,434	2,720,187	1,812,846
1919	1,212	371.78	90	111,658	4,686	2,352.61	269,888	93,160	1,024,976	603,027
1920	1,474	198.27	54	46,147	3,387	1,005.37	116,664	23,942	1,511,676	372,420
1921	2,021	146.50	41	28,983	2,767	903.17	88,022	7,158	867,474	217,000
1922	138	23.04	4	68,197	4,204	1,168.55	164,326	3,569	817,089	800,000
1923	1,796	85.69	28	19,940	2,813	754.90	95,444	12,301	779,681	577,000
	339,557	23,192.52	23,430	793,656	130,796	91,202.91	2,725,648	1,945,397	25,239,991	18,301,665

Custer County

[No production reported for 1920 and 1921]

Year	Short tons	Gold in bullion (fine ounces)	Silver in bullion (fine ounces)	Ore to concentrating mills (short tons)	Short tons	Gold (fine ounces)	Silver (fine ounces)	Copper (pounds)	Lead (pounds)	Zinc (pounds)
1909	5,595	614.33	7,538							
1910	6,800	467.98	4,100	124	31	2.35	299	2,343	7,496	6,796
1911	3,275	244.43	2,020	45	9	0.62	126		5,862	
1912	3,500	562.43	2,785							
1913	4,353	696.41	4,501							
1914	300	59.80	459	19	7					4,470
1915	450	91.91	652	194	65					30,411
1916		1.69	12							
1917		4.96	6	509	206	11.36	3,378	2,049	10,633	
1918	75		690							
1919				572	36		2,164		1,679	
1922				16,665	450				600,000	
1923	500	102.70	1,358	50,000	2,042				2,809,528	
	24,848	2,846.64	24,121	68,119	2,846	14.33	5,967	4,392	3,435,198	41,677

Dolores County

[No production in 1915 and 1918-1923]

Year	Short tons	Gold in bullion (fine ounces)	Silver in bullion (fine ounces)	Ore to concentrating mills (short tons)	Short tons	Gold (fine ounces)	Silver (fine ounces)	Copper (pounds)	Lead (pounds)	Zinc (pounds)
1909	1	52.71	31	3,223	373	267.75	15,390	6,258	150,333	
1910				1,960	95	266.39	16,688	75	1,053	
1911				1,845	90	251.16	14,813		1,003	
1912				1,987	341	205.75	18,225	77	129,879	111,256
1913				6,383	5,063	68.19	63,550	2,282	2,162,302	2,290,892
1914		14.71	2	407	278	3.20	1,298		96,115	148,271
1916		5.30	11							
1917				604	295	11.43	2,732	3,423	138,085	38,840
	1	72.72	44	16,409	6,535	1,073.87	132,696	12,115	2,678,770	2,499,259

Eagle County

[No production in 1921 and 1922]

Year	Short tons	Gold in bullion (fine ounces)	Silver in bullion (fine ounces)	Ore to concentrating mills (short tons)	Short tons	Gold (fine ounces)	Silver (fine ounces)	Copper (pounds)	Lead (pounds)	Zinc (pounds)
1909				5,160	2,488	146.33	6,344	1,318	82,149	728,498
1910				22,613	6,817	205.06	14,148		367,082	4,147,945
1911		5.18	1	27,592	7,480	87.97	7,856		541,833	5,097,597
1912		6.38		28,258	7,467	82.70	7,318	128	537,791	5,601,609
1913				41,040	11,604	269.49	15,096		995,999	6,366,387
1914				45,676	11,737	205.03	14,271	1,258	1,106,086	7,427,029
1915				68,790	18,214	262.65	26,075	741	1,334,524	10,460,085
1916		12.82	5	91,425	54,944	691.02	50,791	14,262	1,463,869	23,849,071
1917		3.93	1	85,883	44,235	631.23	38,585	14,883	1,812,441	20,793,848
1918				68,811	31,894	37.69	38,703	84	2,518,781	14,099,316
1919				18,825	9,003	97.31	28,697		365,358	3,367,548
1920				20,118	15,544	254.07	27,159		282,538	6,546,554
1923				65,707	37,786	227.80	31,335	18,983	336,207	23,600,000
		28.29	7	589,898	259,213	3,198.35	306,378	51,657	11,744,658	132,085,487

El Paso County

[No production in 1909-1912 and 1914-1923]

Year	Short tons	Gold in bullion (fine ounces)	Silver in bullion (fine ounces)	Ore to concentrating mills (short tons)	Short tons	Gold (fine ounces)	Silver (fine ounces)	Copper (pounds)	Lead (pounds)	Zinc (pounds)
1913				84	27				2,037	

Fremont County

[No production in 1909-1912 and 1914-1923]

Year	Short tons	Gold in bullion (fine ounces)	Silver in bullion (fine ounces)	Ore to concentrating mills (short tons)	Short tons	Gold (fine ounces)	Silver (fine ounces)	Copper (pounds)	Lead (pounds)	Zinc (pounds)
1913				25	18		15		4,591	7,161

Mine production of metals from gold and silver mills and from concentrating mills, by counties and years, 1909–1923—Continued

Gilpin County

Year	Ore to gold and silver mills			Ore to concentrating mills (short tons)	Concentrates produced					
	Short tons	Gold in bullion (fine ounces)	Silver in bullion (fine ounces)		Short tons	Gold (fine ounces)	Silver (fine ounces)	Copper (pounds)	Lead (pounds)	Zinc (pounds)
1909	79,778	11,827.22	2,382	15,563	13,988	13,191.98	67,770	157,125	253,051	----------
1910	47,211	5,576.14	1,275	19,575	9,066	6,548.20	47,449	195,060	328,131	----------
1911	29,507	4,321.68	1,125	54,646	17,300	12,262.46	172,413	489,433	1,123,176	23,088
1912	40,771	5,505.58	1,417	62,767	24,095	16,932.60	201,927	494,282	1,209,246	25,377
1913	33,634	4,552.38	1,204	44,428	16,988	9,958.46	167,299	395,163	963,064	8,589
1914	27,749	4,010.39	3,454	15,466	6,870	8,944.55	44,680	203,767	181,700	12,980
1915	31,527	5,642.92	1,683	15,987	9,387	10,865.80	52,388	135,691	230,812	11,000
1916	13,109	2,208.83	954	19,065	5,685	10,010.73	41,611	141,945	223,221	----------
1917	5,912	996.55	308	22,824	5,943	10,942.04	43,303	126,496	385,745	141,490
1918	6,525	852.06	308	14,786	4,782	10,408.79	89,348	368,046	531,830	28,099
1919	7,041	1,041.11	307	5,714	2,544	4,216.04	22,485	76,303	189,539	----------
1920	6,221	674.59	156	2,567	1,002	1,281.55	9,106	11,955	220,694	----------
1921	1,846	625.18	158	1,031	318	263.27	4,667	3,355	22,564	----------
1922	3,781	626.02	190	8,793	582	537.40	13,782	6,452	135,799	----------
1923	3,883	382.84	111	717	384	409.71	5,414	5,890	48,990	----------
	338,495	48,843.49	15,032	303,929	118,934	116,773.58	983,642	2,810,963	6,047,562	250,623

Gunnison County

Year	Short tons	Gold in bullion (fine ounces)	Silver in bullion (fine ounces)	Ore to concentrating mills (short tons)	Short tons	Gold (fine ounces)	Silver (fine ounces)	Copper (pounds)	Lead (pounds)	Zinc (pounds)
1909	8,193	2,925.42	1,851	----------	1,249	2,170.40	9,593	----------	221,074	----------
1910	23,468	4,891.19	2,991	1,204	2,700	6,390.60	31,259	16,948	397,292	----------
1911	10,390	2,751.76	756	417	1,329	3,737.62	11,148	5,238	232,447	48,895
1912	12,793	2,955.14	2,079	529	2,045	2,981.04	11,873	1,917	172,184	230,027
1913	1,146	177.05	165	695	474	107.46	2,066	6,733	49,393	260,547
1914	4,000	242.03	192	439	354	76.43	4,165	10,624	38,654	179,500
1915	2,630	1,370.55	921	----------	230	272.42	6,036	230	12,095	----------
1916	477	117.03	76	4,293	733	221.00	10,883	1,655	189,649	135,439
1917	----------	----------	----------	4,650	2,027	103.89	20,898	16,092	483,981	523,278
1918	280	85.59	64	----------	28	35.11	888	106	2,249	----------
1919	----------	----------	----------	3,390	215	83.40	3,419	3,312	17,455	----------
1920	----------	----------	----------	4,516	603	----------	11,711	----------	534,543	----------
1921	----------	11.85	8	----------	----------	----------	----------	----------	----------	----------
1922	32	64.98	111	----------	----------	----------	----------	----------	----------	----------
1923	3,318	921.59	664	----------	99	190.80	3,820	----------	5,799	----------
	66,722	16,514.18	9,878	20,133	12,086	16,370.17	128,209	62,855	2,356,815	1,377,686

Hinsdale County

[No production in 1914–1917, 1920, and 1921]

Year	Short tons	Gold in bullion (fine ounces)	Silver in bullion (fine ounces)	Ore to concentrating mills (short tons)	Short tons	Gold (fine ounces)	Silver (fine ounces)	Copper (pounds)	Lead (pounds)	Zinc (pounds)
1909		92.25	75	2,213	245	47.06	3,733	13,395	202,777	----------
1910				500	188	24.22	1,210	8,636	108,801	36,439
1911				9,220	1,433	187.88	20,551	36,048	1,204,781	11,926
1912				3,723	810	95.69	7,937	11,861	628,777	39,257
1913				4,270	798	100.03	11,820	4,676	451,482	----------
1918				500	24	29.44	941	51	2,420	----------
1919				700	73	9.00	2,247	3,657	47,628	----------
1922				195	39	5.70	3,213	1,159	1,975	----------
1923										
		92.25	75	21,321	3,610	499.02	51,652	79,483	2,678,641	87,622

Lake County

Year	Short tons	Gold in bullion (fine ounces)	Silver in bullion (fine ounces)	Ore to concentrating mills (short tons)	Short tons	Gold (fine ounces)	Silver (fine ounces)	Copper (pounds)	Lead (pounds)	Zinc (pounds)
1909		2,973.66	842	87,981	38,056	1,112.84	222,419	2,608	3,283,040	19,940,921
1910		3,920.90	1,307	137,772	86,577	873.38	276,594	20,920	6,825,502	39,498,175
1911	50	2,245.21	729	62,540	52,745	1,250.04	184,045	91,108	5,576,244	20,472,073
1912	845	1,743.90	750	80,087	69,230	2,066.98	420,491	128,579	9,753,360	23,690,826
1913	9,749	2,722.59	740	64,995	57,168	2,542.80	389,596	77,312	9,002,597	18,526,720
1914	11,080	6,019.02	1,603	63,670	54,908	2,847.89	256,800	51,434	7,032,368	14,792,668
1915	6,968	4,906.21	1,747	96,764	75,607	2,441.00	353,597	7,345	9,155,629	26,929,262
1916	11,233	2,855.11	716	111,342	93,863	4,264.90	303,992	5,031	10,746,249	31,648,677
1917		1,179.17	562	123,597	84,109	2,820.86	309,960	27,566	9,102,787	29,347,782
1918		780.74	260	99,454	73,879	1,045.54	233,648	----------	5,667,982	30,113,750
1919		1,810.82	587	27,522	19,597	119.70	48,580	2,556	1,155,030	9,904,338
1920	1,479	2,990.85	628	5,532	4,196	517.90	11,101	----------	138,243	1,154,376
1921		608.58	198	----------	----------	----------	----------	----------	----------	----------
1922		2,730.77	750	----------	----------	----------	----------	----------	----------	----------
1923		312.57	87	18,120	3,475	2,080.50	14,260	722	175,419	158,000
	41,404	37,800.10	11,506	979,376	714,410	23,984.33	3,025,083	415,181	77,611,450	266,177,568

La Plata County

[No production in 1917–1920 and 1923]

Year	Short tons	Gold in bullion (fine ounces)	Silver in bullion (fine ounces)	Ore to concentrating mills (short tons)	Short tons	Gold (fine ounces)	Silver (fine ounces)	Copper (pounds)	Lead (pounds)	Zinc (pounds)
1909	1,100	45.67	29	----------	35	85.13	1,595	228	----------	----------
1910		115.18	41	300	10	9.38	147			
1911	3,000	82.29	62		100	147.25	4,112			
1912		237.47	85							
1913		169.43	26							
1914		119.92	24							
1915		126.38	6							
1916		35.33	10							
1921		5.23	38							
1922		18.61	156							
	4,100	955.51	477	300	145	241.76	5,854	228	----------	----------

Mine production of metals from gold and silver mills and from concentrating mills, by counties and years, 1909-1923—Continued

Larimer County

[No production in 1909-1916 and 1918-1923]

Year	Ore to gold and silver mills			Ore to concentrating mills (short tons)	Concentrates produced					
	Short tons	Gold in bullion (fine ounces)	Silver in bullion (fine ounces)		Short tons	Gold (fine ounces)	Silver (fine ounces)	Copper (pounds)	Lead (pounds)	Zinc (pounds)
1917	50	9.00	11							

Mineral County

[No production in 1914 and 1920-1923]

Year	Short tons	Gold in bullion (fine ounces)	Silver in bullion (fine ounces)	Ore to concentrating mills (short tons)	Short tons	Gold (fine ounces)	Silver (fine ounces)	Copper (pounds)	Lead (pounds)	Zinc (pounds)
1909				38,473	7,751	4,350.32	95,333	6,820	7,185,728	1,759,165
1910				37,165	8,752	5,623.68	51,580	17,674	7,188,297	2,421,926
1911				45,528	7,373	7,609.49	57,265	19,649	6,537,493	1,258,561
1912				36,520	5,014	3,322.53	37,340	10,846	4,355,366	308,681
1913				27,169	3,347	1,628.52	81,752	10,847	2,277,226	454,875
1915				17,207	2,274	899.00	12,022	2,800	1,746,106	85,984
1916	200	17.83	2,060	18,226	2,459	776.36	13,494	5,569	1,676,763	240,575
1917				14,335	1,230	322.41	6,251	3,859	756,365	54,971
1918				1,221	111	29.90	586		64,934	
1919				2,226	517	115.10	2,581		362,284	96,274
	200	17.83	2,060	238,070	38,828	24,677.31	358,204	78,064	32,150,562	6,681,012

Montezuma County

[No production in 1909 and 1912-1923]

Year	Short tons	Gold in bullion (fine ounces)	Silver in bullion (fine ounces)	Ore to concentrating mills (short tons)	Short tons	Gold (fine ounces)	Silver (fine ounces)	Copper (pounds)	Lead (pounds)	Zinc (pounds)
1910				128	64	70.52	26	97		
1911		0.77	1							
		0.77	1	128	64	70.52	26	97		

Ouray County

Year	Short tons	Gold in bullion (fine ounces)	Silver in bullion (fine ounces)	Ore to concentrating mills (short tons)	Short tons	Gold (fine ounces)	Silver (fine ounces)	Copper (pounds)	Lead (pounds)	Zinc (pounds)
1909	100,405	117,095.86	57,321		10,259	28,315.82	177,548	505,492	2,355,180	
1910	102,971	76,122.77	47,593	3,535	12,268	28,718.57	188,269	452,668	3,231,405	
1911	102,376	62,389.22	40,315	26,682	17,387	30,252.80	293,843	515,150	2,973,522	
1912	56,593	22,963.84	10,269	30,051	16,234	26,661.95	344,548	353,684	2,530,171	140,667
1913	54,932	21,977.31	8,210	37,308	16,462	23,039.94	378,776	388,649	1,946,899	186,271
1914	48,991	28,407.34	8,813	32,033	12,496	20,691.64	414,251	316,862	1,752,808	36,524
1915	35,411	30,999.03	8,699	32,064	9,150	15,968.45	374,259	236,711	1,530,444	
1916	61,412	14,153.90	97,766	41,924	8,912	8,698.42	588,944	130,825	1,788,046	17,987
1917	47,480	600.05	560	31,624	7,290	3,508.59	677,237	45,518	1,232,726	10,270
1918	36,790	454.13	531	37,418	7,018	3,771.94	572,719	57,306	1,547,311	39,297
1919	35,558	37.00	95	26,128	4,409	4,143.35	446,962	19,132	1,221,698	11,282
1920				38,533	3,334	1,228.20	338,524	28,208	919,806	
1921	31,144	222.00	407	37,575	6,020	3,299.07	651,028	45,232	991,537	
1922	74,373	1,686.25	1,020	48,252	9,997	4,354.97	1,189,427	49,165	1,389,862	
1923	41,379	1,117.00	1,255	45,548	8,376	1,612.67	821,296	40,490	1,475,781	
	829,815	378,225.70	282,854	468,670	149,612	204,266.38	7,457,631	3,185,092	26,887,196	442,298

Park County

[No production in 1921 and 1922]

Year	Short tons	Gold in bullion (fine ounces)	Silver in bullion (fine ounces)	Ore to concentrating mills (short tons)	Short tons	Gold (fine ounces)	Silver (fine ounces)	Copper (pounds)	Lead (pounds)	Zinc (pounds)
1909	735	302.99	82	1,298	147	6.35	4,998			
1910	237	105.77	31							
1911	150	26.10	9							
1912	264	57.64	19							
1913	870	113.23	36							
1914	200	33.46	5							
1915	200	150.39	23							
1916	30	7.05	102							
1917	500	33.21	13	400	70	55.42	2,463	444	19,638	
1918	10	4.31	1	960	101	0.10	6,058	1,656	42,119	
1919				9,405	485	42.19	57,584	7,117	98,247	
1920				996	78	1.30	5,757	1,658	42,422	
1923	100	48.02	67							
	3,296	882.17	388	13,059	881	105.36	76,860	10,875	202,426	

Mine production of metals from gold and silver mills and from concentrating mills, by counties and years, 1909–1923—Continued

Pitkin County

Year	Ore to gold and silver mills			Ore to concentrating mills (short tons)	Concentrates produced					
	Short tons	Gold in bullion (fine ounces)	Silver in bullion (fine ounces)		Short tons	Gold (fine ounces)	Silver (fine ounces)	Copper (pounds)	Lead (pounds)	Zinc (pounds)
1909		36.04	63	65,890	12,884		128,573	25,720	6,485,955	34,741
1910				55,856	13,995	1.35	111,128	24,516	5,908,475	
1911		13.98	620	68,085	17,703	12.23	111,964	7,408	8,022,366	
1912				55,003	14,129		70,444	22,018	5,956,600	
1913		1.02	1,576	71,822	22,100		115,552	35,679	9,400,241	
1914		0.46	153	79,475	25,720		115,933	40,177	13,025,966	
1915				84,309	28,103		164,475	11,537	13,469,845	
1916				79,284	26,428		214,648	9,534	10,063,063	
1917			56	75,399	25,133	2.24	208,953	7,891	7,464,954	281,280
1918				57,213	19,071		210,651		4,845,398	
1919			823	89,900	12,673		185,144		2,161,505	
1920			798	91,333	12,602		208,317		1,655,584	
1921			683	23,697	2,808		47,899		469,845	
1922			490	78,406	10,416		123,656		1,317,127	
1923				18,062	4,774		62,170		710,839	
		71.50	5,262	993,834	248,539	15.82	2,080,507	184,480	91,057,763	316,021

Rio Grande County

[No production in 1909–1911, 1913, 1914, and 1916–1923]

Year	Ore to gold and silver mills			Ore to concentrating mills (short tons)	Concentrates produced					
	Short tons	Gold in bullion (fine ounces)	Silver in bullion (fine ounces)		Short tons	Gold (fine ounces)	Silver (fine ounces)	Copper (pounds)	Lead (pounds)	Zinc (pounds)
1912		21.96								
1915	1,499	723.00		309						
	1,499	714.96		309						

Routt County

[No production in 1909–1916 and 1920–1923]

Year	Ore to gold and silver mills			Ore to concentrating mills (short tons)	Concentrates produced					
	Short tons	Gold in bullion (fine ounces)	Silver in bullion (fine ounces)		Short tons	Gold (fine ounces)	Silver (fine ounces)	Copper (pounds)	Lead (pounds)	Zinc (pounds)
1917	50	34.22	9		7	5.49	229			
1918		3.98	3	136	17	26.49	1,389			
1919				170	17	15.09	1,061			
	50	38.20	12	306	41	47.07	2,679			

Saguache County

[No production in 1910 and 1914–1919]

Year	Ore to gold and silver mills			Ore to concentrating mills (short tons)	Concentrates produced					
	Short tons	Gold in bullion (fine ounces)	Silver in bullion (fine ounces)		Short tons	Gold (fine ounces)	Silver (fine ounces)	Copper (pounds)	Lead (pounds)	Zinc (pounds)
1909	21	14.06	2							
1911		3.66								
1912				8,699	1,096	66.78	9,184	7,084	424,357	510,528
1913				283	34	12.78	415	153	18,254	8,573
1920				8,430	545	126.66	47,697	37,082	113,843	
1921				5,690	217	31.88	30,085	9,770	50,203	
1922				9,200	184	18.50	31,564	190	9,673	
1923				34,170	9,032	183.80	146,329	453,199	2,837,543	
	21	17.72	2	66,472	11,108	440.40	265,574	507,478	3,453,873	519,101

San Juan County

Year	Ore to gold and silver mills			Ore to concentrating mills (short tons)	Concentrates produced					
	Short tons	Gold in bullion (fine ounces)	Silver in bullion (fine ounces)		Short tons	Gold (fine ounces)	Silver (fine ounces)	Copper (pounds)	Lead (pounds)	Zinc (pounds)
1909	127,774	12,817.74	4,091	53,167	28,360	18,220.29	357,374	1,346,718	7,149,924	786,518
1910	131,593	13,953.69	5,752	68,627	32,713	18,645.22	370,265	997,273	8,571,423	3,743,341
1911	62,473	5,751.24	2,357	43,696	17,916	9,558.56	204,788	404,655	6,577,882	2,224,351
1912	44,774	4,661.25	1,876	89,522	20,693	17,406.71	291,480	643,978	7,704,249	2,373,321
1913	52,056	8,790.35	3,137	60,455	19,058	19,289.14	236,782	528,127	6,965,793	1,664,999
1914	55,700	6,053.24	2,629	57,613	16,337	15,362.58	292,988	665,404	3,949,973	971,177
1915	51,842	4,864.87	2,087	90,084	20,758	18,694.53	307,773	892,123	5,267,845	2,259,226
1916	52,195	3,182.00	1,173	82,185	20,809	13,534.63	319,891	951,112	5,635,202	4,014,403
1917	2,000	1,195.48	366	132,438	22,682	11,584.63	377,028	1,090,936	8,576,936	3,270,500
1918		1,651.83	783	128,775	17,936	9,305.64	345,643	822,224	8,673,629	3,410,308
1919		14.14	1	62,050	9,836	5,741.71	163,905	495,351	4,705,805	1,801,664
1920	19,214	197.31	23	180,441	34,734	12,378.52	638,935	1,275,165	15,937,075	11,837,395
1921		18.09	1							
1922		302.08	159	6,050	1,210	395.90	10,518	43,458	853,124	
1923	560	170.32	72	151,927	23,770	11,315.74	452,218	995,438	10,492,183	9,540,000
	600,181	63,623.63	24,507	1,207,030	286,812	181,433.80	4,369,588	11,151,962	101,061,043	47,897,203

Mine production of metals from gold and silver mills and from concentrating mills, by counties and years, 1909–1923—Continued

San Miguel County

Year	Ore to gold and silver mills			Ore to concentrating mills (short tons)	Concentrates produced					
	Short tons	Gold in bullion (fine ounces)	Silver in bullion (fine ounces)		Short tons	Gold (fine ounces)	Silver (fine ounces)	Copper (pounds)	Lead (pounds)	Zinc (pounds)
1909	402,114	80,272.68	189,550	19,024	23,156	27,591.83	1,098,106	357,452	3,928,973	804,296
1910	452,240	87,222.11	304,722	27,364	32,251	33,032.90	781,995	529,633	7,292,248	2,193,981
1911	390,568	91,123.25	319,976	37,438	25,994	27,182.56	585,673	954,801	5,740,535	3,386,088
1912	418,886	72,589.08	305,614	35,600	31,154	43,170.42	777,452	832,675	6,828,462	2,943,783
1913	480,609	62,939.61	377,110	26,659	31,330	39,987.96	621,524	731,980	6,563,375	2,405,750
1914	476,210	68,717.64	598,193	19,091	26,576	33,474.21	668,777	321,657	3,934,168	----
1915	473,986	62,013.15	507,913	9,636	27,284	37,970.25	581,538	555,837	5,095,742	983,496
1916	411,887	61,413.15	385,805	16,260	26,930	38,552.20	417,874	563,150	6,029,774	1,064,091
1917	341,793	61,664.47	344,273	46,922	29,527	35,156.18	416,756	899,627	6,121,542	1,810,245
1918	360,276	70,428.81	294,267	12,769	27,912	32,306.16	508,136	948,944	5,623,388	790,637
1919	376,076	65,588.86	208,426	51,742	26,863	36,203.43	857,101	898,121	7,241,012	515,082
1920	145,189	20,884.28	108,813	228,351	27,649	43,867.05	927,297	938,562	7,308,219	175,617
1921	364,964	22,886.32	20,703	89,008	32,277	47,944.23	1,658,477	893,000	7,704,824	----
1922	338,284	21,559.53	17,315	58,905	30,478	30,297.02	1,581,972	658,053	6,897,419	----
1923	382,197	21,298.96	18,208	100,639	43,534	45,069.56	1,530,259	1,383,442	10,058,040	----
	5,815,309	870,601.90	3,999,888	780,408	442,915	551,787.96	13,012,937	11,466,934	96,367,721	17,073,066

Summit County

Year	Ore to gold and silver mills			Ore to concentrating mills (short tons)	Concentrates produced					
	Short tons	Gold in bullion (fine ounces)	Silver in bullion (fine ounces)		Short tons	Gold (fine ounces)	Silver (fine ounces)	Copper (pounds)	Lead (pounds)	Zinc (pounds)
1909	7,348	330.00	89	13,215	3,606	937.89	61,955	408	3,360,628	2,304,644
1910	85	165.89	79	44,295	17,205	531.53	119,362	15,618	4,776,553	4,708,581
1911		332.53	82	54,289	22,867	558.39	158,423	16,494	5,728,319	7,561,494
1912		130.22	44	41,464	17,264	390.21	76,116	4,514	3,535,366	8,600,499
1913		1,443.38	927	33,801	13,089	429.13	49,634	8,400	2,590,786	6,628,954
1914		1,834.72	876	20,273	7,748	175.98	27,503	2,529	1,175,663	4,834,043
1915	240	2,163.68	931	42,128	13,291	192.49	33,266	3,592	1,606,620	8,488,260
1916	100	1,350.10	636	56,261	21,543	392.57	47,888	6,050	902,933	12,211,023
1917	1,310	209.46	143	44,778	30,229	427.22	46,069	1,223	163,274	15,763,287
1918	4,290	684.43	366	44,319	20,626	428.11	30,590	3,441	337,494	14,189,450
1919				15,210	5,206	47.76	6,928	----	110,844	3,905,311
1920	1,400	236.38	113	32,584	11,625	232.37	13,925	----	201,465	8,335,963
1921	100	63.72	27	8,480	968	168.02	35,275	14,037	215,724	----
1922				2,114	588	8.50	2,040	----	102,000	300,000
1923				43,732	14,420	846.58	52,634	----	3,038,504	7,289,400
	17,783	9,034.51	4,313	495,013	200,275	5,766.75	761,608	76,306	27,846,173	105,120,909

Teller County

Year	Ore to gold and silver mills			Ore to concentrating mills (short tons)	Concentrates produced					
	Short tons	Gold in bullion (fine ounces)	Silver in bullion (fine ounces)		Short tons	Gold (fine ounces)	Silver (fine ounces)	Copper (pounds)	Lead (pounds)	Zinc (pounds)
1909	530,196	446,519.03	23,860	----						
1910	636,632	a 431,052.24	a 26,515	----	6,276	14,868.75	1,868			
1911	723,655	b 424,214.22	b 30,558	----	2,823	10,865.18	1,878			
1912	812,555	c 429,087.95	c 30,562	----	4,804	13,600.40	7,161			
1913	888,822	447,880.88	42,833	----	6,620	14,983.41	5,304			
1914	906,963	492,418.66	53,486	----	8,147	19,249.05	4,662			
1915	907,156	545,955.69	46,118	----	7,352	16,598.06	2,884			
1916	855,461	491,342.74	45,972	53,248	5,275	10,370.25	2,229			
1917	1,078,423	486,083.05	52,817	----	6,054	7,793.10	5,508			
1918	935,489	379,865.76	47,702	----	8,400	11,575.00	2,220			
1919	775,936	276,387.20	31,897	----	3,378	5,255.00	3,524			
1920	448,519	197,547.88	27,712	----	3,495	10,892.60	4,260			
1921	484,110	193,192.76	30,296	----	3,156	14,427.08	7,039			
1922	452,127	189,398.33	20,837	----	1,960	5,890.80	3,622			
1923	382,733	195,740.33	22,264	----						
	10,798,777	5,026,687.62	533,449	53,248	67,740	156,368.68	52,159			

a In addition 13,051.46 ounces of gold was produced from old tailings.
b In addition 6,116.80 ounces of gold and 3,275 ounces of silver was produced from old tailings.
c In addition 7,123.30 ounces of gold and 5,574 ounces of silver was produced from old tailings.

INDEX

www.ingramcontent.com/pod-product-compliance
Lightning Source LLC
Chambersburg PA
CBHW051207200326
41519CB00025B/7037